Linear and Interface Circuits Applications

SECOND EDITION

D.E. Pippenger and E.J. Tobaben
Linear Applications

CONTRIBUTORS
C.L. McCollum
Field Applications Engineering
Linear Product Engineering

McGraw-Hill Book Company
New York   St. Louis   Oklahoma City   San Francisco   Auckland   Bogotá
Hamburg   London   Madrid   Mexico   Montreal
New Delhi   Panama   Paris   São Paulo   Singapore
Sydney   Tokyo   Toronto

Library of Congress Cataloging-in-Publication Data

Pippenger, Dale E.
    Linear and interface circuits applications/D. E. Pippenger and
E. J. Tobaben [and] contributors, C. L. McCollum, Field Applications
Engineering, Linear Product Engineering.—2nd ed.
        p.   cm.
    Written primarily by two members of the Texas Instruments staff.
    Includes index.
    ISBN 0-07-063764-4 (pbk.)
    1. Linear integrated circuits.   2. Interface circuits.
I. Tobaben, E. J.   II. Texas Instruments Incorporated.   III. Title.
TK7874.P568 1988   87-33383
621.395—dc19   CIP

1234567890 EDW/EDW 8921098

ISBN 0-07-063763-6 {I}

ISBN 0-07-063764-4 {P}

Printed and bound by Edwards Brothers

IMPORTANT NOTICE

Texas Instruments (TI) reserves the right to make changes in the devices or the device specifications identified in this publication without notice. TI advises its customers to obtain the latest version of device specifications to verify, before placing orders, that the information being relied upon by the customer is current.

TI warrants performance of its semiconductor products, including SNJ and SMJ devices, to current specifications in accordance with TI's standard warranty. Testing and other quality control techniques are utilized to the extent TI deems such testing necessary to support this warranty. Unless mandated by government requirements, specific testing of all parameters of each device is not necessarily performed.

In the absence of written agreement to the contrary, TI assumes no liability for TI applications assistance, customer's product design, or infringement of patents or copyrights of third parties by or arising from use of semiconductor devices described herein. Nor does TI warrant or represent that any license, either express or implied, is granted under any patent right, copyright, or other intellectual property right of TI covering or relating to any combination, machine, or process in which such semiconductor devices might be or are used.

Contents

Preface   xxi

Section 1
Introduction

INTRODUCTION..   1-1

Section 2
Operational Amplifier and Comparator Theory

OPERATIONAL AMPLIFIER AND COMPARATOR THEORY.....................   2-1
  Operational Amplifier Theory...   2-1
  Major Performance Characteristics..   2-2
    Gain and Frequency Response..   2-2
    Gain-Bandwidth Product..   2-3
    Influence of the Input Resistance...   2-3
    Influence of Input Offset Voltage...   2-3
    Input Offset Compensation..   2-4
    Input Offset Temperature Coefficient.....................................   2-4
    Influence of Input Bias Current..   2-4
    Influence of Output Resistance...   2-5
    Input Common-Mode Range...   2-5
    Common-Mode Rejection Ratio (CMRR)...................................   2-5
    Influence of Voltage and Current Drift....................................   2-5
    Slew Rate..   2-5
    Noise..   2-6
    Phase Margin...   2-6
    Output Voltage Swing (V_{opp})..   2-6
    Feed-Forward Compensation...   2-6
  Basic Operational Amplifier Circuits..   2-7
    Noninverting Operational Amplifier..   2-7
    Inverting Amplifiers...   2-8
    DC Output Offsets...   2-8
    Voltage Follower...   2-9
    Summing Amplifier..   2-9
    Difference Amplifier..   2-9
    Differentiator...   2-9
    Integrator..   2-10
    Active Filters...   2-10
      Unity-Gain Active Filters...   2-10
      Low-Pass Active Filters...   2-11
      Band-Reject Active Filters..   2-11
      Bandpass Active Filters...   2-12
  Choosing the Right Operational Amplifier....................................   2-13
    General Purpose Operational Amplifiers—Bipolar.......................   2-13
    BIFET Operational Amplifiers...   2-14
    LinCMOS Operational Amplifiers...   2-14
    Ultrastable Offsets..   2-14
    Wide Bandwidths..   2-14
    Advantages of LinCMOS Operational Amplifiers..........................   2-15
  Comparators...   2-15
    Comparator Parameters..   2-16
      Source Impedance...   2-16
      Differential Voltage Gain...   2-16
      Output Characteristics..   2-16

Standard Comparators 2-18
Use of Hysteresis 2-18
Application Precaution 2-18

Section 3
Operational Amplifier and Comparator Applications

OPERATIONAL AMPLIFIER AND COMPARATOR APPLICATIONS............... 3-1
  Operational Amplifier Applications 3-1
    General Applications... 3-1
    Optical Sensor to TTL Interface Circuit 3-1
    Bridge-Balance Indicator.. 3-1
    High-Input-Impedance Differential Amplifier....................................... 3-2
    Low Voltage Shunt Limiter 3-3
    Schmitt Triggers 3-3
    PTC Thermistor Automotive Temperature Indicator................................. 3-5
    Accurate 10-Volt Reference.. 3-5
    Precision Large-Signal Voltage Buffer Using an OP-07 3-6
  Active Filter Applications.. 3-6
    Active Filter Introduction 3-6
    Active Band-Pass Filter 3-7
    High- and Low-Pass Active Filters 3-9
    Multiple-Feedback Band-Pass Filters 3-9
    Twin-T Notch Filter.. 3-10
  Audio Op Amp Applications.. 3-11
  Mike Preamp with Tone Control 3-11
    TL080 IC Preamplifier 3-11
    Audio Distribution Amplifier 3-13
    Audio Power Amplifier 3-14
  Oscillators and Generators 3-15
    Audio Oscillator Circuit Description 3-15
    The Basic Multivibrator 3-16
  LinCMOS Op Amp Applications.. 3-17
    Microphone Preamplifier... 3-17
    Positive Peak Detector 3-17
    Instrumentation Meter Driver 3-18
    A Stable Two-Volt Logic Array Power Supply...................................... 3-19
    TLC271 Twin-T Notch Filter 3-20
    Single Supply Function Generator... 3-20
  Comparator Applications... 3-21
    Window Comparator.. 3-21
    Comparator Interface Circuits... 3-21
    LM393 Zero-Crossing Detector 3-23

Section 4
Video Amplifiers

VIDEO AMPLIFIERS 4-1
  Video Amplifier Theory.. 4-1
    Voltage Gain... 4-1
    Common-Mode Output Voltage 4-1
    Output Offset Voltage 4-2
    Wiring Precautions.. 4-2
    Oscilloscope/Counter Preamplifier 4-2
    NE592 Filter Applications... 4-3

MC1445 Balanced Modulator 4-3
MC1445 Frequency Shift Keyer 4-3

Section 5
Voltage Regulators

VOLTAGE REGULATORS 5-1
Basic Regulator Theory 5-1
Voltage Regulator Components 5-1
    Reference Elements.. 5-1
    Sampling Element... 5-1
    Error Amplifier 5-1
    Control Element 5-1
Regulator Classifications 5-2
Series Regulator 5-2
Shunt Regulator 5-2
Switching Regulator.. 5-2
Major Error Contributors 5-3
Regulator Reference Techniques 5-3
    Zener Diode Reference 5-3
    Constant-Current Zener Reference.. 5-3
    Band-Gap Reference.. 5-4
Sampling Element.. 5-5
Error Amplifier Performance.. 5-5
    Offset Voltage 5-5
    Offset Change with Temperature 5-5
    Supply Voltage Variations.. 5-6
Regulator Design Considerations... 5-6
Positive Versus Negative Regulators.. 5-6
Fixed Versus Adjustable Regulators... 5-7
Dual-Tracking Regulators 5-7
Series Regulators 5-7
    Floating Regulator 5-8
Shunt Regulator 5-9
Switching Regulators... 5-9
    Fixed-On-Time, Variable Frequency.. 5-10
    Fixed-Off-Time, Variable Frequency 5-11
    Fixed-Frequency, Variable Duty Cycle 5-11
Regulator Safe Operating Area 5-12
Regulator SOA Considerations.. 5-12
    Input Voltage 5-12
    Load Current 5-12
    Power Dissipation.. 5-12
    Output Voltage of an Adjustable-Voltage Regulator............................... 5-12
    External Pass Transistor.. 5-12
Safe Operating Protection Circuits 5-13
    Reverse Bias Protection 5-13
Current Limiting Techniques 5-13
    Series Resistor 5-13
    Constant-Current Limiting 5-14
    Fold-Back Current Limiting 5-14
Three Thermal Regulators 5-15
Stabilization.. 5-15
Fixed Dual Regulators 5-16
Series Adjustable Regulators... 5-16

Thermal Considerations 5-17
Layout Guidelines 5-17
   Layout Design Factors 5-17
      Input Ground Loop 5-17
      Output Ground Loop 5-17
      Remote Voltage Sense 5-17
      Thermal Profile Concerns 5-18
Input Supply Design 5-18
   Transformer/Rectifier Configuration 5-18
   Capacitor Input Filter Design 5-19
Voltage Regulator Terms and Definitions 5-21

Section 6
Switching Power Supply Design

SWITCHING POWER SUPPLY DESIGN 6-1
Basic Operation of Switching Regulators 6-1
   Advantages of a Switching Regulator 6-2
   Disadvantages of a Switching Regulator 6-2
   Basic Switching Regulator Architecture 6-2
      The Step-Down Regulator 6-2
      The Step-Up Regulator 6-2
      The Inverting Regulator 6-3
   Forward Converters 6-3
      Push-Pull Converter 6-3
      Half Bridge Converter 6-3
      Full Bridge Converter 6-4
   TL593 Floppy Disk Power Supply 6-4
      Transformer Construction 6-5
TL594 12-Volt to 5-Volt Step-Down Regulator 6-5
   The TL594 Control Circuit 6-7
      Reference Regulator 6-7
      Oscillator 6-7
      Dead Time and PWM Comparators 6-8
      Error Amplifiers 6-9
      Output Logic Control 6-9
      The Output Driver Stages 6-9
   Soft-Start 6-9
   Over-Voltage Protection 6-9
Designing a Power Supply (5 Volt/10 Amp Output) 6-10
   Design Objective 6-10
   Input Power Source 6-10
   Control Circuits 6-10
      Oscillator 6-10
      Error Amplifier 6-10
      Current Limit Amplifier 6-10
      Soft-Start and Dead Time 6-12
   Inductor Calculations 6-12
   Output Capacitance Calculations 6-12
   Transistor Power Switch Calculations 6-13
TL497A Switching Voltage Regulator 6-13
   Step-Down Switching Regulator 6-14
   Step-Up Switching Regulator 6-14
   Inverting Configuration 6-14
   Design Considerations 6-17

A Step-Down Switching Regulator Design Exercise with TL497A 6-18
Design and Operation of an Inverting Regulator Configuration 6-19
Adjustable Shunt Regulator TL430-TL431... 6-21
Shunt Regulator Applications (Crowbar) 6-23
Controlling V_O of a Fixed Output Voltage Regulator 6-23
Current Limiter 6-23
Voltmeter Scaler 6-24
Voltage-Regulated, Current-Limited Battery Charger for Lead-Acid Batteries 6-24
    Battery Charger Design 6-24
        Rectifier Section 6-25
        Voltage Regulator Section... 6-25
        Current Limiter Section 6-26
        Series Pass Element.. 6-26
        Design Calculations 6-26
        Power Dissipation and Heat Sinking................................. 6-27
Voltage Supply Supervisor Devices 6-27
    General Operation... 6-28
    TL7700 Series Supervisor Chips 6-28
        Operation During a Voltage Drop....................................... 6-29
    TL7700 Series Application 6-30
uA723 Precision Voltage Regulator.. 6-31
    Typical Applications 6-32
General Purpose Power Supply... 6-34
8-Amp Regulated Power Supply for Operating Mobile Equipment..................... 6-35
±15 Volts at 1.0 Amp Regulated Power Supplies.................................... 6-36
    Positive Supply.. 6-36
    Negative Supply... 6-36
Overvoltage Sensing Circuits 6-38
    The Crowbar Technique 6-38
    Activation Indication Output.. 6-38
    Remote Activation Input.. 6-38

Section 7
Integrated Circuit Timers

INTEGRATED CIRCUIT TIMERS 7-1
RC Time-Base Generator 7-1
A Basic Delay Timer... 7-1
NE/SE555 Timers 7-2
Definition of Block Diagram Functions.. 7-2
    Threshold Comparator 7-2
    Trigger Comparator.. 7-2
    RS Flip-Flop... 7-2
    Discharge Transistor 7-2
    Output Stage.. 7-2
555 Basic Operating Modes 7-3
    Monostable Mode.. 7-4
    Astable Mode... 7-5
    Accuracy... 7-5
TLC555 Timer 7-5
    TLC555 Astable Timing Equations... 7-6
TLC556/NE556 Dual Unit 7-6
uA2240 Programmable Timer/Counter 7-7
Definition of uA2240 Functions 7-7
    Voltage Regulator (V_{reg})... 7-7

Control Logic 7-8
Time Base Oscillator 7-8
RC Input 7-8
Modulation and Sync Input 7-8
Threshold Comparators 7-9
Oscillator Flip-Flop 7-9
Binary Counter 7-9
uA2240 Basic Operating Modes 7-9
Monostable Operating Mode 7-9
Astable Operating Mode 7-10
Accuracy 7-10
General Design Considerations 7-10
Missing Pulse Detector 7-11
NE555 One-Shot Timer 7-13
Oscilloscope Calibrator 7-13
Darkroom Enlarger Timer 7-14
Touch Switch 7-15
Basic Square Wave Oscillator 7-15
Linear Ramp Generator 7-15
Fixed-Frequency Variable-Duty-Cycle Oscillator 7-17
Alternating LED Flasher 7-17
Positive-Triggered Monostable 7-18
Voltage Controlled Oscillator 7-18
Capacitance-to-Voltage Meter 7-20
TLC555 PWM Motor Controller 7-22
Telephone Controlled Night Light 7-23
Programmable Voltage Controlled Timer 7-23
Frequency Synthesizer 7-25
Cascaded Times for Long Time Delays 7-25
uA2240 Operation with External Clock 7-26
uA2240 Staircase Generator 7-28

Section 8
Display Drivers

DISPLAY DRIVERS 8-1
Introduction to Display Driver Technology 8-1
LED Display Drivers 8-3
SN75491 and SN75491A Quad LED Segment Drivers 8-3
SN75492 and SN75492A Hex LED Digit Drivers 8-3
SN75494 Hex LED Digit Driver 8-4
SN75497 7-Channel and SN75498 9-Channel LED Drivers 8-4
Driving LED Displays 8-4
Other Applications 8-7
DC Plasma Display Drivers 8-9
SN75480 DC Plasma Driver 8-9
SN75480 Application 8-11
SN75581 Gas Discharge Source Driver 8-11
SN75581 Application 8-11
SN75584A High-Voltage 7-Segment Latch/Decoder/Cathode Driver 8-11
SN75584A Application 8-15
AC Plasma Display Drivers 8-16
AC Plasma Display Technology 8-16
Construction 8-16

The Functional Cell... 8-17
Control Circuitry Implementation with the SN75500A and SN75501C............. 8-17
The Functional Waveforms.. 8-19
SN75500A AC Plasma Display Axis Driver.. 8-21
SN75501C AC Plasma Display Axis Driver.. 8-24
Functional Adaptation of the SN75500A and SN75501C 8-24
Strobing and Sustaining 8-24
Floating Driver Considerations.. 8-25
Data Coupling Considerations... 8-27
Vacuum Fluorescent Displayer 8-29
Introduction 8-29
The VFD Panel.. 8-29
Construction 8-29
Panel Performance 8-30
VFD Timing Requirements 8-31
Drive Electronics 8-31
The UCN4810A 10-Bit VFD Driver.. 8-31
The TL4810A 10-Bit VFD Driver 8-32
The SN75512A 12-Bit VFD Driver 8-33
The SN75513A 12-Bit VFD Driver 8-33
The SN75514 High Voltage 12-Bit VFD Driver................................... 8-33
The SN75518 32-Bit VFD Driver... 8-34
The SN75501C as a 32-Bit High-Voltage VFD Driver............................ 8-35
VFD Driver Applications 8-35
Driving a Vacuum Fluorescent Character Display................................ 8-35
Driving a Dot Matrix Display 8-36
AC Thin Film Electroluminescent Display Drivers..................................... 8-42
AC TFEL Display Technology 8-42
Factors Affecting TFEL Display Brightness 8-42
AC TFEL Pixel Equivalent Circuit 8-43
Drivers for AC TFEL Panels... 8-43
SN75551 and SN75552 Electroluminescent Row Drivers 8-44
SN75553 and SN75554 Electroluminescent Column Drivers...................... 8-44
Driving AC TFEL Panels 8-44
Practical Refresh Drive Scheme 8-44
Theory of Operation 8-44
Interconnecting the Drivers to the Panel 8-47
Row Driver Operation 8-47
Column Driver Operation... 8-47
Row and Column Driver Requirements 8-47
Row Driver Voltage Supply 8-48
Composite Row Driver.. 8-49
Specifying Driver Requirements.. 8-50

Section 9
Data Transmission

DATA TRANSMISSION... 9-1
General Purpose Data Transmission.. 9-1
General Requirements 9-1
Types of Transmission Lines 9-1
Single Wire and Ground Plane 9-1
Two-Wire Interconnect.. 9-2
Twisted Pair 9-2
Coaxial Lines 9-4

Line Drivers 9-4
    Basic Driver Modes 9-5
Types of Transmission 9-5
    Single-Ended Transmission 9-5
    Single-Ended Application for High-Speed Bus Communication 9-8
    Differential Line Drivers and Receivers 9-11
    Terminating Differential Data Transmission 9-12
    Current-Mode Drivers in Differential Data Line Transmission 9-13
Receiver Performance 9-15
    Input Sensitivity 9-15
    Common-Mode Voltage Range 9-15
    Input Termination Resistors 9-16
    Reference Voltage 9-16
    Input Limitations 9-17
SN55/75107A Series Applications 9-17
    Connection of Unused Inputs and Outputs 9-17
    One-Channel Balanced Transmission System 9-17
    Differential Party-Line Systems 9-17
    Repeaters for Long Lines 9-18
Standard Voltage-Mode Differential Drivers and Receivers 9-19
EIA Standard RS-232-C Circuits and Applications 9-22
Typical Drivers for EIA RS-232-C Applications 9-22
    SN75150 Dual Data Line Driver 9-22
    SN75156/uA9636 Dual Line Driver 9-22
    SN75188/MC1488 Quad Line Driver 9-24
Typical Receivers for EIA RS-232-C Applications 9-24
    SN75152 Dual Data Line Receiver 9-24
    SN75154 Quad Data Line Receiver 9-28
    SN75189, SN75189A, MC1489 and MC1489A Quad Line Receivers 9-29
RS-232-C Applications 9-32
    Interface Using SN75150 and SN75154 9-32
    Typical Interface Using SN75188 and SN75189A 9-33
EIA Standard RS-423-A Circuits and Applications 9-33
RS-423-A Standard 9-33
RS-423-A Devices 9-33
    RS-423-A Drivers 9-33
    RS-423-A Receivers 9-36
Basic RS-423-A Application 9-36
EIA Standard RS-422-A and RS-485 Circuits and Applications 9-37
RS-422-A Standard 9-37
RS-422-A Applications 9-40
    Typical Application 9-40
    Short-Line Application 9-40
EIA RS-485 Standard 9-40
    Unit Load Circuit 9-41
Drivers and Receivers 9-41
    Driver Details 9-42
    Driver Speed Characteristics 9-43
    Receiver Details 9-44
Transceivers 9-45
    SN75176B, SN75177B, SN75178B and SN75179B Transceiver Features . 9-45
    Basic Transceiver Application 9-45
    Long-Line Application 9-46
    SN75179B Application 9-46
IEEE 488-1978 Transmission Systems 9-46

The IEEE-488 Standard... 9-46
  General Information 9-46
  Connectors.. 9-48
  Cable 9-49
  Logic Convention 9-49
  Functions 9-49
  Messages or Commands.. 9-50
Bus Interface Devices 9-51
  Electrical Specifications 9-52
  Driver Requirements... 9-52
  Driver Specifications 9-52
  Receiver Specifications 9-52
  Composite Load Requirements... 9-52
  Device DC Load Line Boundaries.. 9-52
  Device AC Load Line Limit 9-52
  Device Capacitive Load Limit.. 9-52
  Timing Values 9-53
  Data Rates.. 9-53
Interfacing to the IEEE Standard 488 Bus.. 9-53
  SN75160A Octal GPIB Transceiver 9-53
  SN75163A Octal GPIB Transceiver 9-54
  SN75161A and SN75162A Octal GPIB Transceiver............................. 9-54
  MC3446 Quad Bus Transceiver... 9-56
Typical Applications 9-56
A Typical IEEE-488 System Application 9-57
IBM System 360/370 Interfacing Circuits 9-58
  Driver and Receiver Requirements.. 9-59
    Driver Requirements.. 9-59
    Receiver Requirements 9-62
  General Physical and Electrical Requirements.................................... 9-62
    Line Terminations 9-62
    Voltage Levels 9-62
    Cable 9-62
    Ground Shift and Noise... 9-62
    Fault Conditions 9-62
  Electrical Characteristics for Select Out Interface............................... 9-63
    Receiver Requirements 9-63
    Driver Requirements.. 9-63
  IBM System/360 and System/370 Data Line Drivers 9-63
    SN75123 Dual Line Driver 9-63
    SN75126 Quadruple Line Driver 9-64
    SN75130 Quadruple Line Driver 9-65
  IBM System/360 and System/370 Data Line Receivers 9-65
    SN75124 Triple Line Receiver.. 9-65
    SN75125 and SN75127 Seven-Channel Line Receivers 9-66
    SN75128 and SN75129 Octal Line Receivers............................... 9-67
  IBM System 370 Application.. 9-67
    Drivers 9-67
    Cable 9-68
    Receivers 9-68

Section 10
Peripheral Drivers

PERIPHERAL DRIVERS 10-1
Device Considerations and Product Descriptions 10-1

Basic Configurations 10-1
  Typical Requirements 10-1
    Power.. 10-1
    Voltage 10-2
    Current 10-2
    Speed.. 10-2
    Logic 10-2
  Peripheral Driver Devices.. 10-2
    SN75431, '461 and '471 Series... 10-4
    SN75446 and SN75476 Series... 10-4
    SN75436, SN75437A and SN75438 10-5
    SN75435 Quad Driver... 10-6
    SN75440 Quad Peripheral Driver... 10-7
    SN75603, SN75604, and SN75605.. 10-7
    SN75372 Dual and SN75374 Quad Power FET Drivers 10-8
    DS3680 Quad Telephone Relay Driver 10-9
    SN75064/ULN2064 Series Quad Peripheral Drivers 10-10
    SN75068/ULN2068 and SN75069/ULN2069 Quad Darlington Switches 10-12
    SN75074/ULN2074 and SN75075/ULN2075 Quad Darlington Sink or
      Source Drivers... 10-12
    SN75465 and ULN2001A Series Seven Channel Darlington Transistor Arrays..... 10-12
    UDN2841 and UDN2845 Quad High Current Darlington Drivers................ 10-13
Peripheral Driver Applications 10-14
  Peripheral Driver-Opto Applications 10-14
    Driving Tungsten Filament, or Equivalent, Incandescent Lamps 10-14
    Driving Remotely Located LED Devices 10-17
  Relay and Solenoid Driver Applications.................................... 10-17
    Printer Hammer Driver Application 10-18
    Driving a Reversible Solenoid.. 10-18
    Opto Isolated and Time Controlled Reversible Solenoid Drive 10-18
    Power Solenoid Drive... 10-21
  Intelligent Switches 10-21
    Automotive Lights On Warning 10-21
Driving Motors 10-21
  "H" or Bridge Drive... 10-21
  "H" Motor Drive for Negative Supplies.................................... 10-22
  Speed Controlled, Reversible DC Motor Drive with the SN75603 and SN75604...... 10-22
  Driving Power FETs for DC Motor Control 10-23
  Stepper Motor Drive.. 10-23
Driving Data Transmission Lines.. 10-26

Section 11
Data Aquisition Systems

DATA ACQUISITION SYSTEMS.. 11-1
  Typical Systems 11-1
  Microprocessor-Controlled DAS 11-2
  Basic Uses of a DAS.. 11-2
    Data Logging 11-2
    Signal Analysis... 11-2
    Process Control 11-2
  Basic Sampling Concepts.. 11-2
  Types of Signals.. 11-3
    DC Signals... 11-3
    Dynamic Signals 11-3
    Deterministic Signals... 11-3

Random Signals... 11-4
Frequently Used Terms and Concepts................................. 11-5
   The Sampling Theorem 11-5
   Aliasing.. 11-6
   The Nyquist Criterion.. 11-6
   Aliasing Filter... 11-6
   Quantization... 11-7
   Binary Codes.. 11-7
      Straight and Offset Binary Codes........................ 11-7
      One's and Two's Complement Codes.................... 11-8
      Absolute-Value-Plus-Sign Code 11-8
      Gray Code.. 11-8
      BCD Code.. 11-9
Signal Recovery... 11-9
A Data Requisition Example... 11-10
   Summary of Design Considerations............................ 11-12
System Parameters.. 11-12
   Understanding Major System Parameters..................... 11-12
Methods of A/D Conversion.. 11-14
Key Selection Criteria for A/D Converters......................... 11-14
   Number of Bits 11-14
   Conversion Speed.. 11-14
   Conversion Accuracy... 11-14
Conversion Techniques... 11-15
   Single-Slope A/D Converter 11-15
   Dual-Slope A/D Converter 11-16
   Successive-Approximation Converters 11-17
      Binary Bit Weighing 11-19
      Creating a Digital Number 11-19
   Flash A/D Converters 11-20
Device Types 11-21
   ADC0803 and ADC0805 8-Bit Successive-Approximation A/D Converters with
      Differential Inputs 11-21
      Description 11-21
      Principles of Operation...................................... 11-21
   ADC0808, ADC0809 CMOS Analog-to-Digital Converters with
      8-Channel Multiplexers 11-23
      Description 11-23
      Multiplexer 11-24
      Converter... 11-25
ADC0831, ADC0832, ADC0834, ADC0838 2-, 4-, 8-Channel A/D Peripherals with
   Serial Control and Multiplexer Options 11-26
   Description 11-26
   Principles of Operation... 11-33
TL500C thru TL503C Analog-to-Digital Converter Building Blocks TL500C/TL501C... 11-34
   TL502C/TL503C.. 11-34
   General Overall Description 11-35
   Description of TL500C and TL501C Analog Processors............ 11-37
   Description of TL502C/503C Digital Processors 11-37
   Principles of Operation... 11-39
   Auto-Zero Phase... 11-41
   Integrate-Input Phase 11-41
   Integrate-Reference Phase 11-41
   Capacitor Selection Guidelines................................... 11-41
   Bypassing and Stray Coupling.................................... 11-41

TL505C Analog-to-Digital Converter... 11-41
   Description 11-41
   Definition of Terms 11-42
   Principles of Operation... 11-42
TL5071, TL507C Analog-to-Digital Converter.. 11-44
   Description 11-44
   Definition of Terms 11-44
   Principles of Operation... 11-44
TLC532, TLC533 LinCMOS 8-Bit Analog-to-Digital Peripherals with 5 Analog and 6
   Multipurpose Inputs 11-47
   Description 11-47
   Principles of Operation... 11-48
TLC540, TLC541 8-Bit Analog-to-Digital Peripherals with Serial Control and
   11 Inputs 11-51
   Description 11-51
   Principles of Operation... 11-51
TLC548, TLC549, LinCMOS 8-Bit Analog-to-Digital Converters with Serial Control ... 11-53
   Description 11-53
   Principles of Operation... 11-54
TL0808, TL0809 Low-Power CMOS Analog-to-Digital Converter with
   8-Channel Multiplexers 11-55
   Description 11-55
   Principles of Operation... 11-56
   Multiplexer 11-57
   Converter.. 11-57
A/D Converter Applications 11-58
   Interface for ADC0803, ADC0804, and ADC0805 Converters to Zilog Z80A and
   Z80 Microprocessors.. 11-58
      Circuitry 11-59
      Timing Diagram 11-59
      Software 11-59
   Interface for ADC0803, ADC0804, and ADC0805 Converters to the Rockwell
   6502 Microprocessor 11-61
   Interface for ADC0808, ADC0809, TL0808, TL0809, TL520, TL521, and TL522
   Converters to Zilog Z80A and Z80 Microprocessor 11-62
      Timing Diagram 11-63
      Software 11-63
   Interface for ADC0808 and ADC0809 Converters to the Rockwell
   6502 Microprocessor 11-66
   Interface for ADC0808, ADC0809, TL0808, TL0809, TL520, TL521, and TL522
   Converters to the Motorola 6800 Microprocessors............................... 11-68
      Timing Diagram 11-68
      Software 11-68
   Interface for ADC0831, ADC0832, ADC0834, and ADC0838 Converters to Zilog
   Z80A and Z80 Microprocessors... 11-71
      Circuitry—ADC0832, ADC0834, and ADC0838....................... 11-71
      Timing Diagram—ADC0838 Device 11-71
      Software—ADC0832, ADC0834, and ADC0838 Devices 11-72
      Circuitry—ADC0831 Device 11-75
      Timing Diagram—ADC0831 Device 11-75
      Software—ADC0831 Device 11-75
   Interface for ADC0831, ADC0832, and ADC0838 Converters to the Rockwell
   6502 Microprocessor 11-77
   Interface for ADC0831, ADC0832, ADC0834, and ADC0838 Converters to the
   Motorola 6805 Microprocessor 11-80

Circuitry—ADC0832, ADC0834, and ADC0838 Devices 11-80
Timing Diagrams—ADC0838 Device... 11-80
Software—ADC0832, ADC0834, and ADC0838 Devices 11-80
Circuitry—ADC0831 Device 11-83
Timing Diagram—ADC0831 Device 11-83
Software—ADC0831 Device 11-83
Interface for ADC0831, ADC0832, ADC0834, and ADC0838 Converters to Motorola
  6800, 6802, 6809, and 6809E Microprocessors 11-84
Circuitry—ADC0832, ADC0834, and ADC0838 Devices 11-85
Timing Diagram—ADC0838 Device 11-85
Software—ADC0832, ADC0834, and ADC0838 Devices 11-86
Circuitry—ADC0831 Device 11-89
Timing Diagram—ADC0831 Device 11-89
Software—ADC0831 Device 11-89
Interface for ADC0831, ADC0832, ADC0834, and ADC0838 Converters to Intel
  8051 and 8502 Microprocessors.. 11-91
Circuitry—ADC0832, ADC0834, and ADC0838 Devices 11-91
Timing Diagram—ADC0838 Device 11-91
Software 11-93
Circuitry—ADC0831 Device 11-95
Timing Diagram—ADC0831 Device 11-95
Software—ADC0831 Device 11-96
Interface for ADC0831, ADC0832, ADC0834, and ADC0838 Converters to Intel
  8048 and 8049 Microprocessors... 11-97
Circuitry—ADC0838, ADC0834, and ADC0832.. 11-97
Timing Diagram—ADC0838 Device 11-97
Software—ADC0838, ADC0834, and ADC0832 Devices 11-97
Circuitry—ADC0831 Device 11-98
Timing Diagram—ADC0831 Device 11-98
Software—ADC0831 Device 11-98
TL500, TL501, TL502, and TL503 Devices Application Examples 11-100
External Components Selection Guide 11-100
Printed Circuit Board Layout Notes.. 11-101
TL502 and TL503 Control Circuits 11-101
Logic Inputs and Outputs 11-101
Driving a Display Using the TL502 Device 11-104
Calculation of the Current-Limiting Resistor 11-104
Application of the TL503 BCD Outputs.. 11-104
Digital Panel Meter Using the TL501 and TL502 Devices.......................... 11-105
Digital Thermometer... 11-107
Precision Panel Meter Application 11-107
TL505 A/D Converter Applications... 11-109
Functional Description 11-109
External Component Selection Guide 11-109
Digital Panel Meter Using the TL505 and TL502 Devices.......................... 11-110
TL505 to TMS1000 Interface... 11-111
TL507 A/D Converter Applications... 11-112
TL507 Device Description.. 11-112
TL507 Device Inputs and Outputs.. 11-113
Single-Wire Power, Data, and Clock Cycle Transmitter........................... 11-115
TL507 High Voltage Insulation Application and Signal Coupling................... 11-116
TL507 to TMS1000 Interface... 11-116
TL507 to TMS1000C Interface 11-117
Summary 11-118

Interface for TLC532A, TLC533A, TL530, and TL531 Devices to Zilog Z80A and
    Z80 Microprocessors... 11-118
    Input/Output Mapping... 11-120
    Timing Diagram 11-120
    Software 11-120
    Additional Comments.. 11-121
Interface for TLC532A, TLC533A, TL530, and TL531 Devices to Intel 8048 and
    8049 and Intel 8051 and 8052 Microprocessors................................... 11-122
    Hardware—Interface 1.. 11-122
    Timing Diagram—Interface 1 11-122
    Software—Interface 1... 11-122
    Hardware—Interface 2.. 11-126
    Timing Diagram—Interface 2 11-126
    Software—Interface 2... 11-126
    Additional Comments.. 11-129
Interface for TLC532A, TLC533A, and TL530 Devices to the Rockwell
    6502 Microprocessor 11-129
    Principles of Operation... 11-129
Interface for TLC532A, TLC533A, TL530, and TL531 Devices to the Motorola 6800,
    6802, 6809, and 6809E Microprocessors 11-131
    Input/Output Mapping... 11-131
    Timing Diagram 11-132
    Software 11-132
    Considerations for Different Microprocessors.................................... 11-135
    Interface Selection 11-136
    Additional Comments.. 11-136
Interface for the TLC540 Device to Zilog Z80A and Z80 Microprocessor.............. 11-137
    Interface 1 11-137
    Interface 2 11-139
Interface for the TLC540 Device to the Rockwell 6502 Microprocessor Using
    the 6522 VIA 11-140
    Principles of Operation... 11-140
Interface for the TLC540 Device to the Rockwell Microprocessor Using TTL Gates ... 11-142
    Principles of Operation... 11-142
Interface for TLC540 and TLC541 Devices to the Motorola 6805 Microprocessor 11-144
    Principles of Operation... 11-144
Software Interface for TLC540 and TLC541 Devices to Intel 8051 and
    8052 Microprocessors 11-146
    Circuit—Interface 1 (ALE CLOCK) 11-148
    Circuit—Interface 2 (CRYSTAL CLOCK) 11-148
    Timing Diagrams... 11-148
    Software 11-148
    Hardware 11-151
    Timing Diagrams... 11-151
    Software 11-151
Software Interface for TLC540 and TLC541 Devices to Intel 8048 and
    8049 Microprocessors 11-151
    Hardware 11-151
    Timing Diagram 11-151
    Software 11-151
Interface for the TLC549 Device to the Zilog Z80A Microprocessor.................. 11-156
    Principles of Operation... 11-156
Interface for the TLC549 Device to the Rockwell 6502 Microprocessor
    Using TTL Gates 11-157
    Principles of Operation... 11-157

Interface for the TLC549 Device to the Rockwell 6502 Microprocesor Using
    the 6522 VIA... 11-158
    Principles of Operation.. 11-158
Interface for the TLC549 Device to the Motorola 6805 Microprocessor 11-160
    Principles of Operation.. 11-160
Software Interface for the TLC549 Device to Intel 8048 and 8049 Microprocessors..... 11-161
    Hardware... 11-161
    Timing Diagram 11-161
    Software... 11-163
Serial Port Interface for the TLC549 Device to Intel 8051 and 8052 Microprocessors... 11-163
    Hardware... 11-164
    Timing Diagram 11-164
    Software... 11-164
Software Interface for the TLC549 Device to Intel 8051 and 8052 Microprocessors..... 11-166
    Hardware... 11-166
    Timing Diagram 11-166
    Software... 11-167
Interface for TLC1540 and TLC1541 Devices to the Zilog Z80A Microprocessor....... 11-169
    Principles of Operation.. 11-169
Interface for TLC1540 and TLC1541 Devices to the Rockwell 6502 Microprocessor.... 11-170
    Hardware... 11-170
    Timing Diagram 11-170
    Software... 11-171
Interface for TLC1540 and TLC1541 Devices to the Motorola 6802 Microprocessor.... 11-172
    Principles of Operation.. 11-173
Software Interface for TLC1540 and TLC1541 Devices to the Motorola
6805 Microprocessor 11-174
    Hardware... 11-174
    Software... 11-174
Software Interface for TLC1540 and TLC1541 Devices to the Intel 8051 and
8052 Microprocessor 11-176
    Hardware... 11-176
    Timing Diagram 11-176
    Software... 11-176
Hardware Interface for the TLC545 Device to the Zilog Z80A Microprocessor 11-178
    Principles of Operation.. 11-178
Software Interface for the TLC545 Device to the Zilog Z80A Microprocessor 11-179
    Principles of Operation.. 11-179
Software Interface for the TLC545 and TLC546 Devices to Intel 8051 and
8052 Microprocessors 11-181
    Hardware... 11-181
    Timing Diagram 11-181
    Software... 11-181
Serial Port Interface for TLC545 and TLC546 Devices to Intel 8051 and
8052 Microprocessors 11-185
    Hardware... 11-185
    Timing Diagram 11-186
    Software... 11-187
Interface for the TLC545 Device to the Rockwell 6502 Microprocessor Using
    TTL Gates... 11-190
    Principles of Operation.. 11-190
Interface for the TLC545 Device to the Rockwell 6502 Microprocessor Using
    the 6522 VIA.. 11-191
    Principles of Operation.. 11-191
Interface for the TLC545 and TLC546 Devices to the Motorola 6805 Microprocessor .. 11-193

Principles of Operation.. 11-194
Introduction to Data Acquisition for Digital Signal Processors........................ 11-195
Interfacing the TLC32040 to the TMS320 Family of Digital Signal Processors.......... 11-196
   Description 11-196
   Principles of Operation.. 11-196
Interfacing the TLC7524 to the TMS32010 11-199
   Hardware 11-199
   Principles of Operation.. 11-200
Interfacing the TLC0820 to the TMS32010 11-201
   Hardware 11-201
   Principles of Operation.. 11-201
Interfacing the TLC7524 to the TMS32020 11-202
   Hardware 11-202
   Principles of Operation.. 11-202
Interfacing the TLC0820 to the TMS32020 11-205
   Hardware 11-205
   Principles of Operation.. 11-205
Interfacing the TLC1540, TLC1541, TLC540, and TLC541 to the TMS32020
  and TSMS320C25.. 11-207
   Hardware 11-207
   Software 11-209
   Other Considerations 11-209
Interfacing the TLC548/9 to the TMS32020 11-211
   Hardware 11-211
   Principles of Operation.. 11-211

Section 12
Special Functions

SPECIAL FUNCTIONS... 12-1
  The Hall Effect 12-1
    The Hall Effect in Silicon 12-1
  Hall-Effect Devices 12-2
    Switching Devices.. 12-2
    Latching Devices... 12-2
    Linear Hall Devices.. 12-2
    Advantages of Hall-Effect Devices 12-3
    Hall-Effect Device Selection 12-3
  Hall-Effect Applications.. 12-4
    TL3103 Linear Hall-Effect Device in Isolated Sensing Applications.................. 12-4
      Toroid Design 12-5
    TL594 Isolated Feedback Power Supply 12-5
    Tachometer and Direction of Rotation Circuit................................... 12-5
      Tachometer Operation 12-7
      Direction of Rotation Circuit.. 12-8
      LED V_{CC} Supply Control... 12-8
    Angle of Rotation Detector 12-8
    Hall-Effect Compass 12-8
    Security Door Ajar Alarm 12-9
    Multiple Position Control System.. 12-9
    Door Open Alarm 12-10
  Appendix A-1

Preface

This book presents linear and interface circuit applications in a manner that will give the reader a basic understanding of the products and provide simple but practical examples for typical applications. Care has been taken to choose illustrations which are of interest, at least by analogy, to a wide class of readers. This material is written for not only the design engineer but also for engineering managers, engineering technicians, system designers, and marketing or sales people with some technical background. The authors have attempted to avoid lengthy mathematical analyses for technical elegance, so that the important points may be clearly emphasized and not obscured by distracting derivations. In cases where a rigorous derivation has been omitted, an attempt has been made to state the results precisely and to emphasize limitations that are practically significant.

To facilitate their use, the sections have been made basically independent. The primary goal of the book is to assist the user in selecting the proper device for a particular application. To accomplish this, key features of devices are presented along with discussions of device or system theory and requirements.

Potential uses of the devices are demonstrated in circuit applications. These applications are not intended to be a how-to for specific circuits but to be examples of how the device might be used to solve your specific design requirements. In each case, a data book or data sheet should be referred to for complete device characteristics and operating limits. The circuit examples selected for this book have accrued from numerous customer inquiries and related laboratory simulations.

This book has been written primarily by two members of the Linear Applications Staff. They would like to express their appreciation for the helpful inputs and assistance from members of the Linear Applications Lab, Product Engineering Staff, Field Application Engineering, and the European Applications Staff.

Section 1

Introduction

The technology of incorporating microprocessors and other logic circuits on a single integrated circuit chip has heralded the computer age. Accompanying this technology have been linear and interface circuits to provide the variety of functions required for these complex computer devices to communicate with each other and external systems. The broad range of integrated circuits available today can be divided into two general classes; logic and nonlogic. This is the first in a new series of application books that address the nonlogic devices.

The book has been divided into basically independent sections for ease of use. Each section covers a product category, beginning with the basic theory of that product, followed by the key characteristics of devices in that category, and then applications of the devices.

The primary objective is to assist the user in understanding the operating principles and characteristics of the wide variety of devices. By understanding the devices the user can solve one of the most common application problems — selecting the proper device for a particular application. Some theory is discussed; however emphasis is placed on demonstrating operating characteristics of devices and their potential uses in circuits. Obviously presentation of all possible circuits is beyond the scope of any book. The circuit examples selected for this book have accrued from numerous customer inquiries and related laboratory simulations. In many cases they are solutions to actual customer design problems. However, the circuits are presented only as examples to stimulate your thinking on how the devices could be used to solve your specific design requirements. In each case, a data book or data sheet should be referred to for complete device characteristics and operating limits.

As an overview of the device numbering system, Table 1-1 shows the meaning of the various characters in Texas Instruments Linear and Interface circuit device numbers. Texas Instruments devices that are direct alternate sources for other manufacturers' parts carry the original part number including its prefix. Alteration in device characteristics from the original data sheet specifications, generally to improve performance, results in a new number with the appropriate SN55, SN75, TL, or TLC prefix. The type of package is also included in the device number. Table 1-2 lists the package suffixes and their definitions.

Table 1-1. Linear and Interface Circuits

XXX XXXXX XX
└─ Package Type
(See Table 1-2)

ORIGINAL MANUFACTURER	TI PREFIX	DEVICE NUMBER	TEMP* RANGE
TI	TL or TLC	XXXC	COM
		XXXI	IND
		XXXM	MIL
	SN	75XXX	COM
		55XXX	MIL
NATIONAL	LM	1XXX	MIL
		2XXX	IND
		3XXX	COM
	ADC	XXXX	COM
	DS	78XX	MIL
		88XX	COM
RAYTHEON	RC	4XXX	COM
	RM	4XXX	MIL
SIGNETICS	NE	5/55XX	COM
	SA	5/55XX	AUTO
	SE	5/55XX	MIL
	N8T	XX	COM
FAIRCHILD	uA	7XXXC	COM
		7XXXI	IND
		7XXXM	MIL
	uA	9XXX	COM
MOTOROLA	MC	13/33XX	IND
		14/34XXX	COM
		15/35XX	MIL
SPRAGUE	UCN	XXX	COM
	UDN	XXXX	COM
	ULN	XXXX	COM
AMD	AM	XXXXXM	MIL
		XXXXXC	COM
SILICON GENERAL	SG	15XX	MIL
		25XX	IND
		35XX	COM
PMI	OP-	XX	COM

*Temperature ranges:

COM = 0°C to 70°C

IND = −25°C to 85°C

AUTO = −40°C to 85°C

MIL = −55°C to 125°C

Table 1-2. Packages

TYPE	PACKAGE DESCRIPTION
N	Plastic DIP
NE, NG	Plastic DIP, copper lead frame
NF	Plastic DIP, 28 pin, 400 mil
NT	Plastic DIP, 24 pin, 300 mil
P	Plastic DIP, 8 pin
D	Plastic SO, small outline
J	Ceramic DIP
JD	Ceramic DIP, side braze
JG	Ceramic DIP, 8 pin
FE,FG	Ceramic chip carrier, rectangular
FH, FK	Ceramic chip carrier, square
FN	Plastic chip carrier, square
KA	TO-3 metal can
KC	TO-220 plastic, power tab
LP	TO-226 plastic
U	Ceramic flatpack, square
W, WC	Ceramic flatpack, rectangular

Section 2

Operational Amplifier and Comparator Theory

OPERATIONAL AMPLIFIER THEORY

In 1958, the age of the integrated circuit was ushered in by Jack Kilby of Texas Instruments. From the two hand-built circuits which he fabricated, the variety and quantity of integrated circuits have mushroomed at an ever increasing rate. One type of integrated circuit is the operational amplifier that is characterized by its high gain and versatility. Because of its versatility and ease of application, the operational amplifier has become one of the most widely used linear integrated circuits. Operational amplifiers are designed to be used with external components to provide the desired transfer functions.

The rapid evolution and versatility of the operational amplifier is shown by its initial development and use. One of the two hand-built integrated circuits which Jack Kilby built was a phase shift oscillator, the first linear integrated circuit. This was soon followed by the introduction of the uA702 and SN523 operational amplifiers. Even with their lack of short-circuit protection and their requirements for complex compensation they quickly gained acceptance. Among the improved designs which quickly followed was the uA741 single operational amplifier which required no external compensation. Conversely, the uA748 was designed for compensation by external components to change the frequency response for applications requiring wider bandwidth and higher slew rate.

Operational amplifier capabilities and versatility are enhanced by connecting external components to change the operating characteristics. Typical operational amplifier characteristics include frequency response, signal phase shift, gain and transfer function. The external components are placed in one or more feedback networks and/or the circuits that terminate the input.

To adequately evaluate the potential of an operational amplifier for a specific application, an understanding of operational amplifier characteristics is required. Figure 2-1 represents an equivalent operational amplifier circuit and its parameters. The parameters illustrated in Figure 2-1 are as follows:

- Input bias currents (I_{IB1} and I_{IB2}) — the current flowing into both operational amplifier inputs. In an ideal condition, I_{IB1} and I_{IB2} are equal.
- Differential input voltage (V_{DI}) — the differential input voltage between the noninverting (+) input and the inverting (−) input.

- Input offset voltage (V_{IO}) — an internally generated input voltage identified as the voltage that must be applied to the input terminals to produce an output of 0 V.
- Input resistance (R_I) — the resistance at either input when the other input is grounded.
- Output voltage (V_O) — normal output voltage as measured to ground.
- Output resistance (R_O) — resistance at the output of the operational amplifier.
- Differential voltage gain (A_{VD}) or open-loop voltage gain (A_{OL}) — the ratio of the input voltage to the output voltage of the operational amplifier without external feedback.
- Bandwidth (BW) — the band of frequencies over which the gain (V_O/V_{DI}) of the operational amplifier remains within desired limits.

The generator symbol Ⓖ in Figure 2-1 represents the output voltage resulting from the product of the gain and the differential input voltage ($A_{VD} V_{DI}$).

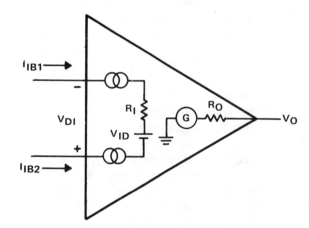

Figure 2-1. Operational Amplifier Equivalent Circuit

An ideal operational amplifier (see Figure 2-2) provides a linear output voltage that is proportional to the difference in voltage between the two input terminals. The output voltage will have the same polarity as that of the noninverting (+) input with respect to the voltage at the inverting (−) input. When the noninverting input is more positive than the inverting input, the output voltage will have a positive amplitude. When the noninverting input is more negative than the inverting input, the output voltage will have a negative amplitude.

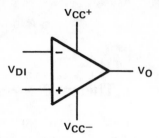

Figure 2-2. Ideal Operational Amplifier

An operational amplifier having no external feedback from output to input is described as being in the open-loop mode. In the open-loop mode, the characteristics of the ideal operational amplifier are as follows:

Differential gain $= \rightarrow \infty$
Common-mode gain $= 0$
Input resistance $= \rightarrow \infty$
Output resistance $= 0$
Bandwidth $= \rightarrow \infty$
Offset and drift $= 0$

MAJOR PERFORMANCE CHARACTERISTICS

The detailed and specific performance characteristics of a particular operational amplifier can be found on the appropriate data sheet. An operational amplifier data sheet will normally provide many nongeneric electrical characteristics. The electrical characteristics provided are for a specified supply voltage and ambient temperature and usually will have minimum, typical, and maximum values.

Major operational amplifier characteristics and their meaning are as follows:

- Input offset current (I_{IO}) — the difference between the two input bias currents when the output voltage is zero.
- Common-mode input voltage range (V_{ICR}) — the range of the common-mode input voltage (i.e., the voltage common to both inputs).
- Output short-circuit current (I_{OS}) — the maximum output current that the operational amplifier can deliver into a short circuit.
- Output voltage swing (V_{OPP}) — the maximum peak-to-peak output voltage that the operational amplifier can produce without saturation or clipping occurring. This characteristic is dependent upon output load resistance.
- Large-signal differential voltage gain (A_{VD}) — the ratio of the ouput voltage swing to the input voltage swing when the output is driven to a specified large-signal voltage (typically ± 10 V).
- Slew rate (SR) — the time rate of change of the closed-loop output voltage with the operational amplifier circuit having a voltage gain of unity (1).
- Supply current (I_{CC}) — The total current that the operational amplifier will draw from both power supplies when unloaded (per amplifier for multiunit packages).
- Common-mode rejection ratio (CMRR) — a measure of the ability of an operational amplifier to reject signals that are present at both inputs simultaneously. The ratio of the common-mode input voltage to the generated output voltage and is usually expressed in decibels (dB).

The preceding paragraphs have discussed basic operational amplifier characteristics. The following paragraphs will provide more detailed information. The specific characteristics that will be discussed are as follows:

- Gain and frequency response
- Difference of input resistance
- Influence of input offset voltage
- Input offset compensation
- Input offset voltage temperature coefficient
- Influence of input bias current
- Influence of output resistance
- Input common-mode range
- Common-mode rejection ratio (CMRR)
- Slew rate
- Noise
- Phase margin
- Output voltage swing
- Feed-forward compensation

Gain and Frequency Response

Unlike the ideal operational amplifier, a typical operational amplifier has a finite differential gain and bandwidth. Because many of the ideal operational amplifier characteristics cannot be achieved, the characteristics of a typical amplifier differ significantly from those of the ideal amplifier. The open-loop gain of a TL321 operational amplifier is shown in Figure 2-3. At low frequencies, open-loop gain is constant. However, at approximately 6 Hz it begins to roll off at the rate of -6 dB/octave (an octave is a doubling in frequency and decibels are a measure of gain

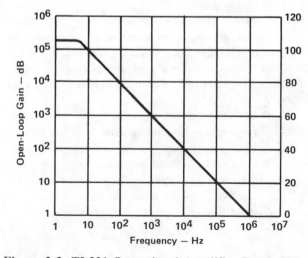

Figure 2-3. TL321 Operational Amplifier Bandwidth

calculated by $20 \log_{10} V_O/V_I$). The frequency at which the gain reaches unity is called unity gain bandwidth and referred to as B1.

When a portion of the output signal is fed back to the input of the operational amplifier, the ratio of the output to input voltage is called closed-loop gain. Closed-loop gain is always less than the open-loop gain. Because gain error is proportional to the ratio of closed-loop gain to open-loop gain, a very high value of open loop gain is desirable.

Gain-Bandwidth Product

When selecting an operational amplifier for a particular application, gain-bandwidth product is one of the primary factors to consider. The product of closed-loop gain and frequency response (expressed as bandwidth, BW, remains constant at any point on the linear portion of the open-loop gain curve (see Figure 2-4).

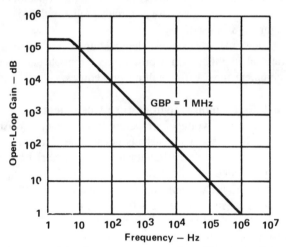

Gain-Bandwidth Product = Closed-Loop Gain X Frequency Response

Figure 2-4. Bandwidth for Operational Amplifier TL321

The bandwidth is the frequency at which the closed-loop gain curve intersects the open-loop curve as shown in Figure 2-4. The bandwidth may be obtained for any desired closed-loop gain by drawing a horizontal line from the desired gain to the roll-off intersection of the open-loop gain curve. In a typical design, a factor of 1/10 or less of the open loop gain at a given frequency should be used. This ensures that the operational amplifier will function properly with minimum distortion. When the voltage gain of an operational amplifier circuit is increased, the bandwidth will decrease.

Influence of the Input Resistance

The influence of the input resistance can be determined with Kirchhoff's law. By applying Kirchhoff's law to the circuit in Figure 2-5, we can use the following equations:

$$I_1 = I_2 + I_3 \text{ or}$$

$$\frac{V_I - V_{DI}}{R1} = \frac{V_{DI} - V_O}{R2} + \frac{V_{DI}}{R_I}$$

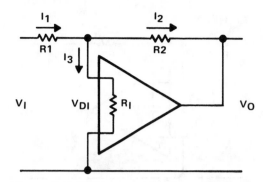

Figure 2-5. Influence of the Input Resistance

If the open-loop gain is infinite, the differential input voltage will be zero and the value of input resistance (if it is not zero) will have no influence. Since $V_{DI} = V_O/A_{VD}$, the following equations apply:

$$\frac{V_I - \dfrac{V_O}{A_{VD}}}{R1} = \frac{\dfrac{V_O}{A_{VD}} - V_O}{R2} + \frac{\dfrac{V_O}{A_{VD}}}{R_I}$$

Therefore:

$$\frac{V_I}{V_O} = \frac{1}{A_{VD}} + \frac{1}{\dfrac{R2}{R1} A_{VD}} + \frac{R1}{R_I A_{VD}} - \frac{1}{\dfrac{R2}{R1}}$$

or:

$$\frac{V_I}{V_O} = -\frac{1}{\dfrac{R2}{R1}} + \frac{1}{\dfrac{R2}{R1} A_{VD}} + \frac{1}{A_{VD}} \left(1 + \frac{R1}{R_I} \right)$$

The previously listed equations indicate that the input resistance (unless it is small relative to R1) will have little or no effect on the ratio of output voltage to input voltage. Therefore, the closed-loop gain for typical applications is independent of the input resistance.

Influence of Input Offset Voltage

The input offset voltage (V_{IO}) is an internally generated voltage and may be considered as a voltage inserted between the two inputs (see Figure 2-6). In addition, it is a differential input voltage resulting from the mismatch of the operational amplifier input stages.

The effect on currents I_1 and I_2 can be determined by the following equations:

$$\frac{V_I - V_{IO}}{R1} = \frac{V_{IO} - V_O}{R2}$$

If the input voltage (V_I) is zero, the equation is as follows:

$$\frac{-V_{IO}}{R1} = \frac{V_{IO} - V_O}{R2}$$

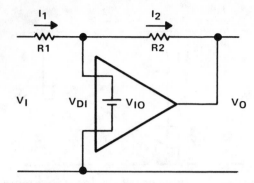

Figure 2-6. Influence of Input Offset Voltage

The output voltage is the output offset voltage (V_{OO}). The following equation can be used to determine V_{OO}:

$$V_{OO} = \left(\frac{R2}{R1} + 1\right) V_{IO}$$

The value of the input offset voltage can be found by dividing the output offset voltage by the closed-loop gain.

Input Offset Compensation

An ideal operational amplifier has zero input offset voltage and no drift. However, because of the mismatch of input transistors and resistors on the monolithic circuit, typical operational amplifiers have a low but definite offset voltage. Most operational amplifiers have provisions for connecting an external potentiometer so that the input offset can be adjusted to zero. The exact method used and total resistance of the null adjustment potentiometer is dependent upon the type of operational amplifier circuit. A general-purpose internally compensated operational amplifier (a uA741) may require a 10 kΩ potentiometer. A BIFET or externally compensated operational amplifier may require a 100 kΩ potentiometer. Recommended input offset voltage null adjustment circuits are usually shown in the data sheet.

Methods of nulling the input offset voltage are shown in Figures 2-7 and 2-8. When the offset null pins (N1 and N2) are connected to the emitter of the constant-current

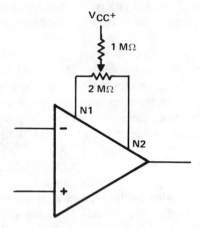

Figure 2-8. Null Pins Connected to Collectors

generators, a circuit similar to that shown in Figure 2-7 is used. When the null pins are connected to the collectors of the constant-current generator, a circuit similar to that shown in Figure 2-8 is used.

Actual resistor values depend upon the type of operational amplifier used. Consult the appropriate data sheet for complete input offset nulling procedures.

Input Offset Voltage Temperature Coefficient

Input offset voltage temperature coefficient (offset voltage drift) is specified in volts per degree Celsius. The amount of drift that occurs with temperature variations is directly related to how closely matched the input characteristics are when the device is manufactured. BIFET input devices (such as the TL080 family) typically have 10 to 12 μV/°C. The LinCMOS™ operational amplifier family has from 0.7 to 5 μV/°C depending upon the bias mode selected.

Influence of Input Bias Current

Both input bias current (I_3) and the normal operating currents (I_1 and I_2) flow through resistors R1 and R2 (see Figure 2-9). A differential input voltage equal to the product of $I_3(R1R2)/R1 + R2$ is generated by this current. The differential input (which is similar to input offset voltage)

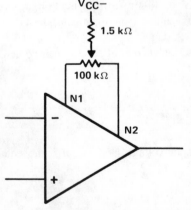

Figure 2-7. Null Pins Connected to Emitters

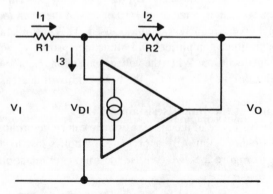

Figure 2-9. Influence of Input Bias Current

LinCMOS is a trademark of Texas Instruments

also appears as a component of the output that is amplified by the system gain. Methods of correcting for the effects of input bias current are discussed later.

Influence of Output Resistance

The influence of output resistance is illustrated by Figure 2-10. Output current can be expressed by the following equation:

$$I_O = I_2 + I_L \text{ and } I_2 + I_1 = \frac{V_O}{\dfrac{R2R_L}{R2 + R_L}}$$

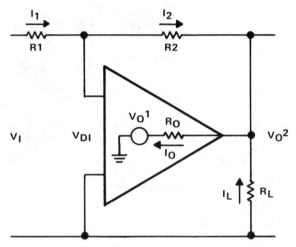

Figure 2-10. Influence of Output Resistance

If V_O1 is the output voltage of the equivalent ideal amplifier, and V_O2 is the output voltage of the actual device, then V_O2 can be determined by the following equation:

$$V_O2 = V_O1 - R_OI_O = V_O1 - \frac{R_OV_O2}{\dfrac{R2R_L}{R2 + R_L}}$$

For the ideal case $V_O1 = V_{DI}A_{VD}$; therefore:

$$V_O2 = V_{DI}A_{VD} - R_O\left[\frac{V_O2}{\dfrac{R2R_L}{R2 + R_L}}\right]$$

Input Common-Mode Range

The input common-mode range may be defined as the maximum range of the input voltage that can be simultaneously applied to both inputs without causing cutoff, clipping, or saturation of the amplifier gain stages. The input stage must be capable of operating within its specifications over the dynamic range of output swing. If it cannot, the amplifier may saturate (or latch-up) when the input limits are exceeded. Latch-up occurs most often in voltage-follower stages where the output voltage swing is equal to the input voltage swing and the operational amplifier is driven into saturation. The specified common-mode voltage range of the input stage must exceed the maximum peak-to-peak voltage swing at the input terminals or the input stage may saturate on peaks. When saturation occurs, an inverting stage no longer inverts. The negative feedback becomes positive feedback and the stage remains in saturation.

Common-Mode Rejection Ratio (CMRR)

The common-mode rejection ratio may be defined as the ratio of the differential signal gain to the common-mode signal gain and is expressed in decibels.

$$\text{CMRR (dB)} = 20 \ (\log_{10}) \ \frac{\dfrac{V_O}{V_I}}{\dfrac{V_O}{V_{CM}}}$$

$$\text{or } \frac{\text{(differential signal gain)}}{\text{(common-mode signal gain)}}$$

An ideal operational amplifier responds only to differential input signals and ignores signals common to both inputs. In a typical circuit, however, operational amplifiers have a small but definite common-mode error. Common-mode rejection is important to noninverting or differential amplfiers because these configurations see a common-mode voltage. Depending upon the type of device, dc rejection ratios may range from 90 dB to 120 dB. Generally, bipolar operational amplifiers have higher rejection ratios than FET-input amplifiers.

Influence of Voltage and Current Drift

Input offset voltage, input bias current, and differential offset currents may drift with temperature. Although it is relatively easy to compensate for the effects of these characteristics themselves, correcting for their drift with temperature variations is difficult. However, there is some limited control offered by the design over any drift characteristics. When drift tendencies are expected to be a design problem, device type, construction, and application should be considered.

Slew Rate

The slew rate may be defined as the maximum rate of change of the output voltage for a step voltage applied to the input (see Figure 2-11). Slew rate is normally measured with the amplifier in a unity gain configuration. Both slew rate and gain bandwidth product are measures of the speed of the operational amplifier.

Slew-rate limiting is accomplished by the limiting of the operational amplifier internal circuit ability to drive capacitive loads. Capacitance limits the slewing ability of the operational

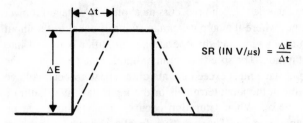

$$SR \text{ (IN V/}\mu s) = \frac{\Delta E}{\Delta t}$$

NOTE: Solid line is a square-wave input. Broken line is slewed output.

Figure 2-11. Effect of Slew Rate

amplifier at high frequencies. When the current available to charge and discharge the capacitance becomes exhausted, slew-rate limiting occurs.

Noise

Although not specifically stated as one of the primary characteristics of the ideal operational amplifier, noise-free operation is desirable. Typical operational amplifiers degrade the input signal by adding noise components. Noise components are usually random and determine the ultimate lower limit of signal-handling capability. Noise is usually specified on the data sheet as equivalent input noise, and like the other input factors, is increased by the gain of the stage. There are several potential sources of noise in an operational amplifier. The most common are thermal noise caused by the two source resistances (this noise exists within an ideal operational amplifier), internal noise current, and noise-voltage generators. Under normal audio applications the noise-voltage will be the dominant source of amplifier noise. As the source resistance is increased, the effect of noise-current increases until (at high source resistance) noise current and the bias compensation resistor noise together are the dominant components of amplifier input noise. In specifications, these two parameters are detailed separately. Noise voltage is specified at a low source resistance. Noise current is specified at a high source resistance. Both V_n and I_n are given in terms of density. These are measured with a narrow-bandwidth filter (1 Hz wide) at a series of points across a useful spectrum of the amplifier. Data is usually given in terms of noise voltage versus frequency. Practical data or curves on data sheets are normally given as the following:

$$V_n = e_n/\sqrt{F \text{ (Hz)}}$$

NOTE: Typically a frequency and source resistance will be given in the test conditions included in the device data sheet.

In general, low-input-current operational amplifiers (FET) or low-bias-current bipolar operational amplifiers will have lower noise current and tend to be quieter at source impedances above 10 kΩ. Below 10 kΩ, the advantage swings to bipolar operational amplifiers which have lower input voltage noise. When the source impedance is below 10 kΩ,

actual source resistance is composed mostly of generator resistance. The noninverting operational amplifier configuration has less noise gain than the inverting configuration for low signal gains and, thus, high signal-to-noise ratio. At high gains, however, this advantage diminishes.

Phase Margin

Phase margin is equal to 180° minus the phase shift at the frequency where the magnitude of the open-loop voltage gain is equal to unity. Phase margin is measured in degrees and must be positive for unconditional stability. Figure 2-12 illustrates a typical circuit used to measure phase margin.

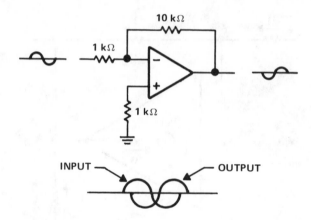

Figure 2-12. Phase Margin Measurement Circuit

If the phase difference between the input and output waveform is 120°, 180° minus 120° phase difference leaves 60° as the phase margin. Phase difference may or may not be given on the data sheet. Phase margin will normally be from 50° to 70° on commercially available operational amplifiers. When phase margin decreases to 45°, the operational amplifier becomes unstable and may oscillate.

Output Voltage Swing (V_{OPP})

V_{OPP} is the peak-to-peak output voltage swing that can be obtained without clipping or saturation. Peak-to-peak swing may be limited by loading effects, operational amplifier frequency capability, load resistance, and power supply used. Load resistances given on the data sheet are usually 2 kΩ or 10 kΩ. With load resistances of 2 kΩ or less, the output decreases due to current limiting. Normally, this will not damage the operational amplifier as long as the specified power-dissipation limits of the package are not exceeded. However, the open-loop gain will be reduced because of excessive loading.

Feed-Forward Compensation

The TL080 through TL084 BIFET operational amplifiers have been developed by Texas Instruments through state-of-the-art semiconductor technology. The BIFET process allows optimum circuit design with the fabrication of bipolar and FET transistors on a common substrate. This process, along with ion-implanted FET inputs, produces an almost ideal operational amplifier family which typically exhibits a higher input impedance than that obtainable with conventional manufacturing technology. High input impedance and the FET's inherently low input bias currents make the BIFET family ideal for numerous instrumentation and audio amplifier applications.

The TL080 provides for externally controlled compensation on pins 1 and 8. This is an advantage over the internally compensated version (TL081) because it allows the user to obtain slew rates from a typical 12 V/μs (for nominal compensation of 12 pF) to 30 V/μs (for a compensation of 3 pF). This increased slew rate is also reflected in the small signal response where the rise time is decreased from 0.1 μs to less than 50 ns. The power bandwidth can be extended to greater than 1 MHz. This greatly increases the potential for large-signal wide-bandwidth applications and for filters with frequencies at or above 1 MHz. The unity gain bandwidth is identical to the normal compensation mode but the first pole frequency is extended above 10 kHz; thus, gain accuracy is maintained at higher frequencies.

In the feed-forward circuit (see Figure 2-13), a 100 kΩ resistor is shown in parallel with a 3 pF capacitor in the negative feedback loop. A 500 pF feed-forward capacitor is connected from pin 1 to the inverting input of the TL080. The high-frequency response increases from approximately 6 kHz to over 200 kHz. Figure 2-14 is the feed-forward compensation curves.

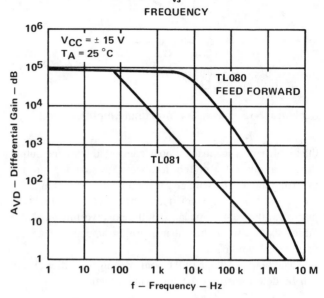

OPEN-LOOP LARGE SIGNAL DIFFERENTIAL VOLTAGE GAIN vs FREQUENCY

Figure 2-14. Feed-Forward Compensation Curves

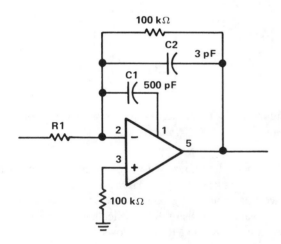

Figure 2-13. Feed-Forward Compensation

BASIC OPERATIONAL AMPLIFIER CIRCUITS

Operational amplifiers, because of their variable characteristics and wide range of adaptablility, can be configured to perform a large number of functions. Operational amplifier applications are frequently limited more by the imagination than by their functional limitations or operating parameters. The basic operational amplifier circuits that are discussed in this section are as follows:

- Noninverting Amplifier
- Inverting Amplifier
- DC Output Offset Amplifier
- Summing Amplifier
- Differentiator
- Active Filter
- Voltage Follower
- Differential Amplifier
- Integrator
- Unity-Gain Active Filter
- Band-Reject Active Filter
- Low-Pass Active Filter

NONINVERTING OPERATIONAL AMPLIFIER

A noninverting amplifier circuit provides an amplified output that is in phase with the circuit input. Figure 2-15 illustrates a basic noninverting operational amplifier circuit. In the circuit shown in the figure, the output is in phase with the input at low frequencies.

In this circuit, the input signal is applied to the noninverting (+) input of the amplifier. A resistor (R1), which is usually equal to the resistance of the input element, is connected between ground and the inverting (−) input of

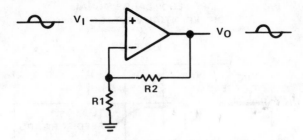

Figure 2-15. Basic Noninverting Amplifier Circuit

the amplifier. A feedback loop is connected from the output of the amplifier, through feedback resistor R2, to the inverting input. The voltage gain of a noninverting amplifier circuit is always greater than unity (1). For practical purposes, the input impedance of the noninverting amplifier circuit is equal to the intrinsic input impedance of the operational amplifier.

The output voltage of the noninverting amplifier circuit can be determined by the following equation:

$$V_O = \left[1 + \frac{R2}{R1}\right](V_I)$$

The voltage gain of the noninverting amplifier circuit can be determined by the following equation:

$$A_V = \frac{V_O}{V_I} = 1 + \frac{R2}{R1}$$

INVERTING AMPLIFIERS

An inverting amplifier circuit, as illustrated in Figure 2-16, provides an output that is 180 degrees out of phase with the input signal.

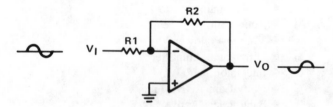

Figure 2-16. Basic Inverting Amplifier Circuit

In this circuit, the input signal is applied through a resistor (R1) to the inverting input (−) of the operational amplifier. The noninverting input (+) is connected to ground. A feedback loop is connected from the output, through feedback resistor R2, to the inverting input. The voltage gain of an inverting amplifier circuit can be less than, equal to, or greater than unity (1). Resistor R1 determines the input impedance of the inverting amplifier circuit. The input impedance is much lower than for a noninverting amplifier circuit.

The output voltage for an inverting amplifier can be determined by the following equation:

$$V_O = \left[-\frac{R2}{R1}\right](V_I)$$

NOTE: The minus sign in the equation indicates the 180° phase reversal.

The voltage, or closed-loop, gain can be determined by the following equation:

$$A_{CL} = \frac{V_O}{V_I} = -\left[\frac{R2}{R1}\right]$$

DC OUTPUT OFFSETS

When the input voltage to an operational amplifier is zero, the ideal output voltage is also zero. However, the ideal condition cannot be realistically achieved because of dc offset. DC offset may be caused by the internal input offset and input bias currents. It may also be caused by an input signal offset voltage. With no signal into the amplifier shown in Figure 2-17, input bias current flows through resistors R1 and R2. Because of the voltage drop across R1 and R2, these input currents will produce an offset voltage. Since the noninverting input is grounded, the voltage appears as input offset and is amplified by the operational amplifier.

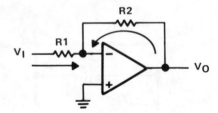

Figure 2-17. Inverting Amplifier with Input Bias Currents

The method commonly used to correct for a dc offset condition is to place an additional resistor (R3) between the noninverting input and ground as shown in Figure 2-18.

The value of resistor R3 is calculated as the parallel combination of R1 and R2 as follows:

$$R3 = \frac{R1R2}{R1 + R2}$$

A voltage is developed across R3 that is equal to the voltage across the parallel combination of R1 and R2. Ideally, the voltages appear as common-mode voltages and are cancelled. However, in a typical operational amplifier, the bias currents are not exactly equal. Because of this difference, a small dc offset voltage remains.

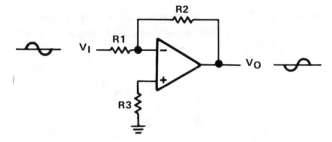

Figure 2-18. Inverting Amplifier with DC Offset Correction

The remaining source of output offset voltage is due to the internal input offset voltage. This may be nulled in several ways. The most common method is to connect a potentiometer across the offset null terminals available on many operational amplifiers. Depending upon the type of circuit and chip construction, the center arm is connected to the V_{CC+} rail or V_{CC-} rail. The terminals on the operational amplifier used for this purpose are usually labeled N1 and N2. For complete information on a specific device, consult the appropriate data sheet.

VOLTAGE FOLLOWER

The voltage or source follower is a unity-gain, noninverting amplifier with no resistor in the feedback loop (see Figure 2-19). The output is exactly the same as the input. The voltage follower has a high input impedance which is equal to the operational amplifier intrinsic input impedance.

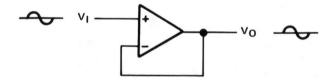

Figure 2-19. Basic Voltage Follower Circuit

The function of the voltage follower circuit is identical to a emitter follower for a bipolar transistor or a source follower on an FET transistor. The main purpose of the circuit is to buffer the input signal from the load. The input impedance is high and the output impedance is low.

SUMMING AMPLIFIER

If several input resistors are connected to the inverting input of the operational amplifier, as shown in Figure 2-20, the result is an amplifier which sums the separate input voltages.

The output voltage of the summing amplifier circuit can be determined by the following equation:

$$V_O = -R4\left(\frac{V_I1}{R1} + \frac{V_I2}{R2} + \frac{V_I3}{R3}\right)$$

If feedback resistor R4 and input resistors R1, R2, and R3 are made equal, the output voltage can be determined by the following equation:

$$V_O = -(V_I1 + V_I2 + V_I3)$$

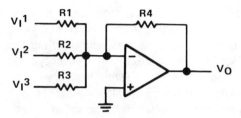

Figure 2-20. Basic Summing Amplifier

DIFFERENCE AMPLIFIER

In a difference amplifier circuit, input voltages V_I1 and V_I2 are applied simultaneously to the inverting and noninverting inputs of the operational amplifier (see Figure 2-21).

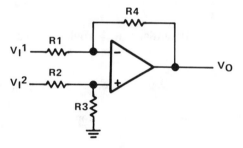

Figure 2-21. Basic Difference Amplifier Circuit

When all four resistors are equal, the output voltage is equal to the difference between V_I2 and V_I1. This circuit is called a unity-gain analog subtractor. Mathematically, the output voltage is stated as follows:

$$V_O = V_I2 - V_I1$$

DIFFERENTIATOR

The operational amplifier differentiator is similar to the basic inverting amplifier circuit except that the input component is a capacitor rather than a resistor (see Figure 2-22).

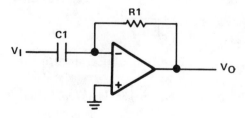

Figure 2-22. Basic Differentiator Circuit

The output voltage of the differentiator circuit can be determined by the following equation:

$$V_O = -R1C1 \frac{\Delta V_I}{\Delta t}$$

In this equation $\Delta V_I / \Delta t$ is the change in input voltage divided by a specified time interval. A problem with the basic differentiator circuit is that the reactance of input capacitor C1 ($1/2\pi fC1$) varies inversely with frequency. This causes the output voltage to increase with frequency and makes the circuit susceptible to high-frequency noise. To compensate for this problem, a resistor is connected in series with the capacitor on the inverting input (see Figure 2-23).

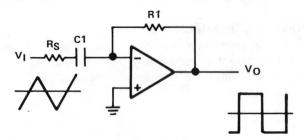

Figure 2-23. Differentiator with High Frequency Noise Correction

However, this circuit functions as a differentiator only on input frequencies which are less than those which can be determined by the following equation:

$$fC1 = \frac{1}{2\pi R1C1}$$

The time constant (R1C1) should be approximately equal to the period of the input signal to be differentiated. In practice, series resistor R_S is approximately 50 Ω to 100 Ω.

INTEGRATOR

An operational amplifier integrator circuit can be constructed by reversing the feedback resistor and input capacitor in a differentiator circuit (see Figure 2-24).

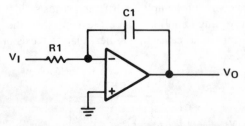

Figure 2-24. Basic Integrator Circuit

The resistor (R1) is the input component and the capacitor (C1) is the feedback component. However, if the low-frequency gain of the circuit is not limited, the dc offset

(although small), would be integrated and eventually saturate the operational amplifier. A more practical integrator circuit is shown in Figure 2-25.

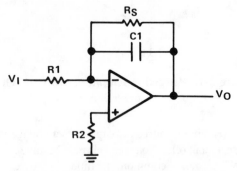

Figure 2-25. Typical Integrator Circuit

In this circuit, a shunt resistor (R_S) is connected across feedback capacitor C1 to limit the low-frequency gain of the circuit. The dc offset (due to the input bias current) is minimized by connecting resistor R2 between the noninverting input and ground. Resistor R2 is equal to the parallel combination of R1 and shunt resistor R_S. The shunt resistor helps limit the circuit low-frequency gain for input frequencies greater than those determined by the following equation.

$$fC1 = \frac{1}{2\pi R_S C1}$$

ACTIVE FILTERS

Filters are often thought of as discrete networks consisting of resistors, capacitors, and inductors (passive components). Because the components are passive, the energy from a passive filter is always less than the energy applied by the input signal. The attenuation (or insertion losses) limit the effectiveness of passive filters and make some applications impractical. However, resistors and capacitors can be combined with operational amplifiers to form active filters which operate without signal loss.

Depending upon the circuit type, low-pass filters as well as high-pass, bandpass, or band-reject filters can be designed with a roll-off characteristic of 6 to 50 dB or greater per octave. Some of the more common active filters that use operational amplifiers are discussed in the following paragraphs.

Unity-Gain Active Filters

The unity-gain active filter is the simplest to design. It combines an operational amplifier connected in a unity gain configuration with RC filter networks. It can be either a low-pass filter [Figure 2-26(a)], or a high-pass filter [Figure 2-26(b)], depending upon the positions of its discrete resistors and capacitors.

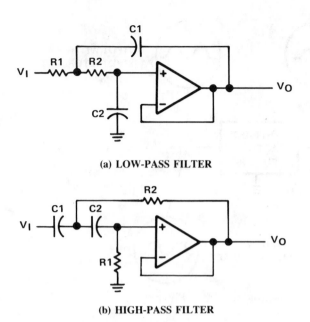

(a) LOW-PASS FILTER

(b) HIGH-PASS FILTER

Figure 2-26. High-Pass and Low-Pass Filter Circuits

The -3 dB (cutoff) frequency of the filter can be determined from the following equations:

Low-frequency cutoff $\quad f_o = \dfrac{1}{2\pi\ C2\sqrt{R1R2}}$

High-frequency cutoff $\quad f_o = \dfrac{1}{2\pi\ R2\sqrt{C1C2}}$

The Q of the circuit can be calculated using the following formulas for a low or high pass filter:

Low-pass filter $\quad\quad Q = 1/2\sqrt{C1/C2}$
High-pass filter $\quad\quad Q = 1/2\sqrt{R1/R2}$

These formulas are valid for a value of Q greater than 10.

Low-Pass Active Filters

Figure 2-27 illustrates the response curve typical of a low-pass active filter using a general-purpose operational amplifier. Outside the passband, the attenuation is computed at 12 dB per octave. However, at high frequencies the attenuation of the filter is less than predicted. In simple theory, the operational amplifier is considered to be perfect, and, for a typical general-purpose operational amplifier, this perfection proves to be acceptable up to 100 kHz. However, above 100 kHz the output impedance and other characteristics of the amplifier can no longer be ignored. The combined effect of these factors causes a loss of attenuation at high frequencies. General-purpose operational amplifiers are most effective in the audio frequency range. For higher frequency applications, a broad band amplifier such as the LM318 or TL291 should be used.

When the frequency spectrum of the input signal is especially wide, the high-frequency rejection characteristic must be considered. This is true when the input to the filter

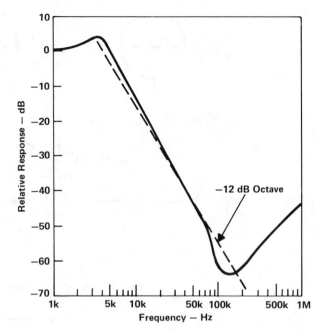

Figure 2-27. Response Curve of a Low-Pass Active Filter

is a rectangular signal. Figure 2-28 shows the response of a low-pass active filter to a 1-MHz square-wave signal.

The high-frequency-cutoff problem is resolved by using a simple RC filter ahead of the active filter. The combination of an RC filter and an active filter having superior low-frequency performance will significantly improve high-frequency cutoff. In addition, an impedance adapter should be inserted between the two filters shown in Figure 2-29.

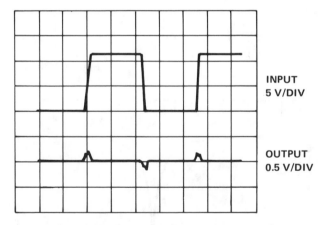

Figure 2-28. Response of a Low-Pass Active Filter to High Frequency Signals

Band-Reject Active Filters

In addition to the previously described functions, an active filter may be used to perform a band-reject function. A filter with a band-reject characteristic is frequently referred to as a notch filter. A typical circuit using a uA741 in unity-gain configuration for this type of active filter is shown in Figure 2-30.

The filter response curve shown in Figure 2-31 is a second-order band-reject filter with a notch frequency of

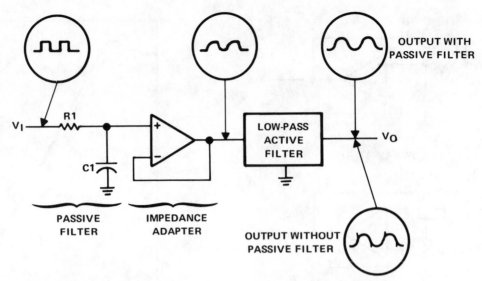

Figure 2-29. Use of a Passive Filter Preceding an Active Filter

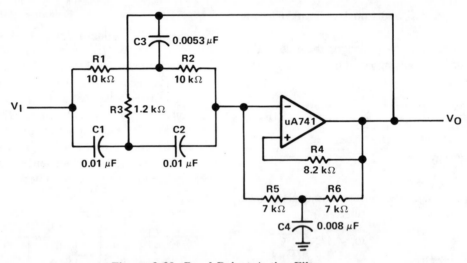

Figure 2-30. Band-Reject Active Filter

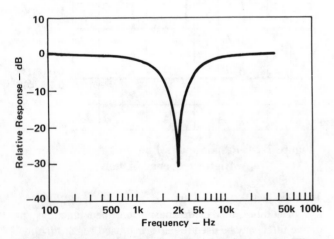

Figure 2-31. Response Curve of Band-Reject Active Filter

3 kHz. The resulting Q of this filter is about 23, with a notch depth of − 31 dB. Although three passive T networks are used in this application, the operational amplifier has become a sharply tuned low-frequency filter without the use of inductors or large-value capacitors.

Bandpass Active Filters

A bandpass filter permits a range of frequencies to pass while attenuating frequencies above and below this range. The center frequency (f_o) is the frequency at which the maximum voltage gain occurs. The bandwidth of this type of filter is the difference between the upper and lower frequencies at the points where the voltage gain is 0.707 times the maximum value, or 3 dB lower than the response at the center frequency. As shown in Figure 2-32 f_L is called the lower 3-dB frequency and f_H is called the upper 3-dB frequency. Bandwidth is determined by the following equation:

$$\text{Bandwidth} = f_H - f_L$$

The bandpass filter bandwidth and center frequency are related to each other by the Q, which is defined as follows:

$$Q = \frac{f_o}{f_H - f_L} \quad \text{or} \quad \frac{f_o}{BW}$$

Bandpass filter responses like those shown in Figure 2-32 can be built with operational amplifiers. The filter circuit shown in Figure 2-33 uses only one operational amplifier and is most often used for Q's of 10 or less.

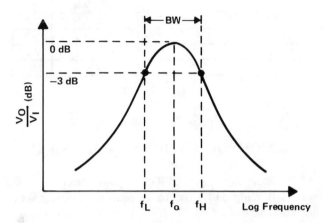

Figure 2-32. Active Bandpass Filter Response Curve

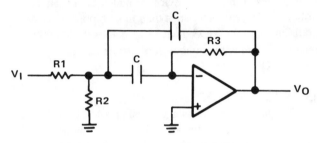

Figure 2-33. Active Bandpass Filter

Given the center frequency (f_0), the Q, and the desired gain (G), choose a convenient value of capacitance C. For typical audio filters, C is often 0.01 to 0.1 μF. The component values are easy to find from the following equations.

$$R1 = \frac{Q}{2\pi f_0 \, GC}$$

$$R2 = \frac{Q}{(2Q^2 - G)2\pi f_0 \, C}$$

$$R3 = \frac{2Q}{2\pi f_0 \, C}$$

Best performance is obtained when the gain is somewhat greater than the square root of the Q. For instance, if the filter is designed for a Q of 16, then the gain should be greater than four.

CHOOSING THE RIGHT OPERATIONAL AMPLIFIER

The operational amplifier, because of its versatility and ease of application, is the most widely used linear integrated circuit today. Because of this useful linear building block, many electronic circuits are much less complex. Due to the popularity of the operational amplifier, many different types are available that offer a variety of features. Which device to use for a specific application is a question that must be answered. If the characteristics of the selected device are not adequate, total system performance may be less than desired. If the selected device is too complex for the job, system cost may be increased unnecessarily. The following paragraphs provide a summary of the various types of operational amplifiers. To assist in the selection of the most effective operational amplifier for a specific application, the features and key applications are presented.

GENERAL-PURPOSE OPERATIONAL AMPLIFIERS — BIPOLAR

Since their inception, many operational amplifier designs have used bipolar transistors as the differential amplifier pair at the operational amplifier inputs. Because these input transistors operate from constant-current sources, an additional pair of matched transistors are used to obtain closely matched base-emitter voltages for a predictable ratio of currents for the constant-current generators. Phase-shift is controlled by frequency compensation that is internal to the amplifier. Amplfier phase-shift must be less than 135° at the frequency where the open-loop gain curve and closed-loop gain curve intersect. In bipolar operational amplifiers, the phase-shift is typically set with an internal capacitance of approximately 30 pF. The output stage should be designed to have a wide range of voltage swing with medium current capability.

The bipolar operational amplifier is usually operated in a class-B configuration. The key features of a bipolar operational amplifier are as follows:

Input impedance of $10^6 \, \Omega$.
Typical slew rates from 0.5 to 1 V/μs.
Typical unity-gain bandwidth of 1 MHz.
Noise levels of approximately 25 to 30 nV/$\sqrt{\text{Hz}}$.

Table 2-1 is a selection guide showing the major parameters to be considered in choosing bipolar operational amplifiers for a particular circuit design.

Table 2-1. Bipolar Operational Amplifier Comparison Chart

PARAMETER	DEVICE									UNITS
	OP-07	741	TL321	SE5534A	LM358	LM318	MC1458	RC4136	RC4558	
V_{IO}	30	1	2	0.5	2	2	1	0.5	0.5	mV
I_{IO}	0.5	20	5	10	5	30	20	5	5	nA
I_B	±1.2	80	45	400	45	150	80	40	40	nA
SR	0.2	0.5	0.5	6	0.5	70	0.5	1	1	V/μs
B1	0.6	1	1	10	1	15	1	3	3	MHz

†Test conditions are V_{CC} = ±15 V. All values are typical.

‡Unity-gain bandwidth

BIFET OPERATIONAL AMPLIFIERS

BIFET operational amplifiers combine JFET input transistors with bipolar transistors in a monolithic integrated circuit. The ion-implantation process used in making BIFET devices results in closely matched transistors. This permits true class-AB operation in the output stage which results in near zero crossover distortion and low total harmonic distortion.

In addition to high input impedance (10^{12} Ω) and input bias currents in the picoampere range, most BIFET operational amplfiers have slew rates of approximately 13 V/μs and a typical unity-gain bandwidth of 3 MHz. However, BIFET operational amplifiers have higher offset voltages and input noise than bipolar operational amplifiers.

Some BIFET operational amplifiers are power-adjustable. This allows the user to select (with an external resistor) the operating current levels. While this causes a tradeoff with power dissipation, it gives greater control over slew rate or signal bandwidth. An example of such a device is the TL066 BIFET operational amplifier. The TL066 can be adjusted for a no-signal supply current of 5 to 200 μA. Slew rate and bandwidth will also change depending upon the level of operating current. Except for the adjustable feature, the TL066 is similar to the TL061. The key application for power adjustable operational amplifiers is in battery-operated and telecommunication equipment where power consumption is an important factor. Table 2-2 is a selection guide listing the major parameters to be considered when choosing a BIFET operational amplifier for a particular application.

Table 2-2. BIFET Operational Amplifier Comparison Chart

PARAMETER	DEVICE SERIES				UNITS
	TL080	TL070	TL060	TL087	
V_{IO}	5	3	3	0.1	mV
I_B	30	30	30	60	pA
NOISE	25	18	42	18	nV/Hz
SR	13	13	3.5	13	V/μs
B1	3	3	1	3	MHz

Test conditions are V_{CC} = ±15 V. All values are typical.

LinCMOS™ OPERATIONAL AMPLIFIERS

The linear silicon-gate CMOS integrated circuit process was first developed by Texas Instruments and designated by the trademark LinCMOS. The LinCMOS technology combines the high speed of the bipolar device with the low power, low voltage, and high input impedance of the CMOS device. The LinCMOS device provides better offset and voltage swing characteristics than most bipolar devices. In addition, the LinCMOS device overcomes the stability and bandwidth limitations imposed on linear designs by metal-gate CMOS.

ULTRASTABLE OFFSETS

The primary disadvantage of using conventional bipolar metal-gate CMOS for linear applications is the unavoidable threshold-voltage shifts that take place with time and with changes in temperature and gate voltage. These shifts (caused by the movement of sodium ions within the device transistor) are frequently more than 10 mV/V of applied gate voltage. However, LinCMOS technology overcomes this problem by replacing the metal gates with phosphorus-doped polysilicon gates that bind the sodium ions. The result is linear integrated circuits with low (2 to 10 mV) input-offset voltages that vary no more than a few microvolts from their original values.

The TLC251 and TLC271 series of general-purpose operational amplfiers have low input offset voltages that typically vary only 0.1 μV per month and 0.7 μV per degree Celsius. The extremely low offset voltage can be reduced even further by using the offset null pins on the device. Unlike metal-gate CMOS devices, the input-offset voltage of LinCMOS devices is not sensitive to input-overdrive voltages.

WIDE BANDWIDTHS

In addition to providing stable offset voltages, LinCMOS technology produces integrated circuits with bandwidths that are two to three times wider than those of metal-gate CMOS devices. This occurs because the silicon gate in LinCMOS transistors is formed during the same processing step that forms the source and drain. As a result, the source, gate, and drain are self-aligned. In contrast, metal gates are formed after the source and drain regions are diffused, necessitating a built-in overlap to ensure source, gate, and drain alignment.

The self-aligned gate of LinCMOS transistors results in a gate-drain capacitance that is approximately one-seventh that of typical metal-gate CMOS integrated circuits. This enhances the bandwidth and speed of LinCMOS devices.

The TLC251 and TLC271 operational amplifiers offer a 2.3 MHz bandwidth, 60-ns rise-time with 25% overshoot, and a slew rate of 4.5 V/μs. These speeds are better than most bipolar operational amplifiers, approach those of BIFET operational amplifiers, and are several times faster than their metal-gate CMOS counterparts.

ADVANTAGES OF LinCMOS OPERATIONAL AMPLIFIERS

The TLC251 and TLC271 series operational amplifiers provide a low input offset voltage (10 mV maximum) that remains highly stable over time and temperature and is not sensitive to input-overdrive voltages. They are also available with tightened, guaranteed input offset voltages.

The TLC251 and TLC271 series operational amplifiers can be adjusted for low-, medium- or high-bias operation. This is accomplished by connecting the bias-select pin to V_{DD} for low bias, to ground for high bias, or to 1/2 V_{DD} for medium bias. By providing a choice of bias conditions, the TLC251 and TLC271 allow users to select between ac performance and power consumption to meet a wide range of circuit requirements. When operated in high-bias with V_{DD} equal to 10 V, these devices draw 100 μA of I_{DD} for 10-mW power dissipation and feature 4.5-V/μs slew rate and 2.3-MHz bandwidth. In the low bias mode with V_{DD} equal to 10 V and I_{DD} equal to 10 μA (100 μW power dissipation), the devices have a slew rate of 0.04 V/μs and a bandwidth of 100 kHz. In low-bias, and at 1 V, the TLC251 consumes just 10 μW making it the ideal choice for battery-operated applications. The bias-select pin can be driven with a logic signal from a microprocessor, allowing the operational amplifier performance to be software-controlled.

Additional features of the TLC251 and TLC271 include a common-mode rejection ratio of 88 dB and a low input-noise voltage of 30 to 70 nV/$\sqrt{\text{Hz}}$ (depending upon whether the device is operating in high, medium, or low bias).

These capabilities make the TLC251 and TLC271 suited for a wide range of applications. These applications include active filters, transducer interfacing, current drivers, voltage-to-current converters, long-interval timers, and many types of amplifiers. The TLC251 and TLC271 series are particularly suited for low-power designs and instrumentation amplifiers that require stable offsets.

When using the TLC251 or TLC271 LinCMOS devices for design, the following characteristics must be considered:
- Supply Voltage, V_{DD}
   TLC251 . . . . 1 V to 16 V
   TLC271 . . . . 4 V to 16 V
- True Single Supply or a Maximum of ±8 V
- Adjustable Supply Current, I_{DD}
   Low Bias = 10 μA typical
   Medium Bias = 150 μA typical
   High Bias = 1000 μA typical

- Extremely Low Input Bias and Offset
   Currents: 1 pA Typical
- Low Input Offset Voltage: 3 mV typical
- Ultra Stable Input-Offset Voltage:
   0.1 μV/Month Typical
- Noise: 30 nV/Hz Typical
- Slew Rate, SR
   High Bias . . . . 4.5 V/μs typical
   Medium Bias . . . . 0.6 V/μs typical
   Low Bias . . . . 0.04 V/μs typical
- Bandwidth, BW
   High Bias . . . . 2.3 MHz
   Medium Bias . . . . 0.7 MHz
   Low Bias . . . . 0.1 MHz

COMPARATORS

A basic comparator is similar to a differential amplifier operating in the open-loop mode. Because of high gain, the output is normally saturated in either the high state or the low state depending upon the relative amplitudes of the two input voltages. With these conditions, the comparator provides a logic-state output which is indicative of the amplitude relationship between two analog input signals.

In typical applications, a comparator provides an indication of the relative state of the two input signals. Figure 2-34 illustrates a basic comparator and its transfer function.

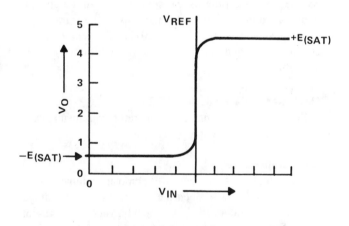

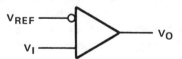

Figure 2-34. Basic Comparator and Transfer Function

In the circuit in Figure 2-34, if a reference voltage is applied to the inverting input and an unknown potential to the noninverting input, the output will reflect the relationship between the two inputs. When V_I is more negative that V_{REF}, the device output will be in saturation at a logic low level. When V_I becomes more positive than V_{REF}, the output will change states and become saturated at a logic high level.

Because comparators are normally used to drive logic circuits, the output must change states as rapidly as possible. High open-loop gain, wide bandwidth and slew rate are key factors in comparator speed. Operation in the open-loop mode (no feedback), with minimum or no frequency compensation, results in maximum gain-bandwidth product for best performance. Most comparators operate in this manner.

The ideal comparator has the same characteristics as the ideal operational amplifier. Those characteristics are as follows:

- Differential Gain = → ∞
- Common-Mode Gain = 0
- Input Impedance = → ∞
- Output Impedance = 0
- Bandwidth = → ∞
- Offset Voltage and Current = 0

Initially operational amplifiers were used in the open-loop mode to perform comparator functions. However, devices designed specifically for this operation resulted in improvements in recovery time, switching speed, and output levels. Since the comparator amplifier stage is usually followed by a TTL logic stage, output logic-state levels normally match those required by TTL loads.

Circuits designed as a comparator use none of the phase/frequency compensation usually required for operational amplifier stabilization with feedback. In fact, these compensation components are detrimental because they slow the response time of the comparator. Although any operational amplifier may be used as a comparator, a compensated device (such as the TL071) will result in longer response times and an output that is not directly TTL compatible.

COMPARATOR PARAMETERS

Some of the common comparator parameters are discussed in the following paragraphs.

Source Impedance

The input bias current of a bipolar comparator is approximatley 10 μA. When the differential input voltage makes the comparator switch, the input bias current is present at one of the inputs and is almost zero at the other. If the source impedances are not negligible, feedback will lower the gain of the comparator and create parasitic oscillations. Figure 2-35 shows this phenomenon occuring with a TL810. The comparator is driven by a ramp voltage at a rate of 1 mV/μs as represented by the center waveform. The upper output waveform shows the response of the comparator with a source impedance of 50 Ω. The lower output waveform represents the response of the same comparator with a source impedance of 10 kΩ. The initial switching occurs sooner with a high source impedance because the bias current characteristics produce an additional offset voltage. However, the subsequent oscillations make this circuit configuration unusable with low-slew-rate input signals.

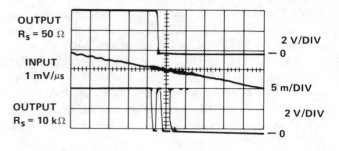

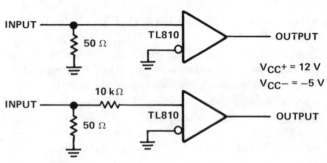

Figure 2-35. Influence of Source Impedance

Differential Voltage Gain

Differential voltage gain (A_{VD}) determines the sensitivity and threshold accuracy of a comparator. In the ideal comparator, the gain would be infinite and an extremely small voltage applied between the two inputs would cause a change in the output. In actual practice, the gain is not infinite and some minimum voltage variation at the input is required to obtain a change in the output. The ratio of the variation of output voltage to that of input voltage is the voltage gain of the comparator. The voltage gain of the comparator may be expressed by the following equation:

$$A_{VD} = \frac{\Delta V_O}{\Delta V_I}$$

The quantity ΔV_O (the difference between the high and low states of the output) is normally set at 2.5 V to ensure matching between the comparator and a TTL load. For example, if the TL810 has a minimum A_{VD} of 12,500, then (for an output swing of 2.5 V) $\Delta V_{I(min)}$ = 2.5 V/12,500, or 0.2 mV.

Output Characteristics

Although some comparators have a full TTL fanout capability of 10 or greater, others have a fanout that is limited to one TTL load. An evaluation of the output circuits should indicate the basic limiting factors and how maximum performance can be obtained.

In the active pull-down mode [see Figure 2-36(a)], the output low-level sink current (I_{OL}) is limited. The emitter of Q2 is clamped at one base-emitter voltage drop or −0.7 V with Q3 providing another base-emitter voltage drop. The resulting low-level output current (I_{OL}) may be calculated from the following equation:

$$I_{OL} = \frac{V_{CC} - 2\,V_{BE}}{1.77\ k\Omega}$$

$$= \frac{-6\ V + 1.4\ V}{1.77\ k\Omega} = -2.6\ mA$$

The resulting value is near the typical value for this device. The minus sign indicates a sink current. The corresponding $V_{OL} = V_E\ (Q2) + V_{CE(SAT)(Q2)}$ or $(-0.7\ V + 0.2\ V) = -0.5\ V$, which is the typical data sheet value.

In a logic low-level ouput state, the TL810 can handle one standard TTL gate with its maximum requirement of -1.6 mA. Increased fanout capability can be obtained by connecting an external resistor between the comparator output and the negative supply.

In the active pull-up mode, the typical high-level output voltage (V_{OH}) is 3.2 V for the TL810. The voltage at the base of the pull-up transistor [Q1, Figure 2-36(b)] is defined by the following equation:

$$V_{OH} + V_Z + V_{BE(Q1)}$$

Where:

$$V_Z = 6.2\ V$$
$$V_{BE(Q1)} = 0.7\ V$$

The base voltage is $(3.2 + 6.2 + 0.7)$, 10.1 V. The resulting base drive is determined by the following equation:

$$I_b = \frac{V_{CC+} - V_B}{R_b}$$

$$= \frac{12\ V - 10.1\ V}{3.9\ k\Omega}\ 0.488\ mA$$

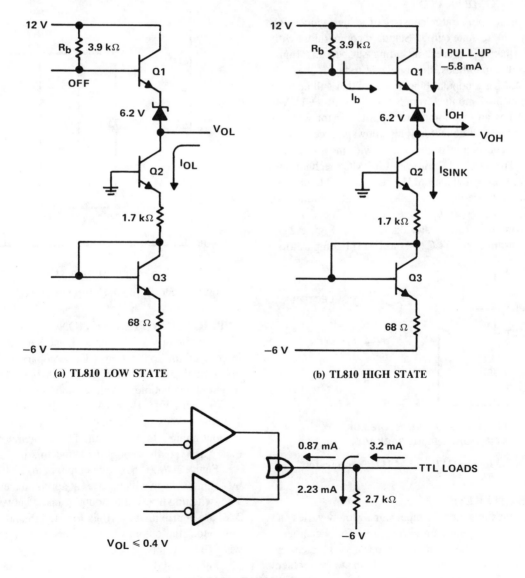

(a) TL810 LOW STATE

(b) TL810 HIGH STATE

$V_{OL} \leqslant 0.4\ V$

(c) TL811 FANOUT OF 2

Figure 2-36. Comparator Output Configurations

Assuming a typical saturated h_{FE} of 12, the resulting pull-up drive capability is (0.488 mA) (12) or 5.8 mA. Only part of the 5.8-mA drive is available to the external circuit. Since the current sink is not turned off during the logic-high output condition, the remainder of the current will be shunted through the pull-down circuit. For TL810, the resulting I_{OH} level available for external drive will be the difference between the pull-up drive of 5.8 mA and the pull-down sink of 2.6 and is 3.2 mA [Figure 2-36(c)]. The 3.2 mA is adequate because the logic high level (I_{OH}) required is only 40 μA per TTL load. Similar calculations for the TL811 comparator yield an I_{OL} level of 0.87 mA and an I_{OH} of 4.3 mA. For example, a fanout capability of 2 requires an I_{OL} level of 3.2 mA. With the TL811 [see Figure 2-36(c)], a 2.7-kΩ resistor is connected between the output and the negative supply. The resulting I_{OL} is 3.2 mA at a 0.4-V maximum V_{OL}. The effective I_{OH} capability, therefore, is reduced to 1.19 mA at a minimum V_{OH} of 2.4 V.

STANDARD COMPARATORS

The typical comparator consists of a high-gain stage followed by a logic-state output. Standard comparators with many performance features, including strobes and high-output current capabilities, are available today.

The LM311 is a popular bipolar device that will operate from single or dual supplies from 5 V to 30 V (or $\pm$ 15 V). The LM311 has an uncommitted output transistor with an available emitter and collector. This allows source or sink output drive. The output is compatible with most standard logic levels. The TLC311, built with LinCMOS technology, is an improved version of the LM311. The LinCMOS process allows common-mode input levels down to and including the negative V_{CC} rail or ground and the input impedance is increased from $10^6 \Omega$ to greater than $10^{12} \Omega$. Figure 2-37 is a basic diagram of the TLC311 and LM311 comparators.

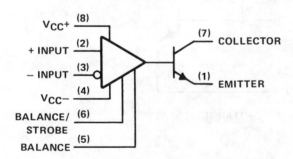

PIN NUMBERS SHOWN ARE FOR
8-PIN DUAL-IN-LINE PACKAGE.

Figure 2-37. TLC311 or LM311 Comparator

USE OF HYSTERESIS

Applications in which the input signal slowly varies can cause the ouput to change proportionally. This becomes a problem when the comparator is used to trigger a logic stage requiring fast rise and fall inputs. One solution to the problem is the introduction of positive feedback. This causes a fast or Schmitt trigger action. This action is accomplished by

feeding a portion of the output signal back to the noninverting input. Depending upon the amount of positive feedback, a new trip-level will be introduced after each transition. The result is two (rather than one) threshold points. These are called the upper threshold point (UTP) and lower threshold point (LTP). the difference between these two points is the hysteresis. A comparator with hysteresis is shown in Figure 2-38.

A typical hysteresis loop diagram for this type of circuit is shown in Figure 2-39.

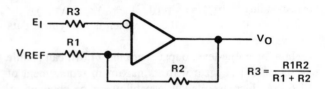

Figure 2-38. Comparator with Hysteresis

$$R3 = \frac{R1R2}{R1 + R2}$$

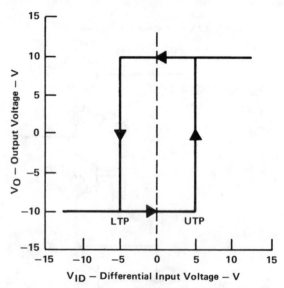

Figure 2-39. Typical Comparator Hysteresis Loop

APPLICATION PRECAUTION

The rise time of the input signal is a critical parameter in comparator applications. The comparator is basically a differential amplifier with very high open-loop gain. The output is compatible in voltage and current with the inputs of TTL circuits. However, this type of logic requires switching times of less than 150 ns to function correctly without going into oscillation. The comparator input signal must vary rapidly enough to avoid this problem.

Figure 2-40(a) shows the output of a TL710 being driven by a ramp voltage which varies at approximately 0.1 mV/μs. The switching times of the output, taken between 0.8 V and 2 V, are approximately 10 μs for the fall and rise times. In this mode, the output of the comparator is not compatible with TTL circuits.

Figure 2-40(b) shows a TL810 under the same conditions as described for the TL710. With a higher gain than the TL710, the switching speed for the TL810 is also higher,

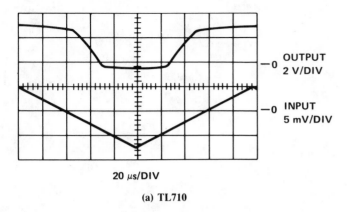

(a) TL710

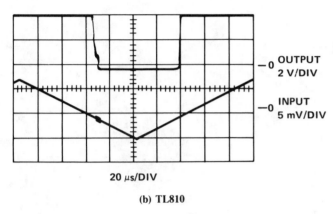

(b) TL810

Figure 2-40. Response of a TL710 and a TL810 to a Ramp Input

When these input conditions are not being met, some positive feedback must be added or a Schmitt trigger configuration must be designed (see Figure 2-41) to accelerate the switching speed. However, the resulting hysteresis makes the comparator less voltage sensitive.

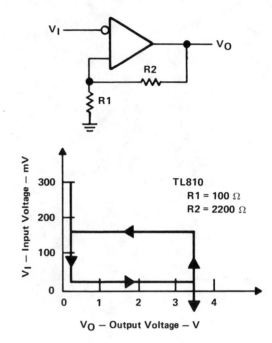

Figure 2-41. Use of Hysteresis to Prevent Oscillations

and the rise time is compatible with TTL circuits. However, some oscillation is present during the periods of switching. This occurs because the input signal remains in the high-gain linear range of the comparator for an excessive period of time. For the output of a comparator to be good, the input must force the output to vary between 0.8 V and 2 V in 150 ns or less.

When the minimum gain (A_{VD}) of the comparator is known, the input signal must vary at a minimum rate determined by the following equation:

$$\frac{2\text{ V} - 0.8\text{ V}}{150\text{ ns} \times A_{VD}}$$

For the TL710, the minimum rate is determined as follows:

$$\frac{2\text{ V} - 0.8\text{ V}}{150\text{ ns} \times 500} = 16\text{ mV}/\mu\text{s}$$

For the TL810, the minimum rate is determined as follows:

$$\frac{2\text{ V} - 0.8\text{ V}}{150\text{ ns} \times 8000} = 1.0\text{ mV}/\mu\text{s}$$

Section 3

Operational Amplifier and Comparator Applications

OPERATIONAL AMPLIFIER APPLICATIONS

GENERAL APPLICATIONS

This section contains information on several specific circuits which will assist the reader in designing circuits which can be applied to a variety of applications. The specific functions which these operational amplifiers perform include: amplification, measurement, control, sensing, and regulation. Although specific applications are given for each operational amplifier circuit, sufficient information is presented to allow the design of circuits for additional applications. The specific applications circuits discussed in this section are as follows:

- Optical Sensor to TTL Interface
- Bridge-Balance Indicator
- High-Input-Impedance Differential Amplifier
- Low-Voltage Shunt Limiter
- Schmitt Triggers
- PTC Thermistor Automotive Temperature Indicator
- Accurate 10-Volt Reference
- Precision Large Signal Voltage Buffer

OPTICAL SENSOR TO TTL INTERFACE CIRCUIT

This optical sensor to TTL interface circuit is designed to detect a low light level at the sensor, amplify the signal, and provide a TTL-level output. When the optical sensor detects low-level light (ON condition), its output is small and must be amplified. Because of this small output, an operational amplifier with very low input bias current and high input resistance must be used to detect the ON condition of the sensor and provide an amplified output.

Figure 3-1 is an optical sensor to TTL interface circuit. When connected as shown in the figure, operational amplifier OP-07 meets all of the previously stated characteristics. When sensor TIL406 is in the ON condition, its output is assumed to be 250 nA (allowing a safety margin). This results in a 250-mV signal being applied to the noninverting input of amplifier OP-07. Because of the circuit configuration, the OP-07 provides a gain of 100 and its output is in positive saturation. The OP-07 output level is applied to a loading network that provides the basic TTL level. Because an optoelectronic circuit may operate at slow speeds, it may be

CONDITION	TIL406 CURRENT	OUTPUT LOGIC
LIGHT ON	≈ 3 μA	0
LIGHT OFF	≈ 3 nA	1

necessary to connect an SN7513 Schmitt-trigger device on the output of the load network to shape the TTL output signal.

BRIDGE-BALANCE INDICATOR

A bridge-balance indicator provides an accurate comparison of two voltages by indicating their degree of balance (or imbalance).

A common bridge, referred to as a Wheatstone bridge, consists of two impedance divider networks as shown in Figure 3-2. Generally one side of the bridge consists of known impedances (one of which is usually adjustable) resulting in a known voltage (E1). The other side of the bridge may consist of combined known and unknown impedances resulting in an unknown voltage E2. The unknown impedance might be varied by a system condition (speed, temperature, etc.). A sensitive voltmeter is used to detect the balance of F1 and E2. For example, the bridge may be used to control motor speed, temperature, or physical position. The accuracy of this type of control is dependent on how close to zero difference, or the null point a deviation can be detected.

Detecting small variations near the null point is difficult with the basic Wheatstone bridge alone. Amplification of voltage differences near the null point will improve circuit accuracy and ease of use. The bridge-balance indicator circuit (Figure 3-3) replaces the meter shown in Figure 3-2 and provides the gain required near the null-point for improved accuracy. An OP-07 operational amplifier is used as the gain element in this circuit.

In this application, the 1N914 diodes in the feedback loop result in high sensitivity near the point of balance (R1/R2 = R3/R4). When the bridge is unbalanced the amplifier's closed-loop gain is approximately R_F/r, where r is the parallel equivalent of R1 and R3. The resulting gain equation is $G = R_F(1/R1 + 1/R3)$. During an unbalanced condition the voltage at point A is different from that at point B. This difference voltage (V_{AB}), amplified by the gain factor G, appears as an output voltage. As the bridge approaches a balanced condition (R1/R2 = R3/R4), V_{AB} approaches zero. As V_{AB} approaches zero the 1N914 diodes in the feedback loop lose their forward bias and their resistance increases, causing the total feedback resistance to increase. This increases circuit gain and accuracy in detecting a balanced condition. Figure 3-4 shows the effect of approaching balance on circuit gain. The visual indicator used at the output of the OP-07 could be a sensitive voltmeter or oscilloscope.

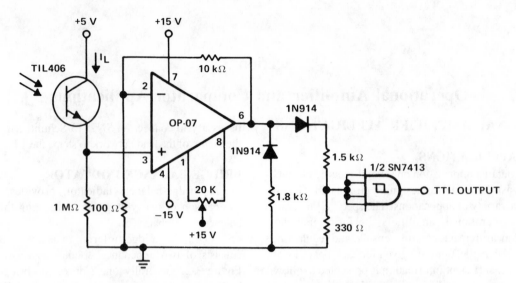

Figure 3-1. Optical Sensor to TTL Interface Circuit

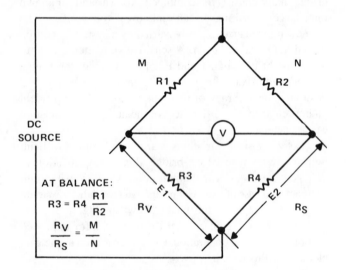

Figure 3-2. Wheatstone Bridge Circuit

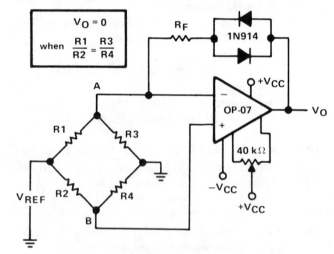

Figure 3-3. Bridge-Balance Indicator Circuit

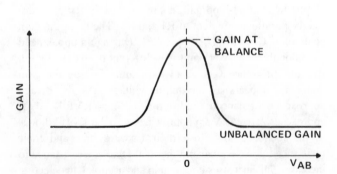

Figure 3-4. Gain as a Function of V_{AB}

HIGH-INPUT-IMPEDANCE DIFFERENTIAL AMPLIFIER

One of the most useful applications of an operational amplifier is the differential-input dc amplifier configuration shown in Figure 3-5.

Operational amplifiers A1 and A2 are connected in a noninverting configuration with their outputs driving amplifier A3. Operational amplifier A3 could be called a subtractor circuit which converts the differential signal floating between points X and Y into a single-ended output voltage. Although not mandatory, amplifier A3 is usually operated at unity gain and R4, R5, R6 and R7 are all equal.

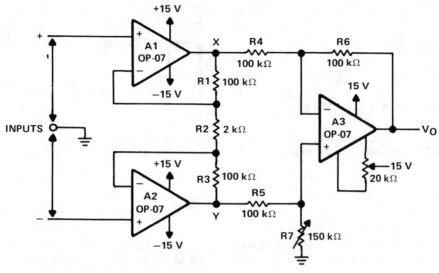

Figure 3-5. High-Input-Impedance Differential Amplifier

The common-mode-rejection of amplifier A3 is a function of how closely the ratio R4:R5 matches the ratio R6:R7. For example, when using resistors with 0.1% tolerance, common-mode rejection is greater than 60 dB. Additional improvement can be attained by using a potentiometer (slightly higher in value than R6) for R7. The potentiometer can be adjusted for the best common-mode rejection. Input amplifiers A1 and A2 will have some differential gain but the common-mode input voltages will experience only unity gain. These voltages will not appear as differential signals at the input of amplifier A3 because, when they appear at equal levels on both ends of resistor R2, they are effectively canceled.

This type of low-level differential amplifier finds widespread use in signal processing. It is also useful for dc and low-frequency signals commonly received from a transducer or thermocouple output, which are amplified and transmitted in a single-ended mode. The amplifier is powered by ±15-V supplies. It is only necessary to null the input offset voltage of the output amplifier A3.

LOW VOLTAGE SHUNT LIMITER

In some circuits, it is necessary to symmetrically limit or clip the peak output voltage of an amplifier stage. Limiting may be required to prevent overdriving a following amplifier stage. This type of circuit is also used in volume compressor and amplitude leveler designs.

The limiting function may be accomplished by several methods. Figure 3-6 shows a simple back-to-back diode limiter.

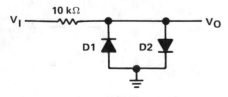

Figure 3-6. Simple Back-to-Back Shunt Limiter

If standard small signal diodes such as 1N914 are used for D1 and D2, ±0.6 V would be available at the limiter output. If germanium diodes are used for D1 and D2, ±0.4 V will be the output voltage. Limiting may also be accomplished utilizing resistor/zener diode networks. The low-voltage shunt limiter shown in Figure 3-7 is useful when the signal level must be limited at a very low level, such as several hundred millivolts. Such levels are, of course, below the range of conventional diodes, so alternate methods are necessary to accomplish limiting at these levels.

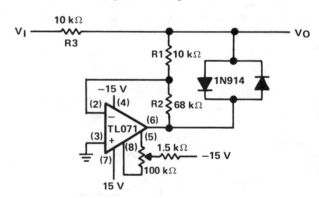

Figure 3-7. Low-Voltage Shunt Limiter

In this circuit, the operational amplifier is used to shift the apparent threshold of a conventional 1N914 silicon diode. The output voltage limit can be adjusted to any fraction of the diode voltage. This circuit divides the signal level, even when below the threshold, because of the feedback through resistor R2. As an example, assume a 10-V peak-to-peak input sine-wave signal at a frequency of 1 kHz. With the values shown the output will be a 150-mV peak-to-peak square wave as shown in the scope photos in Figure 3-8.

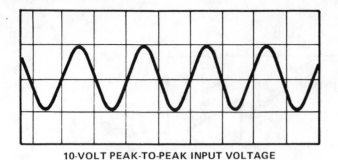

10-VOLT PEAK-TO-PEAK INPUT VOLTAGE

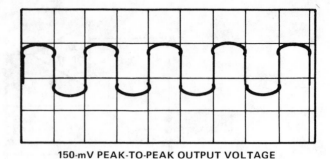

150-mV PEAK-TO-PEAK OUTPUT VOLTAGE

Figure 3-8. Input/Output Voltage Waveforms of Low-Voltage Shunt Limiter

SCHMITT TRIGGERS

A Schmitt trigger may be defined as a comparator with positive feedback, or hysteresis. With an analog input signal, the output is a squarewave or logic output. The hysteresis may be made large or small to prevent input noise, or the unwanted portion of the input signal, from appearing at the output. These types of circuits are very useful as the input section to data line receivers. They may also be used as level detectors, threshold detectors, pulse generators, and pulse shapers.

A voltage-follower Schmitt trigger can be constructed by connecting a positive feedback loop around an operational amplifier as shown in Figure 3-9. The feedback bias is connected to the noninverting input, maintaining it as a voltage level proportional to the output which is latched at either the V_{CC+} or V_{CC-} supply level. An input voltage V_I may be applied to overcome the feedback and force the output to the opposite polarity. The sum of the absolute values of the input voltages required to switch states is the hysteresis. With $V_I = 0$, the voltage at the noninverting input terminal is $\pm V_{O(max)} R1/(R1 + R2)$. An input voltage opposite in polarity and slightly greater in magnitude than $V_{O(max)} R1/(R1 + R2)$ is required to switch states. The hysteresis is therefore $2\,V_{O(max)} R1/(R1 + R2)$.

The hysteresis curve in Figure 3-9 illustrates the transfer characteristics of a typical operational amplifier.

The inverting Schmitt trigger, shown in Figure 3-10, has several advantages over the noninverting Schmitt trigger. The primary advantage is that the signal input current, because it flows only through R3, is independent of the output and feedback. Therefore, the input may be from a current or voltage source and must generate only enough voltage across R3 to result in switching. The input thresholds are $\pm V_{O(max)} R2/(R1 + R2)$. This circuit may be used to convert analog signals to standard logic levels. A positive voltage pulse of sufficient amplitude will switch the output negative. It will remain negative until a negative input pulse is applied causing the output to switch back to the positive state. This is similar to the operation of a set-reset type flip-flop. Operational amplifiers respond slowly for some switching applications. If additional speed is required, use externally compensated amplifiers. Very low value (1 to 2 pF) frequency-compensation capacitors will enhance switching speeds considerably and allow data rates as high as 1 MHz. A device recommended for this application is the TL070.

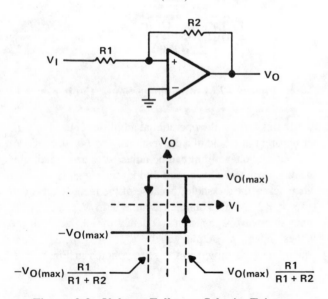

Figure 3-9. Voltage-Follower Schmitt Trigger

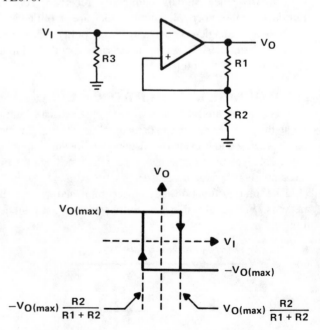

Figure 3-10. Inverting Schmitt Trigger

PTC THERMISTOR AUTOMOTIVE TEMPERATURE INDICATOR

To reach maximum efficiency, present day automobiles require many control methods. For example, temperature control of engine parts and fluids is essential. However, accurate electronic temperature measurements are not simple. Of the variety of thermocouples, resistance sensors, and thermistors available, the positive-temperature-coefficient silicon thermistor is an excellent choice for this application. Planar technology using the spreading-resistance principle allows this integrated circuit to be built. The TSP102 has a positive resistance temperature coefficient of 0.7%/ °C and has very close resistance tracking from unit to unit. Nominally it is about 1 kΩ resistance at 25 °C and changes from 500 Ω at −40 °C to 1900 Ω at 120 °C.

The example circuit (Figure 3-11) is used to indicate two different water temperature trip points by turning on LEDs when the temperatures are reached. The circuit is constructed around the LM2904 dual operational amplfier which was designed mainly for the automotive industry.

The circuit is powered from the 12-V auto system. The 1N5239 zener diode supplies a regulated 9.1 V to operate the circuit. The thermistor is in series with a 10-kΩ resistor from ground to the positive 9.1-V point. The top of the thermistor is tied to both noninverting inputs of the LM2904. The voltage at these inputs will change as the thermistor resistance changes with temperature. Each inverting input on the LM2904 has a reference, or threshold trip point, set by a 10-kΩ resistor and a 2-kΩ potentiometer in series across the 9.1-V regulated voltage. When this threshold is exceeded on the noninverting input of LM2904, the TIL220 LED lights. In this circuit, the FAN ON trip point was set at 70 °C. This occurs with approximately 1.3 V on the inverting input of the top section of the LM2904. The OVERHEAT trip point was set at 95 °C. This condition exists when the bottom section of the LM2904 has approximately 1.44 V on the inverting input. The two trip points can be recalibrated or set to trip at different temperatures by adjusting the 2-kΩ potentiometer in each section. In addition to being used as warning lights as shown here, circuits can be added to turn on the fan motor or activate a relay.

Other types of thermistors and temperature sensors manufactured by other companies can be used with this type of circuit.

ACCURATE 10-VOLT REFERENCE

A stable 10-V reference is a valuable asset for calibrating oscilloscopes and other laboratory equipment. The 10-V reference was selected because it can be used in decade fashion (multiplied or divided by 10). One of the major requirements for a laboratory reference is not only initial accuracy but long-term stability. This requires precision low-drift components. An OP-07 bipolar operational amplifier was chosen because it has low offset and long-term stability. The offset voltage drift is approximately 0.3 µV/ °C. The OP-07 is excellent because of its low noise and high-accuracy amplification of very low-level signals. Figure 3-12 illustrates an accurate 10-V reference circuit.

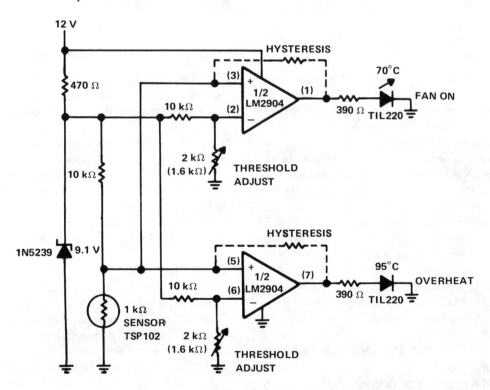

Figure 3-11. PTC Sensor Automotive Temperature Control Circuit

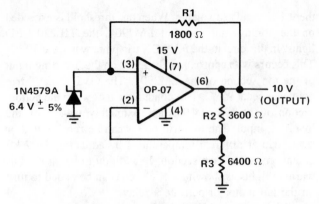

Figure 3-12. Accurate 10-Volt Reference Circuit

The accuracy of the circuit can be enhanced by using precision resistors. The 1N4579A zener diode was chosen because of its 0.0005%/°C temperature coefficient. The resistor values were calculated from the following formulas:

$$R1 = \frac{10\text{ V} - V_Z}{2 \times 10^{-3}}$$

Where:

$$V_Z = \text{Zener Voltage}$$

$$R2 = \frac{10\text{ V} - V_Z}{1 \times 10^{-3}}$$

$$R3 = \frac{V_Z}{1 \times 10^{-3}}$$

Assuming a zener diode voltage of 6.4 V, resistors R2 and R3 total 10 kΩ from the 10-V output to ground. The values of R2 and R3 are calculated to have 6.4 V between their junction and ground. This voltage is applied to the inverting input of the OP-07. Resistor R1 has 0.002 A of current and a 3.6 V drop across it, hence a value of 1800 Ω. This establishes a stable reference at the noninverting input of the OP-07. If the output voltage moves either higher or lower, the operational amplifier holds it at 10 V. By using the recommended components good long-term stability at the desired output voltage can be expected. If other voltages are needed, they can be calculated with the same formulas. However, the output voltage can never be lower than the reference zener diode voltage. To compensate for zener and other component variations, a multiturn potentiometer may be used at the junction of R2 and R3.

PRECISION LARGE-SIGNAL VOLTAGE BUFFER USING AN OP-07

A voltage follower may be defined simply as a unity-gain noninverting amplifier. There is no feedback resistance. A basic voltage follower is shown in Figure 3-13.

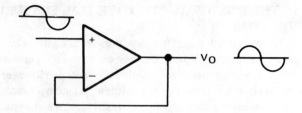

Figure 3-13. Basic Voltage-Follower Circuit

For the basic voltage-follower circuit, the output voltage is an exact reproduction of the input voltage. The input impedance is equal to the operational amplifier intrinsic input impedance. The output impedance for the voltage-follower circuit is, for all practical purposes, equal to the operational amplifier output impedance.

The primary purpose of the voltage-follower circuit is to buffer an input signal from its load. The input impedance is high and the output impedance is low. The function of an operational amplifier voltage follower is identical to the cathode, emitter, and source followers for vacuum tubes, bipolar transistors, and field-effect transistors, respectively. A complete voltage-follower circuit using an OP-07 is shown in Figure 3-14.

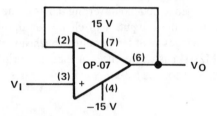

Figure 3-14. Precision Large-Signal Voltage Follower

The circuit in Figure 3-14 is a precision large-signal voltage buffer with a worst-case accuracy of 0.005%. This accuracy is possible because of the ultralow input-offset voltage of the OP-07 and the total absence of external components. With ± 15-V power supplies, an input signal with a ± 14-V swing can be used. Under these conditions, a peak-to-peak output voltage swing of ± 13 V can be expected. The OP-07 is an extremely stable device for use in this type of circuit.

ACTIVE FILTER APPLICATIONS

INTRODUCTION

Filters are frequently thought of as passive networks consisting of resistors, capacitors, and inductors. Therefore the energy out of a passive filter is always less than the energy applied to the input. Attenuation and insertion losses limit the effectiveness of passive filters and make some applications impractical. However, resistors and capacitors can be combined with operational amplifiers to form active filters.

In an active filter the amplifier may add energy to the system, resulting in both filtering and some power gain. Other advantages of active filters include low output impedance, cascaded stages without gain loss, and the capability of generating filtering functions having relatively high Q at low frequencies without inductors. In low-frequency applications, the inductors required for passive filters are generally cumbersome as well as difficult and costly to build. On the other hand, only a few easily used components are needed for active filters in these applications.

Depending on the circuit type; low-pass, high-pass, band-pass, or band-reject filters can be designed with a roll-off characteristic from 6 to 50 dB or greater per octave. An active filter offers several advantages over a passive (LC) filter.

1. No insertion loss. The op amp can provide gain if needed.
2. Cost. Active filter components are more economical than inductors.
3. Tuning. Active filters are easily tuned and adjusted over a wide range without altering the desired response.
4. Isolation. Active filters have good isolation due to their high input impedance and low output impedance, assuring minimal interaction between the filter and its load.

ACTIVE BAND-PASS FILTER

A band-pass filter passes a specific range of frequencies while preventing the passage of all others. Passive LC band-pass filters have been used for many years but the tuning procedures are difficult. These difficulties include removing or adding turns to the inductor and the need for specific capacitor values. Although a passive LC filter uses only inductors and capacitors, an active band-pass filter uses an op amp plus a few resistors and capacitors, but it is physically

smaller and requires less PC board space. The primary disadvantage of using an active filter is that it requires a power supply for the op amp.

The circuit in Figure 3-15 is a two-pole active filter using a TL081 op amp. This type of circuit is usable only for Qs less than 10. The gain of this stage is nominally or slightly larger than the square root of the Q. For example, with a Q of five, the gain chosen would be slightly over two.

The component values for this filter are easily calculated from the following equations. Assume you want to build a filter with a center frequency of 800 Hz. R2 is a potentiometer about twice the calculated value, which is then adjusted to set the resonant frequency precisely. This larger than calculated resistance value is used to compensate for the tolerance of the other resistor and capacitor values. A capacitor value of 0.01 μF to 0.1 μF is often used for filters in the audio range.

$$R1 = \frac{Q}{2\pi fGC}$$

$$R2 = \frac{Q}{(2Q^2 - G)\,2\pi fC}$$

$$R3 = \frac{2Q}{2\pi fC}$$

$$R4 = R3$$

Where

$$f = \text{filter center frequency (800 Hz)}$$
$$Q = 5$$
$$G = 2 \text{ (Gain)}$$
$$C = 0.01 \ \mu F$$

$$R1 = \frac{Q}{2\pi fGC}$$

$$= \frac{5}{(6.28)\,(800)\,(2)\,(0.01 \times 10^{-6})}$$

$$= 49,761 = 50 \ k\Omega$$

$$R2 = \frac{Q}{(2Q^2 - 2)\,2\pi fC}$$

$$= \frac{5}{(50 - 2)\,(6.28)\,(800)\,(0.01 \times 10^{-6})}$$

$$= 2073 = 2.2 \ k\Omega$$

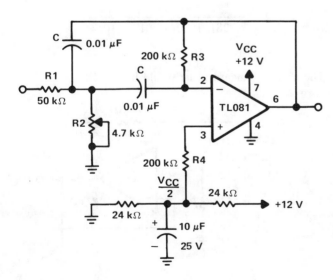

Figure 3-15. Two Pole Active Band-Pass Filter

$$R3 = \frac{2Q}{2\pi fC}$$

$$= \frac{10}{(6.28)(800)(0.01 \times 10^{-6})}$$

$$= 199,045 = 200 \text{ k}\Omega$$

$$R4 = R3 = 200 \text{ k}\Omega$$

Figure 3-16 shows the frequency response of this filter.

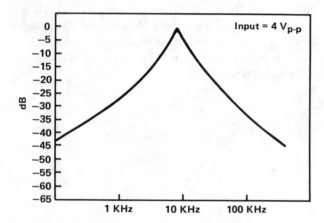

Figure 3-16. Active Band-Pass Filter Response Curve

HIGH- AND LOW-PASS ACTIVE FILTERS

Among the many types of filter circuits, the "Butterworth" filter is often the best overall choice because it has the flattest passband and reasonable overshoot. It also has a characteristic that sets all cascaded sections to the same frequency, which makes voltage control and wide-range tuning easier. Complex filters are normally built using first and second-order networks. A first-order section is not very useful by itself as a filter because all you can control is the center frequency and impedance level. In a second-order section, it is possible to control the impedance level, the center frequency, and another feature called damping, or its inverse, Q. Damping, or Q, sets the peaking or drop of the response near the cutoff frequency.

The simplest second-order low-pass filter is the voltage-controlled-voltage-source (VCVS) circuit shown in (Figure 3-17). In this circuit the capacitors have very little effect at low frequencies, which results in an essentially flat low frequency response. At high frequencies the capacitors separately shunt the signal to low-impedance points, one to ground and one to the output. This two-step shunting causes the response at high frequencies to fall off as the square of the frequency; hence the name, second-order section. The performance starts out flat at low frequencies and falls at 12 dB/octave, or 40 dB/decade past the cutoff frequency.

The most common approach for selecting the values of the two resistors and capacitors in the filter section is to make

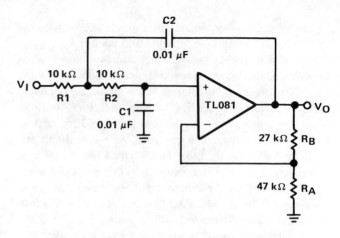

Figure 3-17. Second Order Low-Pass Filter

R1 and R2 equal, and C1 and C2 equal. The cutoff frequency is simply:

$$f_o = \frac{1}{2\pi RC}$$

This is called an "equal-component" low-pass filter. The passband gain is fixed at 1.586 (+4 dB) for a second-order Butterworth response, and is the only gain that will permit this circuit to function properly. The cutoff point will be 3 dB from the passband gain of +4 dB, 4+1 dB.

Since the op amp is in the noninverting mode, the feed back resistor R_B must be 0.586 times the value of the input resistor R_A for a voltage gain of 1.586. To build a low-pass Butterworth filter with a cutoff frequency of 1500 Hz you choose the component values in the following manner. Let $R_A = 47$ kΩ. R_B would be R_A X 0.586 or about 27 kΩ. If we let the capacitor value be 0.01 μF, the resistors will be selected by the formula:

$$R1 = \frac{1}{2\pi f_o C}$$

$$= \frac{1}{(6.28)(1500)(0.01 \times 10^{-6})}$$

$$= 10,617 \ \Omega = R2$$

The nearest standard value would be 10 kΩ.

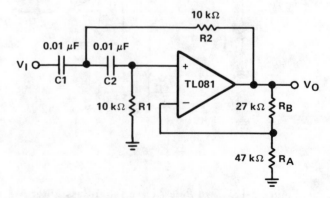

Figure 3-18. Second Order High-Pass Filter

Simply interchanging the position of the resistors and capacitors as shown in Figure 3-18 produces a high-pass active filter with the same cutoff frequency. The passband gain is also 1.586, or +4 dB. Figure 3-19 shows the response curves of both of these filters.

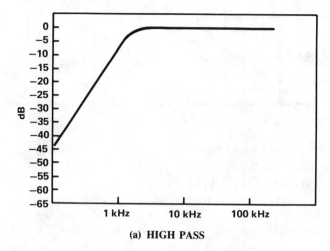

(a) HIGH PASS

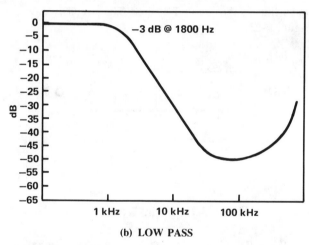

(b) LOW PASS

Figure 3-19. Active Filter Response

MULTIPLE-FEEDBACK BAND-PASS FILTERS

The basic multiple-feedback band-pass filter is useful for Qs up to about 15 with "moderate" gain. Band-pass circuits normally have lower damping and higher Q values than the usual low-pass or high-pass responses. In fact, these circuits are progressively harder to build and tune as the damping goes down and the Q goes up. Experience has shown that a high-performance, high-Q band-pass active filter cannot be built with a single op amp. Component-tolerance problems, sensitivity problems, or severe gain restrictions provide insurmountable barriers as you try to increase the circuit Q of single op amp circuits beyond a certain point. Therefore, single op amp versions of this filter may be used only for low-Q applications (Qs in the 2 to 5 range). Fortunately, Q values of 2 to 5 are ideal for many audio applications, including equalizers and tone controls. Higher Q circuits find use at IF and RF frequencies.

Figure 3-20 shows a single-stage, multiple-feedback band-pass filter where the op amp is connected in the inverting mode. Resistor R3 from the output to the inverting input sets the gain and the current through the frequency-determining capacitor, C1. Capacitor C2 provides feedback from the output to the junction of R1 and R2. C1 and C2 are always equal in value. Resistor R2 may be made adjustable in order to adjust the center frequency which is determined from:

$$fo = \frac{1}{2\pi C} \left[\frac{1}{R3} \times \frac{R1 + R2}{R1R2} \right]^{1/2}$$

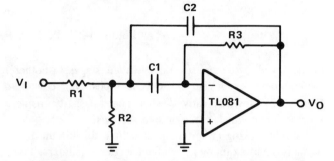

Figure 3-20. Single Stage Feedback Band-Pass Filter

When designing a filter of this type it is best to select a value for C1 and C2, keeping them equal. Typical audio filters have capacitor values from 0.01 μF to 0.1 μF which will result in reasonable values for the resistors. We will design a filter for 10 kHz and assume a Q of 3 and a stage gain of 2. The three resistors values are then determined from the following equations:

$$R1 = \frac{Q}{2\pi fCG} = 2388.5$$

or 2.4 kΩ (nearest standard value)

$$R2 = \frac{Q}{2\pi fC (2Q^2 - G)} = 298.5$$

or 300 Ω (nearest standard value)

$$R3 = \frac{Q}{\pi fC} = 9554$$

or 10 kΩ (nearest standard value)

Where

$$G = 2$$
$$Q = 3$$
$$C = 0.01 \ \mu F$$
$$f = 10 \ kHz$$

As previously stated, a single-stage active filter of this type results in low Qs (2 to 5). Filters which provide a very narrow passband must have a much higher Q than possible with a single section using one op amp. This may

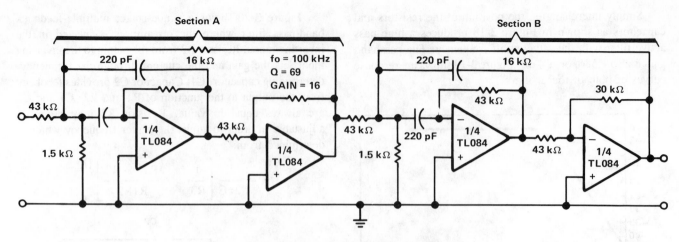

Figure 3-21. Positive Feedback Band-Pass Filter

be achieved by using several cascaded stages. Another method to achieve higher Q is the use of positive feedback. Figure 3-21 shows a positive-feedback band-pass filter using four op amps which make up two sections.

While looking rather complex at first, this circuit may be analyzed by examining each filter section separately. The complete filter is comprised of two identical sections, section "A" and section "B". Each section uses an op amp connected as a multiple-feedback band-pass filter as we have described in the preceding paragraphs. This is followed by a second op amp used as a phase inverter to achieve positive feedback to the input of the first op amp. This stage has a gain of only 0.7. While the 16 kΩ positive feedback resistor gives us a gain of about 10.7, it is reduced to about 7.5 due to the 0.7 gain of the phase inverter stage. The resulting overall gain of section "A" is 4.

When section B is cascaded to section A we have the complete two stage (4 op amp) filter with an overall gain of 16 and Q of 69. The scope photo in Figure 3-22 shows the bandwidth of both stages cascaded. The measured bandwidth with an f_O of 100 kHz is 2.3 kHz at the -3 dB or half-power points.

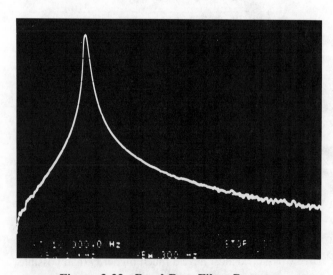

Figure 3-22. Band-Pass Filter Response

TWIN-T NOTCH FILTER

A notch filter is used to reject or block a frequency or band of frequencies. These filters are often designed into audio and instrumentation systems to eliminate a single frequency, such as 60 Hz. Perhaps the best-known passive notch filter is the "twin-T" filter. The circuit is shown in Figure 3-23.

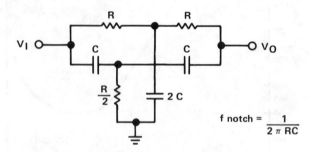

$$f \text{ notch} = \frac{1}{2 \pi RC}$$

Figure 3-23. Twin-T Notch Filter

If the six components are carefully matched, theoretically you can obtain an almost infinite rejection at the null frequency. Commercial grade components (5%—10% tolerance) produce a null depth of at least 30 to 40 dB.

When this twin-T network is combined with a TL081 op amp in a circuit, an active filter can be implemented as shown in Figure 3-24. Notice the added resistor capacitor

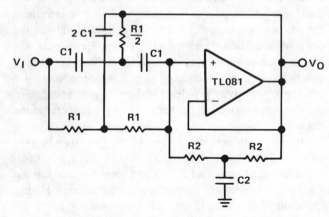

Figure 3-24. Active Twin-T Notch Filter

network (R2, C2), effectively in parallel with the original twin-T network, on the input of the filter. These networks set the Q of the filter. The op amp is basically connected as a unity-gain voltage follower. The Q is found from:

$$Q = \frac{R2}{2R1} = \frac{C1}{C2}$$

Let's now design a 60-Hz notch filter with a Q of 5 using the circuit in Figure 3-24. It is usually best to pick the C1 capacitor value and calculate the resistor R1. Let C1 = 0.22 μF.

$$fn = \frac{1}{2\pi RC}$$

$$R = \frac{1}{2\pi fnC} = \frac{1}{(6.28)\,(60)\,(0.22 \times 10^{-6})}$$

$$R1 = 12{,}063 \text{ or } 12 \text{ k}\Omega$$

Next calculate R2 in the Q network.

$$Q = \frac{R2}{2R1}$$

$$R2 = Q \times 2R1 = 5 \times 24 \text{ k}\Omega = 120 \text{ k}\Omega$$

$$R2 = 120 \text{ k}\Omega$$

Finally, calculate C2 from the equation.

$$Q = \frac{C1}{C2}$$

$$C2 = \frac{C1}{Q} = \frac{0.22 \ \mu F}{5} = 0.044 \ \mu F$$

$$C2 = 0.047 \ \mu F \text{ (nearest standard value)}$$

Standard 5% resistors and 10% capacitors produce a notch depth of about 40 dB as shown in the frequency response curve (Figure 3-25).

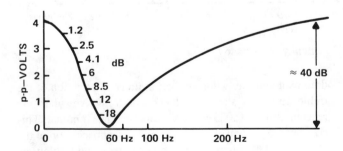

Figure 3-25. 60 Hz Twin-T Notch Filter Response

AUDIO OP AMP APPLICATIONS

MIKE PREAMP WITH TONE CONTROL

Microphones may be classified into two groups: high-impedance (≈ 200 kΩ) with high-voltage output and low-impedance (≈ 200 Ω) with low-voltage output. The output from a high-impedance microphone can be amplified simply and effectively with a standard inverting or noninverting operational amplifier configuration. However, high-impedance microphones are more susceptible to stray RF and 60-Hz noise. They have a fairly flat frequency response but are usually restricted to short cable lengths (10 feet or less). Long cables result in a high frequency roll-off characteristic caused by the cable capacitance.

Low-impedance microphones also have a flat frequency response but their low output levels impose rather stringent noise requirements on the preamp. The preamp shown in Figure 3-26 operates from a low-impedance, unbalanced, two-wire microphone where one of the wires is ground. The circuit consists of the LM318 preamp and the tone control circuitry.

The LM318 op amp is operated as a standard noninverting amplifier. Resistor R1 (47 kΩ) provides an input path to ground for the bias current of the noninverting input. The combination of R2 (560 Ω) and C2 (10 μF) provide a frequency roll-off below 30 Hz. At 30 Hz and above the gain is relatively flat at about 50 dB, set by the ratio R3/R2. R3 (220 kΩ) furnishes negative feedback from the output to the inverting input of the op amp. C3 (1.0 μF electrolytic) ac couples the preamp to the tone control section.

The top half of the tone control section is the bass control. The bottom half controls the treble frequency response. These tone controls (R5 and R8) require audio taper (logarithmic) potentiometers. The 50-kΩ potentiometer on the output can be used to set the output or gain of the preamp. Figure 3-27 shows the bass and treble responses of the circuit.

TL080 IC PREAMPLIFIER

A preamplifier is needed to amplify the signal generated by a tape head or phonograph cartridge. It is also common to include, with the preamplifier, a means of altering the bass and treble frequency response. The "purist" may want the amplifier to be "flat", which means no change from the input's frequency response. This condition should occur with both bass and treble controls at midposition. Sometimes it may be necessary to compensate for the effects of room acoustics, speaker response, etc. Also there is simply a matter of personal taste; one person may prefer music with heavier bass; another may prefer stronger treble.

Active tone control circuits offer some advantages: they are inherently symmetrical in boost and cut operation and have very low total harmonic distortion (THD) because they are incorporated in the negative feedback loop.

The circuit shown in Figure 3-28, is a form of the so-called "Americanized" version of the Baxandall negative-feedback tone control. At very low frequencies the reactance

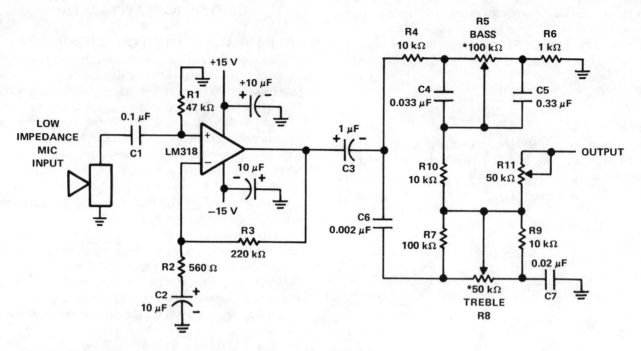

Figure 3-26. Mike Preamp With Tone Control

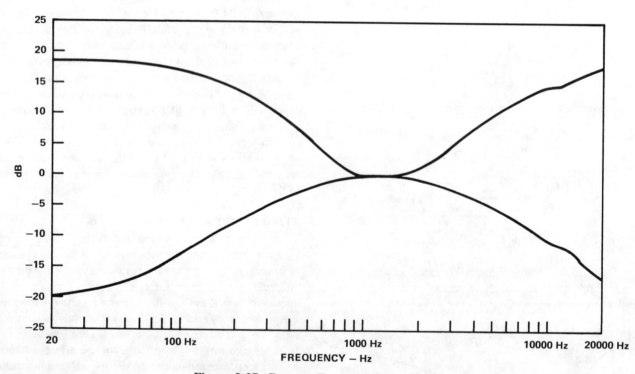

Figure 3-27. Preamp Frequency Response

of the capacitors is large enough that they may be considered open circuits, and the gain is controlled by the bass potentiometer. At low to middle frequencies the reactance of the 0.03 μF capacitors decreases at the rate of 6 dB/octave, and is in parallel with the 100 kΩ bass potentiometer; so the effective impedance is reduced correspondingly, thereby reducing the gain. This process continues until the 10-kΩ resistors, which are in series with the bass pot become dominant and the gain levels off at unity. The action of the treble circuit is similar and becomes effective when the reactance of the 0.003-μF capacitors becomes minimal. This complete tone control is in the negative feedback loop of the second TL080. Figure 3-29 shows the bass and treble tone control response. The response curves were run with 1.0 V equal to "0" dB as the "flat" response line.

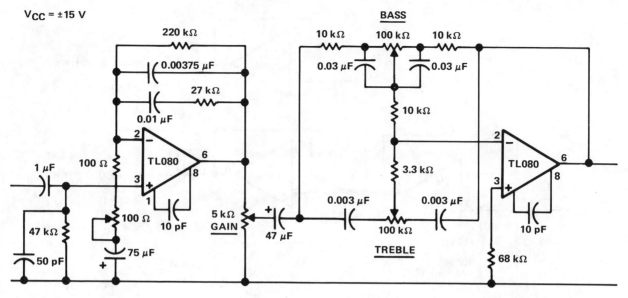

Figure 3-28. IC Preamplifier

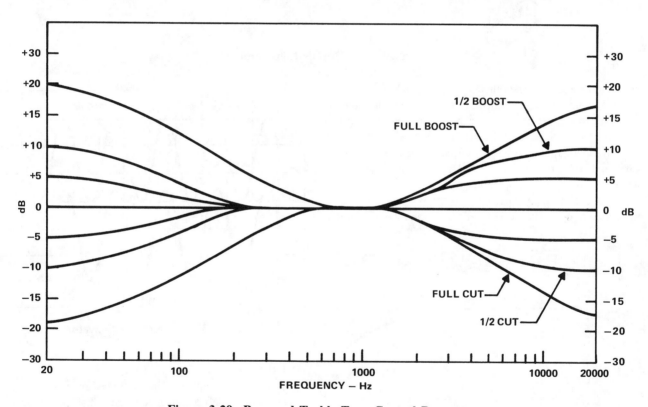

Figure 3-29. Bass and Treble Tone Control Response

The first TL080 is a preamp with a 100-Ω adjustable gain change pot. This gives a gain adjustment of about 6 dB for matching to the output of a particular pick-up or tape head. The negative feedback loop of this TL080 contains the gain setting and frequency compensation components.

AUDIO DISTRIBUTION AMPLIFIER

Sometimes there is a need for a preamplifier to receive the output from a single audio input device such as a microphone and drive several audio power amplifiers. This could be done most easily with a shielded cable to each amplifier from the originating preamplifier. However, if this were done by simply paralleling the shielded cables and connecting them to the preamp, the result could be an oscillating preamp stage or degraded high-frequency response due to the heavy capacitive loading.

A simple solution to this problem is the three channel output distribution amplifier using a single TL084 as shown in Figure 3-30. The first stage is capacitively coupled with a 1.0-μF electrolytic capacitor. The inputs are at 1/2 V_{CC} rail or 4.5 V. This makes it possible to use a single 9-V supply. A voltage gain of 10 (1 MΩ /100 kΩ) is obtained

3-13

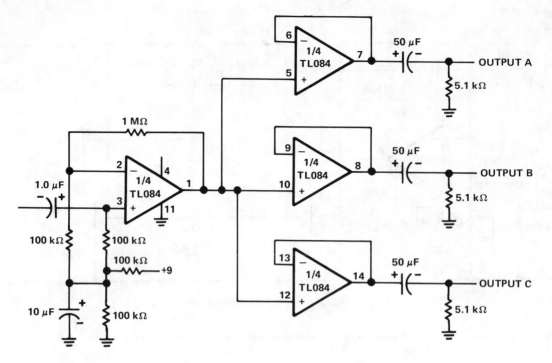

Figure 3-30. Distribution Amplifier

in the first stage, and the other three stages are connected as unity-gain voltage followers. Each output stage independently drives an amplifier through the 50-μF output capacitor to the 5.1-kΩ load resistor.

As shown in the response curve (Figure 3-31), the response is flat from 10 Hz to 30 kHz. Sinewave distortion begins at about 0.45 V$_{PP}$ input with 5.5 V$_{PP}$ output at 1 kHz. The total supply current is about 9 mA at a maximum input of 0.45 V$_{PP}$. The TL070 and TL080 family of op amps

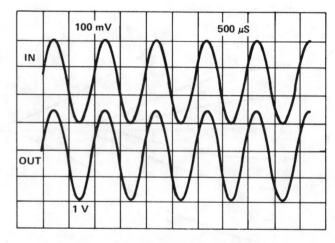

Figure 3-32. Amplifier Waveforms

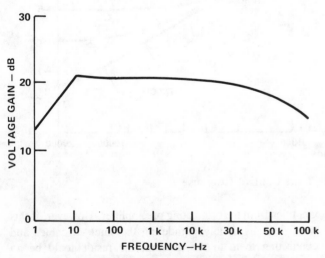

Figure 3-31. Frequency Response Curve

operate in true class-AB and therefore offer zero crossover distortion. The absence of crossover distortion can be noted in the scope photos (Figure 3-32). At a frequency of 1.0 kHz and with 3.0 V$_{PP}$ output the total harmonic distortion is 0.14%.

AUDIO POWER AMPLIFIER

Most audio amplifier circuits today are voltage output devices which apply a voltage to the terminals of the speaker. Large changes in speaker impedance with frequency yield poor frequency response with the high-frequency output falling off rapidly as the speaker impedance increases. It should be noted that the speaker cone displacement is proportional to the current in the voice coil rather than the voltage across the speaker terminals. The current is the primary mover of the voice coil.

The single speaker amplifier circuit shown in Figure 3-33 uses current feedback rather than the more popular voltage feedback. As shown, the feedback loop is from the junction of the speaker terminal and a 0.5-Ω resistor, to the inverting input of the NE5534. Sensing the current through the speaker and feeding it back provides better speaker damping than obtainable with voltage-drive systems.

Note the unusual grounded output pin on the NE5534 op amp. When the input to the amplifier is positive, the power supply supplies current through the TIP32 and the load to ground. Conversely, with a negative input the TIP31 supplies current through the load to ground. The gain in this case is set to about 15 (gain = SPKR 8 Ω /0.5 Ω feedback). The 0.22-μF capacitor across the speaker rolls off its response beyond the frequencies of interest. Using the 0.22-μF capacitor specified, the amplifier output is 3 dB down at 90 kHz where the speaker impedance is about 20 Ω. The Quiescent output stage collector current is determined by the

130-Ω resistors that connect each transistor base to the appropriate supply rail, the output transistor V_{BE}, and the 1-Ω emitter resistors. To set the recommended class "A" output collector current, adjust the value of either 130-Ω resistor. An output current of 50 to 100 mA will provide a good operating midpoint between the best crossover distortion and power dissipation.

The 0.1 μF bypass capacitors on each rail may be mylar or ceramic disk. The 2.0 μF should be a nonpolarized capacitor while the 0.22 μF across the speaker should be mylar.

Figure 3-34 shows the frequency response of the amplifier with a 2.0-μF input capacitor. This response is very flat with the − 3.0-dB point on the low frequency end at 45 Hz. The − 3.0-dB point at the high frequency end occurs at 80 kHz. Total Harmonic Distortion (THD) is 0.01% at 6.25 W rms output into an 8-Ω load with ± 18 V on the supply rails.

This amplifier circuit uses few components, has low total harmonic distortion, excellent frequency response and is easily duplicated. It works well up to 12 W peak output before clipping is noted. The TIP31 and TIP32 output transistors are complementary power transistors in the TO-220 package. Both transistors are rated at 3.0 A continuous collector current.

OSCILLATORS AND GENERATORS

AUDIO OSCILLATOR CIRCUIT DESCRIPTION

A Wein bridge oscillator can be used to produce sinewaves with very low distortion level. The Wein bridge oscillator produces zero phase shift at only one frequency (f = 1/2 πRC) which will be the oscillation frequency. In the configuration shown in Figure 3-35. stable oscillation can occur only if the loop gain remains at unity at the oscillation frequency.

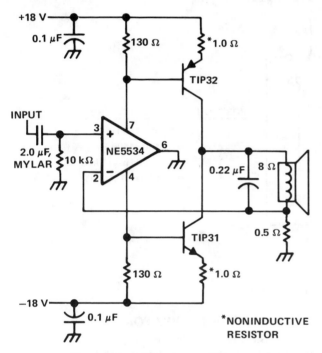

Figure 3-33. Audio Power Amplifier

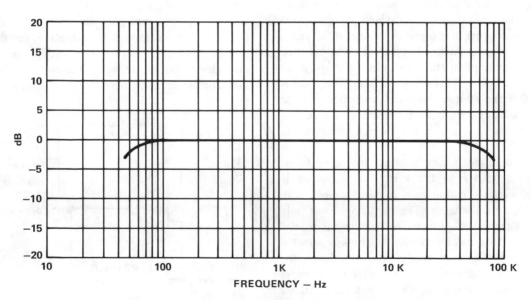

FREQUENCY — Hz

Figure 3-34. Audio Power Amplifier Frequency Response

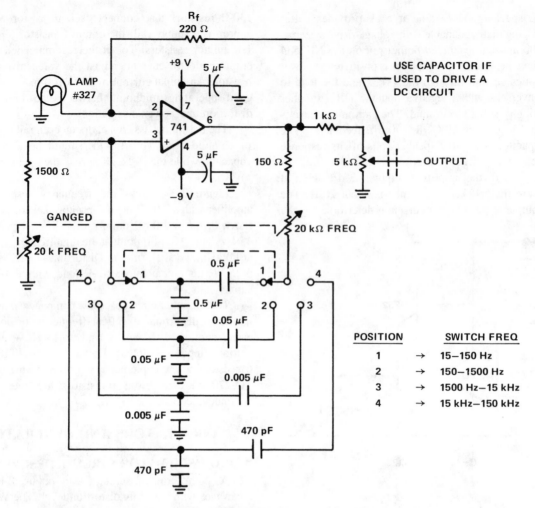

Figure 3-35. Audio Oscillator

POSITION		SWITCH FREQ
1	→	15–150 Hz
2	→	150–1500 Hz
3	→	1500 Hz–15 kHz
4	→	15 kHz–150 kHz

The circuit achieves this control by using the positive temperature coefficient of a small lamp to regulate gain(R_f/R_{LAMP}) as the oscillator attempts to vary its output. This is a classic technique for achieving low distortion that has been used by numerous circuit designers for about 40 years. The smooth limiting action of the bulb, in combination with the Wein network's near-ideal characteristics, yields very high performance. In this circuit a 741 op amp is used with ± 9 V power supplies. The tungsten lamp is a type #327 miniature which has a standard bayonet base. This lamp is rated at 28.0 V and 40 mA. For mass production of these oscillators, the lamps are burned in for a predetermined number of hours to stabilize the characteristics of the filament.

The oscillator shown here has four frequency bands covering about 15 Hz to 150 kHz. The frequency is continuously variable within each frequency range with ganged 20-kΩ potentiometers. The oscillator draws only about 4.0 mA from the 9-V batteries. Its output is from 4 to 5 V with a 10-kΩ load and the R_f (feedback resistor) is set at about 5% below the point of clipping. As shown, the center arm of the 5-kΩ output potentiometer is the output terminal. It should be noted that if you couple the oscillator to a dc type circuit, a capacitor should be inserted in series with the output lead.

THE BASIC MULTIVIBRATOR

A basic multivibrator may be constructed using an operational amplifier and a few external components, as shown in Figure 3-36. When this circuit is turned on, the natural offset of the devices serves as an automatic starting voltage. Assume the output voltage V_O goes positive and the positive feedback through R2 and R1 forces the output to saturate. The high-voltage level at V_O, then charges C through R3 until the voltage at the inverting input exceeds that at the noninverting input. As the inverting input exceeds

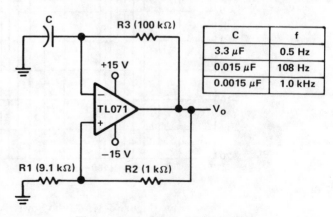

C	f
3.3 μF	0.5 Hz
0.015 μF	108 Hz
0.0015 μF	1.0 kHz

Figure 3-36. Basic Multivibrator

the noninverting input level, the output switches to the negative saturation voltage. This action starts the capacitor discharging toward the new noninverting input level. When the capacitor reaches that level the op amp switches back to the positive saturation voltage, and the process starts again.

With the TL071 the positive and negative output levels are nearly equal, resulting in a 50% duty cycle. The total time period of one cycle will be

$$t_T = 2 (R3)C \ln (1 + 2R1/R2)$$

LinCMOS™ OP AMP APPLICATIONS

MICROPHONE PREAMPLIFIER

It is sometimes necessary to have a microphone preamplifier mounted in the mike head. Obviously, the preamplifier should be as small as possible, battery operated and consume a small amount of power. In the past this was accomplished with bipolar and FET op amps. The primary disadvantage of these circuits is the comparatively large physical size of both the amplifier and power source. Another major factor is relatively large power consumption, which requires frequent battery replacement. The most obvious next choice would be a CMOS op amp. While this approach seems logical at first, it has some disadvantages. A metal-gate CMOS op amp can be operated from a low-voltage supply and has low power consumption; but, it suffers from input-offset voltage instability. This input-offset drift is due primarily to the differential input signals at the op amp input terminals. The LinCMOS op amp overcomes these problems. In addition, it has the advantage of low power comsumption and low voltage operation (down to 1.0 V).

A microphone preamplifier using a LinCMOS op amp is illustrated in Figure 3-37. This unit comes complete with its own battery, and is small enough to be put in a small mike case. The amplifier illustrated was designed to be operated from a 1.5-V mercury cell battery at low supply currents.

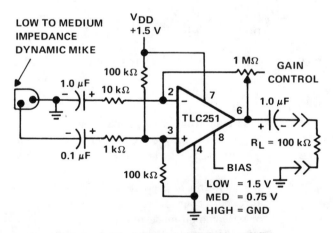

Figure 3-37. Microphone Preamplifier

This preamplifier will operate at very low power levels and maintain a reasonable frequency response as well. The TLC251 operated in the low bias mode (operating at 1.5 V)

LinCMOS is a trademark of Texas Instruments.

draws a supply current of only 10 μA and has a −3–dB frequency response of 27 Hz to 4.8 kHz. With pin 8 grounded, which is designated as the high bias condition, the upper limit increases to 25 kHz. Supply current is only 30 μA under those conditions.

If improved higher frequency performance is desired, the V_{DD} may be increased. For example, when using a 5-V supply the frequency response is from 27 Hz to 11 kHz for the low bias, and from 27 Hz to 220 kHz for the high bias modes respectively. Operating in the high bias mode at 5-V V_{DD} the amplifier requires a supply current of less than 500 μA. Frequency response for the amplifier is shown in Figure 3-38.

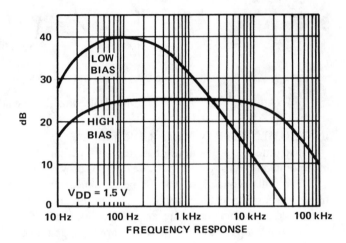

Figure 3-38. Preamplifier Frequency Response

POSITIVE PEAK DETECTOR

Peak detectors measure and hold the maximum value of a fluctuating voltage. The purpose of the circuit in Figure 3-39 is to hold the peak of the input voltage on capacitor C1, and read the value, V_O, at the output of U2. Op amps U1 and U2 are connected as voltage followers. When a signal is applied to V_I, C1 will charge to this same voltage through diode D1. This positive peak voltage on C1 will maintain V_O at this level until the capacitor is reset (shorted). Of course, higher positive peaks will raise this level while lower peaks will be ignored. C1 can be reset manually with a switch, or electronically with an FET that is normally off.

The capacitor specified for C1 should have low leakage and low dielectric absorption. Diode D1 should also have low leakage. The op amps selected for use in a peak detector should be immune to instability due to capacitive loading, and have high output drive and slew rate. They should also have very low input bias currents and extremely high input impedance. The TLC251 meets these requirements well. The TLC251 allows the reading of low-level signals near ground because its input-common-mode range includes the negative connection to the power supply. Peak values of negative polarity signals may be detected by reversing D1.

3-17

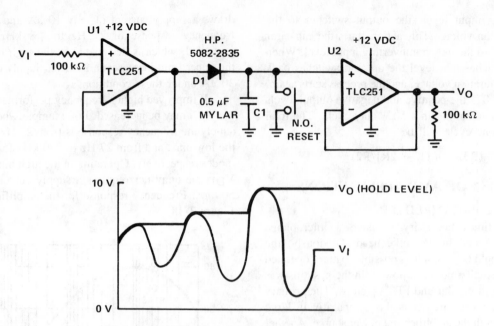

Figure 3-39. Positive Peak Detector and Waveform

INSTRUMENTATION METER DRIVER

Instrumentation amplifier circuitry which has incorporated low-cost general-purpose op amps provides the designer with economical, quality performance options. Improved instrumentation amplifier circuits are possible because of the development of op amps using junction FETs. These op amps have improved input impedance characteristics and ac performance compared with general purpose bipolar devices. Metal gate CMOS op amps have reduced the power required and will operate at voltages as low as 2 V.

Because the input offset voltage of a metal gate CMOS circuit changes with varying differential input voltage levels, there are severe drawbacks in using this technology for op amps. LinCMOS technology overcomes these disadvantages. LinCMOS devices do not have an input offset shift with differential input voltage and can operate satisfactorily down to 1.0 V supply. TLC271 LinCMOS op amps are used in the instrumentation amplifier illustrated in Figure 3-40 because of their unique features. Some of these features are:

- Operate at low voltages
- Input signal operation close to $-V_{CC}$ rail
- Reasonable ac performance at low power
- Provide the high input impedance characteristic of FET input devices
- Offset stability
- High CMRR
- Power/performance adjustment for desired performance levels while maintaining the lowest possible power requirement.

The TLC271 operational amplifiers (the first monolithic devices to combine these characteristics) allow the construction of a ± 5-V instrumentation amplifier with reasonable ac performance. Some of the important features of Figure 3-40 which should be pointed out are:

- Three op amps U1, U2, and U3 are connected

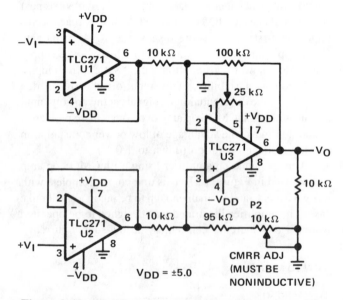

Figure 3-40. TLC Low Power Instrumentation Amplifier

in the basic instrumentation amplifier configuration.

- Operating from ± 5 V, pin 8 of each op amp is connected directly to ground and provides the ac performance desired in this application (high bias mode).
- Two adjustment pots are used. P1 is for offset error correction and P2 allows adjustment of the input common mode rejection ratio.
- The high input impedance allows metering of signals from sources of several megohms without loading. The resulting circuit frequency response is 200 kHz at -3 dB and has a slew rate of 4.5 V/μs. This is a significant improvement over general bipolar performance. The signal response and speed characteristics are

particularly significant in light of the low supply voltage and supply currents. Total supply current is 670 μA per supply.

- Output error voltages of less than 1% are experienced over the 0 °C to 70 °C operating range.

A STABLE TWO-VOLT LOGIC ARRAY POWER SUPPLY

The popularity of logic gate array devices has emphasized the need for closely regulated low-voltage power supplies. Typical power requirements for these devices are +2.0 V at approximately 250 mA per array. A major requirement for many systems is the ability of the power supply to operate over a wide range of input voltages, particularly from 5 V or less to minimize losses. Regulation to within ±5% and good ripple rejection are also desirable in most applications.

Several types of three-terminal adjustable series pass regulators or shunt regulators are capable of providing regulation at 2 V. For example the LM317 will provide 2 V with a minimum number of external components. Good regulation is possible with this device when supply voltages are in excess of 5 V. LM317 operating characteristics will begin to deteriorate at 5 V or less because of insufficient input to output differential voltage.

Adjustable zeners such as the TL431 will operate at low voltage levels. However, they do not produce enough output current to be useful as a regulator in this application.

Combining a TL431, to provide an accurate reference; an op amp, for accurate output level control; with a power pass transistor is very effective (Figure 3-41). However, this configuration puts some rather severe constraints on the op amp used, as shown below:

1. It should be a device capable of operating below 5 V and handling high input, common-mode

voltages (up to 2 V when operating from a single 3.5-V supply)
2. Accommodate varying differential input voltage without adversely affecting input offset stability.
3. Provide adequate output drive current while maintaining a low operating current for maximum circuit efficiency.

The TLC271 is such a device.

In the application discussed here the TLC271 must operate from a single supply that can be as low as 3.5 V. A temperature compensated voltage reference is provided by the TL431 and coupled to the inverting input of the TLC271. The power supply output of 2.0 V is sensed and fed back to the TLC271's noninverting input. This results in a commmon-mode input level of about 2.0 V which can easily be handled by the TLC271 even when operating from a single 3.5-V supply. The TLC271's high-bias, open-loop gain of approximately 96 dB (at a V_{CC} of 3.5 V) provides control of the output drive transistors with minimum error. Most of the no-load supply current required is drawn by the 110 Ω load shunt at the 2-V output (20 mA). The supply's total no-load current with a 15-V input is less than 30 mA.

The circuit's stability is excellent for both input voltage and output current variations. Maximum variation of output voltage, with input voltage swings from 3.5 V to 15 V, is less than 5%. Ripple rejection with a 2-V input swing at 120 Hz was over 60 dB. Variation in output voltage from no load to full load is less than 0.5%. The capability of the TLC271 to operate from a 3.5-V supply and handle 2-V common-mode input signals without input clipping or distortion, coupled with the inherent input offset stability of the Silicon Gate CMOS process, makes this kind of performance possible.

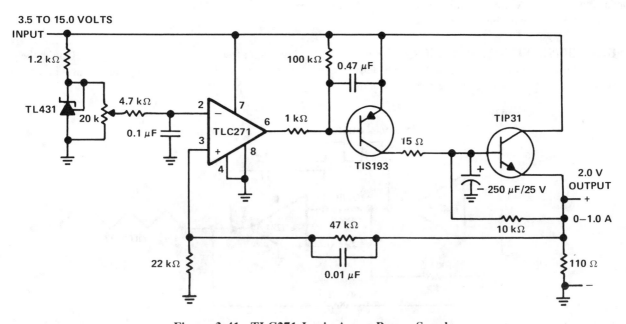

Figure 3-41. TLC271 Logic Array Power Supply

TLC271 TWIN-T NOTCH FILTER

The theory of a bipolar twin-T notch filter was discussed in the section under Active Filter Applications. That twin-T filter required ±15-V power supplies. This filter however, can be powered with a single 5-V supply. Active filters built with LinCMOS op amps will operate well at low frequencies from a single low-voltage supply. They also require minimum space because of their low component count. The 60-Hz filter illustrated in Figure 3-42 has only one op amp, three resistors, and three small capacitors.

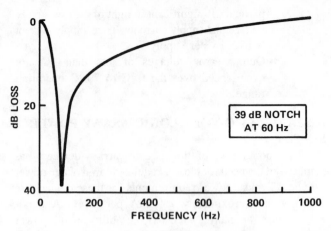

Figure 3-43. TLC271 Twin-T Notch Filter Frequency Response

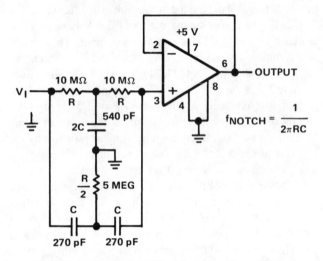

Figure 3-42. Twin-T Notch Filter for 60 Hz

$$f_{NOTCH} = \frac{1}{2\pi RC}$$

If the resistors and capacitors are carefully matched, you theoretically obtain almost complete rejection of the 60-Hz null frequency. The TLC271 has an input bias current of about 1 pA and will not generate adverse offset voltages even though the source impedance is 20 MΩ. Low-level 60-Hz ripple will also be attenuated due to the TLC271's capability of handling input signals at the ground rail while operating from a single supply. As illustrated in Figure 3-43, a 39-dB notch was achieved with this circuit.

SINGLE SUPPLY FUNCTION GENERATOR

A function generator is a circuit that can deliver a number of different waveforms. Some circuits operate at a

fixed frequency while others have the capability of varying their frequencies over a wide range. The example circuit shown in Figure 3-44 has both square-wave and triangle-wave output.

The left section is similar in function to a comparator circuit that uses positive feedback for hysteresis. The inverting input is biased at one-half the V_{CC} voltage by resistors R4 and R5. The output is fed back to the noninverting input of the first stage to control the frequency. The amplitude of the square wave is the output swing of the first stage, which is 8 V peak-to-peak.

The second stage is basically an op amp integrator. The resistor R3 is the input element and capacitor C1, is the feedback element. The ratio R1/R2 sets the amplitude of the triangle wave, as referenced to the square-wave output. For both waveforms, the frequency of oscillation can be determined by the equation:

$$f_o = \frac{1}{4R3C1} \left[\frac{R2}{R1} \right]$$

The output frequency is approximately 50 Hz with the given components. The different waveforms are illustrated in Figure 3-45.

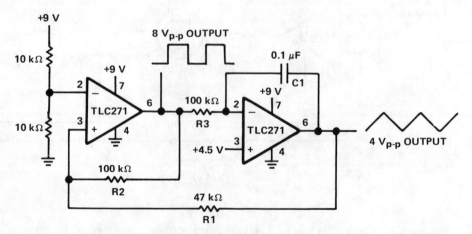

Figure 3-44. Single Supply Function Generator

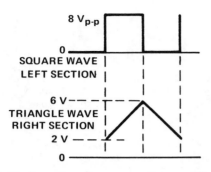

Figure 3-45. Function Generator Output Waveforms

COMPARATOR APPLICATIONS

WINDOW COMPARATOR

As the name implies, a window detector is a specialized comparator circuit designed to detect the presence of a voltage between two prescribed limits that is, within a voltage "window". A window comparator is useful in test and production equipment to select components that are within a specific set of limits.

This circuit is implemented by logically combining the outputs of two single-ended comparators. One indicates an input greater than the lower limit, and the other an input less than the upper limit. If both comparators indicate a true condition, the output is true. If either input is not true, the output is not true. A basic window comparator circuit is illustrated in Figure 3-46.

In this circuit, the outputs of the two comparators are logically combined by the 1N914 diodes. When the input voltage is between the upper limit (V_{UL}) and the lower limit (V_{LL}) the output voltage is zero; otherwise it equals a logic high level. The output of this circuit can be used to drive a logic gate, LED driver or relay driver circuit. The circuit shown in Figure 3-47 shows a 2N2222 npn transistor being driven by the window comparator. When the input voltage to the window comparator is outside the range set by the V_{UL} and V_{LL} inputs, the output changes to positive, which turns on the transistor and lights the LED indicator.

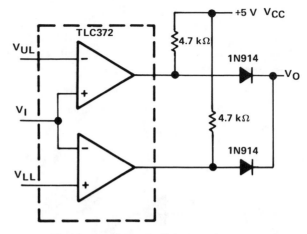

Figure 3-46. Basic Window Comparator

The TLC372 features extremely high input impedance (typically greater than 10^{12} Ω) which allows direct interfacing with high-impedance sources. The outputs are n-channel open-drain configurations, and can be connected to achieve positive-logic wired-AND relationships. While these devices meet the 2000 V ESD (electrostatic discharge) specification, care should be exercised in handling the chips because exposure to ESD may result in a degradation of the device performance. If the input signal will exceed the common-mode voltage, it is good design practice to include protective input circuitry to the comparators. Clamp diodes and/or series resistors, as shown in Figure 3-48, could be used for this purpose.

COMPARATOR INTERFACE CIRCUITS

A comparator is a useful building block in signal conditioning circuits as well as in instrumentation and control circuits. Once the inputs to the comparator have been correctly matched and connected, the comparator must interface with any additional output circuitry to perform functions such as: energizing a relay, lighting a lamp or driving another type of logic circuit.

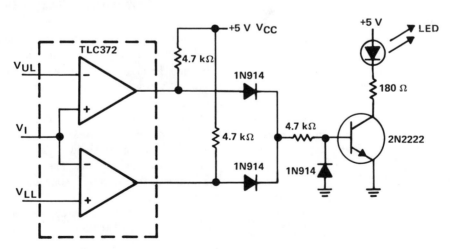

Figure 3-47. Window Comparator with LED Indicator

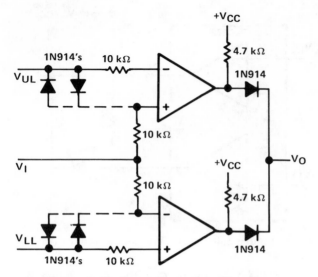

Figure 3-48. Input Protection Circuitry

Figure 3-49 shows three similar output interface circuits which can be used to drive a lamp, a relay, and an LED. All three circuits utilize the popular LM311 comparator and 2N2219/2N3904 family of discrete transistors. A resistor in series with the transistor base should be used in circuits of this type in order to limit the current. Although it is possible to directly drive some small lamps and relays from the output of the comparator, it is advisable to use a buffer transistor as illustrated. The output buffer will minimize loading on the comparator to preserve its gain and drift characteristics, and provide a higher output current.

Figure 3-49(a) is a circuit which can be used to turn on a 28-V lamp. A 100-Ω resistor must be included in series with the lamp. The purpose of this resistor is to limit the cold lamp inrush current. Lower voltage lamps of up to 150-mA current rating can be driven by reducing the lamp supply voltage appropriately.

Figure 3-49(b) illustrates a relay driver circuit. The relay is connected in series with the transistor collector terminal and the 24-V supply. The 1N914 diode across the relay clamps the back EMF voltage generated when the relay is turned off. Lower voltage relays can be driven as well as the relay illustrated here, provided the maximum current is less than 150 mA.

Figure 3-49(c) illustrates an LED driver, using a 2N3904 transistor switch rather than the high-powered 2N2219. The TIL220 is a red LED and has approximately a 1.7-V drop in the forward-bias condition at a forward current of 20 mA with a V_{CC} of 5 V. The value of the current limiting resistor is calculated as shown below.

$$R = \frac{5 \text{ V} - 1.7 \text{ V}}{0.020 \text{ A}} = 165 \text{ ohms}$$

The next higher standard value is 180 Ω which has been used in this circuit.

A comparator output may interface with digital logic. If the comparator operates from a single 5-V supply, there

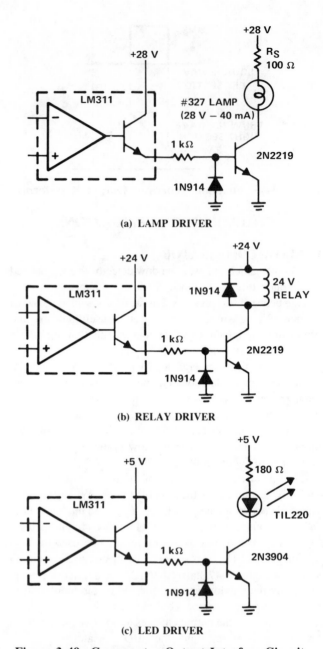

(a) LAMP DRIVER

(b) RELAY DRIVER

(c) LED DRIVER

Figure 3-49. Comparator Output Interface Circuits

is no problem. Interfacing to a comparator that is operating from +15-V supplies may require some level shifting and/or clamping to drive logic circuits. Figure 3-50 illustrates graphically how to interface comparators to different level logic circuits.

Figure 3-50(a) illustrates a comparator with a single 5-V power supply and has a 10-kΩ pull-up resistor on the output. This circuit will drive standard TTL logic circuits or low-level CMOS logic circuits. These logic circuits require a maximum of 0.8 V for low state and a minimum of 2.4 V for high state.

Figure 3-50(b) illustrates a comparator circuit capable of driving high-level CMOS circuits. This circuit operates from dual +15-V supplies and uses a 100-kΩ pull-up resistor on the output. High-level CMOS logic requires a maximum of 4.0 V for a low state and a minimum of 11.0 V for a high state.

Figure 3-50(c) illustrates how a three-state output is produced by following the comparator with a hex-bus driver such as the SN74367. In this circuit the comparator has a 10-kΩ pull-up resistor on its output and sends TTL logic signals to the input of the hex buffer. The output of the hex buffer is controlled by the input control pin's logic level. When this pin is low the device is enabled and the output is TTL logic. When the input control pin is high there is no output from the device and the output looks like a high impedance.

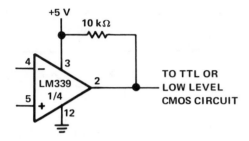

(a) TTL AND LOW-LEVEL CMOS DRIVER

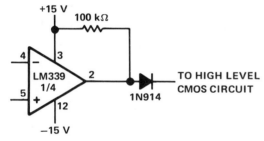

(b) HIGH-LEVEL CMOS DRIVER

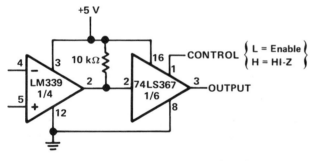

(c) 3-STATE BUS DRIVER (TTL OUTPUT)

Figure 3-50. Comparator Logic Interface Circuits

LM393 ZERO-CROSSING DETECTOR

A zero-crossing detector is sometimes called a zero-level detector or a Schmitt trigger. In operation, a zero-crossing detector determines if an input voltage to the comparator is greater or less than zero. In response to this determination, the output voltage of the comparator can assume only two possible states. The output state may be high or low depending upon which comparator input (plus or minus) is used to detect the incoming signal.

A single comparator may be used as a simple crossover detector, but this can allow several sources of error. These errors may be caused by the input bias and offset currents of the comparator. Temperature may also affect the zero-crossing voltage points. This basic zero-crossing detector also will have another drawback called chatter which is due to noise on the input signal. Chatter can be reduced by adding hysteresis or positive feedback. These provide noise immunity and prevent the output from "chattering" between states as the input voltage passes through zero.

An improved circuit is illustrated in Figure 3-51. This zero-crossing detector of this type uses a dual LM393 comparator, and easily controls hysteresis by the reference levels which are set on the comparator inputs.

The circuit illustrated is powered by ±10-V power supplies. The input signal can be an ac signal level up to +8 V. The output will be a positive going pulse of about 4.4 V at the zero-crossover point. These parameters are compatible with TTL logic levels.

The input signal is simultaneously applied to the noninverting input of comparator A and the inverting input of comparator B. The inverting input of comparator A has a +10-mV reference with respect to ground, while the noninverting input of comparator B has a −10-mV reference with respect to ground. As the input signal swings positive (greater than +10 mV), the output of comparator "A" will be low while comparator "B" will have a high output. When the input signal swings negative (less than −10 mV), the reverse is true. The result of the combined outputs will be low in either case. On the other hand, when the input signal is between the threshold points (±10 mV around zero crossover), the output of both comparators will be high. In this state the output voltage will be one-half the 10 V (V_{CC+}) less the 0.6-V diode drop at the junction of the two 10-kΩ resistors (approximately +4.4 V). This circuit is very stable and immune to noise. If more hysteresis is needed, the ±10-mV window may be made wider by increasing the reference voltages. The 1N914 diode in series with the outputs allows a positive going pulse at the crossover point. This circuit "squares" the input signal into positive rectangular output pulses whose pulse width corresponds to the input zero crossings.

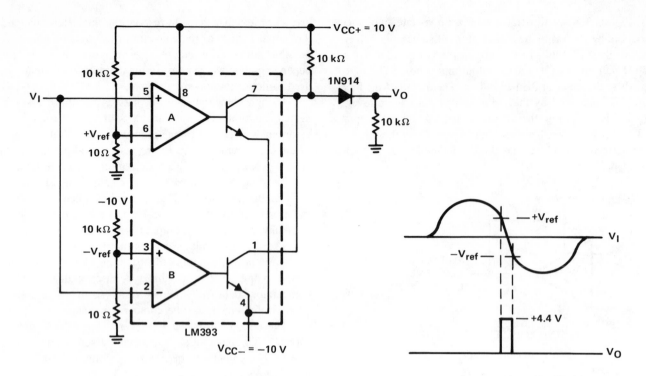

Figure 3-51. LM393 Zero-Crossing Detector

Section 4
Video Amplifiers

VIDEO AMPLIFIER THEORY

The characteristics of an ideal video amplifier are identical to those of an ideal operational amplifier (i.e., infinite input resistance, infinite gain, zero output resistance, and zero offset).

Typical performance differences between operational amplifiers and video amplifiers are bandwidth and gain. The bandwidth averages 100 kHz for typical operational amplifiers. However, video amplifiers have bandwidths as high as 100 MHz. The gain for a video amplifier averages only 40 dB as compared to 100 dB for operational amplifiers. Because their internal phase shift does not permit the use of negative feedback to control gain, most video amplifiers function only in the open-loop mode. Video amplifiers have a limited output voltage swing. For high-frequency operation, the output voltage swing is limited to a few volts. Table 4-1 lists general characteristics of some video amplifiers which are currently available.

The input stage of most video amplifiers consists of a basic emitter-coupled differential transistor pair connected to a constant-current source transistor. Most early video amplifiers consisted of these three transistors combined with a few integrated resistors and diodes. In addition, all component terminals were brought out for external interconnection. A typical video amplifier configuration is illustrated in Figure 4-1. The bias input voltage can be adjusted to provide symmetrical output voltage swing with respect to ground.

Video amplifier characteristics are similar to those of operational amplifiers and comparators. However, the following definitions are specifically applicable to video amplifiers.

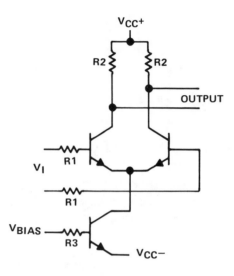

Figure 4-1. Basic Video Amplifier Circuit

VOLTAGE GAIN

Video amplifiers have a differential type of input and output modes. Voltage gain is defined as the ratio between the change in differential output voltage to the change in differential input voltage as stated in the following equation (see Figure 4-2):

$$A_{VD} = \frac{V_{OD}}{V_{ID}}$$

COMMON-MODE OUTPUT VOLTAGE

With the inputs grounded, the outputs of a video amplifier are at dc levels with respect to ground. The average of the two dc output voltages is the common-mode output

Table 4-1. Video Amplifier Selection Guide

DEVICE	CHARACTERISTICS	DESCRIPTION
uA733, TL733	−3 dB bandwidth, 90 MHz	Differential video amplifier. Selectable amplification of 10, 100, or 400.
NE592, SE592	−3 dB bandwidth, 90 MHz	Differential video amplifier. Selectable amplification of 100 or 400. Adjustable gain from 0 to 400. Adjustable passband.
TL592	−3 dB bandwidth, 90 MHz	Differential video amplifier. Adjustable gain from 0 to 400. Adjustable passband.
MC1445	−3 dB bandwidth, 50 MHz	2-Channel-input video amplifier. Gate controlled. 16-dB minimum gain. Broadband noise, typically 25 V.

voltage, V_{OC}, and can be determined by the following equation (see Figure 4-2):

$$V_{OC} = \frac{V_O1 + V_O2}{2}$$

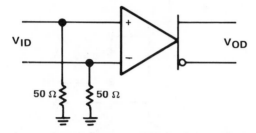

Figure 4-2. Differential Voltage Gain

OUTPUT OFFSET VOLTAGE

While functioning under the same conditions as those shown in Figure 4-3, the difference between the dc levels at the two outputs is defined as the output offset voltage ($V_{OO} = V_O1 - V_O2$). The offset voltage can be compared to the input voltage by dividing the offset voltage by the differential voltage gain of the amplifier.

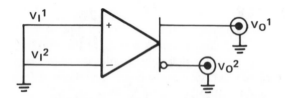

Figure 4-3. Common-Mode And Offset Voltages

Video amplifiers have no provision for adjusting the dc input offset voltage. In circuits where this will cause problems, capacitive coupling is used on both the inputs and outputs to block the dc component and prevent its affecting the amplified signal. Figure 4-4 illustrates the single-ended and differential coupling methods.

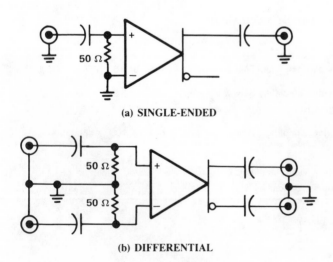

(a) SINGLE-ENDED

(b) DIFFERENTIAL

Figure 4-4. Coupling Methods

WIRING PRECAUTIONS

The mechanical layout of the video amplifier is very important. All leads should be as short as possible. When using a printed circuit board, conductors should be wide and as short as possible. This helps provide low resistance and low inductance connections. In addition, stray signal coupling from the input to the output is minimized.

Grounding is the most important wiring precaution. As with all high frequency circuits, a ground plane and good grounding techniques should be used. The ground plane should connect all areas of the pattern side of the printed circuit board that are not otherwise used. The ground plane provides a low-resistance low-inductance common return path for all signal and power returns. The ground plane also reduces stray signal pick up.

Each power supply lead should have a bypass capacitor to ground as near as possible to the amplifier pins. A 0.1-μF capacitor is normally sufficient. In very high-frequency and high-gain circuits, a combination of a 1-μF tantalum capacitor in parallel with a 470-pF ceramic capacitor is a suitable bypass.

Single point grounding should be used in cases where point-to-point wiring is used or a ground plane is not used. The input signal return, the load signal return, and the power supply common should all be connected at the same physical point. This eliminates ground loops or common current paths which may cause signal modulation or unwanted feedback.

When designing video amplifier circuits, resistor values from $50\,\Omega$ to $100\,\Omega$ should be used for input terminations. Resistors in this range improve circuit performance by reducing the effects of device input capacitance and input noise currents.

OSCILLOSCOPE/COUNTER PREAMPLIFIER

A circuit containing a single NE592A video amplifier (the only active component) can be used to increase the sensitivity of an older oscilloscope or frequency counter.

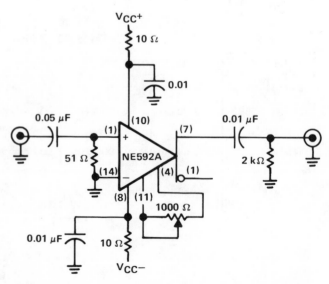

Figure 4-5. Oscilloscope/Counter Preamplifier Circuit

Figure 4-5 shows a circuit which will provide a 20 ± 0.1 dB voltage gain from 500 kHz to 50 MHz. The low-frequency response of the amplifier may be extended by increasing the value of the 0.05-μF capacitor connected in series with the input terminal. This circuit will yield an input-noise level of approximately 10 μV over a 15.7 MHz bandwidth.

The gain can be calibrated by adjusting the potentiometer connected between pins 4 and 11 (gain adjust terminals). These pins go directly to the emitter terminals of the two npn differential input amplifiers. The 1000-Ω potentiometer (a cermet type trimmer) can be adjusted for an exact voltage gain of 10. This preserves the scale factor of the instrument. The usual precautions of short leads and wide area ground planes for low-inductance ground systems, should result in good high-frequency response. A compact assembly package for an oscilloscope/counter preamplifier can be made by forming a small piece of sheet copper or brass into a the shape of a U.

NE592 FILTER APPLICATIONS

The NE592 is a two-stage differential-output wideband video amplifier. It has a voltage gain of 0 to 400 that can be adjusted by one external resistor. The input stage is designed so that by adding a few external reactive elements between the gain-select terminals (pins 4 and 11), the circuit can function as a high-pass, low-pass, or band-pass filter. This feature makes the circuit ideal for use as a video or pulse amplifier in communications, magnetic memories, display and video recorder systems. Figure 4-6 illustrates the basic filter circuit. This circuit has a 50-Ω input termination and a 50-Ω output termination and uses a ±6-V power supply. The 50-Ω output termination allows interfacing to the spectrum analyzer.

Figure 4-7 illustrates the results when five different reactive elements are placed across the gain select terminals (pins 4 and 11). The parametric values and condition of each circuit, as well as the scope photos, are shown.

MC1445 BALANCED MODULATOR

A balanced modulator can be obtained by connecting an MC1445 operational amplifier as shown in Figure 4-8. The internal differential amplifiers are connected in a manner which cross couples the collectors (see Figure 4-9).

When the carrier level is adequate to switch the cross-coupled pair of differential amplifiers, the modulation signal (which has been applied to the gate) will be switched, at the carrier rate, between the collector loads. When switching occurs, it will result in the modulation being multiplied by a symmetrical switching function. If the modulation gate remains in the linear region, only the first harmonic will be present. To achieve good harmonic suppression of the modulation input, the input level must remain in the linear region of the gate.

To balance the MC1445 modulator, equal gain must be achieved in the two separate channels. In Figure 4-10 (the composite gate characteristic for both channels) the equal gain point is at 1.3 V. The midpoint of the linear region of channel B is at 1.2 V. To remain in the linear region, the modulation input must be restricted to approximately 200 mV peak-to-peak. Because the gate bias point is sensitive to the amount of carrier suppression, a high-resolution (10-turn) potentiometer should be used.

In Figure 4-11, the top trace shows the 1 MHz carrier being modulated by a 1 kHz signal. The output is 750 mV peak-to-peak. The bottom trace shows the 600 mV peak-to-peak 1 kHz modulating signal. When functioning under these conditions, a carrier rejection of 38 dB should be obtained.

MC1445 FREQUENCY SHIFT KEYER

To construct a frequency shift keyer with an MC1445, apply a signal to each differential amplifier input pair. When the gate voltage is changed from one extreme to the other, the output may be switched alternately between the two input signals (see Figure 4-12).

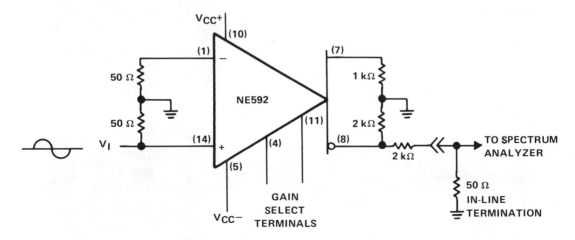

Figure 4-6. Basic Filter Circuit

When the gate level is high (1.5 V), a signal applied between pins 5 and 6 (channel A) will be passed and a signal applied between pins 3 and 4 (channel B) will be suppressed.

The reverse situation will exist when the gate is low (0.5 V). At 0.5 V, a signal applied to pins 3 and 4 (channel B) will pass. The unselected channel will have a gain of one or less.

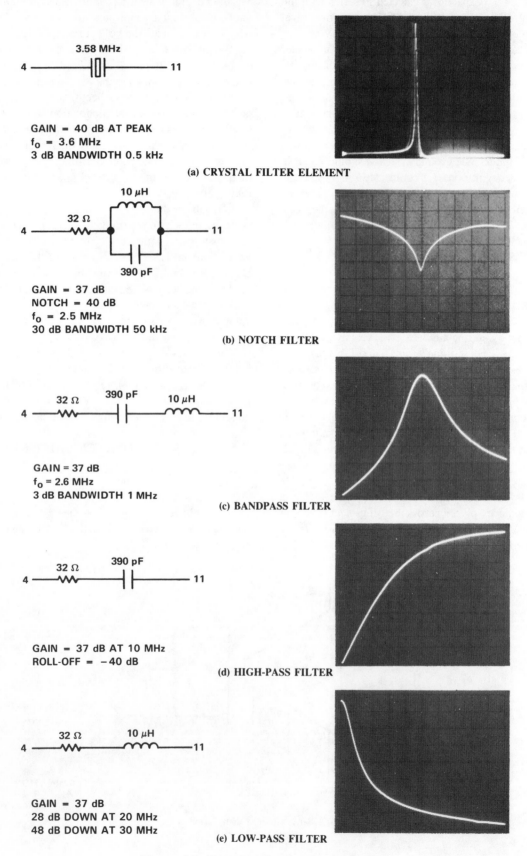

3.58 MHz

4 ——|||——— 11

GAIN = 40 dB AT PEAK
f_O = 3.6 MHz
3 dB BANDWIDTH 0.5 kHz

(a) CRYSTAL FILTER ELEMENT

10 μH

32 Ω

4 ——w——● ●——— 11

390 pF

GAIN = 37 dB
NOTCH = 40 dB
f_O = 2.5 MHz
30 dB BANDWIDTH 50 kHz

(b) NOTCH FILTER

32 Ω   390 pF   10 μH

4 ——w————||———w——— 11

GAIN = 37 dB
f_O = 2.6 MHz
3 dB BANDWIDTH 1 MHz

(c) BANDPASS FILTER

32 Ω   390 pF

4 ——w————||——————— 11

GAIN = 37 dB AT 10 MHz
ROLL-OFF = −40 dB

(d) HIGH-PASS FILTER

32 Ω   10 μH

4 ——w————w———— 11

GAIN = 37 dB
28 dB DOWN AT 20 MHz
48 dB DOWN AT 30 MHz

(e) LOW-PASS FILTER

Figure 4-7. Reactive Component Application

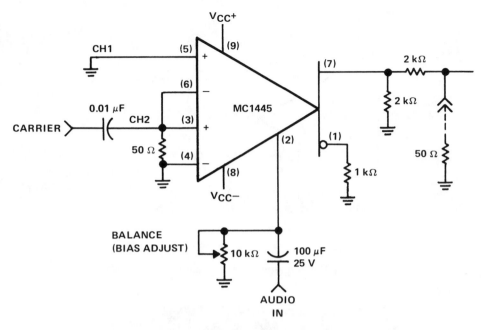

Figure 4-8. Balanced Modulator

In this manner, a binary-to-frequency conversion is obtained that is directly related to the binary sequence which is driving the gate input (pin 2). Figure 4-13 illustrates the waveforms of this basic frequency shift keying (FSK) application using the MC1445. The top trace illustrates a 20-kHz signal applied to channel A and a 4-kHz signal applied to channel B. The bottom trace illustrates a 1-kHz gating signal applied to the gate pin (2). The oscilloscope is triggered by this gate input signal.

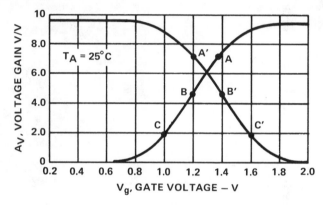

Figure 4-10. Voltage Gain vs Gate Voltage

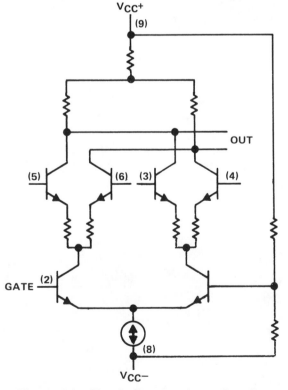

Figure 4-9. Circuit Showing Cross-Coupling

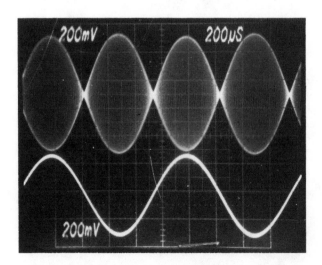

Figure 4-11. Balanced Modulation

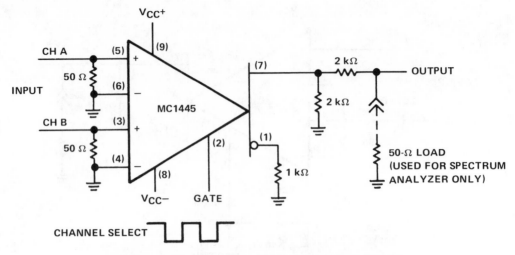

Figure 4-12. Frequency Shift Keying Test Circuit

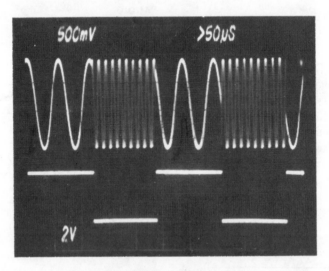

Figure 4-13. FSK Output and Gate
Input Signal Waveforms

Section 5

Voltage Regulators

BASIC REGULATOR THEORY

The function of every voltage regulator is to convert a dc input voltage into a specific, stable, dc output voltage and maintain that voltage over a wide range of load current and input voltage conditions. To accomplish this, the typical voltage regulator (Figure 5-1) consists of:

1. A reference element that provides a known stable voltage level, (V_{REF}).
2. A sampling element to sample the output voltage level.
3. An error-amplifier element for comparing the output voltage sample to the reference and creating an error signal.
4. A power control element to provide conversion of the input voltage to the desired output level over varying load conditions as indicated by the error signal.

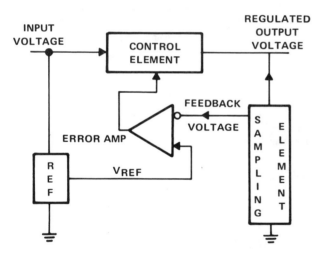

Figure 5-1. Basic Regulator Block Diagram

Although actual circuits may vary, the three basic regulator types are series, shunt, and switching. The four basic functions listed above exist in all three regulator types.

VOLTAGE REGULATOR COMPONENTS

REFERENCE ELEMENTS

The reference element forms the foundation of all voltage regulators since output voltage is directly controlled by the reference voltage. Variations in the reference voltage will be interpreted as output voltage errors by the error amplifier and cause the output voltage to change accordingly. To achieve the desired regulation, the reference must be stable

for all variations in supply voltages and junction temperatures. There are several common techniques which can be used to solve design problems using integrated circuit regulators. Many of these techniques are discussed in the section of the text that outlines error contributions.

SAMPLING ELEMENT

The sampling element monitors the output voltage and converts it into a level equal to the reference voltage. A variation in the output voltage causes the feedback voltage to change to a value which is either greater or less than the reference voltage. This voltage difference is the error voltage which directs the regulator to make the appropriate response and thus correct the output voltage change.

ERROR AMPLIFIER

The error amplifier of an integrated circuit voltage regulator monitors the feedback voltage for comparison with the reference. It also provides gain for the detected error level. The output of the error amplifier drives the control circuit to return the output to the preset level.

CONTROL ELEMENT

All the previous elements discussed remain virtually unaltered regardless of the type regulator circuit. The control element, on the other hand, varies widely, depending upon the type of regulator being designed. It is the element that determines the classification of the voltage regulator; series, shunt, or switching. Figure 5-2 illustrates the three basic

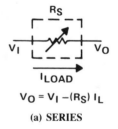

$$V_O = V_I - (R_S)\, I_L$$

(a) SERIES

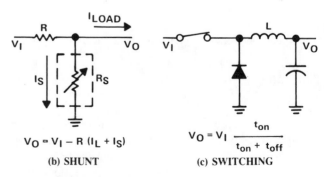

$$V_O = V_I - R\,(I_L + I_S)$$

(b) SHUNT

$$V_O = V_I \frac{t_{on}}{t_{on} + t_{off}}$$

(c) SWITCHING

Figure 5-2. Control Element Configurations

control element configurations, each of which is discussed in detail. These elements contribute an insignificant amount of error to the regulator's performance. This is because the sampling element monitors the output voltage beyond the control element and compensates for its error contributions. However, the control element directly affects parameters such as minimum input-to-output voltage differential, circuit efficiency, and power dissipation.

REGULATOR CLASSIFICATIONS

SERIES REGULATOR

The series regulator derives its name from its control element. The output voltage, V_O, is regulated by modulating an active series element, usually a transistor, that functions as a variable resistor. Changes in the input voltage, V_I, will result in a change in the equivalent resistance of the series element identified as R_S. The product of the resistance, R_S, and the load current, I_L creates a changing input-to-output differential voltage, $V_I - V_O$, that compensates for the changing input voltage. The basic series regulator is illustrated in Figure 5-3, and the equations describing its performance are listed below.

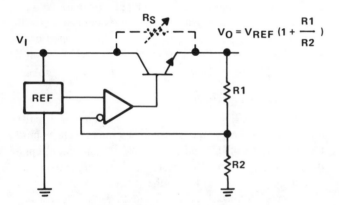

Figure 5-3. Basic Series Regulator

$$V_O = V_I - (V_I - V_O)$$
$$(V_I - V_O) = I_L R_S$$
$$V_O = V_I - I_L R_S$$

The change in R_S for a changing input voltage is:

$$\Delta R_S = \frac{\Delta V_I}{I_L}$$

The change in R_S for a changing load current:

$$\Delta R_S = \frac{\Delta I_L R_S}{I_L + \Delta I_L}$$

Series regulators provide a simple, inexpensive way to obtain a source of regulated voltage. In high-current applications, however, the voltage drop which is maintained across the control element will result in substantial power loss and a much lower efficiency regulator.

SHUNT REGULATOR

The shunt regulator employs a shunt control element in which the current is controlled to compensate for varying input voltage or changing load conditions. The basic shunt regulator is illustrated in Figure 5-4.

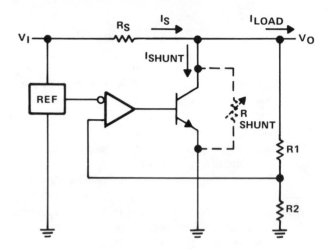

Figure 5-4. Basic Shunt Regulator

The output voltage, V_O as with the series regulator, is held constant by varying the voltage drop across the series resistor, R_S, by varying the current I_S. I_S may vary because of I_L changes or it may vary because of current, $I_{(shunt)}$, through the shunt control element. For example, as I_L increases, $I_{(shunt)}$ decreases to adjust the voltage drop across R_S. In this fashion V_O is held constant.

$$V_O = V_I - I_S R_S$$
$$I_S = I_L + I_{(shunt)}$$
$$V_O = V_I - R_S[I_L + I_{(shunt)}]$$

The change in shunt current for a changing load current is:

$$\Delta I_{(shunt)} = -\Delta I_L$$

The change in shunt current for a changing input voltage is:

$$\Delta I_{(shunt)} = \frac{\Delta V_I}{R_S}$$

$$I_{(shunt)} = \frac{V_O}{R_{(shunt)}}$$

Even though it is usually less efficient than series or switching regulators, a shunt regulator may be the best choice for some applications. The shunt regulator is less sensitive to input voltage transients; does not reflect load current transients back to the source, and is inherently short-circuit proof.

SWITCHING REGULATOR

The switching regulator employs an active switch as its control element. This switch is used to chop the input voltage at a varying duty cycle based on the load requirements. A basic switching regulator is illustrated in Figure 5-5.

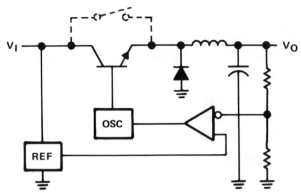

**Figure 5-5. Basic Switching Regulator
(Step-Down Configuration)**

A filter, usually an LC filter, is then used to average the voltage present at its input and deliver that voltage to the output load. Because the pass transistor is either on (saturated) or off, the power dissipated in the control element is minimal. The switching regulator is therefore more efficient than the series or shunt type. For this reason, the switching regulator becomes particularly advantageous for applications involving large input-to-output differential voltages or high load-current requirements. In the past, switching voltage regulators were discrete designs. However, recent advancements in integrated circuit technology have resulted in several monolithic switching regulator circuits that contain all of the necessary elements to design step-up, step-down, or inverting voltage converters. The duty cycle may be varied by:

1. maintaining a constant on-time, varying the frequency
2. maintaining a constant off-time, varying the frequency
3. maintaining a constant frequency, varying the on/off times

MAJOR ERROR CONTRIBUTORS

The ideal voltage regulator maintains constant output voltage despite varying input voltage, load current, and temperature conditions. Realistically, these influences affect the regulator's output voltage. In addition, the regulator's own internal inaccuracies affect the overall circuit performance. This section discusses the major error contributors, their effects, and suggests some possible solutions to the problems they create.

REGULATOR REFERENCE TECHNIQUES
There are several reference techniques employed in integrated circuit voltage regulators. Each provides its particular level of performance and problems. The optimum reference depends on the regulator's requirements.

Zener Diode Reference
The zener diode reference, as illustrated in Figure 5-6, is the simplest technique. The zener voltage itself, V_Z, forms the reference voltage, V_{REF}.

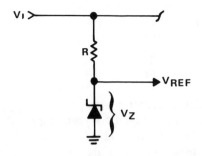

Figure 5-6. Basic Zener Reference

This technique is satisfactory for relatively stable supply-voltage and load-current applications. The changing zener current results in a change in the zener diode's reference voltage, V_Z. This zener reference model is illustrated in Figure 5-7.

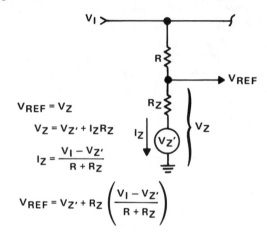

$$V_{REF} = V_Z$$
$$V_Z = V_{Z'} + I_Z R_Z$$
$$I_Z = \frac{V_I - V_{Z'}}{R + R_Z}$$
$$V_{REF} = V_{Z'} + R_Z \left(\frac{V_I - V_{Z'}}{R + R_Z} \right)$$

Figure 5-7. Zener Reference Model

Constant-Current Zener Reference
The zener reference can be refined by the addition of a constant-current source as its supply. Driving the zener diode with a constant current minimizes the effect of zener impedance on the overall stability of the zener reference. An example of this technique is illustrated in Figure 5-8. The reference voltage of this configuration is relatively independent of changes in supply voltage and load current.

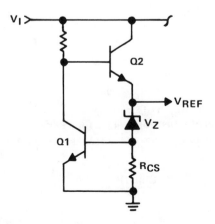

Figure 5-8. Constant-Current Zener Reference

$$V_{REF} = V_Z + V_{BE(Q1)}$$

$$I_Z = \frac{V_{BE(Q1)}}{R_{CS}}$$

In addition to superior supply voltage independence, the circuit illustrated in Figure 5-8 yields improved temperature stability. The reference voltage, V_{REF}, is the sum of the zener voltage (V_Z) and the base-emitter voltage of $Q1[V_{BE(Q1)}]$. A low temperature coefficient can be achieved by balancing the positive temperature coefficient of the zener with the negative temperature coefficient of the base-emitter junction of Q1.

Band-Gap Reference

Another popular reference is the band-gap reference, which developed from the highly predictable emitter-base voltage of integrated transistors. Basically, the reference voltage is derived from the energy-band-gap voltage of the semiconductor material [$V_{go(silicon)} = 1.204$ V]. The basic band-gap configuration is illustrated in Figure 5-9. The reference voltage, V_{REF}, in this case is:

$$V_{REF} = V_{BE(Q3)} + I_2 R_2$$

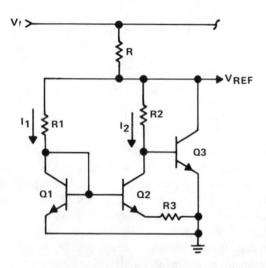

Figure 5-9. Band-Gap Reference

The resistor values of R1 and R2 are selected in such a way that the current through transistors Q1 and Q2 are significantly different ($I_1 = 10I_2$). The difference in current through transistors Q1 and Q2 also results in a difference in their respective base-emitter voltages. This voltage differential [$V_{BE(Q1)} - V_{BE(Q2)}$] will appear across R3. Application of transistors with sufficiently high gain results in current I_2 passing through R3. In this instance, I_2 is equal to:

$$\frac{V_{BE(Q1)} - V_{BE(Q2)}}{R3}$$

$$\therefore V_{REF} = V_{BE(Q3)} + \left[\left(V_{BE(Q1)} - V_{BE(Q2)}\right)\frac{R2}{R3}\right]$$

By analyzing the effect of temperature on V_{REF} it can be shown that the difference between two similar transistors' emitter-base voltages, when operated at different currents is:

$$V_{BE(Q1)} - V_{BE(Q2)} = \frac{kT}{q} \ln \frac{I_1}{I_2}$$

where

$$
\begin{aligned}
k &= \text{Boltzmann's constant} \\
T &= \text{absolute temperature — degrees K} \\
q &= \text{charge of an electron} \\
I &= \text{current}
\end{aligned}
$$

The base-emitter voltage of Q3 can also be expressed as:

$$V_{BE(Q3)} = V_{go}\left[1 - \frac{T}{T_O}\right] + V_{BEO}\left[\frac{T}{T_O}\right]$$

where

$$
\begin{aligned}
V_{go} &= \text{band-gap potential} \\
V_{BEO} &= \text{emitter-base voltage at } T_O
\end{aligned}
$$

V_{REF} can then be expressed as:

$$V_{REF} = V_{go}\left[1 - \frac{T}{T_O}\right] + V_{BEO}\left[\frac{T}{T_O}\right]$$
$$+ \frac{R2}{R3}\ \frac{kT}{q}\ \ln \frac{I_1}{I_2}$$

Differentiating with respect to temperature yields

$$\frac{dV_{REF}}{dT} = -\frac{V_{go}}{T_O} + \frac{V_{BEO}}{T_O} + \frac{R2}{R3}\ \frac{k}{q}\ln \frac{I_1}{I_2}$$

If R2, R3, and I_1 are appropriately selected such that

$$\frac{R2}{R3}\ \ln \frac{I_1}{I_2} = [V_{go} - V_{BEO(Q3)}]\ C$$

where

$$C = \frac{q}{kT_O}$$

and

$$V_{go} = 1.2\ \text{V}$$

the resulting

$$\frac{dV_{REF}}{dT} = 0$$

The reference is temperature-compensated.

Band-gap reference voltage is particularly advantageous for low-voltage applications ($V_{REF} = 1.2$ V) and it yields a reference level that is stable even with variations in supply and temperature.

SAMPLING ELEMENT

The sampling element used on most integrated circuit voltage regulators is an R1/R2 resistor divider network (Figure 5-10), which can be determined by the output-voltage-to-reference-voltage ratio.

$$\frac{V_O}{V_{REF}} = 1 + \frac{R1}{R2}$$

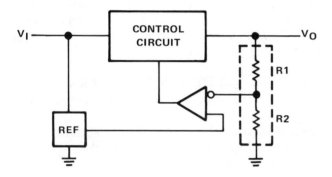

Figure 5-10. R1/R2 Ladder Network Sampling Element

Since the feedback voltage is determined by ratio and not absolute value, proportional variations in R1 and R2 have no effect on the accuracy of the integrated circuit voltage regulator. When proper attention is given to the layout of these resistors in an integrated circuit, their contribution to the error of the voltage regulator will be minimal. The initial accuracy is the only parameter affected.

ERROR AMPLIFIER PERFORMANCE

If a stable reference and an accurate output sampling element exist, the error amplifier becomes the primary factor determining the performance of the voltage regulator. Typical amplifier performance parameters such as offset, common-mode and supply-rejection ratios, output impedance, and temperature coefficient affect the accuracy and regulation of the voltage regulator. These amplifier performance parameters will affect the accuracy of the regulator due to variations in supply, load, and ambient temperature conditions.

Offset Voltage

Offset voltage is viewed by the amplifier as an error signal, as illustrated in Figure 5-11, and will cause the output to respond accordingly.

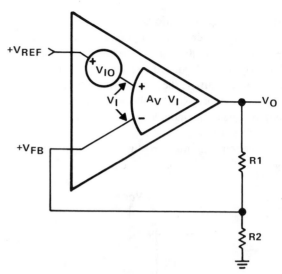

Figure 5-11. Amplifier Model Showing Input Offset Voltage Effect

$$V_O = A_V V_I$$

$$V_I = V_{REF} - V_{IO} - V_{FB}$$

$$V_{FB} = V_O \frac{R2}{R1 + R2}$$

$$V_O = \frac{V_{REF} - V_{IO}}{\dfrac{1}{A_V} + \left[\dfrac{R2}{R1 + R2}\right]}$$

If A_V is sufficiently large

$$V_O = (V_{REF} - V_{IO})\left[1 + \frac{R1}{R2}\right]$$

V_{IO} represents an initial error in the output of the integrated circuit voltage regulator. The simplest method of compensating for this error is to adjust the output voltage sampling element R1/R2.

Offset Change with Temperature

The technique discussed above compensates for the amplifier's offset voltage and yields an accurate regulator, but only at a specific temperature. In most amplifiers, the offset voltage change with temperature is proportional to the initial offset level. Trimming the output voltage sampling element, does not reduce the offset voltage but merely counteracts it. At a different ambient temperature, the offset voltage changes and, thus, error is again introduced into the voltage regulator. Monolithic integrated circuit regulators use technology that essentially eliminates offset in integrated circuit amplifiers. With minimal offset voltage, drift caused by temperature variations will have little consequence.

Supply Voltage Variations

The amplifier's power supply and common-mode rejection ratios are the primary contributors to regulator error which has been introduced by an unregulated input voltage. In an ideal amplifier, the output voltage is a function of the differential input voltage only. Realistically, the common-mode voltage of the input also influences the output voltage. The common-mode voltage is the average input voltage, referenced from the amplifier's virtual ground (see Figure 5-12 and the following equations).

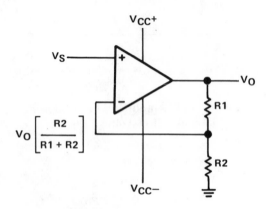

Figure 5-12. Amplifier Model Showing Common-Mode Voltage

$$\text{Virtual Ground} = \frac{V_{CC+} + V_{CC-}}{2}$$

$$V_{I(av)} = \frac{V_S + V_O\left[\dfrac{R2}{R1 + R2}\right]}{2}$$

$$V_{CM} = \frac{1}{2}\left[V_S + V_O\left(\frac{R2}{R1 + R2}\right) - \left(V_{CC+} + V_{CC-}\right)\right]$$

From this relation it can be seen that unequal variations in either power supply bus rail will result in a change in the common-mode voltage.

The common-mode voltage rejection ration (CMRR) is the ratio of the amplifier's differential voltage amplification to the common-mode voltage amplification.

$$CMRR = \frac{A_{VD}}{A_{VCM}}$$

$$A_{VCM} = \frac{A_{VD}}{CMRR}$$

That portion of output which is voltage contributed by the equivalent common-mode input voltage is:

$$V_O = V_{CM}A_{VCM} = \frac{A_{VD}V_{CM}}{CMRR}$$

The equivalent error introduced then is:

$$\text{COMMON-MODE ERROR} = \frac{V_{CM}}{CMRR}$$

The common-mode error represents an offset voltage to the amplifier. Neglecting the actual offset voltage, the output voltage of the error amplifier then becomes:

$$V_O = \left(V_{REF} + \frac{V_{CM}}{CMRR}\right)\left(1 + \frac{R1}{R2}\right)$$

Using constant-current sources in most integrated circuit amplifiers, however, yields a high power-supply (common-mode) rejection ratio. This power-supply rejection ratio is of such a large magnitude that the common-mode voltage effect on V_O can usually be neglected.

REGULATOR DESIGN CONSIDERATIONS

Various types of integrated circuit voltage regulators are available, each having its own particular characteristics, giving it advantages in various applications. The type of regulator used depends primarily upon the designer's needs and trade-offs in performance and cost.

POSITIVE VERSUS NEGATIVE REGULATORS

This classification of voltage regulators is easily understood; a positive regulator is used to regulate a positive voltage, and a negative regulator is used to regulate a negative voltage. However, what is positive and negative may vary, depending upon the ground reference.

Figure 5-13 illustrates conventional positive and negative voltage regulator applications employing a continuous and common ground. For systems operating on a single supply, the positive and negative regulators may be interchanged by floating the ground reference to the load or input. This approach to design is recommended only where ground isolation serves as an advantage to overall system performance.

Figures 5-14 and 5-15 illustrate a positive regulator in a negative configuration and a negative regulator in a positive configuration, respectively.

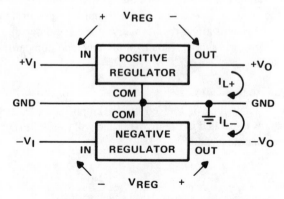

Figure 5-13. Conventional Positive/ Negative Regulator

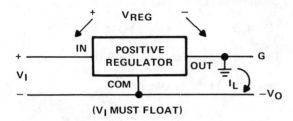

Figure 5-14. Positive Regulator in Negative Configuration (V$_I$ Must Float)

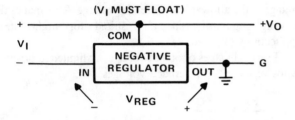

Figure 5-15. Negative Regulator in Positive Configuration (V$_I$ Must Float)

FIXED VERSUS ADJUSTABLE REGULATORS

Many fixed three-terminal voltage regulators are available in various current ranges from most major integrated circuit manufacturers. These regulators offer the designer a simple, inexpensive method to establish a regulated voltage source. Their particular advantages are:

1. Ease of use
2. Few external components required
3. Reliable performance
4. Internal thermal protection
5. Short-circuit protection

There are disadvantages. The fixed three-terminal voltage regulators cannot be precisely adjusted because their output voltage sampling elements are internal. The initial accuracy of these devices may vary as much as $\pm 5\%$ from the nominal value; also the output voltages available are limited.

Current limits are based on the voltage regulator's applicable current range and are not adjustable. Listings of some fixed and variable voltage regulators are given at the end of this chapter. Extended range operation (increasing I$_{LOAD}$) is cumbersome and requires complex external circuitry.

The adjustable regulator may be well suited for those applications requiring higher initial accuracy. This depends on the complexity of the adjustable voltage regulator. Additionally, all adjustable regulators use external feedback, which allows the designer a precise and infinite voltage selection.

The output sense may also be referred to a remote point. This allows the designer to not only extend the range of the regulator (with minimal external circuitry), but also to compensate for losses in a distributed load or external pass components. Additional features found on many adjustable

voltage regulators are: adjustable short-circuit current limiting, access to the voltage reference element, and shutdown circuitry.

DUAL-TRACKING REGULATORS

The dual tracking regulator (Figure 5-16) provides regulation for two power supply buses, usually one positive and one negative. The dual-tracking feature assures a balanced supply system by monitoring the voltage on both power supply buses. If either of the voltages sags or goes out of regulation, the tracking regulator will cause the other voltage to vary accordingly (A 10% sag in the positive voltage will result in a 10% sag in the negative voltage.). These regulators are, for the most part, restricted to applications such as linear systems where balanced supplies offer a definite performance improvement.

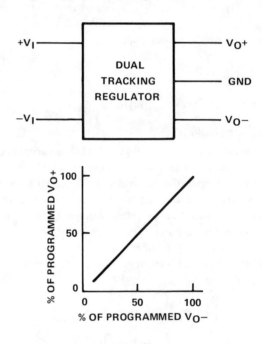

Figure 5-16. Dual Tracking Regulator

SERIES REGULATORS

The series regulator is well suited for medium current applications with nominal voltage differential requirements. Modulation of a series pass control element to maintain a well regulated, prescribed, output voltage is a straightforward design technique. Safe-operating-area protection circuits such as overvoltage, fold-back current limiting, and short-circuit protection are additional functions that series regulators can supply. The primary disadvantage of the series regulator is its power consumption. The amount of power a series regulator (Figure 5-17) will consume depends on the load current being drawn from the regulator and is proportional to the input-to-output voltage differential. The amount of power consumed becomes considerable with increasing load or differential voltage requirements. This power loss limits the amount of power that can be delivered to the load because the amount of power that can be dissipated by the series regulator is limited.

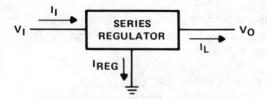

Figure 5-17. Series Regulator

The equations that describe these conditions are listed below. P_{REG} is the power lost in the regulator, I_I is the input current, I_{REG} is the regulator current and I_L is the load current. The differential voltage across the regulator is $(V_I - V_O)$.

$$P_{REG} = V_I I_I - V_O I_L$$

$$I_I = I_{REG} + I_L$$

Since I_L is much greater than I_{REG}

$$I_I = I_L$$

$$P_{REG} = I_L (V_I - V_O)$$

Floating Regulator

The floating regulator (Figure 5-18) is a variation of the series regulator. The output voltage remains constant by changing the input-to-output voltage differential for varying input voltage. The floating regulator's differential voltage is modulated such that its output voltage when referenced to its common terminal $V_{O(reg)}$ is equal to its internal reference (V_{REF}). The voltage developed across the output-to-common terminal is equal to the voltage developed across $R1 (V_{R1})$.

$$V_{O(reg)} = V_{REF} = V_{R1}$$

$$V_{R1} = V_O \left[\frac{R1}{R1 + R2} \right]$$

$$V_O = V_{REF} \left[1 + \frac{R2}{R1} \right]$$

The common-terminal voltage is:

$$V_{COM} = V_O - V_{R1} = V_O - V_{REF}$$

The input voltage seen by the floating regulator is:

$$V_{I(reg)} = V_I - V_{COM}$$

$$V_{I(reg)} = V_I - V_O + V_{REF}$$

$$V_{I(reg)} = V_{DIFF} + V_{REF}$$

Since V_{REF} is fixed, the only limitation on the input voltage is the allowable differential voltage. This makes the floating regulator especially suited for high-voltage applications ($V_I > 40$ V).

Practical values of output voltage are limited to practical ratios of output-to-reference voltages.

$$\frac{R2}{R1} = \frac{V_O}{V_{REF}} - 1$$

The floating regulator exhibits power consumption characteristics similar to that of the series regulator from which it is derived, but unlike the series regulator, it can also serve as a current regulator as shown in Figure 5-19.

$$V_O = V_{REF} \left[1 + \frac{R_L}{R_S} \right]$$

$$V_O = V_L + V_{O(reg)}$$

$$V_{O(reg)} = V_{REF}$$

$$V_L = V_{REF} \left[1 + \frac{R_L}{R_S} \right] - V_{REF}$$

$$V_L = V_{REF} \left[\frac{R_L}{R_S} \right]$$

$$I_L = \frac{V_{LOAD}}{R_L}$$

$$I_{LOAD} = \frac{V_{REF}}{R_S}$$

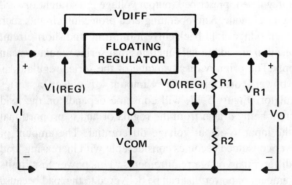

Figure 5-18. Floating Regulator

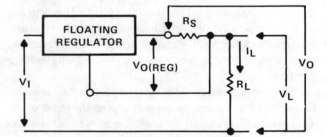

Figure 5-19. Floating Regulator as a Constant-Current Regulator

SHUNT REGULATOR

The shunt regulator, illustrated in Figure 5-20, is the simplest of all regulators. It employs a fixed resistor as its series pass element.

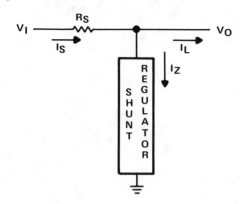

Figure 5-20. Shunt Regulator

Changes in input voltage or load current requirements are compensated by modulating the current which is shunted to ground through the regulator.

For changes in V_I: $\quad \Delta I_Z = \dfrac{\Delta V_I}{R_S}$

For changes in I_L: $\quad \Delta I_Z = - \Delta I_L$

The inherent short-circuit-proof feature of the shunt regulator makes it particularly attractive for some applications. The output voltage will be maintained until the load current required is equal to the current through the series element (see Figure 5-21).

$$I_L = I_S \ (I_Z = 0)$$

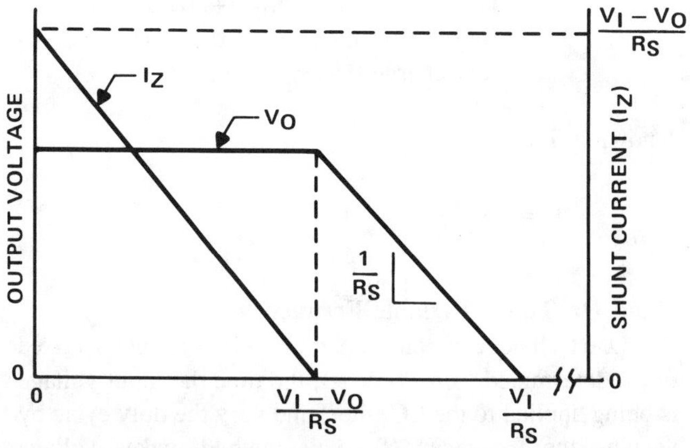

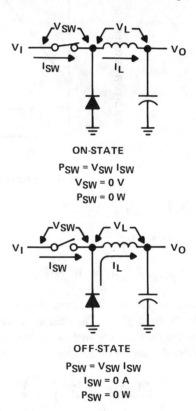

Figure 5-21. Output Voltage vs Shunt Current of a Shunt Regulator

Since the shunt regulator cannot supply any current, additional current required by the load will result in reducing the output voltage to zero.

$$V_O = V_I - I_L R_S$$

The short-circuit current of the shunt regulator then becomes:

$$V_O = 0$$

$$I_{SC} = \frac{V_I}{R_S}$$

SWITCHING REGULATORS

The switching regulator lends itself primarily to the higher power applications or those applications where power supply and system efficiency are of the utmost concern. Unlike the series regulator, the switching regulator operates its control element in an on or off mode. Switching regulator control element modes are illustrated in Figure 5-22.

ON-STATE

$P_{SW} = V_{SW} I_{SW}$
$V_{SW} = 0 \text{ V}$
$P_{SW} = 0 \text{ W}$

OFF-STATE

$P_{SW} = V_{SW} I_{SW}$
$I_{SW} = 0 \text{ A}$
$P_{SW} = 0 \text{ W}$

Figure 5-22. Switching Voltage Regulator Modes

In this manner, the control element is subjected to a high current at a very low voltage or a high differential voltage at a very low current. In either case, power dissipation in the control element is minimal. Changes in the load current or input voltage are compensated for by varying the on-off ratio (duty cycle) of the switch without increasing the internal power dissipated in the switching regulator. See Figure 5-23(a).

For the output voltage to remain constant, the net charge in the capacitor must remain constant. This means the charge delivered to the capacitor must be dissipated in the load.

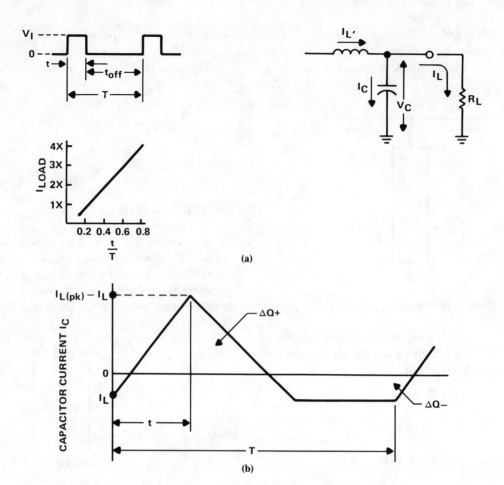

Figure 5-23. Variation of Pulse Width vs Load

$$I_C = I_L' - I_L$$

$$I_C = I_L \text{ for } I_L' = 0$$

$$I_C = I_{L(pk)} - I_L \text{ for } I_L = I_{L(pk)}$$

The capacitor current waveform then becomes that illustrated in Figure 4-23(b). The charge delivered to the capacitor and the charge dissipated by the load are equal to the areas under the capacitor current waveform.

$$\Delta Q+ \ = \ \frac{1}{2} \ \frac{(I_{L(pk)} - I_L)^2}{I_{L(pk)}} \ t \left(\frac{V_I}{V_C}\right)$$

$$\Delta Q- \ = \ I_L \left[T - \frac{1}{2} \ t \left(\frac{V_I}{V_C}\right) \right.$$

$$\left. - \ \frac{1}{2} \ t \left(\frac{I_{L(pk)} - I_L}{I_{L(pk)}}\right)\left(\frac{V_I}{V_C}\right) \right]$$

By setting $\Delta Q+$ equal to $\Delta Q-$, the relationship of I_L and $I_{L(pk)}$ for $\Delta Q = 0$ can be determined;

$$I_L = \frac{1}{2} \ I_{L(pk)} \left[\frac{V_I}{V_C} \ \frac{t}{T} \right]$$

As this demonstrates, the duty cycle t/T can be altered to compensate for input voltage changes or load variations.

The duty cycle t/T can be altered a number of different ways.

> t = t_{on} (inductor charge time)
> T = Total time (t_{on} + t_{off} + t_I) where t_I is the time from t_{off} until the start of the next charge cycle.

Knowing T then:

$$f = \frac{1}{T}$$

Fixed On Time, Variable Frequency

One technique of voltage regulation is to maintain a fixed or predetermined "on" time (t), the time the input voltage is being applied to the LC filter) and vary the duty cycle by varying the frequency (f). This method makes voltage conversion applications design easier (step-up, step-down, invert) since the energy stored in the inductor of the LC filter during the on-time (which is fixed) determines the amount of power deliverable to the load. Thus calculation of the inductor is fairly straightforward.

$$L = \frac{V}{I} \ t$$

where
- L = value of inductance in microhenrys
- V = differential voltage in volts
- I = required inductor current defined by the load in amps
- t = on-time in microseconds

The fixed-on-time approach is also advantageous from the standpoint that a consistent amount of energy is stored in the inductor during the fixed on-time period. this simplifies the design of the inductor by defining the operating parameters to which the inductor is subjected. The operating characteristic of a fixed-on-time switching voltage regulator is a varying frequency, which changes directly with changes in the load. This can be seen in Figure 5-24.

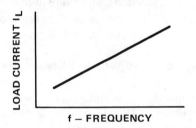

Figure 5-24. Frequency vs Load Current for Fixed On-Time SVR

Fixed Off Time, Variable Frequency

In the fixed-off-time switching voltage regulator, the average dc voltage is varied by changing the on time (t) of the switch while maintaining a fixed off-time (t_{OFF}). The fixed-off-time switching voltage regulator behaves in a manner opposite that of the fixed-on-time regulator.

As the load current increases, the on time is made to increase, thus decreasing the operating frequency; this is illustrated in Figure 5-25. This approach provides the capability to design a switching voltage regulator that will operate at a well defined minimum frequency under full load conditions.

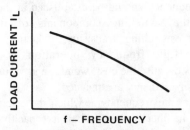

Figure 5-25. Frequency vs Load Current for Fixed Off-Time SVR

The fixed-off-time approach also allows a dc current to be established in the inductor under increased load conditions, thus reducing the ripple current while maintaining the same average current. The maximum current experienced in the inductor under transient load conditions is not as well defined as that found in the fixed-on-time regulator. Thus additional precautions should be taken to ensure that the inductor does not saturate.

Fixed-Frequency, Variable Duty Cycle

The fixed-frequency switching regulator varies the duty cycle of the pulse train in order to change the average power. The fixed-frequency concept is particularly advantageous for systems employing transformer-coupled output stages. The fixed frequency permits efficient design of the associated magnetics. Transformer coupling also has advantages in single and multiple voltage-conversion applications. The fixed-frequency regulator will establish a dc current through the inductor (for increased load conditions) to maintain the required load current, with minimal ripple current. The single-ended and transformer-coupled configurations are illustrated in Figure 5-26.

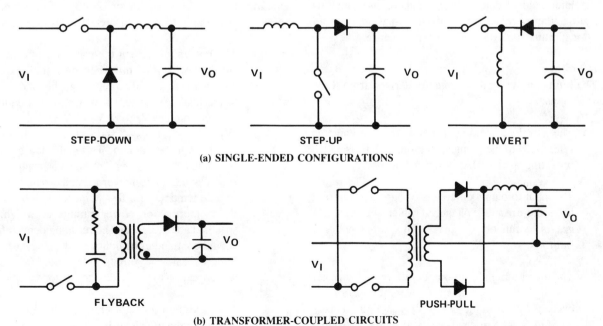

(a) SINGLE-ENDED CONFIGURATIONS

(b) TRANSFORMER-COUPLED CIRCUITS

Figure 5-26. Switching Voltage Regulator Configurations

These types of switching regulators can thus be operated with high efficiency to provide low-voltage, regulated outputs from a high-voltage, unregulated supply or vice versa. The switching frequency should be established at the optimum value for the switching components of the supply (transformer, switching transistor, inductor, and filter capacitor). High frequency operation is distinctly advantageous because the cost, weight, and volume of both L and C filter elements are reduced.

However, the frequency at which the effective series resistance of the filter capacitor equals its capacitive reactance is the maximum allowable frequency.

Operation above 20 kHz is desirable to eliminate the possibility of audible noise. Choosing an operating frequency that is too high will result in power switching transistor losses as well as "catch" diode losses. The higher cost of these high performance components must be balanced against the reduced cost, size, and weight of the L and C components when determining the optimum frequency for a specific application.

REGULATOR SAFE OPERATING AREA

The safe operating area (SOA) is a term used to define the input and output voltage range, and load current range within which any device is designed to operate reliably. Exceeding these limits will result in a catastrophic failure or will render the device temporarily inoperative, depending upon the device and its performance characteristics. Integrated circuit voltage regulators with internal current limiting, thermal and short-circuit protection will merely shut down. External components, such as pass transistors on the other hand, may respond with catastrophic failure.

REGULATOR SOA CONSIDERATIONS

Although particular design equations depend upon the type of integrated circuit voltage regulator used and its application, there are several boundaries that apply to all regulator circuits for safe, reliable performance.

Input Voltage

The limits on the input voltage are derived from three considerations:

V_{Imax}

The absolute maximum rated input voltage as referenced to the regulator's ground. This is a safe operating area (SOA) destruct limit.

$(V_I - V_O)_{min}$

The input-to-output differential voltage also referred to as the dropout voltage, at which the regulator ceases to function properly. This is a functional limit.

$(V_I - V_O)_{max}$

The maximum input-to-output differential voltage. Usually, the regulator's power dissipation is exceeded prior to the $(V_I - V_O)_{max}$ limit. This is an SOA level that can be limited by the allowable Power Dissipation (P_{Dmax}).

Load Current

I_{Lmax}

The maximum load current deliverable from the integrated circuit regulator. If internal current limiting is not provided, external protection should be provided. This is a functional limit that may be further limited by $P_{D\ max}$.

Power Dissipation

P_{Dmax}

The maximum power that can be dissipated within the regulator. Power dissipation is the product of the input-to-output differential voltage and the load current, and is normally specified at or below a given case temperature. This rating is usually based on a 150°C junction temperature limit. The power rating is an SOA limit unless the integrated circuit regulator provides an internal thermal protection.

Output Voltage of an Adjustable-Voltage Regulator

V_{Omin}

The minimum output voltage a regulator is capable of regulating. This is usually a factor of the regulator's internal reference and is a functional limit.

V_{Omax}

The maximum output voltage a regulator is capable of regulating. This is largely dependent on the input voltage and is a functional limit.

External Pass Transistor

For applications requiring additional load current, integrated circuit voltage regulator capabilities may be boosted with the addition of an external pass transistor. When employed, the external pass transistor, in addition to the voltage regulator, must be protected against operation outside its safe operating area. Operation outside the safe operating area is catastrophic to most discrete transistors.

I_{Cmax}

The maximum current the transistor is capable of sustaining. I_{Cmax} now becomes the maximum current the regulator circuit is capable of delivering to the load. Associated with I_{Cmax} is a collector-emitter voltage ($V_{CE} = V_I - V_O$). If the product $(V_I - V_{Omax})I_{Cmax}$ exceeds the SOA then I_{Cmax} will have to be derated. This will then become a functional limit instead of a catastrophic limit. I_{Cmax} is related to power dissipation and junction or case temperature. I_{Cmax} must again be derated if the thermal or power ratings at which it is specified are exceeded. The resulting derated I_{Cmax} should continue to be considered as a catastrophic limit. Actual I_{Cmax} limits and derating information will appear on the individual transistor specification.

V_{CEmax}

The maximum collector-emitter voltage that can be applied to the transistor in the off-state. Exceeding this limit can be catastrophic.

P_{Dmax}

The maximum power that can be dissipated by the transistor. This is usually specified at a specific junction or case temperature. If the transistor is operated at higher temperatures, the maximum power must be derated in accordance with the operating rules specified in the transistor's applicable specification. Prolonged operation above the transistor's maximum power rating will result in degradation or destruction of the transistor.

SAFE OPERATING PROTECTION CIRCUITS

Selection of the proper integrated circuit voltage regulators and external components will result in a reliable design in which all devices can operate well within their respective safe operating areas. Fault conditions (such as a short-circuit or excessive load) may cause components in the regulator circuit to exceed their safe operating area operation. Because of this situation, as well as protection for the load, certain protection circuits should be considered.

Reverse Bias Protection

A potentially dangerous condition may occur when a voltage regulator becomes reverse biased. For example, if the input supply were crowbarred to protect either the supply itself or additional circuitry, the filter capacitor at the output of the regulator circuit would maintain the regulator's output voltage and the regulator circuit would be reverse biased. If the regulated voltage is large enough (greater than 7 V), the regulator circuit may be damaged. To protect against this, a diode can be used as illustrated in Figure 5-27.

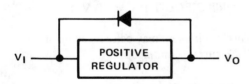

Figure 5-27. Reverse Bias Protection

CURRENT LIMITING TECHNIQUES

The type of current limiting used depends primarily on the safe operating area of the pass element used. The three basic current limiting techniques are series resistor, constant current, and fold-back current limiting.

Series Resistor

This is the simplest method for short-circuit protection. The short-circuit current is determined by the current-limiting resistor R_{CL}, illustrated in Figure 5-28.

$$V_O = V_{O(reg)} - I_L R_{CL}$$

A short-circuit condition occurs when $V_O = 0$, thus:

$$I_{SC} = I_L \text{ @ } (V_O = 0) = \frac{V_{O(reg)}}{R_{CL}}$$

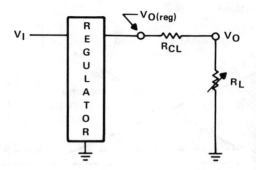

Figure 5-28. Series Resistance Current Limiter

The primary drawback of this technique is error introduced by the voltage dropped across R_{CL} under varying load conditions. The % error, as illustrated by the following equations, depends on the R_{CL} and R_L values.

$$I_L = \frac{V_O}{R_L}$$

$$V_O = \frac{V_{O(reg)}}{1 + \frac{R_{CL}}{R_L}}$$

$$\% \text{ ERROR} = \frac{V_{O(reg)} - V_O}{V_{O(reg)}}$$

$$\% \text{ ERROR} = \frac{R_{CL}}{R_L + R_{CL}}$$

Maintaining R_{CL} at a level which is an order of magnitude less than the nominal load impedance minimizes this effect.

$$R_{CL} = \frac{1}{10} R_L \qquad \% \text{ ERROR} = 9.1\%$$

This also yields a short-circuit current that is an order of magnitude greater than the normal operating load current.

$$I_{L(norm)} = \frac{V_{O(reg)}}{R_{CL} + R_{L(norm)}}$$

$$I_{SC} = \frac{V_{O(reg)}}{R_{CL}}$$

$$I_{SC} = 11 \, I_{L(norm)}$$

This technique is obviously inefficient since it requires using a regulator or pass element with current capabilities in excess (11X) of its normal operating capabilities.

The performance characteristics of a series resistance current limited regulator are illustrated in Figure 5-29.

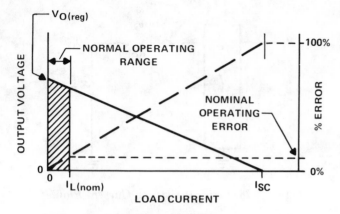

Figure 5-29. Performance Characteristics of a
Series Resistance Current-Limited Regulator

Constant-Current Limiting

Constant-current limiting is the most popular current-limiting technique in low-power, low-current regulator circuits. The basic configuration is illustrated in Figure 5-30.

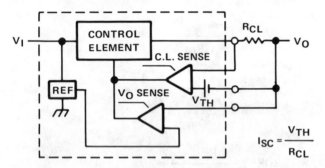

$$I_{SC} = \frac{V_{TH}}{R_{CL}}$$

Figure 5-30. Constant Current Limit Configuration

Note that this method requires access to the control element and remote voltage sense capabilities. By sensing the output voltage beyond the current limiting resistor, the circuit allows the regulator to compensate for the voltage changes across R_{CL}.

If an external pass transistor is used, its base current may be starved to accomplish constant-current limiting, as illustrated in Figure 5-31. Current limiting takes effect as the voltage drop across R_{CL} approaches the potential required to turn on transistor Q1. As Q1 is biased on, the current

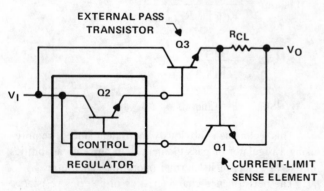

Figure 5-31. Constant Current Limiting for
External Pass Transistor Applications

supplying the base of Q2 is diverted, thus decreasing the drive current to Q3, the regulator's pass transistor. The performance characteristics of a constant-current limited regulator are illustrated in Figure 5-32.

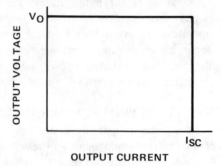

Figure 5-32. Constant Current Limiting

It should be noted that short-circuit conditions are the worst conditions that can be imposed on the pass transistor since it has to survive not only the short-circuit current, but it has to withstand the full input voltage across its collector and emitter terminals.

This normally requires the use of a pass transistor with power handling capabilities much greater than those required for normal operation i.e.:

$$V_I = 20\ V \qquad V_O = 12\ V \qquad I_O = 700\ mA$$

$$\text{NOMINAL } P_D = (20\ V - 12\ V) \times 0.7\ A = 5.6\ W$$

For $I_{SC} = 1\ A(150\%\ I_{OUT})$:

$$\text{SHORT-CIRCUIT } P_D = 20\ V \times 1\ A = 20\ W$$

This requirement may be reduced by the application of fold-back current limiting.

Fold-Back Current Limiting

Fold-back current limiting is used primarily for high-current applications where the normal operating requirements of the regulator dictate the use of an external power transistor. The principle of fold-back current limiting provides limiting at a predetermined current (I_K). At this predetermined current, feedback reduces the load current as the load continues to increase (R_L decreasing) and causes the output voltage to decay.

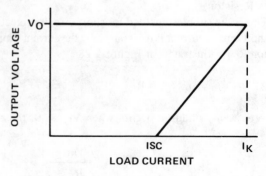

Figure 5-33. Fold-Back Current Limiting

The fold-back current-limiting circuit of Figure 5-34 behaves in a manner similar to the constant-current limit circuit illustrated in Figure 5-31. In Figure 5-33, the potential developed across the current limit sense resistor (R_{CL}) must not only develop the base-emitter voltage required to turn on Q1, but it must develop sufficient potential to overcome the voltage across resistor R1.

$$V_{BE(Q1)} = R_{CL}I_L - \frac{V_O + R_{CL}I_L}{R1 + R2} \times R1$$

$$\therefore I_K = \frac{V_{BE(Q1)}(R1 + R2) + V_O R1}{R_{CL}R2}$$

As the load current requirement increases above I_K, the output voltage (V_O) decays. The decreasing output voltage results in a proportional decrease in voltage across R1. Thus, less current is required through R_{CL} to develop sufficient potential to maintain the forward-biased condition of Q1. This can be seen in the above expression for I_K. As V_O decreases, I_K decreases. Under short-circuit conditions ($V_O = 0$) I_K becomes:

$$I_{SC} = I_K @ (V_O = 0) = \frac{V_{BE(Q1)}}{R_{CL}}\left[1 + \frac{R1}{R2}\right]$$

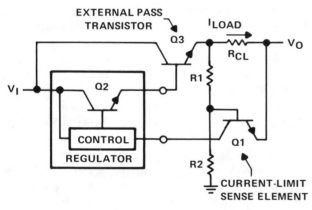

EXTERNAL PASS TRANSISTOR

Figure 5-34. Fold-Back Current Limit Configuration

The approach illustrated in Figure 5-34 allows a more efficient design because the collector current of the pass transistor is less during short-circuit conditions than it is during normal operation. This means that during short-circuit conditions, when the voltage across the pass transistor is maximum, the collector-emitter current is reduced. As illustrated in Figure 5-35, fold-back current limiting fits closer to the typical performance characteristics of the transistor, thus allowing a better design match of the pass transistor to the regulator.

THREE TERMINAL REGULATORS

Three-terminal IC regulators have been especially useful to the designer of small, regulated power supplies or on-card regulators. Three-terminal regulators are popular because they are small and require a minimum number of external components.

STABILIZATION

Mounting and using three-terminal regulators usually presents no problem, however, there are several precautions that should be observed. Positive regulators, in general, use npn emitter follower output stages whereas negative regulators use npn common-emitter stages with the load connected to the collector. The emitter follower output stage configuration is not used in negative regulators because monolithic pnp series-pass transistors are more difficult to make. Due to their output stage configuration, positive regulators are more stable than negative regulators. Therefore, the practice of bypassing positive regulators may be omitted in some applications. It is good practice, however, to use bypass capacitors at all times.

For a positive regulator, a 0.33 μF bypass capacitor should be used on the input terminals. While not necessary for stability, an output capacitor of 0.1 μF may be used to improve the transient response of the regulator. These capacitors should be on or as near as possible to the regulator terminals. See Figure 5-36.

When using a negative regulator, bypass capacitors are a must on both the input and output. Recommended values are 2 μF on the input and 1 μF on the output. It is considered

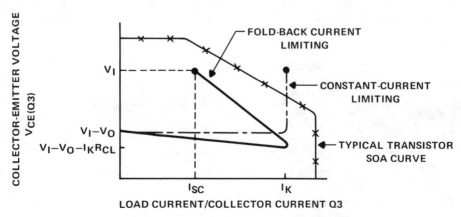

Figure 5-35. Fold-Back Current Limit Safe Operating Area

good practice to include a 0.1-μF capacitor on the output to improve the transient response (Figure 5-37). These capacitors may be mylar, ceramic, or tantalum, provided that they have good high frequency characteristics.

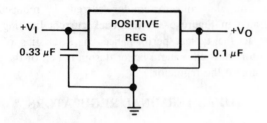

Figure 5-36. Positive Regulator

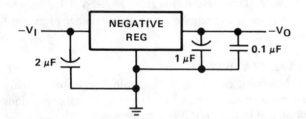

Figure 5-37. Negative Regulator

FIXED DUAL REGULATORS

When building a dual power supply with both a positive and a negative regulator, extra precautions should be taken. If there is a common load between the two supplies, latch-up may occur. Latch-up occurs because a three-terminal regulator does not tolerate a reverse voltage of more than one diode drop. To prevent this latch-up problem, it is good design practice to place reversed-biased diodes across each output of a dual supply. While the diodes should not be necessary if the dual regulator outputs are referenced to ground, latch-up may occur at the instant power is turned on, especially if the input voltage to one regulator rises faster than the other. This latch-up condition usually affects the positive regulator rather than the negative regulator. These diodes prevent reverse voltage to the regulator and prevent parasitic action from taking place when the power is turned on. The diodes should have a current rating of at least half the output current. A recommended circuit for a dual 15 V regulated supply is illustrated in Figure 5-38.

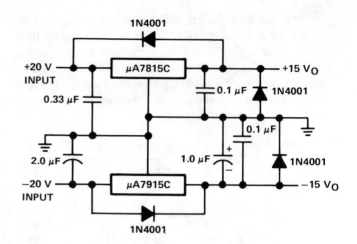

Figure 5-38. Regulated Dual Supply

In Figure 5-38, 1N4001 diodes are placed directly across the regulators, input to output. When a capacitor is connected to the regulator output, if the input is shorted to ground, the only path for discharging the capacitor normally is back through the regulator. This could be (and usually is) destructive to the regulator. The diodes across the regulator divert any discharge current, thus protecting the regulator.

SERIES ADJUSTABLE REGULATORS

Figure 5-39 illustrates a typical circuit for an LM317 adjustable positive regulator with the output adjustable from 1.2 V to 17 V and up to 1.5 A of current. (A typical input supply uses a 25.2-V transformer and a full-wave bridge rectifier.)

Stabilization, as described earlier for fixed three-terminal regulators, is usually not required. Although the LM317 is stable with no output capacitors, like any feedback circuit, certain values of external capacitance can cause excessive ringing. This effect occurs with values between 500 pF and 5000 pF. Using a 10-μF aluminum electrolytic on the output swamps this effect and ensures stability.

C1 is the power supply filter capacitor following the rectifier section and should be connected close to the regulator input for maximum stability. If the input were to be shorted, D1 would divert the discharge current around the regulator, protecting it. Also, with both D1 and D2 in the circuit, when the input is shorted, C2 is discharged through both diodes. In general, a diode should be used in

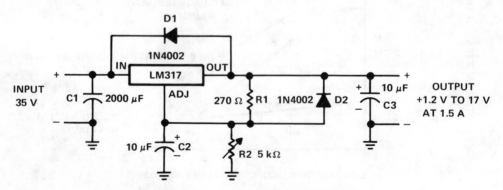

Figure 5-39. Positive Adjustable Series Regulator

the position occupied by D1 on all positive regulators to prevent reverse biasing. This becomes more important at higher output voltages since the energy stored in the capacitors is larger. Bypassing the adjustment terminal (C2) improves ripple rejection. Output capacitor C3 is added to improve the transient response of the regulator.

In both the negative (LM337) and the positive (LM317) series adjustable regulators there is an internal diode from the input to the output. If the total output capacitance is less than 25 μF, D1 may be omitted.

THERMAL CONSIDERATIONS

Like any semiconductor circuit, lower operating temperature greatly improves reliability of a voltage regulator. It is good practice to make the input-to-output drop across a three-terminal regulator as low as possible while maintaining good regulation. Larger voltage drops mean more power dissipated in the regulator. Although most regulators are rated to withstand junction temperatures as high as 150 °C, heat sinking should be provided to maintain the lowest possible temperature.

LAYOUT GUIDELINES

As implied in the previous sections, component layout and orientation plays an important, but often overlooked, role in the overall performance of the regulator. The importance of this role depends upon such things as power level, the type of regulator, the overall regulator circuit complexity, and the environment in which the regulator operates. The general layout rules, as well as remote voltage sensing, and component layout guidelines are discussed in the following text.

LAYOUT DESIGN FACTORS

Most integrated circuit regulators use wide-band transistors to optimize their response. These regulators must be compensated to ensure stable closed-loop operation. This compensation can be counteracted by a layout which has excess external stray capacitance and line inductance. For this reason, circuit lead lengths should be held to a minimum. Lead lengths associated with external compensation or pass transistor elements are of primary concern. These

components especially, should be located as close as possible to the regulator control circuit. In addition to affecting a regulator's susceptibility to spurious oscillation, the layout of the regulator also affects its accuracy and performance.

Input Ground Loop

Improper placement of the input capacitor can induce unwanted ripple on the output voltage. Care should be taken to ensure that currents in the input circuit do not flow in the ground line that is in common with the load return. This would cause an error voltage resulting from the peak currents of the filter capacitor flowing through the line resistance of the load return. See Figure 5-40 for an illustration of this effect.

Output Ground Loop

Similar in nature to the problem discussed on the input, excessive lead length in the ground return line of the output results in additional error. Because the load current flows in the ground line, an error equivalent to the load current multiplied by the line resistance (R3′) will be introduced in the output voltage.

Remote Voltage Sense

The voltage regulator should be located as close as possible to the load. This is true especially if the output voltage sense circuitry is internal to the regulator's control device. Excessive lead length will result in an error voltage developed across the line resistance (R4′).

$$V_O = V_{O(reg)} - (R_2' + R_3' + R_4') I_L + R_2' I_{reg}$$

$$ERROR = I_L(R_3' + R_4') - I_{Ireg} R_2'$$

If the voltage sense is available externally, the effect of the line resistance can be minimized. By referencing the low current external voltage sense input to the load, losses in the output line are compensated. Since the current in the sense line is very small, error introduced by its line resistance is negligible (Figure 5-41).

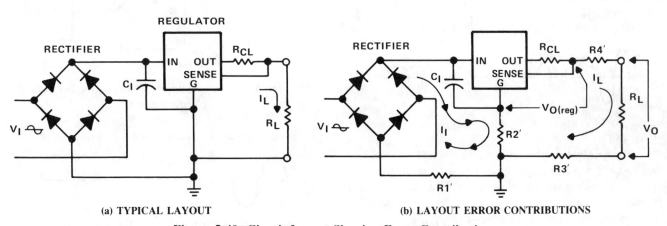

(a) TYPICAL LAYOUT   (b) LAYOUT ERROR CONTRIBUTIONS

Figure 5-40. Circuit Layout Showing Error Contributions

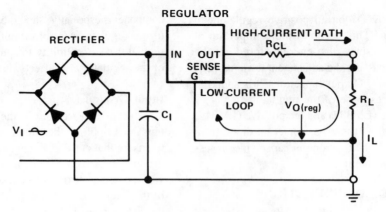

Figure 5-41. Proper Regulator Layout

Thermal Profile Concerns

All semiconductor devices are affected by temperature; therefore, care should be taken to the placement of these devices so that their thermal properties are not additive. This is especially important where external pass transistors or reference elements are concerned.

INPUT SUPPLY DESIGN

When the power source is an ac voltage, the transformer, rectifier, and input filter design are as important as the regulator design itself for optimum system performances. This section presents input supply and filter design information for designing a basic capacitor input supply.

TRANSFORMER/RECTIFIER CONFIGURATION

The input supply consists of three basic sections: (1) input transformer, (2) rectifier, and (3) filter as illustrated in Figure 5-42.

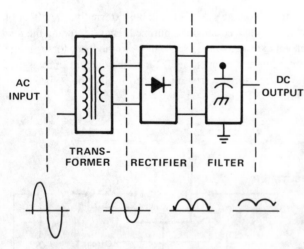

Figure 5-42. Input Supply

The first two sections, the transformer and the rectifier, are partially dependent upon each other because the structure of one depends upon that of the other. The most common transformer configurations and their associated rectifier circuits are illustrated in Figure 5-43.

The particular configuration used depends upon the application. The half-wave circuit [Figure 5-3(a)] is used in low-current applications. This is because the single rectifier diode experiences the total load current and its conversion efficiency is less than 50%. The full-wave configurations [Figures 5-43(b) and 5-43(c)] are used for higher current applications. The characteristic output voltage waveforms of these configurations are illustrated in Figure 5-44.

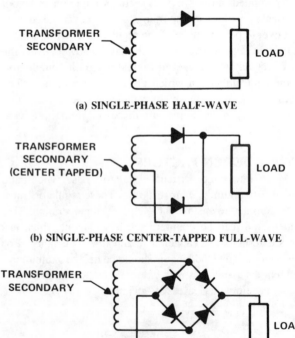

(a) SINGLE-PHASE HALF-WAVE

(b) SINGLE-PHASE CENTER-TAPPED FULL-WAVE

(c) SINGLE-PHASE FULL-WAVE BRIDGE

Figure 5-43. Input Supply Transformer/Rectifier Configurations

Figure 5-44. Rectifier Output-Voltage Waveforms

Before the input supply and its associated filter can be designed, the voltage, current, and ripple requirements of its load must be fully defined. The load, as far as the input supply is concerned, is the regulator circuit. Therefore, the input requirements of the regulator itself become the governing conditions.

Because the input requirements of the regulator control circuit govern the input supply and filter design, it is easiest to work backwards from the load to the transformer primary.

CAPACITOR INPUT FILTER DESIGN

The most practical approach to a capacitor-input filter design remains the graphical approach presented by O.H. Schade[1] in 1943. The curves illustrated in Figures 5-45 through 5-48 contain all of the design information required for full-wave and half-wave rectifier circuits.

[1]O.H. Schade, "Analysis of Rectifier Operation", *Proc. IRE.*, **VOL. 31**, 343, 1943.

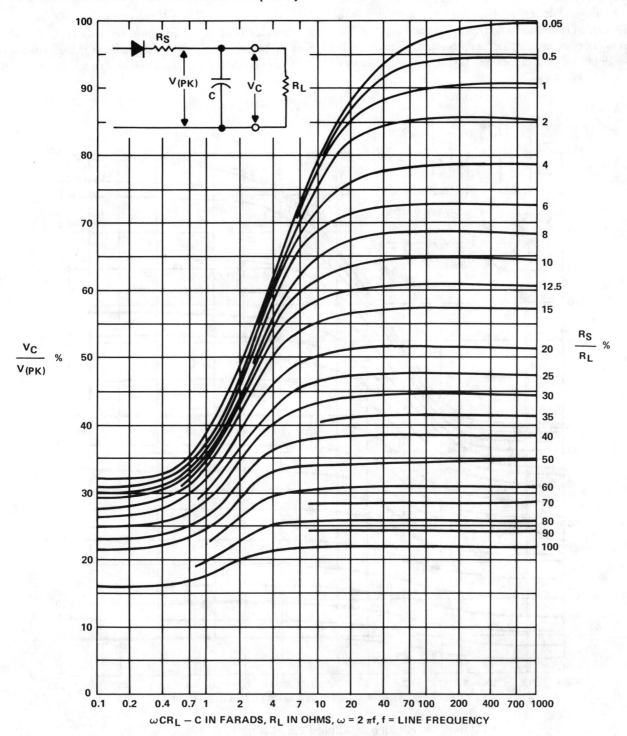

Figure 5-45. Relation of Applied Alternating Peak Voltage to Direct Output Voltage in Half-Wave Capacitor-Input Circuits (From O.H. Schade, *Proc. IRE*, Vol. 31, p. 343, 1943)

Figures 5-45 and 5-46 illustrate the ratio of the dc output voltage developed (V_C), to the applied peak input voltage ($V_{(PK)}$), as a function of ωCR_L, for half-wave and full-wave rectified signals respectively. For a full-wave rectified application, the voltage reduction is less then 10% for $\omega CR_L > 10$ and $R_S/R_L < 0.5\%$. As illustrated, the voltage reduction decreases as ωCR_L increases or the R_S/R_L ratio decreases. Minimizing the reduction rate, contrary to initial impressions, may prove to be detrimental to the optimum circuit design. Further reduction requires a reduction in the series to load resistance ratio (R_S/R_L) for any given ωCR_L. This will result in a higher peak-to-average current ratio through the rectifier diodes (see Figure 5-47). In addition,

and probably of more concern, this increases the surge current experienced by the rectifier diodes during turn-on of the supply. It is important to realize that the surge current is limited only by the series resistance R_S.

$$I_{SURGE} = \frac{V_{SEC(PK)}}{R_S}$$

In order to control the surge current, additional resistance is often required in series with each rectifier. It is evident that a compromise must be made between the voltage reduction and the rectifier current ratings.

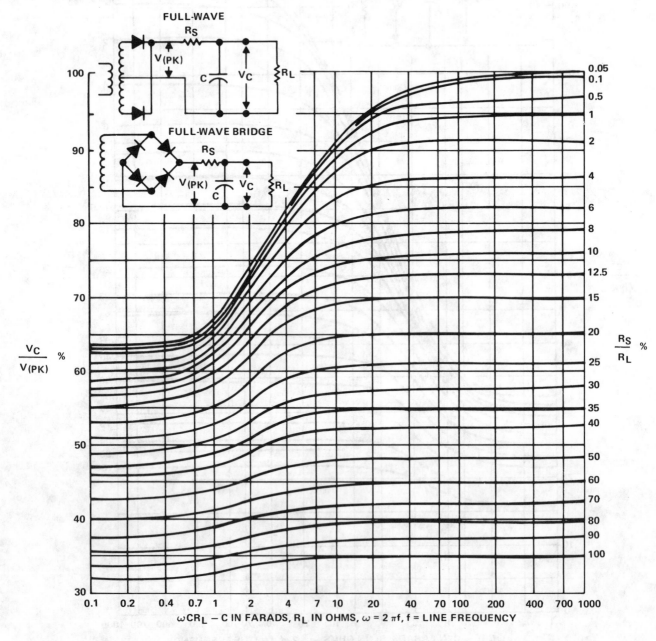

Figure 5-46. Relation of Applied Alternating Peak Voltage to
Direct Output Voltage in Full-Wave Capacitor-Input Circuits
(From O.H. Schade, *Proc. IRE*, Vol. 31, p. 344, 1943)

The maximum instantaneous surge current is $V_{(pk)}/R_S$. The time constant (τ) of capacitor C is:

$$\tau \cong R_S C$$

As a rule of thumb, the surge current will not damage the diode if:

$$I_{SURGE} < I_{F(SURGE)max} \text{ and } \tau < 8.3 \text{ ms}$$

Figure 5-48 illustrates the relationships between the ripple factor r_f, $\omega C R_L$, and R_S/R_L. The ripple factor is the ratio of the rms value of the ripple component of the output voltage, expressed as a percent of the nominal dc output voltage.

VOLTAGE REGULATOR TERMS AND DEFINITIONS

SERIES REGULATORS

Input Regulation

The change in output voltage, often expressed as a percentage of output voltage, for a change in input voltage from one level to another.
NOTE: Sometimes this characteristic is normalized with respect to the input voltage change.

Ripple Rejection

The ratio of the peak-to-peak input ripple voltage, to the peak-to-peak output ripple voltage.
NOTE: This is the reciprocal of ripple sensitivity.

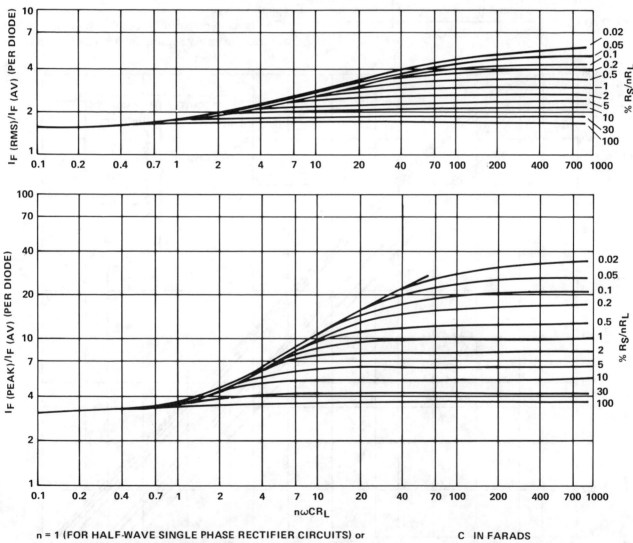

n = 1 (FOR HALF-WAVE SINGLE PHASE RECTIFIER CIRCUITS) or
n = 2 (FOR FULL-WAVE SINGLE PHASE RECTIFIER CIRCUITS)
$\omega = 2\pi f$ = LINE FREQUENCY

C IN FARADS
R_L IN OHMS
R_S = RMS EQUIVALENT SOURCE RESISTANCE

Figure 5-47. Relation of RMS and Peak to Average Diode Current in Capacitor Input Circuits
(From O.H. Schade, *Proc. IRE*, Vol. 31, p. 345, 1943)

Ripple Sensitivity

The ratio of the peak-to-peak output ripple voltage, sometimes expressed as a percentage of output voltage, to the peak-to-peak input ripple voltage.

NOTE: This is the reciprocal of ripple rejection.

Output Regulation

The change in output voltage, often expressed as a percentage of output voltage, for a change in load current from one level to another.

Temperature Coefficient of Output Voltage (αV_O)

The ratio of the change in output voltage, usually expressed as a percentage of output voltage, to a change in temperature. This is the average value for the total temperature change.

$$\alpha_{VO} = \pm \left[\frac{(V_O \text{ at } T_2) - (V_O \text{ at } T_1)}{V_O \text{ at } 25\,^\circ\text{C}} \right] \frac{100\%}{T_2 - T_1}$$

Output Voltage Change with Temperature

The percentage of change in the output voltage for a change in temperature. This is the net change over the total temperature range.

Output Voltage Long-Term Drift

The change in output voltage over a long period of time.

Output Noise Voltage

The rms output voltage, sometimes expressed as a percentage of the dc output voltage, with constant load and no input ripple.

Current-Limit Sense-Voltage

A voltage that is a function of the load current and is normally used for control of the current-limiting circuitry.

Dropout Voltage

The input-to-output differential voltage at which the circuit ceases to regulate against further reductions in voltage.

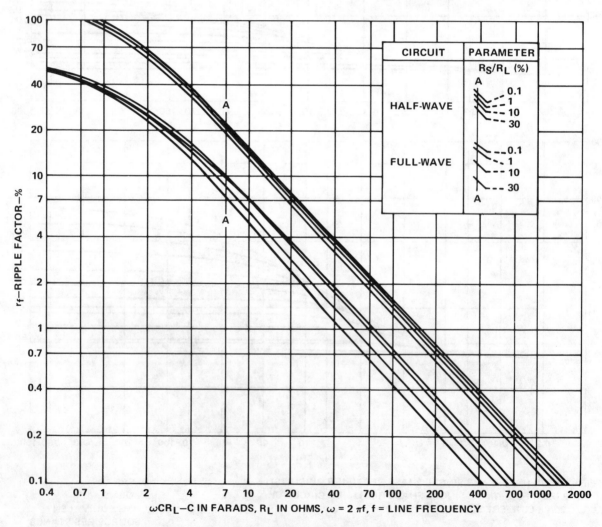

Figure 5-48. Root-Mean-Square Ripple Voltage for Capacitor-Input Circuits
(From O.H. Schade, *Proc. IRE*, Vol. 31, p. 346, 1943)

Feedback Sense Voltage

The voltage that is a function of the output voltage, used for control of the regulator.

Reference Voltage

The voltage that is compared with the feedback sense voltage to control the regulator.

Bias Current

The difference between input and output currents. NOTE: This is sometimes referred to as quiescent current.

Standby Current

The input current drawn by the regulator with no output load and no reference voltage load.

Short-Circuit Output Current

The output current of the regulator with the output shorted.

Peak Output Current

The maximum output current that can be obtained from the regulator.

Shunt Regulators

NOTE: These terms and symbols are based on JEDEC and IEC standards for voltage regulator diodes.

Shunt Regulator

A device having a voltage current characteristic similar to that of a voltage regulator diode. It is normally biased to operate in a region of low differential resistance (corresponding to the breakdown region of a regulator diode) and develops across its terminals an essentially constant voltage throughout a specified current range.

Anode

The electrode to which the regulator current flows within the regulator when it is biased for regulation.

Cathode

The electrode from which the regulator current flows within the regulator when it is biased for regulation.

Reference Input Voltage (V_{ref}) (of an adjustable shunt regulator)

The voltage at the reference input terminal with respect to the anode terminal.

Temperature Coefficient of Reference Voltage (V_{ref})

The ratio of the change in reference voltage to the change in temperature. This is the average value for the total temperature change. To obtain a value in ppm/°C:

$$V_{ref} = \left[\frac{(V_{ref} \text{ at } T_2) - (V_{ref} \text{ at } T_1)}{V_{ref} \text{ at } 25\,°C} \right] \left[\frac{10^6}{T_2 - T_1} \right]$$

Regulator Voltage (V_Z)

The dc voltage across the regulator when it is biased for regulation.

Regular Current (I_Z)

The dc current through the regulator when it is biased for regulation.

Regulator Current near Lower Knee of Regulation Range (I_{ZK})

The regulator current near the lower limit of the region within which regulation occurs; this corresponds to the breakdown knee of a regulator diode.

Regulator Current at Maximum Limit of Regulation Range (I_{ZM})

The regulator current above which the differential resistance of the regulator significantly increases.

Differential Regulator Resistance (r_z)

The quotient of a change in voltage across the regulator and the corresponding change in current through the regulator when it is biased for regulation.

Noise Voltage (V_{nz})

The rms voltage across the regulator with the regulator biased for regulation and with no input ripple.

FIXED OUTPUT VOLTAGE REGULATORS

POSITIVE OUTPUT REGULATORS				
DEVICE SERIES	OUTPUT VOLTAGE TOLERANCE	MINIMUM DIFFERENTIAL VOLTAGE	OUTPUT CURRENT RATING	AVAILABLE VOLTAGE SELECTIONS
LM340	±5%	2 V	1.5 A	8.5 V to 24 V
TL780-00C	±2%	2 V	1.5 A	3.5 V to 15 V
uA7800C	±5%	2 V to 3 V	1.5 A	10.5 V to 24 V
uA78L00C	±10%	2 V to 2.5 V	100 mA	8.2.6 V to 15 V
uA78L00AC	±5%	2 V	100 mA	8.2.6 V to 15 V
uA78M00C	±5%	2 V to 3 V	500 mA	9.5 V to 24 V

NEGATIVE OUTPUT REGULATORS				
DEVICE SERIES	OUTPUT VOLTAGE TOLERANCE	MINIMUM DIFFERENTIAL VOLTAGE	OUTPUT CURRENT RATING	AVAILABLE VOLTAGE SELECTIONS
uA7900C	± 5%	2 V to 3 V	1.5 A	8.5 V to 24 V
MC79L00AC	± 5%	2 V	100 mA	3.5 V to 15 V
uA79M00C*	± 5%	2 V to 3 V	500 mA	7.5 V to 24 V

AVAILABLE OUTPUT VOLTAGES FOR ABOVE REGULATOR SERIES														
DEVICE SERIES	VOLTAGE SELECTIONS													
	2.6	5	5.2	6	8	8.5	9	10	12	15	18	20	22	24
LM340		X		X	X			X	X	X	X			X
MC79L00AC		X							X	X				
TL780-00C		X							X	X				
uA7800C		X		X	X	X		X	X	X	X		X	X
uA78L00C	X	X		6.2	X		X	X	X	X				
uA78L00AC	X	X		6.2	X		X	X	X	X				
uA78M00C*		X		X	X			X	X	X		X	X	X
uA7900C		X	X	X	X				X	X	X			X
uA79M00C*		X		X	X				X	X		X		X

*Also available in Military Temperature Range (M suffix)

VARIABLE OUTPUT VOLTAGE REGULATORS

POSITIVE OUTPUT SERIES REGULATORS				
DEVICE NUMBER	DIFFERENTIAL VOLTAGE		OUTPUT VOLTAGE	OUTPUT CURRENT RATING
	MIN	MAX	MAX	
LM317	1.2 V	37 V	$V_I - 1.2$ V	1.5 A
TL317	1.2 V	32 V	$V_I - 1.2$ V	100 mA
TL783	1.25 V	125 V	125 V	700 mA
uA723C*	3 V	38 V	37 V	25 mA

NEGATIVE OUTPUT SERIES REGULATORS				
DEVICE NUMBER	DIFFERENTIAL VOLTAGE		OUTPUT VOLTAGE	OUTPUT CURRENT RATING
	MIN	MAX	MAX	
LM337	1.2 V	37 V	$V_I + 1.2$ V	1.5 A

POSITIVE SHUNT REGULATORS					
DEVICE NUMBER	SHUNT VOLTAGE		SHUNT CURRENT		TEMP. COEFF.
	MIN	MAX	MIN	MAX	MAX
TL430C	3 V	30 V	2 mA	100 mA	200 ppm/°C
TL431C	2.55 V	36 V	1 mA	100 mA	100 ppm/°C
TL431I**	2.55 V	36 V	1 mA	100 mA	100 ppm/°C

*Also available in Military Temperature Range (M suffix)
**I denotes Industrial Temperature Range

Section 6

Switching Power Supply Design

Modern electronic equipment usually requires one or more dc power sources. The usual method of supplying dc power is a power supply which converts ac power to dc power. The two types of dc power supplies in common use are classified by the type of regulator employed; linear regulator or switching regulator.

Linear power supplies consist of a power transformer, rectifier and filter circuits, and a linear regulator. Switching power supplies don't require line transformers; the ac input is rectified and filtered, chopped by a high frequency transistor switch/transformer combination, then rectified and filtered again.

Switching power supplies have been used for some time in the military and space industry due to their smaller size and higher efficiency. In 1975 switching power supplies were more cost effective than linear power supplies from approximately the 500 W power level. Now the breakeven point is down to approximately 5 W.

BASIC OPERATION OF SWITCHING REGULATORS

Figure 6-1 is a block diagram of a typical switching power supply which consists of four basic circuits:
1. Input rectifier and filter.
2. High frequency inverter.
3. Output rectifier and filter.
4. Control circuit.

The ac line voltage is applied to an input rectifier and filter circuit. The dc voltage output from the rectifier and filter circuit is switched to a higher frequency (typically 25 kHz to 100 kHz) by the transistor switch in the high frequency inverter circuit. This circuit contains either a high frequency transformer or inductor, depending on the output voltage required.

Output from the high frequency inverter circuit is applied to the output rectifier and filter circuit. The circuit is monitored and controlled by the control circuit which attempts to keep the output at a constant level.

The control circuit consists of an oscillator driving a pulse-width modulator, an error amplifier, and a precision voltage reference. The error amplifier compares the input reference voltage with a sample of the voltage from the output rectifier and filter circuit. As the load increases the output voltage drops. The error amplifier senses this drop and causes the pulse-width modulator to remain on for a longer period of time, delivering wider control pulses to the transistor switch.

The width of the pulse determines how long the transistor switch allows current to flow through the high frequency transformer and, ultimately, how much voltage is available at the output. If the load decreases, narrower control pulses are delivered to the switching transistor until the output voltage remains at a constant value.

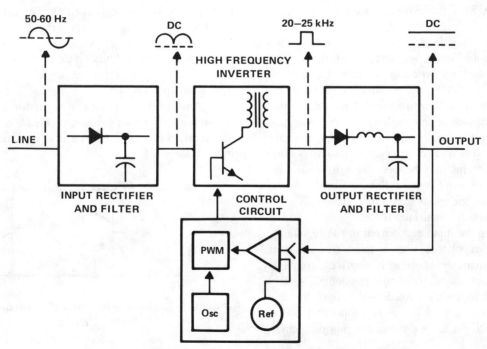

Figure 6-1. Basic Switching Regulator Block Diagram

ADVANTAGES OF A SWITCHING REGULATOR

The primary advantages of switching regulators are higher efficiency and smaller size. Conventional linear series and shunt regulators operate in a continuous conduction mode, dissipating relatively large amounts of power. The efficiency of linear regulators is typically around 40% to 50%. When the input-to-output voltage differential is large, the resultant efficiency is much lower than 40%.

Switching regulators have typical efficiencies of 60% to 90%; much higher than either the linear series or shunt regulator. Switching regulators achieve their higher efficiency as a result of three factors:

1. The power-transistor switch is always turned completely on or off, except when it is switching between these two states, resulting in either low voltage or low current during most of its operation.
2. Good regulation can be achieved over a wide range of input voltage.
3. High efficiency can be maintained over wide ranges in load-current.

Switching regulators use the on-off duty cycle of the transistor switch to regulate the output voltage and current. By using a frequency much higher than the line frequency (typically 20 kHz to 500 kHz) the transformers, chokes, capacitors, and other filter elements can be made smaller, lighter, and less costly. The smaller elements used in switching regulators result in smaller power losses than the larger components used in linear regulators.

The highest cost elements of a switching power supply are the transistor switches. The remaining costs, in descending order, are due to the magnetic components, capacitors, and rectifiers.

DISADVANTAGES OF A SWITCHING REGULATOR

Switching regulators can generate some electromagnetic and radio frequency interference (EMI/RFI) noise due to high switching currents and short rise and fall times. EMI/RFI noise, which is generated at higher frequencies (100 kHz to 500 kHz), is easily filtered. In those applications where a large series impedance appears between the supply and the regulator, the rapid changes in current also generate a certain amount of noise.

These problems may be overcome or significantly reduced by one or more of the following steps:

1. Reducing the series impedance.
2. Increasing the switching time.
3. Filtering the input and output of the regulator.

Switching regulators with a fixed frequency are easier to filter than regulators with a variable frequency because the noise is at only one frequency. Variable frequency regulators with a fixed "on" time increase or decrease the switching frequency in proportion to load changes, presenting a more difficult filtering problem.

BASIC SWITCHING REGULATOR ARCHITECTURE

There are three basic switching regulator configurations from which the majority of present day circuits are derived:

1. Step-down, or "buck", regulator.
2. Step-up, or "boost", regulator.
3. Inverting, or "flyback" regulator (which is a variation of the "boost" regulator).

The Step-Down Regulator

Figure 6-2 illustrates the basic step-down or "buck" regulator. The output voltage of this configuration is always less than the input voltage.

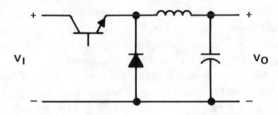

Figure 6-2. Step-Down or "Buck" Switching Regulator Circuit

In the buck circuit, a semiconductor switch is placed in series with the dc input from the input rectifier/filter circuit. The switch interrupts the dc input voltage providing a variable-width pulse to a simple averaging LC filter. When the switch is closed, the dc input voltage is applied across the filter and current flows through the inductor to the load. When the switch is open, the energy stored in the field of the inductor maintains the current through the load.

In the buck circuit, peak switching current is proportional to the load current. The output voltage is equal to the input voltage times the duty cycle.

$$V_O = V_I \times \text{Duty Cycle}$$

The Step-Up Regulator

Another basic switching regulator configuration is the step-up or "boost" regulator (Figure 6-3). In this type of circuit, the output voltage is always greater than the input voltage.

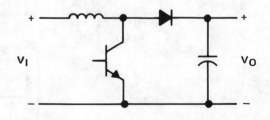

Figure 6-3. Step-Up or "Boost" Switching Regulator Circuit

The boost circuit first stores energy in the inductor and then delivers this stored energy along with the energy from the dc input voltage to the load. When the switch is closed, current flows through the inductor and the switch, charging the inductor but delivering no current to the load. When the switch is open, the voltage across the load equals the dc input voltage plus the charge stored in the inductor. The inductor discharges, delivering current to the load.

The peak switching current in the boost circuit is not related to the load current. The power output of a boost regulator can be determined by the following equation:

$$P_{OUT} = \frac{LI^2f}{2}$$

where:

P_{OUT} = power output
L = inductance
I = peak current
f = operating frequency

The Inverting Regulator

The third switching regulator configuration is the inverting or "flyback" regulator. This circuit is a variation of the step-up or "boost" circuit discussed previously. The flyback circuit is illustrated in Figure 6-4.

Flyback regulators, which evolved from "boost" regulators, deliver only the energy stored by the inductor to the load. This type of circuit can step the input voltage up or down. When the switch is closed the inductor is charged, but no current is delivered to the load because the diode is reverse biased. When the switch is open the blocking diode is forward biased and the energy stored in the inductor is transferred through it to the load.

The flyback circuit delivers a fixed amount of power to the load regardless of load impedance. It is widely used in photo flash, capacitor-discharge ignition circuits, and battery chargers.

To determine the output voltage of an electronic equipment supply, the load (R_L) must be known. If the load is known, the output voltage may be calculated using the following equation:

$$V_O = \sqrt{P_O R_L} = I \sqrt{\frac{LfR_L}{2}}$$

where:

V_O = voltage output
P_O = power out
R_L = load resistance
I = inductor current
f = operating frequency

The inductor current is proportional to the "on time" (duty cycle) of the switch and regulation is achieved by varying the duty cycle. However, the output also depends on the load resistance (which was not true with the step-down circuit).

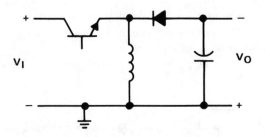

Figure 6-4. Inverting or "Flyback" Switching Regulator Circuit

Transient response to abrupt changes in the load is difficult to analyze. Practical solutions include limiting the minimum load and using the proper amount of filter capacitance to give the regulator time to respond to this change. Flyback type circuits are used at power levels of up to 100 W.

FORWARD CONVERTERS

The forward converter family, which includes the push-pull and half-bridge circuits, evolved from the step-down or "buck" type of regulator. A typical forward converter circuit is illustrated in Figure 6-5.

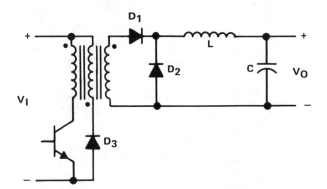

Figure 6-5. Forward Converter Switching Regulator

When the transistor switch is turned on the transformer delivers power to the load through diode D1 and the LC filter. When the switch is turned off diode D2 is forward biased and maintains current to the load.

Without the third winding, and diode D3, the converter would lose efficiency at higher frequencies. The function of this winding is to return energy stored in the transformer to the line and reset the transformer core after each cycle of operation.

This is a popular low power (up to about 200 W) converter and is almost immune to transformer saturation problems.

Push-Pull Converter

The push-pull converter is probably one of the oldest switching regulator type circuits. It was first used in the 1930's with mechanical vibrators functioning as the switch. When transistors became available push-pull converters were used as free-running oscillators in the primary of many automobile communication converters.

Some recreational vehicles still use this free-running type of oscillator converter in dc-to-dc converters as well as in dc-to-ac inverters.

A typical push-pull converter circuit is shown in Figure 6-6.

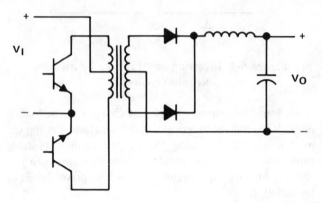

Figure 6-6. Basic Push-Pull Converter Circuit

Half Bridge Converter

The most popular type of high power converter is the half bridge circuit illustrated in Figure 6-7.

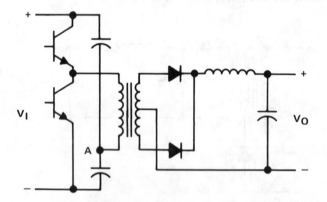

Figure 6-7. Half Bridge Converter Circuit

The half bridge converter has several advantages over the push-pull circuit. First, the midpoint between the capacitors (point A) can be charged to $V_I/2$. This allows the use of transistors with lower breakdown voltage. Second, because the primary is driven in both directions (push-pull), a full-wave rectifier and filter are used which allows the transformer core to be more effectively utilized.

Full Bridge Converter

In contrast to the half bridge, the full bridge (or H-Bridge) converter uses four transistors as shown in Figure 6-8.

In a full bridge circuit the diagonally opposite transistors (Q1/Q2 or Q3/Q4) are turned on during alternate half cycles. The highest voltage any transistor is subjected to is V_I, rather than $2 \times V_I$ as is the case in the push-pull converter circuit. The full bridge circuit offers increased reliability because less voltage and current stress is placed on the transistors.

The disadvantage of this circuit is the space required by the four transistors and the cost of the two additional transistors.

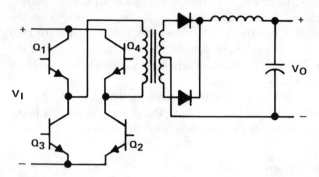

Figure 6-8. Full Bridge Converter Circuit

TL593 FLOPPY DISK POWER SUPPLY

The TL593 incorporates, on a single monolithic chip, all the functions required for a pulse-width modulation control circuit. The TL593 is similar to the TL594, from which it was derived, except that the TL593 includes a current-limit amplifier instead of a second error amplifier.

The current-limit amplifier of the TL593 has an offset voltage of approximately 80 mV in series with the inverting input (pin 15). This makes it easier to design the current-limit portion of the power supply and also requires fewer components. With 80 mV on the inverting input, it is only necessary to apply an 80 mV control voltage to the non-inverting input (pin 16). This is easily accomplished by taking the voltage across a resistor in series with the load.

The floppy disk power supply schematic is shown in Figure 6-9. The power supply uses a pair of TIP34 pnp transistors in a push-pull configuration. The oscillation frequency is set at 25 kHz and −5 V at 500 mA by the .01 μF capacitor on pin 5 and the 5 kΩ resistor on pin 6.

The center connection of the two 5.6 kΩ resistors on pins 13 and 14 establishes a 2.5 V reference voltage on pin 2, which is the inverting input of the voltage control error amplifier. The voltage feedback to pin 1, the non-inverting input, comes from the center connection of the two 5.6 kΩ resistors located on the 5 V/2.5 A power supply output terminal. Because this voltage supplies the logic circuits, it requires closer regulation.

The 24 V winding, on the other hand, is not critical as it furnishes voltage for the stepping motor. The −5 V supply is regulated separately with a uA7905 three-terminal regulator.

In choosing components for this circuit, the same precautions taken in the construction of any switching power supply should be observed; be careful of layout, ground loops, and heatsinking of the power transistors. In the output section, where high frequency rectifiers are needed, either Schottky or fast recovery diodes should be used. For output capacitors, low equivalent series

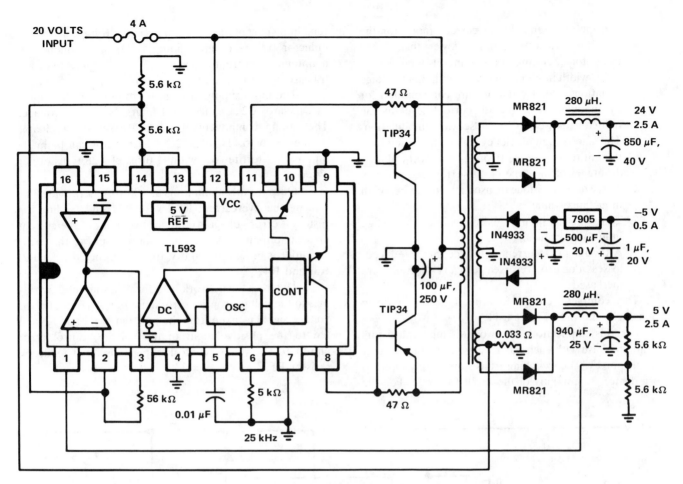

Figure 6-9. TL593 Floppy Disk Power Supply

resistance (ESR) types should be considered. The output ripple depends more on this resistance than on the capacitor value.

TRANSFORMER CONSTRUCTION

The transformer for this circuit was wound on a toroid core. The core used was 3C8 ferrite material (F-42908-TC).

The winding layout is shown in Figure 6-10.

Transformer Winding Data

Primary A + B = 20 turns bifilar #20 HNP
Secondary C + D = 28 turns bifilar #20 HNP over A+B
Secondary E + F = 6 turns bifilar #20 HNP over C+D
Secondary G + H = 10 turns bifilar #26 HNP over E+F

NOTE: All windings to be center tapped.

DC Resistance

Winding 1 - 3 = 0.11 Ω
Winding 4 - 6 = 0.11 Ω
Winding 7 - 9 = 0.025 Ω
Winding 10 - 12 = 0.15 Ω

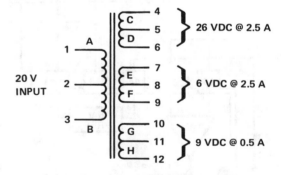

Figure 6-10. Transformer Winding Layout

TL594 12-VOLT TO 5-VOLT STEP-DOWN REGULATOR

The TL594 switching voltage regulator operates as a step-down converter in a discontinuous mode. When the output current falls below a specified minimum value the inductor current becomes discontinuous. The advantages of a step-down converter in this mode of operation are:

1. The ripple voltage at the output can be kept low, even in high current designs.
2. The ratio of peak current in the switching device to output current is determined by the inductor value and is typically low. For a specific output

current requirement, the current rating for the switching transistor can be lower than for a transistor operating in a continuous mode.

3. Pulse-width modulation occurs with input voltage variations. Load variations are compensated for by modulation of the dc current level in the inductor, as well as by pulse-width modulation. This allows high efficiency to be maintained over the entire load range (from I_0 max to I_0 min).

The disadvantages of this type of converter are:

1. The size of the inductor used may result in a high inductance value.
2. Transient response is impaired by high inductance values.
3. Although peak current in the rectifier is reduced, losses due to reverse recovery current are increased.

The complete circuit for the TL594 Step-down Regulator is shown in Figure 6-11.

For this application the two switching transistors operate in phase with each other by grounding the output control, pin 13. The switching transistors supply input to the inductor, L, for part of the oscillator cycle. For the remaining part of the oscillator cycle the voltage across the inductor reverses and diode D1 starts conducting, maintaining current flow in the inductor while the transistors are off (see Figure 6-12).

The input supply through R1 to pin 12 is decoupled by capacitor C2. Capacitor C4 filters the output voltage. The timing components C3 and R6 set the oscillator frequency to 15 kHz. The 2.2 mH inductor can be made on an RM7 ferrite core with 94 turns of #28 transformer wire.

Output current limiting of 500 mA is provided by sensing the overcurrent level with R11 and feeding the resultant error voltage to the positive input of the current error amplifier on Pin 1. The negative input to this error amplifier is biased to 500 mV from reference divider R2, R3, and R4.

This resistor network also furnishes about 2.3 volts bias to the voltage control error amplifier. An output error voltage signal is taken from the junction of R1 and R8 and fed to the positive input of the voltage control error amplifier. The voltage control loop gain is set by feedback resistor R5.

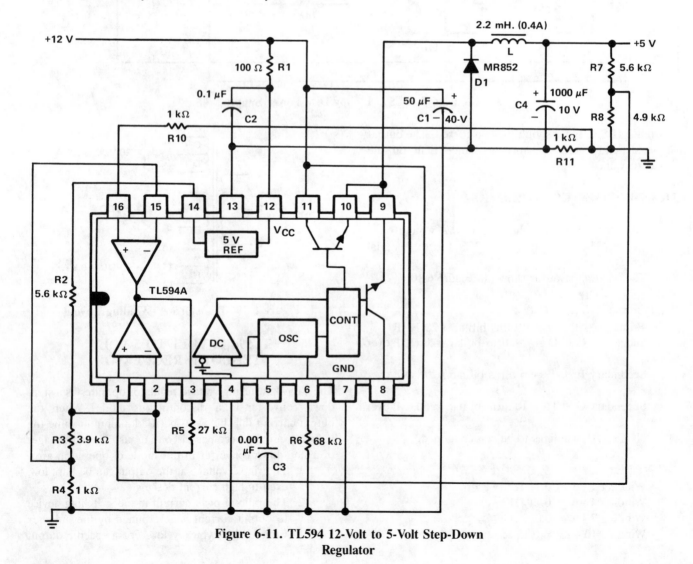

Figure 6-11. TL594 12-Volt to 5-Volt Step-Down Regulator

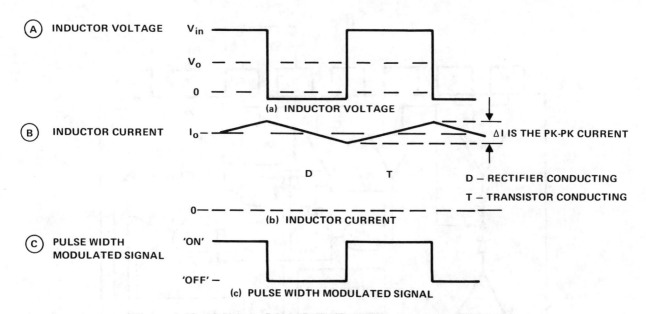

(a) INDUCTOR VOLTAGE

(b) INDUCTOR CURRENT

ΔI IS THE PK-PK CURRENT

D — RECTIFIER CONDUCTING

T — TRANSISTOR CONDUCTING

(c) PULSE WIDTH MODULATED SIGNAL

Figure 6-12. 12-Volt to 5-Volt Series Switching Regulator Waveforms

SPECIFICATIONS

Input Voltage	— 12 V nominal (10 V to 15 V)
Output Voltage	— 5 V ± 10%
Output Ripple	— 50 mV$_{pp}$
Output Current	— 400 mA
Output Power	— 2 W at 5 V output
Short Circuit Protection	— 500 mA constant current
Efficiency	— typically 70%

THE TL594 CONTROL CIRCUIT

The TL594 is a fixed frequency pulse-width-modulation control for switching power supplies and voltage converters. The TL594 includes an adjustable oscillator, a pulse width modulator, and an error amplifier. Additional functions include over-current detection, independent dead-time control, a precision 5 V reference regulator, and output control logic which allows single-ended or push-pull operation of the two switching transistors. Figure 6-13 shows a block diagram of the TL594.

Modulation of the output pulses is accomplished by comparing the sawtooth waveform created by the internal oscillator on timing capacitor C_T to either of two control signals. The output stage is enabled when the sawtooth voltage is greater than the voltage of the control signal. See Figure 6-14.

As the control signals increase the output pulse width decreases. The control signals are derived from two sources: the dead-time control and the error amplifiers. The dead-time comparator has a fixed offset of 10 mV which provides a preset dead time of about 5%. This is the minimum dead time that can be programmed with pin 4 grounded.

The pulse-width-modulation (PWM) comparator generates the control difference signal created by the input from either of the error amplifiers. One error amplifier is used to monitor the output voltage and provide a change in control signal voltage. The other error amplifier monitors the output current and its change in control voltage provides current limiting.

Reference Regulator

The internal 5-volt reference at pin 14 provides a stable reference for the control logic, pulse steering flip-flop, oscillator, dead-time-control comparator and pulse-width-modulation circuitry. It is a band-gap circuit with short circuit protection and is internally programmed to an accuracy of ± 5%.

Oscillator

The internal oscillator provides a positive sawtooth waveform to the dead-time and PWM comparators for comparison with the various control signals. The oscillator frequency is set by an external timing capacitor and resistor on pins 5 and 6. The oscillator frequency is determined by the equation:

$$f_{OSC} = \frac{1}{R_T C_T} \text{ (single-ended applications)}$$

The oscillator frequency is equal to the output frequency only for single-ended applications. The output frequency for push-pull applications is one-half the oscillator frequency as shown by the equation:

$$f_{OSC} = \frac{1}{2 R_T C_T} \text{ (push-pull applications)}$$

There is a frequency variation of ± 5% between devices due to internal component tolerances.

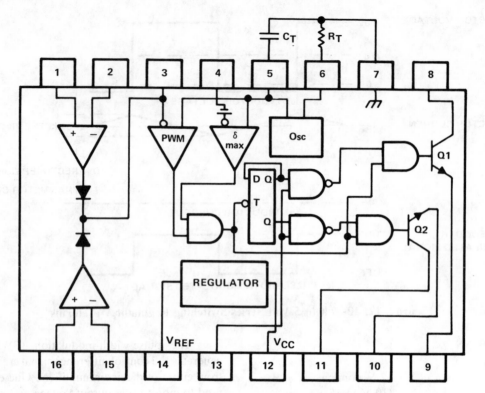

PIN ASSIGNMENT

PIN NO.	FUNCTION	PIN NO.	FUNCTION
1.	ERROR AMP. 1, NONINVERTING INPUT	9.	DRIVE TRANSISTOR 1, EMITTER
2.	ERROR AMP. 1, INVERTING INPUT	10.	DRIVE TRANSISTOR 2, EMITTER
3.	COMPENSATION INPUT	11.	DRIVE TRANSISTOR 2, COLLECTOR
4.	DEAD TIME CONTROL INPUT	12.	INPUT SUPPLY
5.	OSCILLATOR TIMING CAPACITOR	13.	OUTPUT MODE CONTROL
6.	OSCILLATOR TIMING RESISTOR	14.	STABILIZED REFERENCE VOLTAGE
7.	GROUND	15.	ERROR AMP 2, INVERTING INPUT
8.	DRIVE TRANSISTOR 1, COLLECTOR	16.	ERROR AMP 2, NONINVERTING INPUT

Figure 6-13. TL594 Block Diagram

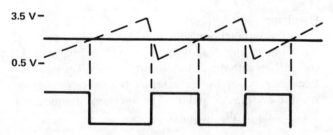

PWM CONTROL RANGE, PIN 3

**RESULTANT OUTPUT PULSE WITH
PIN 3 VOLTAGE AS ABOVE PIN 13 WIRED FOR
SINGLE ENDED OPERATION**

Figure 6-14. Output Pulses vs. Sawtooth Control Voltage

The oscillator charges the external timing capacitor, C_T, with a constant current which is determined by the external timing resistor, R_T. This circuit produces a linear ramp voltage waveform. When the voltage across the timing capacitor reaches 3.0 V, the circuit discharges and the charging cycle is initiated again.

Dead Time and PWM Comparators

Both the dead time and PWM comparator functions use a single logic comparator with parallel input stages. The comparator output is a pulse-width-modulated signal, whose width is determined by comparison with the oscillator ramp waveform. The comparator outputs drive the output control logic.

A fixed 100 mV offset voltage input to the dead-time comparator allows a minimum dead time between output pulses to be maintained when the dead-time control input (pin 4) is grounded (Figure 6-15).

The full range of pulse width control (0% - 90%) is available when the dead time control voltage (pin 4) is between 3.0 V and 0 V. The relationship between control voltage and maximum output pulse width is essentially linear. A typical application for this may be in a push-pull converter circuit where overlap of the conduction times of power transistors must be avoided.

The PWM comparator input is coupled internally to the outputs of the two error amplifiers. This input is accessible on pin 3 for control loop compensation. The

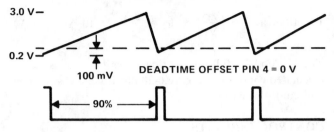

DEADTIME OFFSET PIN 4 = 0 V

OSCILLATOR RAMP, PIN 5

MAXIMUM OUTPUT PULSE WIDTH
SINGLE ENDED OPERATION, PIN 4 = 0 V

Figure 6-15. Deadtime Comparator Operation

output pulse width varies from 90% of the period to zero as the voltage present at pin 3 varies from 0.5 V to 3.5 V (Figure 6-14).

Error Amplifiers

Both error amplifiers are high gain amplifiers which operate as single-ended single-supply amplifiers, in that each output is active high only. This allows each amplifier to pull up independently for a decreasing output pulse width demand. With the outputs ORed together, the amplifier with the higher output level dominates. The open loop gain of these amplifiers is 60 dB. Both error amplifiers exhibit a response time of about 400 ns from their inputs to their outputs on pin 3. Figure 6-16 shows the amplifier transfer characteristics and a Bode plot of the gain curves.

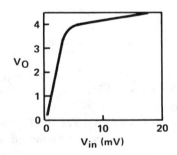

AMPLIFIER TRANSFER CHARACTERISTICS

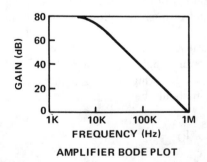

AMPLIFIER BODE PLOT

Figure 6-16. Amplifier Performance Curves

Output Logic Control

The output control logic interfaces the pulse width modulator to the output stages. In the single-ended mode (both outputs conducting simultaneously) the pulse width modulated signal is gated through to both output stages when the output control (pin 13) is connected to ground.

For push-pull operation (each output stage conducting alternately) the output control (pin 13) is connected to the internal reference voltage (pin 14) enabling the pulse steering flip-flop. The flip-flop is toggled on the the trailing edge of the pulse width modulated signal gating it to each of the outputs alternately; therefore, the switching frequency of each output is one-half the oscillator frequency.

The output control (pin 13) must never be left open. It may be connected to the internal voltage reference (pin 14) or ground (pin 7).

The Output Driver Stages

The two identical Darlington output drivers may be operated in parallel or push-pull mode. Both the collector and emitter terminals are available for various drive configurations. $V_{CE(sat)}$ of each output at 200 mA is typically 1.1 V in common-emitter configuration and 1.5 V in common-collector configuration. These drivers are protected against overload but do not have sufficient current-limiting to be operated as current source outputs.

SOFT-START

Use of a soft-start protection circuit is recommended. This circuit prevents current surges during power-up and protects against false signals which might be created by the control circuit when power is applied.

Implementing a soft-start circuit is relatively simple using the dead-time control input (pin 4). Figure 6-17 shows an example.

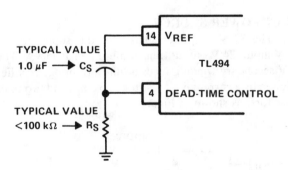

Figure 6-17. Soft-Start Circuit

Initially, capacitor C_S forces the dead-time control input to follow the internal 5 V reference which disables both outputs (100% dead time). As the capacitor charges through R_S, the output pulse width increases until the control loop takes command.

OVER-VOLTAGE PROTECTION

The dead-time control input (pin 4) also provides a convenient input for over-voltage protection, which may be sensed as an output voltage condition, or input voltage protection as shown in Figure 6-18.

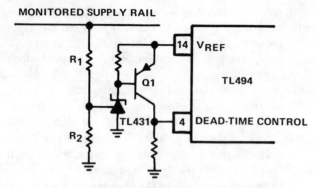

Figure 6-18. Over-Voltage Protection Circuit

A TL431 is used as the sensing element. When the monitored supply rail voltage increases to the point that 2.5 V is developed across R2, the TL431 conducts, Q1 becomes forward biased, and the dead-time control is pulled up to the reference voltage which disables the output transistors.

DESIGNING A POWER SUPPLY
5 VOLT/10 AMP OUTPUT

DESIGN OBJECTIVE
This design uses the TL594 integrated circuit based on the following parameters:

$$V_O = 5 \text{ V}$$
$$V_I = 32 \text{ V}$$
$$I_O = 10 \text{ A}$$
$$f = 20 \text{ kHz Switching Frequency}$$
$$V_R = 100 \text{ mV peak-to-peak } (V_{ripple})$$
$$\Delta I_L = 1.5 \text{ A Inductor Current Change}$$

INPUT POWER SOURCE
The 32 V dc power source for this supply uses a 120 V input, 24 V output transformer rated at 75 VA. The 24 V secondary winding feeds a full-wave bridge rectifier followed by a current limit resistor (0.3 ohm) and two filter capacitors, as shown in Figure 6-19.

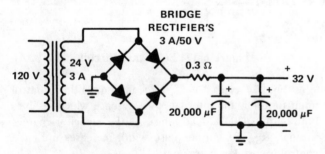

Figure 6-19. Input Power Source

The output current and voltage are determined by the following equations.

$$V \text{ rectifier} = V \text{ secondary} \times \sqrt{2} = 24 \text{ V} \times \sqrt{2} = 34 \text{ V}$$

$$I \text{ rectifier(avg)} \approx \left(\frac{V_O}{V_I}\right) \times I_O \approx \frac{5 \text{ V}}{32 \text{ V}} \times 10 \text{ A} \approx 1.6 \text{ A}$$

The 3 A/50 V full-wave bridge rectifier meets these calculated conditions. Figure 6-20 illustrates the switching and control section.

CONTROL CIRCUITS
Oscillator
The TL594 oscillator frequency is controlled by connecting an external timing circuit consisting of a capacitor and resistor to pins 5 and 6. The oscillator is set to operate at 20 kHz using the component values calculated by the following equations.

$$f = \frac{1}{R_T C_T}$$

where:

$$R_T = \text{Value of timing resistor}$$
$$C_T = \text{Value of timing capacitor}$$

Choose $C_T = 0.001 \text{ } \mu\text{F}$ and calculate RT.

$$R_T = \frac{1}{f \times C_T} = \frac{1}{20 \times 10^3 \times 0.001 \times 10^{-6}}$$
$$= 50 \text{ k}\Omega$$

Error Amplifier
The error amplifier compares a sample of the 5 V output to a reference and adjusts the pulse-width modulator to maintain a constant output as shown in Figure 6-21.

The TL594's internal 5 V reference (pin 14) is divided to 2.5 V by R3 and R4. The output voltage error signal is also divided to 2.5 V by R8 and R9. If the output must be regulated to exactly 5.0 V, a 10 kΩ potentiometer may be used in place of R8 to provide an adjustment control.

To increase the stability of the error amplifier circuit, the output of the error amplifier is fed back to the inverting input through R7, reducing the gain to 100.

Current Limit Amplifier
The power supply was designed for a 10 A load current and an I_L swing of 1.5 A; therefore, the short circuit current should be

$$I_{SC} = I_O + \frac{I_L}{2} = 10.75 \text{ A}$$

The current limit portion of the circuit is shown in Figure 6-22.

Resistors R1 and R2 set a reference of about 1 V on the inverting input of the current limit amplifier. Resistor R11, in series with the load, applies 1 V to the non-

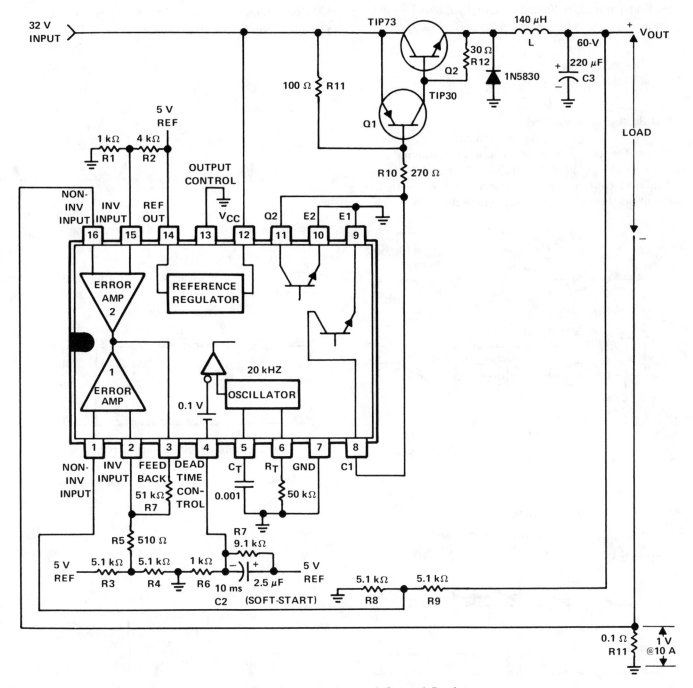

Figure 6-20. Switching and Control Section

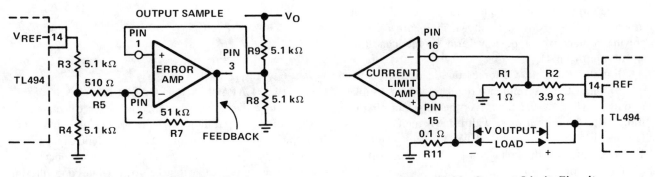

Figure 6-21. Error Amplifier Section

Figure 6-22. Current Limit Circuit

inverting terminal of the current limit amplifier when the load current reaches 10 A. The output pulse width will be reduced accordingly. The value of R11 is calculated as follows:

$$R11 = \frac{1\ V}{10\ A} = 0.1\ \Omega$$

Soft-Start and Dead Time

To reduce stress on the switching transistors at startup, the startup surge which occurs as the output filter capacitor charges must be reduced. The availability of the dead-time control makes implementation of a soft-start circuit, as shown in Figure 6-23, relatively simple.

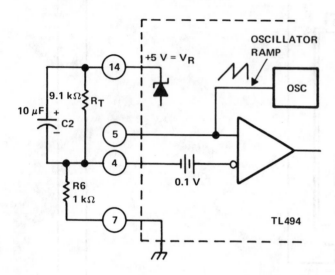

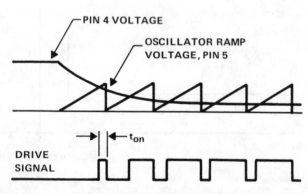

Figure 6-23. Soft-Start Circuit

The "soft-start" circuit allows the pulse width at the output to increase slowly, as shown in Figure 6-23, by applying a negative slope waveform to the dead-time control input (pin 4).

Initially, capacitor C2 forces the dead-time control input to follow the 5 V reference regulator, which disables the outputs (100% dead time). As the capacitor charges through R6, the output pulse width slowly increases until the control loop takes command. With a resistor ratio of 1:10 for R6 and R7, the voltage at pin 4 after startup will be 0.1 × 5 V or 0.5 V.

The soft-start time is generally in the range of 25 to 100 clock cycles. If we select 50 clock cycles at a 20 kHz switching rate, the soft start time is calculated as follows:

$$T = \frac{1}{f} = \frac{1}{20\ kHz} = 50\ \mu s\ \text{per clock cycle}$$

The value of the capacitor is then determined by

$$C2 = \frac{\text{soft start time}}{R6} = \frac{50\ \mu s \times 50\ \text{cycles}}{1\ k\Omega}$$

$$= 2.5\ \mu F$$

This helps to eliminate any false signals which might be created by the control circuit as power is applied.

INDUCTOR CALCULATIONS

The switching circuit used is shown in Figure 6-24.

INDUCTOR CALCULATIONS

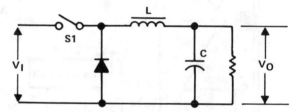

Figure 6-24. Switching Circuit

The size of the inductor (L) required is calculated as follows:

$$d = \text{Duty Cycle} = \frac{V_O}{V_I} = \frac{5\ V}{32\ V} = 0.156$$

$$f = 20\ kHz\ \text{(Design Objective)}$$

$$t_{on} = \text{time on (S1 closed)} = \frac{1}{f} \times d = 7.8\ \mu s$$

$$t_{off} = \text{time off (S1 open)} = \frac{1}{f} - t_{on} = 42.2\ \mu s$$

$$L \approx \frac{(V_I - V_O) \times t_{on}}{\Delta IL} \approx \frac{(32\ V - 5\ V) \times 7.8\ \mu s}{1.5\ A}$$

$$\approx 140.4\ \mu H$$

$$L \approx 140\ \mu H$$

OUTPUT CAPACITANCE CALCULATIONS

Once the filter inductance has been calculated, the value of the output filter capacitor is calculated to meet the output ripple requirements. An electrolytic capacitor can be modeled as a series connection of an inductance, a resistance, and a capacitance. To provide good filtering, the ripple frequency must be far below the frequencies at which the series inductance becomes important; so, the two components of interest are the capacitance and the effective series resistance (ESR). The maximum ESR is calculated according to the relation between the specified peak-to-peak ripple voltage and peak-to-peak ripple current.

$$ESR(max) = \frac{\Delta V_O \text{ (ripple)}}{\Delta I_L} = \frac{0.1 \text{ V}}{1.5 \text{ A}} = 0.067 \ \Omega$$

The minimum capacitance of C3 necessary to maintain the V_O ripple voltage at less than the 100 mV design objective was calculated according to the following equation.

$$C3 = \frac{\Delta I_L}{8 \ f \Delta V_O} = \frac{1.5 \text{ A}}{8 \times 20 \times 10^3 \times 0.1 \text{ V}} = 94 \ \mu F$$

A 220 μF, 60 V capacitor is selected because it has a maximum ESR of 0.074 Ω and a maximum ripple current of 2.8 A.

TRANSISTOR POWER SWITCH CALCULATIONS

The transistor power switch was constructed with a TIP30 pnp drive transistor and a TIP73 npn output transistor. These two power devices were connected in a pnp hybrid Darlington circuit configuration as shown in Figure 6-25.

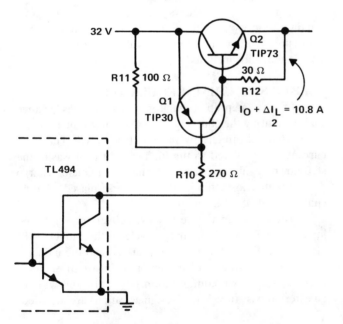

Figure 6-25. Power Switch Section

The hybrid Darlington must be saturated at a maximum output current of $I_O + \Delta I_L/2$ or 10.8 A. The Darlington h_{FE} at 10.8 A must be high enough not to exceed the 250 mA maximum output collector current of the TL594. Based on published TIP30 and TIP73 h_{FE}

specifications, the required power switch minimum drive was calculated by the following equations to be 108 mA.

$$h_{FE}(Q1) \text{ at } I_C \text{ of } 1.2 \text{ A} = 10$$

$$h_{FE}(Q2) \text{ at } I_C \text{ of } 12.0 \text{ A} = 10$$

$$i_B \geq \frac{I_O + \dfrac{\Delta I_L}{2}}{h_{FE}(Q2) \times h_{FE}(Q1)} \geq 108 \text{ mA}$$

The value of R10 was calculated by the following equation.

$$R10 \leq \frac{V_I - (V_{BE}(Q1) + V_{CE}(TL594))}{i_B}$$

$$= \frac{32 - (1.3 + 0.7)}{0.108}$$

$$R10 \leq 277 \ \Omega$$

Used on these calculations, the nearest standard resistor value of 270 Ω was selected for R10. Resistors R11 and R12 permit the discharge of carriers in the switching transistors when they are turned off.

The power supply described demonstrates the flexibility of the TL594 pulse-width-modulation control circuit. This power supply design demonstrates many of the power supply control methods provided by the TL594 as well as the versatility of the control circuit.

TL497A SWITCHING VOLTAGE REGULATOR

The TL497A is a fixed-on-time, variable-frequency voltage regulator controller. The block diagram of the TL497A is shown in Figure 6-26.

The on-time is controlled by an external capacitor connected between the frequency control pin (pin 3) and ground. This capacitor, C_T, is charged by an internal constant-current generator to a predetermined threshold. The charging current and threshold vary proportionately with V_{CC}; thus, the on-time remains constant over the allowable input voltage range.

The output voltage is controlled by two series resistors in parallel with the supply output. The resistance ratios are calculated to supply 1.2 V to the comparator input (pin 1) at the desired output voltage. This feedback voltage is compared to the 1.2 V bandgap reference by the high-gain error amplifier. When the output voltage falls below the desired voltage, the error amplifier enables the oscillator circuit, which charges and discharges C_T.

The npn output transistor is driven "on" during the charging cycle of C_T. The internal transistor can switch currents up to 500 mA. It is current driven to allow operation from either the positive supply voltage or ground. An internal diode matched to the current

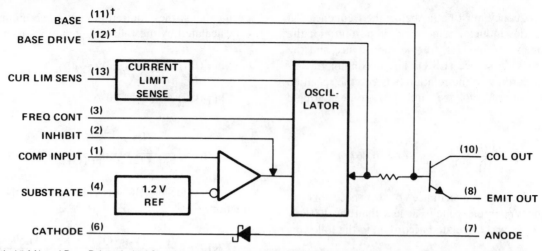

† The Base pin (#11) and Base Drive pin (#12) are used for device testing only. They are not normally used in circuit applications of the device.

Figure 6-26. TL497A Block Diagram

characteristics of the output transistor is included on the chip and may be used for blocking or commutating purposes.

The TL497A also contains current-limiting circuitry which senses the peak currents in the switching regulator and protects the inductor against saturation and the output transistor against overstress. The current limit is adjustable and is set by a single sense resistor between pins 13 and 14. The current-limit circuitry is activated when 0.5 V is developed across current-limit resistor R_{CL}.

The TL497A contains all the active elements required for constructing a single-ended dc-to-dc converter. The output transistor and the rectifier are uncommitted allowing maximum flexibility in the choice of circuit configuration.

The TL497A's primary feature is design simplicity. Using six external components; three resistors, two capacitors, and one inductor, the step-up, step-down, and inverting power supplies shown in Figure 6-27 may be constructed.

STEP DOWN	STEP-UP	INVERTING
POS → POS	POS → POS	POS → NEG
$+V_I > +V_O$	$+V_O > +V_I$	$+V_I > -V_O$

STEP-DOWN SWITCHING REGULATOR

The circuit in Figure 6-28(a) illustrates the basic configuration for a step-down switching regulator.

When switch S1 is closed, the current in the inductor and the voltage across the capacitor start to build up. The current increases while switch S1 is closed as shown by the inductor waveform in Figure 6-28(b). The peak current in the inductor is dependent on the time S1 is closed (t_{on}).

When S1 opens, the current through the inductor is I_{pk}. Since the current cannot change instantaneously, the voltage across the inductor inverts, and the blocking diode (D1) is forward biased providing a current path for the discharge of the inductor into the load and filter capacitor.

The inductor current discharges linearly as illustrated in Figure 6-28b.

For the output voltage to remain constant, the net charge delivered to the filter capacitor must be zero. The charge delivered to the capacitor from the inductor must be dissipated in the load. Since the charge developed in the inductor is fixed (constant on-time), the time required for the load to dissipate that charge will vary with the load requirements. It is important to use a filter capacitor with minimal ESR. Note, however, some ripple voltage is required for proper operation of the regulator.

Figure 6-29 shows a positive, step-down configuration both with and without an external pass transistor. Design equations for calculating the external components are included.

STEP-UP SWITCHING REGULATOR

In the step-up regulator, the formulas change slightly. During the charging cycle (S1 closed) the inductor (L) is charged directly by the input potential. The peak current is not related to the load current as it was in the step-down regulator because during the inductor charge cycle the blocking diode D1 is reverse-biased and no charge is delivered to the load.

The circuit in Figure 6-30(a) delivers power to the load only during the discharge cycle of the inductor (S1 open). The diode (D1) is forward biased and the inductor discharges into the load capacitor. Figure 6-31 shows a positive, step-down configuration both with and without an external pass transistor. Design equations are included.

INVERTING CONFIGURATION

The inverting regulator is similar to the step-up regulator. During the charging cycle of the inductor the load is isolated from the input. The only difference is in the potential across the inductor during its discharge. This can best be demonstrated by a review of the basic inverting regulator circuit (Figure 6-32).

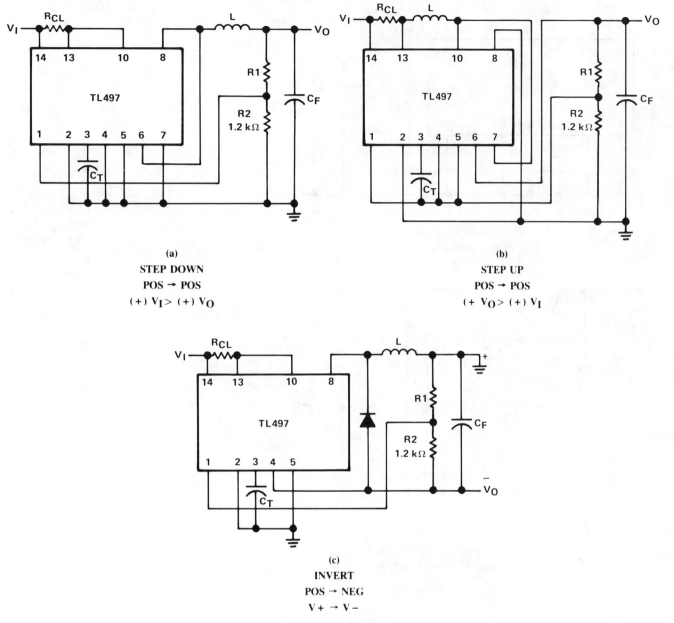

(a)
STEP DOWN
POS → POS
(+) V$_I$ > (+) V$_O$

(b)
STEP UP
POS → POS
(+ V$_O$ > (+) V$_I$

(c)
INVERT
POS → NEG
V + → V −

Figure 6-27. Basic Power Supply Configurations

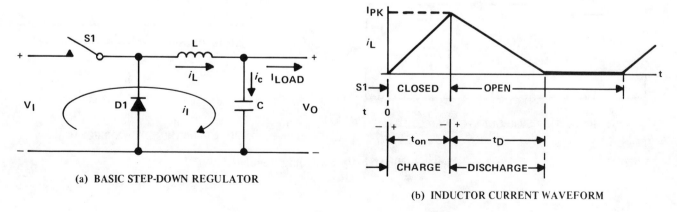

(a) BASIC STEP-DOWN REGULATOR

(b) INDUCTOR CURRENT WAVEFORM

Figure 6-28. Step-Down Switching Regulator

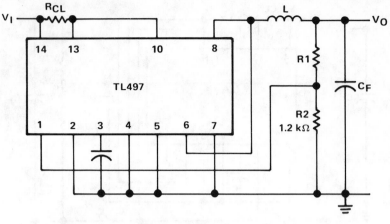

BASIC CONFIGURATION

$I_{PK} < 500$ mA)

DESIGN EQUATIONS

- $I_{PK} = 2\, I_{LOAD}$ max

- $L\,(\mu H) = \dfrac{V_I - V_O}{I_{PK}}\, t_{on}\,(\mu s)$

Choose L (50 to 500 μH), calculate t_{on} (20 to 150 μs)

- $C_T\,(pF) \approx 12\, t_{on}\,(\mu s)$

- $R1 = (V_O - 1.2)\,k\Omega$

- $R_{CL} = \dfrac{0.5\ V}{I_{PK}}$

- $C_F = \dfrac{(I_{PK} - I_{LOAD})^2}{(V_{ripple})\,2\,I_{PK}} \times \dfrac{T_{on}\,V_I}{V_O}$

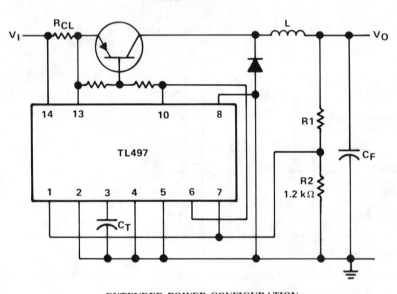

EXTENDED POWER CONFIGURATION
(USING EXTERNAL TRANSISTOR)

Figure 6-29. Positive Regulator, Step-Down Configurations

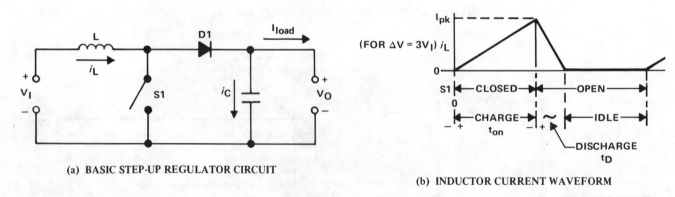

(a) **BASIC STEP-UP REGULATOR CIRCUIT**

(b) **INDUCTOR CURRENT WAVEFORM**

Figure 6-30. Step-Up Switching Regulator

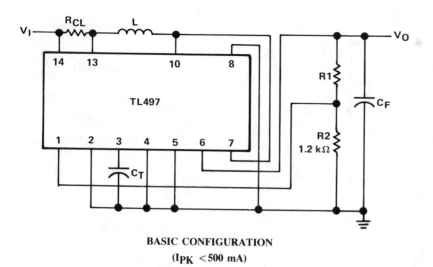

BASIC CONFIGURATION

(I_{PK} < 500 mA)

- $I_{PK} = 2\,I_{LOAD\ max}\left[1 + \dfrac{V_O}{V_I}\right]$

- $L\,(\mu H) = \dfrac{V_I}{I_{PK}}\,t_{on}\,(\mu s)$

Choose L (50 to 500 μH), calculate t_{on} (25 to 150 μs)

- $C_T\,(pF) \approx 12\,t_{on}\,(\mu s)$

- $R1 = (V_O - 1.2)\ k\Omega$

- $R_{CL} = \dfrac{0.5\ V}{I_{PK}}$

- $C_F = \dfrac{(I_{PK} - I_{LOAD})^2}{(V_{ripple})\,2\,I_{PK}} \times \dfrac{T_{on}\,V_I}{V_O}$

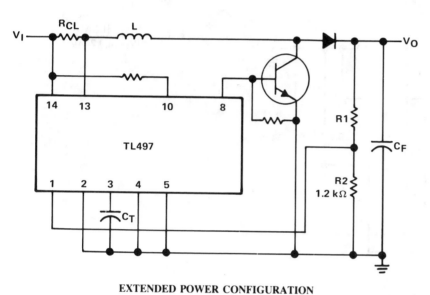

EXTENDED POWER CONFIGURATION
(USING EXTERNAL TRANSISTOR)

Figure 6-31. Positive Regulator, Step-Up Configurations

During the charging cycle (S1 closed) the inductor (L) is charged only by the input potential, similar to the step-up configuration. In the inverting configuration the input provides no contribution to the load current during the charging cycle. The maximum load current for discontinuous operation will be limited by the peak current, as observed in the step-up configuration. The inductor current waveform looks identical to the

waveform demonstrated in the step-up configuration [see Figure 6-30(b)].

Figure 6-33 shows the inverting applications both with and without an external pass transistor. Design equations are also included.

Note that in the inverting configuration the internal diode is not used. An external diode must be used because pin 4 (substrate) must be the most negative point on the chip. The cathode of the internal diode is also the cathode of a diode connected to the substrate. When the cathodes are at the most negative voltage in the circuit, there will be conduction to the substrate resulting in unstable operation.

DESIGN CONSIDERATIONS

An oscilloscope is required when building a switching regulator. When checking the oscillator ramp on pin 3 the oscilloscope may be difficult to synchronize. This is a normal operating characteristic of this regulator and is

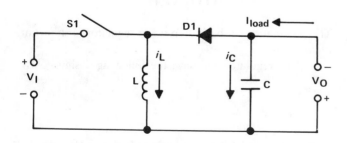

Figure 6-32. Basic Inverting Regulator Circuit

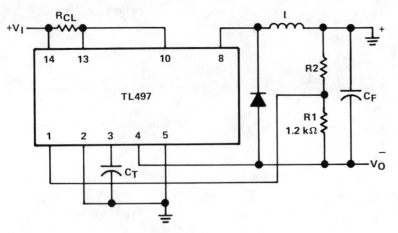

BASIC CONFIGURATION
(I_{PK} < 500 mA)

- $I_{PK} = 2\,I_{LOAD}\,max\left[1 + \dfrac{V_O}{V_I}\right]$

- $L\,(\mu H) = \dfrac{V_I}{I_{PK}}\,t_{on}\,(\mu s)$

Choose L (50 to 500 μH), calculate t_{on} (25 to 150 μs)

- $C_T\,(pF) \approx 12\,t_{on}\,(\mu s)$

- $R2 = (V_O - 1.2)\,k\Omega$

- $R_{CL} = \dfrac{0.5\,V}{I_{PK}}$

- $C_F = \dfrac{(I_{PK} - I_{LOAD})^2}{(V_{ripple})\,2\,I_{PK}} \times \dfrac{t_{on}\,V_I}{V_O}$

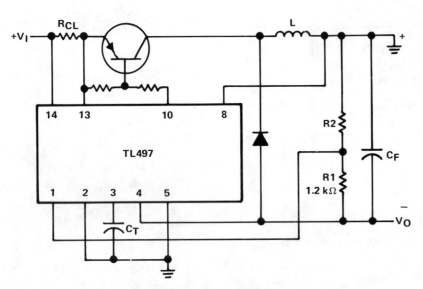

EXTENDED POWER CONFIGURATION
(USING EXTERNAL TRANSISTOR)

Figure 6-33. Inverting Applications

caused by the asynchronous operation of the error amplifier to that of the oscillator. The oscilloscope may be synchronized by varying the input voltage or load current slightly from design nominals.

High frequency circuit layout techniques are imperative. Keep leads as short as possible and use a single ground point. Resistors R1 and R2 should be as close as possible to Pin 1 to eliminate noise pick-up in the feedback loop. The TL497A type of circuits do not need "hi-Q" inductors. They are, in fact, not desirable due to the broad frequency range of operation. If the "Q" is too high, ringing will occur. If this happens a shunt resistor (about 1 kΩ) may be placed across the coil to damp the oscillation.

While not necessary, it is highly desirable to use a toroidal inductor as opposed to a cylindrically wound coil. The toroidal type of winding helps to contain the flux

closer to the core and in turn minimize radiation from the supply. All high current loops should be kept to a minimum length using copper connections that are as large as possible.

A STEP-DOWN SWITCHING REGULATOR DESIGN EXERCISE WITH TL497A

The schematic of a basic step-down regulator is shown in Figure 6-34.

This regulator will have the following design goals:

V_I = 15 V
V_O = 5 V
I_O = 200 mA
V_{ripple} = < 1.0% or 50 mV (1.0% × 5 V)

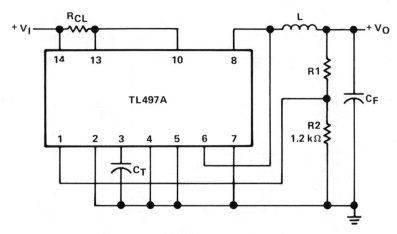

Figure 6-34. Basic Step-Down Regulator

Calculations:

$$I_{PK} = 2 \, I_L \text{ max} = 400 \text{ mA}$$

For design margin, I_{PK} will be designed for 500 mA which is also the limit of the internal pass transistor and diode.

$$\therefore I_{PK} = 500 \text{ mA}$$

The next step will be to select t_{on}. You may select a timing capacitor to match an inductor you may already have. You may also assume an on-time and calculate the inductor value. We will assume an on-time of 20 μs.

$$t_{on} = 20 \text{ μs}$$

$$L(\mu H) = \frac{V_I - V_O}{I_{PK}} \times t_{on} \text{ μs}$$

$$= \frac{15 - 5}{0.5} \times 20 = 400 \text{ μH}$$

$$L = 400 \text{ μH}$$

To set the TL497A for 5 V output:

$$R2 = 1.2 \text{ k}\Omega \text{ (fixed)}$$

$$R1 = (5 - 1.2) \text{ k}\Omega = 3.8 \text{ k}\Omega$$

To set current limiting:

$$R_{CL} = \frac{0.5}{I_L} = \frac{0.5}{500 \times 10^{-3}} = 1 \, \Omega$$

$$R_{CL} = 1 \, \Omega$$

For the on-time chosen, C_T can be approximated:

$$C_T(pF) = 12 \, t_{on} \text{ μs}$$

$$C_T = 240 \text{ pF}$$

or it may be selected from a table in the data sheet.

To determine filter capacitor (C_F) for desired ripple voltage:

$$C_F = \frac{(I_{PK} - I_L)^2}{(V_{ripple}) \, 2 \, I_{PK}} \times \frac{t_{on} \, V_I}{V_O}$$

$$C_F = \frac{(0.5 - 0.2)^2}{(0.05) \, 2 \times 0.5}$$

$$\times \frac{20 \times 10^{-6} \times 15}{5} = 108 \text{ μF}$$

We selected C_F to be 120 μF, the next higher standard value.

Figure 6-35 illustrates the regulator with the calculated values applied to it.

A 150 μF filter capacitor may be used as a prefilter as well as a 0.01 μF disc capacitor to take care of any transients on the incoming V_I rail.

For peak currents greater than 500 mA, it is necessary to use an external pass transistor and diode. Such a technique is illustrated in Figure 6-36 which is an automotive power supply. With a 12 V battery, this step-down regulator supplies 5 V at 2 A output current.

Figure 6-37 illustrates a basic step-up regulator. This design steps up the output voltage from 5 V to 15 V. The equations for determining the values of the external components are provided in Figure 6-31.

DESIGN AND OPERATION OF AN INVERTING REGULATOR CONFIGURATION

Figure 6-38 illustrates a basic inverting regulator designed to have −5 V output with +5 V input using the design equations in Figure 6-33.

Conditions:

$$V_I = 5 \text{ V}$$
$$V_O = -5 \text{ V}$$
$$I_O = 100 \text{ mA}$$
$$V_{ripple} = 1.0\% \text{ or } 50 \text{ mV} \, (1\% \times 5 \text{ V})$$

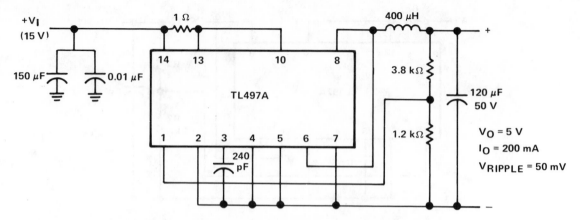

Figure 6-35. 15-Volt to 5-Volt Step-Down Regulator

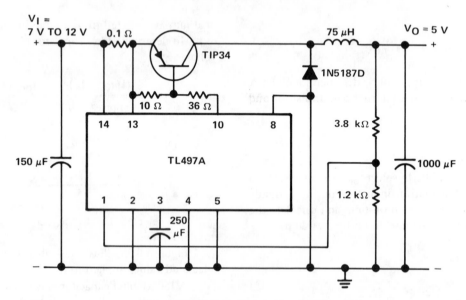

Figure 6-36. Step-Down Regulator

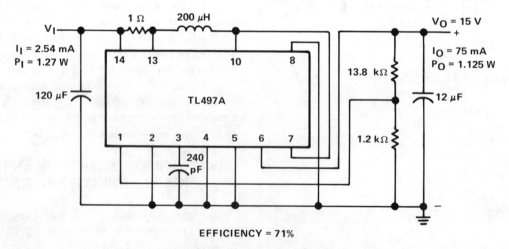

EFFICIENCY = 71%

Figure 6-37. 5-Volt to 15-Volt Switching Regulator

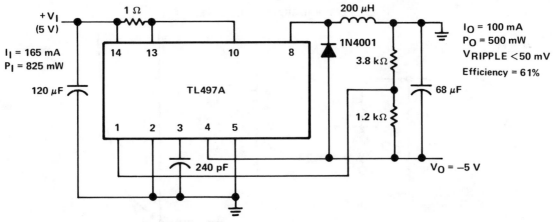

NOTE — Do not use internal diode (Pins 6, 7) on an inverting circuit.

Figure 6-38. +5-Volt to −5-Volt Switching Regulator

Calculations:

$$I_{PK} = 2 I_L(max) \left(1 + \frac{|V_O|}{V_I} \right)$$

$$I_{PK} = 400 \text{ mA (for design margin use 500 mA)}$$

Assume $t_{on} = 20 \ \mu s$

$$C_{T(pF)} = 12 \ t_{on} \ \mu s$$
$$C_T = 240 \text{ pF}$$

$$L = \frac{V_I}{I_{PK}} \ t_{on} = \frac{5}{0.5} \times 20 = 200 \ \mu H$$

To set the output voltage:

$$R2 = 1.2 \ k\Omega$$

$$R1 = (5 - 1.2) \ k\Omega = 3.8 \ k\Omega$$

To set the current limiting:

$$R_{CL} = \frac{0.5}{I_{PK}} = \frac{0.5}{0.5} = 1 \ \Omega$$

$$RCL = 1 \ \Omega$$

$$C_F = \frac{(I_{PK} - I_L)^2}{(V_{ripple}) \ 2 \ I_{PK}} \times \frac{t_{on} \ V_I}{V_O}$$

To determine C_F for desired ripple voltage:

$$\begin{aligned} C_F &= \frac{(I_{PK} - I_L)^2}{(V_{ripple}) \ 2 I_{PK}} \times \frac{t_{on} \ V_I}{|V_O|} \\ &= \frac{(0.5 - 0.1)^2}{(0.05) 2 \times 0.5} \times \frac{20 \times 10^{-6} \times 5}{|-5|} \end{aligned}$$

$$C_F = 64 \ \mu F \text{ (nearest standard value = 68 } \mu F)$$

ADJUSTABLE SHUNT REGULATOR
TL430 - TL431

The TL430 and TL431 are three-terminal 'programmable' shunt regulators. The devices are basically the same except the TL431 contains a diode connected between the emitter and collector of the output transistor. The standard symbol and block diagram are shown in Figure 6-39.

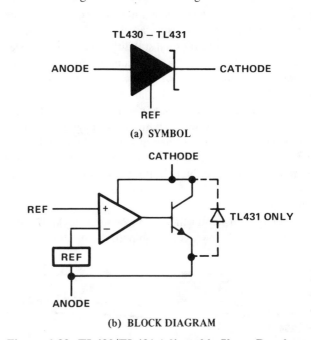

Figure 6-39. TL430/TL431 Adjustable Shunt Regulators

The circuit consists of a bipolar op amp driving an npn transistor. The reference on the TL430 is a band-gap reference (not temperature compensated). The TL431 has a true temperature compensated band-gap reference and is more stable and accurate than other shunt regulators. The TL431 also has a diode across the emitter-collector of the npn output transistor. If the cathode goes negative, the diode conducts around the transistor, emulating the

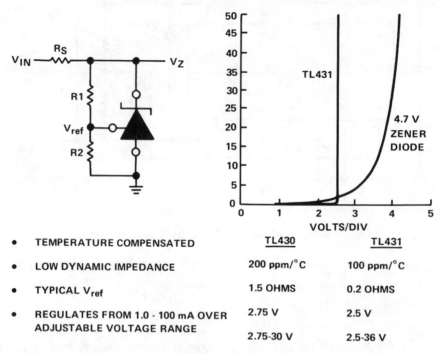

TL430	TL431
• TEMPERATURE COMPENSATED		
• LOW DYNAMIC IMPEDANCE	200 ppm/°C	100 ppm/°C
• TYPICAL V_{ref}	1.5 OHMS	0.2 OHMS
• REGULATES FROM 1.0 - 100 mA OVER ADJUSTABLE VOLTAGE RANGE	2.75 V	2.5 V
	2.75-30 V	2.5-36 V

Figure 6-40. Basic Operating Characteristics

performance characteristics of a normal zener diode. The basic operating characteristics are shown in Figure 6-40.

Their excellent thermal stability make these devices extremely attractive as a replacement for high cost, temperature-compensated zeners. As seen in Figure 6-41, the TL431 offers improved characteristics, even at low voltages. Since the TL431 operates as a shunt regulator, it can be used as either a positive or negative voltage reference. The TL431 has an equivalent full-range temperature coefficient of 50 ppm/°C (typical) and has low output noise voltage. Note in the graph (Figure 6-41) that for a nominal 2.495 V reference the curve is essentially flat from 0°C to 70°C.

Depending upon the zener voltage, the TL431 also has an extremely low dynamic impedance of about 0.2 Ω,

compared to a standard zener diode's dynamic impedance of about 30 to 60 Ω.

A 2.5 V reference voltage is developed across R2 as shown in Figure 6-42. I_{ref}, the current input at the reference terminal, is about 10 μA. To maintain a steady reference, it is advisable to allow 1 mA of current flow through series resistors R1 and R2. This will assure a stable reference voltage independent of I_{ref} variations. The TL431 is available in either the commercial temperature range of 0° - 70°C or the military temperature range of −55° to +125°C.

$$R = \frac{V_I - (V_{be} + V_O)}{I_R}$$

$$R = \frac{32 - (2 + 24)}{10 \text{ mA}} = 600 \ \Omega$$

$$V_O = \left(1 + \frac{R1}{R2}\right) V_{ref}$$

R1 = 21.4 kΩ
R2 = 2.5 kΩ

The circuit in Figure 6-43 uses a TL431 as a regulator to control the base drive to a TIP660 series pass transistor. For good reference stability, a current flow of about 1 mA (I_2) though the resistor divider is recommended. A 2.5 V reference voltage is developed across R2, and R1 will develop a voltage drop of 21.5 V. The Darlington power transistor is used because of the reduced base drive requirement of the TIP660 which has a V_{be} (max) of about 2 V. The h_{FE} at 2.5 A I_C is about 1000, so it would only

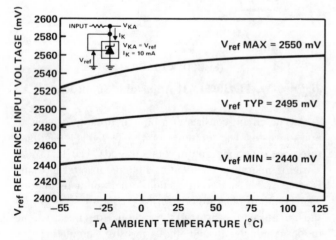

Figure 6-41. Reference Input Voltage Versus Ambient Temperature

require about 2.5 mA of base drive to produce 2.5 A of output current. In calculating the value of the current limit resistor, R3, we assume about 7.5 mA of current through the TL431. The value of R3, therefore, would be 600 Ω and the current about 10 mA, so a 1/2 W resistor will suffice. This is a simple method of designing a medium output current power-supply using only four components plus the series pass transistor.

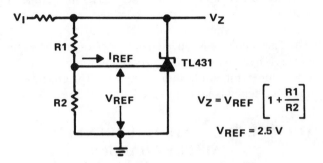

$$V_Z = V_{REF}\left[1 + \frac{R1}{R2}\right]$$

$$V_{REF} = 2.5\ V$$

Figure 6-42. Basic Operational Circuit

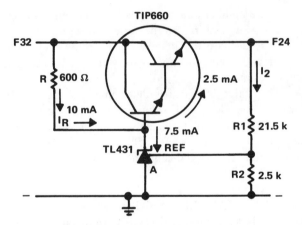

Figure 6-43. Series Regulator Circuit

SHUNT REGULATOR APPLICATIONS CROWBAR

To protect solid-state electronic equipment from overvoltage due to a power-supply component failure, it is sometimes desirable to use a 'crowbar' circuit. When a preset voltage is exceeded, the SCR turns on, shorting the output and blowing the fuse on the input side of the crowbar circuit. The circuit in Figure 6-44 is set to trip when V_O reaches 27 V. When that occurs, the reference voltage should be 2.5 V which turns on the TL431, thus biasing the SCR low. This turns the SCR on and immediately blows the safety fuse on the circuit input, thus protecting the equipment using this power supply.

$$V_L = \left(1 + \frac{R1}{R2}\right) V_{ref}$$

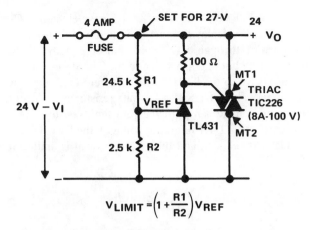

$$V_{LIMIT} = \left(1 + \frac{R1}{R2}\right) V_{REF}$$

Figure 6-44. Shunt Regulator in Crowbar Circuit

CONTROLLING V_O OF A FIXED OUTPUT VOLTAGE REGULATOR

Sometimes it is necessary to have a regulated output voltage different from that for which the regulator is designed. This may be accomplished with any three-terminal regulator, although it should be noted that the lowest obtainable voltage will be 2.5 V for the TL431 plus the voltage of the three-terminal regulator. In the circuit in Figure 6-45, the lowest possible regulated voltage would be 7.5 V (2.5 V for the TL431 + 5 V for the 7805). This particular circuit provides 9 V output using a uA7805 three-terminal regulator.

Note: Minimum $V_O = V_{ref} + 5$ V

$$V_O = \left(1 + \frac{R1}{R2}\right) V_{ref}$$

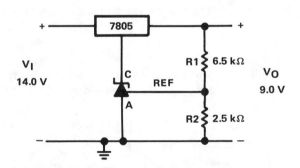

NOTE: MINIMUM $V_O = V_{REF} + 5.0$ V

$$V_O = \left(1 + \frac{R1}{R2}\right) V_{REF}$$

Figure 6-45. Fixed Output Shunt Regulator

CURRENT LIMITER

Figure 6-46 is an example of a current limiter designed to limit the current from a 12 V supply to 1.5 A using a TIP31 npn transistor as the pass element. The value of R1 is calculated from the equation in Figure 6-46. The current through R1 is split almost equally in this circuit, with about 30 mA going to the TL431, and 30 mA for base drive to the TIP31. With a current load of 6 mA, and an R1 value of 128 Ω a 1/2 watt resistor is sufficient. When the voltage across the current limit resistor (R_{CL}) reaches 2.5 V (TL431 reference voltage), the base drive to the TIP31 is reduced and the output current is limited to 1.5 A.

$$R1 = \frac{V_1 - (V_{be} + V_{RCL})}{I_1}$$

$$= \frac{12 - (1.8 + 2.5)}{0.06} = 128 \; \Omega$$

$$R_{CL} = \frac{V_{ref}}{I_L} = \frac{2.5 \; V}{1.5 \; A} = 1.7 \; \Omega$$

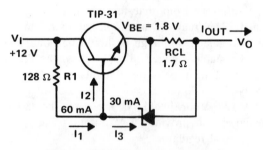

Figure 6-46. Current Limiter

VOLTMETER SCALER

The circuit in Figure 6-47 is a voltmeter scaler (or multiplier) to extend the range of a 0 to 10 V voltmeter to 40 V. Most multiplier circuits extend the range with 0 V being the low reading on any given scale. This circuit actually divides the 40 V total range into 4 separate 10 V scales.

With the selector switch in position #1, the reference input of the TL431 is bypassed and the TL431

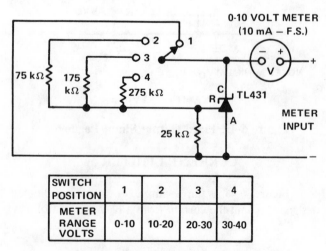

SWITCH POSITION	1	2	3	4
METER RANGE VOLTS	0-10	10-20	20-30	30-40

Figure 6-47. Voltmeter Scaler

does not influence circuit operation. The meter is effectively connected directly to the voltage being measured. This scale would be the normal meter range of 0 to 10 V.

When in position #2, a 75 kΩ and a 25 kΩ resistor are added in series across the anode and cathode of the TL431. The voltmeter will remain near zero until the input reaches 10 V. At this time, there is 2.5 V between the reference terminal and anode which causes the voltmeter to start reading at 10 V. It will continue reading on this scale until it reaches full scale, which is 20 V.

This sequence is repeated in 10 V steps until position #4 is reached. This circuit is very useful when expanded-scale voltmeter multiplication is required. The precision of the scaler depends upon the accuracy of the resistors.

VOLTAGE-REGULATED, CURRENT-LIMITED BATTERY CHARGER FOR LEAD-ACID BATTERIES

There are a number of approaches to recharging lead-acid batteries. Many will return the battery to service, but fail to fully rejuvenate the battery. To keep a battery fully charged, and attain maximum battery life, proper charging techniques must be observed.

The status of a cell is determined by the specific gravity of the electrolyte solution. A specific gravity of 1.280 (obtained by hydrometer reading) indicates a fully charged cell. A reading of 1.250 or better is considered good. A fully discharged cell exhibits a specific gravity of 1.150 or less.

BATTERY CHARGER DESIGN

The battery charger design shown in Figure 6-48 is based on a charging voltage of 2.4 V per cell, in accordance with most manufacturers' recommendations. The battery charger circuit pulses the battery under charge with 14.4 V (6 cells × 2.4 V per cell) at a rate of 120 Hz.

The design provides current limiting to protect the charger's internal components while limiting the charging rate to prevent damaging severely discharged lead-acid batteries. The maximum recommended charging current is normally about one-fourth the ampere-hour rating of the battery. For example, the maximum charging current for an average 44 ampere-hour battery is 11 A.

If the impedance of the load requires a charging current greater than the 11 A current limit, the circuit will go into current limiting. The amplitude of the charging pulses is controlled to maintain a maximum peak charging current of 11 A (8 A average).

The charger circuit is composed of four basic sections:

1. Rectifier
2. Voltage Regulator
3. Current Limiting
4. Series Pass Element

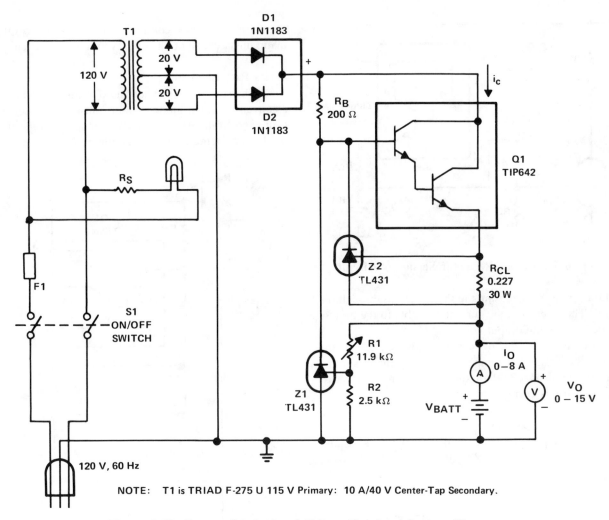

Figure 6-48. Current Limited and Voltage Regulated Battery Charger

NOTE: T1 is TRIAD F-275 U 115 V Primary: 10 A/40 V Center-Tap Secondary.

Rectifier Section

A full-wave rectifier configuration with a center-tapped transformer (Figure 6-49) achieves maximum performance with minimum component count. The breakdown voltage requirement for the diode is:

$$VR > Vsecondary(pk) - V_F(rectifier\ drop)$$
$$\therefore VR > 20 \times 1.414 - 1 = 27.28\ V$$

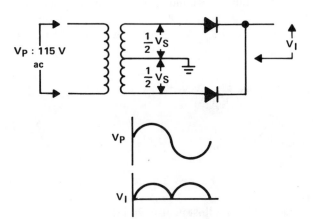

Figure 6-49. Full-Wave Rectifier Section of Circuit

This design is set to current limit at 11 A so a rectifier rating of 25 A is recommended to handle the maximum current drain plus any current surges. A pair of 1N1183 diodes was chosen (35 A/50 V rectifiers).

Voltage Regulator Section

The components which make up the voltage regulator portion of the circuit are: Z1, Q1, R1, R2 and R_B as shown in Figure 6-50.

Z1 is a TL431 programmable shunt regulator which serves as the control element, Q1 is the pass transistor, and R1 - R2 sense the output voltage providing feedback to Z1. R1 and R2 are chosen so that their node voltage is 2.5 V at the desired output voltage. This node voltage is applied to the TL431's error amplifier which compares it to the internal 2.5 V reference.

When the feedback voltage is less than the internal 2.5 V reference, the series impedance (anode-to-cathode) of the TL431 increases, decreasing the shunt current through the TL431. This increases the current available to the base of pass transistor Q1, increasing the output voltage.

When the feedback voltage is greater than the internal 2.5 V reference, the series impedance of the

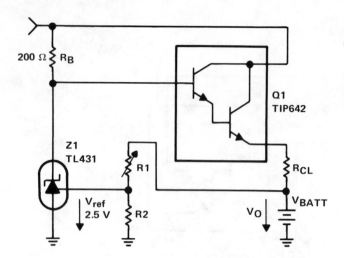

Figure 6-50. Voltage Regulator Section of Circuit

TL431 decreases, increasing the shunt current through the TL431. This decreases the current available to the base of Q1, decreasing the output voltage.

Because the feedback voltage is sensed at the output, the TL431 will compensate for any changes in the base-emitter drop of Q1 or the voltage dropped across R_{CL} for various currents.

Current Limiter Section

The components which make up the current-limit portion of this circuit are: Z2, Q1, and R_{CL} as shown in Figure 6-51.

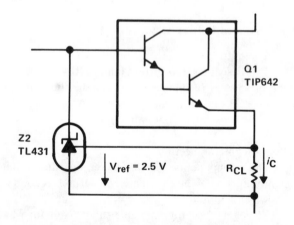

Figure 6-51. Current Limiter Section of Circuit

The value of the current-limit setting resistor, R_{CL}, is chosen so that 2.5 V will be developed across it at the desired limit current. The voltage across R_{CL} is sensed by a TL431 programmable shunt regulator (Z2). When the output current is less than the current limit, V_{ref} is less than 2.5 V and Z2 is a high impedance which does not affect the operation of Q1.

When the output current reaches maximum, V_{ref} is 2.5 V and the impedance of Z2 decreases, decreasing the current available at the base of Q1 and controlling the maximum output current. Under this condition, shunt

regulator Z2 takes control of pass transistor Q1 and maintains a constant current, even into a short circuit.

Series Pass Element

The series pass element used in this configuration is a conventional Darlington power transistor, whose control is derived from either Z1 or Z2 depending on the state of the battery being charged. See Figure 6-52.

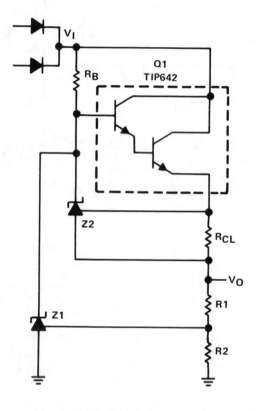

Figure 6-52. Series Pass Element

The performance characteristics of Q1 are important in determining the circuit design and in the choice of the transformer to be used. This relationship is shown in the following section on the design of the battery charger.

Design Calculations

The values of R1 and R2 set the output voltage level at 2.4 V per cell or 14.4 V for 6 cells. For optimum performance of Z1, 1 mA should flow through the R1 and R2 combination.

$$R1 + R2 = \frac{14.4 \text{ V}}{1 \text{ mA}} = 14.4 \text{ k}\Omega$$

$$R2 = \frac{2.5 \text{ V}}{1 \text{ mA}} = 2.5 \text{ k}\Omega$$

$$R1 = 14.4 \text{ k}\Omega - 2.5 \text{ k}\Omega = 11.9 \text{ k}\Omega$$

For ease of final adjustment, a 20 kΩ potentiometer may be used for R1.

Current limiting starts when 2.5 V is developed across R_{CL} at the desired current limit. For a 44 A hour battery, the maximum charge rate is 11 A.

$$R_{CL} = \frac{2.5 \text{ V}}{11 \text{ A}} = 0.227 \ \Omega$$

The average current $= 0.707 \times 11 \text{ A} = 7.777 \text{ A}$ or
$$\approx 8A$$

The average power dissipation $= I^2R = 8^2 \times 0.227$
$$= 14.5 \text{ W}$$

After the pass transistor has been selected, its base drive resistor, R_B, may be calculated. A TIP642 meets the requirements. From the data sheet:

h_{FE} @ 11 A = 500 (min)
$V_{CE} \approx 2$ V
$V_{BE} = 1.6$ V
$P_{max} = 160$ W @ 40°C TC
$I_B = 22$ mA @ 11 A peak collector current

To calculate R_B, assume a worst case or short-circuit condition where:

$$R_B \approx \frac{V_I - V_{ref} - V_{BE(Q1)}}{I_{B(Q1)} + I_{SHUNT(Z2)}}$$

$$R_B \approx \frac{27.28 - 2.5 - 1.6}{0.022 + 0.12} = 163 \ \Omega$$

While R_B must be small enough so it does not limit the base current of Q1 at the desired I_{CHG} of 8 A, however, it must be large enough to limit the current during short circuit conditions. This value should be less than the sum of the base drive current required by Q1 and $I_{SHUNT(max)}$ Z2.

$$R_B \approx \frac{(V_I - 14.4 \text{ V} - 2.5 \text{ V} - V_{BE(Q1)})}{I_{CHG}/h_{FE(Q1)}}$$

$$= \frac{27.28 - 14.4 - 2.5 - 1.6}{8/500}$$

$$R_B \approx \frac{8.78}{0.016} = 548.7 \ \Omega$$

A value of R_B within this range assures sufficient drive to Q1 for a charging rate of 8 A, yet allows total control of Q1 by Z2 during short-circuit conditions. R_B was selected to be 200 Ω.

Power Dissipation and Heat Sinking

To determine the power dissipation in the 1N1183 rectifier and the TIP642 Darlington, the RMS currents and voltages must be calculated. The voltage and current paths are shown in Figure 6-53.

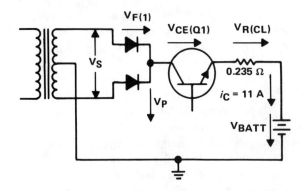

Figure 6-53. Voltage and Current Path

$$V_{CE(Q1)} = V_I - V_{BATT} - V_{RCL}$$
$$= 27.78 - 14.4 - 2.5 = 10.88 \text{ V}$$
$$V_{CE(Q1)} = 10.9 \text{ V}$$

The transistor power dissipation is:

$$P_{Q1} = I_{(RMS)} \times V_{CE(RMS)}$$
$$= (11 \text{ A} \times 0.707)(10.9 \text{ V} \times 0.707)$$
$$P_{Q1} = (7.78 \text{ A}) (7.7 \text{ V})$$
$$P_{Q1} = 59.9 \text{ W}$$

The rectifier power dissipation is:

$$P_{(RECT)} = I_{(RMS)} \times V_F = (7.78 \text{ A}) (1.3 \text{ V})$$
$$P_{(RECT)} = 10.1 \text{ W total}$$

If the pass transistor and rectifiers are mounted on separate heat sinks, the sinks must be capable of dissipating the heat transferred by each device and maintain a surface temperature which satisfies the temperature requirement for each device. Mounted separately, the respective heat sink requirements are as follows:

PASS TRANSISTOR	RECTIFIERS
$R_{\theta CA} \leq \dfrac{150°C - 25°C}{59.9 \text{ W}}$	$R_{\theta CA} < \dfrac{140°C - 25°C}{10.1 \text{ W}}$
$R_{\theta CA} \leq 2.08°C/W$	$R_{\theta CA} < 11.4°C/W$

Depending on the mass of the heat sink and the type of cabinet, forced air cooling may be required.

VOLTAGE SUPPLY SUPERVISOR DEVICES

Voltage supply supervisor devices deliver a digital output signal (high or low) if supply voltage (V_{CC}) falls below a predefined value. The digital output signal remains in its high or low state for a certain period of time (t delay) after V_{CC} returns to normal. These devices are used to sequentially initialize digital systems for proper operation at power-on or following a V_{CC} interruption.

The versatility, few external components, and accurate threshold voltage of the TL7700 series make these devices easy to use in digital systems requiring V_{CC} line supervision.

GENERAL OPERATION

At power-on, digital systems must normally be forced into a definite initial state. In simple microcomputer and microprocessor applications an RC network connected to the RESET input pin will generally suffice. However, in more complex systems a discrete component design as illustrated in Figure 6-54 may be used.

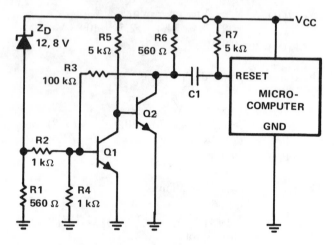

Figure 6-54. Discrete Solution of a Voltage Supply Supervisor

In this circuit, after V_{CC} reaches a specific value, defined by Z_D, the input voltage divider, and V_{BE}, the collector of Q2 becomes high and coupling capacitor C1 provides enough power to the RESET input pin of the digital system to execute the reset function.

The major deficiency with this type of circuit is that after power-on and the system is operating, low V_{CC} conditions and short drops in V_{CC} may not be recognized. A small decrease of V_{CC} below the recommended supply voltage can destroy the content of the memory and registers without activating the reset circuit. This may have catastrophic consequences.

Moreover, the circuit in Figure 6-54 contains an excessive number of components, one being Z_D, which has to be specially selected and is therefore relatively expensive.

Several features are provided in larger computers to prevent some of the problems just mentioned. In some cases the content of the memory is protected by a battery back-up. However, for most applications and in small microcomputer systems, these solutions are too expensive and generally not required. After any serious voltage drop, it is usually sufficient to force the microcomputer into a defined initial condition.

To implement this function, while preventing the problems previously mentioned, a chip with the following features is required:

1. Accurate detection of a serious voltage drop.
2. Generation of a continuous reset signal while the supply voltage is not in the operational range to prevent undefined operations.
3. Maintenance of the reset signal for a certain time after the supply voltage has returned to its nominal value to ensure a proper reset.

TL7700 SERIES SUPERVISOR CHIPS

A functional block diagram of the chip is illustrated in Figure 6-55.

The most critical element of this chip is the reference voltage source, which consists of a very stable, temperature-compensated bandgap reference. An external capacitor (typically 0.1 μF) must be connected to the Reference (REF) voltage output to reduce the influence of fast transients in the supply voltage. The voltage at the SENSE INPUT pin is divided by resistors R1 and R2 and compared with the reference voltage. The divider is adjusted to achieve high accuracy at the probing operation during manufacture of the chip.

When the sensed input voltage is lower than the threshold voltage, the thyristor is triggered discharging the timing capacitor C_T. It is also possible to fire the thyristor with a TTL logic level (active low) at the $\overline{RESIN}$ input.

The thyristor is turned off again when the voltage at the SENSE INPUT (or $\overline{RESIN}$ input) increases beyond the threshold, or during short supply voltage drops when the discharge current of the capacitor becomes lower than the hold current of the thyristor.

Capacitor C_T is recharged by a 100 μA current source; the charge time is calculated as follows:

$$t_d \text{ (internal time delay)} = C_T (1.3 \times 10^4)$$

A second comparator forces the output into the active state as long as the voltage at the capacitor is lower than the reference voltage. Figure 6-56 is a graph plotting C_T versus t_d.

The SENSE INPUT pin is connected to V_{CC} in typical applications. Figure 6-57 shows the timing of the Supply Voltage and $\overline{RESET}$ signals.

The minimum supply voltage for which operation is guaranteed is 3.6 V. Between POWER-ON (0 V) and 3.6 V, the state of the outputs is not defined. In practical applications this is not a limitation because the function of the reset inputs of the other devices is not guaranteed at such supply voltages.

Above 3.6 V capacitor C_T is discharged and the outputs stay in the active state. When the input voltage

Note: SENSE INPUT pin connected to V_{CC}

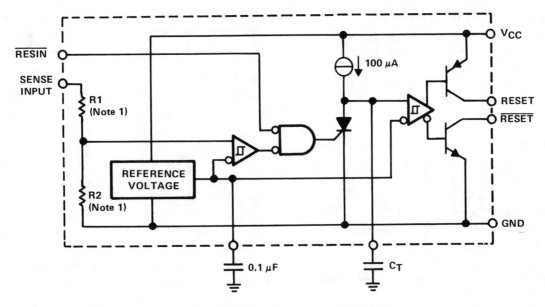

Figure 6-55. TL7700 Series Function Block Diagram

exceeds the threshold voltage, V_S, the thyristor is turned off and capacitor C_T is charged. After a delay of t_d, the voltage passes the trigger level of the output comparator and the outputs become inactive. The microcomputer is then set to a defined initial state and starts operation.

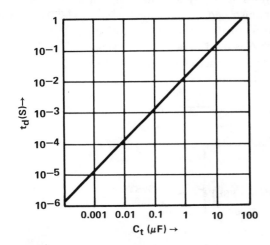

Figure 6-56. Graph for Calculation of C_T

Operation During a Voltage Drop

The thyristor is triggered when the supply voltage drops below the minimum recommended value. After the supply returns to its required value the output stays in the active state for the duration of t_d.

The delay time, t_d, is determined by the requirements of the computer system to be controlled. Typically, in TTL systems, a reset time of 20 to 50 ns is sufficient. Microcomputers usually require a reset signal which lasts several machine cycles. The duration of the reset signal is dependent on the type of microcomputer, but is typically 10 to 200 μs. In most practical applications, t_d is determined by the characteristics of the power supply.

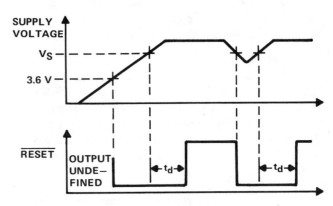

Figure 6-57. Timing Diagram

During and shortly after power-on make sure voltage fluctuations do not repetitively reset the system. Delay times of 10 to 20 ns will usually prevent this problem.

Four versions of this device are available:

	Threshold Voltage	V_{CC}
TL7702A	2.53 V	3.6 – 18.0 V
TL7705A	4.55 V	5.0 V
TL7712A	10.8 V	12.0 V
TL7715A	13.5 V	15.0 V

The TL7702A may be used in applications where V_{CC} voltages up to 18 V are used. The required trigger level (2.5 V) may be set with a resistor divider network at the SENSE INPUT pin. The TL7705A, TL7712A, and TL7715A have an internal resistor divider network and operate on 5 V, 12 V, and 15 V, respectively.

TL7700 SERIES APPLICATIONS

Since, for most applications, the devices are already adjusted to the appropriate voltage levels these chips are easy to use. Figure 6-58 illustrates an undervoltage protection circuit for a TMS9940 microcomputer system with a 5 V power supply.

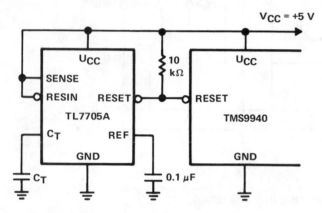

Figure 6-58. TL7705A in 5-Volt Microcomputer Application

External components are the 0.1 μF bypass capacitor at the REF terminal, which reduces transients from the supply voltage, and the C_T capacitor, which sets the time delay t_d. The TL7705A devices do not have internal pull-up (or pull-down) resistors. An external 10 kΩ pull-up resistor is connected from the RESET pin to the 5 V V_{CC} to produce a high level. A similar application is illustrated in Figure 6-59.

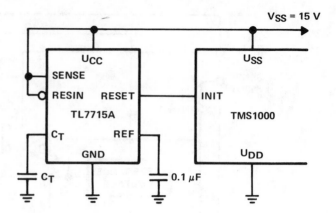

Figure 6-59. TL7715A in TMS1XXXNLP Application

This circuit utilizes a TL7715A as a protection device for a TMS1000 microcomputer system. The C_T and reference bypass capacitors are also used in this application. Note, however, the absence of the pull-up resistor used in Figure 6-58. This circuit has a required internal pull-down resistor at the INIT INPUT pin on the TMS1000 microcomputer chip.

In large systems, where several supply voltages are required (e.g., TMS8080, TMS9900), it is necessary to supervise all supply voltages that may cause dangerous conditions if a power failure or transient occurs. The circuit illustrated in Figure 6-60 uses two TL7712A devices to check the positive and negative 12 V supplies. A TL7705A is used to check a 5 V supply.

The outputs of the two TL7712A's are fed to the $\overline{\text{RESIN}}$ input of the TL7705A. The output of this device, a

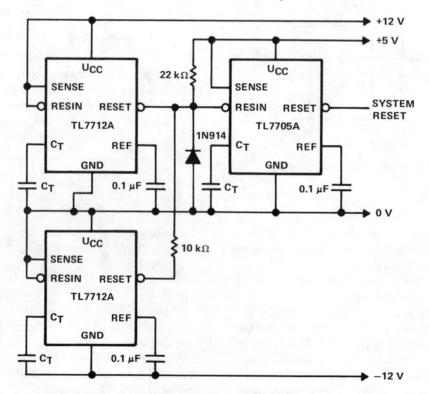

Figure 6-60. Voltage Supervision of a Multiple Power Supply

system reset signal, becomes active when any one of the three supply voltages fail.

The supply voltage supervisor devices were designed to detect very short voltage drops of 150 ns. In applications where this sensitivity is not required, the circuit may be delayed by adding an RC network ahead of the SENSE INPUT pin (Figure 6-61).

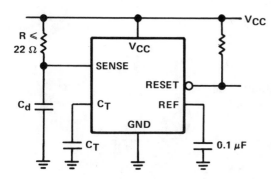

Figure 6-61. Delayed Triggering

To avoid influence on the threshold voltage of this input, the resistor should be less than 22 Ω. The capacitor C_d is then calculated to the required delay time ($C_d = t/R$).

Another application for the TL7705A is in battery-buffered memory systems. After a line-voltage failure the content of the memory has to be protected against spikes on the write line. It is usually sufficient to switch the chip-select line into the inactive state; however, some memories also require that the write line be disabled. See Figure 6-62.

A switch, formed by transistor Q1 and diode D1, is inserted into the chip-select line of the memory. Under normal operation (line voltage present) the RESET output of the TL7705A is turned off (high), transistor Q2 is turned on, and transistor Q1 draws its base current through transistor Q2 and resistor R1.

When the chip-select line is switched from high to low transistor Q1 conducts and the CS input of the memory goes low. Because of the small dc load of resistor

R2 the saturation voltage of the transistor is very small (typically 40 mV).

When the chip-select line is switched high again transistor Q1 is turned off and diode D1 conducts, charging the circuit capacitance.

In case of a power failure the TL7705A is triggered and its RESET output becomes low, turning off transistor Q2 and the base current to transistor Q1. In this way the CS input of the memory is separated from the chip-select line. In some cases it is also recommended that memory be disabled during the system reset with the $\overline{\text{RESIN}}$ input. This protects the memory content against spikes on the write line during this time.

uA723 PRECISION VOLTAGE REGULATOR

The uA723 monolithic integrated circuit voltage regulator is used extensively in power supply designs. The device consists of a temperature compensated reference amplifier, an error amplifier, a 150 mA series-pass transistor, and current-limiting circuitry. See Figures 6-63 and 6-64 for the functional diagram and schematic.

Additional external npn or pnp pass elements may be used when output currents exceeding 150 mA are required. Provisions are made for adjustable current limiting and remote shutdown. In addition, the device features low standby current drain, low temperature drift and high ripple rejection. The uA723 may be used with positive or negative supplies as a series, shunt, or floating regulator.

When using an external series pass device, the 3-dB bandwidth of the uA723 must also be taken into consideration. Adequate uA723 compensation may be provided by connecting a 100 to 500 pF capacitor from the compensation terminal to the inverting input. Extra capacitance may be required at both the input and output of any power supply due to the inductive effects of long lines. Adding output capacitance provides the additional benefit of reducing the output impedance at high frequencies.

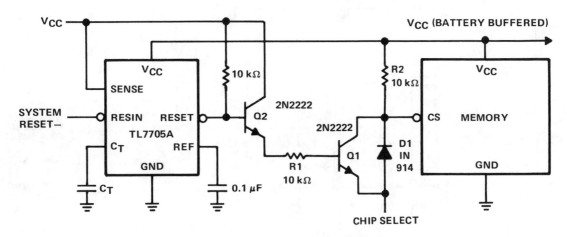

Figure 6-62. Circuit Diagram for Memory Protection

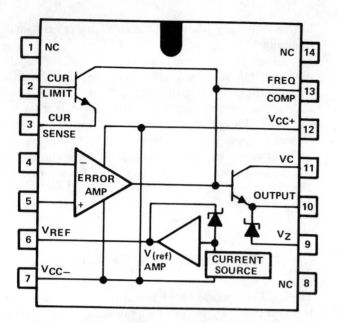

Figure 6-63. uA723 Functional Block Diagram

TYPICAL APPLICATIONS

The required output voltage and current limits for the applications shown in Figure 6-65 can be calculated from the equations given in Table 6-1. In all cases the resulting resistor values are assumed to include a potentiometer as part of the total resistance. Table 6-2 affords a quick reference for many standard output voltage requirements.

Table 6-1. Formulas for Output Voltages

Outputs from +2 to +7 V
[Figures 6-65 (a), (e), (f)]

$$V_O = V_{(ref)} \times \frac{R2}{R1 + R2}$$

Outputs from +7 to +37 V
[Figures 6-65 (b), (d), (e), (f)]

$$V_O = V_{(ref)} \times \frac{R1 + R2}{R2}$$

Outputs from −6 to −250 V
[Figure 6-65 (c)]

$$V_O = -\frac{V_{(ref)}}{2} \times \frac{R1 + R2}{R1}$$

$$R3 = R4$$

Current Limiting $\qquad I_{(limit)} \approx \dfrac{0.65 \text{ V}}{R_{SC}}$

Foldback Current Limiting
[Figure 6-65 (f)]

$$I_{(knee)} \approx \frac{V_O R3 + (R3 + R4)0.65 \text{ V}}{R_{SC} \, R4}$$

$$I_{OS} \approx \frac{0.65 \text{ V}}{R_{SC}} \times \frac{R3 + R4}{R4}$$

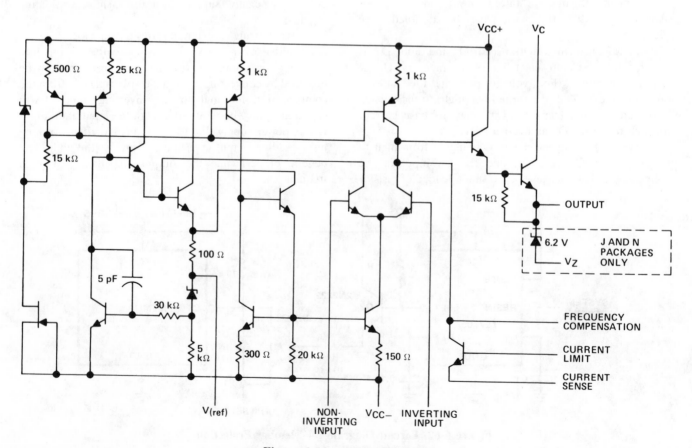

Figure 6-64. uA723 Schematic

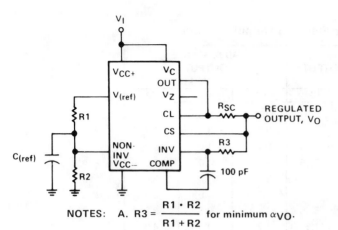

NOTES: A. $R3 = \dfrac{R1 \cdot R2}{R1 + R2}$ for minimum α_{VO}.

B. R3 may be eliminated for minimum component count. Use direct connection (i.e., $R_3 = 0$).

(a) BASE LOW-VOLTAGE REGULATOR
(V_O = 2 to 7 Volts)

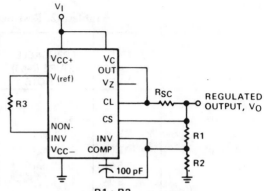

NOTES: A. $R3 = \dfrac{R1 \cdot R2}{R1 + R2}$ for minimum α_{VO}.

B. R3 may be eliminated for minimum component count. Use direct connection (i.e., $R_3 = 0$).

(b) BASIC HIGH-VOLTAGE REGULATOR
(V_O = 7 to 37 VOLTS)

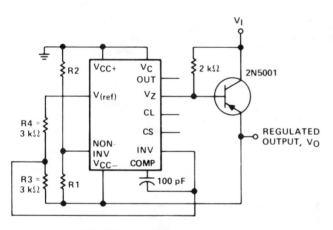

(c) NEGATIVE-VOLTAGE REGULATOR

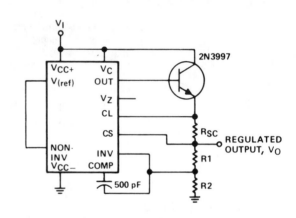

(d) POSITIVE-VOLTAGE REGULATOR
(EXTERNAL N-P-N PASS TRANSISTOR)

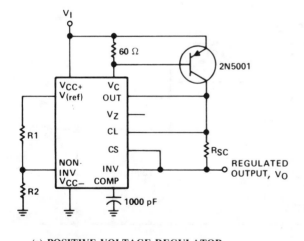

(e) POSITIVE-VOLTAGE REGULATOR
(EXTERNAL P-N-P PASS TRANSISTOR)

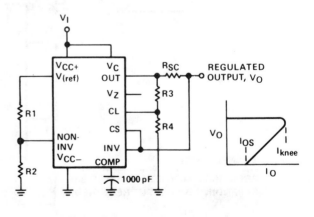

(f) FOLDBACK CURRENT LIMITING

Figure 6-65. Typical Applications

Table 6-2. Resistor Values for Standard Output Voltages

OUTPUT VOLTAGE (V)	APPLICABLE FIGURE (6-65) SEE NOTE 1	FIXED OUTPUT ± 5% kΩ		ADJUSTABLE OUTPUT ± 10% kΩ		
		R1	R2	R1	P1	R2
+5.0	a, e, f	2.15	4.99	0.75	0.5	2.2
+6.0	a, e, f	1.15	6.04	0.5	0.5	2.7
+9.0	b, d, e, f	1.87	7.15	0.75	1.0	2.7
+12.0	b, d, e, f	4.87	7.15	2.0	1.0	3.0
+15.0	b, d, e, f	7.87	7.15	3.3	1.0	3.0
−9.0	c ⎫ see	3.48	5.36	1.2	0.5	2.0
−12.0	c ⎬ note 2	3.57	8.45	1.2	0.5	3.3
−15.0	c ⎭	3.57	11.5	1.2	0.5	4.3

NOTES: 1. To make the voltage adjustable, the R1/R2 divider shown in the figures must be replaced by the divider shown here.

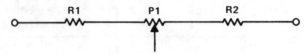

2. For negative output voltages less than 9 V, $V_{CC}+$ and V_C must be connected to a level large enough to allow the voltage between $V_{CC}+$ and $V_{CC}−$ to be greater than 9 V.

GENERAL PURPOSE POWER SUPPLY

The general purpose power supply shown in Figure 6-66 may be used for supply output voltages from 1 to 35 V.

The line transformer should be selected to give about 1.4 times the desired output voltage from the positive side of the filter capacitor, C1, to ground. R1 discharges the carriers in the base-emitter junction of the TIP31 when the drive is reduced. Its value is determined as follows:

$$R1 = \frac{\text{TIP31 Voltage (at point of conduction)}}{\text{Leakage Current of TIP31 and uA723 Output}}$$

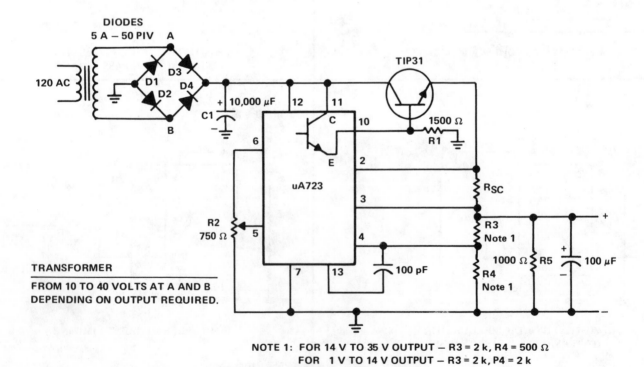

Figure 6-66. General Purpose Power Supply

where: TIP31 voltage at point of conduction is 0.35 V, leakage current (collector-base) of the TIP31 plus the collector-emitter leakage of the uA723 output transistor (worst case = 200 µA).

therefore: $R1 = \dfrac{0.35 \text{ V}}{0.0002 \text{ A}} = 1750 \ \Omega$ maximum

$R1 = 1.5 \text{ k}\Omega$ (standard value)

Potentiometer R2 sets the output voltage to the desired value by adjusting the reference input voltage. It is connected between pin 6 (7.15 V reference) and ground. The center arm of R2, connected to pin 5, will select any point between zero and the 7.15 V reference.

Resistors R3 and R4 are connected in series across the supply output. The junction of these two resistors is connected to the inverting input (pin 4) of the error amplifier establishing an output voltage reference. This voltage reference is compared to the selected voltage at the non-inverting input to the error amplifier (pin 5) to set the level of output voltage regulation. The values for R3 and R4 are listed in Note 1 of Figure 6-66.

R_{SC} is the current limit set resistor. Its value is calculated as:

$$R_{SC} = \dfrac{0.65 \text{ V}}{I_L}$$

For example, if the maximum current output is to be 1 A, $R_{SC} = 0.65/1.0 = 0.65 \ \Omega$.

The 1 kΩ resistor, R_S, on the output is a light-load resistor designed to improve the no-load stability of the supply. The 100 µF electrolytic capacitor improves the overall output ripple voltage. A 100-pF capacitor from the compensation terminal (pin 13) to the inverting input

(pin 4) allows for gain variations in the uA723 error amplifiers and for parasitic capacitances.

The output voltage and current of this supply must be restricted to the specifications of the TIP31 series pass transistor. Since it is rated at two watts in free air at 25°C, sufficient heat sinking is necessary.

8-AMP REGULATED POWER SUPPLY FOR OPERATING MOBILE EQUIPMENT

It is often necessary to operate or test equipment used in automotive applications. This supply, as shown in Figure 6-67, provides up to 8 A at 13.8 V.

The uA723 is used as the control element, furnishing drive current to series-pass transistors which are connected in a Darlington configuration. Two 2N3055 npn transistors are used as the pass transistors, so proper heat sinking is necessary to dissipate the power.

This supply is powered by a transformer operating from 120 VAC on the primary and providing approximately 20 VAC on the secondary. Four 10-A diodes with a 100 PIV rating are used in a full-wave bridge rectifier. A 10,000 µF/36 VDC capacitor completes the filtering, providing 28 VDC.

The dc voltage is fed to the collectors of Darlington-connected 2N3055's. Base drive for the pass transistors is from pin 10 of the uA723 through a 200 Ω current limiting resistor, R1. The reference terminal (pin 6) is tied directly to the non-inverting input of the error amplifier (pin 5), providing 7.15 V for comparison.

The inverting input to the error amplifier (pin 4) is fed from the center arm of a 10 kΩ potentiometer connected across the output of the supply. This control is set for the desired output voltage of 13.8 V. Compensation of the error amplifier is accomplished with a 500-pF capacitor connected from pin 13 to pin 4.

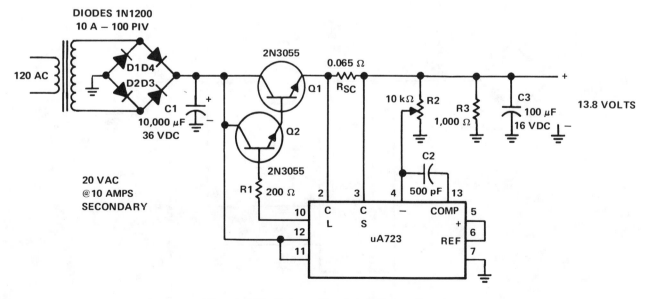

Figure 6-67. 8-Amp Regulated Power Supply

The 1 kΩ resistor on the output is a light load to provide stability when the supply has a no load condition. The 100 μF/16 VDC electrolytic capacitor completes the filter action and reduces the ripple voltage.

The current output of the supply is sampled through resistor R_{SC} between the output transistor and the output terminal. The resistor value for a 10 A maximum current is calculated from the formula:

$$R_{SC} = \frac{0.65 \text{ V}}{I_{(load\ max)}} = \frac{0.65}{10} = 0.065\ \Omega$$

If the power supply should exceed 8 A or develop a short circuit, the uA723 regulator will bias the transistors to cutoff and the output voltage will drop to near zero until the short circuit condition is corrected. This circuit features a no-load-to-full-load (8 A) voltage regulation of no more than 0.2 VDC variation (better than 2% regulation).

± 15 VOLTS @ 1.0 AMP REGULATED POWER SUPPLIES

When working with operational amplifiers, a common requirement is plus and minus supplies in the 15 V range. A positive 15 V supply is shown in Figure 6-68 and a negative 15 V supply is shown in Figure 6-69.

Positive Supply

The positive supply, shown in Figure 6-68, receives + 20 VDC from the rectifier/filter section. This is applied to pins 11 and 12 of the uA723 as well as to the collector of the 2N3055 series-pass transistor. The output voltage is sampled through R1 and R2 providing about 7 V with respect to ground at pin 4.

The reference terminal (pin 6) is tied directly to pin 5, the non-inverting input of the error amplifier. For fine trimming of the output voltage, a potentiometer may be installed between R1 and R2. A 100-pF capacitor from pin 13 to pin 4 furnishes gain compensation for the amplifier.

Base drive to the 2N3055 pass transistor is furnished by pin 10 of the uA723. Since the desired output of the supply is 1 A, maximum current limit is set to 1.5 A by resistor R_{SC} whose value is calculated as:

$$R_{SC} = \frac{0.65 \text{ V}}{I_{(max\ limit)}} = \frac{0.65}{1.5} = 0.433\ \Omega$$

A 100-μF electrolytic capacitor is used for ripple voltage reduction at the output. A 1 kΩ output resistor provides stability for the power supply under no-load conditions. The 2N3055 pass transistor must be mounted on an adequate heat sink since the 3.5 W, 25°C rating of the device would be exceeded at 1 A load current.

Negative Supply

The negative 15 V version of this power supply is shown in Figure 6-69.

The supply receives − 20 V from the rectifier/filter which is fed to the collector of the Darlington pnp pass transistor, a TIP105. A different uA723 configuration is required when designing a negative regulator.

The base drive to the TIP105 is supplied through resistor R5. The base of the TIP105 is driven from pin 9 (V_Z terminal), which is the anode of a 6.2 V zener diode that connects to the emitter of the uA723 output control transistor.

The method for providing the positive feedback required for foldback action is shown in Figure 6-69. This technique introduces positive feedback by increased current flow through resistors R1 and R2 under short-circuit conditions. This forward biases the base-emitter junction of the 2N2907 sensing transistor, which reduces base drive to the TIP105.

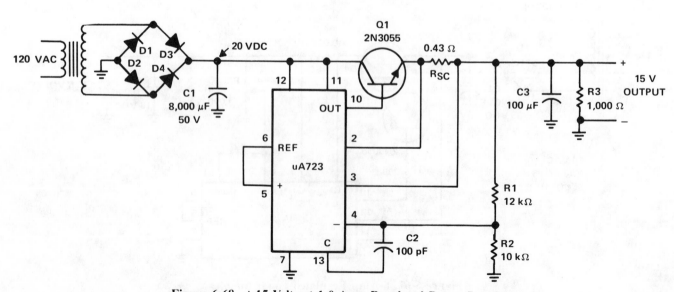

Figure 6-68. + 15-Volts at 1.0-Amp Regulated Power Supply

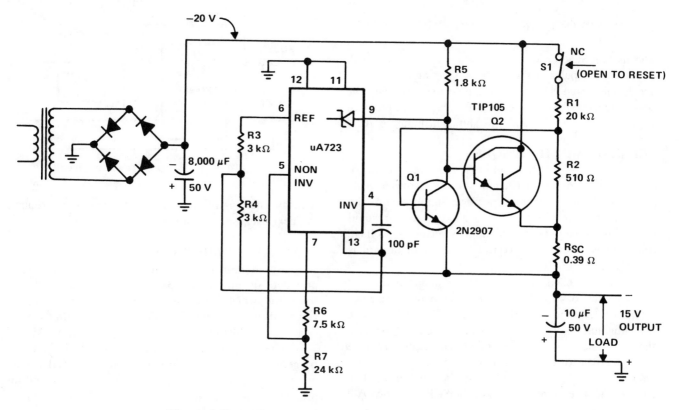

Figure 6-69. −15-Volts at 1.0-Amp Regulated Power Supply

The final percentage of foldback depends on the relative contributions of the voltage drop across R2 and R_{SC} to the base current of the 2N2907 sensing transistor. From the start of base-emitter conduction of the sense transistor to the full shut-off of the TIP105 pass transistor requires a 2 µA base current.

The latch condition, or 100% positive feedback, is generated by any change in the input voltage which increases the voltage drop across R2 turning on the sense transistor (2N2907). It can only be reset by breaking the positive feedback path with switch S1. This allows the series pass device to once more be driven in a normal fashion.

R3 and R4 are equal in value and divide the 7.15 V reference in half. The resulting 3.6 V reference is tied to the inverting input of the error amplifier. R6 and R7 are connected in series across the output of the power supply. The junction of R6 and R7 furnishes 3.6 V to the non-inverting input of the error amplifier. At this point the output is regulated at −15 V with respect to ground.

Resistors R1 and R2 are calculated as follows:

$$R1(k\Omega) = V_I - *V_{SENSE(V)}$$
$$= 20 - 0.5$$
$$= 19.5\ k\Omega$$

$$R1 = 20\ k\Omega\ \text{(standard value)}$$

$$R2(k\Omega) = *V_{SENSE(V)} = 0.5\ k\ \text{or}$$
$$510\ \Omega\ \text{(standard value)}$$

$$\text{Resistor } R5 = (V_I - V_O - V_{BEQ2} - V_{R_{SC}})$$
$$\times \frac{(\text{min beta Q2})}{I_M\ (\text{max load current})}$$

$$R5 = (20 - 15 - 2.8 - 0.4)$$
$$\times \frac{1000}{1} = 1800\ \Omega$$

$$R5 = 1.8\ k\Omega$$

The current sense resistor R_{SC} is calculated as follows:

$$R_{SC} = \frac{V_O}{I_M}\left(\frac{V_{SENSE}}{V_I - V_{SENSE}}\right) = \frac{15}{1}\left(\frac{0.5}{20 - 0.5}\right) = 0.384\ \Omega$$

$$R_{SC} = 0.39\ \Omega$$

Foldback limiting, as used in this circuit, is advantageous where excessive pass transistor power dissipation is a problem. The TIP105 can tolerate only 2 W dissipation in free air at 25°C ambient, so adequate heat sinking is necessary.

---

*V_{SENSE} is defined as the base to emitter voltage needed to start turn-on of the 2N2907. From the data sheet this is about 0.5 V.

OVERVOLTAGE SENSING CIRCUITS

The use of SCR crowbar overvoltage protection (OVP) circuits is a popular method for providing protection from accidental overvoltage stress for a power supply load. The sensing function for this type of OVP circuit can be provided by a single IC, the MC3423, as shown in Figure 6-70.

THE CROWBAR TECHNIQUE

One of the simplest and most effective methods of obtaining overvoltage protection is to use a crowbar SCR placed across the equipment's dc power supply bus. As the name implies, the SCR is used much like a crowbar would be, to short the input of the dc supply when an overvoltage condition is detected. A typical circuit configuration is shown in Figure 6-71.

The MC3423 operates from a V_{CC} minimum of 4.5 V to a maximum of 40 V. The input error amplifier has a 2.6 V reference between the non-inverting input and V_{EE}. The inverting input is V_{sense1} (Pin 2) and is the point to which the output sense voltage is applied. This is usually done through a resistor voltage divider which sets the trip point (V_{ref}) at 2.6 V. The output of the device, Pin 8, then triggers the gate drive terminal of the SCR. A basic OVP circuit is shown in Figure 6-72.

When V_{CC} rises above the trip point set by R1 and R2, an internal current source (Pin 4) begins charging capacitor C1 which is also connected to Pin 3. When triggered, Pin 8 supplies gate drive through the current limit resistor (RG) to the gate of the SCR. The minimum value of RG is given in Figure 6-73.

The value of capacitor C determines the minimum duration of the overvoltage condition necessary to trip the OVP. The value of C can be determined from Figure 6-74.

If the overvoltage condition disappears before C is charged, C discharges at a rate which is 10 times faster

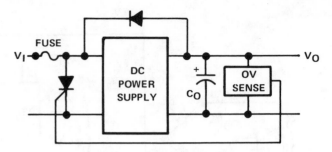

Figure 6-71. Typical Crowbar Circuit

than the charging rate, and resets the timing feature until the next overvoltage condition occurs.

ACTIVATION INDICATION OUTPUT

An additional output for use as an OV indicator is provided on the MC3423. This is an open-collector transistor which saturates when the OVP circuit is activated. It will remain in a saturated state until the SCR crowbar pulls the supply voltage, V_{CC}, below 4.5 V.

This output may also be used to clock an edge-triggered flip-flop whose output inhibits or shuts down the power supply when the OVP trips. This method of protection reduces or eliminates the heat sinking requirements for the crowbar SCR.

REMOTE ACTIVATION INPUT

Another feature of the MC3423 is its remote activation input, Pin 5, which has an internal pull-up current source. This input is CMOS/TTL compatible and, when held below 0.8 V, the MC3423 operates normally. However, if it is raised above 2 V, the OVP is activated regardless of whether an overvoltage condition is present. This feature may be used to accomplish an orderly and sequenced shutdown of system power supplies during a system fault condition.

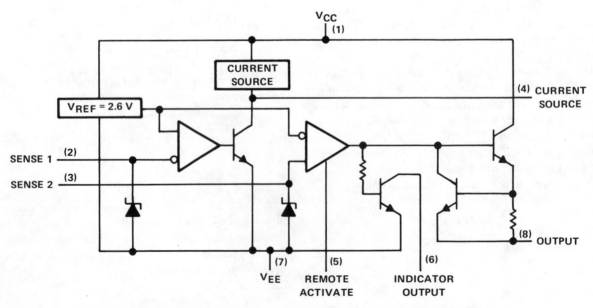

Figure 6-70. MC3423 Overvoltage Crowbar Sensing Circuit Block Diagram

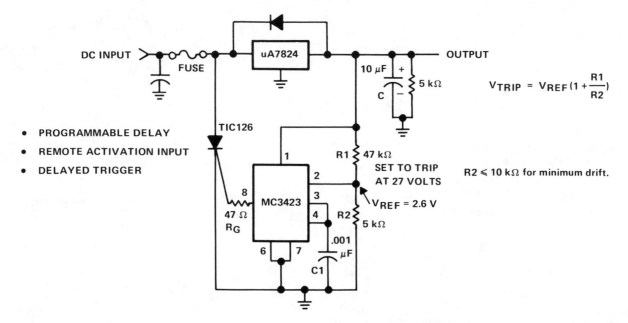

$$V_{TRIP} = V_{REF}\left(1 + \frac{R1}{R2}\right)$$

R2 ≤ 10 kΩ for minimum drift.

- PROGRAMMABLE DELAY
- REMOTE ACTIVATION INPUT
- DELAYED TRIGGER

Figure 6-72. Overvoltage Protection Circuit

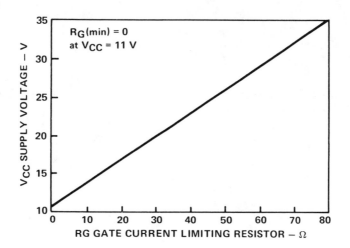

Figure 6-73. Minimum RG vs Supply Voltage

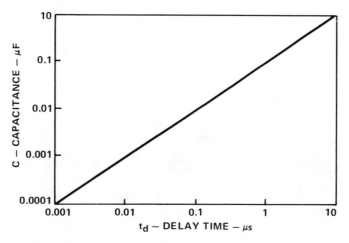

Figure 6-74. Capacitance vs Minimum Overvoltage Duration

Section 7

Integrated Circuit Timers

Timing may be accomplished by a variety of methods; mechanical, thermal, chemical, electronic, or a combination of these. Regardless of the method, a timer depends upon a time base generated internally or applied from an external source. The spring-driven clock, for example, generates its own internal time base, whereas the electric clock uses the period of the ac line voltage as an external time base. The first consideration in the design of a solid state timer is the generation of a suitable time base.

RC TIME-BASE GENERATOR

A simple time base circuit may be established by the use of a resistor, a capacitor, and sensing network as shown in Figure 7-1.

The time, t, for the capacitor C to charge to voltage V_C in this circuit is:

$$t = \text{time in seconds}$$
$$\ln = \text{natural log}$$
$$R = \text{resistance in ohms}$$
$$C = \text{capacitance in farads}$$
$$V_C = \text{capacitor voltage in volts}$$

$$t = RC \ln \frac{V_{CC}}{V_{CC} - V_C}$$

For one RC time constant $(t = RC)$ V_C equals 63.21% of V_{CC}.

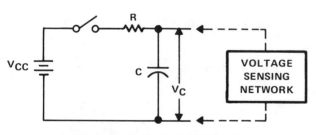

Figure 7-1. Basic RC Circuit

A BASIC DELAY TIMER

A basic delay circuit may be implemented with an RC network, a discharge switch, a voltage reference, a threshold comparator and an output switch. The timer circuit and its typical waveforms are illustrated in Figure 7-2.

The circuit operates as follows: Initially the timer output (V_O) is high (near V_{CC}) and switch S is closed, shorting capacitor C to ground. When switch S is opened the timing period begins and V_O is still high.

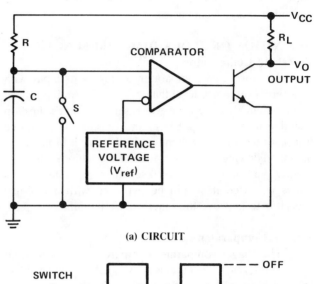

(a) CIRCUIT

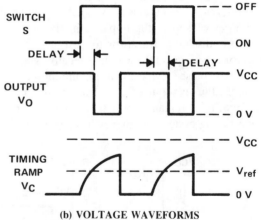

(b) VOLTAGE WAVEFORMS

Figure 7-2. Basic Delay Timer

When switch S is opened capacitor C starts charging, through resistor R, toward the V_{CC} rail. The voltage V_C rises non-linearly, forming a timing ramp (Figure 7-2). V_O is delayed from changing state until V_C reaches the reference voltage, V_{ref}, which is some fraction of V_{CC}. When the timing ramp voltage reaches V_{ref} the comparator output goes high, turning on the output transistor, thus switching the output low. The output will remain low until the switch S is closed or power is turned off.

The addition of another comparator, an RS type flip-flop and a discharge transistor to the structure shown in Figure 7-2 would allow a trigger pulse to start the timing function. Many other performance features including oscillation would be possible. NE555 type IC timers have all of these functions, except the timing resistor and capacitor, included on a single chip.

NE/SE555 TIMERS

The NE555 was the first monolithic IC timer with multi-functional capabilities (introduced in 1972) and has been accepted as the standard for basic timing and oscillator functions. The NE555 (Figure 7-3) is a general-purpose bipolar IC capable of monostable and astable pulse generating modes covering a wide range of pulse durations and/or frequencies.

DEFINITION OF BLOCK DIAGRAM FUNCTIONS
Threshold Comparator

The threshold comparator compares its input with an internal reference level that is 2/3 V_{CC}. An input level greater than the reference will reset the timer's flip-flop resulting in a low output level and causing the discharge transistor to turn on. The internal reference is brought out on a pin allowing external control of the reference level to modify the timing period or reset the comparator. If this pin is not used it should be bypassed with a 0.01 μF capacitor to improve the timer's noise immunity.

Trigger Comparator

The trigger comparator compares its input with an internal reference level that is 1/3 V_{CC}. An input level less than 1/3 V_{CC} will set the flip-flop causing the output to go high and the discharge transistor to turn off. The trigger comparator functions on the leading (negative going) edge of the input pulse. Typically a minimum input pulse duration of 1.0 μs is required for reliable triggering. If the trigger input level remains lower than 1/3 V_{CC} for longer than one timing cycle, the timer will retrigger itself after

the first output pulse. Propagation delay in the trigger comparator can delay turn off up to several microseconds after triggering. An output pulse duration of 10 μs or greater will prevent double triggering due to these effects.

RS Flip-Flop

The RS flip-flop receives its reset input from the threshold comparator, its set input from the trigger comparator and an additional reset from an external source. The external reset input overrides all other inputs and can be used to initiate a new timing cycle. When this input is at a logic low level the timer output is low and the discharge transistor is on, resulting in a reset condition. The reset input is TTL compatible. When not used it is recommended that the reset pin be connected to the V_{CC} rail to prevent false resetting.

Discharge Transistor

The discharge transistor will be on when the device output is low. Its open-collector output is used to discharge the external timing capacitor during the reset phase of operation.

Output Stage

The device output stage is driven by the flip-flop output. It is an active-pull-up, active-pull-down circuit with a 200 mA sink or source capability.

555 BASIC OPERATING MODES

There are two basic operating modes for 555 timer circuits: the monostable (one-shot) mode for timing functions, and the astable (free-running) mode for oscillator functions. There are many variations of the two basic modes, allowing numerous applications.

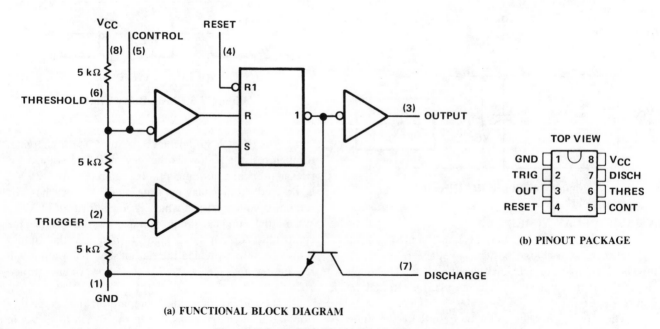

(a) FUNCTIONAL BLOCK DIAGRAM

Figure 7-3. NE555 Timer Block Diagram

Monostable Mode

Figure 7-4 illustrates the circuit and waveforms for the 555 connected in its most basic mode of operation — a triggered monostable.

With the trigger input terminal held higher than $1/3$ V_{CC}, the timer is in its standby state and the output is low. When a trigger pulse appears with a level less than $1/3$ V_{CC}, the timer triggers and its timing cycle starts. The output switches to a high level near V_{CC} and concurrently C_t begins to charge toward V_{CC}. When the V_C voltage crosses $2/3$ V_{CC}, the timing period ends with the output falling to zero. The circuit is now ready for another input trigger pulse. The output pulse duration (T) is defined as $(1.1) \times R_t C_t$. In the monostable mode T is used to represent the on time which is the time base in this mode. With few restrictions, R_t and C_t can have a wide range of values. Assuming zero capacitor leakage current, there is no theoretical upper limit on T while the short pulse durations are limited by internal propagation delays to about 10 µs.

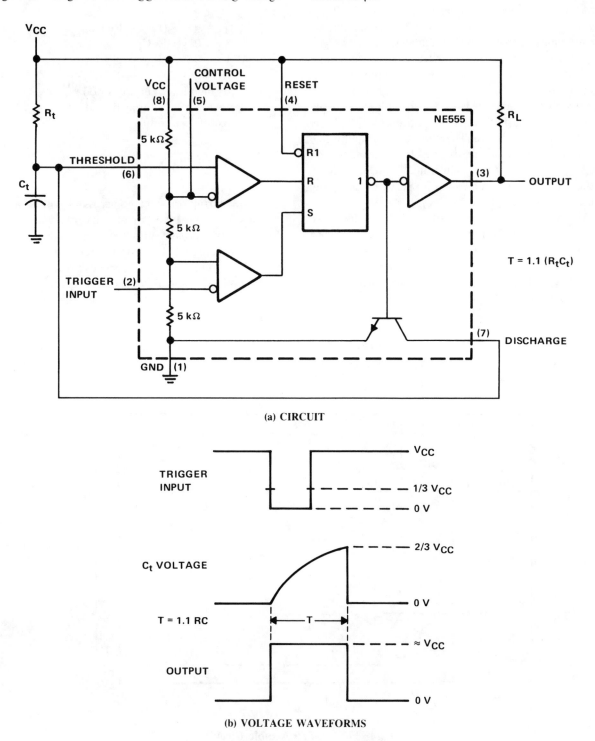

(a) CIRCUIT

(b) VOLTAGE WAVEFORMS

Figure 7-4. 555 Triggered Monostable Circuit

A reasonable lower limit for R_t, for 15 V operation, is about 1 kΩ for the NE555 and is limited only by power dissipation considerations. The upper R_t limit, for a V_{CC} of 15 V, is about 10 MΩ. Allowing for only 0.25 μA input leakage and 0.25 μA capacitor charging current the total current through R_t at the threshold level would be 0.5 μA. R_t max is equal to the voltage across R_t at threshold, (which for a 15 V supply is 5 V), divided by 0.5 μA. This yields an R_t max of 10 MΩ. However, lower values should be used if accurate timing is required.

A practical minimum value for C_t is about 100 pF. Below this value stray capacitance becomes a limiting factor for timing accuracy. The maximum value for C_t is limited by capacitor leakage. Low leakage capacitors are available in values up to about 10 μF and are preferred for long timing periods. Capacitor values as high as 1000 μF could possibly be used if the leakage current is low enough for the application. The real limitation on C_t is leakage current and not capacitance. The ultimate criterion for the selection of R_t and C_t is the degree of accuracy desired. Staying within the limitations illustrated in the 555 device data sheet charts is recommended for relatively accurate designs.

As given in Figure 7-4(b), the pulse duration T is slightly more than an RC time constant (T = 1.1 RC). This is a result of a threshold level that is 66.7% of V_{CC} while one RC level is 63.2% of V_{CC}.

In a typical application input leakage currents may also lead to some slight differences between actual values of "T" and calculated values. Operation at high speeds (very short pulse durations) will result in variations from the calculated values due to internal propagation delays.

Astable Mode

Figure 7-5(a) illustrates the 555 connected as an astable timer. Like the monostable circuit, the astable circuit requires only a few external components.

Figure 7-5(b) shows the timing diagram. The timing calculations are as follows.

$$t_1 = 0.693 (R_A + R_B) C_t$$
$$t_2 = 0.693 R_B C_t$$
$$T = 0.693 (R_A + 2R_B) C_t$$

where: t_1 is high-level output period
   t_2 is low-level output period
   T is total period ($t_1 + t_2$)

$$f = \frac{1}{T} = \frac{1.44}{(R_A + 2R_B)C_t}$$

On startup, the voltage across C_t will be near zero which causes the timer to be triggered via pin 2. This forces the output high, turning off the discharge transistor and allows charging of C_t through R_A and R_B. C_t charges toward V_{CC} until its voltage reaches a level of 2/3 V_{CC}, at which point the upper threshold is reached, causing the output to go low and the discharge transistor to turn on.

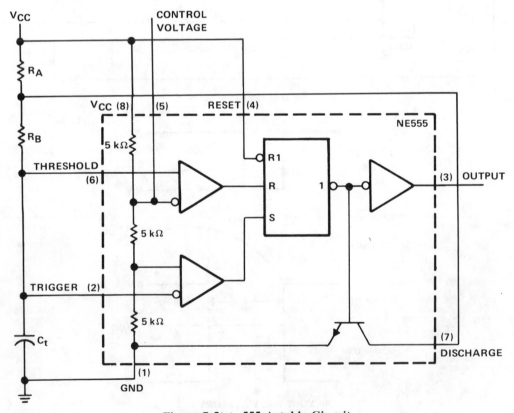

Figure 7-5(a). 555 Astable Circuit

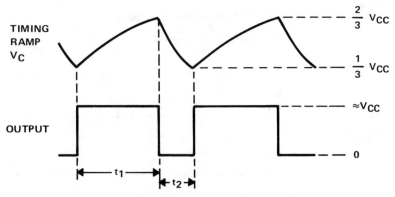

Figure 7-5(b). NE555 Astable Timing Diagram

Capacitor C_t then discharges toward ground through R_B until its voltage reaches 1/3 V_{CC}, the lower trigger point. This retriggers the timer, beginning a new cycle. The timer threshold input therefore oscillates between the 2/3 V_{CC} and 1/3 V_{CC} comparator threshold levels.

The frequency of operation is simply the reciprocal of T as stated above. The duty cycle for either the high or low output state is simply that period (t_1 or t_2) divided by the total period. For reliable operation, the upper frequency limit of the bipolar NE555 is about 100 kHz. Device upper frequency limitations are due to internal propagation delays. Low frequencies are not limited by the 555 devices but are limited by the leakage characteristics of C_t.

Specific duty cycles may be required in some applications. Duty cycle can be controlled (within limits) by adjusting the resistance ratios of R_A and R_B in Figure 7-5(a). As R_B becomes large with respect to R_A, the duty cycle approaches 50% (square wave operation). Conversely, as R_A becomes large with respect to R_B, the duty cycle increases toward 100%. R_A must not be allowed to reach zero. Practical duty cycles range from 49.8% to 99% or in terms of resistor ratios R_A may be 1/100 of R_B or R_B may be 1/100 of R_A.

Accuracy

Although the 555 is a simple device, it performs accurately. In the monostable mode there is a typical initial error of only 1% due to process imperfections (R_t and C_t errors must be considered separately). For astable operation the error is somewhat greater, typically about 2%. Drift with temperature is typically 55 ppm/°C (or 0.005%/°C) for the monostable mode, and is about 150 ppm/°C for the astable mode.

TLC555 TIMER

In 1984, TI introduced a LinCMOS version of the 555 timer, the TLC555. It features the performance characteristics of a bipolar 555 and, in addition, some important improvements. Table 7-1 is a performance comparison of major parametric differences in the NE555 and TLC555.

Due to its high-impedance inputs, the TLC555 is capable of producing accurate time delays and oscillations using less expensive, smaller timing capacitors than the NE555. A duty cycle of 50% in the astable mode is easily achieved using only one resistor and one capacitor as illustrated in Figure 7-6.

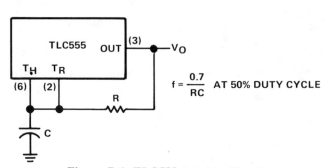

Figure 7-6. TLC555 Astable Circuit

While the complementary CMOS output is capable of sinking over 100 mA and sourcing over 10 mA, the TLC555 exhibits greatly reduced supply current spikes during output transitions (see Table 7-1). This minimizes the need for the large decoupling capacitors required by the bipolar 555. The TLC555 also provides very low power consumption (typically 1 mW at V_{DD} = 5 V) for supply voltages ranging from 2 V to 18 V.

TLC555 Astable Timing Equations

At astable operating frequencies above one megahertz, the propagation delays of the TLC555 must be accounted for in the equations for calculating the astable frequency (see Figure 7-7). The on-state resistance (R_{Don}) of the NMOS discharge transistor (typically 10 Ω) is also included to give greater accuracy. Besides the low to high and high to low progagation delays, (t_{PLH} and t_{PHL} respectively), two additional times, T_C and T_D, must also be included along with the maximum charge and minimum discharge voltages, (V_h and V_l respectively).

As the capacitor is charging, it continues to charge up to V_h after crossing the 2/3 V_{DD} level for a time equal to the t_{PHL} propagation delay. T_D is the length of time it takes to discharge from V_h to 2/3 V_{DD}.

Table 7-1. Performance Comparison

SPECIFICATION	NE555	TLC555	UNITS
QUIESCENT CURRENT	6.0	0.3	mA
BIAS CURRENT	500,000	10	pA
MAXIMUM FREQUENCY	100	2000	kHz
CURRENT SPIKES	≥ 200	3	mA

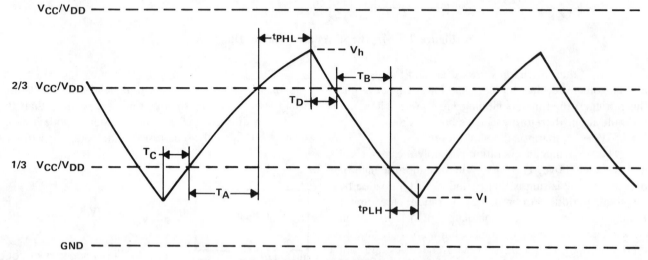

Figure 7-7. TLC555 Astable Timing Diagram

Likewise, during the discharge part of the cycle the capacitor continues to discharge down to V_1 after crossing the 1/3 V_{DD} level for a time equal to t_{PLH}. T_C is the length of time required to charge back up from V_1 to 1/3 V_{DD}.

$$R_T = R_A + R_B$$
$$R_T' = R_B + R_{Don}$$
$$T_A = 0.693\, R_T C_t$$
$$T_B = 0.693\, R_T' C_t$$
$$\therefore V_h = (2/3)V_{DD}\{1 - EXP[-(0.693 + t_{pHL}/R_T C_T)]\}$$
$$\qquad + (1/3)V_{DD}$$
$$\therefore V_1 = (2/3)V_{DD}\, EXP[-(0.693 + t_{pLH}/R_T' C_T)]$$
$$\therefore T_C = -R_t C_t \ln\{1 - [(1/3)V_{DD} - V_1]/(V_{DD} - V_1)\}$$
$$\therefore T_D = (-R_T' C_t)\ln[(2/3)V_{DD}/V_h]$$
$$\therefore T_{period} = T_A + T_B + T_C + T_D + t_{PLH} + t_{PHL}$$
$$\therefore f_{astable} = 1/T_{period}$$

Notice that for t_{PLH}, t_{PHL}, and R_{Don} equal to zero the above equations reduce to the standard equation as printed in the data book for the NE555:

$$(NE555)\ T_{period} = T_A + T_B = 0.693(R_A + 2R_B)(C_t)$$

It should also be noted that a 10% increase in astable frequency will also occur if a 0.1 μF capacitor is connected from the control voltage pin (pin 5) to ground.

TLC556/NE556 DUAL UNIT

The NE556 bipolar timer contains two NE555 timers in the same package, with common power-supply and ground pins. The LinCMOS version of the dual 555 timer is the TLC556. See Figure 7-8 for the pinout of these devices.

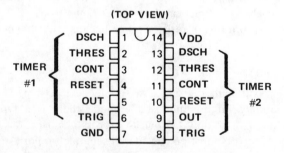

Figure 7-8. 556 Dual Timer Pinout

The TLC556 dual timer is identical to the NE556 electrically and functionally. It is fabricated using the LinCMOS process and has the same advantages noted for the TLC555.

uA2240 PROGRAMMABLE TIMER/COUNTER

The uA2240 programmable timer/counter is a special class of timer IC. This device includes a timing section made up of a timer similar to a 555 type, an eight-bit counter, a control flip-flop, and a voltage regulator. Timing periods from microseconds to days may be programmed with an accuracy of 0.5%. Two timing circuits cascaded could generate time delays of up to three years. The 2240 may be operated in either the monostable or astable mode with programmable timing in either mode. A functional diagram and pin-out are shown in Figure 7-9.

DEFINITION OF uA2240 FUNCTIONS
Voltage Regulator (V_{REG})

The on-chip regulator provides voltage to the binary counter and control logic and through pin 15 to external circuits. If V_{CC} (pin 16) is 15 V, V_{REG} is typically 6.3 V. For a V_{CC} of 5 V, V_{REG} is typically 4.4 V. For a V_{CC} supply of less than or equal to 4.5 V, pin 16 (V_{CC}) should be shorted to pin 15 (V_{REG}). The minimum supply voltage for this condition is 4 V. When supplying external circuitry via pin 15 (regulator output), the output current should be 10 mA or less.

Control Logic

The control logic block provides trigger and reset signals to the binary counter and time-base generator flip-flops. Trigger and reset inputs are high impedance and TTL compatible, requiring only about 10 μA of input current. Therefore, they may be controlled by TTL, low level MOS or CMOS logic, and respond to positive input transitions. The reset input (pin 10) terminates the timing cycle by returning the counter flip-flops to zero and disabling the time-base oscillator. A logic-high reset stops

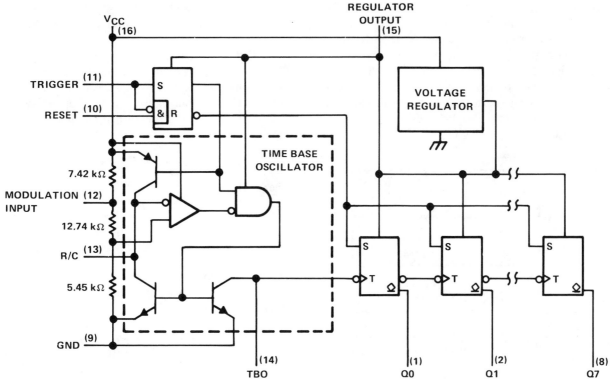

◊ open-collector outputs

(a) FUNCTIONAL BLOCK DIAGRAM

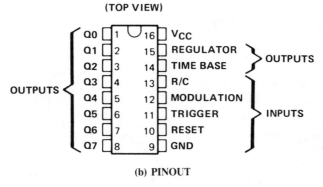

(b) PINOUT

Figure 7-9. uA2240 Functional Block Diagram and Pinout

the timer and sets all outputs high. When powered on, the uA2240 comes up in the reset state. A trigger input (pin 11) initiates the timing cycle by enabling the time-base oscillator and setting all outputs low. Once triggered, the circuit is immune to further trigger inputs until the timing cycle is completed or a reset signal is applied. However, the trigger takes precedence if both trigger and reset are applied simultaneously.

Time-Base Oscillator

The heart of a uA2240 is its time-base oscillator that is made up of threshold comparators, a flip-flop, discharge and time-base output transistors, RC input, and a modulation and sync input.

RC Input

A resistor and capacitor connected in series between V_{CC} and ground provide an exponential ramp at the RC input (pin 13). The comparator thresholds are designed to detect at levels allowing a 63% charge time (1 RC time interval). Thus, the time-base output (TBO) pulse will have a period T = 1 RC. Figure 7-10 shows the recommended range of timing component values. The timing capacitor leakage currents must be low to allow it to charge to the comparator threshold levels with a large value (1 MΩ or greater) timing resistor.

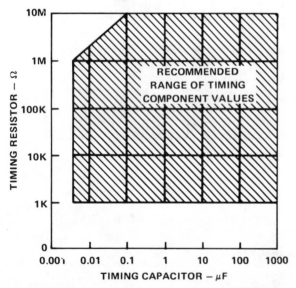

Figure 7-10. Recommended Range of Timing Component Values

Modulation and Sync Input

The MOD input (pin 12) is nominally at 73% of the V_{CC} level due to internal biasing. This level may be varied by connecting a resistor from pin 12 to ground or V_{CC}, or by applying an external voltage. This change, or modulation, of voltage on pin 12 changes the upper threshold for the time-base comparator resulting in a change, or modulation, of the time-base period T. Figure 7-11 illustrates the effects of an externally applied modulation voltage on the time-base period.

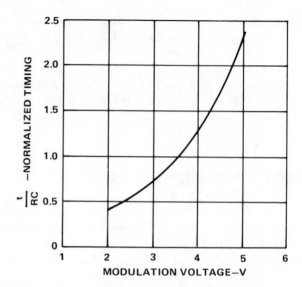

Figure 7-11. Normalized Change in Time Base Period as a Function of Modulation Voltage at Pin 12

The MOD input may also be used to synchronize the time-base oscillator with an external clock. Synchronization is achieved by setting the time-base period (T) to be an integer multiple of the sync pulse period (T_S). This is accomplished by choosing R and C timing components so that:

$$T = RC = T_S \times M$$

where: M is an integer from 1 to 10.

Figure 7-12 gives the typical pull-in range for harmonic synchronization versus the ratio of time-base period to sync pulse period (T/T_S).

Threshold Comparators

The two levels of threshold are set at 27% and 73% of V_{CC}. Charging and discharging of the timing capacitor

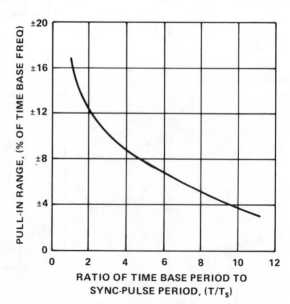

Figure 7-12. Typical Pull-In Range for Harmonic Synchronization

occurs between these two levels. When charging from the 27% level toward V_{CC} to the second threshold at 73%, the percentage interval to be changed is $73 - 27$ or 46%. The actual percentage of the range from 27% to the V_{CC} rail is 73% so the charge range to be covered is 0.46/0.73 or 63%, exactly one RC time constant. The resulting time base $T = RC$.

Oscillator Flip-Flop

Comparator outputs feed the oscillator flip-flop which controls the discharge and time-base output (TBO) transistors. Once triggered (see Figure 7-13) the oscillator continues to run until reset. Output pulses from the TBO are internally connected to the counter input for automatic triggering. The TBO output is an open-collector transistor and requires a pull-up resistor (typically 20 kΩ) to be connected to the V_{REG} output (pin 15). Grounding the TBO output (pin 14) will disable the counter section.

NOTE: When using a high supply voltage ($V_{CC} > 7$ V) and a small timing capacitor ($C < 0.1$ μF) the pulse width of the TBO output may be too narrow to trigger the counter section. Connecting a 300 pF capacitor from pin 14 to ground will widen the TBO output pulse width and allow a proper trigger time. This capacitor is also recommended to improve noise immunity.

Binary Counter

The uA2240 has an on-chip 8-bit binary counter with outputs that are buffered open-collector type stages. Each output is capable of sinking 2 mA at 0.4 V V_{OL}. At turn on, or in the reset condition, all counter outputs are high, or in the off state. Following a trigger input (Figure 7-14) the outputs will change states according to the sequence shown. The outputs may be used individually, or can be

connected together in a wired-OR configuration for special programming.

Combining counter outputs in a wired-OR configuration results in the addition of the time delays associated with each output connected together. As an example pin 5 alone results in a timing cycle (T_O) that is equal to 16T. Similarly connecting Q_0 (pin 1), Q_4 (pin 5), and Q_7 (pin 8) together will yield $T_O = (1 + 16 + 128)T = 145$ T. A proper selection of counter output terminals will allow programming of T_O from 1 T to 255 T.

uA2240 BASIC OPERATING MODES
Monostable Operating Mode

Figure 7-15 illustrates the 2240 used in the monostable mode.

In the circuit, Figure 7-15(a), R_t and C_t set the time base, T, for the desired time period, T_O. Programming of various output times may be accomplished by connecting the desired counter output pins together. The timer output appears across R_L. The output pulse width, T_O, is equal to the number of timing pulses, n, multiplied by R_tC_t or $T_O = nR_tC_t$.

As shown in the timing diagram, Figure 7-15(b), the output is high (at V_{CC}) prior to triggering. When a trigger pulse is received, the output falls low and the timing cycle is initiated. The time-base oscillator will now run until the counter reaches the count programmed by the selector switches or jumpers. When this count is reached, the output rises from the low level to V_{CC}. This rise in level is fed to the reset input, which stops the oscillator and resets the counter. The timer is now in its standby state, awaiting the next trigger pulse.

R1 is a load resistor for the time-base output. The 270 pF capacitor on the time-base pin is a noise bypass to

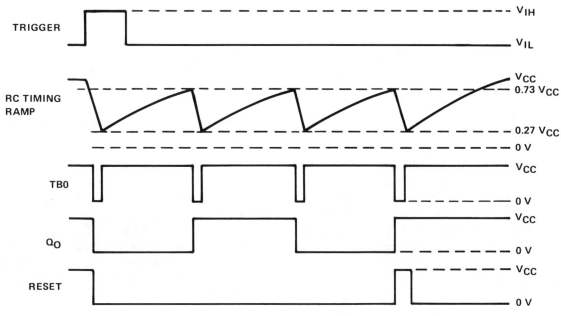

Figure 7-13. Timing Diagram

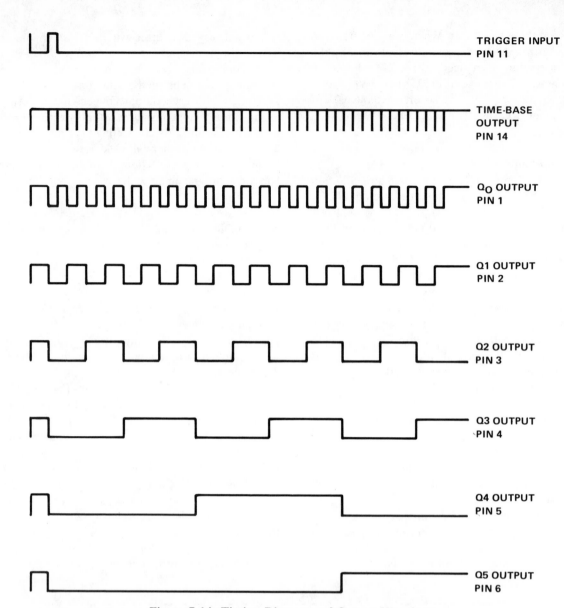

Figure 7-14. Timing Diagrams of Output Waveforms

ensure noise immunity within the time-base oscillator. The R_t and C_t value ranges appear in Figure 7-15. The maximum oscillator frequency should be limited to about 100 kHz.

Astable Operating Mode

The astable mode circuit is shown in Figure 7-16(a). This circuit is similar to the monostable circuit with the exception that the reset is not connected to the output. This allows the oscillator to continue oscillating once started by a trigger pulse. With a single counter output connected to R_L and the output bus, the frequency of oscillation will be:

$$f_o = \frac{1}{2nR_tC_t}$$

where: n is the counter tap selected (1, 2, 4, 8, etc)

The factor 2 is required because the basic timing taps are multiples of 1/2 cycle.

This circuit will not self-start on power-up. A pulse applied to the trigger input will start the synchronous oscillations. The oscillator may be stopped by applying a reset pulse, which causes the output to go high. It will remain in this state until triggered again. If automatic power-up oscillation is desired, connect the trigger input to pin 15 (regulator output). The timing component ranges shown in the monostable circuit, Figure 7-15, are applicable to the astable mode circuit.

Accuracy

The 2240 timer is somewhat more complex than the 555 family. Timing accuracy is good, typically 0.5% at a V_{CC} of 5 V. The maximum operating frequency is about 130 kHz. The trigger and reset inputs have a threshold sensitivity of 1.4 V.

GENERAL DESIGN CONSIDERATIONS

Several precautions should be taken with respect to the V_{CC} supply. The most important is good power supply

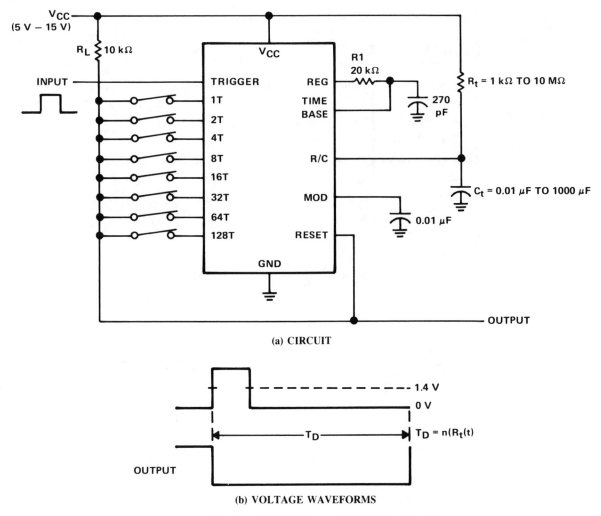

(a) CIRCUIT

(b) VOLTAGE WAVEFORMS

Figure 7-15. uA2240 Monostable Mode Circuit

filtering and bypassing. Ripple on the supply line can cause loss of timing accuracy. A capacitor from V_{CC} to ground, ideally directly across the device, is necessary. The capacitance value will depend on the specific application. Values of from 0.01 μF to 10 μF are not uncommon. The capacitor should be as close to the device as physically possible.

If timing accuracy is to be maintained, stable external components are necessary. Most of the initial timing error is due to the inaccuracies of the external components. The timing resistors should be the metal film type if accuracy and repeatability are important design criteria. If the timing is critical, an adjustable timing resistor is necessary. A good quality multi-turn pot might be used in series with a metal film resistor to make up the R portion of the RC network.

The timing capacitor should also be high quality, with very low leakage. Do not use ceramic disc capacitors in the timing network under any circumstance. Several acceptable capacitor types are silver mica, mylar, polystyrene, and tantulum. If timing accuracy is critical over temperature, timing components with a small positive temperature coefficient should be chosen. The

most important characteristic of the capacitor is low leakage. Obviously any leakage will subtract from the charge count causing the actual time to be longer than the calculated value.

One final precaution should be observed. Make certain that the power dissipation of the package is not exceeded. With extremely large timing capacitor values, a maximum duty cycle which allows some cooling time for the discharge transistor may be necessary.

MISSING PULSE DETECTOR

Figure 7-17 illustrates an NE555 timer utilized as a missing pulse detector. This circuit will detect a missing pulse or abnormally long spacing between consecutive pulses in a train of pulses. The timer is connected in the monostable mode. In addition, a 2N2907 is connected with the collector grounded and the emitter tied to pins 6 and 7. This outboard switch is in parallel with the internal discharge transistor. The transistor base is connected to the trigger input of the NE555.

For this application, the time delay should be set slightly longer than the timing of the input pulses. The timing interval of the monostable circuit is continuously

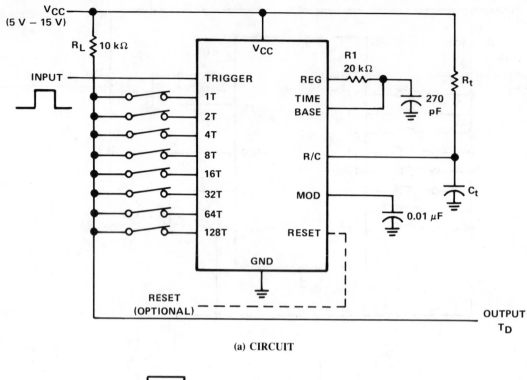

(a) CIRCUIT

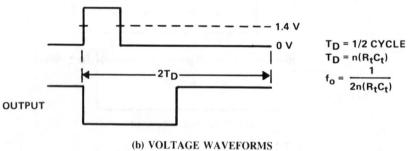

$$T_D = 1/2 \text{ CYCLE}$$
$$T_D = n(R_t C_t)$$
$$f_o = \frac{1}{2n(R_t C_t)}$$

(b) VOLTAGE WAVEFORMS

Figure 7-16. uA2240 Astable Mode Circuit

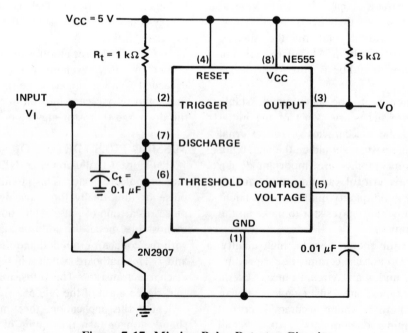

Figure 7-17. Missing Pulse Detector Circuit

retriggered by the input pulse train (V_I). The pulse spacing is less than the timing interval, which prevents V_C from rising high enough to end the timing cycle. A longer pulse spacing, a missing pulse, or a terminated pulse train will permit the timing interval to be completed. This will generate an output pulse (V_O) as illustrated in Figure 7-18. The output remains high on pin 3 until a missing pulse is detected at which time the output goes low.

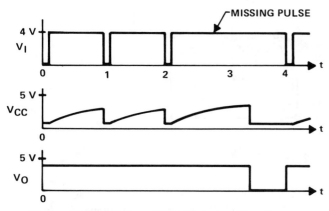

Figure 7-18. Missing Pulse Detector Waveforms

The NE555 monostable circuit should be running slightly slower (lower in frequency) than the frequency to be analyzed. Also, the input cannot be more than twice this free-running frequency or it would retrigger before the timeout and the output would remain in the low state continuously. The example in Figure 7-17 operates in the monostable mode at about 8 kHz so pulse trains of 8 to 16 kHz may be observed.

NE555 ONE-SHOT TIMER

Figure 7-19 shows an NE555 connected in its most basic mode of operation, a triggered monostable one-shot. This simple circuit consists of only the two timing components R_t and C_t, the NE555, and bypass capacitor C2. While not essential for operation, C2 is recommended for noise immunity.

During standby, the trigger input terminal is held higher than 1/3 V_{CC} and the output is low. When a trigger pulse appears with a level less than 1/3 V_{CC}, the timer is triggered and the timing cycle starts. The output rises to a high level near V_{CC}, and at the same time C_t begins to charge toward V_{CC}. When the C_t voltage crosses 2/3 V_{CC}, the timing period ends with the output falling to zero, and the circuit is ready for another input trigger. This action is illustrated in Figure 7-20.

Due to the internal latching mechanism, the timer will always time out when triggered, regardless of any subsequent noise (such as bounce) on the trigger input. For this reason the circuit can also be used as a bounceless switch by using a shorter RC time constant. A 100 kΩ resistor for R_t and a 1 μF capacitor for C_t would give a clean 0.1 s output pulse when used as a bounceless switch.

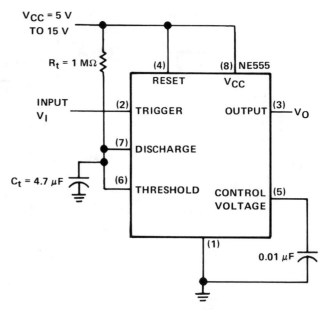

Figure 7-19. NE555 One-Shot Timer

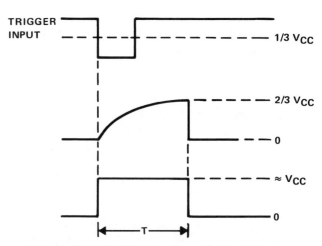

Figure 7-20. NE555 One-Shot Timing Diagram

OSCILLOSCOPE CALIBRATOR

The oscilloscope is one of the most useful electronic instruments. It is capable of displaying both low and high frequency signals on its screen, in a form which lends itself to circuit analysis. Calibrated amplifiers and time base generators are usually incorporated.

The calibrator shown in Figure 7-21 can be used to check the accuracy of the time-base generator as well as to calibrate the input level of the amplifiers.

The calibrator consists of an NE555 connected in the astable mode. The oscillator is set to exactly 1 kHz by adjusting potentiometer P1 while the output at pin 3 is being monitored against a known frequency standard or frequency counter. The output level likewise is monitored from potentiometer P2's center arm to ground with a standard instrument. P2 is adjusted for 1 V peak-to-peak at the calibrator output terminal. The circuit may be supplied with at least 8 V and is regulated to 5 V to supply the NE555.

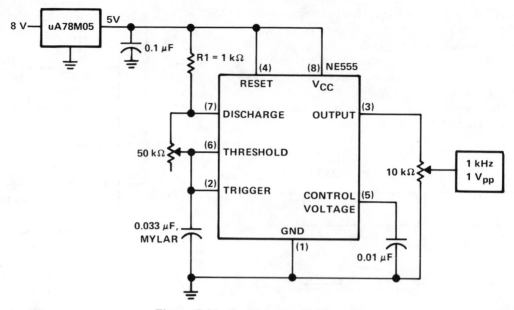

Figure 7-21. Oscilloscope Calibrator

During operation, the calibrator output terminal will produce a 1 kHz square wave signal at 1 V peak-to-peak with about 50% duty cycle. For long term oscillator frequency stability, C1 should be a low leakage mylar capacitor.

DARKROOM ENLARGER TIMER

An enlarger timer is essential for consistent quality and repeatability. The timer controls the exposure time of the paper in the enlarger.

The NE555 circuit illustrated in Figure 7-22 is a basic one-shot timer with a relay connected between the output and ground. It is triggered with the normally open momentary contact switch which when operated, grounds the trigger input (pin 2). This causes a high output to energize K1 which closes the normally open contacts in the lamp circuit. They remain closed during the timing interval, then open at time out. Timing is controlled by a 5 MΩ potentiometer, R_t. All timer driven relay circuits should use a reverse clamping diode, such as D1, across the coil. The purpose of diode D2 is to prevent a timer output latch up condition in the presence of reverse spikes across the relay.

With the RC time constant shown, the full scale time is about 1 minute. A scale for shaft position of the 5 MΩ potentiometer can be made and calibrated in seconds. Longer or shorter full scale times may be achieved by changing the values of the RC timing components.

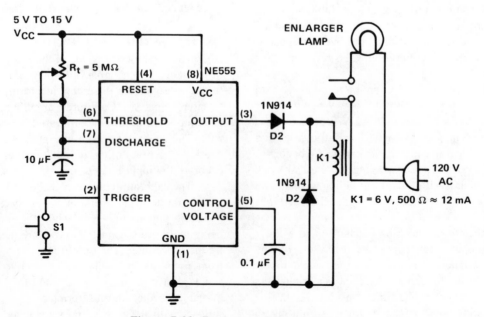

Figure 7-22. Darkroom Enlarger Timer

TOUCH SWITCH

An interesting type of circuit that uses an IC timer is shown in Figure 7-23. This circuit is a "touch switch", a device in which switching action is accomplished without conventional mechanical movement of a lever or push button. A circuit such as this would also be useful in a burglar alarm.

The circuit is basically an NE555 monostable, the only major difference being its method of triggering. The trigger input is biased to a high value by the 22 MΩ resistor. When the contact plates are touched, the skin resistance of the operator will lower the overall impedance from pin 2 to ground. This action will reduce the voltage at the trigger input to below the $1/3 \ V_{CC}$ trigger threshold and the timer will start. The output pulse width will be $T = 1.1 \ R1C1$, in this circuit about 5 seconds. A relay connected from pin 3 to ground instead of the LED and resistor could be used to perform a switching function. Diodes should be used with the relay as shown in Figure 7-22.

The contact strips or plates may be copper, brass, or any conducting material arranged for convenient finger contact.

BASIC SQUARE WAVE OSCILLATOR

A basic square wave oscillator is shown in Figure 7-24. The NE555 is connected in the astable mode and uses only three timing components (R_A, R_B, and C_t). A 0.01 μF bypass capacitor is used on pin 5 for noise immunity.

Operating restrictions of the astable mode are few. The upper frequency limit is about 100 kHz for reliable operation, due to internal storage times. Theoretically there is no lower frequency limit, only that imposed by R_t

and C_t limitations. There are many variations of this astable circuit, but it is shown here in its simplest form.

Oscillators such as this are useful in test equipment or as a signal generator for testing other circuits.

The frequency for the circuit in Figure 7-24 may be calculated as follows.

$$f = \frac{1.44}{(R_A + 2R_B)C_t}$$

$$= \frac{1.44}{(4.7K + 2M)(0.0047 \ \mu F)}$$

$$= \frac{1.44}{9.42209 \times 10^{-3}}$$

$$f = 152.8 \ Hz$$

LINEAR RAMP GENERATOR

A very useful modification to the standard monostable configuration is to make the timing ramp a linear waveform, rather than an exponential one. The linear charging ramp is most useful where linear control of voltage is required. Some possible applications are a long period voltage controlled timer, a voltage to pulse width converter, or a linear pulse width modulator.

One of the simplest methods to achieve a linear ramp monostable is to replace the timing resistor, R_t, with a constant current source as illustrated in Figure 7-25. Q1 is the current source transistor, supplying constant current to the timing capacitor C_t. When the timer is triggered, the clamp on C_t is removed and C_t charges linearly toward V_{CC} by virtue of the constant current supplied by Q1. The threshold at pin 6 is $2/3 \ V_{CC}$; here, it is termed V_C. When

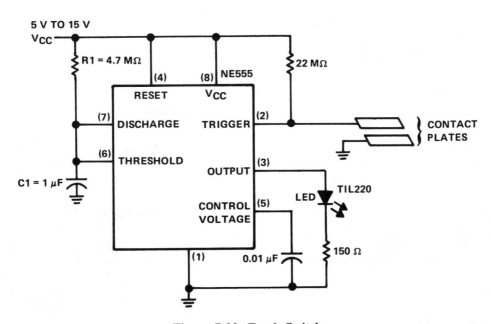

Figure 7-23. Touch Switch

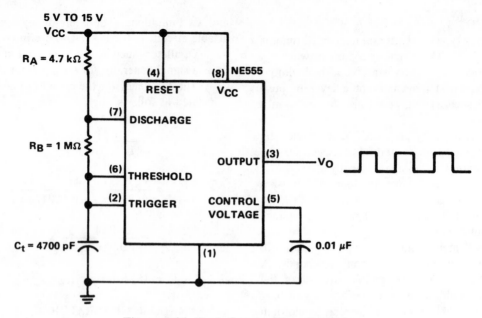

Figure 7-24. Basic Square Wave Oscillator

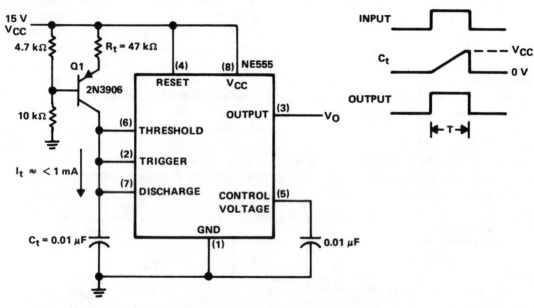

Figure 7-25. Linear Ramp Generator

the voltage across C_t reaches V_C volts, the timing cycle ends. The timing expression for output pulse width T is:

$$T = \frac{V_C C_t}{I_t}$$

where: V_C is the voltage at pin 5

I_t is the current supplied by Q1

I_t can be approximated for 15 V V_{CC} as:

$$\frac{\text{Voltage across } R_t}{I_t} = \frac{4.2 \text{ V}}{R_t \ 47 \text{ k}\Omega}$$

Then T is: T $\approx 0.24 \ V_C R_t C_t$

$\approx 0.24 \ (10)(4.7 \text{ k}\Omega)(0.01 \ \mu F)$

≈ 1 ms

The ramp frequency f_0 is then:

$$f_0 = \frac{1}{2T} = \frac{1}{0.002} = 0.5 \text{ Hz}$$

In general, I_t should be 1 mA or less, and C_t can be any value compatible with the NE555.

FIXED-FREQUENCY VARIABLE-DUTY-CYCLE OSCILLATOR

In a basic astable timer configuration timing periods t_1 and t_2, as shown in Figure 7-26, are not controlled independently. This makes it difficult to maintain a constant period (T) if either t_1 or t_2 is varied. Figure 7-26 illustrates this relationship.

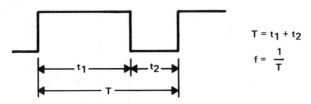

$$T = t_1 + t_2$$
$$f = \frac{1}{T}$$

Figure 7-26. Astable Mode Output Timing

A number of methods have been developed to maintain a constant period T for this versatile timer. One method, as illustrated in Figure 7-27, employs a circuit that adds two diodes to the basic astable mode circuit.

In this circuit the charge (R_{AB}) and discharge (R_{BC}) resistances are determined by the position of the common wiper arm (B) of the potentiometer, making it possible to adjust the duty-cycle by adjusting t_1 and t_2 proportionately without changing the period T.

At startup, the voltage across C_t is less than the trigger level voltage (1/3 V_{DD}), causing the timer to be triggered via pin 2. The output of the timer (pin 3) goes

high, turning off the discharge transistor (pin 7) and allowing C_t to charge through diode D1 and resistance R_{AB}.

When capacitor C_t charges to the upper threshold voltage (2/3 V_{DD}), the flip-flop is reset and the output (pin 3) goes low. Capacitor C_t then discharges through diode D2, resistance R_{BC}. When the voltage at pin 2 reaches 1/3 V_{DD}, the lower threshold or trigger level, the timer triggers again and the cycle is repeated.

In this circuit the oscillator frequency remains fixed and the duty cycle is adjustable from < 0.5% to > 99.5%.

ALTERNATING LED FLASHER

It is often desirable to have two LEDs turn on and off alternately. Advertising signs frequently use this type of operation. Alternating LED action is easily accomplished using the circuit shown in Figure 7-28 with the TLC555 operating in the astable mode.

The timing components are R1, R2, and C_t. C1 is a bypass capacitor used to reduce the effects of noise. At startup, the voltage across C_t is less than the trigger level voltage (1/3 VDD), causing the timer to be triggered via pin 2. The output of the timer (pin 3) goes high, turning LED1 off, LED2 on, the discharge transistor (pin 7) off, and allowing C_t to charge through resistors R1 and R2.

When capacitor C_t charges to the upper threshold voltage (2/3 V_{DD}), the flip-flop is reset and the output (pin 3) goes low. LED1 is turned on, LED2 is turned off, and capacitor C_t discharges through resistor R2 and the

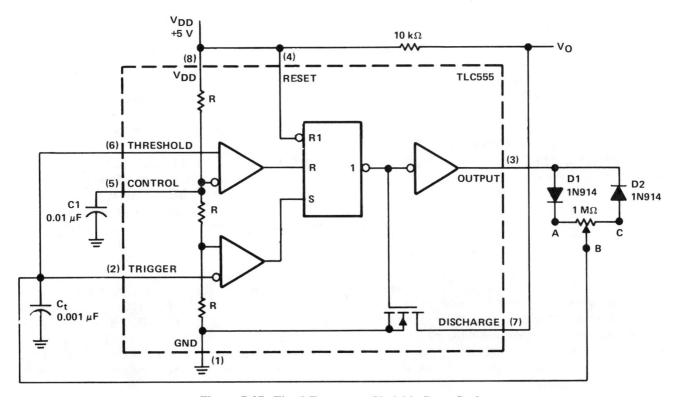

Figure 7-27. Fixed-Frequency Variable-Duty Cycle Astable Oscillator

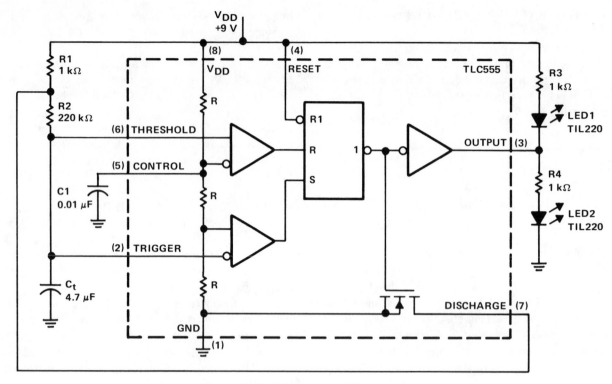

Figure 7-28. Alternating LED Flasher

discharge transistor. When the voltage at pin 2 reaches 1/3 V_{DD}, the lower threshold or trigger level, the timer triggers again and the cycle is repeated.

The totem-pole output at pin 3 is a square wave with a duty cycle of about 50%. The output alternately turns on each LED at slightly less than one blink per second.

If the unit is battery operated, the TLC555 uses minimum current to produce this function. With a 9-V battery the circuit draws 5 mA (no load) and 15 mA when turning on an LED. Most of the ON current is LED current.

POSITIVE-TRIGGERED MONOSTABLE

The standard 555 ordinarily requires a negative-going trigger pulse to initialize operation; however, a positive-going trigger pulse may be used to start the timing cycle with the circuit shown in Figure 7-29.

In this design the trigger input (pin 2) is biased to 6 V (1/2 V_{DD}) by divider R1 and R2. The control input (pin 5) is biased to 8 V (2/3 V_{DD}) by the internal divider circuit. With no trigger voltage applied, point 'A' is at 4 V (1/3 V_{DD}). To turn the timer on, the voltage at point 'A' has to be greater than the 6 V present on pin 2.

A positive 5 V trigger pulse (V_I) applied to the control input (pin 5) is ac coupled through capacitor C1, adding the trigger voltage to the 8 V already on pin 5 which results in 13 V with respect to ground. The pulse width of the input trigger (V_I) is not critical; the main criterion is that the width of the input trigger pulse (following C1) should be less than the desired output pulse

width. The output pulse width is determined by the values of R_t and C_t.

When the voltage at point 'A' is increased to 6.5 V, which is greater than the 6 V on pin 2, the timer cycle is initialized. The output of the timer (pin 3) goes high, turning off the discharge transistor (pin 7) and allowing C_t to charge through resistor R_t. When capacitor C_t charges to the upper threshold voltage of 8 V (2/3 V_{DD}), the flip-flop is reset and the output (pin 3) goes low. Capacitor C_t then discharges through the discharge transistor.

The timer is not triggered again until another trigger pulse is applied to the control input (pin 5).

VOLTAGE-CONTROLLED OSCILLATOR

There are many different types of voltage controlled oscillator (VCO) circuits that can use an IC timer. VCO circuits are valuable for instrumentation, electronic music applications, and function generators. The output is a rectangular pulse stream whose frequency is related to the external control voltage. A VCO circuit using the TLC555 is shown in Figure 7.30.

At startup, the voltage at the trigger input (pin 2) is less than the trigger level voltage (1/3 V_{DD}), causing the timer to be triggered via pin 2. The output of the timer (pin 3) goes high, allowing capacitor C_t to charge very rapidly through diode D1 and resistor R1. The charge time of C_t is extremely short and may essentially be neglected.

When capacitor C_t charges to the upper threshold voltage (2/3 V_{DD}), the flip-flop is reset, the output (pin 3)

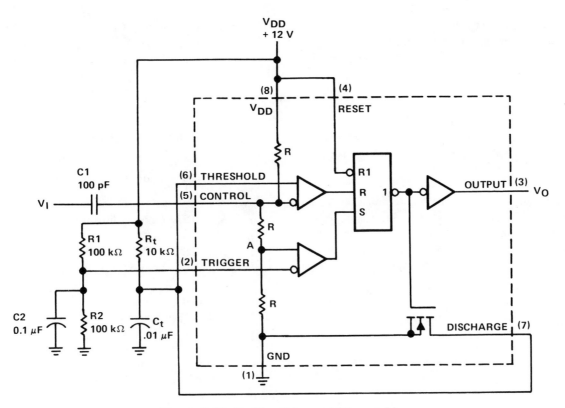

Figure 7-29. Positive-Triggered Monostable

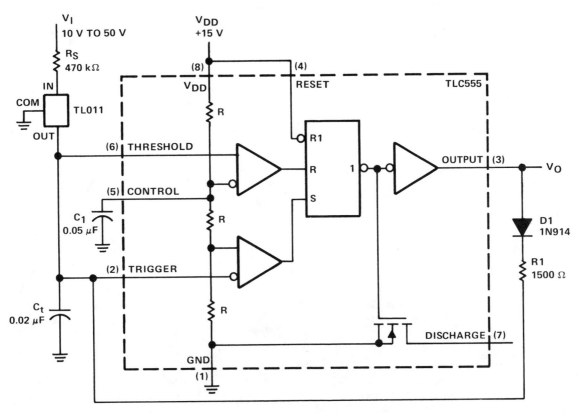

Figure 7-30. Voltage-Controlled Oscillator

goes low, and capacitor C_t discharges through the current mirror, TL011. When the voltage at pin 2 reaches 1/3 V_{DD}, the lower threshold or trigger level, the timer triggers again and the cycle is repeated.

The input voltage (V_I) determines the constant current output of the current mirror, which is used as a voltage to current converter and sets the discharge rate of capacitor C_t. The discharge time of C_t determines the frequency of the oscillator. As the input voltage is varied from 10 V to 50 V, the output frequency varies at a linear rate.

As an example, assume an application calls for an output midrange frequency of 500 Hz. Since T = 1/f, the time between output pulses will be 2 ms. The charge time, which will be less than 1 μs, may be neglected. The discharge current of C_t for a specific input control voltage is:

$$I_{DISCHARGE} = \frac{V_I}{R_S} = \frac{10 \text{ V}}{470 \text{ k}\Omega} = 20 \text{ μA at 10 V input}$$

$$I_{DISCHARGE} \text{ at midrange} = 50 \text{ μA at 25 V input}$$

$$I_{DISCHARGE} = \frac{V_I}{R_S} = \frac{50 \text{ V}}{470 \text{ k}\Omega} = 100 \text{ μA at 50 V input}$$

With an input voltage of 10 V to 50 V, the TL011 current will vary linearly from 20 μA to 100 μA. Figure 7-31 shows the voltage to frequency conversion obtained with two different values for capacitor C_t. With C_t = 0.001 μF, a frequency range of 3.3 kHz to 10 kHz is obtained. When a value of 0.02 μF is used, a frequency range of 187 Hz to 1 kHz is obtained.

Since the capacitor, C_t, discharges from 10 V to 5 V (2/3 V_{DD} to 1/3 V_{DD}), the capacitor value may be calculated for 500 Hz as follows:

$$C = \frac{IT}{V_C} = \frac{(50 \text{ μA}) (0.002)}{5}$$

$$I = \text{midrange discharge current}$$

$$T = \frac{1}{f} = \text{midrange output pulse}$$

$$V_C = 5 \text{ V (charge-discharge)}$$

$$C = \frac{(50 \times 10^{-6}) (0.002)}{5} = 0.02 \text{ μF}$$

Note that the current mirror is sinking current during both the charge and discharge of C_t. However, the small discharge current is easily overcome during the charge cycle by the lower impedance, high-current charge path from the output pin.

For linear ramp applications, the output is obtained across C_t.

CAPACITANCE-TO-VOLTAGE METER

This circuit performs a function that is somewhat different than a frequency-to-voltage converter. The circuit can be more easily analyzed by examining the individual sections as shown in Figure 7-32.

Timer U1 operates as a free-running oscillator at 60 Hz, providing trigger pulses to timer U2 which operates in the monostable mode. Resistor R1 is fixed and

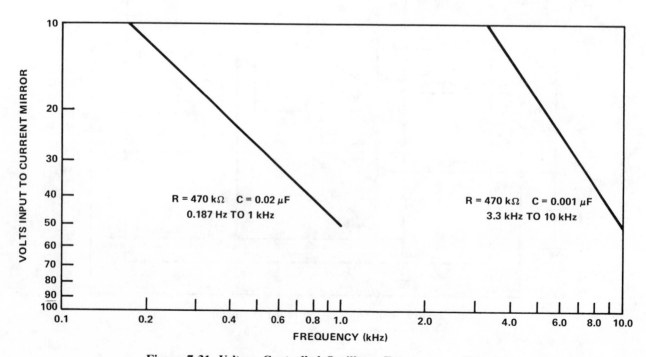

Figure 7-31. Voltage-Controlled Oscillator Frequency vs Voltage

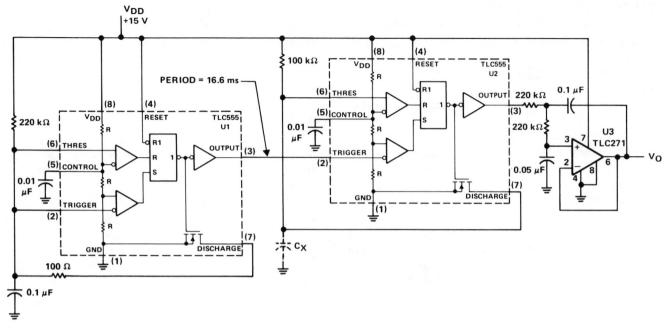

Figure 7-32. Capacitance-to-Voltage Meter Converter

capacitor Cx is the capacitor being measured. While the output of U2 is 60 Hz, the duty cycle depends on the value of Cx. U3 is a combination low-pass filter and unity-gain follower whose dc voltage output is the time-averaged amplitude of the output pulses of U2, as shown in Figure 7-33 (a and b).

Figure 7-33(a) shows when the value of Cx is small the duty cycle is relatively low. The output pulses are narrow and produce a lower average dc voltage level at the output of U3. As the capacitance value of Cx increases, the duty cycle increases making the output pulses at U2 wider and the average dc level output at U3 increases.

As an example, the graph in Figure 7-34 illustrates capacitance values of 0.01 μF to 0.1 μF plotted against the output voltage of U3. Notice the excellent linearity and direct one-to-one scale calibration of the meter. If this does not occur with your design, the 100 kΩ resistor, R1, can be replaced with a potentiometer which can be adjusted to the proper value for the meter being used.

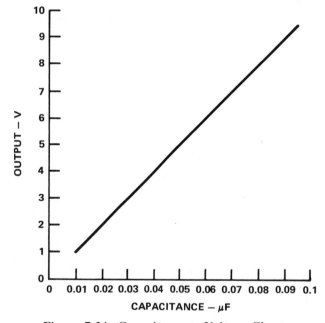

Figure 7-34. Capacitance-to-Voltage Chart

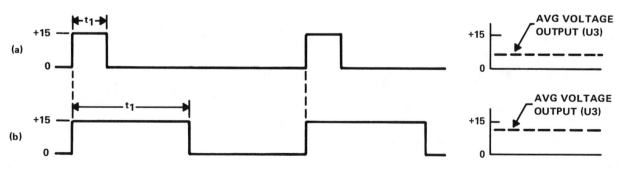

Figure 7-33. U2 Duty Cycle Change

TLC555 PWM MOTOR CONTROLLER

The speed of a dc motor is proportional to the applied voltage, but the torque diminishes at low voltages. Low speed performance is usually erratic when analog controllers are used, especially under changing load conditions. Pulse-width-modulated (PWM) controllers offer superior control and operate efficiently at low speeds.

The PWM controller shown in Figure 7-35 uses complementary half-H peripheral drivers (SN75603, SN75604) with totem-pole outputs rated at 40 V and 2.0 A. These drivers effectively place the motor in a full bridge configuration which has the ability to provide bidirectional control.

Timer U1 operates in the astable mode at a frequency of 80 Hz. The 100 Ω discharge resistor results in an 8 μs trigger pulse which is coupled to the trigger input of timer U2. Timer U2 serves as the PWM generator. Capacitor C1 is charged linearly with a constant current of 1 mA from the IN5297, which is an FET current regulator diode.

Motor speed is controlled by feeding a dc voltage of 0 to 10 V to the control input (pin 5) of U2. As the control voltage increases, the width of the output pulse (pin 3) also increases. These pulses control the on/off time of the two motor drivers. Note that the trigger pulse width of timer U1 limits the minimum possible duty cycle from U2. Figure 7-36 illustrates the analog control voltage versus drive motor pulse width.

Figure 7-37 illustrates the output waveforms of U1 with respect to the output of U2. The maximum duty cycle that may be achieved is about 98% at which time the control voltage is 12.5 V.

CAUTION: Careful grounding is required to prevent motor-induced noise from interfering with the proper operation of the U1 and U2 timer circuits. Supply lines must be bypassed and decoupled at each timer to prevent transients from causing circuit instability. Separate power supplies should be used for the motor and the timer circuits.

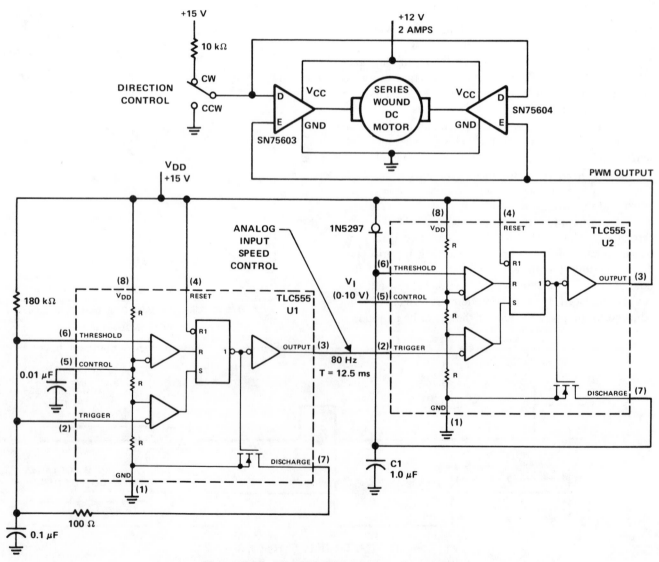

Figure 7-35. TLC555 PWM Motor Controller

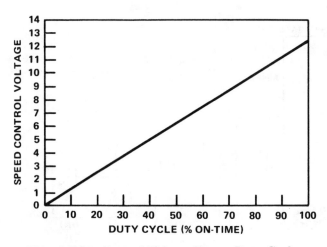

Figure 7-36. Control Voltage Versus Duty Cycle

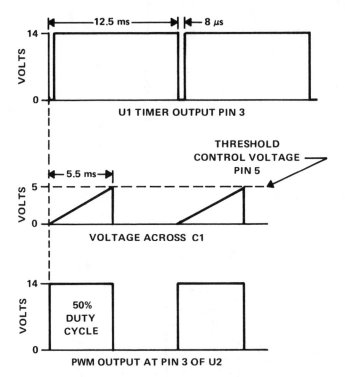

Figure 7-37. PWM Controller Waveforms

TELEPHONE CONTROLLED NIGHT LIGHT

This application is a useful addition for the bedroom telephone stand. When the telephone rings, or when the handset is lifted, the night light is turned on and remains on while the conversation takes place. When the handset is replaced in the cradle, the light remains on for about 11 s. This circuit is illustrated in Figure 7-38.

Operationally the circuit may be divided into four sections.

1. U1 — TCM1520 Ring Detector
2. U2 — TLC555 Timer
3. U3 — TLC271 End-of-Call Detector
4. U4 — MOC3011 Light Switch

During standby conditions, the −48 V dc bias on the phone line maintains the output of U3 in a high state.

Also the timer discharge transistor is on, preventing the charging of timer capacitor C1. When the ac ring signal is applied to the phone line, it is processed by the ring detector U1, producing a negative output pulse at pin 2 for each ring. These pulses are applied to the trigger input (pin 2 of U2) and trigger U2 causing its output to go high and the discharge transistor to turn off. The high output of U2 activates opto-isolator U4 which turns on the night light. The discharge transistor, being off, allows C1 to begin charging. Each ring retriggers the timer and discharges C1 preventing it from reaching 2/3 V_{DD} threshold level. Thus the night light will remain on while the phone is ringing and for about 11 s after the last ring. After 11 s C1 will be charged to the U2 threshold level (2/3 V_{DD}) resulting in the U2 output returning to a low level and its discharge output turning on discharging C1. At this time the lamp will turn off if the phone is not answered.

When the phone is answered, a 1 kΩ load is placed across the phone line (tip and ring in Figure 7-38). This removes the differential input to operational amplifier U3, causing its output to go low and capacitor C1 starts discharging through R1. As long as the voltage across C1 remains low, timer U2 cannot start its cycle and the lamp will remain on.

When the phone is hung up, the low impedance is removed from the phone line and the differential voltage across the line causes the U3 output to go high. This allows C1 to start charging, initiating the timing that will turn off the night light.

PROGRAMMABLE VOLTAGE CONTROLLED TIMER

The uA2240 may easily be configured as a programmable voltage controlled timer with a minimum number of external components. The basic programmable timer circuit is shown in Figure 7-39, including the functional clock diagram of the uA2240. Counter outputs Q2 through Q6 are not shown in Figure 7-39.

A useful feature of the uA2240 timer is the modulation input (pin 12), which allows external adjustment of the input threshold level. A variable voltage is applied to pin 12 from the arm of a 10 kΩ potentiometer connected from V_{CC} to ground. A change in the modulation input voltage will result in a change in the time base oscillator frequency and the period of the time base output (TBO). The time-base period may be trimmed via the modulation input to supply the exact value desired, within the limits determined by the values of R1 and C1. In this application, the basic time-base period set by R1 and C1 is 0.5 ms.

The effect of the voltage modulation input on the time-base oscillator period is shown in Figure 7-40. Although the voltage that can be safely applied to pin 12 may range from V_{CC} to ground, oscillator operation may cease near V_{CC} (within 0.5 V). Note that the chart also shows TBO operation inhibited with modulation inputs below approximately 2 V. With a modulation input voltage

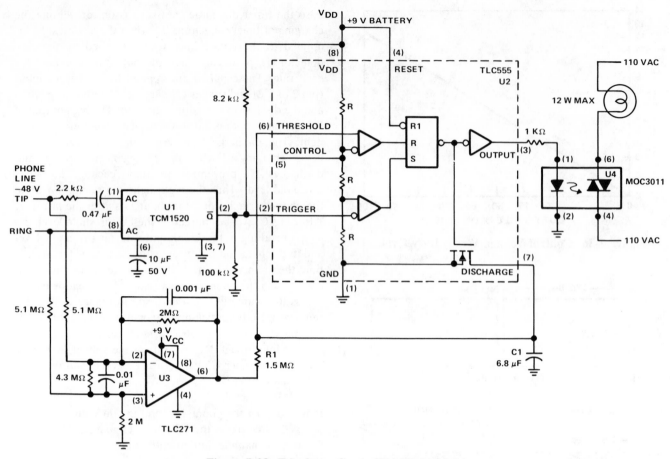

Figure 7-38. Telephone Controlled Night Light

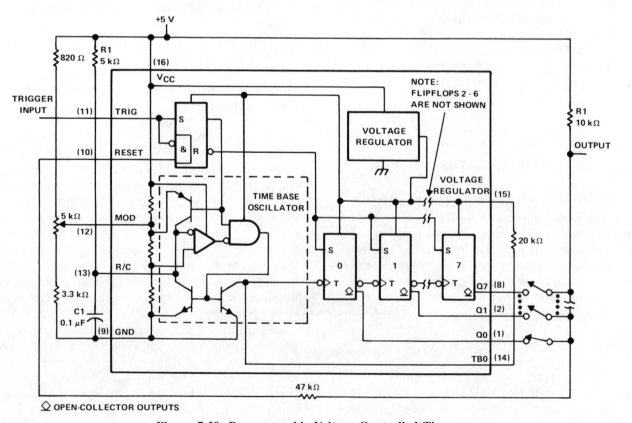

Figure 7-39. Programmable Voltage Controlled Timer

of 3.5 V, the Q5 connection gives an active low output for 32 cycles of the time base oscillator, or 16 ms in the case of the circuit shown in Figure 7-39.

The TBO has an open-collector output that is connected to the regulator output via a 20 kΩ pull-up resistor. The output of the TBO drives the input to the 8-stage counter section.

At start-up, a positive trigger pulse starts the TBO and sets all counter outputs to a low state. The trigger pulse duration must be at least 2 μs with a voltage level exceeding 2 V. Once the uA2240 is initially triggered, any further trigger inputs are ignored until it is reset.

The binary outputs are open-collector stages that may be connected together to the 10 kΩ pull-up resistor to provide a "wired-OR" output function. The combined output is low if any single connected counter output is low. In this application, the output is connected to the reset input through a 47 kΩ resistor.

This circuit may be used to generate 255 discrete time delays that are integer multiples of the time-base period. The total delay is the sum of the number of time-base periods, which is the binary sum of the Q outputs connected. For example, if only Q5 is connected to the reset input, each trigger pulse generates an active-low output for 32 periods of the time-base oscillator. If Q0, Q2 and Q3 are connected together, the binary count is: $1 + 4 + 8 = 13$. Thus, 255 discrete time delays are available with the eight counter outputs, Q0 through Q7. Delays from 200 μs to 0.223 s are possible with this configuration.

FREQUENCY SYNTHESIZER

The uA2240 may easily be connected to operate as a programmable voltage controlled frequency synthesizer as shown in Figure 7-41. The uA2240 consists of four basic circuit elements: (1) a time-base oscillator, (2) an eight-bit counter, (3) a control flip-flop, and (4) a voltage regulator.

The basic frequency of the time-base oscillator (TBO) is set by the external time constant determined by the values of R1 and C1 ($1/R1C1 = 2$ kHz). The basic frequency may be changed by varying the modulation input voltage supplied to terminal 12. With a modulation voltage range of 2 to 4.5 V, the basic frequency multiplier is changed from 0.4 to 1.75 nominally with a V_{CC} of 5 V. See Figure 7-40.

The open-collector output of the TBO is connected to the regulator output via a 20 kΩ pull-up resistor, and drives the input to the eight-bit counter. Each counter output is an open-collector stage that may be connected to the 10 kΩ load pull-up resistor to provide a "wired-OR" function. The combined output is low if any single counter output is low.

At power-up, a positive trigger pulse is detected across C2 which starts the TBO and sets all counter outputs to a low state. Once the uA2240 is initially triggered, any further trigger inputs are ignored until it is reset. In this astable operation, the uA2240 will free-run from the time it is triggered until it receives an external reset signal.

Up to 255 discrete frequencies may be synthesized by connecting different counter outputs. For example, connecting the counter outputs of Q0, Q2, and Q3 supplies the output waveform shown in Figure 7-42. if the modulation voltage is set at 3.5 V, the TBO frequency is 2 kHz, then the output frequency will be 125 Hz. By adjusting the modulation voltage, the TBO frequency may be trimmed from 800 Hz to 3.5 kHz. A wide range of basic time-base frequencies are also available to the designer through the selection of different values of R1 and C1. This includes time-base frequencies as high as 100 kHz. The RC and modulation inputs, pins 12 and 13, have high speed capability and sensitivity for accurate, repeatable performance. It is essential, therefore, that high frequency layout and lead dress techniques be used to avoid noise problems which could result in undesirable jitter on the output pulse.

CASCADED TIMERS FOR LONG TIME DELAYS

Two uA2240 timers may be cascaded to provide long time delays as shown in Figure 7-43. Each uA2240 counter consists of a time-base oscillator, an eight-bit counter, a control flip-flop, and a voltage regulator. The frequency of the time-base oscillator (TBO) is set by the time constant of an external resistor and capacitor.

The open-collector output of the TBO drives the internal eight-bit counter if the TBO output is connected to the regulator output via a pull-up resistor. Otherwise, an external source can be connected to the TBO terminal to supply the input to the eight-bit counter. The open-collector counter outputs Q0 through Q7 provide a "wired-OR" function. The combined output is low if any of the connected counter outputs is low.

The trigger and reset inputs to the internal control flip-flop determine the uA2240 operational state. Once the positive trigger pulse starts the TBO and resets all counter outputs to a low level, further trigger signals are ignored until a reset pulse is received. The reset input inhibits the TBO output and sets all counter outputs to a high level.

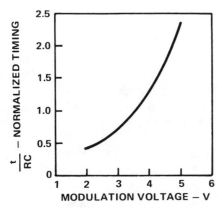

Figure 7-40. Modulation Voltage Control of Time-Base Period

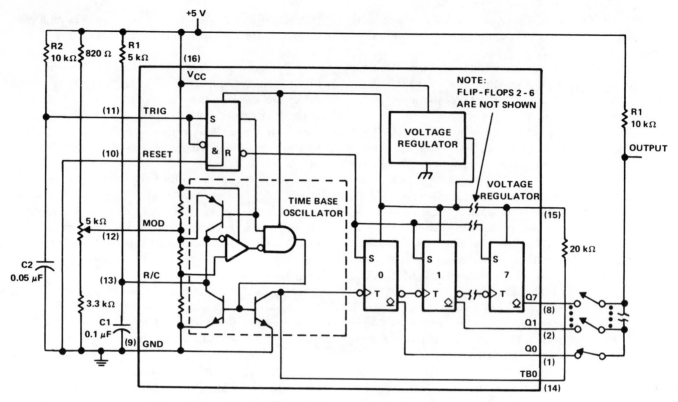

Figure 7-41. Programmable Voltage Controlled Frequency Synthesizer

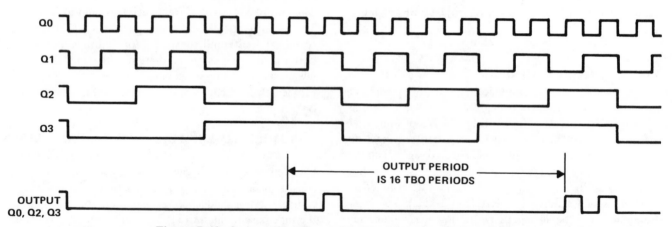

Figure 7-42. Output Waveform with Q0, Q2, and Q3 Connected

In this application, the TBO frequency of U1 is set at 2 kHz by the time constant of R1 and C1. This provides a circuit time-base period of 0.5 ms that drives the internal eight-bit counter of U1.

When the trigger switch is momentarily closed, the trigger input starts the U1 TBO and sets all outputs low. The U1 TBO output is connected to its regulator output through the 30 kΩ pull-up resistor. At the end of 256 U1 time-base periods, the U1 Q7 counter output generates a negative-going transition that supplies the active time-base (clock) input for U2. This clock input has a period of 128 ms and is active until a reset is generated by a high U2 output. The U2 time-base oscillator is inhibited by connecting its trigger input (pin 13) to ground through a 1 kΩ resistor.

The U2 counter outputs are connected together resulting in a continuous low U2 output until its final count of 256. At this time, all U2 outputs are high. This ends the output pulse period and resets both uA2240 counters. Thus, the output period is low for about 33 s (256 × 128 ms). If the values of R1 and C1 are changed to 4.8 MΩ and 100 µF, respectively, the output period duration will be about 1 year.

uA2240 OPERATION WITH EXTERNAL CLOCK

The uA2240 programmable timer/counter may be operated by an external clock as shown in Figure 7-44. The internal time-base oscillator is disabled by connecting the trigger input (pin 13) to ground through a 1 kΩ resistor. An external clock is applied to the time-base output

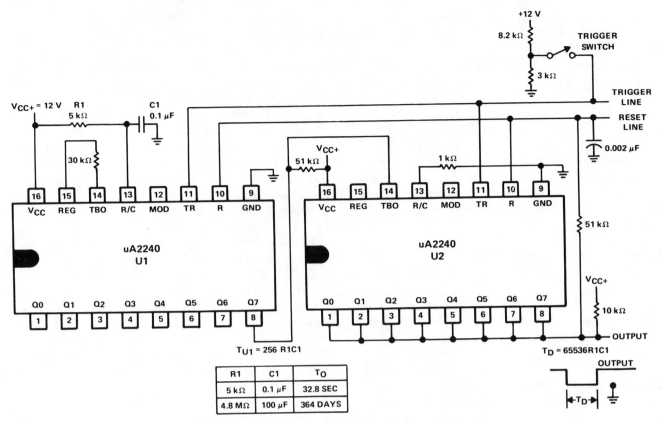

Figure 7-43. Cascaded Operation for Long Delays

R1	C1	T_O
5 kΩ	0.1 μF	32.8 SEC
4.8 MΩ	100 μF	364 DAYS

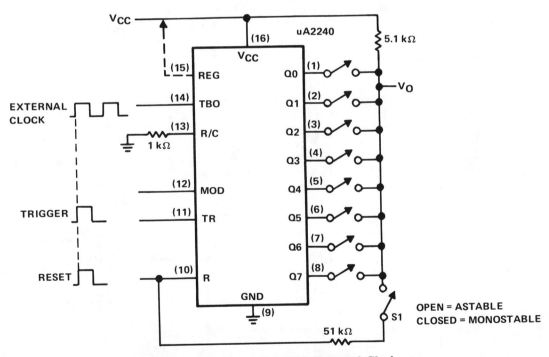

Figure 7-44. Operation with External Clock

(TBO), pin 14. For proper operation, the minimum clock amplitude and pulse-duration are 3 V and 1 μs respectively.

In this application the uA2240 consists of an eight-bit counter, a control flip-flop, and a voltage regulator. The external clock triggers the eight-bit counter on the negative-going edge of the clock pulse. The open-collector counter outputs are connected together to the 5.1 kΩ pull-up resistor to provide a "wired-OR" function where the combined output is low if any one of the outputs is low.

This arrangement provides time delays or frequency outputs that have a period equal to integer multiples of the external clock time-base period.

In the astable mode, the uA2240 will free-run from the time it is triggered until it receives an external reset signal. If the monostable mode is selected, one or more of the counter outputs is connected to the reset terminal, and provides the reset at the end of the pulse delay period.

For operation with a supply voltage of 6 V or less, the internal time-base can be powered down by open-circuiting pin 16 and connecting pin 15 to V_{CC}. In this configuration, the internal time-base does not draw any current, and the overall current drain is reduced by approximately 3 mA.

uA2240 STAIRCASE GENERATOR

The uA2240 timer/counter combined with a precision resistor ladder network and an operational amplifier form the staircase generator shown in Figure 7-45. The uA2240 consists of a time-base oscillator, an eight-bit counter, a control flip-flop, and a voltage regulator.

In the astable mode, once a trigger pulse is applied, the uA2240 operates continuously until it receives a reset pulse. The trigger input (pin 11) is tied to the time base output (pin 14) resulting in automatic starting and continuous operation. The frequency of the time-base oscillator (TBO) is set by the time constant of R1 and C1 ($f = 1/R1C1$). for this example, a 10 kΩ resistor and a 0.01 µF capacitor form the timing network. The total ramp generation time is 25.6 ms for an output frequency of 39.1 Hz. The open-collector TBO output is connected to the regulator output via a 20 kΩ pull-up resistor, and drives the input to the eight-bit counter.

The counter outputs are connected to a precision resistor ladder network with binary weighted resistors. The current sink through the resistors connected to the counter outputs correspond to the count number. For example, the current sink at Q7 (most significant bit) is 128 times the current sink at Q0 (least significant bit).

The positive bias of approximately 0.5 V applied to the non-inverting input of the operational amplifier generates a current feedback at the inverting input that supplies the current sink for the open-collector pull-up resistors. As the count is generated by the uA2240 eight-bit counter, the current sink through each active binary weighted resistor decreases the positive output of the operational amplifier in discrete steps.

The feedback potentiometer is set at a nominal 10 kΩ to supply a maximum output voltage range. A 12 V supply was used to allow a 10 V output swing. Operation from a single 5 V supply will require an adjustment of the gain (feedback resistor) and reference voltage. With a feedback resistance of 4.5 kΩ and a reference of 0.4 V, the output will allow a 3.6 V output change from 0.4 V to 4.0 V in 14.1 mV steps.

The staircase waveform is shown in Figure 7-46. With a 0.5 V input reference on pin 3 of the TLC271, the output will change from 10.46 V maximum, in 256 steps of 38.9 mV per step, to a 0.5 V minimum. Each step has a pulse duration of 100 µs and an amplitude decrease of 38.9 mV. The period of the staircase waveform is 25.6 ms and the waveform output is repeated until a reset is applied to the uA2240.

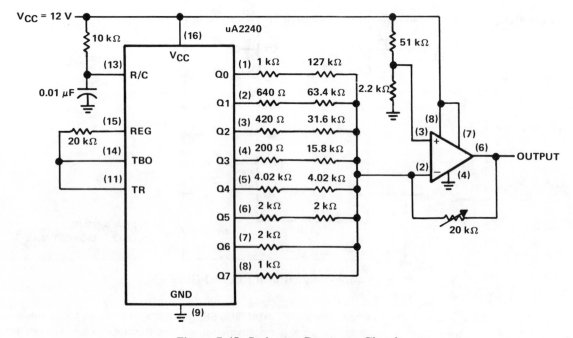

Figure 7-45. Staircase Generator Circuit

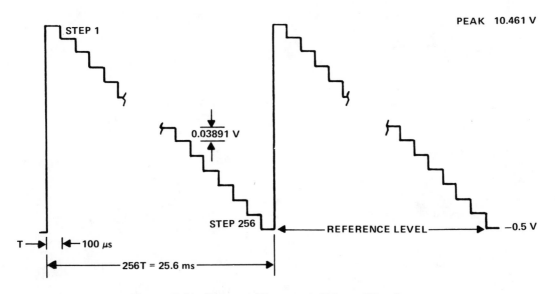

Figure 7-46. Staircase Generator Voltage Waveforms

Section 8

Display Drivers

INTRODUCTION TO DISPLAY DRIVER TECHNOLOGY

Visual displays exist in a wide range of complexities from the simple incandescent bulb or light emitting diode (LED) indicators to complex full digital read-out or large flat-screen display panels. Until recently, many discrete transistors were required to drive the more complex visual display systems.

As the complexity of display systems grew, alternatives to discrete transistor drivers became imperative. An increasing number of monolithic integrated circuits are becoming available for driving many types of display systems, some of which were not considered practical until recently. Monolithic ICs that have been developed to drive digital and more complex displays include:

> Light Emitting Diode Display Drivers
> DC Plasma (Gas Discharge) Display Drivers
> AC Plasma Display Drivers
> Vacuum Fluorescent Display Drivers
> Electroluminescent Display Drivers
> Electrophoretic Display Drivers

Reliability of the higher voltage display drivers (all except LED drivers) has been poor in the past. In fact, ac plasma and electroluminescent display drivers were considered impractical because their complexity and high voltage required the use of many discrete power transistors and logic circuits. "BIDFET"* technology has made monolithic IC drivers possible by providing the voltage capability and complexity required for 64 power output transistors and control logic on a single chip.

---

*Bipolar, double diffused, N-Channel and P-Channel MOS transistors on the same chip — a patented process.

BIDFET is a rugged, low-cost wafer processing technology which merges precision control, self-isolated CMOS logic and high-voltage interface circuitry on a common monolithic substrate. It is manufactured using standard junction isolation techniques (see Figure 8-1). Many multitechnology processes have been developed, but BIDFET is the only merged process which alleviates the high voltage limitations of conventional integrated circuits while retaining their LSI logic capabilities. BIDFET devices are being produced with working voltages to 250 V and breakdown voltages exceeding 300 V. This has been achieved by replacing the conventional bipolar output stage with a Double-Diffused MOS (DMOS) transistor structure. The two output stage structures are compared in the following paragraphs.

First let us consider the limitations imposed by operating a bipolar switch within the Reverse Bias Safe Operating Area (RBSOA).

Figure 8-2 shows the typical load line characteristics of a switch operated within the device's RBSOA and $V_{CES(SUS)}$ ratings. If the load line penetrates the RBSOA, a destructive condition occurs. Thus, reliable operation may be limited to the $V_{BR(CEO)}$ rating of the switch.

With common topologies used in conventional integrated circuits, breakdown voltages [$V_{BR(CEO)}$] are limited by the thickness of the epitaxial layer. Practical limitations on the thickness of conventional junction isolated integrated circuits limits $V_{BR(CEO)}$ to 70 V.

The DMOS structure, on the other hand, is a surface (lateral) device whose breakdown characteristic is limited only by bulk junction avalanching and horizontal topology (channel length). Breakdown ratings are governed by doping levels and surface area, which is an economic consideration, instead of a physical limitation. Unlike NPN transistors,

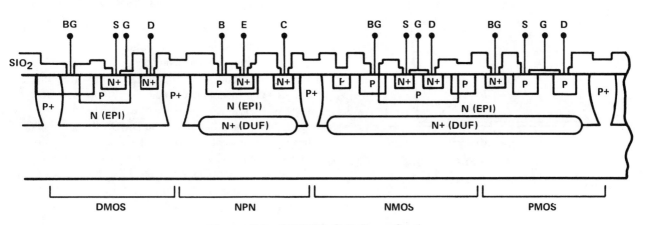

Figure 8-1. BIDFET Cell Cross Section

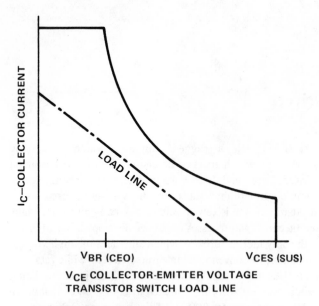

Figure 8-2. RBSOA Curve and Load Line

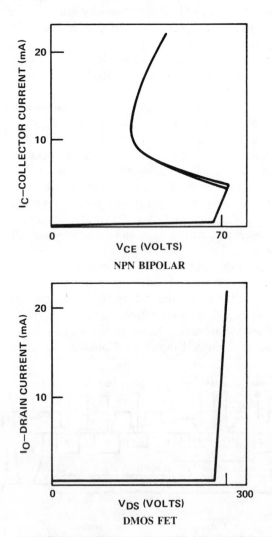

Figure 8-3. Breakdown Characteristics

DMOS can operate safely to its breakdown voltage limit without risk of destructive secondary breakdown or sacrifice of reliability. Figure 8-3 shows the superior breakdown characteristics of the DMOS structure as compared to the bipolar structure.

Characteristic waveforms (Figure 8-4) show the virtual independence of the output current to the output voltage of a DMOS structure. The Early voltage, as well as the stored charge characteristics of the DMOS structure, is superior to that of the bipolar transistor in switching applications.

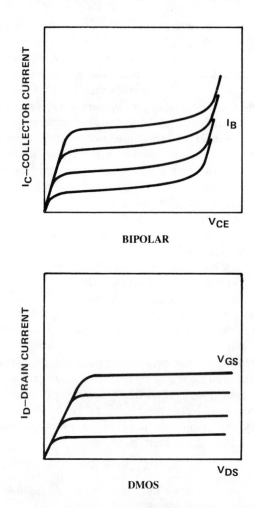

Figure 8-4. Characteristic Waveforms

Using the variety of structures offered by the BIDFET process, design engineers can optimize high voltage circuits by partitioning them and using the optimum technologies for the various sections. The bipolar structure offers durability and is very forgiving of input conditions. CMOS allows increased circuit complexity while requiring minimal power consumption and bar area. The DMOS structure's benefits have been presented herein. The result is a high-voltage interface circuit which is capable of performing data registration, manipulation or decoding functions to reduce the requirements on system electronics.

LED DISPLAY DRIVERS

LED displays have generally been fairly simple systems. Applications vary from one or two discrete LEDs to small arrays and digital displays. A variety of multichannel drivers have evolved that allow easy implementation of the drives required. Some examples of these devices follow:

SN75491 AND SN75491A QUAD LED SEGMENT DRIVERS (Figure 8-5)

Features

- 50 mA source or sink capability
- '491 rated for 10 V operation
- '491A rated for 20 V operation
- MOS compatible inputs
- Low standby power
- High gain Darlington circuits

SN75492 AND SN75492A HEX LED DIGIT DRIVERS (Figure 8-6)

Features

- 250 mA sink capability
- '492 rated for 10 V operation
- '492A rated for 20 V operation
- MOS compatible inputs
- Low standby power
- High-gain Darlington circuits

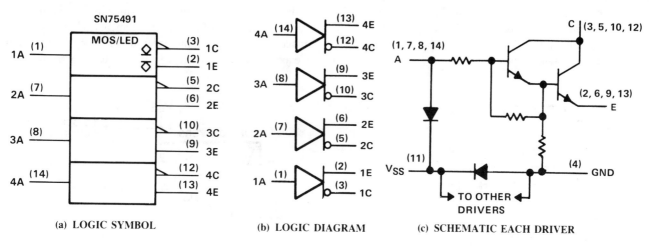

(a) LOGIC SYMBOL   (b) LOGIC DIAGRAM   (c) SCHEMATIC EACH DRIVER

Figure 8-5. SN75491 and SN75491A

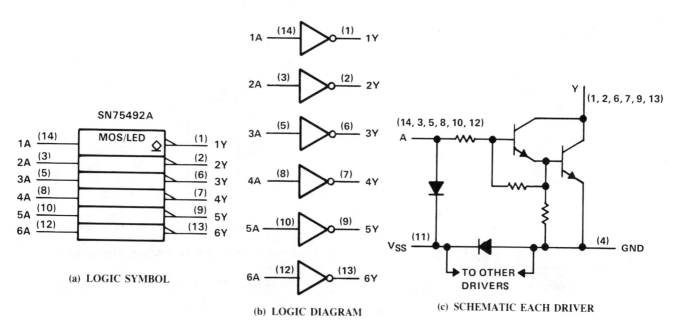

(a) LOGIC SYMBOL   (b) LOGIC DIAGRAM   (c) SCHEMATIC EACH DRIVER

Figure 8-6. SN75492 and SN75492A

SN75494 HEX LED DIGIT DRIVER

The SN75494 (Figure 8-7) provides the same basic function as the SN75492 but with some key differences:

- Operates from lower V_{CC} voltages; as low as 3.2 V.
- In addition to V_{SS} and ground the '494 provides access to predriver collectors via its V_{CC} pin thus allowing good saturation of output transistors for low-voltage applications.
- An enable input is provided to allow display blanking.

SN75497 7-CHANNEL and SN75498 9-CHANNEL LED DRIVERS (Figure 8-8)

Features:

- 100 mA output sink capability
- Low-voltage operation (2.7 V min)
- MOS and TTL compatible inputs
- Low standby power

DRIVING LED DISPLAYS

Light emitting diodes exist in a variety of sizes and operating current requirements. Typical digital displays require 10 to 20 mA forward current for normal operating brightness levels.

Digital displays are usually driven using time-multiplex techniques to minimize the number of drivers required. A segment-address selection and digit-scan method of multiplexing is used in the application (Figure 8-9) for a 12-digit display.

The TIL804 used in this application is a 12-digit numeric LED display with right-hand decimal points. It consists of common-cathode red LED digits. Twenty tab connections allow control of the 84 segments and 12 decimal points. The segments, and decimal points, are addressed by the "0" outputs of a TMS1200 microcomputer chip. Each digit of the display is connected in a common-cathode configuration and the anodes of like-position segments of all digits are connected together for multiplex drive. Normal operation of each digit is 8.3% (1/12) duty cycle. For an average

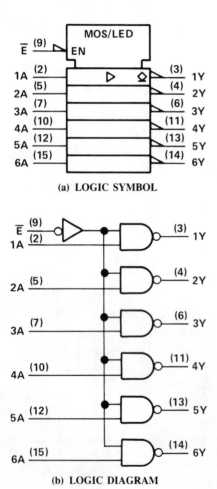

(a) LOGIC SYMBOL

(b) LOGIC DIAGRAM

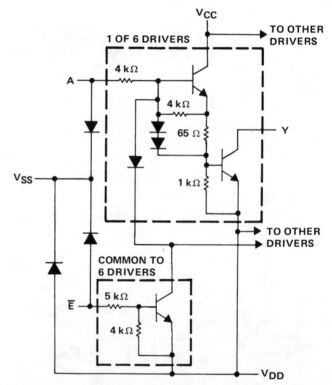

NOTES: 1. THE V_{SS} TERMINAL MUST BE CONNECTED TO THE MOST POSITIVE VOLTAGE THAT IS APPLIED TO THE DEVICE.
2. RESISTOR VALUES SHOWN ARE NOMINAL AND IN OHMS.

(c) SCHEMATIC EACH DRIVER

Figure 8-7. SN75494

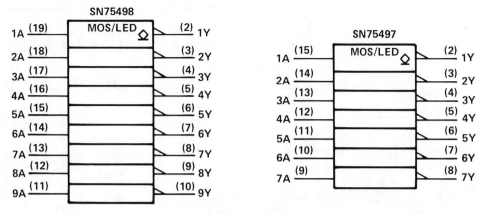

(a) LOGIC SYMBOLS

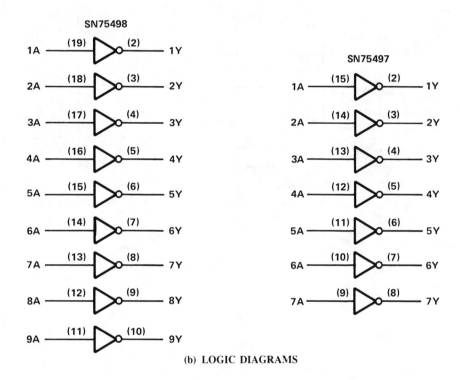

(b) LOGIC DIAGRAMS

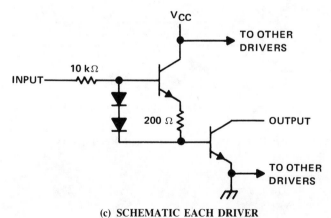

(c) SCHEMATIC EACH DRIVER

Figure 8-8. SN75497 and SN75498

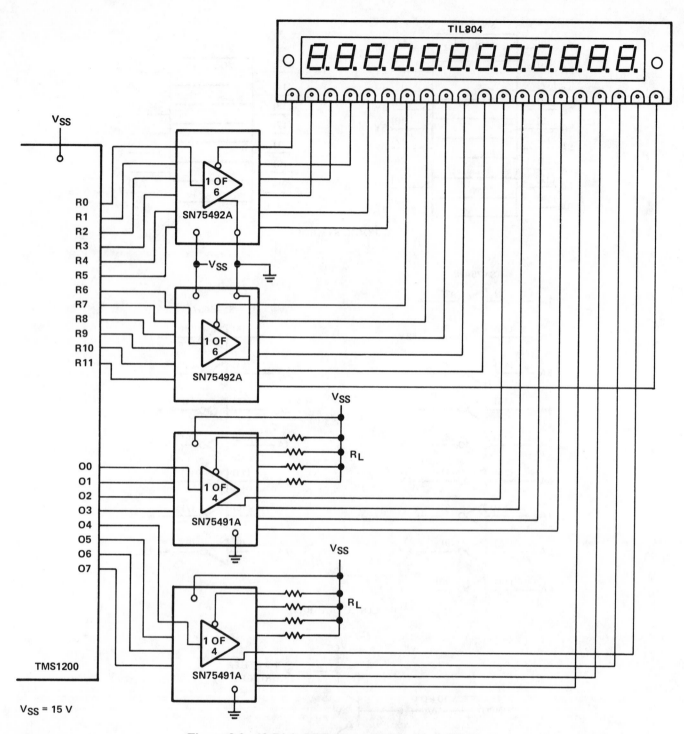

Figure 8-9. 12-Digit LED Numeric Display and Drive

forward current of 5 mA a pulse current of 60 mA will be required. System operation may be from a single 15 V supply as shown, or the SN75491A source drivers may be operated from a separate supply of typically 5 V. The current limit resistors, R_L may be calculated as follows:

$$R_L = \frac{V_{SS} - V_{CE(ON)}(SN75491A) - V_{OL}(SN75492A) - V_F(LED)}{I_{FM}(LED)}$$

Typical Parameters

$$V_{SS} = 12 \text{ V}$$
$$V_{CE(ON)} = 1.2 \text{ V}$$
$$V_{OL} = 1.2 \text{ V}$$
$$V_F = 1.7 \text{ V}$$
$$I_{FM} = 60 \text{ mA}$$
$$R_L = 130 \ \Omega$$

OTHER APPLICATIONS

Although the multichannel drivers shown here were designed specifically for driving light emitting diodes, they are basically Darlington transistor drivers and as such they have many other uses. The circuits shown in Figure 8-10 through 8-17 are examples of the variety of applications that may be implemented with these devices.

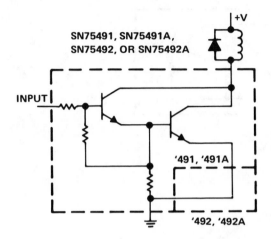

Figure 8-10. Quad or Hex Relay Driver

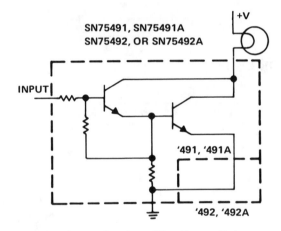

Figure 8-11. Quad or Hex Lamp Driver

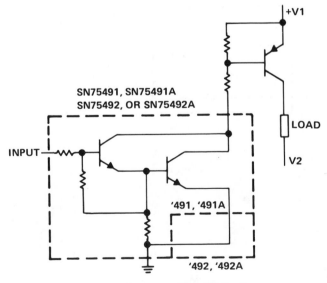

Figure 8-12. Quad or Hex High-Current P-N-P Transistor Driver

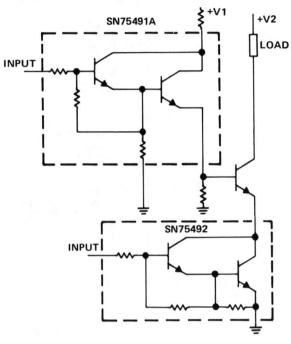

Figure 8-13. Base/Emitter Select N-P-N Transistor

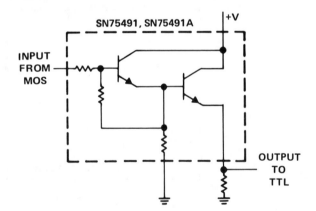

Figure 8-14. MOS to TTL Level Shifter

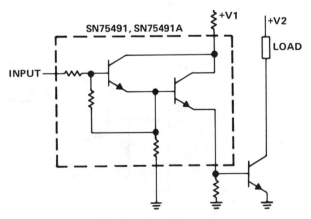

Figure 8-15. Quad High-Current N-P-N Transistor Driver

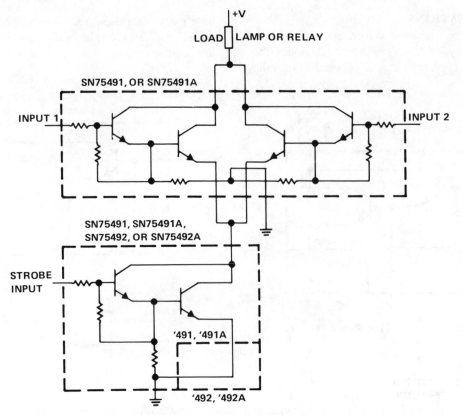

Figure 8-16. Strobed NOR Driver

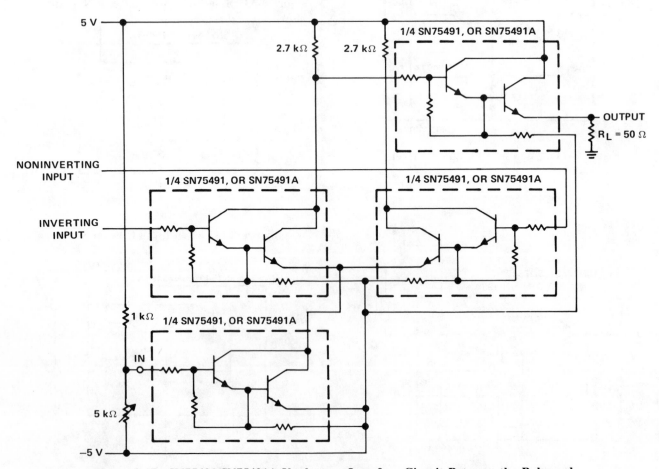

Figure 8-17. SN75491/SN75491A Used as an Interface Circuit Between the Balanced 30-MHz Output of an RF Amplifier and a Coaxial Cable

DC PLASMA DISPLAY DRIVERS

DC plasma has been popular for many years as the media for digital displays. DC plasma provides a relatively bright display ideal for panel meters and various types of test equipment. Numerical 7-segment displays are used in most cases, and therefore the display drivers for dc plasma are configured specifically for this application. DC plasma displays require a rather high voltage to fire (ionize) the gas and maintain conduction. Firing voltages are generally in excess of 100 V and operating current levels are from 0.2 mA to 3 or 4 mA depending on the display's physical size and the brightness required. Multiplexing may be used in some applications to minimize the number of parts required.

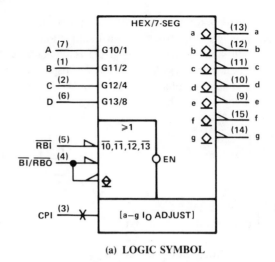

(a) LOGIC SYMBOL

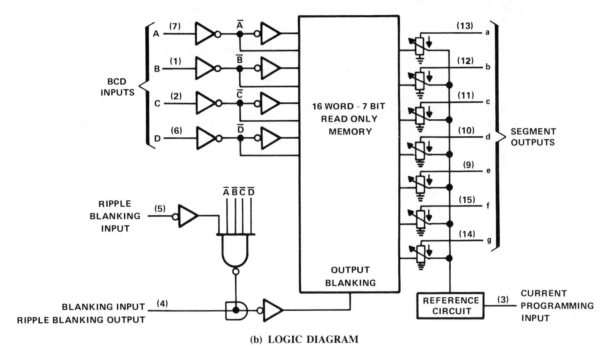

(b) LOGIC DIAGRAM

SN75480 DC PLASMA DRIVER

The SN75480 (Figure 8-18) is a bipolar high-voltage 7-segment decoder/cathode driver for dc plasma displays. Off-state output voltage capability is 80 V minimum. It is designed to decode four lines of hexadecimal input and drive a seven-segment Panaplex II* type gas-filled (dc plasma) display tube. The SN75480 employs a 112 bit read only memory to provide input decoding and output segment on or off control in accordance with the function table (Figure 8-19).

Segment drive outputs are constant-current sinks with adjustable operating levels. The current sink level is adjustable by connecting an external program resistor (R_p) from V_{CC} to the current programming input in accordance with the curve shown in Figure 8-18(c). Output current may be adjusted from nominally 0.2 mA up to 1.5 mA in order to drive various tube types and permit multiplex operation.

**(c) SEGMENT OUTPUT CURRENT
vs PROGRAMMING RESISTOR**

Figure 8-18. SN75480 DC Plasma Driver

*Trademark of Burroughs Corporation

Output sink currents for other segments are proportional to the b-segment current to provide even illumination of all segments. Each sink output (Figure 8-18) is regulated to ensure a constant brightness level across the display even with fluctuating supply voltages. Typical on-state currents are within 1% for voltage changes from 3 to 50 V. Off-state voltages can vary from 3 to 80 V. The blanking input (pin 4) provides unconditional blanking of all outputs, while the A through D inputs (pins 1, 2, 6 and 7) and the ripple blanking input (pin 5) into the blanking circuit, allow simple leading or trailing zero blanking.

SEGMENT IDENTIFICATION

DECIMAL OR FUNCTION	INPUTS					BI/RBO	SEGMENT OUTPUTS							DISPLAY
	RBI	D	C	B	A		a	b	c	d	e	f	g	
0	H	L	L	L	L	H	ON	ON	ON	ON	ON	ON	OFF	*0*
1	X	L	L	L	H	H	OFF	ON	ON	OFF	OFF	OFF	OFF	*1*
2	X	L	L	H	L	H	ON	ON	OFF	ON	ON	OFF	ON	*2*
3	X	L	L	H	H	H	ON	ON	ON	ON	OFF	OFF	ON	*3*
4	X	L	H	L	L	H	OFF	ON	ON	OFF	OFF	ON	ON	*4*
5	X	L	H	L	H	H	ON	OFF	ON	ON	OFF	ON	ON	*5*
6	X	L	H	H	L	H	ON	OFF	ON	ON	ON	ON	ON	*6*
7	X	L	H	H	H	H	ON	ON	ON	OFF	OFF	OFF	OFF	*7*
8	X	H	L	L	L	H	ON	ON	ON	ON	ON	ON	ON	*8*
9	X	H	L	L	H	H	ON	ON	ON	ON	OFF	ON	ON	*9*
10	X	H	L	H	L	H	ON	ON	ON	OFF	ON	ON	ON	*A*
11	X	H	L	H	H	H	OFF	OFF	ON	ON	ON	ON	ON	*b*
12	X	H	H	L	L	H	ON	OFF	OFF	ON	ON	ON	OFF	*C*
13	X	H	H	L	H	H	OFF	ON	ON	ON	ON	OFF	ON	*d*
14	X	H	H	H	L	H	ON	OFF	OFF	ON	ON	ON	ON	*E*
15	X	H	H	H	H	H	ON	OFF	OFF	OFF	ON	ON	ON	*F*
BI	X	X	X	X	X	L	OFF	OFF	OFF	OFF	OFF	OFF	OFF	
RBI	L	L	L	L	L	L[†]	OFF	OFF	OFF	OFF	OFF	OFF	OFF	

H = high level, L = low level, X = irrelevant
[†]BI/RBO is wired AND logic serving as a blanking input (BI) and/or ripple-blanking output (RBO). When RBI and inputs A, B, C, and D are all low, all segment outputs go off and RBO goes to a low-level (response condition).

Figure 8-19. SN75480 Function Table

SN75480 Application

In the basic application circuit, Figure 8-20, a serial count from a controller or microprocessor is fed to the "A" input of an SN7493A binary counter. The SN7493A output is a BCD representation of the decimal number to be displayed. To hold the last input until time for an update, an SN7475 4-bit latch is used. The BCD input is decoded by the SN75480 to provide proper segment drive resulting in a displayed hexadecimal value in accordance with the function table (Figure 8-19).

SN75581 GAS DISCHARGE SOURCE DRIVER

The SN75581 (Figure 8-21) is a high-voltage (150 V), 7-channel, source output monolithic BIDFET integrated circuit designed to drive a dot matrix or segmented display. Its output characteristics make this driver compatible to several display types including vacuum fluorescent and dc plasma displays.

All device inputs are diode-clamped pnp inputs and, when left open, assume a high logic level. The nominal input threshold is 1.5 V. Outputs are open source DMOS transistors for excellent high voltage characteristics and reliability.

The device consists of a 7-bit shift register, seven latches, and seven output AND gates. Serial data is entered into the shift register on the low-to-high transition of the Clock input. When the Latch Enable input is high, data is transferred from the shift registers to the latch outputs. When Latch Enable makes a high-to-low transition, the shift register is cleared. Taking the Output Enable input high enables all outputs simultaneously. The Serial Output is not affected by the Output Enable input. See Figures 8-22 and 8-23.

SN75581 Application

A typical SN75581 gas discharge display driver application is shown in Figure 8-23. Serial data is clocked into the SN75581 under microprocessor control. When the latch enable input is high, the data is transferred from input shift registers to the latches. The latches store the data until output enable goes high, which enables all outputs (pins 9 through 16). The outputs are connected to a seven segment display (a through g). Individual segments (a through g) will fire if a data bit turns on the corresponding DMOS output transistor connected to that segment. Serial output is not utilized.

SN75584A HIGH-VOLTAGE 7-SEGMENT LATCH/DECODER/CATHODE DRIVER

The SN75584A is a BIDFET integrated circuit and has high voltage DMOS outputs. The SN75584A (Figure 8-24) is designed to decode four lines of BCD data and drive a seven segment (plus decimal point) gas display tube such as the Beckman and Panaplex II displays. Latches are provided to store the four data and decimal point inputs while the enable input is at a low logic level voltage.

This circuit employs a read only memory to provide output decoding for the BCD input digits 0 to 9. For input data codes greater than 9, the segment outputs are blanked. Each sink output is a constant current, regulated to ensure uniform brightness of the display even with fluctuating supply voltage. On-state output current is essentially constant over the output voltage range of 4 V to 100 V. Each current sink is normalized to the "b" segment output current as required for even illumination of all segments. Output currents may be varied from 0.1 mA to 4 mA for driving various displays.

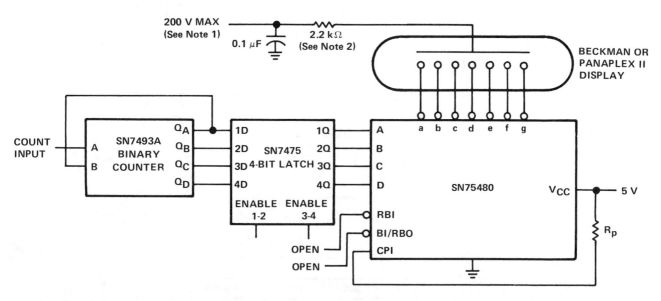

NOTES: 1. This voltage will be different for different displays. It must be set to ensure that the on-state and off-state voltages do not exceed 55 V and 80 V at the SN75480.

2. In all applications peak transient segment current must be limited to 50 mA. This may be accomplished by connecting a 2.2 kΩ resistor as shown or current limiting in an anode driver in multiplex applications.

Figure 8-20. Basic Digital Display Drive

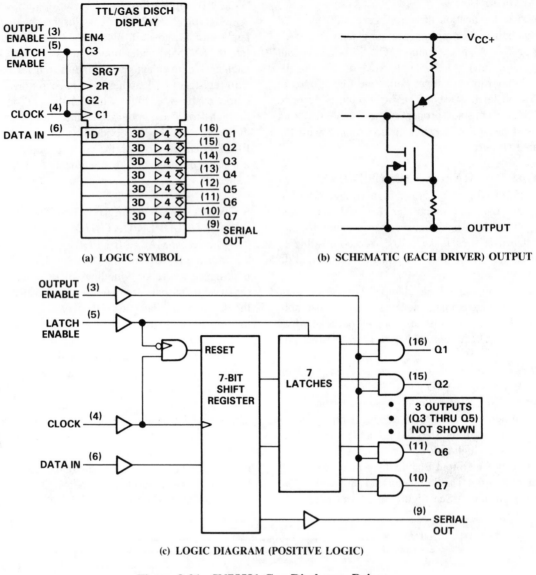

(a) LOGIC SYMBOL (b) SCHEMATIC (EACH DRIVER) OUTPUT

(c) LOGIC DIAGRAM (POSITIVE LOGIC)

Figure 8-21. SN75581 Gas Discharge Driver

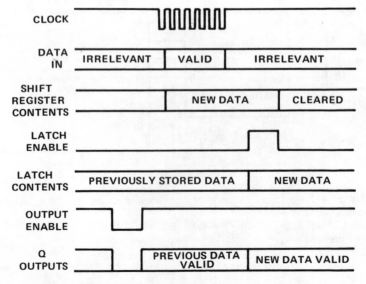

Figure 8-22. SN75581 Typical Operating Sequence

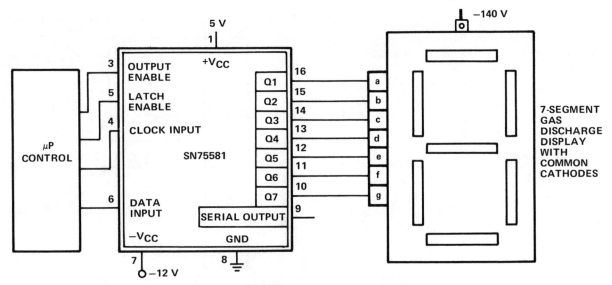

Figure 8-23. Typical SN75581 Application

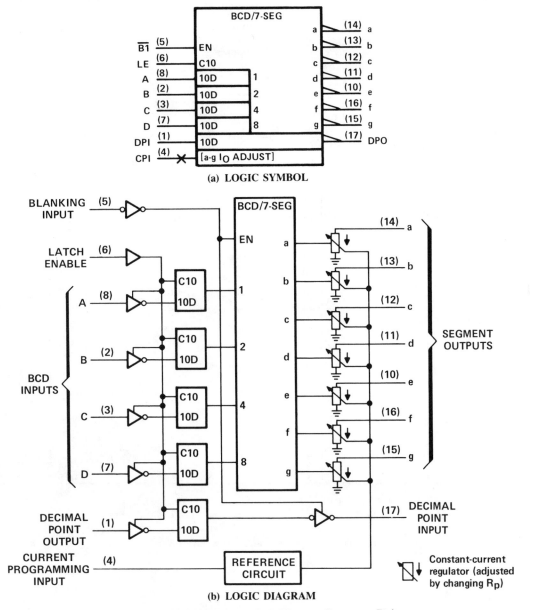

(a) LOGIC SYMBOL

(b) LOGIC DIAGRAM

Figure 8-24. SN75584A DC Plasma Segment Driver

The output current is adjusted by connecting an external programming resistor (R_p) from the current programming input to ground. Figure 8-25 shows the "b" segment output current versus programming resistance. $I_{O(b)}$ in mA = $3.616/R_p$ in kΩ. The blanking input provides unconditional blanking of all segment outputs including the decimal point output. The enable input allows data to be stored internally while input data is changing. When enable is at a high level voltage, the outputs will reflect conditions on the A, B, C, D, and DP inputs in accordance with the function table in Figure 8-26. A transition from a high level voltage to a low level voltage at enable will cause the input data set up prior to the transition to be latched. In the latched state, the A, B, C, D, and DP inputs are in a high-impedance state to minimize input loading.

Thermal protection circuitry will blank the display, regardless of input conditions, whenever junction temperature exceeds approximately 150°C.

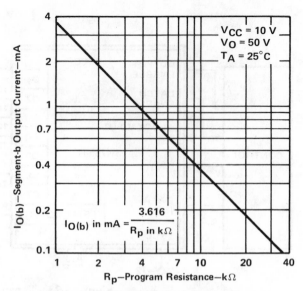

Figure 8-25. SN75584A Segment Output Current vs Program Resistance

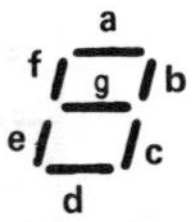

DECIMAL OR FUNCTION	DP INPUT†	BCD INPUTS†				BI	SEGMENT OUTPUTS							DP OUTPUT	DISPLAY
		D	C	B	A		a	b	c	d	e	f	g		
0	X	L	L	L	L	H	ON	ON	ON	ON	ON	ON	OFF	X	
1	X	L	L	L	H	H	OFF	ON	ON	OFF	OFF	OFF	OFF	X	
2	X	L	L	H	L	H	ON	ON	OFF	ON	ON	OFF	ON	X	
3	X	L	L	H	H	H	ON	ON	ON	ON	OFF	OFF	ON	X	
4	X	L	H	L	L	H	OFF	ON	ON	OFF	OFF	ON	ON	X	
5	X	L	H	L	H	H	ON	OFF	ON	ON	OFF	ON	ON	X	
6	X	L	H	H	L	H	ON	OFF	ON	ON	ON	ON	ON	X	
7	X	L	H	H	H	H	ON	ON	ON	OFF	OFF	OFF	OFF	X	
8	X	H	L	L	L	H	ON	ON	ON	ON	ON	ON	ON	X	
9	X	H	L	L	H	H	ON	ON	ON	OFF	ON	ON	X	
10	X	H	L	H	L	H	OFF	OFF	OFF	OFF	OFF	OFF	OFF	X	
11	X	H	L	H	H	H	OFF	OFF	OFF	OFF	OFF	OFF	OFF	X	
12	X	H	H	L	L	H	OFF	OFF	OFF	OFF	OFF	OFF	OFF	X	
13	X	H	H	L	H	H	OFF	OFF	OFF	OFF	OFF	OFF	OFF	X	
14	X	H	H	H	L	H	OFF	OFF	OFF	OFF	OFF	OFF	OFF	X	
15	X	H	H	H	H	H	OFF	OFF	OFF	OFF	OFF	OFF	OFF	X	
BI	X	X	X	X	X	L	OFF	OFF	OFF	OFF	OFF	OFF	OFF	OFF	
DP	H	X	X	X	X	H	X	X	X	X	X	X	X	ON	
DP	L	X	X	X	X	H	X	X	X	X	X	X	X	OFF	

H = high level, L = low level, X = irrelevant
†Table is valid for the indicated BCD and decimal point inputs while enable is high. See description.

Figure 8-26. SN75584A Function Table

SN75584A Application

A basic circuit (Figure 8-27) that automatically counts 0 through 9, resets and counts through again is used to demonstrate the SN75584A's ease of operation and drive capabilities.

The display is a 3-digit display with decimal points requiring a 180 V supply and typical segment operating currents of 1 mA.

The TLC555 provides clock pulses at about 1.6 Hz to an SN7493A 4-bit binary counter which in turn drives the A, B, C and D inputs of the SN75584A display driver in accordance with its count sequence. Sequencing through the binary input codes results in the output segment drivers cycling from 0 through 9. For all input codes over 9 all segment outputs are blanked off. Thus the display will be blanked off for about four seconds before the count sequence

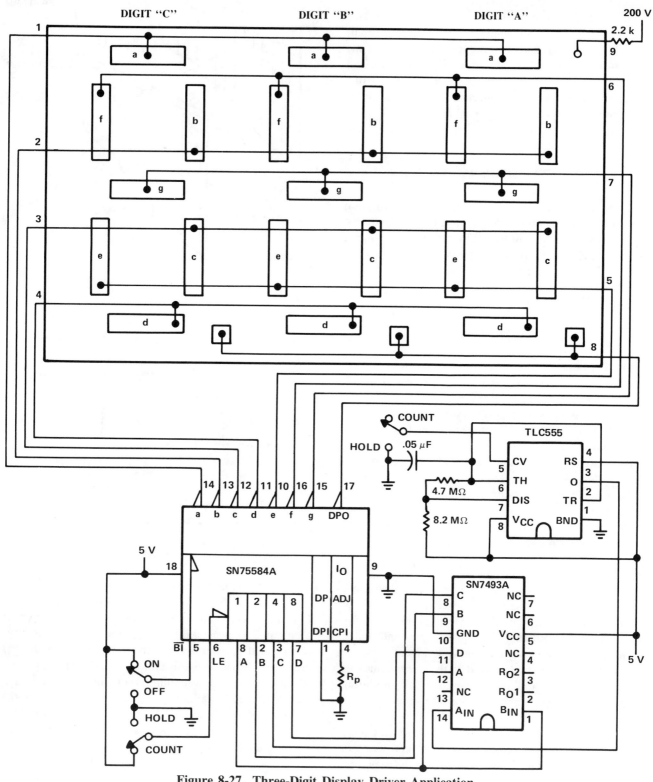

Figure 8-27. Three-Digit Display Driver Application

starts over with zero. Blanking of all output drives (including decimal point) is accomplished by switching pin 5, the blanking input, to ground. If the latch enable input, pin 6, is connected to 5 V the SN75584A continues to sequence. Switching the latch enable to ground prevents further transferring of data to the 584A outputs and the segment outputs are held (latched) in the mode they were in just prior to switching. Setting pin 6 high again allows data present in the 584A memory to again control the outputs. The driver output may also be held by simply stopping the clock. This is accomplished by switching pin 5, control voltage input, of the TLC555 to ground. In this application, pin 4 of the SN75584A, (the current programming input) is connected through a 3.3 kΩ resistor to ground allowing about 1 mA of segment drive current. The decimal point input, pin 1, is grounded which leaves the decimal point on at all times except when blanked. A TTL logic high level at pin 1 could turn off the decimal point output.

AC PLASMA DISPLAY DRIVERS

The persistent interest in flat panel information displays has stimulated development of the gas discharge display. Their safety, thinness, durability, and compact screen size have been the primary advantages over conventional display technologies. The following sections discuss ac plasma gas discharge displays and display drivers.

AC PLASMA DISPLAY TECHNOLOGY
The ac plasma display is an X-Y matrix gas discharge display. The basic display element is the gas discharge that occurs at the intersection of selected electrodes when the applied voltage between the electrodes exceeds the breakdown voltage of the media gas with which the display is filled. When the breakdown voltage of the gas is exceeded, the gas is ionized and the discharge that occurs emits a visible spot of light at the intersection of the selected electrodes. Once initiated, the display element can be maintained active without further selective control. The data retention property of the ac plasma display eliminates the necessity of a memory map for simple information displays.

Construction
The simple construction techniques employed are another feature encouraging the development of the ac plasma display panel. The panel envelope is essentially two flat pieces of ordinary glass spaced apart and sealed around the peripheral edges as shown in Figure 8-28.

The electrodes are deposited on the internal surfaces of the glass plates and then covered by an insulating dielectric layer prior to their joining. The space between the glass plates is evacuated and filled with a media gas under low pressure (approximately one-fifth atm.). Unlike the dc plasma display panel where the electrodes are immersed in the media gas, the electrodes of the ac plasma

panel are isolated from the media gas by the dielectric layer. This dictates ac operation utilizing capacitive coupling to the insulated ac plasma display cell.

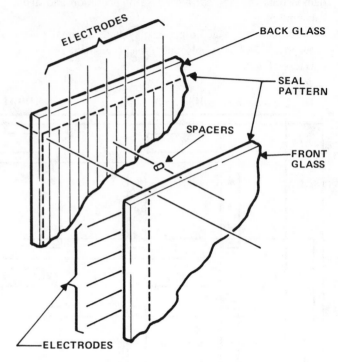

QUADORTHOGONAL EXPLOSION

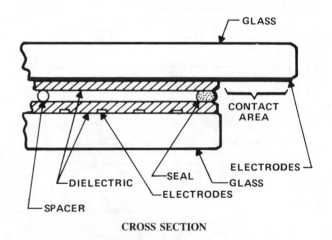

CROSS SECTION

Figure 8-28. Panel Construction

Early panels utilized a third piece of perforated glass, which defined the individual display cell or pixel. Current panels, however, use an open-cell structure which eliminates the masking glass. Individual cells thus constructed are defined by properly ratioed media gas pressure, electrode width and resolution, glass spacing, and excitation and sustaining potentials. Individual cells are defined as the area located at the intersection of the mutually perpendicular electrodes of the front and back plates. The parallel electrodes of each plate of the panel are usually divided, every other electrode exiting from opposing edges of the plate, to allow easier access for the mechanical interface required to connect the electrodes to the control circuitry. The simple construction of the ac plasma panel yields a rugged sandwich containing only a few cubic centimeters of inert gas. There is no danger of implosion as found with conventional vacuum tube displays, and no danger of contact with high voltages through the glass faceplate. In addition to the low cost of the ac plasma display panel, the construction techniques yield one of the safest display panels in today's marketplace.

The Functional Cell

Light is emitted from the ac plasma cell as a result of the energy release that occurs when the media gas is ionized. This is accomplished by simply applying sufficient potential across the cell to break down the media gas. Since the actual cell is only capacitively coupled to the electrode potential, the voltage waveform, frequency and amplitude are interdependent for reliable plasma display operation.

When ionization of the gas occurs, a charge buildup is created by the high electron and ion currents present in the ionization discharge sequence. This charge buildup, or wall charge as it is commonly identified, plays an important role in the ability of the ac plasma panel to maintain display information without further selective control. Figure 8-29 shows a typical applied ac waveform and the wall charge waveforms of an active and extinguished cell. These relationships are observed at all cell locations during that period of time in which no panel information is being altered (written or erased). This is normally identified as the sustain mode.

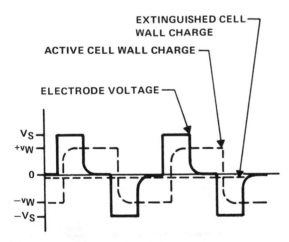

Figure 8-29. Cell Waveforms

A typical write-sustain cycle waveform is shown in Figure 8-30. The electrode potential and resulting wall charge are plotted to show their interdependence in writing (initializing) and sustaining a selected cell.

Prior to the cycle in which the selected cell will be written, the cell being exercised exhibits zero wall charge. Therefore, the potential seen by the cell (the cell voltage) is solely the potential applied to the electrodes (the electrode voltage). When the electrode voltage exceeds the breakdown voltage of the media gas, the gas ionizes and the wall charge is deposited such that its polarity opposes the applied cell voltage. Once created, the wall charge remains even after the ionization discharge extinguishes and the electrode potential decays to zero. In the next half-cycle, the electrode potential reverses, thus the wall charge that opposed the electrode voltage in the previous half-cycle now is additive. The cell voltage is therefore the sum of the wall voltage and the electrode voltage. This allows the electrode voltage to be reduced and still create a cell voltage of sufficient amplitude to cause the cell to fire. Stable operation exists when the sum of the electrode potential and wall charge create a cell voltage (V_{W2}) which is sufficient to create ample excess charge ($2 V_W$) to cause the wall voltage to invert. Thus the cell can be maintained indefinitely by an alternating electrode potential which is actually less than the potential required to fire the cell. The ability to do this is attributed chiefly to the nonlinear charge transfer characteristics of the ac plasma cell, as illustrated in Figure 8-31. Since the sustaining electrode potential is less than the required firing potential of the extinguished cell, application of this voltage will have no effect on cells which have not been fired previously. With this in mind, operation of the ac plasma display is relatively simple. A background signal is applied to the entire display panel. This is usually called the sustain signal. Select circuitry superimposes the write pulse required to initially fire a cell on the X and Y-axis electrodes common to the cell to be written into. Once fired, the background signal will maintain the integrity of a cell until the cell is extinguished by another selective control signal, normally called the erase pulse. This is accomplished by application of a single pulse to the cell electrodes whose amplitude (V_{W1}) and duration are only sufficient to create enough excess charge to counterbalance the wall charge. Thus the wall charge is removed and the sustain chain sequence is broken. An illustration of this operation is shown in Figure 8-32.

Control Circuitry Implementation with the SN75500A and SN75501C

Actual circuitry may vary based on the specific application being addressed so long as the net result, the differential electrode potential waveforms, yields reliable control. One such approach utilizes the principle of the H-Bridge. This reduces the requirements on the power supply and enables the two drivers to share in the generation of the functional waveforms. Additional approaches will be presented primarily to illustrate the variety of acceptable drive schemes.

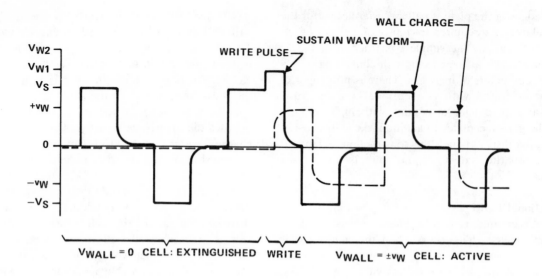

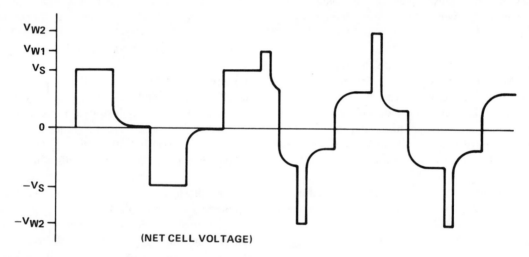

(NET CELL VOLTAGE)

Figure 8-30. Write-Sustain Cycle Cell Waveforms

The high voltage BIDFET drivers used in these approaches are the SN75500A and SN75501C ac plasma display drivers. Designed specifically for ac plasma display applications, the SN75500A and SN75501C mark a significant milestone in the development of the ac plasma display technology. Prior to their development, the high voltage requirements of the plasma panel prevented the use of conventional integrated circuits as electrode drivers. This meant that the electrode interface must employ discrete components for proper control. This is quite prohibitive on larger panels without a more sophisticated drive scheme. The SN75500A and SN75501C drivers each provide circuitry required for active control of 32 electrodes. Interface to a 256-line by 256-line panel can now be achieved with a total of 16 integrated circuits (eight SN75500A and eight SN75501C). Use of these devices allows a significant reduction in the complexity and cost of the system electronics required to operate the ac plasma display.

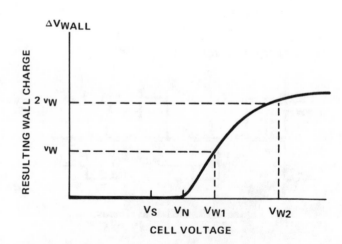

Figure 8-31. Charge Transfer Characteristics of AC Plasma Cell

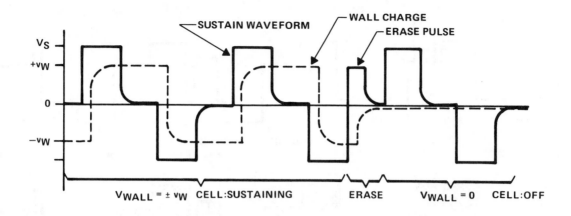

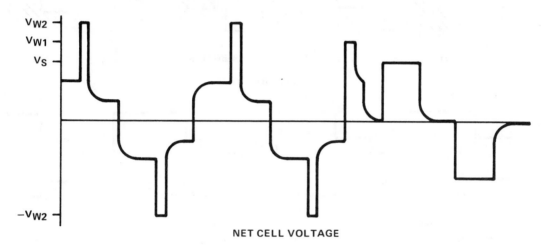

NET CELL VOLTAGE

Figure 8-32. Sustain-Erase Cycle Cell Waveforms

The Functional Waveforms

The functional waveforms developed are the primary waveforms which provide the basic functions required in the operation of an ac plasma panel display: sustain, write, erase, and blanking. These waveforms are only one of several approaches, all of which may provide satisfactory operation. The main intent of this application section is to present the methods used to develop the waveforms and the implementation of the SN75500A and SN75501C drivers.

Figure 8-33(a) illustrates the basic waveforms for write, sustain, erase, and blanking. This is the net differential waveform created between the X and Y-axis electrodes by excursions on the X-axis or Y-axis or both. Figure 8-33(b) shows the composition of the differential waveform and identifies the origin of the components.

The first sequence of waveforms to consider is the sustain waveform. Review of Figure 8-33 shows the sustain waveform to be composed of two parts: the base pulse applied to the X-axis and a negative excursion on the Y-axis. It is important to note that this basic waveform is found in all other waveforms which require the retention of the panel data (selective write, selective erase). Additionally, this signal is nonselective, it must appear at all electrode nodes where the cell data is to be maintained.

Any cell not experiencing this differential signal will fail to retain its active status. The first component of the sustain waveform, the base pulse, is generated *external* of the X-axis drivers, and is applied to all electrodes addressed along the X-axis. Thus it is identified as the X-axis sustain. Additionally, since the drivers for all the X-axis electrodes share common circuitry, the X-axis sustain is commonly called the bulk sustain. The second component of the sustain waveform, the negative pulse appearing on the Y-axis is not considered a bulk-sustain signal since it is created by the Y-axis drivers and each electrode addressed along the Y-axis is driven by its associated driver output circuitry. Supplemental pulses thus created are identified with their axis, Y-axis sustain. Together, the bulk (X-axis) sustain minus the Y-axis sustain combine to compose the basic sustain waveform. The SN75501A ac plasma driver was specifically designed to provide the Y-axis driver function. Additional control of the output circuitry allows all outputs (32) to be switched low, independent of the select control circuitry employed in selective output operations such as write and erase. All the outputs of the SN75501A switch low when the sustain input is taken low, thus all electrodes addressed along the Y-axis experience the Y-axis sustain signal.

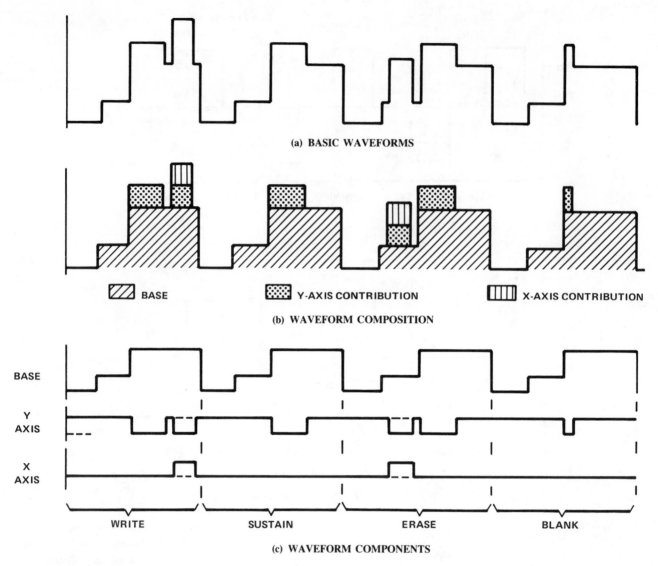

(a) BASIC WAVEFORMS

BASE   Y-AXIS CONTRIBUTION   X-AXIS CONTRIBUTION

(b) WAVEFORM COMPOSITION

BASE

Y
AXIS

X
AXIS

WRITE   SUSTAIN   ERASE   BLANK

(c) WAVEFORM COMPONENTS

Figure 8-33. Functional Waveforms

The blanking waveform is also a nonselective waveform as it is used to blank the entire panel. Composed of the same components as the basic sustain waveform, it is similarly created.

The write and erase operations are selective operations. In other words, the pedestals superimposed on the basic sustain waveform, which create the write and erase waveforms, appear only at the pixels (electrode intersections) whose information is to be altered. This is accomplished by superimposing half the required pedestal on each of the associated X and Y-axis electrodes. This is illustrated in Figure 8-34. Note the importance of timing in these selective operations.

All other nodes experience either a standard sustain waveform or a half-select waveform. Standard sustain waveforms appear at all nodes where neither of the associated electrodes (X-axis or Y-axis) exhibit a half-select pulse. A half-select waveform appears at all nodes where only one of the associated electrodes exhibits a half-select pulse. The pedestal created by a single electrode excursion in a half-select waveform is insufficient to ionize

the media gas and initiate a write-sustain sequence. The half-select pulses appearing on the X and Y-axis are created by the X and Y-axis drivers respectively. The SN75500A X-axis and SN75501C Y-axis drivers are designed to provide these functions. Both devices contain circuitry for selective control of their outputs. The specifics of this circuitry are discussed in their respective sections. Let it be sufficient to say at this time that with proper data controls established, the selected outputs of the SN75500A switch positive and the selected outputs of the SN75501C switch negative when the strobe input of both devices is pulled low. The determination of a write or erase operation depends on the timing of the strobe. Figure 8-35 illustrates use of the sustain and strobe inputs in the generation of the basic operational waveforms.

As mentioned previously, a variety of approaches for driving the ac plasma panel can be employed. The previous approach utilized the half-select principle for selective operations with an X-axis bulk sustain and a Y-axis supplemental sustain. Additionally, the X-axis drivers floated on the bulk sustain signal while the Y-axis remained ground based. The following example incorporates split sustain and uses a common driver for both axes. Selective operations are performed using the blanking principle.

A split axis sustain merely says both axes employ an externally generated bulk sustain signal of opposite polarities. Figure 8-36 illustrates how the basic sustain waveform is created using this approach. The write technique for selective operations superimposes a full-select pedestal of opposite polarity on both axes. Locations along the selected electrode which are not to be altered are not affected by a single pulse on either axis. Only the pixel at the intersection of the selected electrodes is affected. Both axes require a selective pulse of the opposite polarity for the write operation. Figure 8-37 illustrates the array of waveforms created using this approach and their origin.

SN75500A AC PLASMA DISPLAY AXIS DRIVER

The operation of the SN75500A on the horizontal or vertical electrodes is primarily dependent on the panel's application. The outputs of the SN75500A are normally low and switch high selectively when the strobe input is low. The SN75500A thus provides the positive select pulses. The logic symbol, functional block diagram, and output structure of the SN75500A is shown in Figure 8-38. Selection of the outputs (32) is accomplished with the select and data inputs. The 32 outputs of the SN75500A

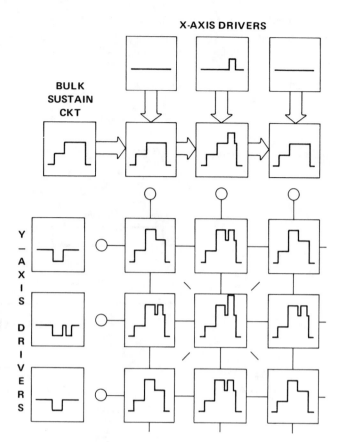

Figure 8-34. Write Waveform Array and Origin

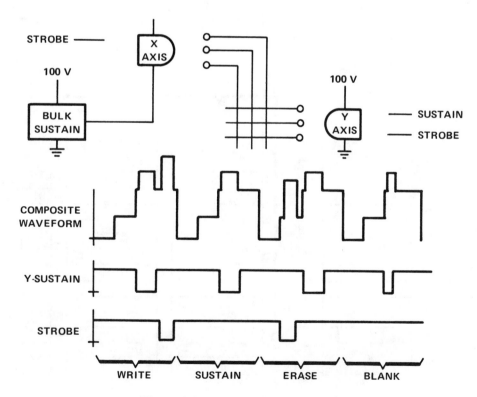

Figure 8-35. Control Signal Timing

are divided into four sections (1QX, 2QX, 3QX, 4QX) of eight outputs each. Only one of the four sections can be activated at one time (eight outputs). All other outputs (24) remain low. For this reason most systems use the SN75500A to scan the electrodes along which the information is written.

Selection of the specific section is determined by the select inputs S0 and S1 (Table 8-1). When selected, the state of the eight outputs of the section is determined by the data stored in the 8-bit storage register. Data is shifted into the storage register in a serial fashion on the positive transition of the clock. The maximum guaranteed data

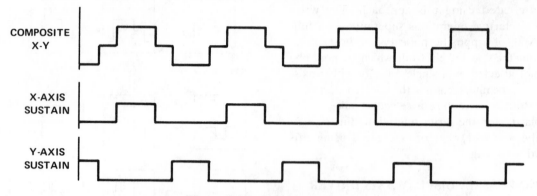

Figure 8-36. Split Axis Sustain

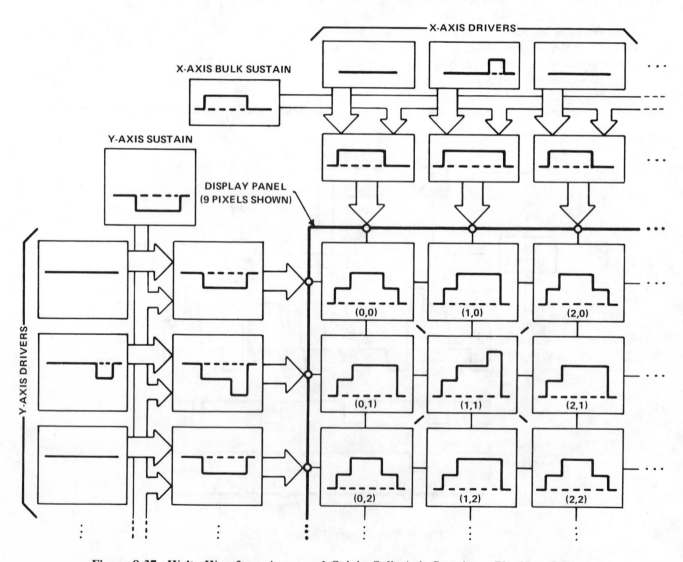

Figure 8-37. Write Waveform Array and Origin Split Axis Sustain — Blanking Select

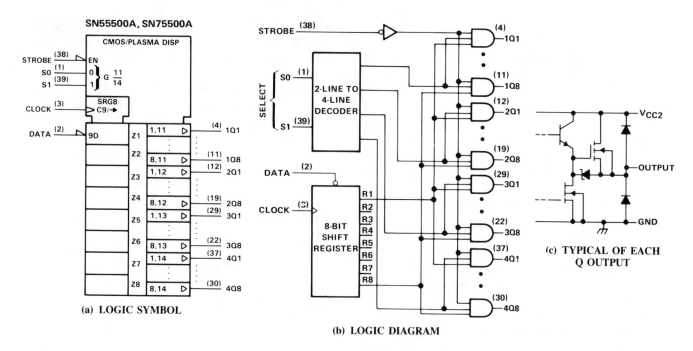

(a) LOGIC SYMBOL

(b) LOGIC DIAGRAM

(c) TYPICAL OF EACH Q OUTPUT

Figure 8-38. SN55500A and SN75500A

rate is 4 MHz. Data is shifted into the Q1 register and progresses to the Q8 register. A logic zero previously entered in the serial data input stream determines which of the outputs will switch high when the SN75500A is strobed (pulsed low). All outputs of the SN75500A contain clamp diodes to the V_{CC2} and GND supply inputs. This allows it to be operated on a base waveform where required. These diodes play an important role as discussed in the following

Table 8-1. Select Input Truth Table

S0	S1	Outputs Enabled
0	0	1Q1 thru 1Q8
1	0	2Q1 thru 2Q8
0	1	3Q1 thru 3Q8
1	1	4Q1 thru 4Q8

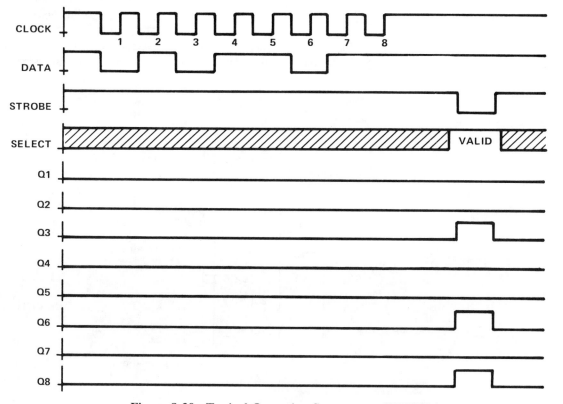

Figure 8-39. Typical Operating Sequence — SN75500A

section on Functional Adaptation of the SN75500A and SN75501C. Push-pull circuitry on each output provides active switching between the GND and V_{CC2} supplies. All inputs of the SN75500A are CMOS compatible ($V_{TH} = 5$ V) and assume a logical 1 if left open. A typical operating sequence is shown in Figure 8-39.

SN75501C AC PLASMA DISPLAY AXIS DRIVER

The SN75501C is designed to generate the negative select pulses. In addition, the internal control circuitry provides for all outputs to be switched low independent of the stored data when the sustain input is taken low. This feature is provided primarily for the purpose of creating a supplemental sustain signal. Unlike the SN75500A, the SN75501C can operate on all 32 of its outputs at one time. The logic symbol, functional block diagram, and output structure of the SN75501C are shown in Figure 8-40. Control of the output gates is established in the internal 32-bit shift register. A logic zero previously entered in the serial data input stream determines which of the outputs will switch low when the strobe input is taken low. Data enters the serial shift register on the positive transition of the clock input. The maximum data rate is 4 MHz. All outputs of the SN75501C contain clamp diodes to the V_{CC2} and GND supply terminals. This allows it to be operated on a base waveform where required. The push-pull output circuitry provides positive switching of all outputs between the V_{CC2} and GND supplies. The CMOS compatible inputs ($V_{TH} = 5V$) assume a logical 1 if left open. A typical operating sequence is shown in Figure 8-41.

FUNCTIONAL ADAPTATION OF THE SN75500A AND SN75501C

In the previous text, functional waveforms were discussed. It is the intent of this section to discuss the adaptation of the SN75500A and SN75501C to these drive techniques and to identify specific considerations which must be observed for satisfactory operation.

Strobing and Sustaining

The output gate circuitry for the SN75500A [Figure 8-38(c)] and SN75501C [Figure 8-40(c)] is virtually identical. Both devices contain a pair of DMOS output transistors for active control of the output. The lower DMOS transistor receives its drive from the low voltage supply, V_{CC1} (12 V). Thus the power consumption of the gate drive circuitry is minimal. The gate drive of the upper DMOS structure however experiences the full high voltage bias (100 V). To minimize its power dissipation, the gate circuitry incorporates a dynamic drive scheme which coincides with the dynamic output current requirements of the panel. In other words, the drive circuitry initially provides a large gate drive capable of saturating the upper DMOS transistor during the transition period of the output when the current demand is large. As the panel capacitance is charged and the current requirements decrease, so does the gate drive. This kick current occurs every time the output of the drivers is required to switch high.

The kick current is approximately 60 mA with a duration of 300 ns. If operated in a system whose V_{CC2} is 100 V, the resulting power consumption is 1.8 μJ. The average power dissipation then depends on the frequency at which the kick circuit is employed. In a system operating at a 50 kHz (period = 20 μs), this equates to an average dc power of 90 mW.

This causes a 7°C rise in the chip temperature over the ambient. The kick circuit being discussed here is activated every time the output is strobed regardless of whether or not the data causes the output to switch. For this reason, a selective strobe architecture is preferable in

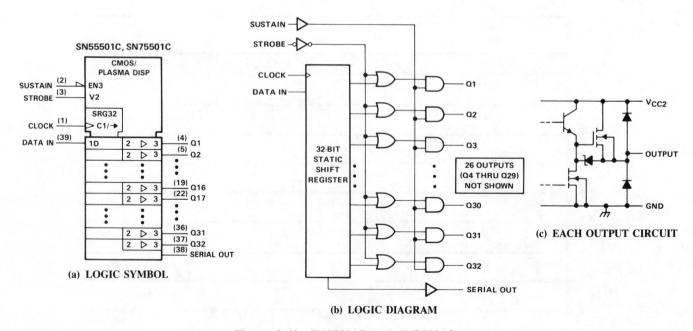

(a) LOGIC SYMBOL

(b) LOGIC DIAGRAM

(c) EACH OUTPUT CIRCUIT

Figure 8-40. SN55501C and SN75501C

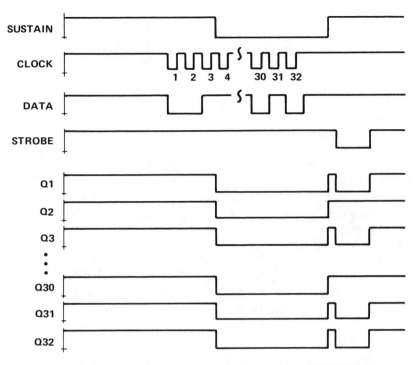

Figure 8-41. Typical Operating Sequence — SN75501C

a system that updates panel information every cycle. Thus, each driver experiences only the additional power dissipation while it is actively performing the panel operations (writing or erasing) thus reducing the duty cycle and effective dc power. In a 512-line panel (16 drivers: 8/side), this represents an 88% reduction in power due to data operations. Panels which fill an entire line or column with data to be written in parallel also reduce the effective dc power. With a maximum data rate of 4 MHz, it requires 64 μs to enter 256 serial data bits (8 drivers × 32 bits). If the information is then written on the panel in the fifth cycle, the effective power resulting from the data operations on the panel is reduced by 80%.

These strobing techniques affect only the kick circuit power associated with the data operations of the devices. In the case of the SN75500A, this is its only contribution. With the SN75501C, however, the kick circuit not only applies to strobed response of the driver but also to the sustain response. The sustain feature of the SN75501C is employed as in the initial example, the SN75501C output is pulsed twice in each cycle and the average power increases to 180 mW. This causes a 14°C rise in the chip temperature over the ambient.

Floating Driver Considerations

In most applications, one or both axis drivers will be required to operate (float) on a base waveform. Output clamp diodes have been provided on each output to accommodate this requirement. Figure 8-42 shows the output structure which is common to both driver circuits, SN75500A and SN75501C.

In the case of the SN75501C, the output is normally high, therefore Q1 is normally on. A base pulse applied to the SN75501C is therefore applied to the V_{CC2} terminal. As shown in Figure 8-43, positive excursions on the V_{CC2} terminal are passed through the output transistor Q1 while negative excursions are coupled through the catch diode D_1. Since the data retention and decode circuitry receive their bias from the V_{CC1} supply (12 V), variations in the V_{CC2} supply will be reflected at all outputs but will not affect the stored data present in the SN75501C's register. If the outputs of the SN75501C are not required to switch below ground, the SN75501C may be operated ground-

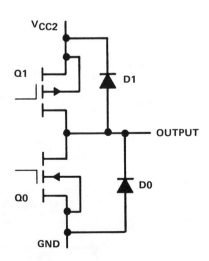

Figure 8-42. SN75500A and SN75501C Output Structure

based even though the V_{CC2} incorporates a base signal. Figure 8-44 illustrates a typical application which behaves in this manner. The only limitations are:

1. V_{CC2} must never be less than GND or greater than 100 volts above GND
2. The strobed data pulse occurs while V_{CC2} is at any potential other than GND.
3. The strobed data output pulse is always required to switch to 0 volts (GND).

If the data pulse is required to occur at various levels on the base pulse or switch to a potential other than GND (0V), the SN75501C must float with the base pulse waveform. This can be accomplished with floating V_{CC2} and V_{CC1} supplies referenced to the base waveform or by utilizing a capacitive storage bridge like that shown in Figure 8-45.

The SN75500A output is normally low, thus the output transistor Q0 of Figure 8-42 is normally on. A base pulse applied to the SN75500A is therefore applied to the

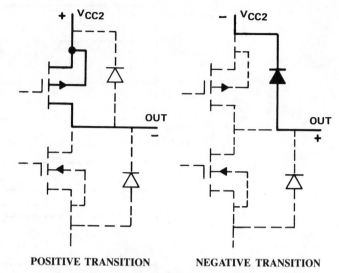

POSITIVE TRANSITION   **NEGATIVE TRANSITION**

Figure 8-43. Floating SN75501C Current Paths

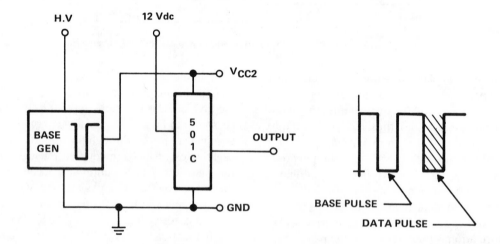

Figure 8-44. SN75501C with Pulsed V_{CC2}

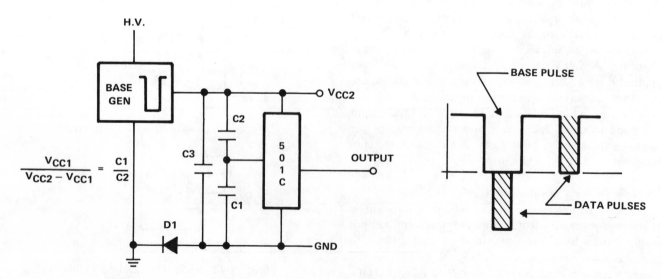

$$\frac{V_{CC1}}{V_{CC2} - V_{CC1}} = \frac{C1}{C2}$$

Figure 8-45. Floating SN75501C Fixed Bias

GND terminal. As shown in Figure 8-46, negative excursions on the GND terminal are passed through the output transistor Q0 while positive excursions are coupled through the catch diode D0. When utilizing this feature of the SN75500A, caution must be taken to assure proper retention of the data and to prevent excessive power consumption. When the lower clamp diode D0 is forward biased as is the case during the positive transition of the GND terminal, the SN75500A substrate also becomes forward biased. In this condition the output current demand is supplied in part from the high-voltage supply. Even though the current is small (typically 200 mA) when drawn from the high-voltage supply (100 V), it represents a significant power dissipation.

$$200 \text{ mA} \times 100 \text{ V} \times 1 \text{ } \mu s = 20 \text{ } \mu J$$

The average power consumed depends on the operating frequency of the system.

$$\text{at 50 kHz} : 20 \text{ } \mu J \times 50 \text{ kHz} = 1 \text{ W}$$

The SN75500A is designed such that the data registration and retention circuitry are powered from the V_{CC1} supply (12 V). The high-voltage supply, V_{CC2} is required only to provide pull-up of the SN75500A outputs when a positive output pulse is desired. It is most beneficial therefore to blank the V_{CC2} input except for the period of time the outputs of the SN75500A are switched high (strobed). Utilization of this feature (strobing the high voltage) allows lower power dissipation and improved performance with the SN75500A. Figure 8-47 shows a typical circuit which provides for strobed application of the V_{CC2} supply and a regulated 12 V supply for V_{CC1}. Figure 8-48 shows a circuit using an FET to strobe V_{CC2}. It is also possible to use a simple series impedance to limit the current from the high-voltage supply.

Data Coupling Considerations

If operated ground-based, all inputs are CMOS compatible ($V_{TH} = 5$ V). If, however, either driver is floated on a base, signal entry of data into that device must be given due consideration.

One approach that is not necessarily cost effective, but simple, employs the use of coupling transformers on all data lines: clock, serial data in, strobe, sustain, select. In this configuration, data can be entered into the device regardless of the state of the base signal. This provides the maximum utilization of the drivers by minimizing the time required to shift data into the drivers' storage registers. Another approach requires gating of the data but creates a totally solid-state interface. In this approach, data is gated into the drivers when they reside at ground. Figure 8-49 in which a base pulse is employed, illustrates the timing of the gated clock which registers the data in the respective driver. Considering 4 MHz as the maximum data rate, 8 μs is required to fill the 32-bit shift register of the SN75501A. The frequency and duty cycle of the base pulse then decide whether the data can be entered in a single cycle. The limiting factor in most applications is the speed of the source from which the data is received.

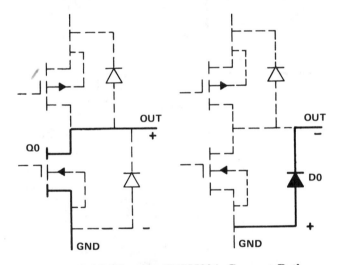

Figure 8-46. Floating SN75500A Current Paths

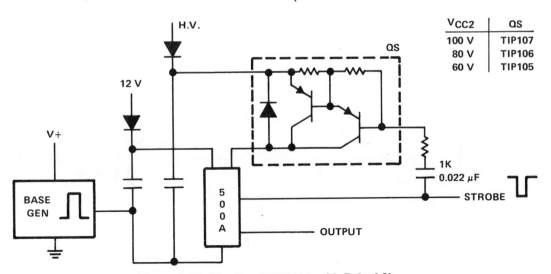

V_{CC2}	QS
100 V	TIP107
80 V	TIP106
60 V	TIP105

Figure 8-47. Floating SN75500A with Pulsed V_{CC2}

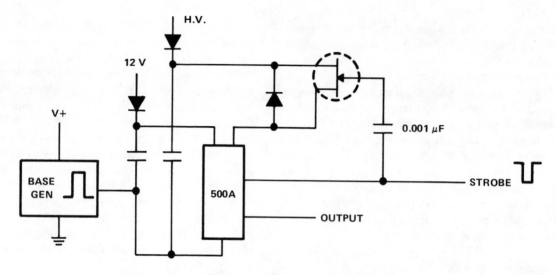

Figure 8-48. Floating SN75500A with FET Pulsed V$_{CC2}$

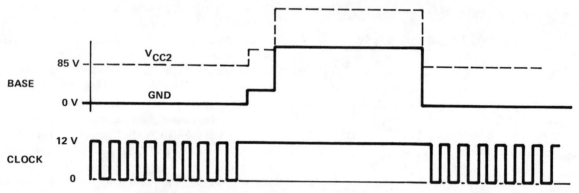

Figure 8-49. Gated Clock for Floating Applications

Figure 8-50 shows a circuit for use with positive base signals. When the base pulse is at ground, the input signal operates as a pull-up to the input of the CMOS buffer through the blocking diode. As the ground bias of the buffer circuit rises with its associated plasma driver, the diode becomes reverse biased and the resistive termination of the input maintains a logic zero at its input. With this circuit, transitions in the floating bias applied to the buffer gate are not detectable at its output, providing the input signal is low.

Figure 8-51 shows a similar interface circuit for use with negative base signals. Note the reversal of the blocking diode. This circuit uses a normally high (logic 1) input signal. While ground based, the buffer gate will respond to logic zero transitions at its input. As the ground and V$_{CC}$ inputs of the buffer circuit go below ground, the blocking diode becomes reverse biased and the terminating resistor establishes a logic 1 (high) on the gate input. Thus, provided that the input is at a logic 1, transitions in the ground and V$_{CC}$ bias inputs concurrent

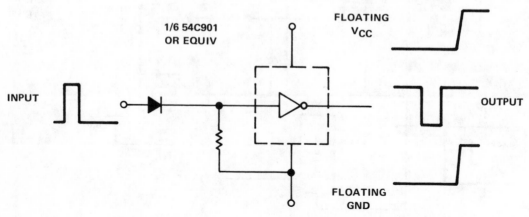

Figure 8-50. Positive Base Signal Data Buffer Circuit

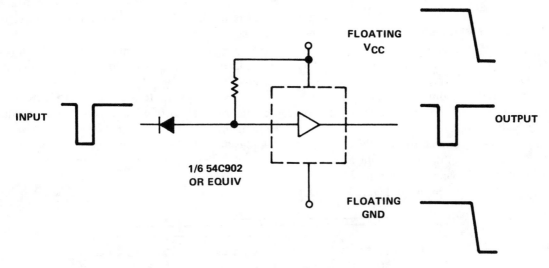

Figure 8-51. Negative Base Signal Data Buffer Circuit

with the respective plasma driver will not affect the output of the buffer. The use of the inverting or noninverting buffer gate is determined primarily on the preferred status of the output signal during the period of time the driver is floating. Both circuits of Figures 8-50 and 8-51 create normally high (logic 1) signals. This is the preferred state of the positive edge triggered clock circuits of the SN75500A and SN75501C plasma drivers.

VACUUM FLUORESCENT DISPLAYS

INTRODUCTION

A vacuum fluorescent display (VFD) operates on the same basic principles as a triode vacuum tube. It is composed of a grid, cathode, and an anode that has been coated with phosphor. The anode will emit light when electrons from the cathode strike it with sufficient energy. The blue-green color light emitted is bright, yet clear and pleasing to the human eye.

The VFD comes in a wide variety of sizes in dot matrix, segmented, or dot character arrangements. Its flat panel construction, relatively low operating voltage, and low power consumption make VFDs an attractive option in many display applications.

THE VFD PANEL
Construction

The VFD is contained in a glass envelope which provides the vacuum environment required for proper operation (Figure 8-52).

The cathode is an oxide-coated tungsten filament that emits free electrons when heated by a current. The display is controlled by the bias conditions placed on the grid and anodes. Column or character selection is provided by the grid while individual segment control is governed by the anodes. Electrons from the cathode pass through the grid when it is at a positive potential and are blocked when it is at a negative potential. Electrons passing through the grid are attracted by anodes with

positive potentials and are repelled by anodes with negative potentials (Figure 8-53). The applied voltages vary depending on the display format. Table 8-2 shows the range of element voltages and currents required by the majority of VFDs.

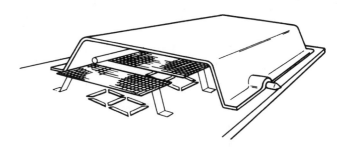

Figure 8-52. Vacuum Fluorescent Display Construction

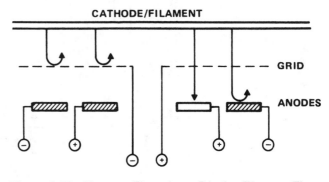

Figure 8-53. Vacuum Fluorescent Display Electron Flow

Table 8-2. VFD Operating Ranges

Parameter	Min	Max
Cathode Voltage	2.4 V ac	10 V ac
Grid/Anode Voltages	24 V p-p	70 V p-p
Cathode Current	20 mA ac	250 mA ac
Grid Current	2.5 mA p-p	30 mA p-p
Anode Current	2.5 mA p-p	30 mA p-p
Power Dissipation	14 mW/Char.	125 mW/Char.

Panel Performance

Higher resolution VFDs pose problems. As higher resolution is pursued the mechanical and electrical properties of the VFD require special considerations. The grid is a fine screen mesh with practical limitations on grid spacing and durability. Additionally, as neighboring display elements are brought closer, the fields created by their grids affect the display site being activated. The fringe field effect causes a nonuniformity in the luminance of the display. Most character displays employ a common grid for each character or column of characters. Thus the character spacing prevents this problem (Figure 8-54). Dot matrix displays, however, have no such space, thus grid/anode multiplexing is required so that no two adjacent cells are on at the same time. Figure 8-55 shows several arrangements of anode/grid connections used in dot matrix VFDs.

Except for the format shown in Figure 8-55(a), the configurations shown in Figure 8-55 provide good illumination. The particular style chosen depends on the application. Table 8-3 shows the total control pin count and cycle time required to complete one frame for various panel sizes.

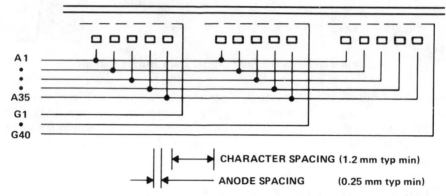

Figure 8-54. A 40: 5 × 7 Dot Character VFD

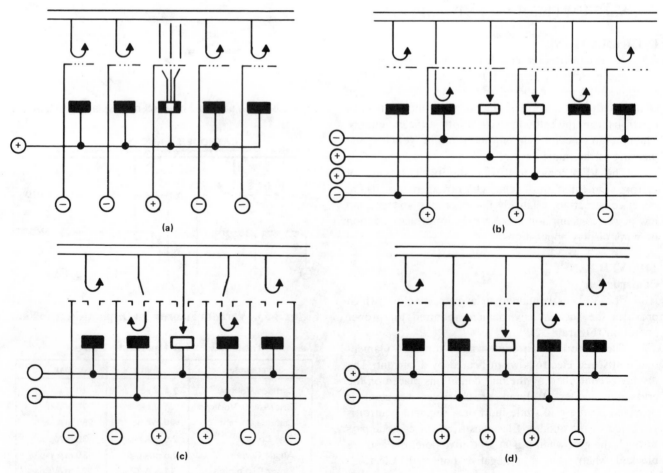

Figure 8-55. Grid-Anode Configurations of Dot Matrix VFDs

Table 8-3. VFD Configuration Comparison

Parameter	Panel size	128 X 64				128 X 128				256 X 64			
	Config.	A	B	C	D	A	B	C	D	A	B	C	D
Anode Lines		64	256	128	128	128	512	256	256	64	256	128	128
Grid Lines		128	64	128	128	128	64	128	128	256	128	256	256
Total Lines		192	320	256	256	256	576	384	384	320	384	384	384
Cycles/Frame		128	64	128	128	128	64	128	128	256	128	256	256

VFD Timing Requirements

The VFD from a timing standpoint, is very forgiving. A typical cycle is shown in Figure 8-56.

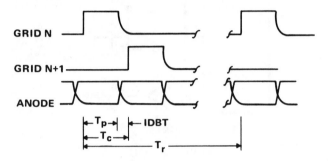

Figure 8-56. VFD Timing Diagram

The actual timing is governed largely by the application and panel size. The minimum panel refresh rate with undetectable flicker is 60 Hz. This defines the time (T_r) allotted to write the entire display one time (one frame). The amount of time allowed for each character or grid operation (T_c) is determined by the number of character or grid control lines (N) external to the display ($T_c = T_r/N$). To prevent noticeable degradation in the display intensity, each display site should be exercised at a minimum duty cycle of 1:150 ($T_p = T_r \times$ DUTY CYCLE). Neglecting any inter-digit-blanking time (IDBT), a display could contain as many as 150 grid or character control lines. However, to prevent ghosts, some time must be allotted for inter-digit-blanking. The allowable dead-time between strobes (IDBT) is the difference between the character cycle time T_c and the character or grid strobe width T_p. As the number of grid or character control lines approach 150, the IDBT approaches zero (0).

$$IDBT = \frac{T_r}{N} - \frac{T_r}{150}$$

Conventional drivers with resistive pull-down require an IDBT as large as 30 μs (20 μs typically). Thus the maximum allowable number of grid or character control lines must realistically be less than 150. Most displays limit this parameter to 128 (a convenient binary number < 150).

For a panel refresh rate of 100 Hz:
$T_r = 10$ ms
For an 80 character display
$T_c = 10$ ms/80 $= 125$ μs
For a 1:100 character duty cycle
$T_p = 10$ ms/100 $= 100$ μs
The allowable IDBT is:
$IDBT = T_c - T_p = 25$ μs

Some applications may be limited by the excessive IDBT requirements of the resistive pull-down driver:
For a 128 grid display
$T_c = T_r/128$
For a 1:150 duty cycle operation
$T_p = T_r/150$
For a driver requiring 20 μs IDBT
$IDBT = T_c - T_p = 20$ μs
$T_r = 17.5$ ms
The maximum panel refresh rate is:
$R = 1/T_r = 57$ Hz (MARGINAL)

DRIVE ELECTRONICS

The following discussion on drive techniques will be limited primarily to the interface drivers for the display, and their requirements and performance characteristics. Because the characteristics of the various display grid-anode configurations change (see Figure 8-55), interface-driver circuit requirements differ. Texas Instruments offers a selection of VFD drivers with a variety of features.

The UCN4810A 10-Bit VFD Driver

Figure 8-57 shows the block diagram, logic symbol and output circuit of the UCN4810A. The UCN4810A is a 10-bit active-high VFD driver. Input data is stored in a 10-bit serial shift register on the positive transition of the clock. Parallel data is presented to the output buffers through a 10-bit parallel D-type latch. Data at the respective output of the shift register will be transferred through the 10-bit latch while the strobe (latch enable) input is high. Data present at the latch inputs during a negative transition of the strobe will be stored regardless of subsequent changes, if the strobe input is low. A blanking control is provided which inhibits all output gates and assures they are low when the blanking input is high. All outputs [Figure 8-57(c)] are capable of sourcing 40 mA

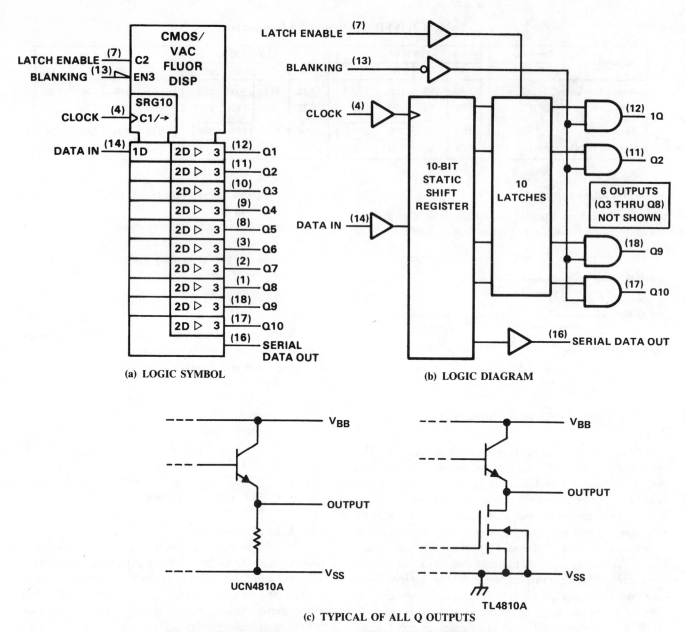

(a) LOGIC SYMBOL

(b) LOGIC DIAGRAM

(c) TYPICAL OF ALL Q OUTPUTS

Figure 8-57. UCN4810A and TL4810A VFD Display Drivers

each from a V_{BB} supply voltage of 60 V, if the maximum allowable package power limitation of 1.3 W is not exceeded. This limits the duty cycle of the on time for various load requirements (Figure 8-58). All inputs are CMOS compatible but require the addition of a pull-up resistor to V_{DD} when driven by standard TTL logic.

The TL4810A 10-Bit VFD Driver

The TL4810A can be used as a direct replacement for the UCN4810A. The TL4810A utilizes an active totem-pole output to improve the sink current capability (1 mA minimum versus 400 μA) without sacrificing the resulting power consumption as conventionally experienced in a passive pull-down structure. The totem-pole output is composed of an npn emitter follower (source) and double-diffused MOS (DMOS) (sink) transistors [Figure 8-57(c)].

This improvement decreases the inter-digit-blanking time required and the overall device power consumption.

Unlike most VFD Drivers which are limited to an 85% duty cycle at 50°C, the TL4810A will sustain 25 mA in as many as six outputs per output load at a 100% duty cycle over its entire operating temperature range of 0°C to 70°C.

All device inputs are diode-clamped and compatible with standard MOS, CMOS and DMOS logic. Designed to control 10 VFD inputs, the TL4810A provides a positive edge triggered 10-bit serial shift register and a serial data output of the display information. A 10-bit D-type latch accepts parallel data from the serial shift register when the strobe (latch enable) input is high. The data stored in the latch circuitry when the strobe input is taken low remains unaltered regardless of subsequent changes in the data

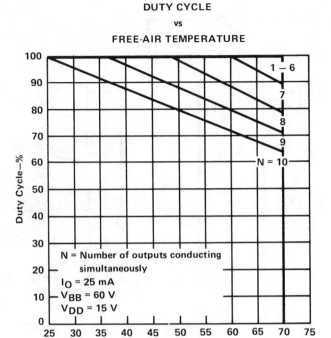

DUTY CYCLE
vs
FREE-AIR TEMPERATURE

N = Number of outputs conducting
simultaneously
I_O = 25 mA
V_{BB} = 60 V
V_{DD} = 15 V

Figure 8-58. UCN4810A Operational Duty Cycle

present in the serial shift register. The latched information is then transferred to the output through the gated output buffers when the blanking output is low. A logic high on the blanking input causes all outputs to go low. All outputs are capable of sourcing 40 mA at 60 V.

The SN75512A 12-Bit VFD Driver

Figure 8-59 shows a logic symbol, logic diagram and output circuit of the SN75512A. The SN75512A is a 12-bit VFD driver with totem-pole outputs. The totem-pole outputs minimize the required IDBT to less than 1 μs. Input data is stored in the 12-bit serial shift register on the positive transition of the clock input. Parallel data is presented to the output buffers through a 12-bit D-type latch. Data at the respective output of the serial shift register is transferred through the 12-bit latch while the latch enable is high. Data present at the latch inputs during the negative transition of the latch is stored regardless of subsequent changes, providing the latch input remains low. The active-low strobe input enables all output gates. Each output is capable of sourcing 25 mA at a supply voltage of 60 V, providing the maximum package power dissipation of 1150 mW is not exceeded. Based on the maximum allowable voltage drop across the output at 25 mA sink current, the total package capabilities are as shown in Table 8-4. All inputs of the SN75512A are TTL compatible. A serial data-out is also available for cascading additional drivers.

The SN75513A 12-Bit VFD Driver

Figure 8-60 shows the logic symbol, logic diagram and driver output circuit of the SN75513A. The SN75513A

is a 12-bit VFD driver with totem-pole outputs which minimize the required IDBT to less than 1 μs. Input data is shifted into a 12-bit serial shift register on the positive transition of the clock. Data appearing at the corresponding outputs of the shift register is presented directly to the output gates and is transferred to the Q outputs when the strobe input is low. Data in the shift register can be cleared with the reset input. A logic 0 on the reset input clears the shift register contents to a logic 0. Each output is capable of supplying 25 mA of source current at a V_{CC2} supply voltage of 60 V, providing the absolute maximum package power limitation of 1150 mW is not exceeded. Table 8-4 reflects the derating resulting from this consideration. All inputs are TTL compatible and assume a logic high if left open. A serial data output allows cascading of several devices without additional circuitry.

The SN75514 High Voltage 12-Bit VFD Driver

The SN75514 (Figure 8-61) is similar to the SN75512A except it is capable of operating at up to 125 instead of 60 V. Although not limited to high density applications, the SN75514 was designed to accommodate the specific requirements of large, high resolution (dot matrix) VF displays: high anode driver voltages, high-speed data reception and short inter-digit-blanking times (higher duty cycle).

As with the SN75512A, the 12-bit serial-in, parallel-out shift register is loaded on the positive edge of clock transition. However, the SN75514 combines the latch enable and output strobe functions of the 60 V device in one control line (strobe). Valid data is transferred from the shift register to the parallel latch outputs when the active-high strobe line is taken low. This is truly an edge-triggered data transfer as changes in the shift register contents while the strobe is held low will have no effect on the latch contents. With the strobe held low, high-voltage output lines go low (disabled state) and remain low until the strobe goes high. Upon raising the strobe line to a logic one, the latch contents are presented to the display through the high-voltage outputs (enabled state).

Another difference between SN75514 and the SN75512A lies in the high voltage supply requirements. Whereas a single supply of up to 70 V (V_{CC2}) was sufficient with the lower voltage device, the SN75514 requires two high-voltage supplies to deliver the maximum 25 mA per channel current drive capability. V_{CC2} (130 V, absolute maximum) provides the actual current to the display load through a high-voltage transistor in the output totem-pole structure. However the bias of that DMOS switch requires a slightly higher voltage (V_{CC2} + 10 V) to be applied to the V_{CC3} input. Often, these two voltages can be obtained from one supply and a few passive components. In addition, it is possible to operate the device with V_{CC2} and V_{CC3} in common, with the only penalty being reduced current source capabilities on the high voltage outputs.

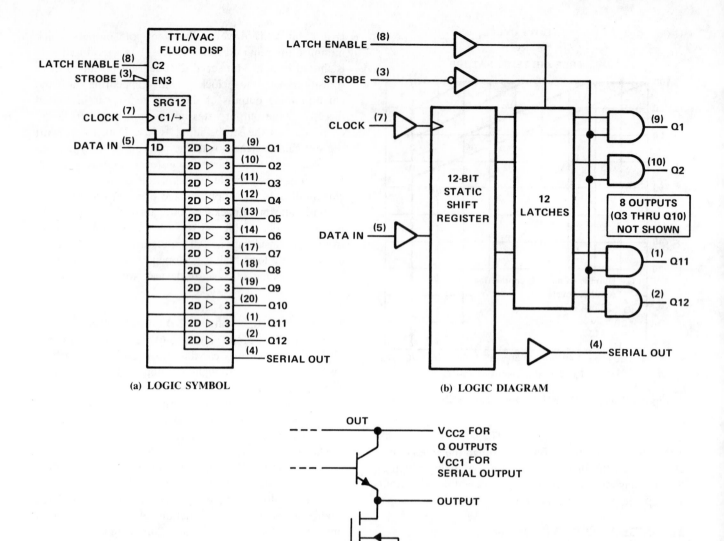

(a) LOGIC SYMBOL

(b) LOGIC DIAGRAM

(c) TYPICAL OF ALL OUTPUTS

Figure 8-59. SN75512A 12-Bit VFD Driver

The inputs are directly compatible with CMOS logic and can be interfaced with TTL with the addition of pull-up resistors.

The SN75518 32-Bit VFD Driver (Figure 8-62)

The most cost-effective driver for large displays with many anodes and grids is the SN75518. It is comprised of the same elements as the SN75512A, except it is 32 bits wide rather than only 12: a 32-bit serial-in, parallel-out shift register, 32 bits of parallel latches and 32 DMOS, totem-pole output stages capable of operation to 70 V (absolute maximum). The logical relationships of input clocking, latch enable and output strobe are also identical to the SN75512A. However, the SN75518 has been designed for direct CMOS input compatibility with TTL signals requiring the addition of resistive pull-ups. Power dissipation cannot exceed the 1650 mW limit of the 40-pin plastic package.

Table 8-4. SN75512A—SN75513A Operational Duty Cycle

Number of Outputs on I_O = 25 mA	Max. Allowable Duty Cycle at Ambient Temperature				
	25°C	40°C	50°C	60°C	70°C
12	77%	68%	62%	55%	48%
11	84	75	67	60	52
10	92	82	74	66	58
9	100	91	82	73	61
8		100	93	83	73
7			100	94	82
6				100	96
5					100
1	100	100	100	100	100

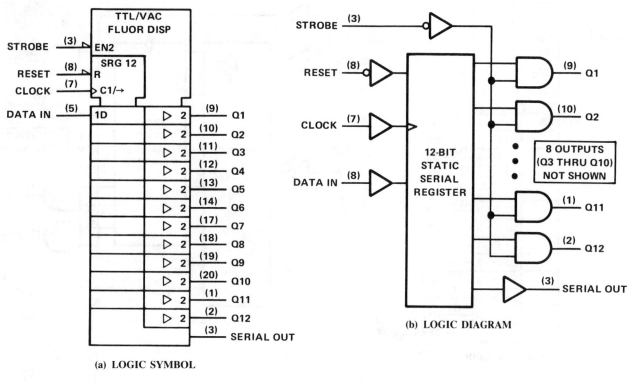

(a) LOGIC SYMBOL

(b) LOGIC DIAGRAM

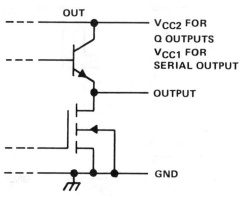

(c) TYPICAL OF ALL OUTPUTS

Figure 8-60. SN75513A 12-Bit VFD Driver

The SN75501C as a 32-Bit High-Voltage VFD Driver

Although designed originally for ac plasma applications, the SN75501C can be used to drive VFDs. The SN75501C is a 32-bit high-voltage display driver. A logic diagram of the SN75501C is shown in Figure 8-63(b). When used as a VFD driver, the strobe input is grounded and the sustain input is operated as an active high strobe input. The 32-bit serial shift register, capable of 4 MHz operation, registers the data on the positive edge of the clock. A logical "1" stored in the register will cause the respective output to pulse high when the sustain input is pulsed high. The outputs of the SN75501C are totem-pole outputs and a serial data output is provided for use in cascading multiple drivers. The use of the SN75501C as a VFD driver should be limited to applications where the IDBT (sustain high) inputs are 74 μs or greater.

VFD DRIVER APPLICATIONS

Driving a Vacuum Fluorescent Character Display

The following application uses a 5×7 40-digit VFD by Noritake. Each character is written in a single cycle since all 35 anodes (A1 through A35) of the 5×7 matrix are pinned out. The characters (1 through 40) are scanned by selective control of their respective grid, each of which is pinned out (G1 through G40). Respective anodes of all characters (A1 of Char 1 through Char 40) are connected. This format is common to most dot character or segment character displays (Figure 8-64). Multirow displays require additional control. This is usually provided through parallel access to the anodes of each additional row (Table 8-5).

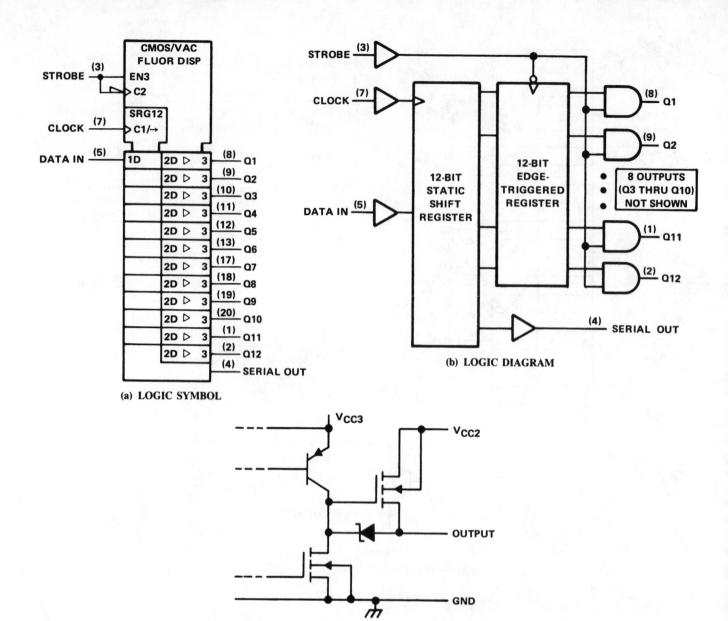

(a) LOGIC SYMBOL

(b) LOGIC DIAGRAM

(c) TYPICAL OF ALL Q OUTPUTS

Figure 8-61. SN75514 12-Bit VFD Driver

A typical driver scheme for a single line 40:5 × 7 dot character VFD is shown in Figure 8-65.

Whether or not the anode drivers require latched outputs depends on the circuit timing. Figure 8-66 shows a typical timing diagram for the display as shown in Figure 8-65. The drivers remain inactive for 183 μs (IDBT) prior to each character registration. This is more than sufficient time to load the 35 bits of data required for each character. With a 1 MHz data rate, the SN75513A requires only 35 μs to load this information. Modification of the timing to take advantage of the latch capability of the SN75512A for this particular application (1 line — 40:5 × 7 character VFD) will produce only minor improvement in display aesthetics. This is not the case for larger displays. Take for example, a six-line display of similar format (Table 8-5). A six-line display (DC40066A) requires control of 210 anodes (6 × 5 × 7). Thus, unless received in parallel format, this requires

210 μs loading time. With a latched driver however, this presents no problem as new data can be entered independent of the IDBT. Figure 8-67 illustrates a typical timing diagram incorporating this design.

The latch function virtually extends the time allotted for data registration in the anode drivers to the full character cycle time. For a 100 Hz panel refresh rate (T_r = 10 ms) and 1:150 minimum character duty cycle, the minimum character cycle time is 67 μs (T_p = T_rXDC = 10 ms/150). The larger the panel the more complex the anode/grid configuration, the more beneficial the latch feature (as that available with the SN75512A) becomes.

Driving a Dot Matrix Display

As discussed in a preceding section of Panel Performance, several variations of grid/anode configurations exist. The following will present the panel

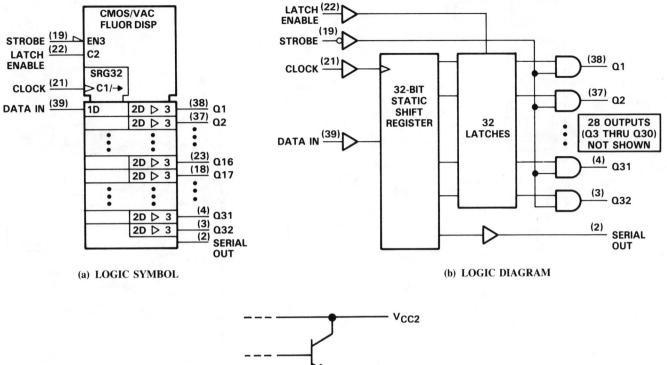

(a) LOGIC SYMBOL

(b) LOGIC DIAGRAM

(c) TYPICAL OF ALL Q OUTPUTS

Figure 8-62. SN75518 VFD Driver

requirements and suggested drive techniques for the displays shown in Figures 8-55(b), (c), and (d).

Figure 8-68 illustrates the VFD grid/anode arrangement for a DM256X64A. This is a 256 × 64 dot matrix VFD by Noritake whose grid/anode configuration is as illustrated in Figure 8-55. Figure 8-69 shows the required timing of the anode and grid signals to properly operate the DM256X64A. As can be seen in Figures 8-68 and 8-69, the active columns are composed of anodes which are between the activated grids. When grids 1 and 2 are activated, columns 2 and 3 are between them, and columns 1 and 4 are outside them. The purpose of this arrangement is to eliminate fringing effects of neighboring grids and thus achieve uniform intensity. Analysis of this configuration also shows requirements on panel drive electronics which are common to the previous examples. If the total panel refresh rate is held to 100 Hz, the total panel period (T_r) is 10 ms. With 128 write cycles required, each write cycle is 78 μs (T_c). Maintaining the 1:150 duty cycle, each strobe signal is 66 μs (T_p). This allows only 12 μs dead time or IDBT which dictates the use of an active pull-down driver. The time between the strobe signals of a particular anode group is 90 μs. This opens options in the VFD driver architecture. Each group of anode drivers requires 64 bits of data and two groups (A & D or B & C) must be loaded during each column write cycle (128 bits). If the SN75512A is used, its latch feature allows use of total strobe cycle period (156 μs), and a 1 MHz data rate allows the data to be received in a serial format. If the SN75513A is used, the A(B) group and D(C) group data must be loaded in parallel (64 μs), since it must be loaded during the dead time between strobe signals. Also available is the SN75501C. Since the strobe signal duty cycle is less than 50%, the SN75501C can also be used. With a 4 MHz data rate, all 128 bits of data can be registered serially in a 32 μs dead time between strobes.

Figures 8-70 through 8-73 illustrate the dot matrix pinout and timing diagrams for the anode/grid configurations shown in Figures 8-55(c) and 8-55(d). Table 8-6 identifies applicable anode and grid drivers and the number required for each of the configurations presented.

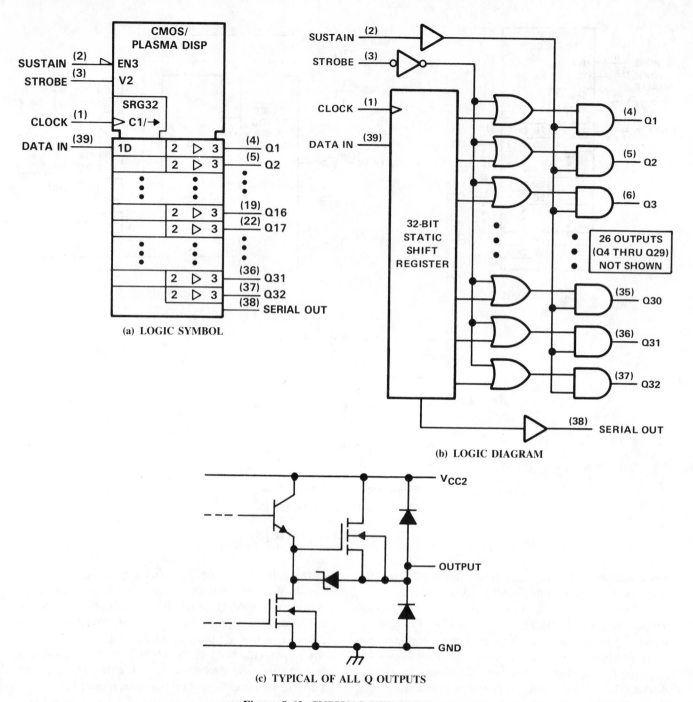

(a) LOGIC SYMBOL

(b) LOGIC DIAGRAM

(c) TYPICAL OF ALL Q OUTPUTS

Figure 8-63. SN75501C VFD Driver

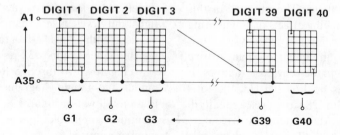

Figure 8-64. A 40: 5 × 7 Dot Character VFD Configuration

Table 8-5. Dot Character VFD Driver Requirements

| Control Pins | | | | | Required Drivers | |
Anode	Grid	Rows	Char	Matrix	10 Bit	12 Bit
35	10	1	10	5 × 7	5	4
	40		40	5 × 7	8	7
60	10		10	5 × 12	7	6
	40		40	5 × 12	10	9
70	10	2	10	5 × 7	8	7
	40		40	5 × 7	11	10
140	10	4	10	5 × 7	15	13
	40		40	5 × 7	18	16
210	10	6	10	5 × 7	22	19
	40		40	5 × 7	25	22

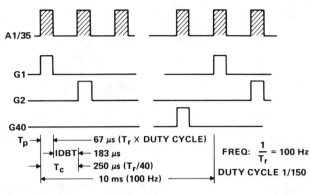

Figure 8-66. A 40 Character VFD Timing Diagram

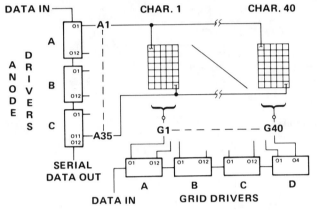

Figure 8-65. A 40: 5 × 7 Dot Character VFD Drive Scheme

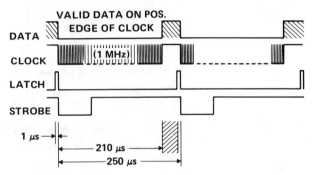

Figure 8-67. Data Registration of an SN75512 12-Bit VFD with Latch

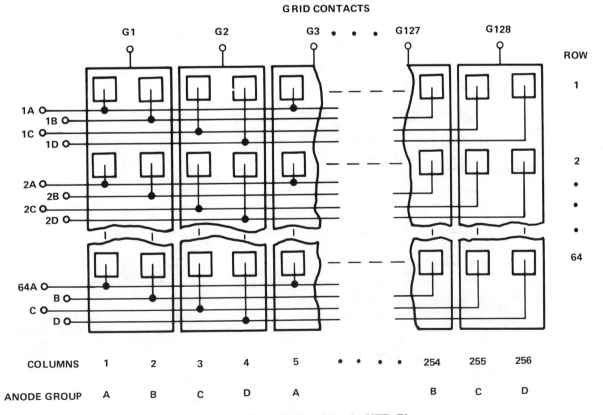

Figure 8-68. A 256 × 64 Dot Matrix VFD Pinout

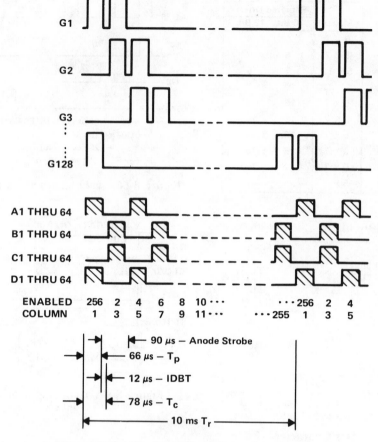

Figure 8-69. Timing Diagram for VFD of SN75512

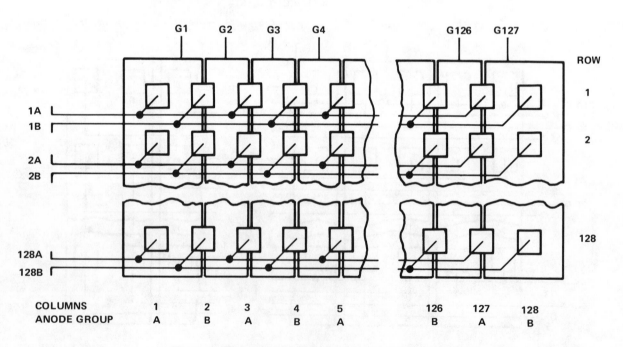

Figure 8-70. A 128 × 128 Dot Matrix VFD Pinout

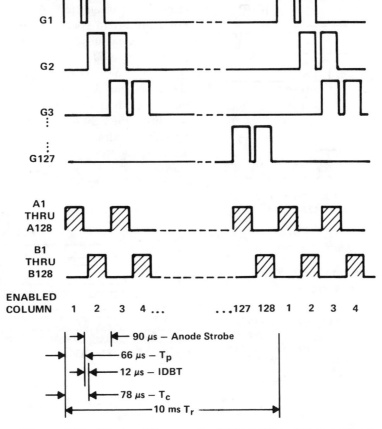

Figure 8-71. Timing Diagram for VFD 1128 × 128 Dot Matrix

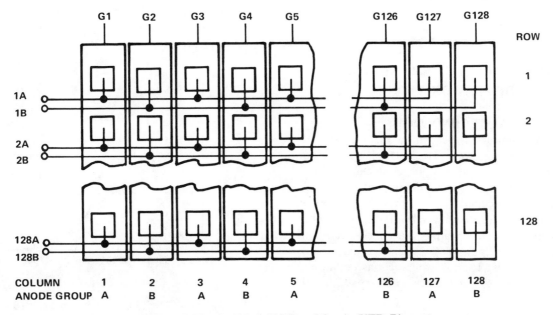

Figure 8-72. A 128 × 128 Dot Matrix VFD Pinout

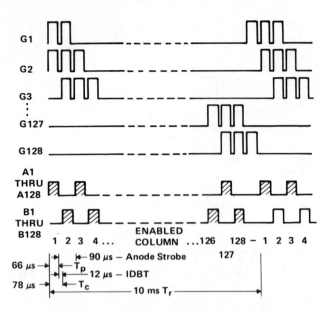

Figure 8-73. Timing Diagram for VFD of 128 × 128 Dot Matrix Pinout

Table 8-6. Dot Matrix VFD Driver Requirements

Display Size	Format Figure	Control Lines Anode	Grid	No. Drivers Required '4810	'512	'513	'518
128 × 64	8-68	256	64	26 + 7	22 + 6	24 + 6	8 + 2
	8-70	128	128	14 + 13	12 + 11	12 + 11	4 + 4
	8-72	128	128	14 + 13	12 + 11	12 + 11	4 + 4
128 × 128	8-68	512	64	52 + 7	44 + 6	NA 6	16 + 2
	8-70	256	128	26 + 13	22 + 11	NA 11	8 + 4
	8-72	256	128	26 + 13	22 + 11	NA 11	8 + 4
256 × 64	8-68	256	128	26 + 13	22 + 11	24 + 11	8 + 4
	8-70	128	256	14 + 26	12 + 22	12 + 22	4 + 8
	8-72	128	256	14 + 26	12 + 22	12 + 22	4 + 8

AC THIN FILM ELECTROLUMINESCENT DISPLAY DRIVERS

AC TFEL DISPLAY TECHNOLOGY

Due to technical advances in the past decade, AC Thin Film Electroluminescent (TFEL) Display technology has matured to the extent that the production of cost-effective reliable displays is a reality. TFEL displays' thinness, durability, compact screen size, and low power requirements are the primary advantages over other display technologies. X-Y matrix panels have been built with line resolutions up to 100 lines per inch and as large as 500 lines per axis. The following sections outline the technology, operation and drive circuitry of TFEL displays.

The AC TFEL display is a solid state device. It is available in several different formats, i.e., segmented character, dot matrix character, large X-Y dot matrix and custom graphic shapes. The basic mode of operation applies an alternating voltage across any two crossing electrodes. When this voltage exceeds the threshold

voltage, light is emitted from the luminescent layer. The threshold is considered to be the potential at which any visible light is emitted. A bright yellow light, with a relatively broad spectrum, is emitted. TFEL devices exhibit a view angle of up to 180 degrees and contrast ratios of up to 60:1.

The typical TFEL device is a sandwich structure as shown in Figure 8-74. All layers are transparent except for possibly the back electrodes. If a reflecting back electrode is utilized, the light output will be nearly doubled but the optimum contrast will be lost. A back cover which is sealed to the front glass at the edges of the display is not shown.

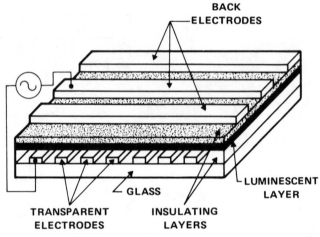

Figure 8-74. TFEL Construction

Mn doped ZnS is most often used for the luminescent layer. It is believed that the Mn centers emit light when excited by high energy electrons. The typical emission spectrum of a Mn doped ZnS TFEL display is shown in Figure 8-75.

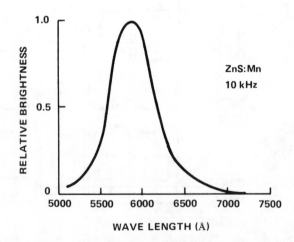

Figure 8-75. TFEL Emission Spectrum

Factors Affecting TFEL Display Brightness

The brightness of TFEL devices is very dependent on the polarity, pulse amplitude, frequency and pulse width of the drive voltages. The brightness of a TFEL display is very dependent on the polarity of drive voltages, as can be seen in Figure 8-76. Little light is emitted during the second of two emissions if the threshold voltage is exceeded twice in one polarity. If the next pulse is of opposite polarity however, the effective field is increased and electrons are accelerated in the opposite direction (faster than before) resulting in a greater light emission. The dependence of TFEL brightness on voltage and its characteristic hysteresis is shown in Figure 8-77(a).

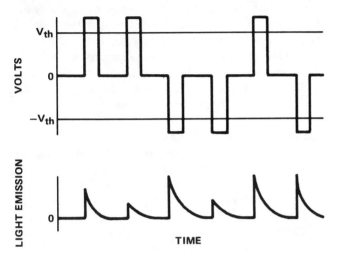

Figure 8-76. Brightness Dependence on Polarity

Simply stated, once the applied voltage pulse exceeds the threshold voltage, the polarity of the next pulse must be reversed to obtain a consistent brightness.

Most TFEL devices are built without the hysteresis feature. The brightness can be altered by operating along the B-V curve as shown in Figure 8-77(b).

Figure 8-77(c) shows the relationship of brightness to pulse width. Once the threshold voltage is exceeded, increasing the width of the excitation pulse causes a dramatic increase in brightness.

Figure 8-78 shows the B-V characteristics of a TFEL panel at different operating frequencies. At frequencies less than 500 Hz, brightness is linear with variations in excitation frequency.

Inasmuch as many applications typically employ a low-frequency refresh drive scheme (60 — 500 Hz), the panel brightness is usually directly proportional to the refresh rate.

AC TFEL Pixel Equivalent Circuit

Figure 8-79 is an approximate model for the TFEL pixel. The pixel is highly capacitive due to its physical construction of conductors separated by insulators. The capacitor shunted by a variable resistor represents the luminescent layer where the energy is converted to light.

DRIVERS FOR AC TFEL PANELS

With the pixels organized in X-Y dot matrix configurations most TFEL panels require row (or X axis) drivers and column (or Y axis) drivers. Slightly different control schemes are used for each type of driver resulting in the row and column drivers having somewhat different input logic characteristics.

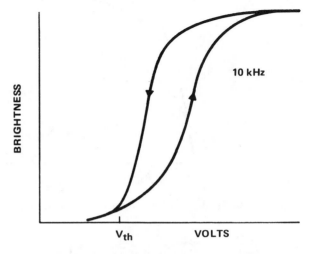

(a) BRIGHTNESS VOLTAGE HYSTERESIS

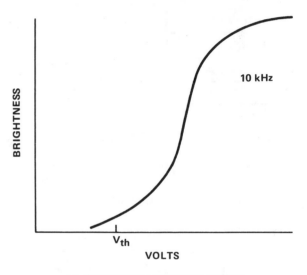

(b) BRIGHTNESS VOLTAGE CURVE

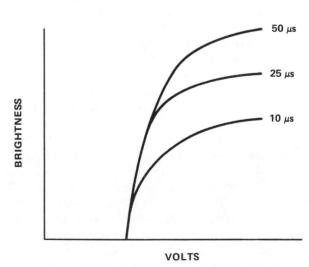

(c) BRIGHTNESS DEPENDENCE ON PULSE WIDTH

Figure 8-77. TFEL Device Characteristics

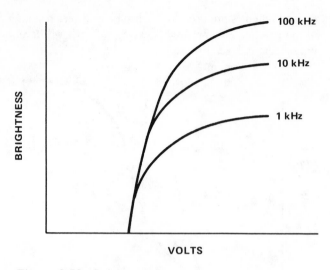

Figure 8-78. Brightness Dependence on Frequency

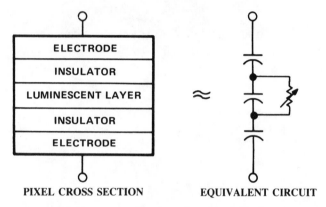

PIXEL CROSS SECTION   EQUIVALENT CIRCUIT

Figure 8-79. Pixel Equivalent Circuit

SN75551 and SN75552 Electroluminescent Row Drivers

The SN75551 and SN75552 (Figure 8-80) are monolithic BIDFET integrated circuits designed to drive the row electrodes of an electroluminescent display. All inputs are CMOS compatible and all outputs are high-voltage open-drain DMOS transistors. The SN75552 output sequence has been reversed from the SN75551 for ease in printed circuit board layout.

The devices consist of a 32-bit shift register, 32 AND gates, and 32 output OR gates. Typically, a composite row drive signal is externally generated by a high-voltage switching circuit and applied to the Substrate Common terminal. Serial data is entered into the shift register on the high-to-low transition of the clock input. A high Enable input allows those outputs with a high in their associated register to be turned on causing the corresponding row to be connected to the composite row drive signal. When the Strobe input is low, all output transistors are turned on. The Serial Data output from the shift register may be used to cascade additional devices. This output is not affected by the Enable or Strobe inputs.

SN75553 and SN75554 Electroluminescent Column Drivers

The SN75553 and SN75554 (Figure 8-81) are monolithic BIDFET integrated circuits designed to drive the column electrodes of an electroluminescent display. The SN75554 output sequence has been reversed from the SN75553 for ease in printed circuit board layout.

The devices consist of a 32-bit shift register, 32 latches, and 32 output AND gates. Serial data is entered into the shift register on the low-to-high transition of the clock input. When high, the Latch Enable input transfers the shift register contents to the outputs of the 32 latches. When Output Enable is high, all Q outputs are enabled. Serial data output from the shift register may be used to cascade shift registers. This output is not affected by the Latch Enable or Output Enable inputs.

DRIVING AC TFEL PANELS

The refresh approach is presently the most popular drive scheme for X-Y matrix panels. Due to the large capacitance of electroluminescent (EL) devices and the energy lost in charging and discharging the capacitive pixel elements, drive frequency has a large effect on power requirements. The refresh approach operates a panel typically at a frequency of 60 to 500 Hz. Most drive schemes operate the EL cells in binary fashion, either off or on. Brightness control or "Grey Scaling" is not incorporated into most simple refresh approaches.

Practical Refresh Drive Scheme

Figure 8-82 shows a block diagram of the drive scheme for a typical EL panel and Figure 8-83 shows the associated waveforms. The Texas Instruments SN75551 and SN75552 EL drivers are used to drive the rows and the SN75553 and SN75554 EL drivers are used to drive the columns. The column drivers are referenced to ground while the substrate common pin of the row drivers is connected to the composite row drive signal and therefore all row signals are relative to the composite signal.

Theory of Operation

Suppose no light is emitted from a display element when driven by a 200 V pulse followed by a −140 V pulse, but light is emitted when the initial 200 V pulse is followed by a −200 V pulse. Then selective operation can be achieved by applying a pulse train (200 V, −140 V) to the selected row, and coincident to the negative pulse, applying −60 V to the selected columns and 0 V to the nonselected columns. Light will be emitted at the intersection of the selected row and columns. This is due to the selected column potential adding to the selected row potential to effectively create the 200 V, −200 V pulse train. Each row sees a positive pulse (200 V refresh pulse) at the beginning of each scan. The delay between this initial positive pulse and the following negative pulse has little effect upon the brightness.

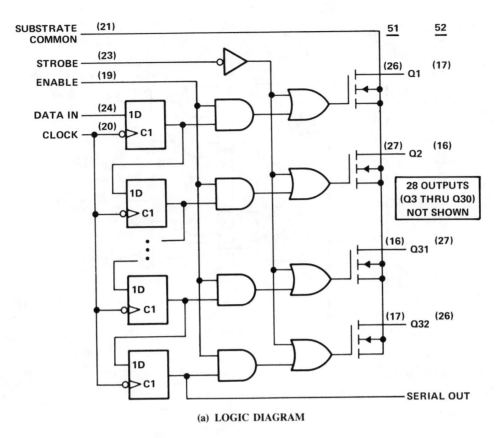

(a) LOGIC DIAGRAM

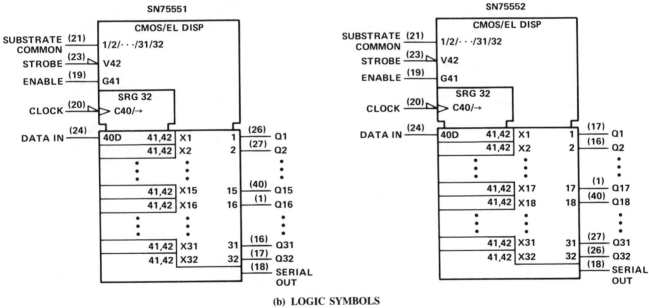

(b) LOGIC SYMBOLS

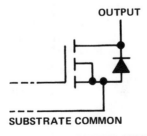

(c) TYPICAL EACH Q OUTPUT STRUCTURE

Figure 8-80. SN75551 and SN75552 Electroluminescent Row Drivers

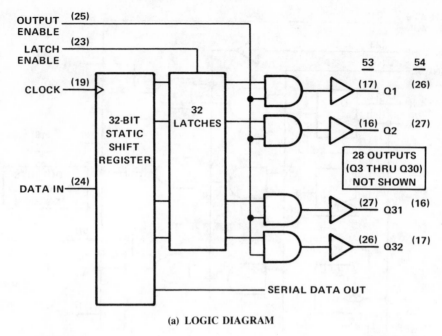

(a) LOGIC DIAGRAM

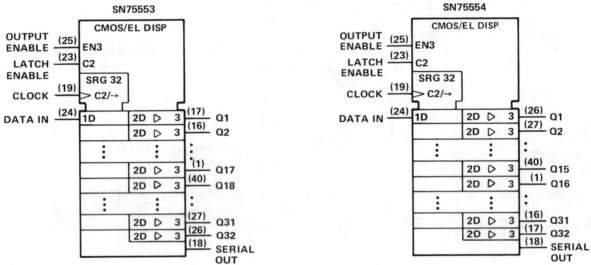

(b) LOGIC SYMBOLS

(c) TYPICAL OF ALL Q OUTPUTS

Figure 8-81. SN75553 and SN75554 Electroluminescent Column Drivers

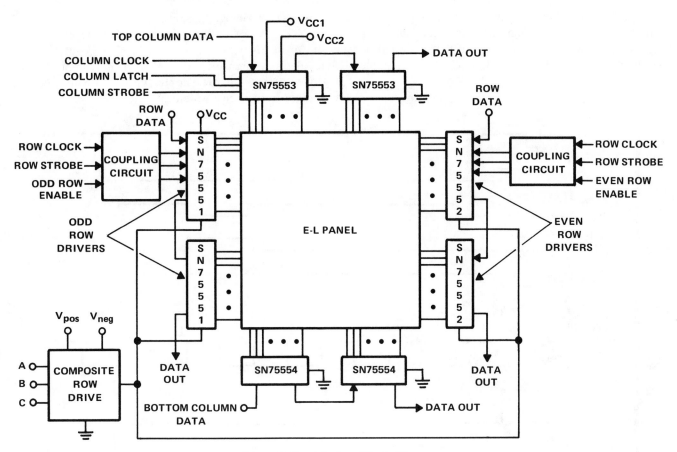

Figure 8-82. Display Block Diagram

Interconnecting the Drivers to the Panel

On most X-Y matrix panels, row electrodes are brought out alternately on opposite sides of the display. The SN75551 and SN75552 devices are identical except that the high-voltage outputs are pinned out clockwise on one and counterclockwise on the other to aid interconnections along the edges of the display. In much the same way, column electrodes are brought out alternately on the top and bottom of the display with the SN75553 and SN75554 column drivers varying only in opposite high voltage output pinouts.

Row Driver Operation

At the beginning of each scan a refresh pulse is applied to all rows. The positive pulse is applied to all rows due to the output structure of the row drivers [Figure 8-80(c)]. When the composite row drive signal goes positive, the clamp diode of each row driver output is forward biased and pulls all of the rows up to follow the composite row drive. Next, the row drivers are strobed, turning on each output's DMOS FET. As the composite row drive returns to ground, the diodes are reverse biased and the rows are pulled to ground with the capacitor discharge current flowing through the DMOS transistors. After a delay to allow the rows to return to ground potential, strobe is removed leaving the rows floating. Each row will continue to float until it is selected. When a

row is selected, the associated DMOS FET is turned on to allow the selected row to follow the composite row drive negative transition. Successive rows are selected by clocking one bit of data into the register of the first row driver on each side of the display. (See Figure 8-82) The odd side of the drivers is enabled first, followed by the even side. Then the row drivers are clocked one time and the next even/odd pair is selected.

Column Driver Operation

Before a row is selected, all the data for that line must be clocked into the column drivers and latched. The data for the next lines can be clocked into the column registers as soon as the previous data is latched into the output latches. (See Figure 8-83). The column drivers must be enabled during the time the selected row composite signal goes negative.

Row and Column Driver Requirements

The data signals to the row drivers must be coupled through proper isolation circuitry since the devices are referenced to the composite row drive waveform. An optical isolator can be used as shown in Figure 8-84. The row strobe, row clock and row enable would likely be coupled using an optical isolator. The opposing odd and even row enable signal could be implemented as shown in Figure 8-85. The row data could be generated as shown in Figure 8-86.

The column drivers are referenced to ground. Therefore no special coupling is required for column data signals. It should be noted, however, that even and odd column data must be clocked, latched, and enabled simultaneously.

Row Driver Voltage Supply

The V_{CC} for the row drivers can be created, as shown in Figure 8-87, by using a 12 V regulator. The regulator reference is connected to the composite row drive so that the row supply, V_{CC}, is floating on the composite row drive waveform.

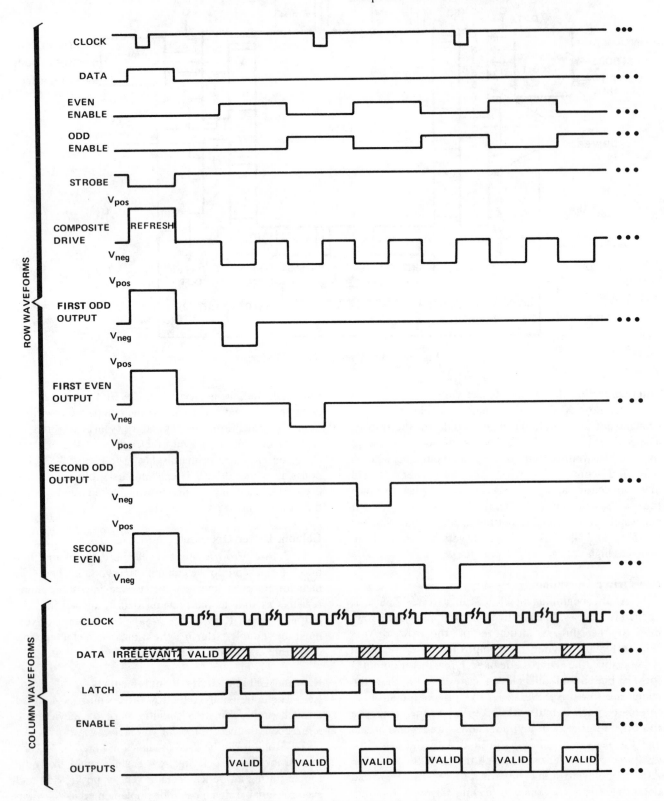

Figure 8-83. Refresh Operation Waveforms

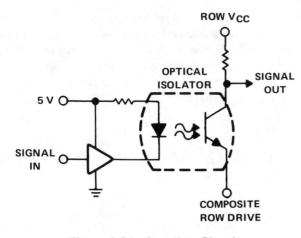

Figure 8-84. Coupling Circuit

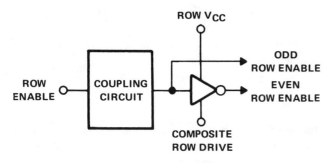

Figure 8-85. Odd and Even Enable Circuits

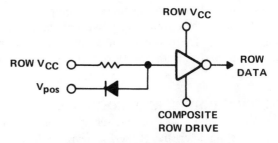

Figure 8-86. Row Data Circuit

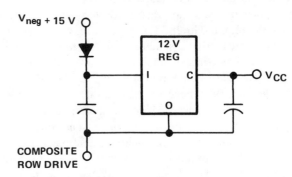

Figure 8-87. Row V$_{CC}$

COMPOSITE ROW DRIVER

The composite row drive can be generated using four high voltage FETs as shown in Figure 8-88. Q3 pulls the output up to V$_{pos}$ and Q4 pulls the output down to

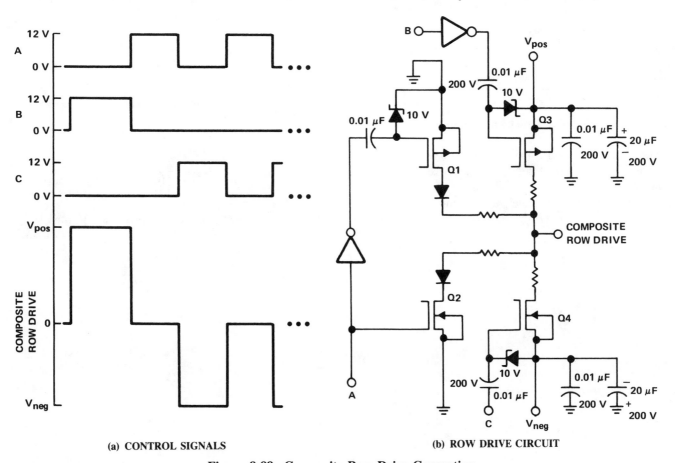

(a) CONTROL SIGNALS   (b) ROW DRIVE CIRCUIT

Figure 8-88. Composite Row Drive Generation

V_{neg}. Q1 and Q2 are used to bring the output back to ground: Q1 when the output is negative and Q2 when the output is positive. The entire circuit is controlled by using inputs A, B, and C with appropriate non-overlapping control signals.

SPECIFYING DRIVER REQUIREMENTS

When designing an EL drive scheme, it is essential to model the characteristics of the EL device, taking into consideration the drive scheme. The EL designer needs to pay close attention to the current levels required to drive his display and make sure that the drivers can sink and/or source the required levels. The drive current requirements for the refresh drive scheme can be determined by examining worst-case voltage excursions and estimated display capacitance. As reported in *Proceedings of the SID*, vol 23/2, 1982, page 87, the equivalent circuit shown in Figure 8-89 can be used in the analysis of the drive requirements. For purposes of discussion, some simplifying assumptions are made. First, each display element (pixel) is purely capacitive with $C = C_e$, C_e being the same for all pixels. Next, all points that are at the same potential are treated as if they were connected together.

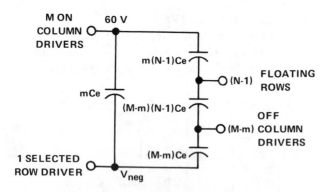

Figure 8-89. Panel Equivalent Circuit

The equivalent circuit shown in Figure 8-89 is for a panel having N rows and M columns. This circuit ignores the negligible capacitance between adjacent rows and between adjacent columns. The panel is driven one row at a time by selecting one of the N rows and m of the M elements in that row. If one row and m columns are selected, the m elements are turned on and the m columns are capacitively coupled to the N minus 1 floating rows. Each of these N minus 1 floating rows is coupled to M minus m off columns which are again coupled to the one selected row. Consider the following example:

Row switch voltage = 200 V
Pull down time = 10 μs
Number of rows = 240
Number of columns = 320
C_e = 4 pF

The worst-case capacitance is when all columns are on. This reduces the equivalent circuit to:

$$MC_e = 320 \times 4 \text{ pF} = 1280 \text{ pF}.$$

Therefore, to satisfy this requirement, the row driver must be able to sink at least:

$$i = C\frac{dv}{dt} = (1280 \text{ pF} \times 200 \text{ V})/10 \text{ μs} = 25.6 \text{ mA}.$$

During the refresh, the row driver would be required to source current through the clamp diode. Since all the N rows are pulled up, the equivalent circuit is as shown in Figure 8-90. The worst case is when all the rows are at zero volts and all are switched to V_{pos}. Suppose V_{pos} is 200 V and a 10 μs rise time is required, then the clamp diode must be able to deliver at least:

$$i = C\frac{dv}{dt} = (4 \text{ pF} \times 320 \times 200)/10 \text{ μs} = 25.6 \text{ mA}.$$

The current requirements for the column drivers can also be calculated as above.

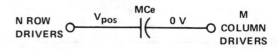

Figure 8-90. Refresh Equivalent Circuit

Section 9

Data Transmission

GENERAL PURPOSE DATA TRANSMISSION

The ever increasing use of computers, microprocessors, and elaborate logic control systems constantly present designers with the problem of transmitting digital data from one system to another. Digital communications between components of systems ranging from machine controls in a factory to consumer appliances and automotive systems have made data line drivers and receivers as prevalent as power supplies.

Many systems use a central computer or controller with several remote terminals or control points. Remote stations may be a few feet or thousands of feet from the central computer. Even small independent systems will usually have one or more pieces of peripheral hardware such as a printer located at some, usually short, distance away. The need for dedicated data transmission products became more apparent as the industry matured. At first, simple single-ended data transmission was used; products were basic and often tailored for each specific system. Then widespread data transmission required system compatibility, and industry standards were adopted. Before addressing the standards and their applications, we will look at the basic line circuit requirements and general purpose data line drivers and receivers.

GENERAL REQUIREMENTS

Basic requirements for a good data transmission system include:

Speed: Data capabilities of up to 10 MHz are desirable for most systems. Frequency of operation is often expressed as data frequency (bits per second). Each cycle has both logic levels, so bps is twice the frequency in hertz. Data transmission speeds in this text are generally expressed in bps.

Power Supplies: Single 5 V supplies are the most popular although dual ± 5 V, ± 9 V and ± 12 V are sometimes required.

Logic Compatibility: TTL compatibility is understood although many devices are also compatible with low-power/low-level CMOS.

Drivers: Capable of driving low-impedance transmission lines. Able to withstand line voltages up to the V_{CC} supply levels.

Receivers: Good input sensitivity, typically less than 500 mV. Provide immunity to noise either with common-mode rejection capabilities or threshold hysteresis.

Lines: Uniform impedance characteristics over the length of the line allowing proper termination for high speed transmission without error-generating reflections.

TYPES OF TRANSMISSION LINES

Transmission lines are often the key to successful data communications. Popular line types include coax cable, twisted pairs, shielded twisted pairs, flat or ribbon cable and discrete wires for short line applications.

Single Wire and Ground Plane

A single wire connection between a driver and receiver (see Figure 9-1) may be satisfactory in some simple system applications. For example, the output of one gate is driving the input of another located on the same PC board or a board nearby. Distances between drivers and receivers must be short and the system relatively free from switching transients or similar electrical noise. Distance limitations, dependent on system noise levels and operating speeds, are approximately three to six inches for most applications.

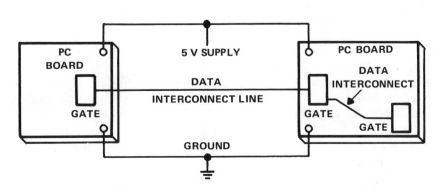

Figure 9-1. Single Wire and Ground Plane

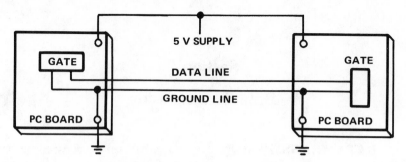

Figure 9-2. Double Wire

Two Wire Interconnect

Nontwisted-dual-wire interconnects (see Figure 9-2) are seldom used. The second wire serves as ground and application limitations are similar to those for single wires. A variation of the two wire interconnect is found in ribbon cable (Figure 9-3). Two adjacent wires are used, one as the signal lead and the other as ground.

Because of the uniform characteristics of ribbon cable, grounding every other wire results in a uniform impedance for all signal wires. Proper termination will then allow relatively high data rates without ringing or bad reflections.

Twisted Pair

When two wires are uniformly twisted, the result is a definite impedance characteristic that is uniform along the length of the line (Figure 9-4).

System noise is reduced by the mutual coupling of the twisted wires. It should be noted that grounding both ends of the ground wire may result in little magnetic field protection due to ground loop noise currents. Grounding the high impedance (receiver) end only, as shown, will result in good noise reduction. Twisted pair lines are also used in balanced transmission applications where both wires serve as signal leads (Figure 9-5).

As data rates approach 100 kHz or greater and line lengths are longer, over 50 ft, twisted pair lines have proven to be advantageous over the other types. Following are a few advantages of twisted pair lines:

1. Cancellation of noise due to mutual coupling of adjacent lines in the twisted pair.
2. Both wires in the pair are equally affected by electromagnetically and electrostatically coupled noise resulting in a net common-mode noise voltage with respect to ground. This is easily rejected by balanced differential receiver inputs.
3. Voltage differences between driver/receiver locations will appear as common-mode signals and be rejected by the receiver.
4. Uniform impedance characteristics along the line make it easy to terminate.
5. Twisted pair lines are low cost with long life, and are mechanically very rugged.

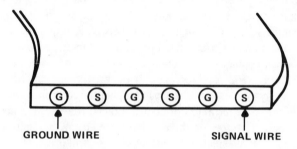

Figure 9-3. Ribbon Cable

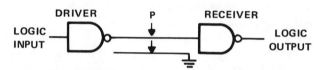

Figure 9-4. Twisted Pair Lines

The voltage across an impedance terminating a transmission line is a function of the real and imaginary components of the impedance, the characteristic impedance of the line, and the incident power. When the line terminating impedance is purely resistive (in practice, the reactive components of impedance can usually be neglected) the transmission line may be described by the following equations (see Figure 9-6):

$$P_R = P_I \left(\frac{R_L - Z_O}{R_L + Z_O} \right)^2 \qquad (1)$$

$$P_L = P_I - P_R = P_I \left[1 - \left(\frac{R_L - Z_O}{R_L + Z_O} \right)^2 \right] \quad (2)$$

$$V_L = \sqrt{P_L R_L} = \sqrt{I_L^2 R_L^2} \qquad (3)$$

Where:

P_I = Incident Power
P_R = Reflected Power
P_L = Power Delivered to R_L
R_L = Load Resistance
Z_O = Line Characteristic Impedance

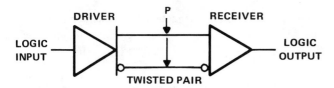

Figure 9-5. Balanced (Differential) Driven Twisted Pair

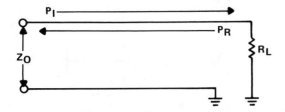

Figure 9-6. Line Incident (P_I) and Reflected (P_R) Power

When $R_L = Z_O$, the numerators of the fractional terms in equations 1 and 2 become zero and the reflected power is zero. With reflections reduced to zero, a major source of signal distortion and noise is eliminated.

In line circuits design R_L is actually a lumped value representing the combination of termination resistor (R_T) and the input resistance of a line receiver, R_{IN}, (Note Figure 9-7). When $R_{IN} >> R_T$, large induced power transients are shunted to ground by R_T, decreasing the maximum power absorbed by the input of the receiver.

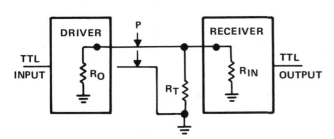

Figure 9-7. Basic, Single Ended, Line Terminations

Figure 9-7 illustrates the basic resistive termination of a single-ended data transmission line. The values of R_O and R_T have a definite effect on circuit performance as is seen in Figure 9-8.

The line used is a 100 Ω twisted pair line. Case A is correctly terminated and results in best performance. Irregularities in other case examples are a result of reflections on the data line due to mismatched termination values.

Figure 9-9 shows the results of a Case C operation at a line length of 50 ft and data rate of 5.4 Mbps. Although the bit rate achievable with this type of mismatched termination is high, reflections result in ringing on the line and false triggering of the receiver.

Noise sources and interference that result in lost or improper data are produced in many ways. Induced pickup on the line, as seen by the receiver, may result in distortion

and general interference. Balanced or differential line techniques greatly reduce the effects of common-mode noise but proper line terminations are critical to good operation.

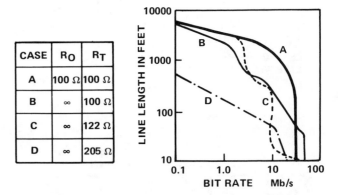

CASE	R_O	R_T
A	100 Ω	100 Ω
B	∞	100 Ω
C	∞	122 Ω
D	∞	205 Ω

Figure 9-8. Effects of Line Terminations

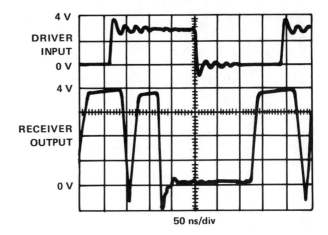

Figure 9-9. Case C Voltage Waveform

Noise induced into the line from electromagnetic or electrostatic sources is an energy source that will generate interference voltage amplitudes proportional to the line to ground impedance. In short lines with low interference energy levels, the typical termination for balanced lines (Figure 9-10) is directly from Line A to Line B. Line termination value (R_T) should equal the characteristic impedance of the line (Z_O) assuming a receiver input resistance (R_{IN}) that is much larger than Z_O. As can be seen from the equivalent circuit in Figure 9-10, the resistance path from each line to ground is very high. With relatively high line to ground impedance the resulting induced noise voltage levels will be high.

The termination technique shown in Figure 9-11 splits the line termination by placing a resistor from each line to ground with a value of $1/2\ Z_O$. Now the resistance to ground is low, approximately equal to R_{T1} or R_{T2}, and the resulting common-mode voltage is much lower than that generated in the circuit shown in 9-10. A typical twisted pair line has a characteristic impedance of about 120 Ω and typically $R_{T1} = R_{T2} = 62$ Ω.

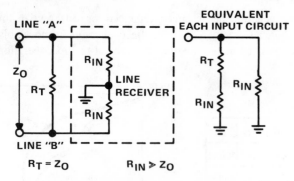

Figure 9-10. Line-to-Line Termination

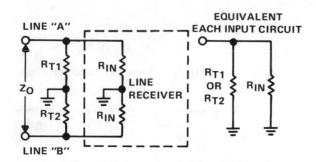

Figure 9-11. Termination to Ground

Noise induced into the balanced twisted pair lines may result in significant levels of noise voltage, but it is common to both lines. This type of noise voltage is referred to as common-mode noise having the same amplitude, at any instant, on both lines. Differential line drivers and receivers will reject common-mode voltages. Further discussion of this will occur later in the text. Twisted pairs are available in ribbon cable (Figure 9-3) as well as discrete pairs and shielded pairs. Shielding yields better noise immunity, if grounded properly, but results in much higher distributed capacitance which significantly attenuates the transmitted signal at high frequencies. If a shielded twisted pair is used (Figure 9-12) grounding the shield at the receiver end only will provide the least signal attenuation and the best rejection of unwanted signals.

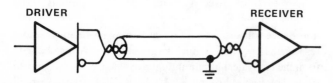

Figure 9-12. Shielded Twisted Pair Transmission

Coaxial Lines

Popular 50 to 200 Ω coaxial cables [Figure 9-13(a)] have a very uniform and definite impedance and offer the best transmission line characteristics for single-ended applications. Primary advantages are low loss and good shielding against magnetically induced interference. As with the shielded twisted pair lines, the coax line's shield should be grounded at the receiver end only. If the shield is grounded

at both ends, very little magnetic field protection is possible due to mutual coupling of noise currents in the shield [Figure 9-13(b)].

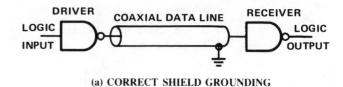

(a) CORRECT SHIELD GROUNDING

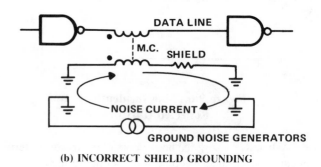

(b) INCORRECT SHIELD GROUNDING

Figure 9-13. Coaxial Data Lines

LINE DRIVERS

Probably the most common single-ended application is the transmission of data from one logic gate directly to another. Standard TTL gates can operate at frequencies up to 20 MHz. However, interconnects must be short (a few inches), and special care taken to assure adequate noise margin and minimum line reflections. The higher speeds of ECL and Schottky TTL gates place even more emphasis on properly terminated, well-shielded lines.

Driving directly from a TTL type gate has several limiting factors. Output currents are too small for long lines terminated in their characteristic impedance.

Another factor to consider in the use of gates is the environmental noise level. It can be seen in Figure 9-14 that negative-going noise peaks greater than 0.4 V and positive-going noise peaks greater than 0.4 V may result in false triggering. The guaranteed noise margin is therefore ±0.4 V when using standard TTL gates.

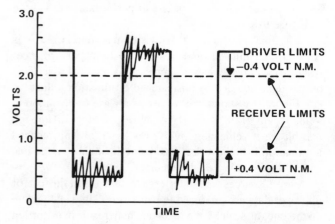

Figure 9-14. TTL Gate Noise Margins

Basic Driver Modes

Data line drivers operate as voltage-mode or current-mode devices. Voltage-mode drivers are active switches used to transfer a voltage from a supply line to the transmission line directly (Figure 9-15) or to switch the transmission line to ground, (Figure 9-16) removing voltage that has been supplied through a termination resistor. In each case the primary drive current required is determined by the termination resistance and voltage drive levels required. Drive current requirements typically range from 25 mA to as high as 100 mA, depending on the application. The three basic types of voltage-mode driver outputs are active pull-up (Figure 9-15), active pull-down (Figure 9-16), and a combination referred to as "totem-pole" (Figure 9-17).

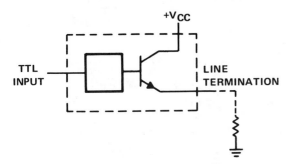

Figure 9-15. Active Pull-Up Configuration

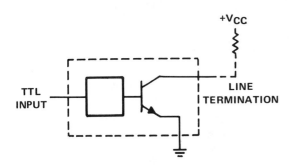

Figure 9-16. Active Pull-Down Configuration

Current-mode drivers are also active switches but they switch a constant current onto the data line. Most current-mode drivers are current sinks. An open-collector transistor (Figure 9-18) is used to switch the data transmission line to a current generator operating from a negative supply rail. Thus the line voltage generated will be negative and is a product of the output current and the line termination resistance. Some advantages of this mode of operation are:

- Output may be shorted without damage to the circuit.

- Line RFI radiation levels are low due to the fact that there are no large surge currents.

- Low impedance data lines can be driven without additional supply power.

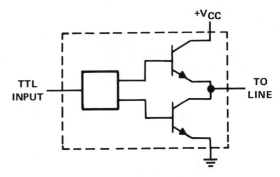

Figure 9-17. Totem-Pole Configuration

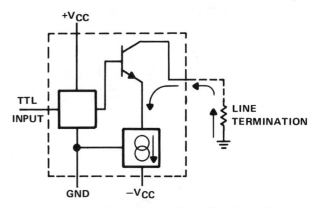

Figure 9-18. Current-Mode Configuration

TYPES OF TRANSMISSION

Whether the driver is voltage-mode or current-mode, it may drive single-ended data transmission lines or dual differential (balanced) transmission lines. Single-wire, dual-wire, twisted pair and coax lines are used for single-ended applications. Discrete single-wire or dual-wire lines are seldom employed, although they can be used where data rates are low and line lengths are short. Normally coax lines or more economical twisted pair lines are used in single-ended transmission. For differential data transmission, a twisted pair line is normally used.

Single-Ended Transmission

Numerous integrated circuit devices are available for driving single-ended data transmission lines. Some are general purpose and others have been designed to meet specific industrial standards. Tables 9-1, 9-2, and 9-3 list drivers, receivers and transceivers designed for single-ended transmission.

Advantages and disadvantages of single-ended drivers:

Advantages
Simplicity: easy to understand and minimum connections.
Low cost.
Usually only one power supply required.

Disadvantages
Radiates RFI easily.
Poor noise immunity.
Coax improves noise but is expensive.
Limited line lengths and data rates due to susceptibility to interference signals.

Table 9-1. Single-Ended Drivers

TYPE NUMBER	CIRCUITS/PACKAGE	DESCRIPTION	APPLICATION
SN55/75121	2	Emitter-follower outputs to 75 mA	General purpose
SN75361A	2	Totem-pole outputs to 100 mA	General purpose
SN75451B	2	Open-collector to 300 mA	General purpose
DS78/8831	4	3-state diode clamped outputs	General purpose
DS78/8832	4	3-state outputs to 40 mA	General purpose
SN75150	2	Totem-pole dual polarity	RS-232-C
SN75156	2	Totem-pole dual polarity	RS-232-C/423-A
uA9636	2	Totem-pole dual polarity	RS-232-C/423-A
SN75186	4	Totem-pole dual polarity	RS-232-C
SN75187	4	Totem-pole dual polarity	RS-423-A
SN75123	2	Emitter-follower output to 60 mA	IBM 360
SN75126	4	Emitter-follower output to 60 mA	IBM 360/370
SN75130	4	Emitter-follower to 60 mA with fault flag	IBM 360/370

Table 9-2. Single-Ended Receivers

TYPE NUMBER	CIRCUITS/PACKAGE	DESCRIPTION	APPLICATION
SN55/75122	3	Input hysteresis	General purpose
SN55/75140	2	100 mV sensitivity, common reference	General purpose
SN55/75141	2	100 mV sensitivity, common reference, diode protected	General purpose
SN55/75142	2	100 mV sensitivity, internal reference	General purpose
SN55/75143	2	100 mV sensitivity, internal reference, diode protected	General purpose
SN75152	2	Adjustable hysteresis	General purpose MIL-STD 188 & RS-232-C
SN75154	4	Adjustable hysteresis	RS-232-C
SN75189	4	Response control	RS-232-C
SN75189A	4	Wide hysteresis, response control	RS-232-C
SN75157	2	Fixed hysteresis	RS-232-C/423-A
uA9637AC	2	Fixed hysteresis	RS-232-C/423-A
AM26LS32AC	4	± 7 V CMR with ± 200 mV sensitivity	RS-422-A/423-A
AM26LS33AC	4	± 15 V CMR with ± 500 mV sensitivity	RS-422-A/423-A
MC3486	4	± 200 mV sensitivity ± 7 V CMIV	RS-422-A/423-A
SN75173	4	± 200 mV sensitivity 12 kΩ min R_{IN}	RS-422-A/423-A
SN75175	4	± 200 mV sensitivity 12 kΩ min R_{IN}	RS-422-A/423-A
SN75124	3	Hysteresis, independent strobes	IBM 360
SN75125	7	7 kΩ min input resistance	IBM 360/370
SN75127	7	7 kΩ min input resistance	IBM 360/370
SN75128	8	Strobe active high	IBM 360/370
SN75129	8	Strobe active low	IBM 360/370

Table 9-3. Single-Ended Transceivers

TYPE NUMBER	CIRCUITS/PACKAGE	DESCRIPTION	APPLICATION
AM26S10M/C	4	High speed, 100 mA, inverting driver	General purpose
AM26S11M/C	4	High speed, 100 mA, noninverting driver	General purpose
N8T26	4	3-state outputs, 40 mA I_{OL}	General purpose
N8T26A	4	3-state outputs, 48 mA I_{OL}	General purpose
SN75136	4	3-state outputs, 40 mA I_{OL}	General purpose
SN55/75138	4	Open-collector drive to 100 mA	General purpose
SN75163A	4	High speed, receiver hysteresis, 48 mA driver	General purpose
SN75160A	8	High speed, receiver hysteresis, 48 mA driver	IEEE 488
SN75161A	8	For single controller MGMT transceiver	IEEE 488
SN75162A	8	For multiple controller MGMT transceiver	IEEE 488
MC3486	4	3-state outputs, receiver hysteresis, 48 mA driver	IEEE 488

The output voltage from a single-ended driver, although typically at TTL compatible levels, may produce high-level output voltages to help override high noise conditions. An application example (Figure 9-19) uses the SN75361A driver. The SN75361A is a TTL to MOS driver that can provide up to 23 V output swing. Figure 9-19 illustrates an application to provide up to 11 V, V_{OH}, and 0.5 V, V_{OL}, for a resulting output peak-to-peak swing of 10.5 V.

A good receiver to use in this application would be one with 12 V or greater input capability and with adjustable hysteresis to take maximum advantage of noise rejection capabilities. The dual channel SN75152 is such a device.

Features of the SN75152 include:

± 25 V common-mode input voltage range.
Adjustable hysteresis.
Adjustable threshold.
Standard TTL output.

The receiver circuit is illustrated in Figure 9-20 with a receiver input signal of 0 to 10.5 V. A threshold center of about 5.0 V may be set using one external resistor from the 12 V rail to the inverting input of the receiver (point D). Since the input resistance is about 9 kΩ and we want a 5.0 V

reference at that location, the reference pull-up resistor R_r may be calculated as follows:

$$R_r = \left(\frac{V_{CC+}}{V_r} - 1 \right) r_{in}$$

Where:

$$\begin{aligned} R_r &= \text{the reference resistor} \\ V_{CC+} &= 12 \text{ V} \\ V_r &= 5.0 \text{ V} \\ r_{in} &= 9 \text{ k}\Omega \\ R_r &= \left(\frac{12}{5.0} - 1 \right) 9 \text{ k}\Omega, \ R_r = 12.6 \text{ k}\Omega \end{aligned}$$

Standard values of 12.0 kΩ and 620 Ω in series may be used.

With 0 V, V_{IL}, and 10.5 V, V_{IH}, levels at point C and a 5 V reference at point D, the threshold hysteresis is selected to be ± 3.5 V. A 620 Ω resistor from the hysteresis adjust pin to the -12 V rail will give a ± 3.5 V hysteresis. The resulting high level, positive going, threshold (V_{TH}) will be 5 V + 3.5 V, or 8.5 V. The noise margin, when in the low input level state will be about $V_{TH} - V_{IL} = 8.5$ V $- 0$ V $= 8.5$ V. The noise margin when in the high input level state will be $V_{IH} - V_{TL} = 10.5$ V $- 1.5$ V $= 9$ V.

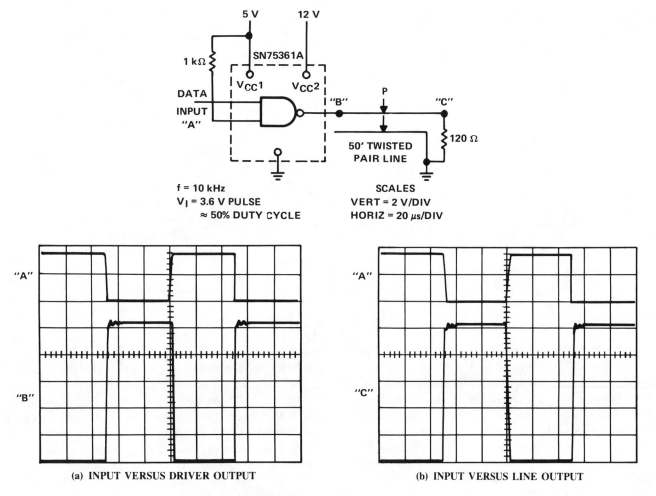

(a) INPUT VERSUS DRIVER OUTPUT

(b) INPUT VERSUS LINE OUTPUT

Figure 9-19. High Level Driver

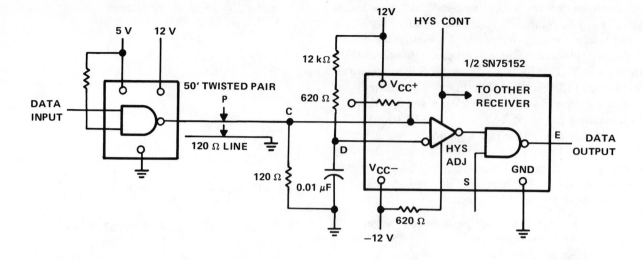

RECEIVER INPUT WAVEFORM POINT C

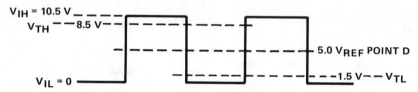

Figure 9-20. Single-Ended Transmission with Receiver Adjusted for High Noise Immunity

Single-Ended Application for High-Speed Bus Communication

In applications where communication between several pieces of equipment over a single data line or bus is required, and the data rates are as high as 1 Mbps, the following device characteristics will be desired:

Propagation delays of ≤25 ns (accomplished by using high-speed Schottky or similar technology) allowing operation at frequencies up to 2 Mbps.

Three-state or open-collector outputs for party-line (data bus) operation. This allows several stations to be connected to a single data bus.

Relatively high output current (≥40 mA) to charge the line capacitance quickly for high-speed requirements.

Low off-state leakage current (≤200 μA) for minimum loading of the bus line allowing several devices to be connected to a single bus without adversely loading the line.

Several transceiver devices are available for this type of application. Two examples are the AM26S10/S11 quad Schottky bus transceivers and the SN75163A octal low-power Schottky bus transceiver. The AM26S10 and AM26S11 have 100 mA open-collector outputs and typical propagation times of 12 ns. The SN75163A outputs may be full totem-pole or open-collector. An output sink current of 48 mA, and typical propagation delays of 14 ns allow operating speeds up to 2 Mbps. Figures 9-21 and 9-22 illustrate the AM26S10 and SN75163A functional diagrams.

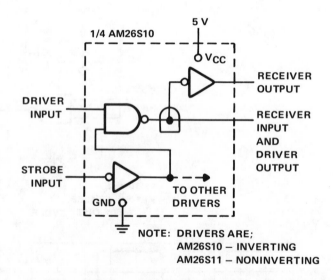

NOTE: DRIVERS ARE;
AM26S10 — INVERTING
AM26S11 — NONINVERTING

Figure 9-21. AM26S10 Functional Diagram

Figure 9-23 is an example of a 4 line interconnect between several terminal stations. The bus lines are 100 Ω twisted pairs but could be coax. Only the extreme ends of the lines are terminated to 5 V which provides matched termination and the pull-up required for the open-collector driver outputs. Typical V_{OL} levels from the driver outputs are 0.35 V and V_{OH} levels are 4.95 V. Receiver low level requirements of 0.8 V or less and high levels greater than 2.25 V result in error-free operation even over relatively long lines (greater than 200 ft).

The SN75121 driver and SN75122 receiver can be used in another type of popular single-ended application, party-

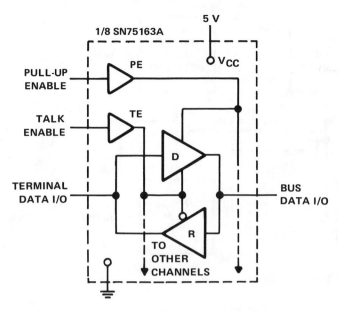

Figure 9-22. SN75163A Functional Diagram

line operation over low-impedance coax lines. Some aircraft and military applications require 50 Ω or 75 Ω coax lines. Because the SN75121 drivers have emitter-follower outputs, which result in high output in the off condition, use in party-line or bus applications is possible. Figure 9-24 illustrates the SN75121 and SN75122 in a party-line application.

SN75122 receivers have a typical 0.6 V input threshold hysteresis. Typical negative going threshold is 1.1 V and the positive going threshold is 1.7 V. SN75121 driver outputs are typically 3.7 V V_{OH} and 0 V V_{OL} when operating into a 75 Ω load. The resulting noise margins are 1.7 V and -2.6 V.

Combining a quad SN75365 driver with two SN75152 dual receivers provides for four line independent communication with very good noise immunity (see Figure 9-25). Connecting the SN75365 V_{CC2} pin to the 5 V supply provides an output V_{OH} of 4.5 V and a V_{OL} of 0.5 V. The inverting inputs of the SN75152 are referenced to the midpoint between the driver V_{OH} and V_{OL} levels (2.5 V). This allows for a wide hysteresis range to be set by connecting 1.5 kΩ resistors from the hysteresis adjust pins to $-V_{CC}$. The resulting hysteresis is ± 1.7 V from the reference for a high level threshold of 4.2 V and a low level threshold of 0.8 V. The resulting noise margins are ± 3.7 V. For additional noise immunity, 125 Ω coax, RG-63B/U, is recommended for the transmission line.

Another receiver that may be used for this application is the SN75154. It may be operated from a single 5 V supply rather than the 12 V and -12 V supplies required by the SN75152. Also the SN75154 is a quad receiver requiring only one package instead of two. There is some reduction in noise margin as it would be 1.7 V and -3.0 V rather than ± 3.7 V.

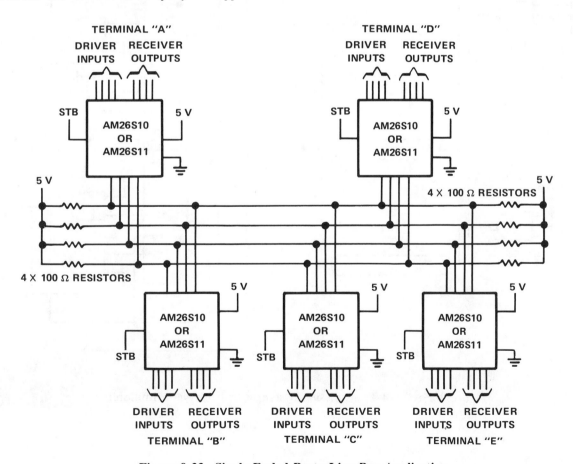

Figure 9-23. Single-Ended Party-Line Bus Application

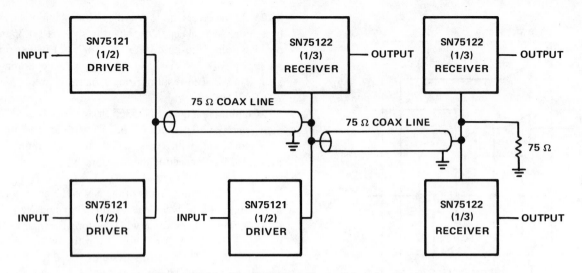

Figure 9-24. Single-Ended Party-Line Application

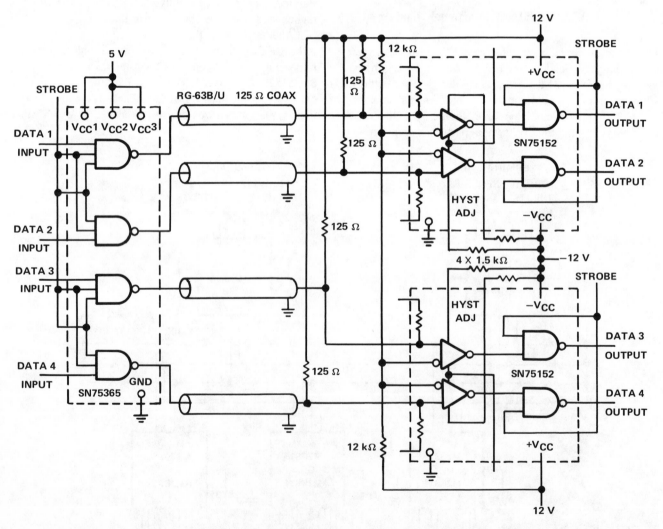

Figure 9-25. Quad Transmission System with High Noise Immunity

Differential Line Drivers and Receivers

The ability to transmit data from one location to another without errors requires immunity to noise. At high data rates, on long lines or under noisy conditions, differential data transmission has an advantage because it is more immune to noise interference than single-ended transmission. Figure 9-26 illustrates the basic sources of noise voltage impressed on a differential, or balanced, data transmission line.

Voltages induced onto the data lines by ground noise or switching transients appear as common-mode signals at the receiver input. Since the receiver has a differential input it responds only to the differential data signal (see Figure 9-26). Differential drivers and receivers can operate safely within specified common-mode voltage ranges. Differential line drivers are designed for general purpose applications as well as specific standards. Tables 9-4 through 9-6 list drivers, receivers and transceivers designed for differential data transmission.

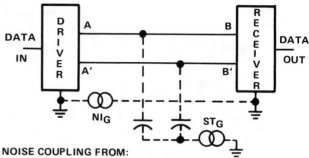

NOISE COUPLING FROM:
NI$_G$, GROUND NOISE CURRENTS
ST$_G$, SWITCHING TRANSIENTS BY CAPACITIVE AND INDUCTIVE COUPLING.

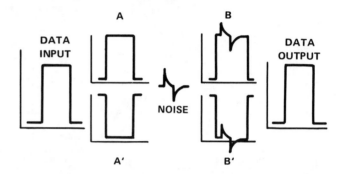

Figure 9-26. Basic Noise Sources and Results

Table 9-4. Differential Drivers

TYPE NUMBER	CIRCUITS/PACKAGE	DESCRIPTION	APPLICATION
DS78/8831	2	Totem-pole output to 40 mA, and diode clamped	General purpose
SN78/8832	2	Totem-pole output to 40 mA	General purpose
SN55/75109A	2	Current-mode drive 6 mA	General purpose
SN55/75110A	2	Current-mode drive 12 mA	General purpose
SN75112	2	Current-mode drive 27 mA	General purpose
SN55/75113	2	3-state output to 40 mA	General purpose
SN55/75114	2	Totem-pole output to 40 mA	General purpose
SN55/75183	2	Totem-pole output to 40 mA	RS-422-A
SN55/75158	2	Totem-pole output to 40 mA	RS-422-A
SN75159	2	3-state output to 40 mA	RS-422-A
uA9638C	2	Totem-pole output to 50 mA	RS-422-A
AM26LS31C	4	Totem-pole output to 20 mA	RS-422-A
MC3487	4	3-state output to 48 mA	RS-422-A
SN55/75151	4	3-state, individual enable 40 mA	RS-422-A
SN55/75153	4	3-state, common enable 40 mA	RS-422-A
SN75172B	4	3-state, common enable 60 mA	RS-422-A
SN75172B	4	3-state, common enable, 60 mA	RS-485
SN75174B	4	3-state, dual enable, 60 mA	RS-422-A
SN75174B	4	3-state, dual enable, 60 mA	RS-485

Table 9-5. Differential Receivers

TYPE NUMBER	CIRCUITS/PACKAGE	DESCRIPTION	APPLICATION
SN55/75107A	2	Totem-pole output, V_{ID} ± 25 mV	General purpose
SN55/75108A	2	Open-collector output V_{ID} ± 25 mV	General purpose
SN55/75107B	2	107A with input power off protected	General purpose
SN55/75108B	2	108A with input power off protected	General purpose
SN75207/207B	2	107A & B with V_{ID} ± 10 mV	General purpose
SN75208/208B	2	107A & B with V_{ID} ± 10 mV	General purpose
SN55/75115	2	Split totem-pole output V_{ID} ± 500 mV	General purpose
SN55/75182	2	Totem-pole output V_{ID} ± 500 mV	General purpose
SN55/75157	2	Input hysteresis V_{ID} ± 7 V CMR	RS-422-A
uA9637AAC	2	Input hysteresis V_{ID} ± 200 mV ± 7 V CMR	RS-422-A
AM26LS32AC	4	Input hysteresis V_{ID} ± 200 mV ± 7 V CMR	RS-422-A
AM26LS33AC	4	Input hysteresis V_{ID} ± 500 mV ± 15 V CMR	RS-422-A
MC3486	4	3-State output V_{ID} ± 200 mV ± 7 V CMR	RS-422-A
SN75173A	4	Input hysteresis V_{ITH} ± 200 mV ± 12 V CMR	RS-422-A
SN75173A	4	Input hysteresis V_{ITH} ± 200 mV ± 12 V CMR	RS-485
SN75175A	4	Input hysteresis V_{ITH} ± 200 mV ± 12 V CMR	RS-422-A
SN75175A	4	Input hysteresis V_{ITH} ± 200 mV ± 12 V CMR	RS-485

Table 9-6. Differential Transceivers

TYPE NUMBER	CIRCUITS/PACKAGE	DESCRIPTION	APPLICATION
SN55/75116	2	3-state split totem-pole 40 mA driver, ± 0.5 threshold	General purpose
SN55/75117	2	3-state totem-pole, 40 mA driver threshold	General purpose
SN55/75118	2	'116 with 3-state receiver output	General purpose
SN55/75119	2	'117 with 3-state receiver output	General purpose
SN75176B	2	60 mA 3-state driver, 200 mV sensor with hysteresis	RS-485
SN75177B	2	'176 configured as bus repeater	RS-485
SN75178B	2	'177 with inverted enable	RS-485
SN75179	2	Full duplex '176 type	RS-485

Advantages and disadvantages of differential (balanced) data transmission, relative to single-ended transmission are:

Advantages
High common-mode noise voltage rejection.
Reduced line radiation — less RFI.
Improved speed capabilities.
Drive longer line lengths.

Disadvantages
Slightly higher costs (sometimes).
Must be used with twisted pair or other types of balanced transmission lines.

Terminating Differential Data Transmission Lines

As with single-ended data transmission lines, it is necessary to terminate balanced lines properly to prevent ringing and errors in transmitted data. To properly absorb a signal on the line, a termination value equal to the characteristic impedance of the line must be used at the receiving end of the line [See Figure 9-27(a)]. If stray signals and noise are introduced to the line from radiation or other sources, it may be desirable to terminate both the driver and receiver ends of the line to prevent reflections, ringing or oscillations. When multiple drivers are involved the line must be terminated at the extreme ends only as shown in Figure 9-27(b). In party-line systems, termination at each location of a driver or receiver should be avoided because these extra terminations will only adversely load the line. Circuits (a) and (b) use line-to-line termination with a single resistor at each termination point. Line-to-line termination is often the desirable method because it requires minimum power from the drivers. The circuit in Figure 9-27(c) uses line-to-ground termination where the value of R_T is $Z_O/2$. With each line terminated to ground with $Z_O/2$, the total termination line-to-line is still Z_O. However more output drive power will be required from the data transmitters when this type of termination is used.

A problem arises with direct line-to-line termination [Figure 9-27(a), Figure 9-27(b)] if noise energy is induced into the transmission line. Noise induced into the data line is common-mode and the noise voltage level, V_N, is a product of the noise current, I_N, and resistance, R_G, from each line to ground. The lowest resistance path to ground is the receiver input resistance, which for the SN75115 is typically about 12 kΩ. If line termination is to ground [Figure 9-27(c)], then the resistance is $Z_O/2$ which may be around 50 Ω. Line noise voltage levels are therefore greatly reduced with the latter termination, resulting in much better common-mode rejection capability. In noisy or long line applications, line-to-ground terminations are preferred.

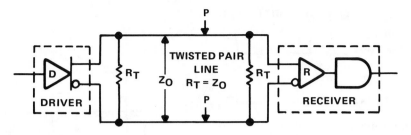

(a) DEDICATED SINGLE DRIVER; LINE TERMINATED AT RECEIVER

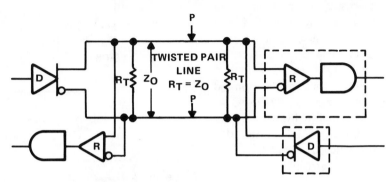

(b) MULTIPLE DRIVERS; LINE TERMINATED AT EXTREME ENDS ONLY

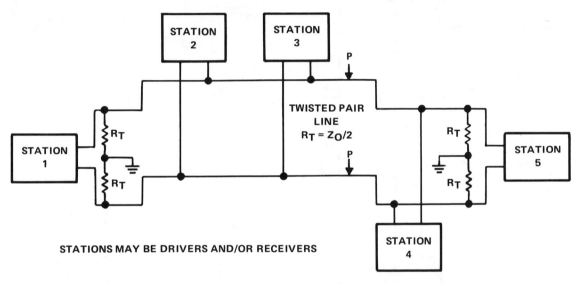

STATIONS MAY BE DRIVERS AND/OR RECEIVERS

(c) MULTIPLE STATIONS; EACH LINE TERMINATED TO GROUND; TERMINATE EXTREME ENDS ONLY

Figure 9-27. Terminating Balanced Twisted Pair Lines

Current-Mode Drivers in Differential Data Line Transmission

The following problems, typical with voltage-mode drive, can be minimized with current-mode drive.

1. Amount of power required to drive properly terminated low-impedance lines.
2. Large output surge currents during pulse transitions (Figure 9-28) which result in radiation and possible cross talk.

The data line current spikes, illustrated in Figure 9-28, result from charging the distributed capacitance of the line. At low data rates these high speed current surges will result

in radiation of RFI. At higher data rates, the surge time remains the same but becomes a large percentage of the overall duty cycle, resulting in excessive power dissipation that can, at some point, lead to deterioration of driver performance or permanent damage.

Examples of current mode drivers include the following dual channel devices:

-55 to $125\,^{\circ}\text{C}$ DEVICE	0 to $70\,^{\circ}\text{C}$ DEVICE	TYPICAL CONSTANT CURRENT OUTPUT LEVEL
SN55109A	SN75109A	6 mA
SN55110A	SN75110A	12 mA
	SN75112	27 mA

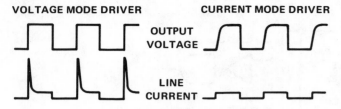

Figure 9-28. Line Current Comparisons

Figures 9-29 and 9-30 show their basic input and output circuit configurations. Driver inputs are basic TTL and followed by a converter circuit (transistors Q2 and Q3) to provide differential control of the output. The differential voltage applied to the bases of Q13 and Q14 (output transistors) is large enough to switch them completely on and off alternately. When on, the output transistor (Q13 or Q14) will sink an amount of current supplied by the constant current generator consisting of Q15, Q16, Q17, Q18, Q19, and R9. Thus, an output when switched on will be a fixed current sink at the level set by the current generator.

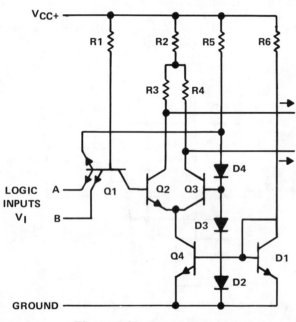

Figure 9-29. Input Circuit

Output current sink is to the device's − 5 V rail which generates an output voltage that is a product of this current and the output terminal resistance to ground. Figure 9-31 shows a typical SN75110A input-to-output transfer curve for a 50 Ω termination. In this example 600 mV is generated by each output for a total differential swing of 1.2 V. Outputs may be shorted to ground or either supply rail without damage to the device. The recommended operating output common-mode range is 10 V to − 3 V. Exceeding these levels will result in output saturation and possible improper operation. Figure 9-32 shows the logic diagram and function table for a typical current-mode driver.

The receivers recommended for use with current-mode drivers are those with very good sensitivity levels (less than or equal to 25 mV).

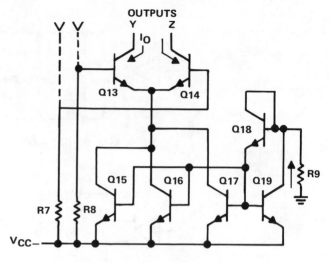

Figure 9-30. Output Circuit

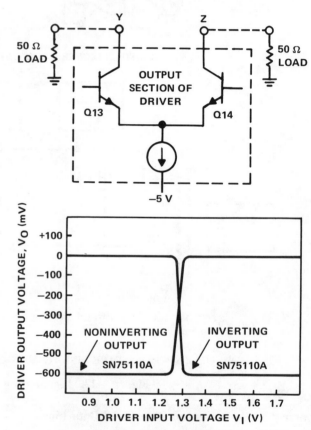

Figure 9-31. Typical Voltage Plot for SN75110A Driver

Following is a list of recommended receivers:

− 55 to 125°C Device	0 to 70°C Device	Input Sensitivity	Receiver Output Type
SN55107A	SN75107A	± 25mV	Totem-pole
SN55107B	SN75107B	± 25mV	Totem-pole
SN55108A	SN75108A	± 25mV	Open-collector
	SN75207	± 10mV	Totem-pole
	SN75207B	± 10mV	Totem-pole
	SN75208	± 10mV	Open-collector
	SN75208B	± 10mV	Open-collector

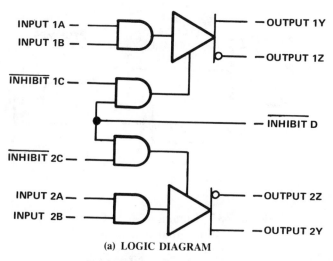

(a) LOGIC DIAGRAM

LOGIC INPUTS		INHIBITOR INPUTS		OUTPUTS	
A	B	C	D	Y	Z
X	X	L	X	OFF	OFF
X	X	X	L	OFF	OFF
L	X	H	H	ON	OFF
X	L	H	H	ON	OFF
H	H	H	H	OFF	ON

H = high level, L = low level, X = irrelevant

(b) FUNCTION TABLE

Figure 9-32. Current-Mode Driver Logic Diagram and Function Table

Figures 9-33 and 9-34 show typical receiver equivalent input and output circuits.

The receiver input is differential and provides a common-mode input voltage range of ±5 V absolute maximum and ±3 V recommended operating maximum. The outputs differ slightly as is shown in Figure 9-34. SN75107A type devices have standard TTL totem-pole outputs while the SN75108A type devices have open-collector outputs. Open-collector outputs allow wired-OR connection of multiple outputs to a single line.

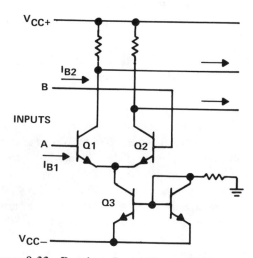

Figure 9-33. Receiver Input Stage, Input Current Source, and Bias Circuit

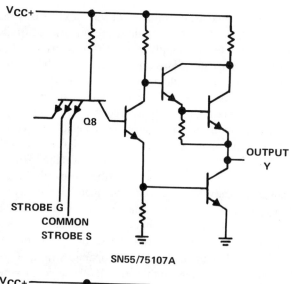

SN55/75107A

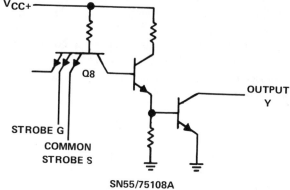

SN55/75108A

Figure 9-34. Output Gates of SN55/75107A and SN55/75108A Compared

RECEIVER PERFORMANCE

The SN55/75107A receiver circuit has a nominal propagation delay of 17 ns, making it ideal for use in high-speed systems. The receiver delay is almost completely insensitive to overdrive voltages of 10 mV. The circuit responds to input signals with repetition rates as high as 20 Mbps.

Input Sensitivity

The input sensitivity is defined here as the differential dc voltage required at the inputs of the receiver to force the output to the logic gate threshold voltage level. The input sensitivity of the receivers is nominally 3 mV. This feature is particularly important when data is transmitted over a long line and the pulse is deteriorated due to line effects. A receiver with this sensitivity also finds many other applications such as comparators, sense amplifiers, and level detectors.

Common-Mode Voltage Range

The common-mode voltage range or CMVR is defined here as that voltage applied simultaneously to both input terminals which, if exceeded, does not allow normal operation of the receiver.

The recommended operating CMVR is ±3 V, making it useful in all but the noisiest environments. In extremely noisy environments, common-mode voltage can easily reach ±10 V to ±15 V if precautions are not taken to reduce ground and power supply noise, as well as cross talk problems. When the receiver must operate in such conditions, input attenuators should be used to decrease the system common-mode noise to a tolerable level at the receiver inputs. Differential noise is also reduced by the same ratio.

These attenuators have been intentionally omitted from the receiver input terminals so the designer may select resistors which will be compatible with the particular application or environment. Furthermore, the use of attenuators adversely affects the input sensitivity, the propagation time, the power dissipation, and in some cases the input impedance (depending upon the selected resistor values), therefore reducing the versatility of the receiver.

The ability of the receiver to operate with a ±15 V common-mode voltage at the inputs has been demonstrated using the circuit shown in Figure 9-35. Dividers with three different values presenting a 5-to-1 ratio were used so as to operate the differential inputs under ±3 V common-mode voltage. Careful matching of the two attenuators is needed to balance the overdrive at the input stage. The resistances used were:

Attenuator 1: R1 = 2 kΩ, R2 = 0.5 kΩ
Attenuator 2: R1 = 6 kΩ, R2 = 1.5 kΩ
Attenuator 3: R1 = 12 kΩ, R2 = 3.0 kΩ

Table 9-7 shows some of the typical switching results obtained under such conditions.

Table 9-7. Typical Propagation Delays for Receiver with Attenuator Test Circuit Shown in Figure 9-35

DEVICE	PARAMETERS	INPUT ATTENUATOR	TYPICAL (ns)
SN55/75107A	$t_{pd(1)}$	1	20
		2	32
		3	42
	$t_{pd(0)}$	1	22
		2	31
		3	33
SN55/75108A	$t_{pd(1)}$	1	36
		2	47
		3	57
	$t_{pd(0)}$	1	29
		2	38
		3	41

NOTE: Output load R_L = 390 Ω.

Input Termination Resistors

To prevent reflections, the transmission line should be terminated in its characteristic impedance. Matched termination resistances normally in the range of 25 to 200 Ω are required not only to terminate the transmission line in a desired impedance, but also to provide a necessary dc path

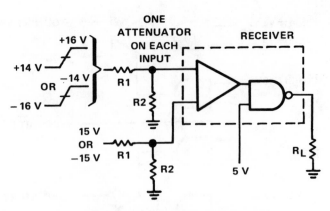

Figure 9-35. Common-Mode Circuit for Testing Input Attenuators

for the receiver input bias. Careful matching of the resistor pairs should be observed or the effective common-mode rejection ratio will be reduced.

The input circuit of the receivers must meet the requirements for low input currents (30 μA typical) and high input impedance. These requirements provide for low loading on the lines, an important consideration for "party-line" applications.

Reference Voltage

The receiver can be used as a single-ended line receiver or comparator by referencing one input as shown in Figure 9-36. The operating threshold voltage level is established by (and is approximately equal to) the applied reference input voltage, V_{ref}, selected within the operating range.

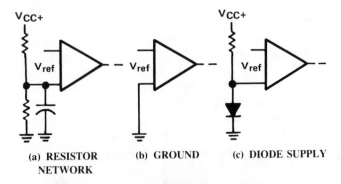

Figure 9-36. Some Methods of Referencing Receiver Inputs

A simple method of generating the reference voltage is the use of a resistor voltage divider from either the V_{CC+} or V_{CC-} supply as in Figure 9-36(a). The reference can also be obtained by a diode [Figure 9-36(c)] or a reference supply or just ground [Figure 9-36(b)]. The bias current required at the referenced input is low (nominally 30 μA). Therefore, voltage dividers of this type may normally be operated with very low current requirements and may be used also to supply a number of paralleled reference inputs. In noisy environments, the use of a filter capacitor may be recommended as indicated in Figure 9-36(a).

Input Limitations

Figure 9-37 shows the "safe operating region" of voltages at the two receiver inputs. Coordinates within the shaded area are considered safe operating conditions. As an example:

1. If B is 2 V, A may be any value from −3 V to 3 V.
2. If B is −2 V, A may be any value from −5 V to 3 V.
3. If A is 0 V, B may be any value from −5 V to 3 V.

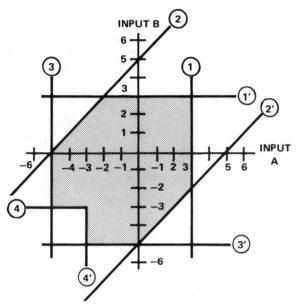

LINES	DESCRIPTION
1 AND 1'	MAXIMUM POSITIVE INPUT SIGNAL IS +3.0 V.
2 AND 2'	MAXIMUM DIFFERENTIAL SIGNAL $\|A-B\|$ IS 5.0 V.
3 AND 3'	MAXIMUM NEGATIVE INPUT SIGNAL IS −5.0 V.
4 AND 4'	MAXIMUM COMMON MODE IS ±3.0 V.

Figure 9-37. "Safe Operating Region" for Receiver Inputs

SN55/75107A SERIES APPLICATIONS

Connection of Unused Inputs and Outputs

In any application of SN55/75107A series devices, there are two important precautions to be observed. First, when only one receiver in a package is being used, at least one of the differential inputs of the other receiver should be terminated at some voltage between 3 V and −3 V, preferably at ground. Failure to do so causes improper operation of the unit being used because of common bias circuitry for the current sources of the two receivers. Second, when only one driver in a package is used, the outputs of the other driver must either be tied to ground or inhibited. Otherwise, excessive power dissipation may result.

There are some other equally basic but less critical recommendations concerning unused driver gate inputs. Where these inputs are left open-circuited, the propagation delay to a logic low level is increased by about one nanosecond per input. In high-speed systems this could be a significant factor.

For faster switching times, unused gate inputs can be tied to a positive voltage source of about 2.4 V. This disposition greatly reduces the distributed capacitance associated with the emitter. It also ensures that no additional degradation occurs in the propagation delay times.

An alternative solution is to tie the unused gate inputs to V_{CC} through a resistor. The value of the resistor should not be less than 1 kΩ or the transient voltages could damage the inputs. The upper resistor value is determined by the voltage drop caused by logic high level input currents and by the minimum supply voltage. The best solution is to join the unused input to the used input on the same gate. This method provides the fastest switching speeds since the stray capacitances are driven by the preceding gate, and it also provides protection against excessive supply surges since the 54/74 TTL outputs at a logic high level are typically 3.5 V and are current-limited.

One-Channel Balanced Transmission System

SN55107A series dual line circuits are designed for use in high-speed data transmission systems that utilize balanced, terminated transmission media such as twisted pair lines. Such a system operates in the balanced mode, so that noise induced on one line is also induced in the other. The noise appears as common-mode at the receiver input terminals, where it is rejected. The ground connection between the line driver and line receiver is not part of the signal circuit, so that system performance is not affected by circulating ground currents and ground noise. Figure 9-38 illustrates the ability of the receivers to reject common-mode noise.

The unique output circuit of the driver allows terminated transmission lines to be driven at normal line impedances. High-speed operation of the system is ensured since line reflections are virtually eliminated when terminated lines are used. Cross talk is minimized because of the low signal amplitude, low line impedances, and because the total current in a line pair remains constant.

A basic balanced transmission system using SN55107A series devices is shown in Figure 9-39. Data is impressed on the twisted pair line by unbalancing the line voltages by means of the driver output current. Line termination resistors labeled R_T are required only at the extreme ends of the line. For short lines, termination resistors only at the receiver end may prove adequate. Signal phasing depends on the driver output and receiver input polarities on the line.

Differential Party-Line Systems

When several stations must communicate with each other it is generally more economical for all stations in the system to share a single transmission line rather than to use many dedicated lines. The strobe feature of the receivers and the inhibit feature of the drivers allow the SN55107A series dual line circuits to be used in party-line (also called data bus) applications. Examples are shown in Figures 9-40, 9-41, and 9-42. In each of these examples, one driver is enabled and

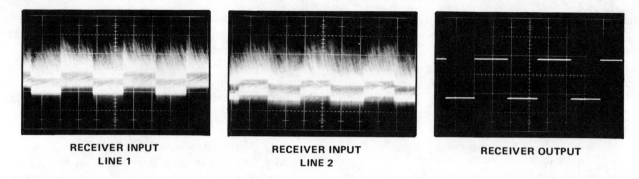

RECEIVER INPUT
LINE 1

RECEIVER INPUT
LINE 2

RECEIVER OUTPUT

**Figure 9-38. Oscilloscope Displays of Typical Noisy Inputs and Noise-Free Output of SN75108 Line Receiver
(Test Frequency: 10 kHz. Input Signal: 50 mV Peak-to-Peak. Input Noise: 4.0 V Peak-to-Peak)**

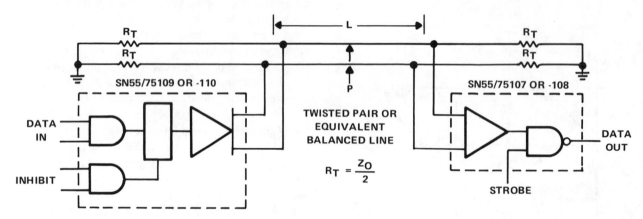

Figure 9-39. Use of SN55107 Series Devices in a Typical Single-Driver, Single-Receiver Transmission System

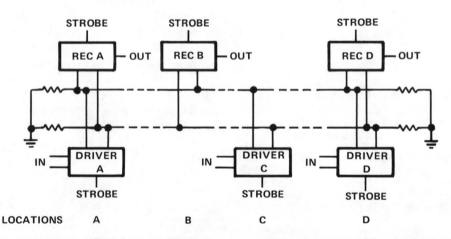

**Figure 9-40. Simple Party-Line System with Driving and Receiving Stations Scattered Along the Receiving
Stations Scattered Along the Line**

transmits to the receivers on the line, while all other drivers are disabled. Data from the various drivers can therefore be time multiplexed on the transmission line.

Figure 9-43 illustrates an eight-channel party-line application similar to that in Figure 9-42, including the clocking arrangements. This method uses two twisted pair lines: one for data transmission and the other for clocking and control information. Details of the clocking and control circuits appear in Figures 9-44 and 9-45. Careful matching of the line delays is necessary to ensure synchronized clocking.

Repeaters for Long Lines

In some cases, the driven line may be so long that the noise level on the line reaches the common-mode limits or the attenuation becomes too large for good reception. In such a case, a simple application of a driver and receiver as a repeater [(shown in Figure 9-46(a)] restores the signal level and allows an adequate signal level at the receiving end. If multi-channel operation is desired, then proper gating for each channel must be sent through the repeater station using another repeater set as in Figure 9-46(b). In most cases, two twisted pair lines suffice, one for the information and the other for the clock pulses.

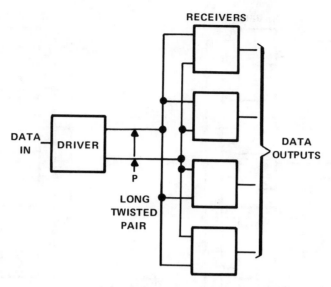

Figure 9-41. Party-Line Concept of One Driver Transmitting to One of Many Receivers

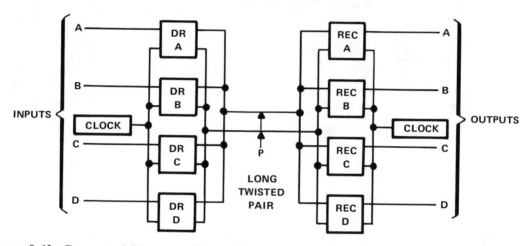

Figure 9-42. Conceptual Diagram of Four Transmission Channels Sharing the Same Party Line

STANDARD VOLTAGE-MODE DIFFERENTIAL DRIVERS AND RECEIVERS

Voltage-mode drivers are not without positive features to make them desirable in differential data line transmission. These features include the following:

- Will operate from single (usually 5 V) supplies.

- Exhibit high common-mode voltage capabilities.

- Less circuit complexity allows four drivers or four receivers in one package.

The following discussion includes only devices in the SN75113 through SN75119 family but is equally applicable to many of the other popular voltage-mode data line drivers and receivers.

SN75113 and SN75114 dual differential line drivers are almost identical. The SN75113 has three-state outputs

allowing operation with other drivers on a party line. These drivers are designed to work with the SN75115 dual receiver and each device operates from a single 5 V supply. For dedicated single drive applications, the SN75114 may be preferred for its triple input gate control, but if multiple drivers are on the line a three-state output SN75113 would be used (see Figure 9-47). A capacitor may be connected in series with Z_O to reduce power dissipation. Also, the SN75115 receiver outputs are three-state. This allows a wired-OR connection of several receivers by using a common pull-up resistor.

Differential voltage levels at the most remote receiver will be dependent on the driver output voltage swing ($V_{OD} = V_{OH} - V_{OL}$), termination loading, and line attenuation. Assuming a worst-case input voltage swing (V_I) requirement at the receiver of 500 mV, the maximum theoretical line length may be derived as follows: $V_{OD} = V_{OH} - V_{OL}$ where V_{OH} and V_{OL} are determined by the dc loading. A typical 120 Ω line terminated rail to

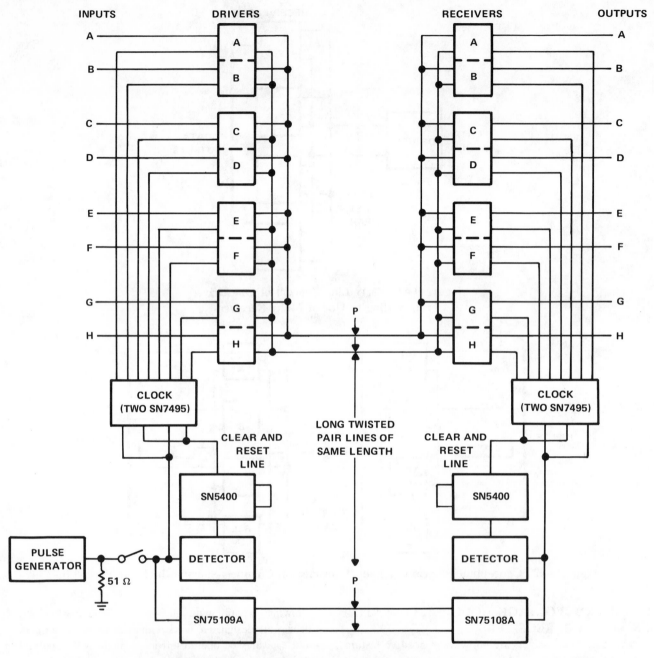

Figure 9-43. A Multi-Channel System with Clocking
(See Details of Detector, Clock, and Test Circuit in Following Figures)

rail as illustrated in Figure 9-47 will allow a V_{OH} of 3.2 V, a V_{OL} of 0.17 V and a resulting V_{OD} of 3.03 V. The maximum attenuation allowable is:

$$A_{MAX} = 20 \log_e (V_{OD} - V_{ID})$$
$$= 20 \log_e (3030-500)$$
$$A_{MAX} = 20 \log_e 2530 = 68 \text{ dB}.$$

Termination at the two extreme ends of the line will result in an additional 6 dB loss, therefore the maximum allowable line loss is 62 dB. For a clock frequency of 1.8 MHz (see Figure 9-48) the attenuation of a typical twisted pair transmission line (AWG 22 solid wire with 0.060 inch plastic cover, twisted 4.5 times per foot) will be 1.25 dB per 100 feet.

Line length is therefore 62 dB divided by 1.25 dB per 100 feet which yields 4960 feet maximum. This calculated value correlates closely with the measured line length versus frequency curves, Figure 13 - page 289, in the Line Driver and Line Receiver Data Book, 1981.

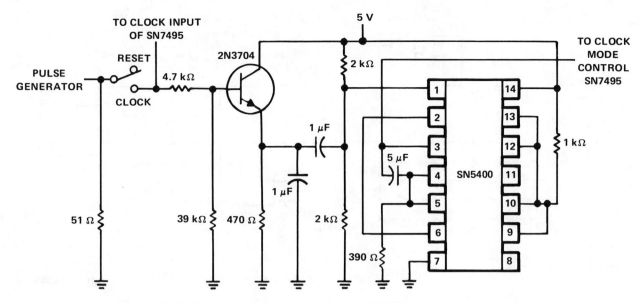

Figure 9-44. Detector Circuit for Clearing SN7495 Clock in Figure 9-43

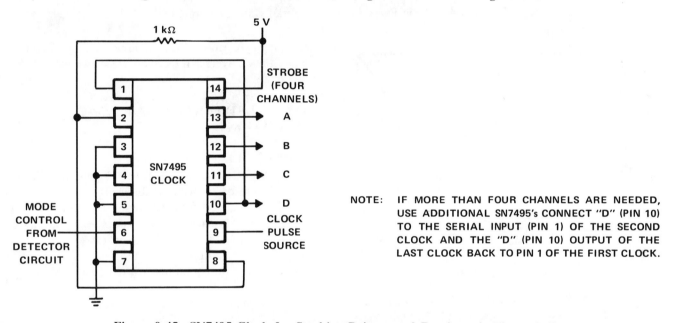

NOTE: IF MORE THAN FOUR CHANNELS ARE NEEDED, USE ADDITIONAL SN7495's CONNECT "D" (PIN 10) TO THE SERIAL INPUT (PIN 1) OF THE SECOND CLOCK AND THE "D" (PIN 10) OUTPUT OF THE LAST CLOCK BACK TO PIN 1 OF THE FIRST CLOCK.

Figure 9-45. SN7495 Clock for Strobing Drivers and Receivers in Figure 9-43

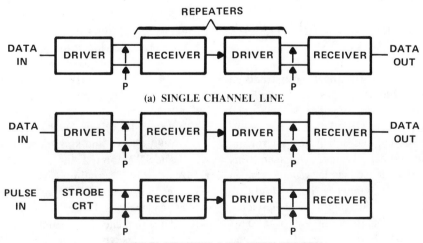

(a) SINGLE CHANNEL LINE

(b) MULTI-CHANNEL LINE WITH STROBE

Figure 9-46. Driver-Receiver Repeaters

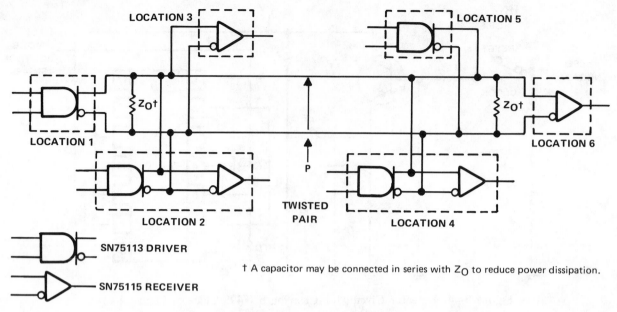

LOCATION 3
LOCATION 5
LOCATION 1
LOCATION 6
$Z_O\dagger$
$Z_O\dagger$
LOCATION 2
LOCATION 4
P
TWISTED
PAIR

SN75113 DRIVER

SN75115 RECEIVER

† A capacitor may be connected in series with Z_O to reduce power dissipation.

Figure 9-47. Basic Party-Line Differential Data Transmission

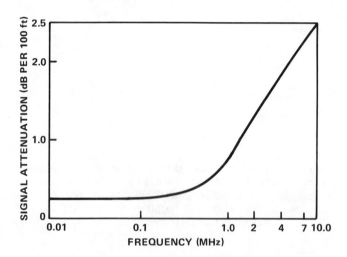

Figure 9-48. Signal Attenuation with Frequency in Twisted Pair Transmission Line (AWG 22 Solid Wire with 0.060-in Plastic Cover, Twisted 4.5 Times Per Foot)

EIA STANDARD RS-232-C CIRCUITS AND APPLICATIONS

The Electronic Industries Association (EIA) introduced the RS-232 standard in 1962 for the purpose of standardizing the interface between data terminal equipment (DTE) and data communication equipment (DCE). See Figure 9-49. Although emphasis was then, and still is, on interfacing between a modem unit and data terminal equipment, this standard is also applicable to serial binary interfacing between various types of data terminal equipment. The revised EIA standard RS-232-C introduced in August 1969 is widely accepted for single-ended data transmission over short distances with low data rates. Several types of drivers and receivers have been developed for use in RS-232-C

interfacing applications. Basic requirements for RS-232-C type data line drivers and receivers are shown in Table 9-8 and Table 9-9.

TYPICAL DRIVERS FOR EIA RS-232-C APPLICATIONS

SN75150 Dual Data Line Driver

The SN75150 dual line driver (Figure 9-50) is an RS-232-C type driver with the following features:

- Withstands sustained output short circuit to any low impedance voltage between -25 V and 25 V.
- Operates from ± 12 V supplies.
- 2 μs max transition time through the 3 V to -3 V transition region under a full 2500 pF load.
- TTL compatible inputs.
- Common strobe input.
- Slew rate controllable with an external capacitor at the output.

SN75156/uA9636 Dual Line Driver

The SN75156/uA9636 (Figure 9-51) is a dual line driver having the following features:

- Wide supply range (± 7.5 V to ± 15 V).
- Low supply current required (4.5 mA max./channel).
- Wave shaping with external resistor.
- TTL or low level CMOS compatible inputs.
- Source and sink output current limiting.

The SN75156 is a single-ended line driver designed for EIA RS-232-C requirements and also EIA RS-423-A, European standard CCITT recommendations V.10, V.28, X.26, and the U.S. Federal Standard FIPS 1030. This device maintains regulated high and low output levels of 5.5 V and -5.5 V, respectively, over a wide range of power supply voltages. The output transition time for both drivers can be

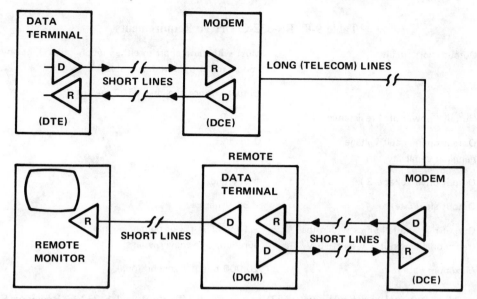

Figure 9-49. RS-232-C Drivers and Receivers in a Basic Data Communications System

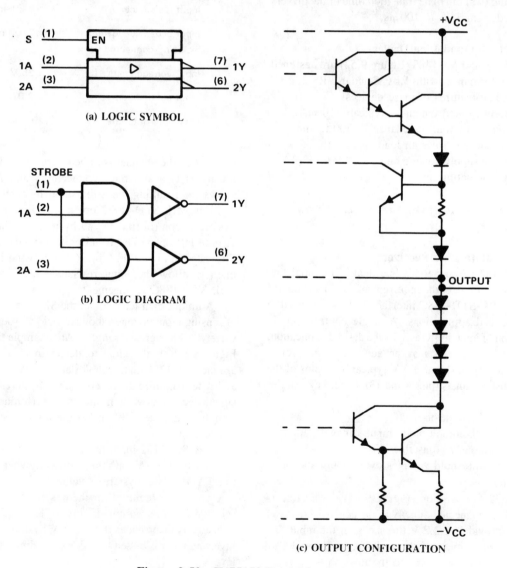

(a) LOGIC SYMBOL

(b) LOGIC DIAGRAM

(c) OUTPUT CONFIGURATION

Figure 9-50. SN75150 Dual Line Driver

Table 9-8. RS-232-C Driver Requirements

Output short circuit	Must withstand short to either supply ground, or any conductor in the interface cable.
	Output current shall be <500 mA.
Output "power off" resistance	>300 Ω
Output open circuit voltage handling capability	0 to ±25 V
Output drive voltage 3 kΩ to 5 kΩ load	>5 V and <15 V
Output slew rate	<30 V per μs
Output rise and fall times within the transitional limits of 3 V and −3 V	≤1 ms or ≤4% of nominal signal duration; (whichever is less)
Maximum data rate	20,000 bits per second (bps)

adjusted by means of an external resistor (Figure 9-52) at the wave shaping (ws) pin. The transition time of the drivers can be adjusted from 1 μs to 100 μs.

SN75188/MC1488 Quad Line Driver

The SN75188 and MC1488 (Figure 9-53) are designed for use in conformance with EIA Standard RS-232-C. Following are key features of these devices:

- Current limited output — typically 10 mA.
- Power off output impedance — 300 Ω min.
- Slew rate control with load capacitor.
- Flexible supply voltage range (±9 V to ±12 V).
- TTL compatible inputs.

TYPICAL RECEIVERS FOR EIA RS-232-C APPLICATIONS

SN75152 Dual Data Line Receiver

The SN75152 (Figure 9-54) is a dual differential line receiver designed to meet the requirements of EIA standard RS-232-C or MIL-STD-188C interfacing. It is also used in other single-ended applications. A single control (pin 1) allows selection of input hysteresis for the desired application. In addition, the hysteresis may be adjusted to any range between ±0.3 V typical and ±5 V typical by means of the hysteresis adjust terminals (pins 4 and 13). Other key features include:

- Independent strobes.
- ±25 V common-mode input voltage range.
- Continuously adjustable hysteresis.
- Stable threshold voltages over supply and/or temperature.

For RS-232-C operation (Figure 9-55) the hysteresis control (pin 1) is connected directly to the negative V_{CC} rail (pin 9). This provides ±2.2 V thresholds at the input. To provide the correct input resistance, R_T(pin 6 channel 1 or pin 11 channel 2) is connected to the inverting line input (pin 5 channel 1 or pin 12 channel 2). Hysteresis adjust pins are left open and the strobes are connected to a logic high

Table 9-9. RS-232-C Receiver Requirements

Input resistance	>3 kΩ and <7 kΩ
Input capacitive loading including connected cable	<2500 pF
Input voltage limits	±25 V
Input open circuit voltage	<2.0 V
Maximum data rate	20,000 bps

level when the channel is to be operational. Figure 9-55(a) illustrates typical circuit connections and Figure 9-55(b) shows the resulting input threshold hysteresis curve.

For MIL-STD-188C operation [Figure 9-56(a)], the hysteresis control (pin 1) and hysteresis adjust (pins 4 and 13) are left open. The resulting thresholds are ±0.3 V as illustrated in Figure 9-56(b). R_T, pins 6 and 11, are also left open to allow the input resistance to be about 9 kΩ, a MIL-STD-188C requirement.

A unique characteristic of the SN75152 is its capability of sensing input voltages above or below ground with a wide variety of hysteresis ranges. An example is the circuit, Figure 9-57(a), designed to detect input signals centered around −3.0 V with thresholds at −1.5 V and −4.5 V. The ability to maximize the receiver input hysteresis allows clean signal reception with minimum interference from noise signals. Figure 9-57(b) shows the input hysteresis for this application.

The SN75152 input hysteresis may be any value from ±0.3 V to ±5.0 V with thresholds anywhere from −25 V to 25 V (less the hysteresis value).

The value required for the hysteresis adjust resistor is selected from the graph, Figure 9-58. A 1.7 kΩ threshold adjust resistor provides the ±1.5 V hysteresis desired. The inverting input is biased to −3 V providing a basic reference level.

A typical application, combining the SN75150 dual driver and SN75152 dual receiver, is shown in Figure 9-59.

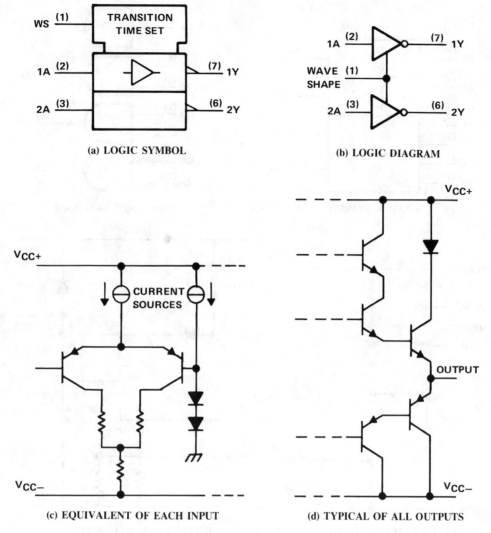

(a) LOGIC SYMBOL

(b) LOGIC DIAGRAM

(c) EQUIVALENT OF EACH INPUT

(d) TYPICAL OF ALL OUTPUTS

Figure 9-51. SN75156 and uA9636 Dual Channel Line Drivers

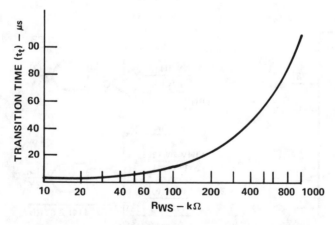

Figure 9-52. SN75156 Transition Time vs R_{WS}

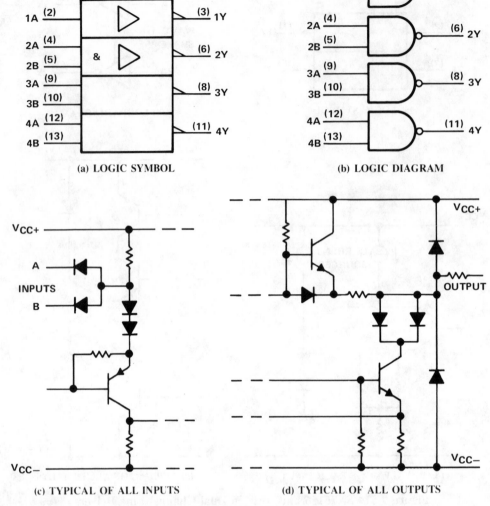

(a) LOGIC SYMBOL

(b) LOGIC DIAGRAM

(c) TYPICAL OF ALL INPUTS

(d) TYPICAL OF ALL OUTPUTS

Figure 9-53. SN75188/MC1488 Quad Drivers

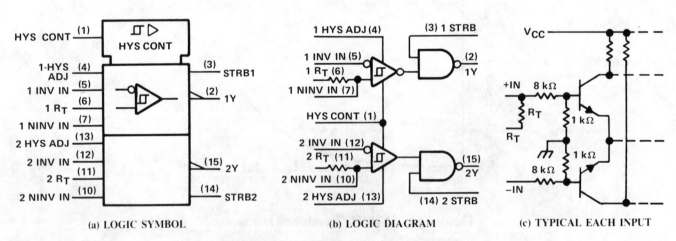

(a) LOGIC SYMBOL

(b) LOGIC DIAGRAM

(c) TYPICAL EACH INPUT

Figure 9-54. SN75152 Dual Line Receiver

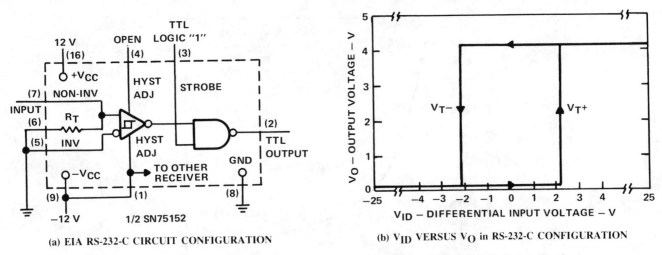

(a) EIA RS-232-C CIRCUIT CONFIGURATION

(b) V_{ID} VERSUS V_O in RS-232-C CONFIGURATION

Figure 9-55. SN75152 Circuit Connections and Results for EIA RS-232-C Application

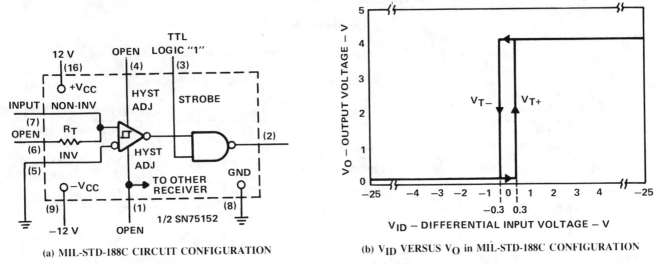

(a) MIL-STD-188C CIRCUIT CONFIGURATION

(b) V_{ID} VERSUS V_O in MIL-STD-188C CONFIGURATION

Figure 9-56. SN75152 Circuit Connections and Results for MIL-STD-188C Application

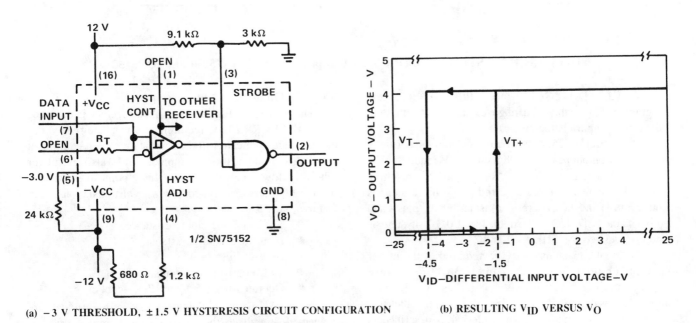

(a) −3 V THRESHOLD, ±1.5 V HYSTERESIS CIRCUIT CONFIGURATION

(b) RESULTING V_{ID} VERSUS V_O

Figure 9-57. Circuit Connections and Results for a Positive-Going Threshold of −1.5 V and a Negative-Going Threshold of −4.5 V

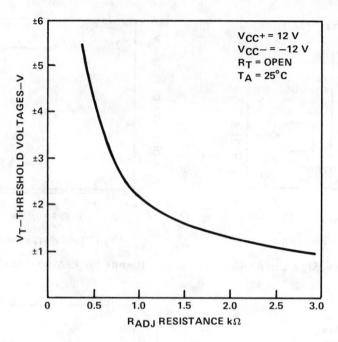

Figure 9-58. Threshold Voltage Versus Hysteresis Adjust Resistance

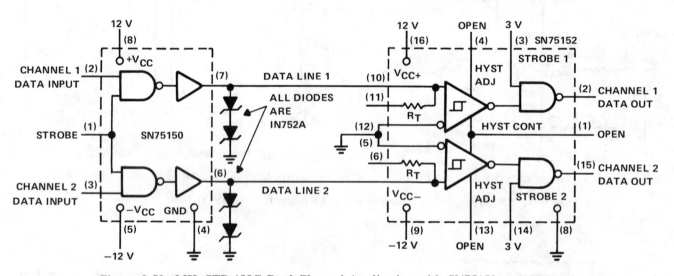

Figure 9-59. MIL-STD-188C Dual Channel Application with SN75150 and SN75152

As a MIL-STD-188C interface, this circuit meets paragraph 7.2 of the the standard governing the driver output and receiver input characteristics.

Back-to-back 1N752A zener diodes clamp the normally ± 9 V driver output peaks to within the ± 6 V required. The SN75152 input resistor terminal "R_T" is left open resulting in the 9 kΩ input resistance required. With the hysteresis control (pin 1) and the hysteresis adjust terminals (pins 4 and 13) open, the SN75152 will exhibit the ± 0.3 V hysteresis condition desired for MIL-STD-188C.

Additional capacitance may be required at the driver outputs to ensure proper rise (t_r) and fall (t_f) times. Wave shaping capacitors must be within 8 inches of the driver output (as close as possible) and provide t_r and t_f values that are to $\leq 15\%$ and $\geq 5\%$ of a unit interval as described in Figure 9-60.

SN75154 Quad Data Line Receiver

The SN75154 (Figure 9-61) quad receiver is designed to satisfy EIA RS-232-C requirements as a data receiver in the interface applications between data terminal equipment and data communications equipment. It is also useful for relatively low frequency, short line, point-to-point data receiver applications and for logic level translations. Key features include:

- 3 kΩ to 7 kΩ input resistance.
- Input threshold adjustable.
- Built-in hysteresis.
- TTL compatible outputs.
- Operates from 5 V or 12 V supplies.

In normal operation, the threshold control terminals are connected directly to the V_{CC1} terminal, pin 15, even if power is being applied via the alternate V_{CC2} terminal.

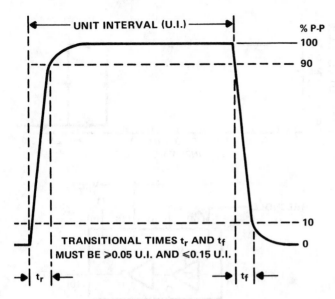

Figure 9-60. MIL-STD-188C Wave Shape Requirements

pin 16. This provides a wide hysteresis loop, [Figure 9-62(a)]. In this operation, if the input voltage goes to zero, the output voltage will remain at the low or high level as determined by the previous input.

For fail-safe operation, the threshold terminals are left open [Figure 9-62(b)]. This reduces the hysteresis loop by causing the negative-going threshold voltage to be above zero, (typically 1.4 V). The positive-going threshold voltage remains above zero, at typically 2.2 V, as it is unaffected by the disposition of the threshold terminals. In the fail-safe mode, if the input voltage goes to zero or an open-circuit condition, the output will go to the high level regardless of the previous input condition.

Figure 9-62(c) shows the specific condition of near zero negative-going threshold voltage achieved by connecting the

threshold terminal (T) to R1 (pin 9). This places a 5 kΩ resistor between T and V_{CC1}. By using an external resistor, and connecting it between the threshold control pin and V_{CC1}, the negative-going threshold may be adjusted to various levels as shown in Figure 9-62(d).

SN75189, SN75189A, MC1489, and MC1489A Quad Line Receivers

The SN75189 and SN75189A receivers (Figure 9-63) are designed to conform with EIA standard RS-232-C. Following are their key features:

- Satisfies EIA RS-232-C requirements.
- Built-in input hysteresis.
- Response control provides: input threshold shifting, input noise filtering.
- Operates from a single 5 V supply.
- TTL compatible outputs.

The SN75189 and SN75189A are quadruple line receivers designed to interface between data terminal equipment and data communication equipment in accordance with EIA standard RS-232-C. Each receiver has a response control terminal allowing adjustment of threshold levels and noise filtering.

Without level shifting the SN75189 will have about 240 mV hysteresis with a positive-going threshold of 1.3 V and a negative-going threshold 1.06 V. Under the same conditions the SN75189A will have 930 mV hysteresis with typical thresholds at 1.9 V and 0.97 V. With wider hysteresis the SN75189A provides more immunity to noise than the SN75189. The threshold level may be adjusted by connecting a resistor from the receiver response control to the positive or negative V_{CC} rail. Figure 9-64 illustrates the effects of different resistor values on threshold level.

For normal EIA RS-232-C applications no bias resistor is used. A frequency compensating capacitor can be used to filter out noise spikes. Figure 9-65 shows the capacitor

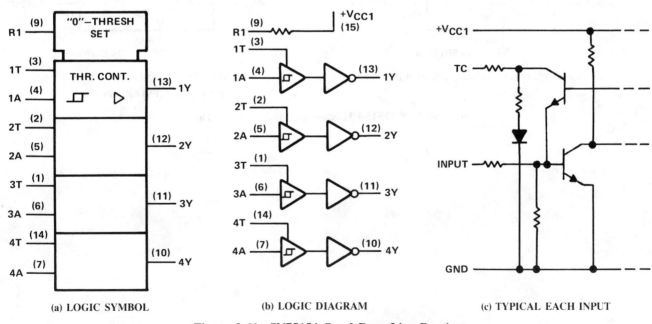

(a) LOGIC SYMBOL   (b) LOGIC DIAGRAM   (c) TYPICAL EACH INPUT

Figure 9-61. SN75154 Quad Data Line Receiver

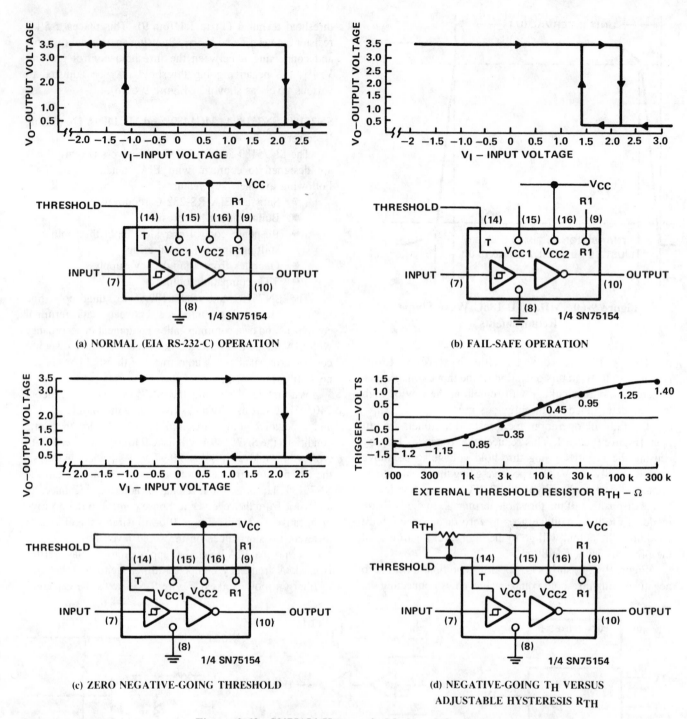

(a) NORMAL (EIA RS-232-C) OPERATION

(b) FAIL-SAFE OPERATION

(c) ZERO NEGATIVE-GOING THRESHOLD

(d) NEGATIVE-GOING T_H VERSUS
ADJUSTABLE HYSTERESIS R_{TH}

Figure 9-62. SN75154 Hysteresis Modes of Operation

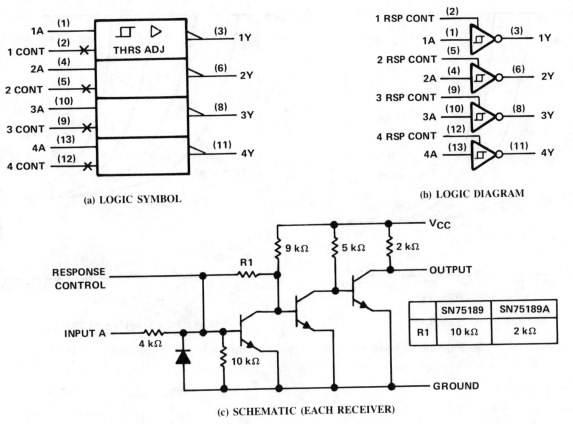

(a) LOGIC SYMBOL

(b) LOGIC DIAGRAM

(c) SCHEMATIC (EACH RECEIVER)

	SN75189	SN75189A
R1	10 kΩ	2 kΩ

Figure 9-63. SN75189/189A Quad Receiver

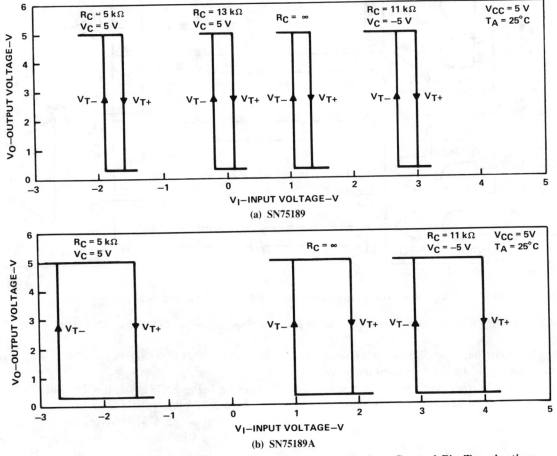

(a) SN75189

(b) SN75189A

Figure 9-64. Output Voltage Versus Input Voltage for Various Control Pin Terminations

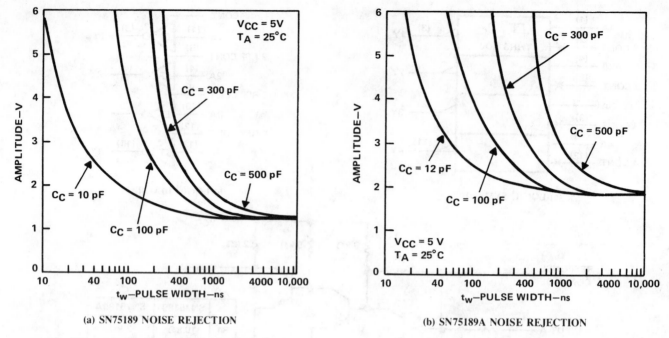

(a) SN75189 NOISE REJECTION   (b) SN75189A NOISE REJECTION

Figure 9-65. Noise Rejection Versus Compensation Capacitance for SN75189 and SN75189A

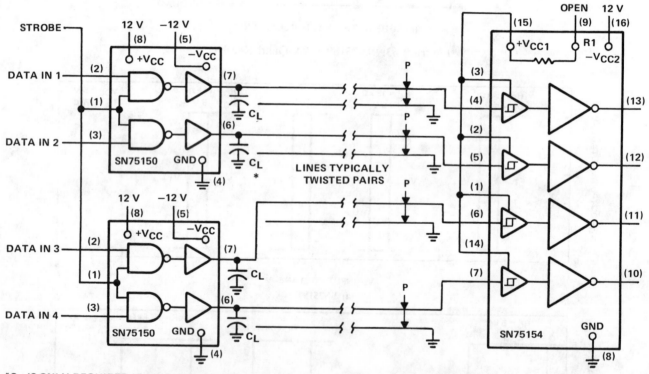

*C_L IS ONLY REQUIRED IF LINE CAPACITANCE IS INSUFFICIENT FOR PROPER WAVE SHAPING.

Figure 9-66. Basic RS-232-C Data Line Interface Using SN75150 and SN75154

value required to eliminate a noise spike knowing its amplitude and pulse width. These curves show the maximum input pulse amplitude that will not cause a change in the output level, for various pulse widths and response control capacitor values.

RS-232-C APPLICATIONS

Interface Using SN75150 and SN75154

Figure 9-66 illustrates a basic EIA RS-232-C interface configuration using SN75150 drivers and SN75154 receivers. Note that capacitive loads may be necessary at the driver outputs to ensure compliance with the standard's wave shaping requirements. The RS-232-C defined transitional region is from -3 V to 3 V. Rising or falling transition times shall not be less than 0.2 μs. Transition speed maximum is therefore 6 V/0.2 μs or 30 V/μs. The maximum transitional time shall not exceed 1 ms. Figure 9-67 illustrates the permissible operating areas for driver transitions versus pulse rates. Figure 9-68 shows the SN75150 output transition times as a function of total load capacitance.

Typical Interface Using SN75188 and SN75189A

Figure 9-69 shows a typical EIA RS-232-C application using the SN75188 quad driver and the SN75189A quad receiver. In this application it is desirable to operate at data rates up to 4800 bps. Noise pulses of up to 4 V amplitude with pulse widths as wide as 300 ns may be experienced. From Figure 9-65 a C_C of 390 pF is chosen to provide the necessary noise rejection. Compensation of 390 pF will not adversely affect the circuit response to the desired signal.

The line used in this application has a capacitance of 230 pF. From Figure 9-70 we can see that the resulting slew rate would be a little over 40 V/μs. The EIA RS-232-C maximum is 30 V/μs. It is desirable to operate within the limit and therefore 20 V/μs is chosen. From Figure 9-70, 20 V/μs will require a total load capacitance of 500 pF. An external capacitor of 270 pF is connected directly from the driver output to ground to properly limit the output slew rate.

EIA STANDARD RS-423-A CIRCUITS AND APPLICATIONS

RS-423-A STANDARD

This standard specifies the electrical characteristics of unbalanced voltage-mode digital interface circuits normally used for the interchange of serial binary signals between Data Terminal Equipment (DTE) and Data Circuit-Terminating Equipment (DCE). Values are given for both the driver and receiver characteristics. Devices designed to meet EIA RS-423-A are used for low speed data communication or control functions. They may be used, under certain conditions, with drivers and receivers of other digital interface standards such as EIA RS-232-C and MIL-STD 188C. Figure 9-71 shows the basic unbalanced digital interface configuration typical of EIA RS-423-A systems. RS-423-A allows one driver and up to 10 receivers on a single data line. Basic requirements for RS-423-A type data line drivers and receivers are in Table 9-10 and Table 9-11.

RS-423-A DEVICES

RS-423-A Drivers

The SN75156 and uA9636 are good examples of EIA RS-423-A drivers. These parts are shown in Figures 9-51 and 9-52 and discussed in the section on RS-232-C drivers

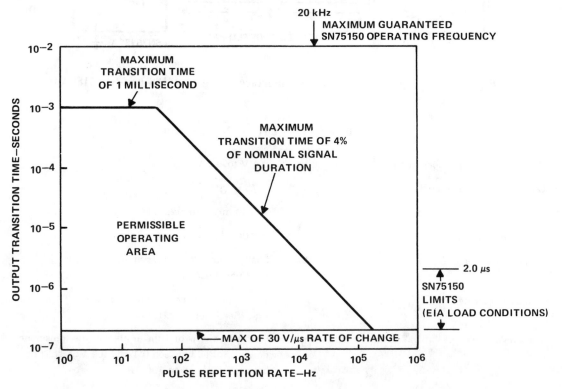

Figure 9-67. Pulse Repetition Rate Versus Output Transition Time of SN75150

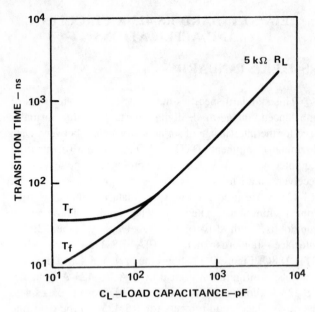

Figure 9-68. Output Transition Times Versus
Total Load Capacitance

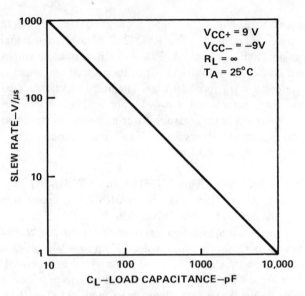

Figure 9-70. Slew Rate Versus Load Capacitance

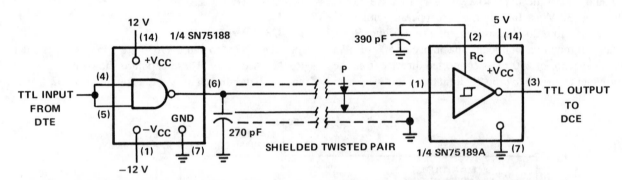

Figure 9-69. Typical EIA RS-232-C Interface Using SN75188 and SN75189A

Table 9-10. RS-423-A Driver Requirements

Output drive voltage open circuited output	V_O	± 4 V to ± 6 V.
Output drive voltage terminated in 450 Ω	V_t	$\geq 0.9\, V_O$.
Output short circuit current	I_S	< 150 mA with output in either logic state.
Output leakage current (with power off)	I_X	$< 100\ \mu A$ with output voltage of -6 V to 6 V.
Output slew rate	S_R	Shall not exceed 15 V/μs at any point during the transitional period.
Output monoticity		The output shall be monotonic between 0.1 and 0.9 V_{SS} (See Figure 9-72).
Transitional time	t_r	For pulse widths of 1 ms or greater, the transitional time measured between 0.1 and 0.9 V_{SS} shall be between 100 μs and 300 μs.
Ringing (overshoot or undershoot)		After completing a transition from one logic state to the other, the signal voltage should not vary more than 10% of V_{SS} from the steady state value until the next transition occurs (See Figure 9-72).

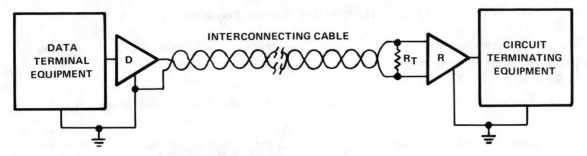

Figure 9-71. Basic RS-423-A Unbalanced Digital Interface

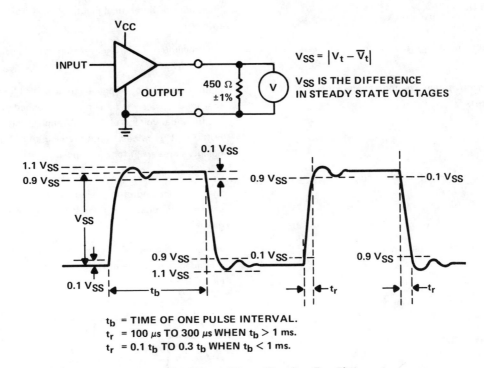

$$V_{SS} = |V_t - \overline{V}_t|$$

V_{SS} IS THE DIFFERENCE
IN STEADY STATE VOLTAGES

t_b = TIME OF ONE PULSE INTERVAL.
t_r = 100 μs TO 300 μs WHEN t_b > 1 ms.
t_r = 0.1 t_b TO 0.3 t_b WHEN t_b < 1 ms.

Figure 9-72. Signal Wave Shaping Requirement

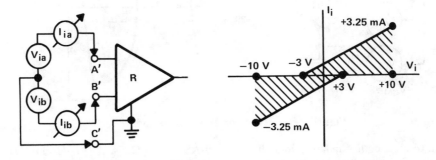

Figure 9-73. Receiver Input Current-Voltage Measurement

Table 9-11. RS-423-A Receiver Requirements

Input current at either input with the other input grounded and an applied input voltage of -10 V to 10 V.	I_{IN}	The input current shall remain within the shaded area of the graph in Figure 9-73.
Input resistance	R_{IN}	4 kΩ minimum under power on or power off conditions.
Input sensitivity over an input common-mode range of -7 V to 7 V. Referred to as differential input threshold voltage. (Figure 9-74)	V_{TH}	The differential input required to ensure the receiver will correctly assume the intended binary output state is 200 mV. The receiver must also maintain correct operation for differential input voltages from 200 mV to 6 V.
Differential input voltage maximum	V_{ID} (max)	The maximum V_{ID} without resulting in damage to the receiver is 12 V.
Input balance with any common-mode voltage from -7 to 7 V		Input voltage balance characteristics shall be such that the receiver will remain in its intended state with a differential voltage of 400 mV applied through 500 Ω ($\pm 1\%$) to each input (See Figure 9-75). V_{ID} polarity is reversed for the opposite binary state.
Multiple receivers and total loading		Up to 10 receivers may be connected on a single line. Total loading of multiple receivers and added fail safe circuits must have a resistance of 400 Ω or greater.
Recommended grounding		The signal ground wire, or common, should be grounded only at the driver end of the data line (See Figure 9-77).

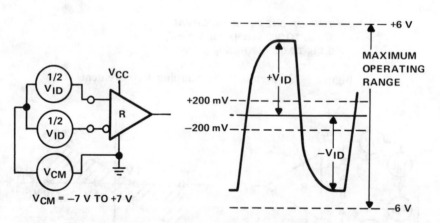

Figure 9-74. Receiver Input Sensitivity Test

as they were designed to meet both standards. In addition these drivers meet CCITT recommendations V.10, V.28 and X.26 as well as the Federal Standard FIPS 1030.

RS-423-A Receivers

Receivers designed to meet RS-423-A requirements normally have differential inputs. Thus they are generally compatible with EIA standard RS-422-A as well.

The SN75157 and uA9637A are examples of devices that fit both EIA standards RS-422-A and RS-423-A as well as Federal Standards 1020 and 1030. Figure 9-76 illustrates the logic diagram and symbol, package pinouts and equivalent input circuit for these devices.

BASIC RS-423-A APPLICATION

Figure 9-77 is an example of a Data Terminal with a remote monitor and printer using EIA RS-423-A circuits to interface data and control signals. The SN75156 dual driver and SN75157 dual receiver are used to provide the RS-423-A interfacing.

In this application, the data rate is to be 5 kilobaud resulting in a pulse width, t_b, of 200 μs. From Figure 9-72, the required transitional times must be between 0.1 and 0.3 t_b or between 20 and 60 μs. A driver waveshaping resistor of 330 kΩ (selected from Figure 9-52) provides about 40 μs transitional times.

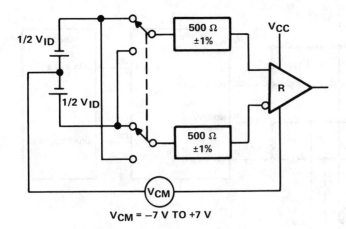

Figure 9-75. Receiver Input Balance Test

$V_{CM} = -7\ V\ TO\ +7\ V$

In this type of application, the data transmission line ground lead is connected to ground at the driver end of the line only. Each line is terminated in its characteristic impedance at the most remote end of the line. In this example 120 Ω resistors are used.

Although this application requires only two receiver stations per data line, as many as 10 receiver stations per line may be employed in an RS-423-A system.

EIA STANDARD RS-422-A AND RS-485 CIRCUITS AND APPLICATIONS

High speed data transmission between computer system components and peripherals over long distances, under high noise conditions, usually proves to be very difficult if not impossible with single-ended drivers and receivers. Recommended EIA standards for balanced digital voltage interfacing provide the design engineer with a universal solution for long line system requirements. As was seen with general purpose balanced data transmission, the ability to cope with common-mode voltages allows a differential (balanced) receiver to receive and reproduce signals that otherwise would be unusable.

A comparison of the single-ended EIA RS-232-C and RS-423-A standards with the differential EIA RS-422-A and RS-485 standards is shown in Table 9-12. The balanced system of data transmission incorporates a differential driver transmitting on balanced interconnecting lines to a receiver with differential inputs.

RS-422-A STANDARD

The balanced voltage digital interface circuit is normally used for data, timing or control lines where the signaling

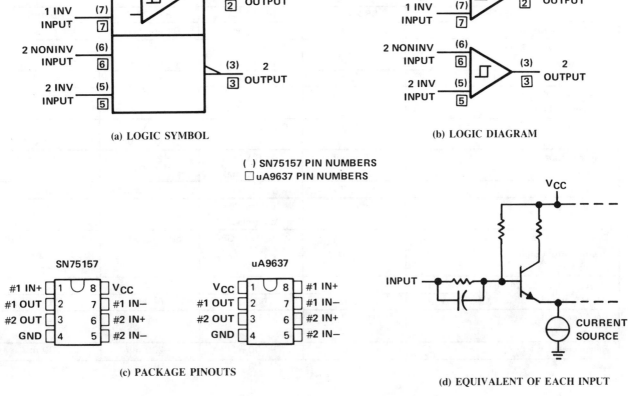

(a) LOGIC SYMBOL

(b) LOGIC DIAGRAM

( ) SN75157 PIN NUMBERS
☐ uA9637 PIN NUMBERS

(c) PACKAGE PINOUTS

(d) EQUIVALENT OF EACH INPUT

Figure 9-76. SN75157 and uA9637 Dual RS-423-A Receivers

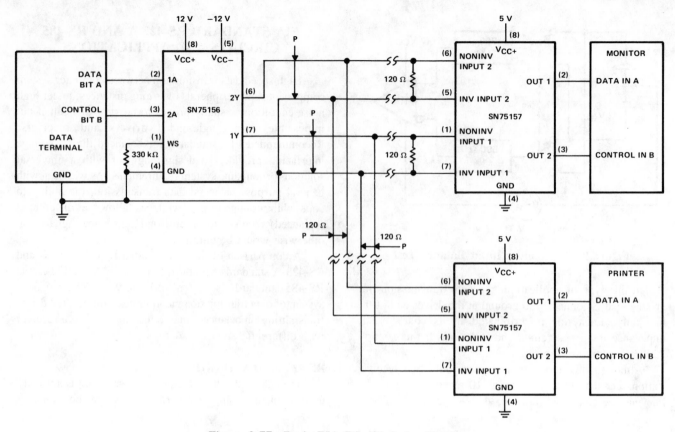

Figure 9-77. Basic EIA RS-423-A Application

Table 9-12. Popular General Purpose EIA Line Circuit Standards

PARAMETER		RS-232-C	RS-423-A	RS-422-A	RS-485
Mode of operation		Single-ended	Single-ended	Differential	Differential
Number of drivers and receivers allowed		1 Driver 1 Receiver	1 Driver 10 Receivers	1 Driver 10 Receivers	32 Drivers 32 Receivers
Maximum cable length (ft)		50	4000	4000	4000
Maximum data rate bits per second		20k	100k	10M	10M
Maximum common-mode voltage		± 25 V	± 6 V	6 V -0.25 V	12 V -7 V
Driver output		± 5 V min ± 15 V max	± 3.6 V min ± 6.0 V max	± 2 V min	± 1.5 V min
Driver load		3 kΩ to 7 kΩ	450 Ω min	100 Ω min	60 Ω min
Driver slew rate		30 V/μs max	Externally controlled	NA	NA
Driver output short circuit current limit		500 mA to V_{CC} or GRD	150 mA to GRD	150 mA to GRD	150 mA to GRD 250 mA to -8 V or 12 V
Driver output resistance (High Z state)	Power on	NA	NA	NA	120 kΩ
	Power off	300 Ω	60 kΩ	60 kΩ	120 kΩ
Receiver input resistance Ω		3 kΩ to 7 kΩ	4 kΩ	4 kΩ	12 kΩ
Receiver sensitivity		± 3 V	± 200 mV	± 200 mV	± 200 mV

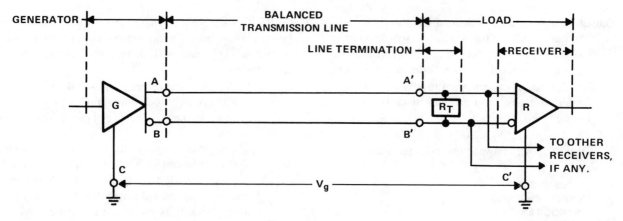

LEGEND: R_T = Terminating resistance, V_g = Ground voltage difference
           A, B = Generator-Line Interface,   A', B' = Load-Line interface
           C   = Generator ground   C'   = Load ground.

Figure 9-78. Balanced Digital Interface

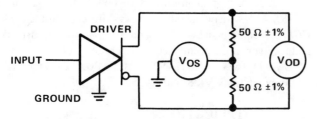

V_{OD} = DIFFERENTIAL OUTPUT VOLTAGE
V_{OS} = DRIVER OUTPUT OFFSET VOLTAGE

Figure 9-79. Driver Output Voltage Test Circuit

rates are from 100 kbps to 10 Mbps. RS-422-A specifications do not place restrictions on minimum or maximum operating frequencies but rather on the relationship of transitional speeds to a unit interval.

Although single-ended transmission circuits are normally used at lower frequencies, differential transmission on balanced lines may be preferred under the following conditions:

A. Interconnecting lines too long for effective unbalanced operation.

B. Transmission lines exposed to large electrostatic or electromagnetic noise levels.

C. When simple inversion of the signals may be desired (obtained by reversing the balanced lines).

A basic balanced digital interface circuit (seen in Figure 9-78) consists of three parts:

A. The generator (G) or data line driver.

B. A balanced transmission line.

C. The loads, where a load may consist of one or more receivers (R) and the line termination resistor (R_T).

The RS-422-A type of driver has a balanced (differential) output voltage source with an impedance of 100 Ω or less. Its output differential voltage is in the range of 2.0 V minimum to 6.0 V maximum. Additionally, the output voltage of either output, with respect to ground, shall not exceed 6.0 V.

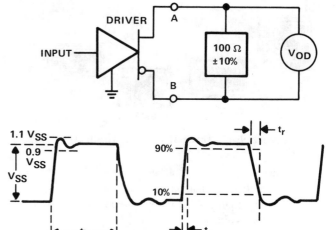

t_b = time of one unit interval
t_r = transitional times between the 10% and 90% of V_{SS} levels.
t_r ≤ 0.1 t_b when t_b is greater than 200 ns.
t_r ≤ 20 ns when t_b is less than 200 ns.
V_{SS} = difference between output steady state levels ($V_{OD} - \overline{V}_{OD}$)

Figure 9-80. Driver Transitional Characteristics

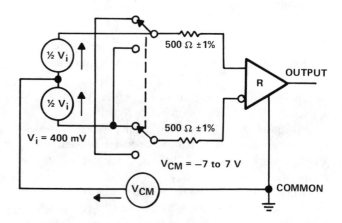

Figure 9-81. Receiver Input Balance Test

Output voltage balance is defined as follows:

The output differential voltage (V_{OD}) shall not be less than 2.0 V with two 50 Ω ($\pm 1\%$) termination resistors connected in series between the outputs. The difference between opposite polarity differential output voltages must be less than 0.4 V (see Figure 9-79).

The driver output offset voltage (V_{OS}), measured from the junction of the two 50 Ω terminators and

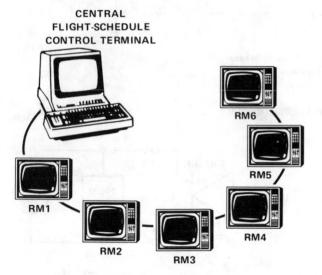

Figure 9.82. Basic RS-422-A Application

driver ground, shall not exceed 3.0 V (either polarity). The magnitude of the change in V_{OS} must be less than 0.4 V for opposite polarity differential output voltages.

Drive output current, with either output shorted to ground, shall not exceed 150 mA. Output off-state leakage

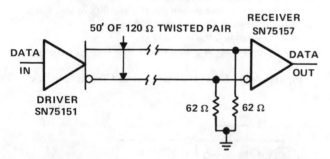

Figure 9-83. Short Line RS-422-A Application

current, with any voltage between −0.25 V and 6.0 V applied to either output, shall not exceed 100 μA.

As illustrated in Figure 9-80, output voltage transition times (t_r) are the transitional times between 0.1 and 0.9 of V_{SS} and must occur within 10% of a unit interval (t_b) or 20 ns, whichever is greater.

Ringing and resulting overshoot and undershoot shall not, as illustrated in Figure 9-80, exceed 10% of V_{SS} where V_{SS} is defined as the difference between the two steady-state values of the output.

RS-422-A receiver characteristics for balanced-line applications are basically the same as those for the RS-423-A single-ended applications. Basic receiver input requirements are as follows:

Differential data input threshold sensitivity of ± 200 mV over a common-mode (V_{CM}) range of −7 to 7 V. Input impedance of greater than or equal to 4 kΩ.

The receiver's input voltage-current characteristics shall be balanced such that its output remains in the intended binary state with a differential input of 400 mV applied, (through 500 Ω $\pm 1\%$ to each input terminal as illustrated in Figure 9-81), and V_{CM} is varied between −7 V and 7 V.

An RS-422-A receiver shall not be damaged, when powered-on or powered-off under the following conditions:

1. Driver output off (open circuit).
2. A short across the data line.
3. A short from either, or both, lines to ground.

RS-422-A drivers and receivers are compatible with CCITT recommendation V.11 and X.27. RS-422-A drivers and receivers are not intended for operation with RS-232-C, MIL-STD-188C, MIL-STD-188-100, or CCITT recommendations V.28 and V.25.

RS-422-A APPLICATIONS

Typical Application

A typical application of an RS-422-A is its use in communicating data from a central computer to multiple remote monitors or stations, such as airport arrival and departure monitors (see Figure 9-82).

In this application, a single twisted pair line is used to connect the central control terminal with several remote monitors distributed throughout the airport. Line termination would be at the most remote end from the control terminal (remote monitor 6). To minimize line noise it may be desirable to use two terminating resistors, with values of $R_T/2$, one from each line to ground.

Short-Line Application

Even in less complex applications the good noise rejection capability of RS-422-A type circuits may be advantageous. Figure 9-83 illustrates a simple, short-line, dedicated single driver and single receiver application with very good noise handling capability.

This combination of RS-422-A driver and receiver will provide −0.25 to 6.0 V common-mode capability and operate at speeds up to 20 Mbps. For improved negative noise voltage rejection capability it would be necessary to use a driver with more negative common-mode capability. The SN75172B driver is such a device with −7 to 7 V common-mode range. The SN75175 receiver also has a −7 to 7 V input capability, allowing very good overall noise performance for the pair.

EIA RS-485 STANDARD

EIA standard RS-485, introduced in 1983, is an upgraded version of EIA RS-422-A. Increasing use of

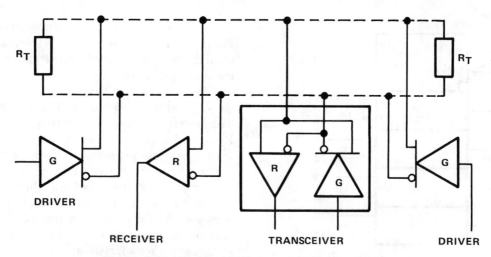

Figure 9-84. Multipoint Balanced Digital Interface

balanced data transmission lines in distributing data to several system components and peripherals over relatively long lines brought about the need for multiple driver/receiver combinations on a single twisted pair line.

EIA RS-485 takes into account RS-422-A requirements for balanced-line transmission plus additional features allowing for multiple drivers and receivers. Figure 9-84 illustrates an application similar to that of Figure 9-78, but with multiple drivers and receivers.

Standard RS-485 differs from the RS-422-A standard primarily in the features that allow reliable multipoint communications. For the drivers these features are:

- One driver can drive as many as 32 unit loads and a total line termination resistance of 60 Ω or more (one unit load is typically one passive driver and one receiver).
- The driver output, off-state, leakage current shall be 100 μA or less with any line voltage from −7 V to 7 V.
- The driver shall be capable of providing a differential output voltage of 1.5 V to 5 V with common-mode line voltages from −7 V to 12 V.
- Drivers must have self protection against contention (multiple drivers contending for the transmission line at the same time). That is, no driver damage shall occur when its outputs are connected to a voltage source of −7 V to 12 V whether its output state is a binary 1, binary 0 or passive.

For receivers these features are:

- High receiver input resistance, 12 kΩ minimum.
- A receiver input common-mode range of −7 V to 12 V.
- Differential input sensitivity of ±200 mV over a common-mode range of −7 V to 12 V.

Unit Load Concept

The maximum number of drivers and receivers that may be placed on a single communication bus depends on their loading characteristics relative to the definition of a "UNIT LOAD" (U.L.). RS-485 recommends a maximum of 32 unit loads (U.L.) per line.

One U.L. is defined (at worst case) as a load that allows 1 mA of current under a maximum common-mode voltage stress of 12 V (for a more detailed definition refer to the EIA Standard RS-485, page 4). The loads may consist of drivers and/or receivers but do not include the line termination resistors which may present an additional load as low as 60 Ω total.

Example: Early production SN75172 drivers and SN75173 receivers.

- Driver output leakage current in the off state with 12 V output voltage is 0.1 mA max.
- Receiver input current at a V_{in} of 12 V is 1 mA max.
- The driver represents 0.1 mA/1.0 mA or 0.1 U.L.
- The receiver represents 1.0 mA/1.0 mA or 1 U.L.
- As a pair they represent 1.1 mA/1.0 mA or 1.1 U.L. Therefore, 32/1.1 or 29 pairs would represent the maximum recommended 32 unit loads.

Example: 1984 production SN75172B drivers and SN75173A receivers.

- Driver I_O off at 12 V is 0.1 mA max
- Receiver I_{IN} at 12 V is 0.6 mA max

Although there were no changes in the driver loading, the receiver now represents only 0.6 U.L. A driver-receiver pair represents 0.7 U.L. and therefore 32 U.L./0.7 U.L. per pair or 45 of these driver-receiver stations could be handled on one twisted pair data transmission line.

DRIVERS AND RECEIVERS

Driver types SN75172B and SN75174B and receivers SN75173A and SN75175A were specifically designed for RS-422-A and RS-485 applications. These parts were designed to be direct plug-in replacements for popular RS-422-A circuits (see Figures 9-85, -86, -87 and -88).

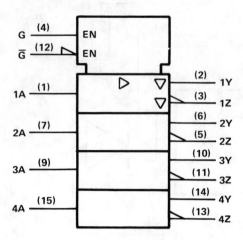

QUAD DIFFERENTIAL LINE DRIVER

<u>AM26LS31</u> MEETS EIA STANDARD
RS-422-A AND FEDERAL STANDARD 1020

<u>SN75172B</u> MEETS EIA STANDARDS RS-422-A AND
RS-485. ALSO MEETS CCITT RECOMMENDATIONS
V.11 AND X.27

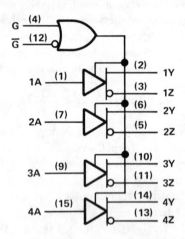

**Figure 9-85. Quad Differential Line Driver
AM26LS31 and SN75172B**

Converting from an RS-422-A system is easy if the AM26LS31 or MC3487 drivers and AM26LS32 or MC3486 receivers were previously being used. Conversion from a dedicated station-to-station line to a party line does not require rewiring of printed circuit boards, only the substitution of RS-485 circuits in the boards.

Driver Details

SN75172B series driver inputs are TTL compatible as previously mentioned. Figure 9-89 shows some additional features.

Using a pnp input transistor provides a relatively high input impedance and allows the input to be both TTL and low-level CMOS compatible. The maximum input high-level current of 20 μA and maximum low-level current of -360 μA are well within the capability of low-power Schottky or low-power CMOS gates. Input diodes will clamp any negative voltages at or below -1.5 V.

The basic driver output circuit configuration in Figure 9-90 is a totem-pole with both source and sink current limiting. High-speed Schottky output pull-up and pull-down transistors provide up to 60 mA of output drive current.

The pull-up output transistor receives its current from the V_{CC} rail through a resistor that is connected across the base-emitter junction of a current sensing transistor. Excessive high-level output current will turn on the current sensing transistor providing shut down drive to an output predrive transistor (not shown).

The pull-down transistor has a different type of current sensing. Its base is connected to the base of a current-sensing transistor. As the output sink transistor current reaches operating limits its V_{BE} rises to a level high enough to turn on the current-sense transistor. The current-sense transistor draws (by bypassing) base drive current from the output thus limiting the output sink current.

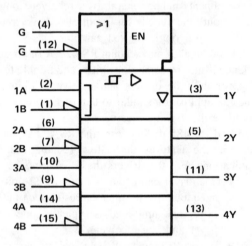

QUAD DIFFERENTIAL LINE RECEIVER

<u>AM26LS32A</u> MEETS EIA STANDARDS RS-422-A AND
RS-423-A

<u>SN75173A</u> MEETS EIA STANDARDS RS-422-A, RS-423-A
AND RS-485. ALSO MEETS CCITT RECOMMENDATIONS
V.10; V.11, X.26 AND X.27

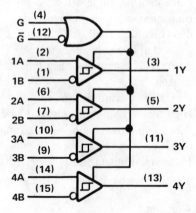

**Figure 9-86. Quad Differential Line Receiver
AM26LS32A and SN75173A**

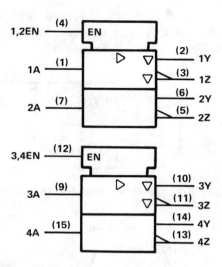

QUAD DIFFERENTIAL LINE DRIVER

<u>MC3487</u> MEETS EIA STANDARDS RS-422-A AND FEDERAL STANDARD 1020.

<u>SN75174B</u> MEETS EIA STANDARDS RS-422-A AND RS-485. ALSO MEETS CCITT RECOMMENDATIONS V.11 AND X.27

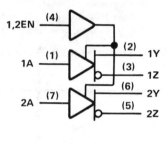

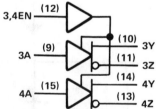

Figure 9-87. Quad Differential Line Driver MC3487 and SN75174B

The differential outputs will typically provide 3.5 V V_{OH} and 1.1 V V_{OL} levels for a 2.6 V differential drive (V_{OD}). If operating into a typical 120 Ω line-to-line termination, Figure 9-91, the resulting drive current is about 22 mA. If termination is to ground, resistor values are half of the line impedance, Figure 9-92. Output high-level currents would typically be: $V_{OH}/R_L/2$ or 3.5 V/62 Ω = 56 mA.

Driver Speed Characteristics

The SN75172B and SN75174B drivers exhibit a differential output transition time of 75 ns maximum. This particular parameter is key to the maximum speed at which the driver may operate in accordance with EIA RS-485 specifications.

EIA RS-485 defines the device transition time, "t_r" relative to one unit interval, "t_b". See Figure 9-93.

If transition time for the device is known then the maximum operating data rate for RS-485 conditions can be calculated as follows:

$$t_r = 0.3\ t_b \quad \text{and} \quad t_b = \frac{t_r}{0.3}$$

The data rate is

$$f_b = \frac{1}{t_b} \quad \text{or} \quad f_b = \frac{0.3}{t_r}$$

For example, the SN75172B or 174B drivers have a specified maximum t_r of 75 ns. At the worst-case level the maximum data rate would be:

$$f_b = \frac{0.3}{t_r} = \frac{0.3}{75 \times 10^{-9}}$$

$$f_b = 4 \text{ Mbps}$$

QUAD DIFFERENTIAL LINE DRIVER

<u>MC3486</u> MEETS EIA STANDARDS RS-422-A, RS-423-A AND FEDERAL STANDARDS 1020 AND 1030

<u>SN75175A</u> MEETS EIA STANDARDS RS-422-A RS-423-A AND RS-485. ALSO MEETS CCITT RECOMMENDATIONS V.10, V.11, X.26 AND X.27

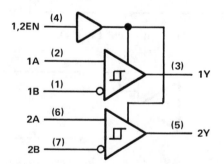

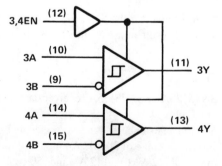

Figure 9-88. Quad Differential Line Receiver MC3486 and SN75175A

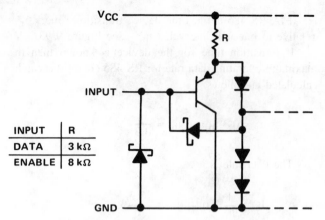

Figure 9-89. Equivalent Driver Input Circuit

INPUT	R
DATA	3 kΩ
ENABLE	8 kΩ

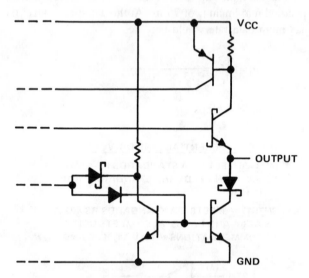

Figure 9-90. Equivalent Each Driver Output

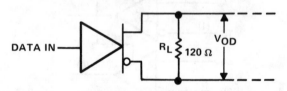

Figure 9-91. Line-to-Line Termination

A typical t_r value for the SN75172B and SN75174B is 50 ns and would result in a max f_b of $0.3/50 \times 10^{-9}$ or 6 Mbps.

Receiver Details

The SN75173A or SN75175A receiver data input circuit is shown in Figure 9-94.

Series input resistance and active impedances provide a typical input resistance of 20 kΩ. An input common-mode voltage range of 12 V to −12 V and 50 mV of input threshold hysteresis provide excellent noise immunity. Differential input threshold voltage is ±200 mV or less allowing the reception of signals that have been attenuated over long line lengths.

The receiver output has typical TTL output characteristics. The equivalent output circuit of the SN75173A/175A is shown in Figure 9-95.

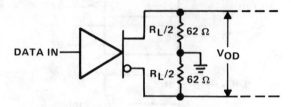

Figure 9-92. Line-to-Ground Termination

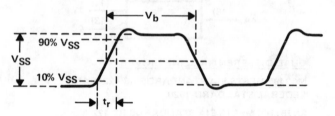

V_{SS} = difference in steady-state voltage levels

t_b = time duration of one unit interval, measured between successive 0.5 V_{SS} levels.

t_r = transitional time from 10% of V_{SS} to 90% of V_{SS}

t_r is also defined as being less than, or equal to, 0.3 t_b

Figure 9-93. Driver Output Waveform

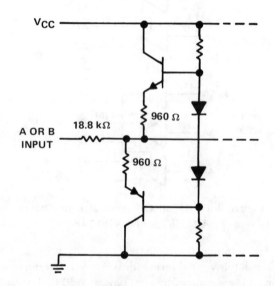

Figure 9-94. Equivalent Receiver A or B Input

In addition, the output has a high impedance when disabled or powered down. Output leakage current, when disabled, is less than 20 μA. High level output current is limited to 85 mA maximum when the output is shorted to ground.

The SN75173A and SN75175A are built with low-power Schottky technology resulting in propagation delay times of only about 20 ns (see Figure 9-96).

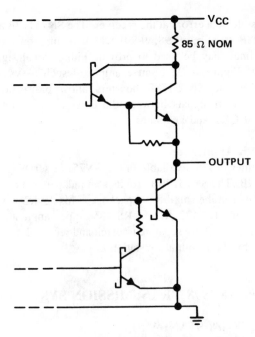

Figure 9-95. Equivalent Receiver Output

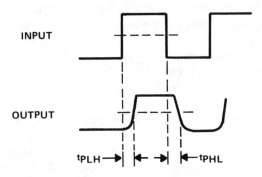

Figure 9-96. Receiver Propagation Delays

TRANSCEIVERS

Some EIA RS-485 type circuit applications require only one driver and one receiver at each location. Point-of-sale terminals is an example. In such applications a transceiver would be desirable. The SN75176B, SN75177B, SN75178B and SN75179B are types of devices used for this purpose. See Figures 9-97, -98, -99 and -100.

SN75176B, SN75177B, SN75178B, and SN75179B Transceivers Features

- Meet EIA Standard RS-485
- Meet EIA Standard RS-422-A, and CCITT V.11 and X.27
- 3-State Drivers (Except SN75179B)
- Wide Positive and Negative Bus Voltage Ranges
- Thermal Shutdown Protection
- Positive and Negative Driver Current Limiting
- Driver Output Current Capability ±60 mA
- Receiver Input Resistance . . . 20 kΩ Minimum

- Receiver Input Sensitivity ±200 mV
- Receiver Input Hysteresis 50 mV Typical
- Operate from Single 5 V Supply

In addition to the features listed, the SN75176B also provides individual driver and receiver enable pins and can be used as a basic transceiver for EIA RS-485 or EIA RS-422-A applications.

Basic Transceiver Application

Figure 9-101 shows transceivers distributed along a transmission line. Each station may transmit or receive data

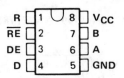

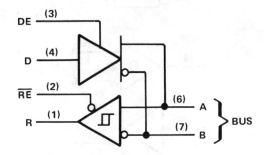

Figure 9-97. SN75176B

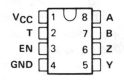

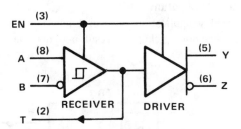

Figure 9-98. SN75177B

which is typical of point-of-sale terminals communicating with a central computer. The central computer handles the protocol conversion which is generally over a control line that is separate from the data bus.

In multipoint communication the transmission line is taken directly to each station, eliminating the use of long stubs or feeder lines. This is necessary to prevent adverse loading

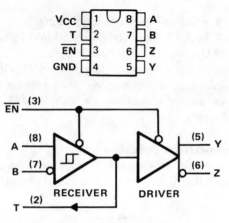

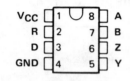

Figure 9-99. SN75178B

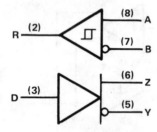

Figure 9-100. SN75179B

on the bus. Use of 3-inch or longer stubs may present a load to the main bus which will reduce signal amplitudes and can result in reflections and unwanted ringing.

The SN75176B represents only 0.7 unit loads and as many as 45 pairs could be connected to a single data transmission line.

Long-Line Application

In long-line applications of several thousand feet, signal attenuation and noise pickup may result in signal deterioration and possible data errors at the receiver. The SN75177B and SN75178B (Figures 9-98, 9-99) which require only one control line, may be used to provide bidirectional signal boosting. Figure 9-102 is an example of such a system. SN75177B and SN75178B boosters allow bidirectional transmission of signals between the CPU and a remotely connected CRT and keyboard.

SN75179B Application

Another device available in the SN75172 series is the SN75179B. The SN75179B provides an independent driver and receiver in the small 8-pin package. This device would be used in RS-422-A or RS-485 type applications (Figure 9-103) where separate transmit and receive lines are required by the terminal or station.

IEEE 488-1978 TRANSMISSION SYSTEMS

THE IEEE-488 STANDARD

The information in this section is intended as an overview of the IEEE-488 Interface System Standard. Brief definitions of some of the interface functions are presented to help you understand the basics of the system.

General Information

The IEEE Std 488-1978 defines a standard digital interface for programmable instrumentation dealing with systems that use a byte-serial bit-parallel method of transferring data between various instruments and/or system components. This interface system is optimized for communication over relatively short data lines and allows for party-line bus operation. The IEEE Std 488-1978 is often referred to by other names or abbreviations. Some examples are: ''GPIB'' (general purpose interface bus), ''HPIB'' (Hewlett-Packard interface bus), ''ASCII bus'' (American standard code for information interchange bus) and sometimes simply the ''488 interface''. Because the 488 interface system is easy to use, this standard has been widely accepted. As a result, the system designer may select various test instruments and control devices, from different manufacturers, that are 488 interface bus compatible and

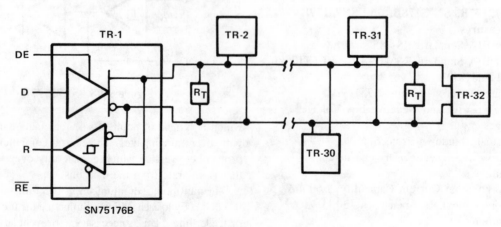

Figure 9-101. SN75176B Transceivers on Multi-Station Bus

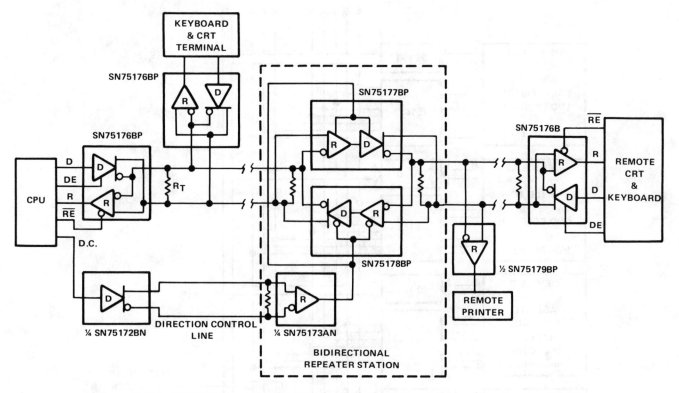

Figure 9-102. SN75177B and SN75178B Allow Bidirectional Long Line Communication

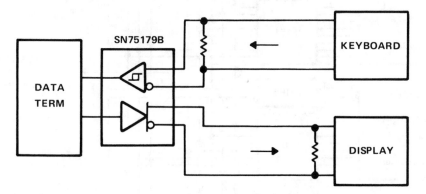

Figure 9-103. SN75179B Independent Driver and Receiver

know that they can work together. The situation was quite different in the mid 1960s when the Hewlett Packard Company began working on an interface standard for its future instruments. This original effort evolved into the IEEE Std 488-1975 which was revised to become the IEEE Std 488-1978 as we know it today.

The IEEE-488 standard defines an instrument interface system that allows up to 15 instruments within a localized area to communicate with each other over a common bus. Each device has a unique address, read from external switches at power-on, to which it responds. Information is transmitted in byte-serial, bit-parallel format and may consist of either data or interface control instructions.

Data may be sent by any one device (the Talker) and received by one or more of the other devices (Listeners).

Instructions such as select range, select function, or measurement data for processing or printing may be sent in this way. One of the devices on the bus, designated the Controller, may send interface control messages. It can assign an instrument to the bus as a listener or talker by sending its unique talk or listen address. The controller can also set the instrument for remote or local control.

The bus itself consists of 24 lines: eight lines carry data, eight are control lines, and eight are system ground lines. The block diagram in Figure 9-104 illustrates a basic IEEE-488 interface. The 16 signal lines shown allow for eight data lines, three hand shaking lines and five control lines.

1. The Data Lines: The eight data input/output lines carry the basic data, or information, to and from the instruments.

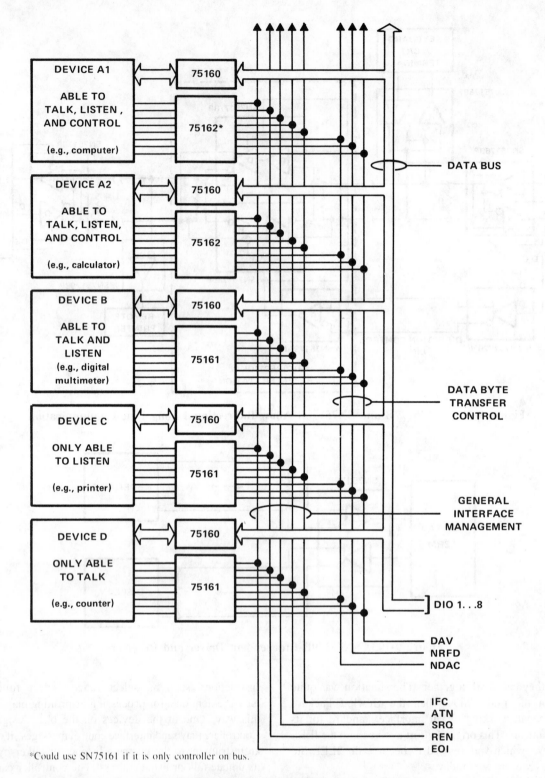

Figure 9-104. Typical IEEE-488 Interface System

2. The Handshaking Lines: The three data-transfer control lines are identified as "Data Valid" (DAV), "Not Ready For Data" (NRFD), and "Data Not Accepted" (NDAC).

3. The Management Lines: The five general management lines are identified as "Interface Clear" (IFC), "Attention" (ATN), "Service Request" (SRQ), "Remote Enable" (REN), and "End or Identify" (EOI).

Connectors

The pin connection assignments, physical dimensions, and mechanical mounting of the connector are given in the mechanical specifications of the IEEE-488 standard. Figure 9-105 illustrates the 24 pin interface connector and identifies each pin assignment. This type of connector is available from sources of Microribbon (Amphenol or Cinch series 57) or Champ (Amp) type connectors.

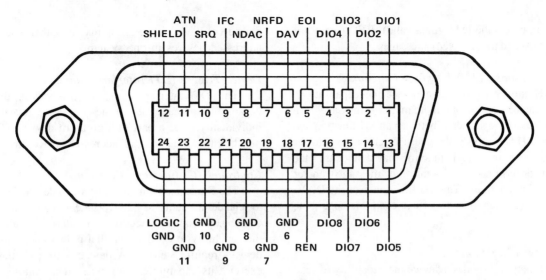

Figure 9-105. Interface Connector

The IEEE Std 488-1978 is very similar to its international counterpart, IEC publication 625-1. The IEC document specifies a connector system identical to that used with the EIA RS-232-C data communications interface. It should be noted, therefore, that component damage due to incompatible voltage levels is possible if data communication (EIA RS-232-C) and instrument (IEEE-488 or IEC 625-1) interfaces are interconnected. If interconnection between IEEE-488 and IEC 625-1 interfaces is required, mechanical differences in the connectors may be accommodated by use of special adapters.

Cable

The cable consists of at least 24 conductors: 16 signal lines and 8 logic ground returns. All lines are contained in a braided shield of 36 AWG wire or equivalent, providing 85% coverage. Line capacitance at 1 kHz on any signal line, with all other lines (signal, ground, and shield) grounded shall not exceed 150 pF per meter. Each signal line DAV, NRFD, NDAC, IFC, ATN, EOI, REN, and SRQ shall be twisted with a ground line or isolated in some other manner so as to minimize cross talk. A cable meeting these requirements is available from several major cable manufacturers. The total cable length in a system is limited to 2 meters times the number of instruments and shall not exceed a maximum of 20 meters without the aid of a booster or bus extender. The instruments may be unevenly spaced along the line but caution should be taken if any individual cable length exceeds 4 meters. Cables may be connected in a linear (daisy-chain) configuration, a star configuration, or combinations of these.

For best operation, the instruments should be at the same frame potential. Differences in frame potentials may result in excessive ground currents or possible system failure.

The maximum resistance of the cable conductors, per meter of cable length, is:

1. Each signal line (data or control) — 0.14 Ω.
2. Individual signal ground return — 0.14 Ω.
3. Common logic ground return — 0.085 Ω.
4. Overall shield — 0.0085 Ω.

Logic Convention

Many data transmission systems use positive logic on the bus lines. However, the IEEE-488 bus uses negative logic. That is, a voltage greater than 2 V on the signal line is considered a logic zero or false, while a voltage of 0.8 V is referred to as a logic one or true.

Three of the signal lines SRQ, NRFD, and NDAC must be driven from open-collector drivers. If any one of the line drivers is on, that line will be in a logic low or true condition. This is referred to as a "wired-OR" configuration and is necessary for these handshaking lines. Other bus lines may be driven by either three-state or open-collector drivers depending on the system application.

Bus lines are connected to three different types of devices. They are:

1. The Talker, which is any device that, when addressed, can transmit device dependent data over the bus.
2. The Listener, which is any device that, when addressed, can receive device dependent data from the bus.
3. The Controller, which can transmit data, receive data, and issue commands. The Controller manages the bus by designating listeners and talkers and by programming the other devices.

Functions

Instrument interface functions are covered in detail in the IEEE-488 standard. Each function may also have some options. Some of the basic functions are defined as follows:

Source Handshake (SH) Interface Function

The SH function provides a device the capability to guarantee the transfer of multiline messages. Asynchronous transfer of each multiline message is accomplished with an interlocked handshake sequence between the SH function and one or more of the AH (acceptor handshake) functions. The SH function controls the initiation and termination of a multiline message byte. To effect each message byte transfer, the SH function uses the DAV (data valid), RFD (ready for data), and DAC (data accepted) messages.

Acceptor Handshake (AH) Interface Function

The AH function provides a device the capability to guarantee proper reception of remote multiline messages. An interlocked handshake sequence between an SH function and one or more AH functions guarantees asynchronous transfer of each message byte. An AH function may delay either the initiation of, or termination of, a multiline message transfer until prepared to continue. The AH function uses the DAV, RFD, and DAC messages to effect each message byte transfer.

Talker (T) Interface Function

The T function provides a device the capability to send device dependent data (including status data during a serial poll sequence) over the interface to other devices when it is addressed. The normal T function uses a 1 byte address. The T with address extension is called the Extended Talker Function (TE) and uses a 2 byte address. Only one of the functions, T or TE, needs to be implemented in a specific device.

Listener (L) Interface Function

The L interface function provides a device the capability to receive device dependent data (including status) from other devices when addressed to listen. The normal L function uses a 1 byte address while the extended listener (LE), with address extension, uses a 2 byte address.

Service Request (SR) Interface Function

The SR interface function provides a device the capability to request service asynchronously from the interface controller. The SR interface function also synchronizes the content of the RQS (request service) message of the status byte present during a serial poll. Thus the SRQ (service request) message can be removed from the interface once the RQS message is received.

Remote Local (RL) Interface Function

The RL interface function provides a device the capability to select between the two sources of input information by indicating to the device which one is to be used, local (from the front panel) or remote (from the interface).

Parallel Poll (PP) Interface Function

The PP interface function provides a device the capability to present a PPR (parallel poll reply) message to the controller without being addressed to talk. A parallel poll enables the transfer of status data from multiple devices concurrently.

Device Clear (DC) Interface Function

The DC interface function provides a device the capability to be cleared (initialized) either individually or with a group of devices. The group may be either a subset or all addressed devices in one system.

Device Trigger (DT) Interface Function

The DT interface function provides a device the capability to have its basic operation started either individually or as a part of a group of devices. The group may be either a subset or all addressed devices in one system.

Controller (C) Interface Function

The C interface function provides a device the capability to send device addresses, universal commands and addressed commands to other devices over the interface. It also provides the capability to conduct parallel polls to determine which devices require service. A device with a C function can exercise its capabilities only when it is sending the ATN message over the interface.

If more than one device on the interface has a C function, then all but one of them shall be in the CIDS (controller idle state) at any given time. The device containing the C function which is not in the CIDS is called the controller-in-charge. IEEE-488 protocol allows devices with a C function to take turns as the controller-in-charge of the interface.

Only one of the devices with a C function that is connected to an interface can exist in the SACS (system control active state). It remains in this state throughout operation of the interface and has the capability to send the IFC and REN messages over the interface at any time whether or not it is the controller-in-charge. This device is called the system controller.

Messages or Commands

All communication between an interface function and its environment is accomplished through messages sent or received. Figure 9-106 illustrates the partitioning of an instrument or apparatus and shows the message flow between device functions and interface functions. Details on message conventions are given in the IEEE Std-488 and will not be covered in this text.

Single-Line Messages

Five interface signal lines are used to manage an orderly flow of information across the interface. Each line carries a specific message and is referred to as a single-line message (or command). These lines are the general management lines and are identified as follows:

1. ATN (attention) is used by a controller to specify how data on the Data I/O signal lines are to be interpreted and which devices must respond to the data. The ATN command causes all devices to cease their activity and listen to the controller. ATN commands are issued only by the controller.

2. IFC (interface clear) is used by a controller to place the entire interface system in a known state. This clears the interface bus and idles all the devices. The IFC command causes all data

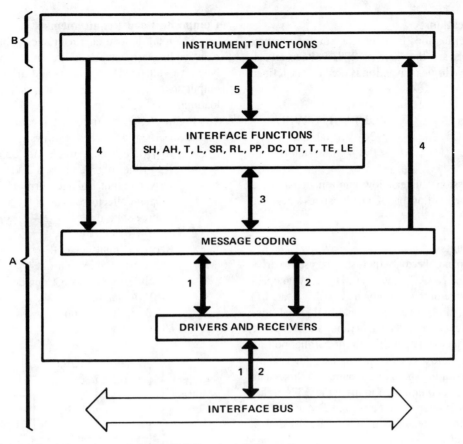

NOTES: A. Capability defined by IEEE Standard 488
B. Capability defined by designer
1. Interface bus data lines
2. Interface bus handshake and management lines
3. Remote interface messages to and from interface functions
4. Device dependent messages to and from device functions
5. Local messages between device functions and interface functions

Figure 9-106. Functional Partitioning Within an Instrument or Apparatus

transmission and all polls, to stop, and unaddresses all devices.

3. SRQ (service request) This command is given by a device, not a controller, to indicate the need for service and request interruption of current activity. The controller will then do a serial poll to determine which device has requested service and switch to a service routine.

4. REN (remote enable) is used by the controller to select device programming data from either the front panel (local) or the interface bus (remote).

5. EOI (end or identify) is used by a talker to indicate the end of a message string or byte sequence. EOI is used by a controller, in conjunction with ATN, to perform a polling sequence.

Multiline Messages

Multiline messages are interpreted as device dependent or interface messages, based on the state of ATN. Four classes of interface messages that may be accepted when the ATN command is true are:

1. The Universal commands are obeyed by all devices whether they have been told to listen or not.
2. Addressed commands are obeyed by those devices addressed to listen.
3. Unaddress commands remove the talker and all listeners from the bus.
4. Address commands designate the talkers and listeners.

BUS INTERFACE DEVICES

Interfacing between the instrument (or apparatus) interface functions and the system bus is accomplished with line driver and receiver circuits generally configured as transceivers. Since IEEE-488 systems consist of eight data lines and eight control lines, interfacing to the bus is often accomplished with octal transceivers. Details concerning the characteristics of these devices are covered in the following section on Electrical Specifications.

Electrical Specifications

Electrical specifications for driver output and receiver input voltage levels are the same as TTL voltage levels. However, a negative logic convention is used and is defined as follows:

LOGIC STATE	VOLTAGE LEVEL
ZERO	>2.0 V (high state)
ONE	<0.8 V (low state)

For this discussion, current flow into a node carries a positive sign and current flow out of a node carries a negative sign.

Driver Requirements

Drivers with open-collector outputs shall be used to drive the SRQ, NRFD, and NDAC signal lines. Drivers with open-collector or three-state outputs may be used to drive data I/O (lines 1—8), DAV, IFC, ATN, REN, and EOI signal lines with the following exception: Data I/O (lines 1—8) shall be driven by open-collector output drivers for parallel polling applications. Three-state driver outputs are used for systems where maximum operating speeds are required. A three-state driver output is also recommended to drive the ATN signal line if the controller is intended to be used in a system where other devices are implemented with three-state drivers on the DIO, DAV, and EOI signal lines.

Driver Specifications

Driver outputs shall meet the following specifications:
Low state output voltage (three-state or open-collector output) <0.5 V at 48 mA sink current.

The driver shall be capable of sinking 48 mA continuously.

High state output voltage (three-state) >2.4 V at −5.2 mA.

High state output voltage (open-collector) dependent on resistive load termination.

Receiver Specifications

Required input signal levels, necessary to provide nominal noise margins, shall be as follows:
Low state input voltage <0.8 V
High state input voltage >2.0 V
Preferred input signal characteristics, necessary to provide added noise margin and improved reliability, are as follows:
Receiver input hysteresis, V_t pos − V_t neg, shall be >0.4 V

Low state negative going threshold voltage V_t neg >0.8 V

High state positive going threshold voltage V_t <2.0 V

Composite Load Requirements

The total dc load characteristics primarily result from resistive line terminations and receiver input characteristics. Negative voltage clamp circuits and driver high-impedance output state characteristics result in some slight additional loading.

Figure 9-107 illustrates a typical circuit configuration whose specifications are as follows:
R_{L1} (line to V_{CC}) is 3 kΩ ±5%
R_{L2} (line to ground) is 6.2 kΩ ±5%
Driver output leakage current:
  open-collector driver is 0.25 mA maximum
  three-state driver is ±40 μA maximum
  at V_o = 2.4 V
Receiver input current:
  −1.6 mA maximum at V_o = 0.4 V
Receiver input leakage current:
  40 μA maximum at V_o = 2.4 V
  1.0 mA maximum at V_o = 5.25 V
V_{CC}: 5.0 V ±5%

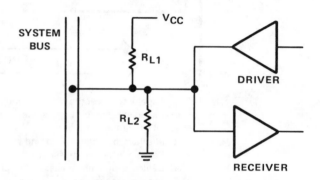

Figure 9-107. Typical Signal Line I/O Circuit

Device DC Load Line Boundaries

The load conditions of a device assume that the driver, receiver and termination network are internal to the device and that the driver is in its high-impedance state. The signal line interface to the device shall have a dc load characteristic that falls within the unshaded area illustrated in Figure 9-108. The load line boundary limits are defined as follows:
1. If I <0 mA, V is <3.7 V.
2. If I >0 mA, V is >2.5 V.
3. If I > −12.0 mA, V is > −1.5 V.
4. If V <0.4 V, I is < −1.3 mA.
5. If V >0.4 V, I is > −3.2 mA.
6. If V <5.5 V, I is <2.5 mA.
7. If V >5.0 V, I is >0.7 mA.

Device AC Load Line Limit

The small-signal load impedance shall be <2.0 kΩ at 1 MHz.

Device Capacitive Load Limit

Each device shall present an internal capacitive load of <100 pF to each of its signal lines.

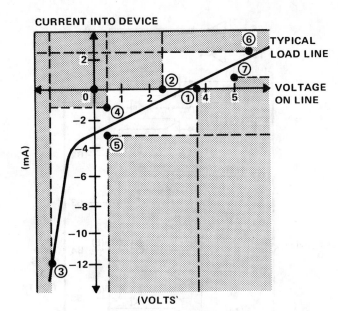

Figure 9-108. DC Load Line Permissible Operating Area

Timing Values

Definite relationships between critical signal inputs and outputs are required for successful interconnection of devices. Consideration of propagation delays, transition times and system response times is necessary for reliable operation.

Table 9-13 lists several of the required limits on these timing values. The longer time values shown allow for the inherently long propagation delays of the transmission lines as well as device circuit delays. The IEEE-488 Standard has a detailed discussion of these timing requirements.

Table 9-13. Time Values

FUNCTION	DESCRIPTION	VALUE
SH	Settling time for multiline messages	$> 2 \mu s$
SH,AH,T,L,LE,TE	Response to ATN	< 200 ns
AH	Interface message accept time	> 0
T,TE,L,LE,C,RL	Response to IFC or REN false	$< 100 \mu s$
PP	Response to ATN with EOI	< 200 ns
C	Parallel poll execution time	$> 2 \mu s$
C	Controller delay to allow current talker to see ATN message	> 500 ns
C	Length of IFC or REN false	$> 100 \mu s$
C	Delay for EOI	$> 1.5 \mu s$

Data Rates

Open-collector drivers will operate at a maximum of 250,000 bytes per second. Three-state drivers with equivalent standard loads will operate at a maximum of 500,000 bytes per second. With special precautions (three-state drivers, short signal lines and minimum capacitive loading) data rates up to 1 megabyte per second are possible.

INTERFACING TO THE IEEE STANDARD 488 BUS

The General-Purpose Interface Bus (GPIB) defined by the IEEE Standard 488 has received wide acceptance, and many commercially available instruments now use it. This allows a user to purchase instruments from several different manufacturers and connect them together using off-the-shelf cable. The TI SN75160 family of bus transceivers is designed to provide the interface between the bus and the bus controller of the instrument. These transceivers may be used with the TMS9914A or any other 488 bus controller. The SN75160 family of transceivers provides the simplest method of interfacing because each part is tailored to either the 8-line data bus or the 8-line control bus, and they require no extra logic or complicated board layout.

SN75160A Octal GPIB Transceiver

The SN75160A is an 8-channel, bidirectional, GPIB transceiver which is designed to meet the IEEE Standard 488-1978. This transceiver features driver outputs that can be operated in either the open-collector or three-state mode. See Figure 9-109 for the pinout and function tables.

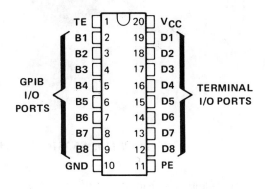

FUNCTION TABLES

EACH DRIVER

INPUTS			OUTPUT
D	TE	PE	B
H	H	H	H
L	H	X	L
H	X	L	Z‡
X	L	X	Z‡

EACH RECEIVER

INPUTS			OUTPUT
B	TE	PE	D
L	L	X	L
H	L	X	H
X	H	X	Z

H = high level, L = low level, X = irrelevant, Z = High-impedance state.

‡ This is the high-impedance state of a normal 3-state output modified by the internal resistors to V_{CC} and ground.

Figure 9-109. SN75160A Pinout and Function Tables

If the Talk Enable (TE) input is high and the Pull-up Enable (PE) is low, the output ports have an open-collector characteristic, and when the PE input is high the outputs are three-state. Taking the TE low places these ports in the high-impedance state. The driver outputs are designed to handle up to 48 mA sink current loading. An active-turn-off feature has been incorporated into the bus-terminating resistors so that the device exhibits a high impedance to the bus when $V_{CC} = 0$. See Figure 9-110 for the logic symbol and functional block diagram.

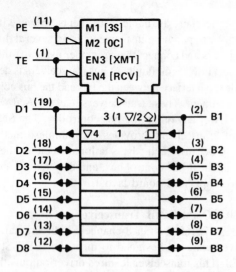

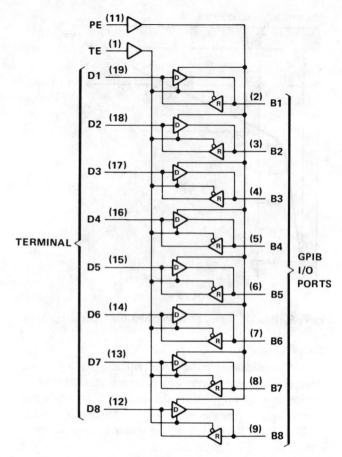

This symbol is in accordance with ANSI/IEEE Std 91-1984 and IEC Publication 617-12.

Figure 9-110. SN75160A Logic Symbol and Logic Diagram (Positive Logic)

Some additional features of the SN75160A are as follows:

PNP inputs for low drive requirements suitable for CMOS as well as TTL levels.

Totem-pole outputs on driver and receiver (open-collector option on driver).

Receiver hysteresis — 650 mV typical.

Low power dissipation — 66 mW per channel.

Fast propagation times — 22 ns maximum.

Figure 9-111 shows the totem-pole outputs on the driver and receiver sections. Notice the open-collector driver output option on the output port.

SN75163A Octal GPIB Transceiver

The SN75163A octal transceiver is functionally similar to the SN75160A but is not dedicated to a specific standard as is the SN75160A. The SN75163A does not have the built-in terminating resistors and as a result may be used to drive low impedance coax lines or other lines with impedances that differ from the IEEE-488 specification. Otherwise, the SN75163A shares the same features as the SN75160A.

SN75161A and SN75162A Octal GPIB Transceiver

The SN75161A and SN75162A eight-channel GPIB transceivers are monolithic, high-speed, low-power Schottky devices designed to meet the requirements of IEEE-488. Each transceiver is designed to handle the bus-management, data-transfer and control signals of a single- or multiple-controller instrumentation system. When combined with the SN75160A, the SN75161A or SN75162A provides the complete 16-wire interface for the IEEE-488 bus. The pinouts and table of abbreviations are given in Figure 9-112.

The SN75161A and SN75162A each feature eight driver-receiver pairs connected in a front-to-back configuration to form input/output (I/O) ports at both the bus and instrument sides. The direction of data through the driver-receiver pairs is determined by the DC, TE, and SC (on SN75162A) enable signals. The SC input on the SN75162A allows the REN and IFC transceivers to be controlled independently. The driver outputs (GPIB I/O ports) feature active bus-terminating resistor circuits designed to provide a high impedance to the bus when $V_{CC} = 0$. See Figure 9-113 for the logic symbols for each device and Figure 9-114 for the functional block diagram of each device.

Table 9-14 shows the receive/transmit functions of the SN75161A and SN75162A.

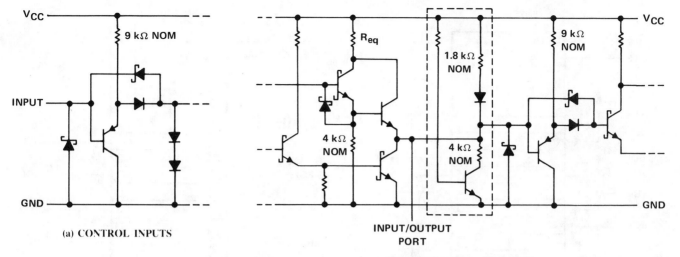

(a) CONTROL INPUTS

INPUT/OUTPUT
PORT

Driver output R_{eq} = 30 Ω NOM
Receiver output R_{eq} = 110 Ω NOM
Circuit inside dashed lines is on the driver outputs only.

(b) INPUT/OUTPUT PORTS

Figure 9-111. Equivalent Schematics of SN75160A Input and Output Sections

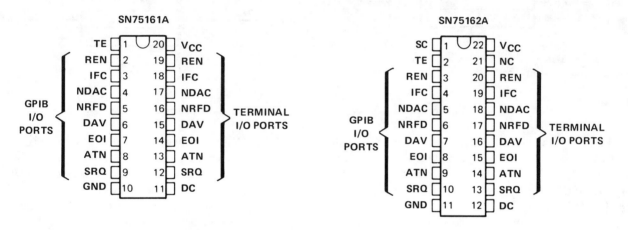

NC — No internal connection.

TABLE OF ABBREVIATIONS

NAME	IDENTITY	CLASS
DC	Direction Control	
TE	Talk Enable	Control
SC	System Control (SN75162 only)	
ATN	Attention	
SRQ	Service Request	
REN	Remote Enable	Bus Management
IFC	Interface Clear	
EOI	End or Edentify	
DAV	Data Valid	
NDAC	Not Data Accepted	Data Transfer
NRFD	Not Ready for Data	

Figure 9-112. SN75161A and SN75162A Pinouts and Table of Abbreviations

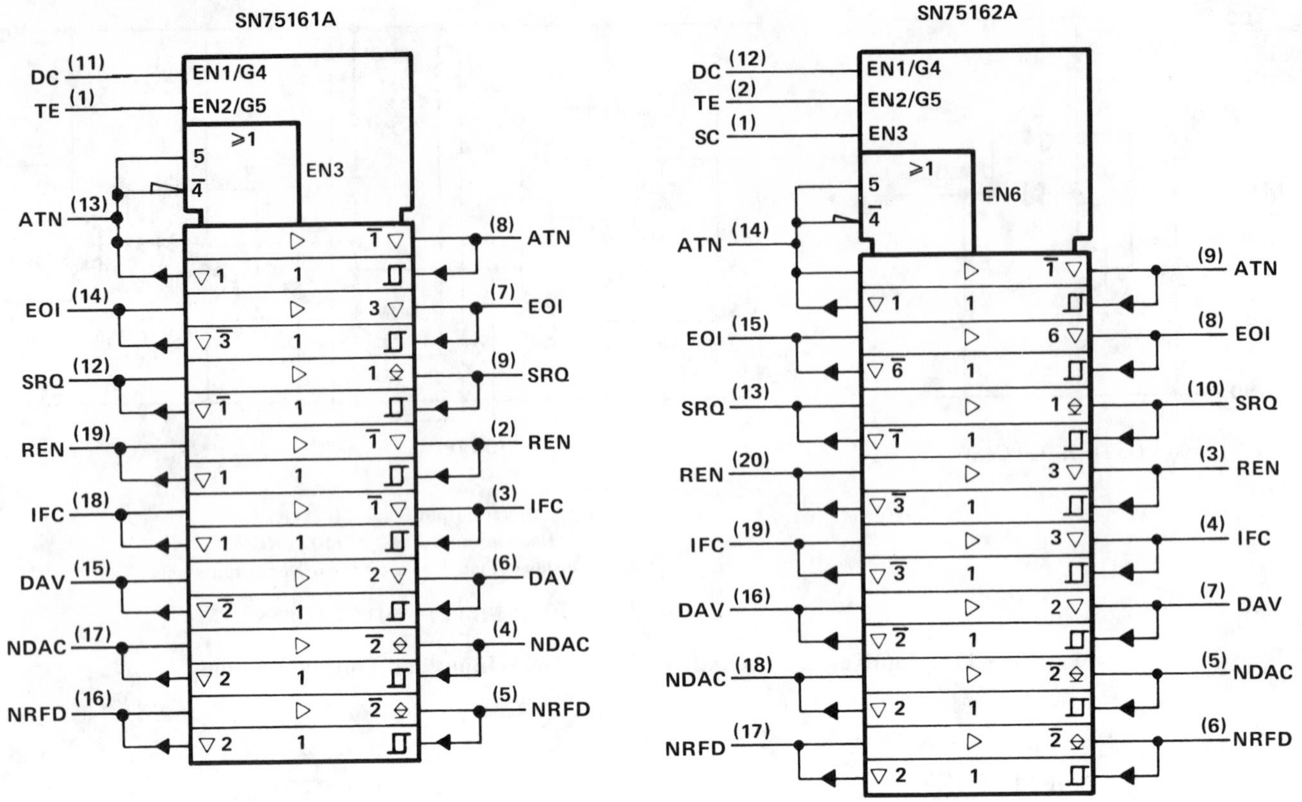

▽Designates 3-state output, ⇔ designates open-collector outputs. These symbols are in accordance with IEEE Std 91-1984 and IEC publication 617-12.

Figure 9-113. Logic Symbols for SN75161A and SN75162A

Figure 9-115 shows the equivalent schematics of the inputs and outputs of the SN75161A and the SN75162A.

MC3446 Quad Bus Transceiver

This device is a quad, single-ended line transceiver designed for bidirectional flow of data and instructions. See Figure 9-116 for pinout and function table.

Each driver output is tied to the junction of an internal voltage divider that sets the no-load output voltage and provides bus termination. The driver outputs are guaranteed to be off during power up and power down if either input is high. The receivers feature 950 mV typical hysteresis for noise immunity. The MC3446 is designed to meet IEEE Standard 488 — 1975 requirements.

Equivalent schematics of the device inputs and outputs are shown in Figure 9-117. The MC3446 drivers feature open-collector outputs for party-line operation. Like the SN75160 family, the driver outputs stay off during power up and power down sequencing.

TYPICAL APPLICATIONS

To use the SN75160 family of bus transceivers effectively, a bus controller or interface device should be used. The Texas Instruments TMS9914A is designed to perform the interface function between the IEEE 488 — 1975/1978 GPIB and a microprocessor. IEEE 488 — 1975/1978 standard protocol is handled automatically in Talker, Listener, or Controller operational modes. The

TMS9914A is used when an intelligent instrument is required to communicate with the IEEE-488 bus. It performs the interface function between the microprocessor and bus and relieves the processor of the task of maintaining IEEE-488 protocol. By utilizing the interrupt capabilities of the device the bus does not have to be continually polled, and fast responses to changes in the interface configuration can be achieved. A block diagram showing the TMS9914A in a typical application is given in Figure 9-118.

The TMS9914A input/output pins are connected to the IEEE-488 bus via bus transceivers. The direction of data flow is controlled by the TE and Controller outputs generated in the TMS9914A. The SN75160A, SN75161A, and SN75162A are designed specifically as bus interfaces. The TE and Controller signals are routed within the devices so that the transceivers on particular lines are controlled as required by the TMS9914A. Other transceivers may be used but they may require additional external logic, particularly around the EOI line transceiver.

Communication between the microprocessor and TMS9914A is carried out via memory-mapped registers. There are 13 registers within the TMS9914A: six are read only registers and seven are write only registers. They are used both to pass control data to, and get status information from, the instrument.

The three least significant address lines from the MPU are connected to the register select lines RS0, RS1, and RS2 and determine the particular register selected. The high order

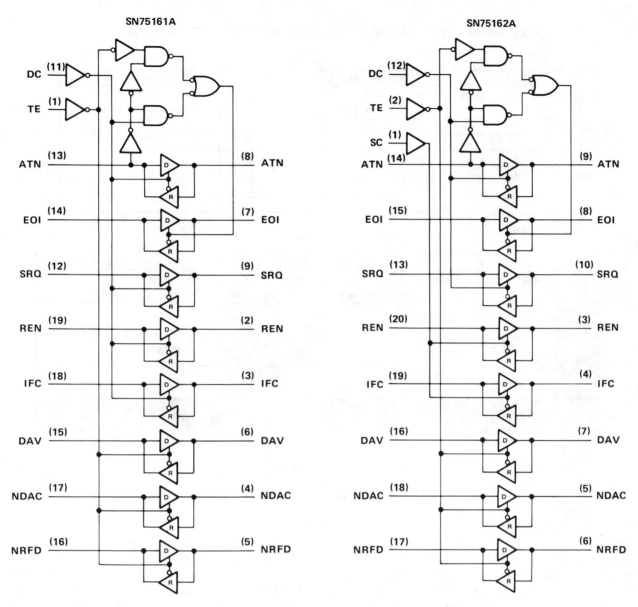

Figure 9-114. Logic Diagrams (Positive Logic) of SN75161A and SN75162A

address lines are decoded by external logic to cause the CE input to the TMS9914A to be pulled low when any one of eight consecutive addresses is selected. Thus, the internal registers appear to be situated at eight consecutive locations within the MPU address space. Reading or writing to these locations transfers information between the TMS9914A and the microprocessor. Note that reading and writing to the same location will not access the same register within the TMS9914A since they are either read only or write only registers. For example, a read operation with RS2 − RS0 = 011 gives the current status of the GPIB interface control lines, whereas a write to this location loads the auxiliary command register.

Each device on the bus is given a 5-bit address enabling it to be addressed as a talker or listener. This address is set on an external DIP switch (usually at the rear of an instrument) before power-on and is both read by the microprocessor and written into the register as part of the initialization procedure. The TMS9914A responds by causing

a MA (My Address) interrupt and entering the required addressed state when this address is detected on the GPIB data lines.

A TYPICAL IEEE-488 SYSTEM APPLICATION

Figure 9-119 shows a possible transceiver connection between the TMS9914A and a remote station. This station could be a printer, digital multimeter, frequency counter, or other equipment of this type. The remote stations may talk only, listen only, or talk and listen. In the case of the SN75162A the remote station may, in addition to being able to talk and listen, be able also to assume control of the system. This is done by the System Control pin (SC) on the SN75162A.

The SN75160A is a 20-pin device used to communicate with the IEEE-488 data lines [DIO(1-8)] in all applications. Its mode of operation is controlled by the Talk Enable (TE) output of the TMS9914A. This active high signal becomes true whenever there is an interface function of the

Table 9-14. SN75161A and SN75162A Receive/Transmit Function Tables

SN75161A

CONTROLS			BUS-MANAGEMENT CHANNELS				EOI	DATA-TRANSFER CHANNELS		
DC	TE	ATN†	ATN†	SRQ	REN	IFC (Controlled by DC)		DAV	NDAC	NRFD (Controlled by TE)
H	H	H	R	T	R	R	T	T	R	R
H	H	L	R	T	R	R	R	T	R	R
L	L	H	T	R	T	T	R	R	T	T
L	L	L	T	R	T	T	T	R	T	T
H	L	X	R	T	R	R	R	R	T	T
L	H	X	T	R	T	T	T	T	R	R

SN75162A

SC	DC	TE	ATN†	ATN†	SRQ (Controlled by DC)	REN	IFC (Controlled by SC)	EOI	DAV	NDAC	NRFD (Controlled by TE)
	H	H	H	R	T			T	T	R	R
	H	H	L	R	T			R	T	R	R
	L	L	H	T	R			R	R	T	T
	L	L	L	T	R			T	R	T	T
	H	L	X	R	T			R	R	T	T
	L	H	X	T	R			T	T	R	R
H						T	T				
L						R	R				

H = high level, L = low level, R = receive, T = transmit, X = irrelevant

Direction of data transmission is from the terminal side to the bus side, and the direction of data receiving is from the bus side to the terminal side. Data transfer is noninverting in both directions.

†ATN is a normal transceiver channel that functions additionally as an internal direction control or talk enable for EOI whenever the DC and TE inputs are in the same state. When DC and TE are in opposite states, the ATN channel functions as an independent transceiver only.

TMS9914A not sending the NUL message on DIO(1-8), that is when the device is in TACS, CACS, SPAS, or PPAS. The Pull-Up Enable (PE) input of the SN75160A is an active high input which selects whether the DIO(1-8) lines are driven by open-collector or totem-pole outputs. A totem-pole driver output is required for faster data rates. Open-collector outputs must be used if parallel polling is required. If only one of these features is desired the PE input may be hard wired. Otherwise, it must be derived from ATN and EOI, as shown in Figure 9-119.

The SN75161A is a 20-pin device used with the IEEE-488 interface management lines. It may be used for a talker/listener device or for a controller which does not pass control. The direction of the handshake line transceivers NRFD, NDAC, DAV are again controlled by the TE signal. However, the SRQ, ATN, REN, and IFC transceivers are controlled by the DC input of the SN75161A, which connects to the Controller Active (CONT) output of the TMS9914A. CONT is low whenever the TMS9914A is an active controller, that is, when it is not in DIDS or CADS. The SN75161A also includes the logic to control the direction of the EOI transceiver. This is dependent on the TE signal when ATN is false (high) and on the DC signal when ATN is true (low).

The SN75162A is a 22-pin device which may be used to interface with the IEEE-488 interface management lines in all applications including devices which pass control. The SN75162A has a separate pin to control the direction of the REN and IFC transceivers, but is otherwise identical to the SN75161A. This input is the System Controller (SC) input which may be hard wired or switchable to determine whether the instrument in question is a controller. The REN and IFC outputs of the TMS9914A are controlled by the auxiliary commands "sre" and "sic". These should never be used by the host MPU unless it is the system controller.

IBM SYSTEM 360/370 INTERFACING CIRCUITS

The purpose of the interface is to provide a ready physical connection to System/360 and System/370 control units. Information, in the form of data, status and sense information, control signals, and I/O addresses is transmitted over the time- and function-shared lines of this interface. The I/O interface (channel to control unit) is the communication link between a channel, or line, and the various I/O control units in the IBM System/360 and System/370. Information on the complete design of this interface is available from:

IBM SYSTEM PRODUCTS DIVISION
PRODUCT PUBLICATIONS DEPT. B98
PO BOX 390
POUGHKEEPSIE, N.Y. 12602

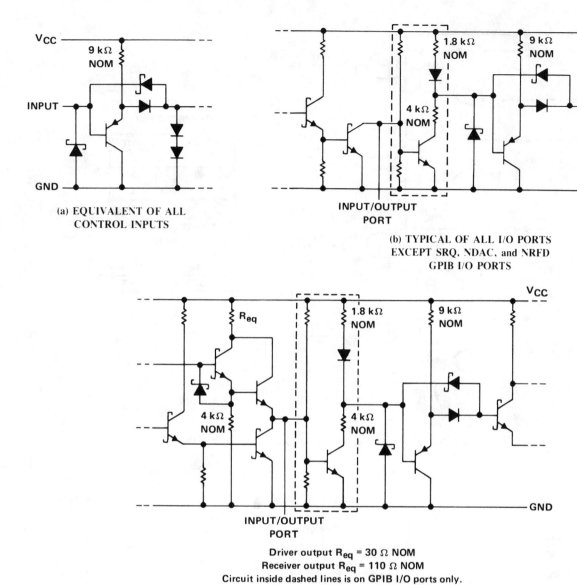

(a) EQUIVALENT OF ALL
CONTROL INPUTS

(b) TYPICAL OF ALL I/O PORTS
EXCEPT SRQ, NDAC, and NRFD
GPIB I/O PORTS

Driver output R_{eq} = 30 Ω NOM
Receiver output R_{eq} = 110 Ω NOM
Circuit inside dashed lines is on GPIB I/O ports only.

(c) TYPICAL OF SRQ, NDAC, and NRFD
GPIB I/O PORTS

Figure 9-115. Input/Output Sections of SN75161A and SN75162A

The design of the interface provides several important features:

1. A degree of consistency in input/output programming for a wide range of control units.
2. Ready physical connection to System/360 and System/370 control units designed by any manufacturer.
3. Ability to physically accommodate future control units designed to meet the parameters of this interface specification.
4. An interlocked interface operation that is, in most cases, time independent: this feature increases the variety of control units that may be attached.
5. An operation applicable to both multiplex and burst mode operations as well as many control operations and channel-to-channel transmissions.

6. Up to eight control units may be serviced per set of lines.

DRIVER AND RECEIVER REQUIREMENTS

Driver Requirements

A driver at one extreme end of the data line must be able to drive up to 10 receivers. Up to 10 drivers must be able to be wire-OR connected to drive one receiver. The receiver would be located at one extreme end of the data line.

In the logic zero state:

1. The driver output voltage must not exceed 0.15 V at a load of 240 μA. (See Figure 9-120 for the definition of driver current polarity.)

In the logic one state:

1. The driver output voltage must be 3.11 V or greater at a load of 59.3 mA.

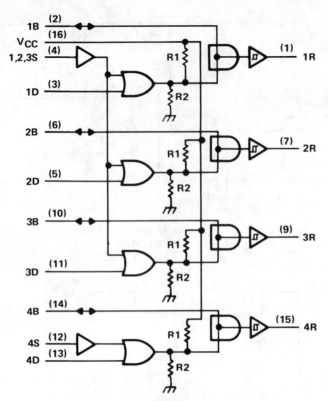

FUNCTION TABLE (TRANSMITTING)			
INPUTS		**OUTPUT**	
S	**D**	**B**	**R**
L	H	H	H
L	L	L	L

FUNCTION TABLE (RECEIVING)			
INPUTS			**OUTPUT**
S	**B**	**D**	**R**
H	H	X	H
H	L	X	L

R1 = 2.4 kΩ NOM, R2 = 5 kΩ NOM

Figure 9-116. MC3446 Logic Diagram and Function Tables

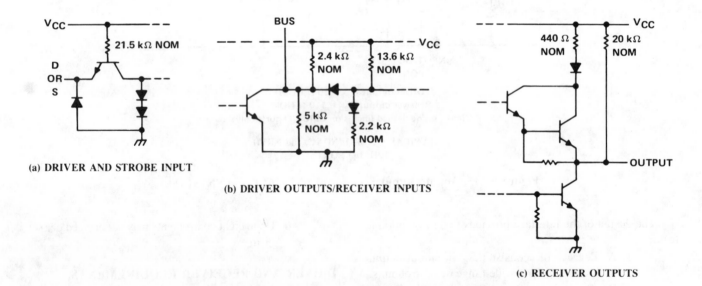

(a) DRIVER AND STROBE INPUT

(b) DRIVER OUTPUTS/RECEIVER INPUTS

(c) RECEIVER OUTPUTS

Figure 9-117. MC3446 Input and Output Equivalent Circuits

2. The output voltage must not exceed 5.85 V at a load of 30 μA.

3. The output voltage must not exceed 7.0 V at a load of 123.0 mA during an overvoltage internal to the driver.

Drivers must be designed to ensure that no spurious noise is generated on the line during a normal power-up or power-down sequence. For the driver this may be accomplished by one of the following methods:

1. Sequencing the power supplies.
2. Building noise suppression into the circuit.
3. Providing an externally controlled gate. (See Figure 9-121.)

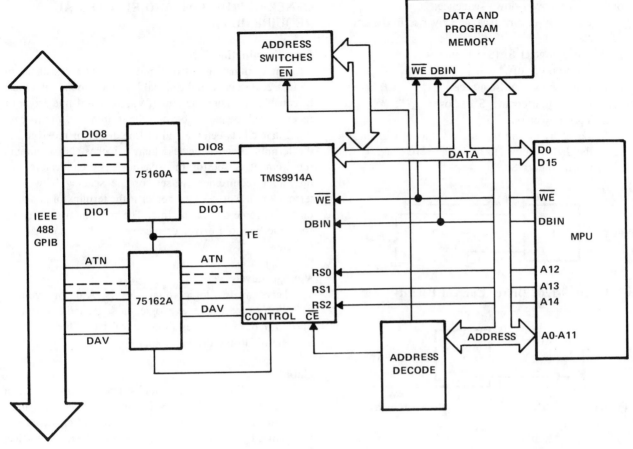

Figure 9-118. Typical TMS9914A Application Block Diagram

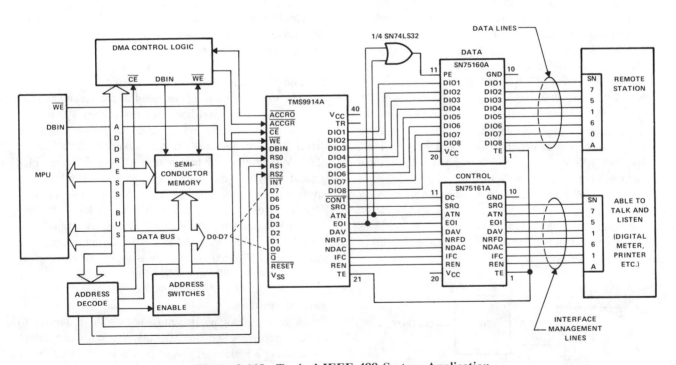

Figure 9-119. Typical IEEE-488 System Application

For a normal power-down sequence:
1. Logically ensure that the driver is in the zero state.
2. Close contact S. (See Figure 9-121.)
3. Turn power off.

For a normal power-up sequence:
1. Ensure that contact S is closed.
2. Turn power on.
3. Logically ensure that the input level will cause the driver output to be in the zero state.
4. Open contact S.

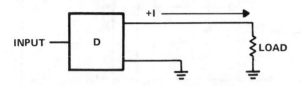

Figure 9-120. Driver Current Polarity

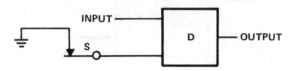

Figure 9-121. Driver Control Gate Switch (S)

Receiver Requirements

If there are multiple receivers on a single data line they must be spaced a minimum of 3 feet apart.

An input voltage (relative to receiver circuit ground) of 1.7 V or greater is interpreted as a logic one; an input of 0.70 V or less is interpreted as a logic zero.

The receiver should not be damaged by:
1. A dc input of 7.0 V with the receiver power on.
2. A dc input of 6.0 V with the receiver power off.
3. A dc input of −0.15 V with power on or off.

To reduce the loading effect on the line, the receiver input resistance must be greater than 7400 Ω with any input voltage from 0.15 V to 3.9 V, and the negative receiver input current must not exceed −0.24 mA at an input voltage of 0.15 V. (See Figure 9-122 for the definition of receiver current polarity.)

Receivers must be designed to ensure that no spurious noise is generated on the line during a normal power-up or power-down sequence.

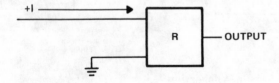

Figure 9-122. Receiver Input Current Polarity

GENERAL PHYSICAL AND ELECTRICAL REQUIREMENTS

Line Terminations

The terminating network must present a resistance of 95 Ω ±2.5% between the signal line and ground, and must be capable of dissipating 390 mW. An end-of-line driver or receiver may be placed beyond the terminator. In this case the distance between the end-of-line driver or receiver and the terminator must be less than 6 inches. No minimum requirement is set for the spacing between drivers. No minimum requirement is set for the spacing between a terminator and driver or receiver if the terminator is placed on the extreme end of the line. The maximum stub length from the line to a driver or receiver on the circuit card is 6 inches.

Voltage Levels

There are two logic voltage levels. A dc line voltage of 2.25 V or more denotes a logic one state, and a dc voltage of 0.15 V or less denotes a logic zero state. These voltages are relative to the driver ground.

Cable

All data lines must have a characteristic impedance of 92 Ω ±10% and, with the exception of "select out", must be terminated at each extreme end in their characteristic impedance by a terminating network. For "select out/select in" see Select Out Circuitry in following paragraphs. Cable length may be limited under special conditions, but is never to exceed a maximum line resistance of 33 Ω. The 33 Ω limit includes all contact resistance, internal line resistance and interunit line resistance.

Ground Shift and Noise

The maximum noise (measured at the receiver input) coupled onto any signal line must not exceed 400 mV.

The maximum allowable ground shift, between an active driver and any receiver on the same interface line, is 150 mV. Therefore the maximum shift (coupled noise plus ground shift) allowed on any line is 550 mV. The line logic levels and receiver threshold levels allow for a 550 mV shift. That is, a worst-case 550 mV shift during a logic one state of 2.25 V (minimum) still guarantees a receiver input of at least 1.7 V. (See Figure 9-123.) Also during a logic zero state of 0.15 V (maximum) there is a guaranteed receiver input of less than 700 mV. (See Figure 9-124)

Fault Conditions

A grounded signal line must not result in damage to drivers, receivers, or line terminators. With one driver transmitting a logic one, loss of power in any other driver or receiver on the line must not result in any damage to circuits on that line. Data transmission must not be affected by a power-off condition of any driver or receiver on the line.

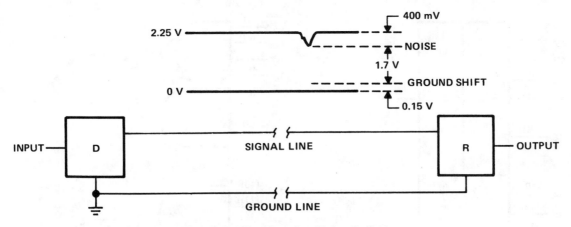

Figure 9-123. Negative Noise Spikes

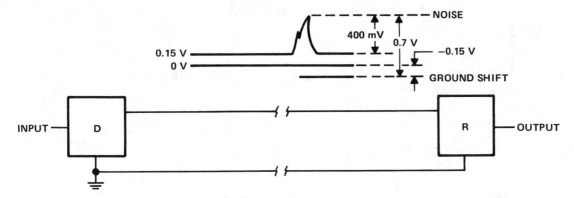

Figure 9-124. Positive Noise Spikes

ELECTRICAL CHARACTERISTICS FOR SELECT OUT INTERFACE

General

The "Select" line has a single-driver to single-receiver configuration, with only the receiver end of the line terminated in its characteristic impedance. A dc line voltage of 1.85 V or greater denotes a logic one state, and a dc line voltage of 0.15 V or less denotes a logic zero state. These voltages are relative to the driver ground. Because of the nature of the select out/select in line, negative noise tolerance has been neglected. All data line requirements not covered here are also applicable to the Select line.

Receiver Requirements

The Select line receiver must meet all the requirements given for the data line receivers.

Driver Requirements

The Select line driver must be capable of withstanding an output short to ground, while in the logic one or logic zero state, without damage to the driver circuit.

For the Select line logic zero state:
1. The driver output voltage must not exceed 0.15 V at a load of 1.0 mA.

For the Select line logic one state:
1. The driver output voltage must exceed 3.7 V at a load of 41 mA.

2. The driver output must not exceed 5.8 V at a load of 0.3 mA.
3. The driver output must not exceed 7.0 V at a load of 72 mA during an overvoltage internal to the driver.

IBM SYSTEM/360 AND SYSTEM/370 DATA LINE DRIVERS

SN75123 Dual Line Driver

The SN75123 dual line driver is designed to meet IBM System/360 requirements. It is also compatible with standard TTL logic and supply voltage levels.

The low-impedance emitter-follower outputs of the SN75123 will drive terminated lines such as coaxial or twisted pair. Having the outputs uncommitted allows wired-OR logic to be performed in party-line applications. Output short-circuit protection is provided by an internal clamping network which turns on when the output voltage drops below approximately 1.5 V. All of the inputs are conventional TTL configuration and the gating can be used during power-up and power-down sequences to ensure that no noise is introduced onto the line.

Figure 9-125 illustrates the logic symbol, logic diagram, package pin-out and equivalent driver output circuit. Features of the SN75123 include the following:

1. Meets IBM System/360 interface requirements.
2. Operates from a single 5 V supply.

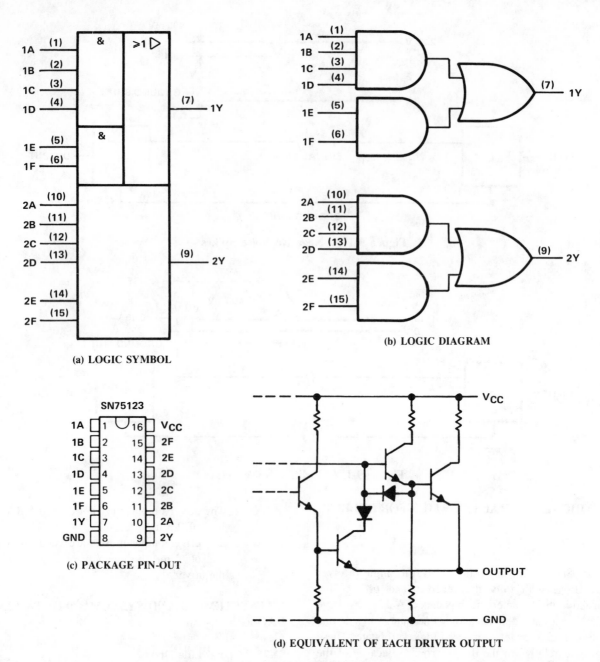

(a) LOGIC SYMBOL

(b) LOGIC DIAGRAM

SN75123

(c) PACKAGE PIN-OUT

(d) EQUIVALENT OF EACH DRIVER OUTPUT

Figure 9-125. SN75123 Dual Line Driver

3. Plug-in replacement for the Signetics N8T23 dual driver.
4. 3.11 V output at $I_{OH} = -59.3$ mA.
5. Uncommitted emitter output structure for party-line operation.
6. Short-circuit protection.
7. Multiple input AND gates.

SN75126 Quadruple Line Driver

The SN75126 quadruple line driver is designed to meet the IBM System/360 and System/370 I/O specifications GA22-6974-3. The output voltage is 3.11 V minimum (at $I_{OH} = -60$ mA) over the recommended ranges of supply voltage (4.5 V to 5.5 V) and temperature (0 °C to 70 °C). This device is compatible with standard TTL logic and supply levels. Fabrication techniques employ low-power Schottky technology to achieve fast switching and low power dissipation. The data bus is not disturbed during power-up or power-down sequencing. Fault flag circuitry is designed to sense a line short on any Y output line, output a low logic level, and reduce the output current to a safe level.

The SN75126 is designed for use with the SN75125 or SN75127 seven-channel receivers, or the SN75128 or SN75129 eight-channel receivers. Figure 9-126 illustrates the logic symbol, logic diagram, package pin-out, and equivalent driver output circuit. Features of the SN75126 include:

1. Meets IBM 360/370 I/O interface specifications GA22-6974-3.

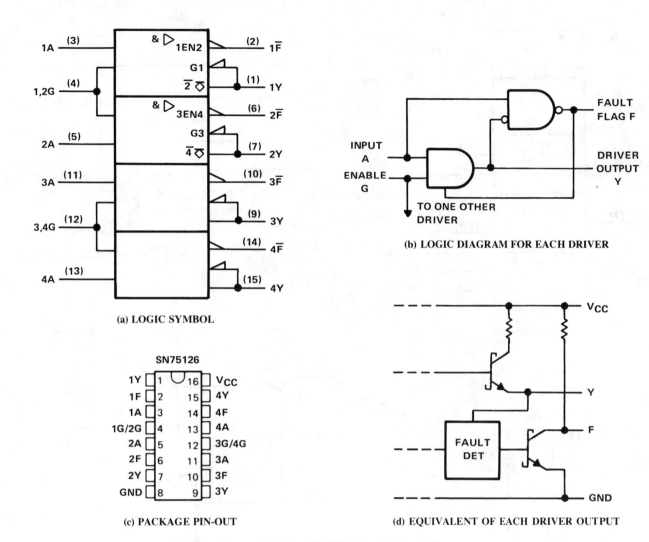

(a) LOGIC SYMBOL

(b) LOGIC DIAGRAM FOR EACH DRIVER

(c) PACKAGE PIN-OUT

(d) EQUIVALENT OF EACH DRIVER OUTPUT

Figure 9-126. SN75126 Quad Line Driver

2. Output voltage of 3.11 V minimum at $I_{OH} = -60$ mA.
3. Overload protection with foldback current limiting.
4. High-speed, low-power Schottky circuitry.
5. Functionally interchangeable with the MC3481.

SN75130 Quadruple Line Driver

The SN75130 quadruple line driver is designed to meet the IBM 360/370 I/O specifications GA22-6974-3. The output voltage is 3.11 V minimum (at $I_{OH} = -60$ mA) over the recommended ranges of supply voltage (4.5 V to 5.5 V) and temperature (0 °C to 70 °C). This device is compatible with standard TTL logic and supply voltages. Fabrication techniques employ low-power Schottky technology to achieve fast switching and low power dissipation. The data bus will not be disturbed during power-up or power-down sequencing. Fault flag circuitry is designed to sense a short on any of the Y output lines, output a logic zero level, and reduce the output current on the shorted line to a safe level. Figure 9-127 illustrates the SN75130 logic symbol, logic

diagram, package pin-out and equivalent output circuitry. SN75130 features include the following:

1. Meets IBM 360/370 I/O interface specification GA22-6974-3.
2. Output voltage of 3.11 V minimum at $I_{OH} = -60$ mA.
3. Overload protection with foldback current limiting.
4. Common enable and common fault flag.
5. High-speed, low-power Schottky circuitry.
6. Functionally interchangeable with MC3485.

IBM SYSTEM/360 AND SYSTEM/370 DATA LINE RECEIVERS

SN75124 Triple Line Receiver

The SN75124 is designed to meet the IBM System 360 interface specifications. It is also compatible with standard TTL logic and supply voltage levels. The inputs have built-in hysteresis to provide increased noise margin for single-ended applications. An open line will affect the receiver input

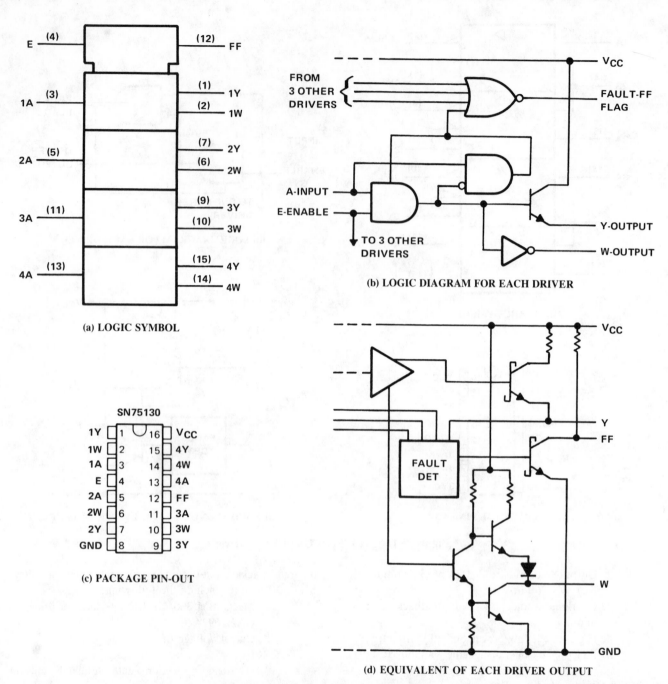

(a) LOGIC SYMBOL

(b) LOGIC DIAGRAM FOR EACH DRIVER

SN75130

(c) PACKAGE PIN-OUT

(d) EQUIVALENT OF EACH DRIVER OUTPUT

Figure 9-127. SN75130 Quad Line Driver

the same way as a low-level input voltage and the receiver input can withstand a level of -0.15 V with power on or off. The S input must be a logic high to enable the receiver input. Two of the receivers have A and B inputs which, if both are high, will hold the output low. The third receiver has only an A input which, if high, will hold the output low. Figure 9-128 shows the logic symbol, logic diagram, package pin-out and equivalent receiver input circuit. Features of the SN75124 include the following:

1. Meets IBM System/360 interface requirements.
2. Operates from a single 5 V supply.
3. TTL compatible output.
4. Plug-in replacement for the N8T24 triple receiver.

5. Built-in input threshold hysteresis.
6. High speed — Typical propagation delay time of 20 ns.
7. Independent channel strobes.

SN75125 and SN75127 Seven-channel Line Receivers

The SN75125 and SN75127 are single-ended, seven-channel line receivers designed to satisfy the requirements of the IBM System 360/370 interface specifications. Special low-power design and Schottky-clamped transistors allow for low supply current requirements while maintaining fast switching speeds. The SN75125 and SN75127 are characterized for operation from 0 °C to 70 °C. These receivers are identical in performance and differ only in their

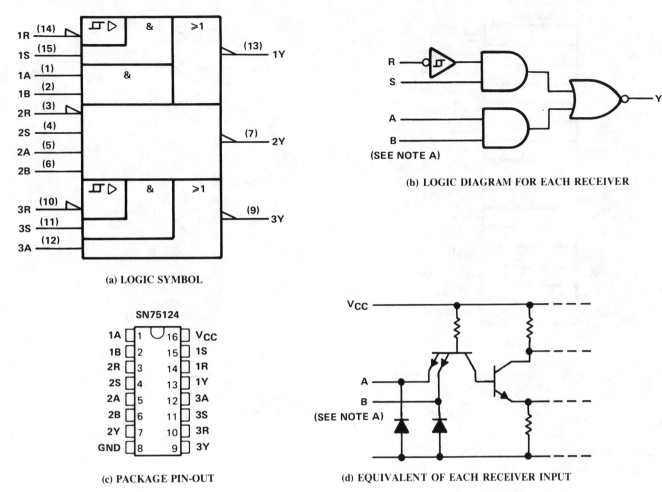

(a) LOGIC SYMBOL

SN75124

1A	1	16	V_{CC}

(c) PACKAGE PIN-OUT

(b) LOGIC DIAGRAM FOR EACH RECEIVER

(d) EQUIVALENT OF EACH RECEIVER INPUT

NOTE A: Channel 3 has only one data input.

Figure 9-128. SN75124 Triple Line Receiver

package pin-outs as shown in Figure 9-129. Figure 9-129 also illustrates the logic symbol, logic diagram and equivalent input circuit for these devices. Additional key features include the following:

1. Meet IBM 360/370 I/O specifications.
2. Input resistance of 7 kΩ to 20 kΩ.
3. TTL compatible outputs.
4. Operate from a single 5 V supply.
5. High speed and low propagation delays.
6. Low-to-high/high-to-low propagation delay ratios are specified.

SN75128 and SN75129 Octal Line Receivers

The SN75128 and SN75129 are eight-channel line receivers designed to satisfy IBM 360/370 system I/O interface requirements. Both devices feature common strobes for each group of four receivers. The SN75128 has an active-high strobe; the SN75129 has an active-low strobe. Low-power Schottky-diode-clamped transistors allow low supply current requirements while maintaining fast switching speeds and high current TTL outputs. These receivers are characterized for operation from 0 °C to 70 °C. Figure 9-130 illustrates the device package pin-outs, logic symbol, logic

diagram, and equivalent input circuit. SN75128 and SN75129 features include the following:

1. Meet IBM 360/370 I/O specifications.
2. Input resistance from 7 kΩ to 20 kΩ.
3. Outputs compatible with TTL.
4. Operate from a single 5 V supply.
5. High speed and low propagation delay.
6. Low-to-high and high-to-low propagation delay ratios specified.
7. Common strobe for each group of four receivers.

IBM SYSTEM 370 APPLICATION

A typical application is shown in Figure 9-131. An output from a 370-type computer provides addressing and control information to eight remote test stations and an input is used to receive data back on status and test results.

Drivers

The SN75126 quad line driver is selected to provide sufficient drive to meet 370 requirements and a fault flag output. The fault flag output can be used to warn the host computer of line shorts and minimize driver power dissipation. The driver power supply must be 5.0 V ±10%

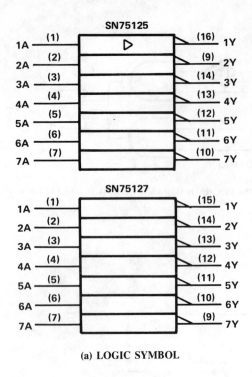

(a) LOGIC SYMBOL

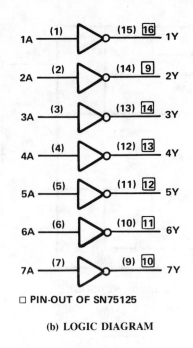

□ PIN-OUT OF SN75125

(b) LOGIC DIAGRAM

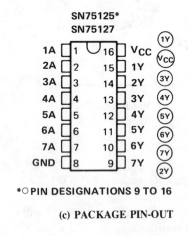

*○ PIN DESIGNATIONS 9 TO 16

(c) PACKAGE PIN-OUT

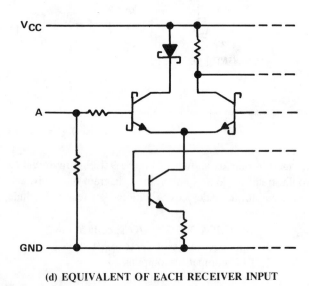

(d) EQUIVALENT OF EACH RECEIVER INPUT

Figure 9-129. SN75125/SN75127 Seven-Channel Receivers

and output drive will meet the 3.11 V minimum at an I_{OH} of −60 mA. Fault flag outputs are standard TTL levels and may be used to interface with special fault detection circuitry or directly with the 370 as illustrated.

Cable

The cable selected is type 62AU coax (Belden 9269). The signal conductor is #22 solid copper with a dc resistance of 0.0412 Ω per foot. Shield coverage is 95% and has a dc resistance of 0.0026 Ω per foot. Allowing 3 Ω for contact resistance and internal connections we are left with 30 Ω maximum line resistance. The resulting overall line length for the system is therefore 30/0.0412 or 728 feet maximum.

It should be remembered that stubs from the main line must be kept as short as possible (less than six inches) and spacing between receiver stations must be in excess of 3 feet.

Receivers

The receivers selected for this application are SN75128 octal receivers with strobes. Receiver input resistance is guaranteed to be greater than 7 kΩ and as many as 10 receivers may be connected to a single line. In this application there are eight receivers on the clock/control line resulting in a total (worst-case) receiver dc loading of 875 Ω.

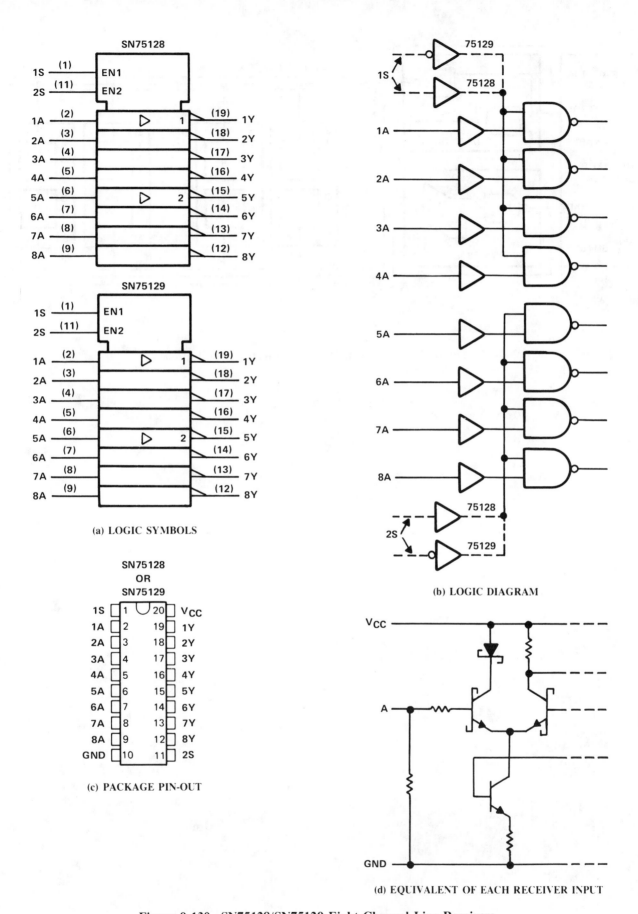

(a) LOGIC SYMBOLS

(b) LOGIC DIAGRAM

(c) PACKAGE PIN-OUT

(d) EQUIVALENT OF EACH RECEIVER INPUT

Figure 9-130. SN75128/SN75129 Eight-Channel Line Receivers

9-69

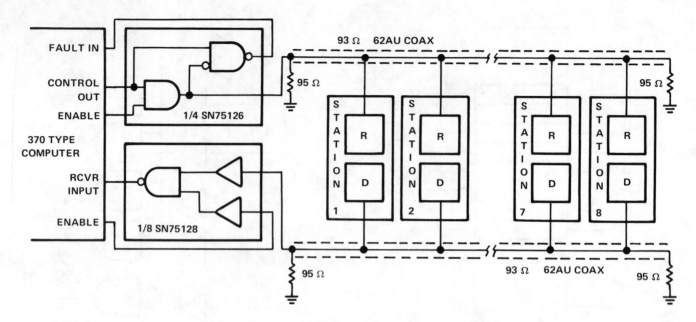

Figure 9-131. Typical IBM 370 Interface Application

Section 10

Peripheral Drivers

Peripheral drivers are general-purpose integrated circuits that can be used to interface between TTL, MOS, and CMOS logic levels and higher voltage, higher current components. Higher voltage and higher current components include lamps, relays, solenoids, data transmission lines, and motors (Figure 10-1).

requiring additional drive power. These peripheral drivers usually include output transistors to provide the required drive capability, preceded by logic level shifting circuitry. Level shifting is accomplished by integrated resistors and diodes or logic gates. Figure 10-2 illustrates the two basic configurations of peripheral drivers. One uses resistor and/or

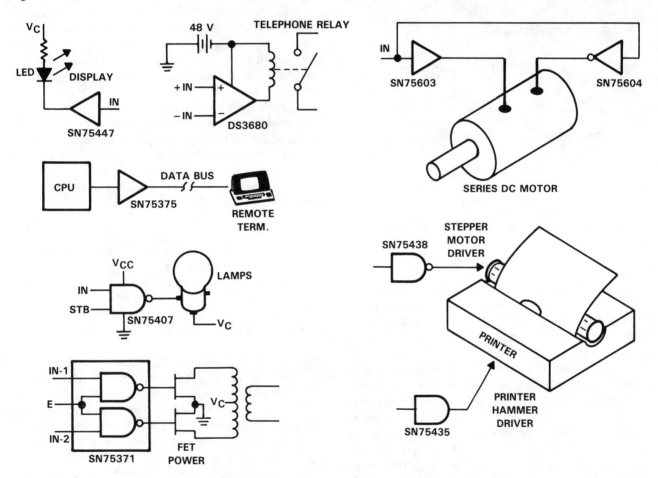

Figure 10-1. Typical Uses for Peripheral Drivers

DEVICE CONSIDERATIONS AND PRODUCT DESCRIPTIONS

BASIC CONFIGURATIONS

In the mid 1960s, integrated circuit logic gates were combined with a variety of discrete small signal transistors and power transistors to provide the desired interfacing. Integrated circuit devices were developed to allow direct, single IC interfacing from logic levels to components

diode level shifting [Figure 10-2(a)], and the other uses logic gate level shifting [Figure 10-2(b)].

TYPICAL REQUIREMENTS
Power

Peripheral drivers are used in situations which require them to dissipate a significant amount of power. As a result, most IC peripheral drivers are designed to handle at least 1 W. Packages with copper lead frames are often used to improve power handling capability. The small dual-in-line 8-pin package with a copper lead frame will typically handle over 1.4 W at 25 °C. The 14- and 16-pin packages with copper lead frames are rated at 2 W or greater.

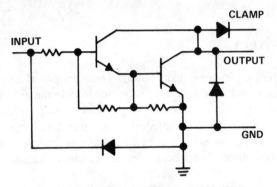

(a) RESISTOR/DIODE INPUT

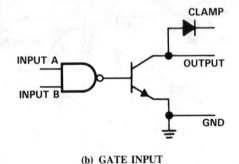

(b) GATE INPUT

Figure 10-2. Basic Peripheral Driver Input Configurations

Voltage

Voltage capability ranges from a minimum of 15 V to 100 V. Care must be exercised in the selection of drivers for switching applications. The switching voltage output limits are generally less than the maximum dc standoff voltage [$V_{(BR)CER}$]. For example, a typical peripheral driver with a standoff voltage rating of 30 V may not be suitable for use as a 24-V relay driver, even if an output clamp diode is used. The high level output voltage after switching (V_{OH}) is typically 20 V for this device. The output voltage swing of 24 V or 25 V would result in secondary breakdown and device latch-up. This is a destructive condition which could result in device failure. The correct device for this application would have a V_{OH} rating of 30 V. Gate controlled input devices require a 5-V supply in addition to any higher level supply required for the output circuit.

Current

Peripheral drivers are designed to operate at output current levels of 100 mA to 2 A. As with output voltage, device output current selection should be done with care. Most devices will have a continuous output current rating and a peak current rating. Peak current ratings are specified for a maximum on time of 10 ms and a duty cycle of 50% or less. NOTE: The peak current level of integrated circuit

drivers should not be exceeded, no matter how short the time or how low the duty cycle. Although average power dissipation may be within limits if the on times and duty cycles are very short, chip surface metal migration may occur with any current above the rated peak level. Metal migration results in destruction of the chip surface metal. This is associated with the output emitter contacts, and eventually causes device failure. For example, short current spikes associated with charging a capacitive load could seem to have no immediate effect on the device. However, if the peak current level is exceeded for even a short time, a small amount of deterioration occurs and the device will exhibit long-term failure.

Speed

Peripheral drivers can be used as switches and are therefore operating in a dc or very low frequency mode. Other applications, such as memory clock drivers require speeds as high as 10 MHz. At low or high data rates or dc operation, the device must never exceed its voltage, current or power limits.

High-speed operation can, in some applications, result in excessive power dissipation. High-speed power dissipation can (and should) be limited to improve operating reliability. Turn-on transients are one form of excessive power that can be controlled to some extent. Figure 10-3 illustrates the effects of transients on power dissipation as frequencies are increased.

Logic

Generally these devices are used to interface between logic level signals and circuits requiring more drive power. Most peripheral driver inputs are compatible with TTL voltage levels. The input resistance of some devices is high allowing compatibility with low level CMOS, MOS, and low power Schottky TTL, as well as standard TTL.

PERIPHERAL DRIVER DEVICES

A wide variety of peripheral drivers is available for today's design engineer. The first monolithic IC peripheral driver, the SN75450B which was introduced by Texas Instruments in 1968 (Figure 10-4), was very basic and versatile, allowing many application options. This versatility has accounted for its continued popularity. The SN75450B's schematic and logic symbol diagrams are illustrated in Figures 10-5 and 10-6. Note that the output transistor's emitter, collector, and base leads are pinned out separately to allow for various methods of interconnections.

Key design features of the SN75450B include the ability to switch load currents of 300 mA. It has a dc off-state transistor collector output voltage capability of 30 V and can switch off 300 mA with inductive loads while operating from collector supply voltages of up to 20 V. The SN75450B also features fast (less than 30 ns) switching speeds for use in high speed logic interfacing applications. As shown in Figure 10-5, its TTL compatible inputs are diode clamped for protection from negative voltage transients.

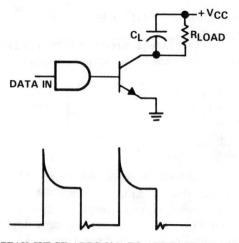

(a) PEAK SURGE ADDS 80% TO AVERAGE DC POWER

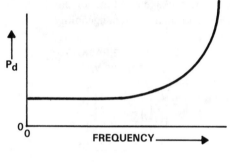

(b) PEAK SURGE ADDS 12% TO AVERAGE DC POWER

Figure 10-3. Transient Effects on Power Dissipation

TOP VIEW

```
         ┌────┐
G    [ 1    14 ] VCC
1A   [ 2    13 ] 2A
1Y   [ 3    12 ] 2Y
1B   [ 4    11 ] 2B
1C   [ 5    10 ] 2C
1E   [ 6     9 ] 2E
GND  [ 7     8 ] SUB
         └────┘
```

Figure 10-4. SN75450B Peripheral Driver

It was noted that in most of the applications using the SN75450B device, the emitters were tied together and grounded. Also the gate outputs were tied to the transistor base inputs. Implementing this configuration yielded the next device in this series — the SN75451B. Other devices with different logic functions, but basically the same operating characteristics, led to the series of drivers summarized in Table 10-1.

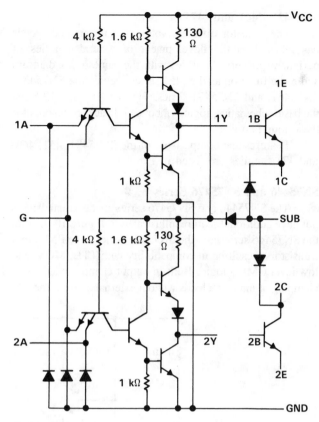

Resistor values shown are nominal.

Figure 10-5. SN75450B Schematic

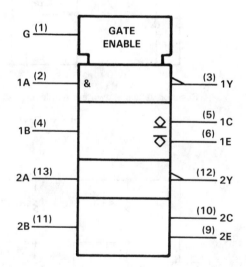

Figure 10-6. SN75450B Logic Symbol

Table 10-1. Summary of Series 55450B/75450B

DEVICE	CIRCUIT LOGIC	PACKAGES
SN75450B	AND	J,N
SN75451B	AND	JG,P,D
SN75452B	NAND	JG,P,D
SN75453B	OR	JG,P,D
SN75454B	NOR	JG,P,D

10-3

SN75431, '461 and '471 Series

The need for different voltages, currents and speeds has resulted in the development of several families of peripheral drivers. Families with the same block diagram and basic function as the '451 series include the SN75431, SN75461, and SN75471 series. Figures 10-7 and 10-8 list the basic schematics, logic symbols, and package pinouts for these four families.

Features and comparisons of the SN75431, 451, 461, and 471 families are listed in Table 10-2.

SN75446 and SN75476 Series

The SN75446 and SN75476 series are also dual drivers but have additional features illustrated in Figure 10-9. Both the SN75446 series and the SN75476 series have PNP input transistors, resulting in compatibility with TTL, MOS, and low level CMOS logic. Built-in output clamp diodes allow clamping of inductive loads without external diodes. The 446

Table 10-2. SN75431 Series Device Features

FAMILY	OFFSTATE VOLTAGE (V)	SWITCH LOAD VOLTAGE (V)	OUTPUT CURRENT (mA)	TYPICAL t_{pd} (ns)
SN75431-434	15	15	300	15
SN75451-454	30	20	300	20
SN75461-464	40	30	300	35
SN75471-474	70	55	300	35

series provides 350 mA continuous output and 70-V standoff with 50-V inductive switching capability. The 476 series provides 300 mA continuous output current with 70-V standoff and 55-V inductive switching capability. These devices are especially suited for driving inductive loads such as relays, solenoids, printer hammers, and motors. Figure 10-10 shows the package and function similarities between these devices and their interchangeability.

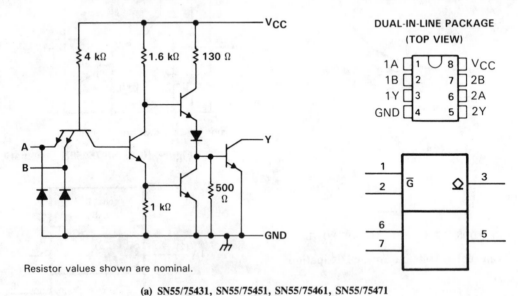

Resistor values shown are nominal.

(a) SN55/75431, SN55/75451, SN55/75461, SN55/75471

(b) SN55/75432, SN55/75452, SN55/75462, SN55/75472

Figure 10-7. SN75431/451/461 and 471 Series (AND, NAND) Schematics and Diagrams

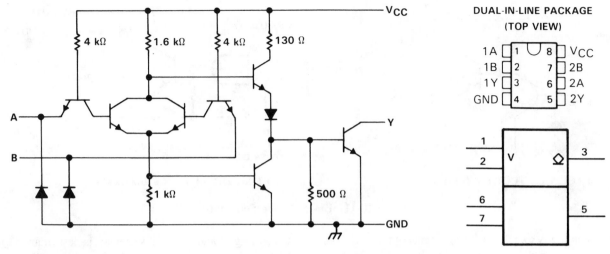

Resistor values shown are nominal.

(a) SN55/75433, SN55/75453, SN55/75463, SN55/75473

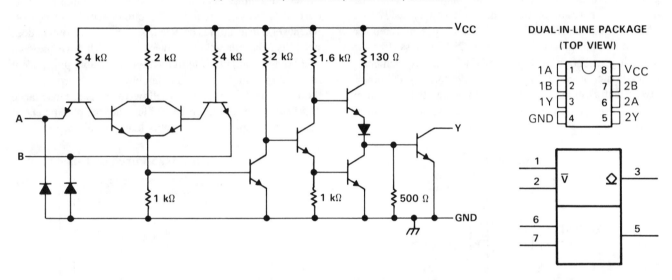

(b) SN55/75434, SN55/75454, SN55/75464, SN55/75474

Figure 10-8. SN75431/451/461 and 471 Series (OR, NOR) Schematics and Diagrams

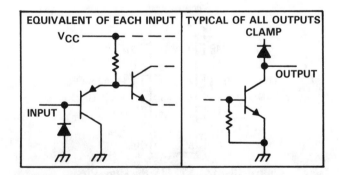

Figure 10-9. Schematics of Device Inputs and Outputs

SN75436, SN75437A, and SN75438

The SN75436, SN75437A and SN75438 are quad, gate controlled, peripheral drivers. They are designed for driving loads requiring relatively high power (15 to 35 W). Each device features four open-collector output drivers with common enables. These devices are designed for use as relay drivers, printer hammer or other types of solenoid drivers, lamp drivers, motor drivers, data line drivers, and memory drivers. The basic device schematic diagrams, Figure 10-11, show some of the special features of this series. PNP transistors provide high-impedance inputs for TTL and CMOS compatibility. Low power logic control circuitry results in less than 26 mW standby power. Open-collector output transistors provide low resistance saturated outputs

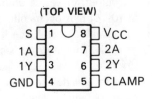

Figure 10-10. SN75446 and SN75476 Series Package Pinouts

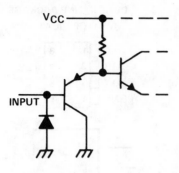

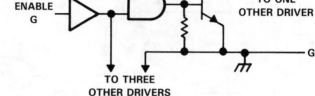

(a) EQUIVALENT OF EACH INPUT   (b) FUNCTIONAL BLOCK DIAGRAM

Figure 10-11. Basic Device Schematics

resulting in low V_{SAT} levels. The outputs of these devices do not have spurious transients during power-up or power-down sequencing. The device package, Figure 10-12, provides four heat sink pins to help conduct heat from the device. As a result of this, and their copper leads, the SN75436, SN75437A and SN75438 packages are rated for 2 W continuous power dissipation at 25 °C or less (free-air operating temperature). Inputs and outputs are conveniently located on opposite sides of the package for easy PC-board layout and assembly. Table 10-3 compares the maximum output current, output saturation voltage, and switching voltage for these three devices.

sequencing. Device standby power is less than 53 mW allowing cool operation and good long-term reliability.

One unique feature is that each driver output is protected against load shorts with its own latching over-current shutdown circuitry. The output will be turned off whenever a load short is detected. A short on one output does not affect the other three drivers. The latch for shutdown will hold the output off until the input or enable pin is taken low and then high again. A delay circuit is incorporated in the over-current shutdown to allow for a load capacitance of 500 μF at 35 V. Figure 10-15 illustrates the recommended maximum supply voltage versus load capacitance.

(TOP VIEW)

```
        ┌──┬──┐
   Y1 □ │1 ⌒16│ □ A1
   D2 □ │2   15│ □ A2
   Y2 □ │3   14│ □ G
HEATSINK{□│4   13│□}HEATSINK
AND GND {□│5   12│□}AND GND
   Y3 □ │6   11│ □ VCC
   D2 □ │7   10│ □ A3
   Y4 □ │8    9│ □ A4
        └─────┘
```

Figure 10-12. Package Pinout

Table 10-3. Device Selection Guide

Feature	436	437A	438
Output current	500 mA	500 mA	1000 mA
Maximum V_{SAT}	0.5 V	0.5 V	1.0 V
Max switching voltage	50 V	35 V	35 V

SN75435 Quad Driver

The SN75435 consists of four peripheral drivers, each with up to 20 W output drive capability. It features (Figures 10-13 and 10-14) four open-collector drivers with a common enable input that, when taken low, disables all four outputs (Table 10-4). Output on resistance is less than 1 Ω at an output current of 500 mA. The standard 2-W DIP (Figure 10-13) is used for this device. Output clamp diodes for transient protection are built in. The SN75435 is also free of spurious transients during power-up and power-down

Table 10-4. SN75435 Function Table

INPUTS					OUTPUTS			
1A	2A	3A	4A	G	1Y	2Y	3Y	4Y
L	L	L	L	X	H	H	H	H
X	X	X	X	L	H	H	H	H
H	L	L	L	H	L	H	H	H
L	H	L	L	H	H	L	H	H
L	L	H	L	H	H	H	L	H
L	L	L	H	H	H	H	H	L
H	H	H	H	H	L	L	L	L

(TOP VIEW)

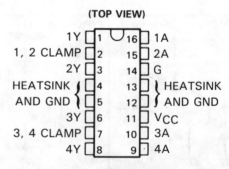

Figure 10-13. SN75435 Package Pinout

The SN75435, Figure 10-16, has high-impedance PNP inputs to provide both TTL and low level CMOS compatibility. Inputs are also diode clamped for negative voltage input transient protection. Although very well suited for driving solenoids, relays, memory systems, and LED

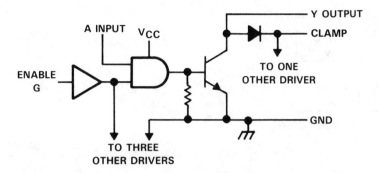

Figure 10-14. SN75435 Basic Logic Diagram

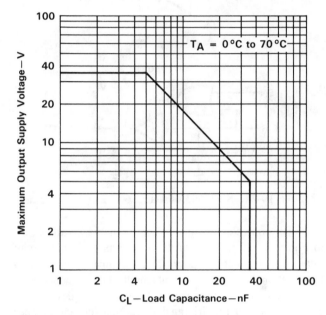

Figure 10-15. Recommended Maximum Supply Voltage
vs Load Capacitance

circuits, this device is particularly well suited for driving lamps and motors.

SN75440 Quad Peripheral Driver

The SN75440 quadruple peripheral driver is designed for use in systems requiring high current, high voltage, and high load power. The package Figure 10-17 allows easy heat sinking and will provide 2-W power handling capability at 25 °C. The device standby power is only 21 mW. Each

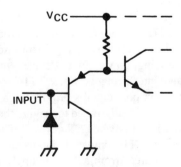

Figure 10-16. SN75435 Equivalent Schematic
of Each Input

device has four noninverting open-collector outputs with 600 mA sink capability and will switch inductive loads with a supply voltage of 35 V. The SN75440 also features an enable input control on pin 14, Figure 10-17, for enabling or disabling all four outputs. The open-collector outputs, Figure 10-18, are diode protected for transient protection and have a low on-state resistance of less than 1.5 Ω. The outputs are free from spurious transitions (glitching) during power-up or power-down sequencing.

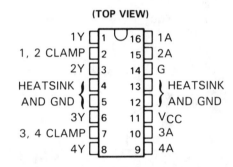

Figure 10-17. SN75440 Package Pinout

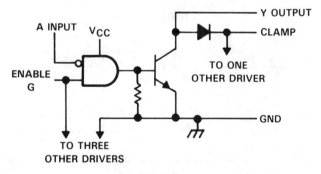

Figure 10-18. SN75440 Logic Diagram (Each Driver)

PNP inputs, Figure 10-19, have low level input currents of less than 10 μA. Functional relationships between the input and output logic functions are given in Table 10-5. Applications include driving relays, lamps, solenoids, motors, LEDs, data transmission lines, printer hammers, and other systems with high drive power requirements.

SN75603, SN75604, and SN75605

The SN75603, SN75604, and SN75605 are power peripherals with three-state outputs having the capability to

Table 10-5. SN75440 Logic Functions

INPUTS		OUTPUT
A	G	Y
L	H	L
H	X	H
X	L	H

H = high level, L = low level,
X = irrelevant

sink or source currents up to 2 A while switching bidirectional loads at voltages of 8 V to 40 V. They come in a straight KH, or formed KV, 5-lead power package illustrated in Figure 10-20. They are rated at 6.25 W at or below 125 °C. The devices have built-in input and output transient diodes (Figure 10-21) and thermal shutdown protection. Table 10-6 provides the input-output functions for all three devices. Note that the SN75603 and SN75604 are both in their high impedance output state if their enable input is low. The SN75605 uses an exclusive-OR input control and if either the enable or the direction control inputs are low the output is in the high impedance state.

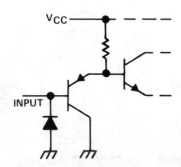

Figure 10-19. SN75440 Equivalent Schematic of Each Input

Table 10-6. SN75603/604 and 605 Function Table

INPUTS		OUTPUT		
EN	DIR	SN75603	SN75604	SN75605
L	L	Z	Z	L
L	H	Z	Z	Z
H	L	L	H	Z
H	H	H	L	H

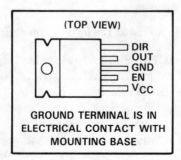

(TOP VIEW)

DIR
OUT
GND
EN
V_{CC}

GROUND TERMINAL IS IN ELECTRICAL CONTACT WITH MOUNTING BASE

Figure 10-20. SN75603/604 and 605 Package Pinout

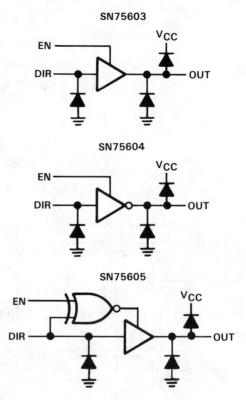

SN75603

SN75604

SN75605

Figure 10-21. SN75603/604 and 605 Logic Diagrams

The equivalent input and output schematics are shown in Figure 10-22. PNP input transistors provide high input impedance with both TTL and CMOS compatibility. Internal propagation delays are such that simultaneous conduction of sink and source outputs cannot occur. These devices are especially suited for driving bidirectional dc and stepper motors as well as reversible solenoids and relays.

The SN75603 and SN75604 are designed to be used together as complementary half-H drivers. Their direction controls are complementary allowing the pair to function as full-H drivers with the direction control function implemented by a single logic control line.

The SN75605 (a functional equivalent to the Sprague UDN2949) provides a high impedance output when either the enable or the direction control is low. Note: If both of the inputs are low at the same time, the output will be in the low state. If both inputs are high, then the output will be high.

SN75372 Dual and SN75374 Quad Power FET Drivers

The SN75372 and SN75374 are designed to drive capacitive type loads at relatively high data rates. Primary uses include driving power FET devices in switching applications and interfacing between TTL and MOS or CMOS. Their totem-pole outputs provide high speed sink and source capability ideally suited to driving the capacitive input characteristics of power FET devices. The SN75372 is a dual driver with TTL compatible inputs and totem-pole outputs that can source, from a V_{CC2} supply level of up to 24 V and sink currents of 100 mA minimum. Even when

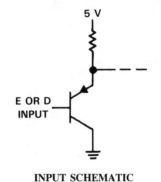

INPUT SCHEMATIC

OUTPUT SCHEMATIC

Figure 10-22. Equivalent SN75603/604 and 605 Input and Output Schematics

operating into a 390-pF load device propagation delays are typically less than 35 ns with transition times accounting for 25 ns of that. The result is a very adequate speed for driving power FETs. Figures 10-23 and 10-24 show the basic circuit configurations for the SN75372 and SN75374. Both devices have input and output transient protection diodes. These devices have transient overdrive protection to minimize power dissipation. The typical standby power is 22 mW for the dual SN75372 and 38 mW for the quad SN75374.

The dual SN75372 comes in the 8-pin DIP package, Figure 10-25, and has a common enable. The quad SN75374 comes in a 16 pin, Figure 10-26, and has 2 enables for each pair of drivers. An additional feature of the SN75374, shown in both Figures 10-24 and 10-26, is the availability of the pre-driver supply rail (V_{CC3} on pin 9). With V_{CC3} at 3 to 4 V above the value of V_{CC2}, it is possible to drive the output level very close to the V_{CC2} rail.

When driving capacitive loads at relatively high data rates, the package power dissipation will become significant and it is desirable to know what to expect. Figures 10-27 and 10-28 illustrate the total package power dissipation that may be expected versus operating frequency with several different capacitive loads for the SN75372 and SN75374.

DS3680 Quad Telephone Relay Driver

The DS3680 quad relay driver is a monolithic integrated circuit designed to interface from TTL to telephone relay systems or other −48 V systems. It is capable of sourcing 50 mA from standard −52 V battery power. To reduce the effects of noise and IR drop between logic ground and battery ground, these drivers are designed to operate with a common- mode input range of ± 20 volts referred to battery ground. Each driver in the package has common-mode input

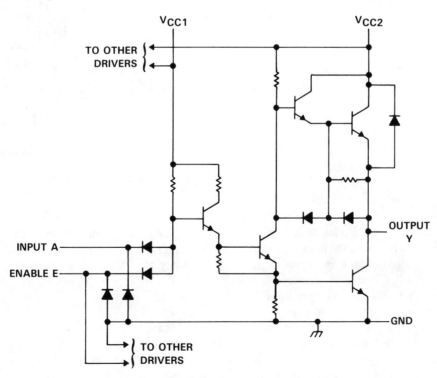

Figure 10-23. SN75372 Schematic (Each Driver)

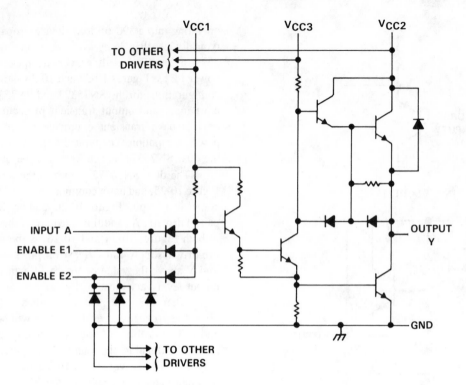

Figure 10-24. SN75374 Schematic (Each Driver)

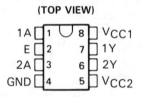

Figure 10-25. SN75372 Package Pinout

(TOP VIEW)

1A	1	8	VCC1
E	2	7	1Y
2A	3	6	2Y
GND	4	5	VCC2

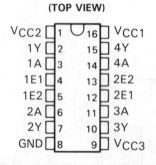

Figure 10-26. SN75374 Package Pinout

(TOP VIEW)

VCC2	1	16	VCC1
1Y	2	15	4Y
1A	3	14	4A
1E1	4	13	2E2
1E2	5	12	2E1
2A	6	11	3A
2Y	7	10	3Y
GND	8	9	VCC3

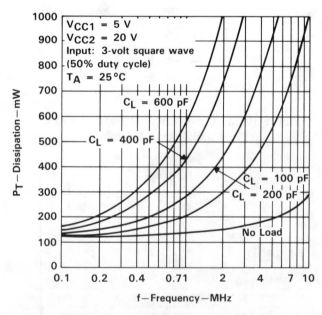

Figure 10-27. Total Dissipation Both 372 Drivers vs Frequency

voltage independent of the other drivers. High input impedance with low input current (typically less than 100 μA) results in minimum loading of the driving circuit. Built-in driver output clamp diodes eliminate the need for external networks to limit high voltage kickback levels present when switching inductive loads. A fail-safe feature incorporated in the DS3680 ensures that if either input is open, the driver output will be off. Figures 10-29 and 10-30 illustrate each driver's logic symbol and schematic diagram respectively. Figure 10-31 shows the convenience of the package pinout with inputs on one side and outputs on the other. Ground is located on a corner pin to implement easy board layouts.

Some peripheral drivers are basically Darlington transistor arrays designed to have logic compatible inputs. They are often used in high current applications where the control logic is provided externally. The following devices are of this type.

SN75064/ULN2064 Series Quad Peripheral Drivers

The SN75064, SN75065, SN75066, SN75067, ULN2064, ULN2065, ULN2066, and ULN2067 are quad high current, high voltage Darlington switches. Each device

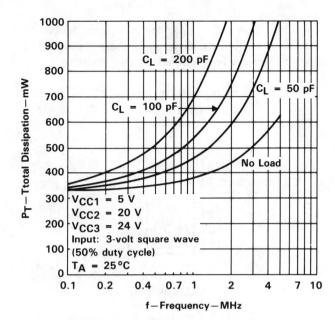

Figure 10-28. Total Dissipation All 374 Drivers vs Frequency

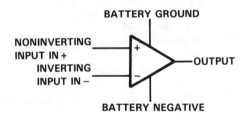

Figure 10-29. DS3680 Logic Diagram (Each Driver)

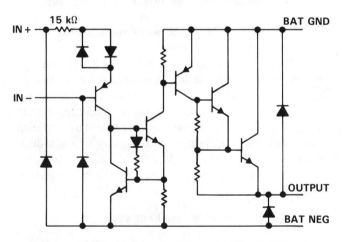

Figure 10-30. Schematic Diagram (Each Driver)

has four Darlington transistor drivers with common-cathode clamp diodes for switching inductive loads (Figure 10-32). Each of the drivers has 0.5 A output current capability, and their inputs and outputs may be paralleled for even higher current handling. Connected as common emitter circuits, these devices provide sink current drive for switching a variety of applications including relays, printer hammers, lamps, display circuits, memory circuits, and data

transmission lines. The NE packages (Figure 10-32) are rated at 2 W. Output loads may be operated from voltages up to 50 or 80 V, depending on the device type.

The SN75064, ULN2064, SN75065, and ULN2065 are intended for use with TTL and 5-V MOS logic. The SN74066, ULN2066, SN75067 and ULN2067 are intended for use with PMOS and higher voltage CMOS logic. The SN75 series devices have a slightly higher maximum $V_{CE(sat)}$ specification and feature economical pricing. The ULN series devices have slightly lower maximum $V_{CE(sat)}$ and are recommended where output low level voltages are critical.

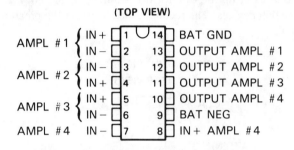

Figure 10-31. DS3680 Package Pinout

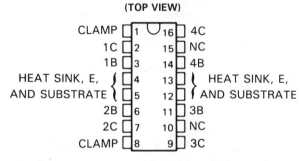

NC—No internal connection

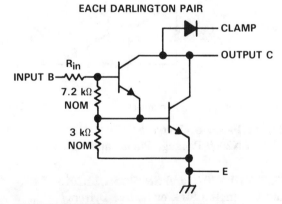

SN75064, SN75065 R_{in} = 350 Ω NOM
ULN2064B, ULN2065B

SN75066, SN75067 R_{in} = 3 kΩ NOM
ULN2066B, ULN2067B

Figure 10-32. SN75064 Series and ULN2064 Series Package Pinout and Schematic

SN75068/ULN2068 and SN75069/ULN2069 Quad Darlington Switches

The SN75068/ULN2068 and SN75069/ULN2069 are quad high voltage, high current Darlington drivers (Figure 10-33). Their outputs have common cathode clamp diodes for switching inductive loads. A third transistor at the input acts as a preamplifier providing high current gain and input compatibility with low power TTL and 5-V CMOS signals. With a maximum input current of only 250 μA at a V_{in} of 2.4 V, these devices may be operated directly from low power sources such as CMOS microprocessors or computers. Their outputs can sink up to 1.5 A and switch voltages of up to 50 V for the 068 devices and 80 V with the 069 devices. The recommended V_{CC} for the preamp is 5 V.

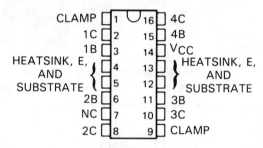

NE DUAL-IN-LINE PACKAGE
(TOP VIEW)

NC—No internal connection

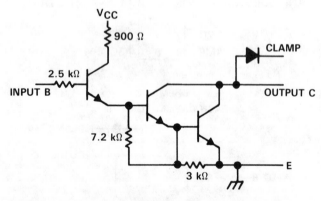

SCHEMATIC
(EACH DRIVER)

Resistor values shown are nominal.

Figure 10-33. SN75068, SN75069, ULN2068, and ULN2069 Package Pinout and Schematic

SN75074/ULN2074 and SN75075/ULN2075 Quad Darlington Sink or Source Drivers

The SN75074/ULN2074 and SN75075/ ULN2075 are quad, high current, high voltage Darlington transistor switches (Figure 10-34). They feature output voltage operation up to 80 V and output current capabilities of up to 1.5 A. These devices are unique in that they feature uncommitted collectors and emitters allowing sink or source applications. Inputs with respect to the emitters are compatible with TTL logic levels. These devices are particularly useful in H-drive applications because of the ability to perform both sink and source functions.

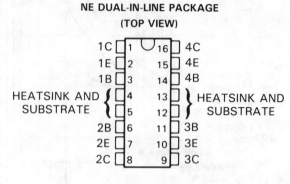

NE DUAL-IN-LINE PACKAGE
(TOP VIEW)

SCHEMATIC
(EACH DRIVER)

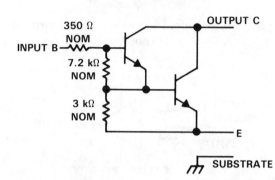

Figure 10-34. SN75074, SN75075, ULN2074, and ULN2075 Package Pinout and Schematic

SN75465 and ULN2001A Series Seven Channel Darlington Transistor Arrays

The SN75465 series and the ULN2001A series are high voltage, high current Darlington npn transistor arrays with common cathode clamp diodes for switching inductive loads (Figure 10-36). Each Darlington pair has a collector current rating of 500 mA. Darlington drivers may also be paralleled for higher current capability when using the (N) plastic dual-in-line package (Figure 10-35.) Total package substrate current for the N package is 2.5 A. The (J) ceramic dual-in-

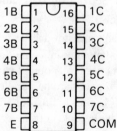

J OR N
DUAL-IN-LINE PACKAGE
(TOP VIEW)

Figure 10-35. SN75465 and ULN2001A Series Package Pinout

line package is limited to 500 mA total current. The SN75465 series devices have a 100 V maximum collector-emitter voltage rating while the ULN2001A series have a 50 V maximum.

The SN75465 and ULN2005A have 1.05-kΩ series base input resistors and are especially designed for standard TTL or other sources with TTL equivalent or greater drive current capability.

The SN75466 and ULN2001A are general-purpose arrays with no input resistors or diodes for level shifting. They have a guaranteed minimum h_{FE} of 1000. These devices will work well as switches or linear amplifiers.

The SN75467 and ULN2002A are especially designed for use with 14 V to 25 V input signals, typical of PMOS logic devices. Each driver has a zener diode and resistor in series with its input to limit the input current for high voltage applications.

The SN75468 and ULN2003A have 2.7-kΩ resistors in series with their inputs allowing them to be compatible with TTL and 5-V CMOS logic.

The SN7469 and ULN2004A have 10.5-kΩ resistors in series with their inputs. They operate from medium logic voltage levels of 6 V to 15 V, typical of some CMOS and PMOS logic.

UDN2841 and UDN2845 Quad High Current Darlington Drivers

The UDN2841 and UD2845 are quad peripheral drivers designed for high voltage (up to 50 V), high current (to 1.5 A) switching applications. These devices are designed to provide a solution to the unique application requirements resulting from loads that must operate from a negative supply rail. Such systems include electronic-discharge printers, bipolar and unipolar dc motors, telephone relays, matrix displays, PIN diodes and other high current loads operating from negative power supplies.

The UDN2841 is intended for current sink applications where, for example, the load is connected between ground and the driver's collector. The emitter is connected to the negative supply and the driver acts as a sink to that supply (Figure 10-37).

The UDN2845 is intended for use as a sink and source combination allowing bidirectional switching where both ends of the load are floating. An application example would be driving a reversible dc solenoid.

Both the UDN2841 and UDN2845 drivers feature a pnp input transistor that acts as a signal level translator and provides a high input impedance for compatibility with

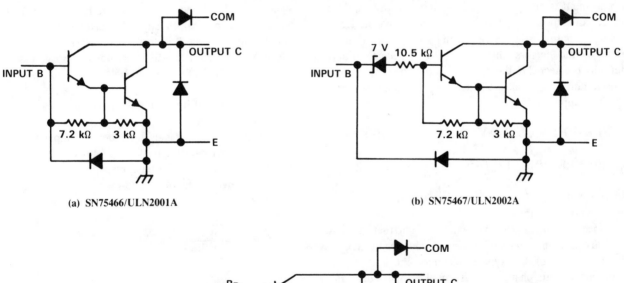

(a) SN75466/ULN2001A

(b) SN75467/ULN2002A

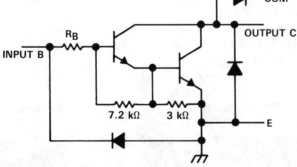

(c) SN75465, ULN2005A, SN75468, ULN2003A, SN75469, AND ULN2004A

Figure 10-36. Schematics of Darlington Pairs

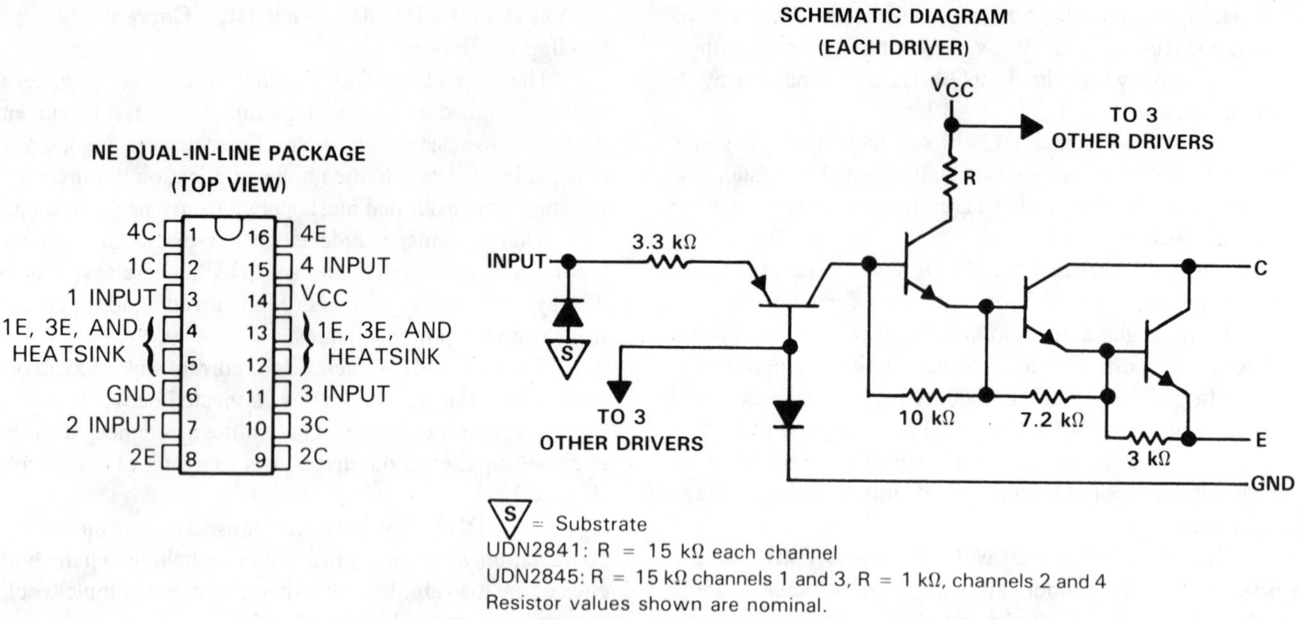

NE DUAL-IN-LINE PACKAGE
(TOP VIEW)

S = Substrate
UDN2841: R = 15 kΩ each channel
UDN2845: R = 15 kΩ channels 1 and 3, R = 1 kΩ, channels 2 and 4
Resistor values shown are nominal.

**Figure 10-37. UDN2841 and UDN2845 Package
Pinout and Schematic**

standard TTL as well as low power 5-V CMOS. Following the pnp input is an npn transistor acting as a preamplifier and providing the gain required to adequately drive the 1.5 A Darlington output pair. Drivers 2 and 4 have uncommitted emitters and collectors while drivers 1 and 3 have emitters internally connected to the substrate. For proper operation, these emitters and the substrate must be connected to the most negative supply voltage.

PERIPHERAL DRIVER APPLICATIONS

PERIPHERAL DRIVER-OPTO APPLICATIONS

**Driving Tungsten Filament, or Equivalent,
Incandescent Lamps**

Incandescant lamps have a characteristic that can pose a serious problem to monolithic integrated circuit drivers. The resistance of the typical incandescent bulb varies significantly with changes in filament temperature. For example, a typical general-purpose bulb, type 1815, is rated at 14 V and 200 mA operating current. One might assume that a monolithic driver rated at 300-mA continuous and 500-mA peak surge would be adequate for driving a lamp requiring only 200-mA drive current. The problem becomes evident if the turn-on surge current is monitored. Type 1815 bulbs have a (cold) turn-on peak surge current of up to 10 times their steady-state value, possibly as high as 2.7 A. Figure 10-38 illustrates a test of surge current for a type 1815 bulb. To overcome the problem of high turn-on surge current the designer has two alternatives: Select a driver with a 3 A surge capability, or do something to limit the surge current.

Several methods can be employed to limit surge currents when using peripheral drivers. One method uses keep-alive resistors as shown in Figure 10-39. These resistors

maintain the off-state current at about 50% of the normal steady-state level and result in a standby light intensity of about 10% of normal. Steady-state current for the particular bulbs being used was 250 mA and the warm-up current was 120 mA. With the keep-alive resistor, the surge current is limited to only 500 mA or twice the normal steady-state value.

Figure 10-40 illustrates the limiting of base drive to limit surge current.

The h_{FE} for the particular device output transistors used was about 84. If a peak surge current of 500 mA is selected, the value of the base resistor can be determined by the following equation. The resulting limitation in the output surge current is shown in Figure 10-41.

$$R = \frac{V_{OH} - V_{BE}}{I_{B(limit)}}$$

Where:

$$V_{OH} = 3.5 \text{ V (typical)}$$
$$V_{BE} = 0.7 \text{ V (typical)}$$

$$I_{B(limit)} = \frac{I_{C(limit)}}{h_{FE}}$$

$$= \frac{500 \text{ mA}}{84} = 5.95 \text{ mA}$$

Therefore:

$$R = \frac{3.5 \text{ V} - 0.7 \text{ V}}{0.00595 \text{ A}} = 470 \text{ } \Omega$$

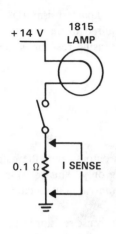

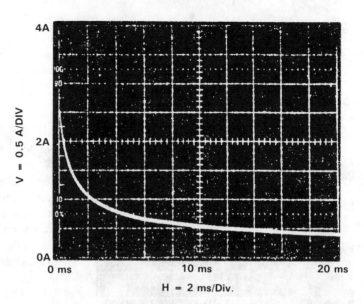

Figure 10-38. Results of Lamp Surge Current Test

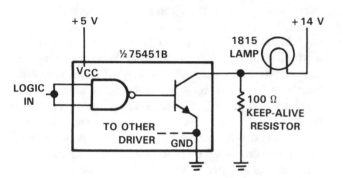

Figure 10-39. Limiting Surge Current with Keep-Alive Resistors

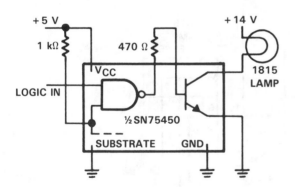

Figure 10-40. Base Drive Current Limiting

This method is not the best solution because of the dependency on h_{FE} and V_{BE}. A resistor value would need to be calculated for each device. A similar method that is much less susceptible to variations in output device parameters is illustrated in Figure 10-42. The current limiting resistor in series with the output transistor's emitter is small enough to be of little significance at the steady-state on level,

but will limit the peak levels. The resistor value required may be calculated from the following equation.

$$R = \frac{V_{OH} - V_{BE}}{I_{OP}}$$

Where:

$$V_{OH} = 3.5 \text{ V} , V_{BE} = 0.7 \text{ V and}$$
$$I_{OP} = 500 \text{ mA (the desired peak limit)}$$

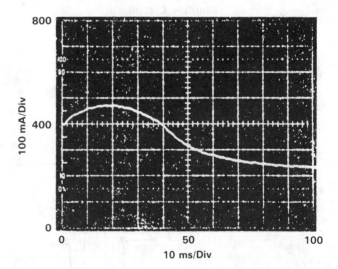

Figure 10-41. Base Drive Limited Surge Current

In this application the value of R is 5.6 Ω. The resulting surge current limitation is shown in Figure 10-43.

Consistency of performance can be improved by using both drivers in a dual peripheral driver package. The circuit in Figure 10-44 uses one transistor (Q2) to sense the output level and clamp it at the desired peak value by controlling

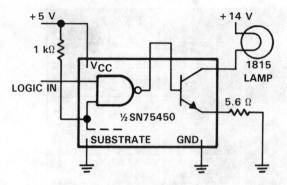

Figure 10-42. Emitter Resistor Surge Limiting Circuit

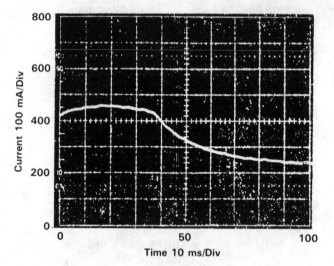

Figure 10-43. Output Current vs Time for Emitter Limiting Circuit

the amount of base drive to the output transistor (Q1). A V_{BE} of only about 0.75 V is required to turn on the sensing transistor. The value of the sensing resistor is therefore:

$$R = \frac{V_{BE}\,(on)}{I_{OP}}$$

For I_{OP} = 0.5 A, R = 1.5 Ω.

The 68-Ω resistor in the gate output limits the gate output current level during clamping only. This resistor must be small enough to not limit the transistor output current during normal operation.

When current through the sense resistor is high enough to cause a 0.75 V base-emitter voltage on Q2, Q2 will start to turn on. The collector of Q2 is connected to the base of Q1 and will clamp its base drive, limiting I_{OUT} peak at the selected value, as shown in Figure 10-45.

Another dual stage circuit that will provide consistent performance is the delayed turn-on configuration shown in Figure 10-46.

In this application, limited lamp current is allowed for a period of 150 to 200 ms while the lamp warms up and increases its resistance. After this delay full power is applied, but the bulb resistance is higher and only a small surge current is experienced. A 27-Ω resistor is used in series with the initial switch, limiting the peak current to under 500 mA. The second driver channel has an RC delay network at its input consisting of a 5.6-kΩ resistor and a 62-μF capacitor. The resulting RC time constant is 354 ms. For a logic input voltage of 3.5 V, with the gate input threshold level of 1.4 V,

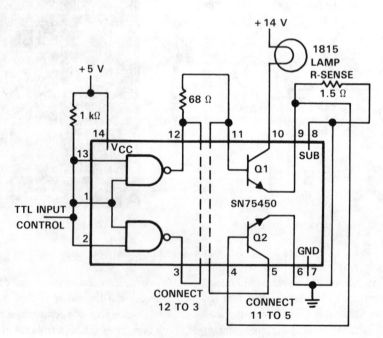

Figure 10-44. Transistor Current Sensing and Peak Limiting

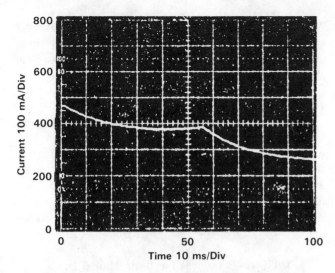

Figure 10-45. Output Current vs Time for Current Sensing Circuit

the second channel will be turned on when its input voltage reaches 1.4 V or 40% of the logic "1" level. A 40% amplitude level will occur at 0.5 RC time constants. Thus the delayed full power turn-on will occur at 0.5×354 ms or at about 177 ms as illustrated in Figure 10-47. In this application, the initial surge current is limited to 425 mA and the delayed turn-on surge is only 300 mA.

Driving Remotely Located LED Devices

There are times when it is necessary to drive an LED indicator from a remote location. This requires sending the control signal from one location to another, over a data transmission line, and then providing sufficient drive power to drive the LED. Figure 10-48 illustrates a method often used for this type of remote application.

The SN75372 dual FET driver has the speed and power capability to drive a data transmission line and the LED directly without additional circuitry. As shown in Figure 10-49, the LED may be connected between a supply and the signal line (driven in the sink mode) or between ground and the signal line (driven in the source mode).

In calculating the values for current limiting resistors (R), consideration of the SN75372's output voltages V_{OH} and V_{OL} must be made. For example:

$$R_H = \frac{V_S - V_{OL} - V_{LR} - V_F}{I_O}$$

$$R_H = 158.5 \ \Omega$$

Where:

V_S = Supply voltage
   = 5 V
V_{OL} = Driver low level output voltage
   = 0.2 V
V_{LR} = Voltage drop due to line resistance.
   = 0.03 V (100 ft line at 20 mA)
V_F = Forward voltage drop of LED
   = 1.6 V (at 20 mA)
I_O = Driver output current
   = 20 mA

$$R_L = \frac{V_{OH} - V_{LR} - V_F}{I_O}$$

$$R_L = 78.5 \ \Omega$$

Where:

V_{OH} = Driver high level output voltage
V_{OH} = $V_{CC2} - 1.8$ V (at 20 mA)
V_{OH} = 3.2 V

Standard resistor values of 150 Ω and 75 Ω may be used.

RELAY AND SOLENOID DRIVER APPLICATIONS

When peripheral drivers are used to drive inductive loads such as relays and solenoids, special attention should be given to the device's switching voltage, as well as current and power capabilities. Most peripheral drivers have their

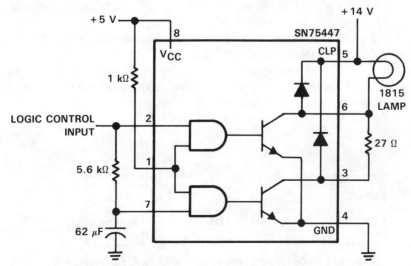

Figure 10-46. Two Stage, Full Power Delayed, Surge Limiting Circuit

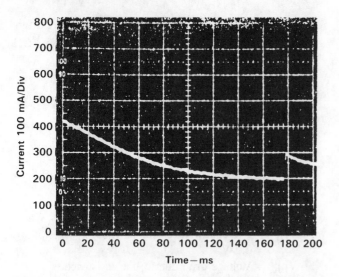

Figure 10-47. Delayed Turn-on Surge Current vs Time

maximum switching voltage specified and care must be taken to avoid exceeding this parameter. Often clamp diodes are used to prevent excessively high voltages at the driver output when driving inductive loads. Figure 10-50 illustrates the basic application and the typical waveform of a diode clamped output. Many peripheral drivers now have built-in inductive clamp diodes.

Printer Hammer Driver Application

In this application, the hammer solenoids require operation from a 30-V supply with peak driver currents of 300 mA. More than one hammer may be actuated at a time. The SN75447 peripheral driver was selected for this application (Figure 10-51). This dual driver will switch up to 50 V (after conducting 300 mA) without latching or breaking down. In addition, the output transient suppression diodes are included in the driver. Since both outputs of the dual driver may be on at the same time, its power handling capabilities must be compared with actual worst case operating conditions. Maximum 5 V supply power dissipation for the SN75447 is equal to 5.25 V times 18 mA (the maximum specified chip supply voltage and current) or 94.5 mW. Output power will be the product of the worst case V_{OL} and the peak (I_{OL}) expected. For this device the output power dissipation, with both outputs on, will be 2(300 mA)(0.65 V) or 390 mW. Power handling capability for this device is 1380 mW at 25 °C. The package has a derating factor of 11.1 mW/°C which yields an 880-mW power capability at the maximum operating temperature of

70 °C. Since the total (chip supply power of 94.5 mW and the output power of 390 mW) is 484.5 mW, it is well within the 880 mW capability at 70 °C, making the SN75447 a good choice.

Driving a Reversible Solenoid

There are times when the solenoid or relay to be driven is bidirectional and therefore must be driven in both directions. Some peripheral drivers are specifically suited for those applications. Bidirectional drive is most easily accomplished with drivers having a combination of both sink and source outputs or totem-pole outputs. The devices in Table 10-7 are examples of drivers suited for bidirectional drive.

Table 10-7. Devices for Bidirectional Drive

DEVICE	V_{CEmax} (V)	I_{Cmax} (A)	$V_{CE(sat)}$ @ I_{Cmax} (V)	t_{PHL} (μs)	t_{PLH} (μs)
SN75603	40	2.0	2.0	1.8	1.4
SN75604	40	2.0	2.0	1.8	1.4
SN75605	40	2.0	2.0	1.8	1.4
UDN2845	50	1.5	1.6	1.5	1.0
ULN2075	80	1.5	1.5	1.5	1.0
ULN2074	50	1.5	1.5	1.5	1.0

Several of these devices are suitable for applications where the supply voltage is over 40 V. In applications where the supply voltage is negative, the UDN2845 will meet the circuit requirements. In Figure 10-52, the basic circuit configuration for driving a bidirectional or reversible solenoid, requiring a negative 48 V supply and 0.7 A, is shown. The logic input control signals could be derived directly from a microprocessor or other types of logic control devices. Input control voltage levels for the UDN2845 are standard TTL or low-level CMOS. Input current requirements are less than 500 μA. Diodes are used to clamp the voltage overshoot and undershoot levels thus protecting the driver outputs from the excessive voltage peaks that can occur when switching inductive loads.

Opto Isolated and Time Controlled Reversible Solenoid Drive

In this application, it is necessary to provide reversible solenoid drive with one position actuated for about 56 minutes out of an hour time period and the other position actuated

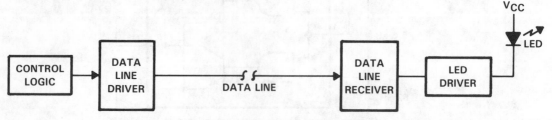

Figure 10-48. Block Diagram of a Typical Remote LED Drive Scheme

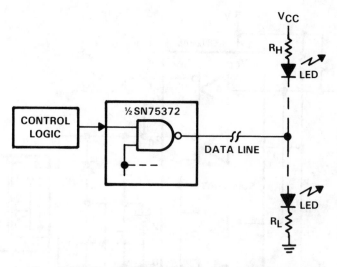

Figure 10-49. SN75372 as a Remote LED Driver

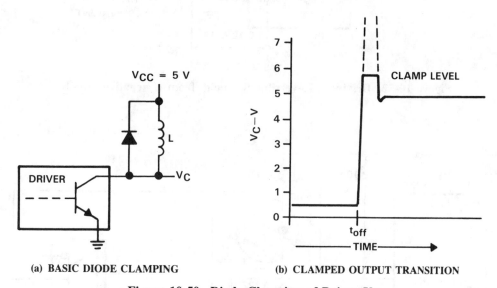

(a) BASIC DIODE CLAMPING   (b) CLAMPED OUTPUT TRANSITION

Figure 10-50. Diode Clamping of Driver V_C

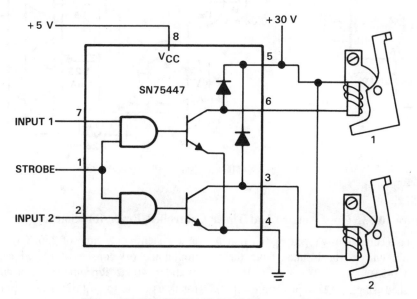

Figure 10-51. SN75447 Dual Hammer Driver

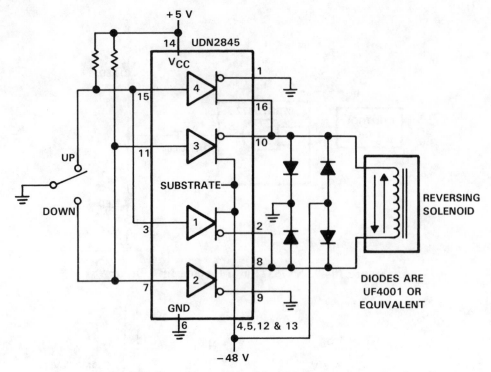

Figure 10-52. Driving a Reversible Solenoid from a Negative Supply

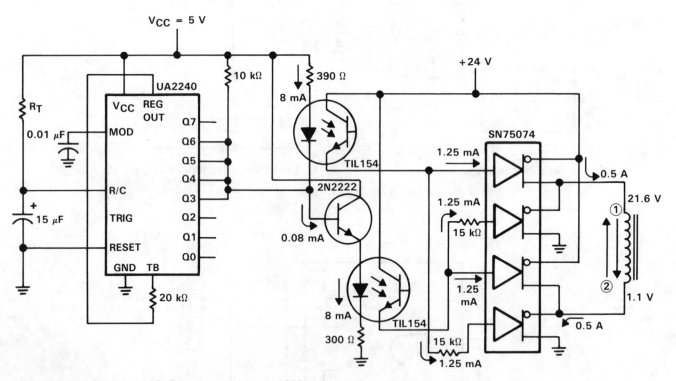

R_T = 937.5 kΩ for exactly 15 μF to generate 1 hour in 256 steps or t_{Bo} = 14.0625 seconds.
Solenoid position ① ~ 56 min
Solenoid position ② ~ 4 min

Figure 10-53. Opto Isolated and Timer Controlled Reversible Solenoid Driver

for about 4 minutes. This is a continuously repeating function that requires the timer circuitry to be isolated from the solenoid supply (Figure 10-53).

Timing is accomplished using a μA2240 programmable timer connected in the astable mode with a time base of about 14.06 s. A time base (t_b) of exactly 14.0625 s will result

in a system cycle time of 256 X t_b or 3600 s (1 hr). The timing network consists of a 15-μF capacitor C_T and an R_T of about 940 kΩ. R_T will need to be adjusted to yield a value for t_b as close to 14.0625 s as is practical. Timer outputs Q3, Q4, Q5 and Q6 are wire-ORed to yield the proper output sequence. The result for an accurate time base will be an output that is low for 56.25 minutes and high for 3.75 minutes. The total time may be adjusted by R_T but the ratio of time high to time low will remain the same. Opto isolation with TIL154 optocouplers provides 2500 V isolation between the controller and the solenoid circuits. The SN75074 driver requires two control inputs, one inverted from the other. One TIL154 is used to couple, in-phase, with the drivers for the forward direction. The inverted drive is done using a 2N2222 NPN transistor operating as an emitter follower driver between the timer and the second TIL154 optocoupler. Output drivers for the reverse direction receive their drive signals inverted from those for the forward direction, allowing the SN75074 to function as a bidirectional solenoid driver. The many applications for a bidirectional solenoid or relay driver include: Fluid Flow Valve Control, Car Electric Door Locks, Relay Switching, and Position Controls.

Power Solenoid Drive

When relatively high power solenoids must be driven, the SN75603, SN75604, and SN75605 are particularly well suited. Individually as drivers, with up to 70 W or more output drive capability, or two in combination for bidirectional drive, this series is a good selection for applications requiring relatively high drive currents.

Figure 10-54 shows the SN75603 and inverting SN75604 in a basic drive configuration for a reversing power solenoid. These devices will sink or source up to 2.0 A from supply rails of 8 V to 40 V. A typical application would be a power solenoid required for operation of a fluid flow control valve.

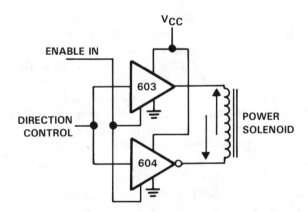

Figure 10-54. Driving a Reversible Power Solenoid

INTELLIGENT SWITCHES
Automotive Lights On Warning

The SN75604, with input control logic but requiring only one supply rail, may be used in special applications requiring input logic where only one supply voltage is available. An example is illustrated in Figure 10-55, a "Lights On" sensor and alarm driver.

In this application, the device V_{CC} and enable inputs are connected to a voltage lead from the light switch. The

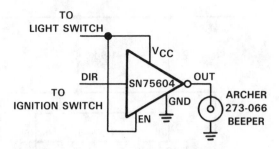

Figure 10-55. Automotive "Lights On" Sensor and Alarm Driver

direction control input is connected to a lead from the ignition switch. Table 10-8 is the truth table for this circuit configuration.

Table 10-8. Lights On Alarm Circuit Truth Table

Light Switch	Ignition Switch	Alarm
ON	ON	OFF
OFF	ON	OFF
ON	OFF	ON
OFF	OFF	OFF

Only operation of the lights without the ignition will result in the alarm sounding. The beeper used in this application is an Archer 273-066 that will operate from 3 V to 28 V. At a typical 12-V level, it will produce a pulsating tone of about 95 dB at 30 cm. The alarm "on" current is about 12 mA when operating from a 12-V supply.

DRIVING MOTORS

"H" OR BRIDGE DRIVE

When it is necessary to switch both ends of a motor winding for forward and reverse operation, a basic "H" or bridge drive is normally used. Figure 10-56 illustrates the basic "H" configuration.

The drivers, or switches, at each end of the motor must be capable of providing both source and sink current to the motor. A single integrated circuit driver capable of providing

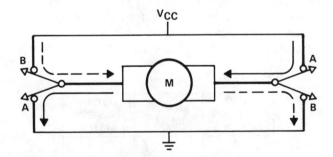

Current B - -→ Flows through the motor left to right
Current A ←— Flows through the motor right to left

Figure 10-56. Basic H Motor Drive Configuration

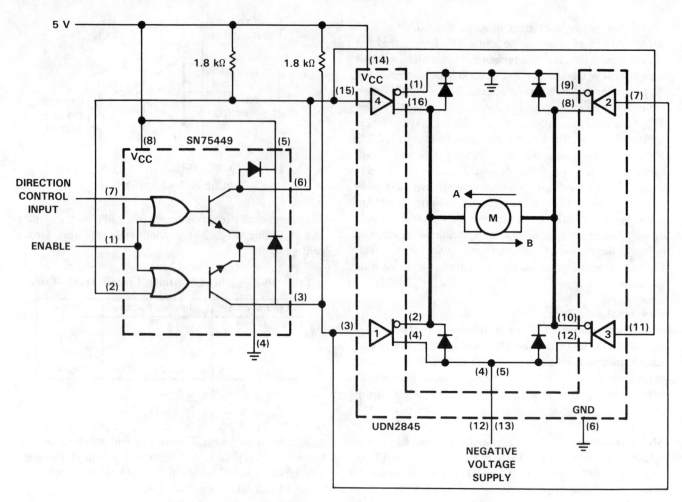

Figure 10-57. UDN2845 Provides Full-H DC Motor Drive

both the source and sink drive is often referred to as a half "H" driver. A circuit with two sink and two source drivers in a single package is referred to as a full "H" driver.

"H" MOTOR DRIVE FOR NEGATIVE SUPPLIES

The UDN2845 provides, in one package, the sink and source drivers needed in a full "H" drive of a dc motor using a negative supply, as illustrated in Figure 10-57. The dc motor in this example is a Pittman model 13104B827. It requires a 300 mA steady state current at 12 V. Turn-on surge currents as high as 1.5 A can be expected. Table 10-9 shows the motor drive conditions for various combinations of input logic.

Table 10-9. Motor Control Truth Table

INPUT DIRECTION CONTROL	ENABLE INPUT	OUTPUT DRIVE TO MOTOR
X	H	OFF
0	L	→ B
1	L	A ←

SPEED CONTROLLED, REVERSIBLE DC MOTOR DRIVE WITH THE SN75603 AND SN75604

For continuous current applications of up to 2 A, the complementary, half "H" SN75603 and SN75604 drivers are recommended. Figure 10-58 illustrates a reversible dc motor drive application with adjustable speed control. The "D" inputs for these drivers are complementary and may be tied together and driven from the same logic control for bidirectional motor drive. The enables are tied together and driven by a pulse-width-modulated generator providing "on" duty cycles of 10% to 90% for speed control. A separate enable control is provided through an SN7409 logic gate. Table 10-10 is the truth table for this motor controller application.

Definitions for the terms used in the truth table (Table 10-10) are as follows:

EN	Enable
DC	Direction Control
SP.C	Speed Control
A	Direction of Current — Right to Left
B	Direction of Current — Left to Right
H	Logic 1 Voltage Level
L	Logic 0 Voltage Level
N	Speed Control Set for Narrow Pulse Widt
W	Speed Control Set for Wide Pulse Width
X	Irrelevant

Table 10-10. Truth Table for Motor Control Circuit (Figure 10-58)

EN	DC	SP.C	MOTOR DIRECTION	MOTOR SPEED
L	X	X	OFF	OFF
H	L	N	A	SLOW
H	L	W	A	FAST
H	H	N	B	SLOW
H	H	W	B	FAST

DRIVING POWER FETs FOR DC MOTOR CONTROL

When motors require discrete drivers for high power levels, it may be desirable to use power FET drivers. This is particularly true where pulse-width-modulated speed control and high-speed switching are required. For the application shown in Figure 10-59, an IRF151 power FET was selected for its speed, low on-state resistance and resulting high efficiency. A natural characteristic of this type of power FET is its large channel area and relatively high gate input capacitance. The IRF151 has a typical on-state resistance of 0.055 Ω and a maximum gate input capacitance of 4000 pF. If the power FET is driven from a typical open-

collector peripheral driver, as shown in Figure 10-59, a long turn-on time will result. The long turn-on time is due to the product of FET input capacitance and static pull-up resistance. This approach is inadequate for efficient high frequency applications. The preferred method for high speed FET driver applications uses an active pull-up as well as an active pull-down, or totem-pole driver such as the SN75372. Figure 10-60 illustrates the use of an SN75372 to drive the FET gate input and the resulting fast gate switching speed.

STEPPER MOTOR DRIVE

Drive circuitry for stepper motors must have characteristics peculiar to the type of motor being driven. Before looking at drivers for stepper motors, a look at the basic stepper motor and its requirements is in order.

Basic stepper motor action may be illustrated (see Figure 10-61) as a permanent magnet rotor that rotates to align itself with magnetic fields. The magnetic fields are generated sequentially by stator coils located around the rotor. If voltage is applied to coil A (A to A') the rotor is attracted to and aligns itself with coil A. If the voltage is switched sequentially to coil B (B to B'), then coil C (C to C'), coil D (D to D'), coil A (A' to A), coil B (B' to B),

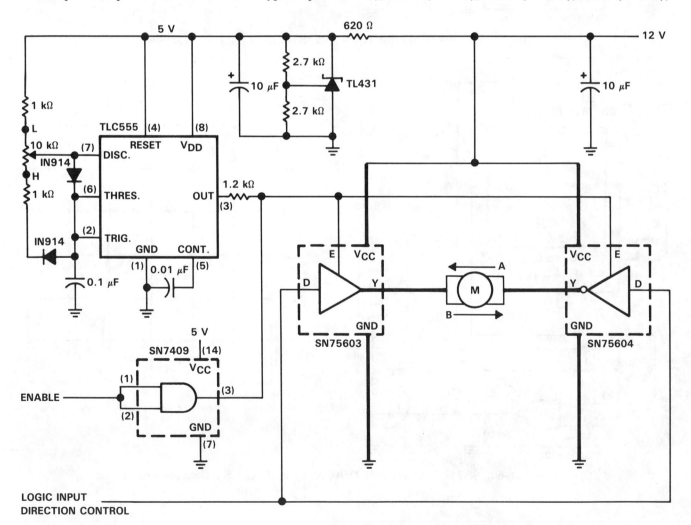

Figure 10-58. SN75602 and SN75604 Devices in a Bidirectional Motor Control Application with Speed Control

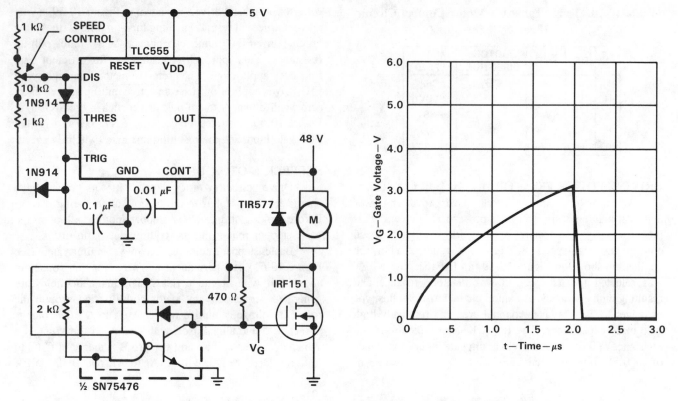

Figure 10-59. Open-Collector Drive and Resulting FET Gate Response

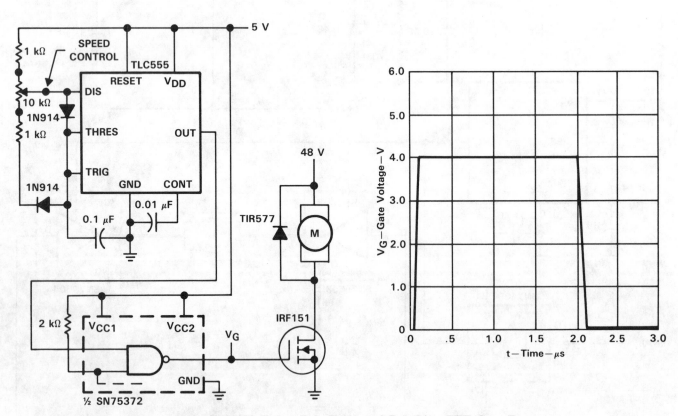

Figure 10-60. SN75372 Totem-Pole Drive and Resulting FET Gate Response

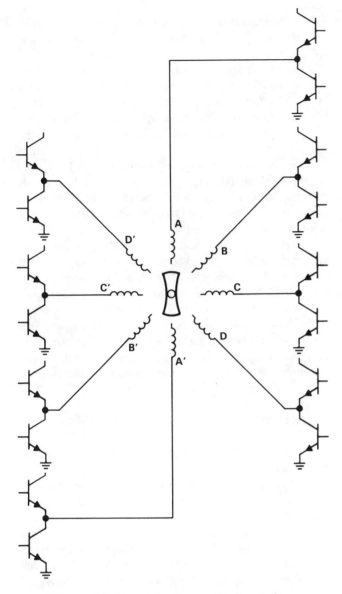

Figure 10-61. Basic Stepper Motor Action

coil C (C′ to C), coil D (D′ to D) and then repeat coil A (A to A′), the rotor will have followed the magnetic field step-by-step for a full 360 degrees of rotation.

The magnetic field in a stator coil will hold the rotor in a fixed position as long as voltage is applied to that coil. Switching the voltage from coil to coil causes the rotor to move, following the field. The rate at which the coils are sequentially switched will determine the rotor speed. In a basic two-pole rotor stepper configuration, with eight stator positions, (8 x 1500) or 12,000 steps per minute will be required to rotate at 1500 rpm. In this typical application a clock rate of 200 Hz is required to provide 12,000 steps per minute. Since the rotor follows the field, simply reversing the timing sequence will result in reversing the direction of rotation.

The basic drive mode just described is sometimes called the wave-mode and requires only one phase (or coil) to be on at a time. However most wave-mode stepper motor drives

are not this simple. Other more common drives include the normal-mode, with two phases on at a time, and the "half-step" mode having one phase on then two phases on then one phase, and so on. The waveforms for these drive methods are illustrated in Figure 10-62.

A multiple-pole wave-mode motor is often used to provide short rotations per step where rotor position is more critical. Figure 10-63 shows a stepper motor configuration having eight stator poles and seven permanent magnet rotor poles. Opposing stator coils (180 degrees apart) are connected in series and are phased to provide rotor attraction at one coil and repulsion at the other. Shifting the drive from stators 1 and 5 to stators 2 and 6 (clockwise) results in a rotor pole being attracted to stator 2 and repelled from stator 6, resulting in a slight counterclockwise rotation. Sixty-four steps are required to complete one full revolution of the rotor, each step being 5.625°.

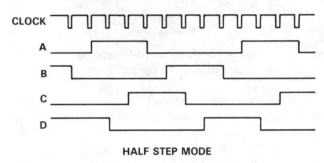

HALF STEP MODE

Provides 2 phases on, 1 phase on, 2 phases on, 1 phase on type of drive sequence.

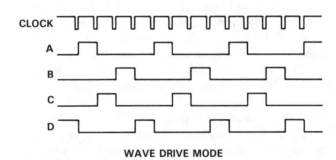

WAVE DRIVE MODE

Provides 1 phase on at a time.

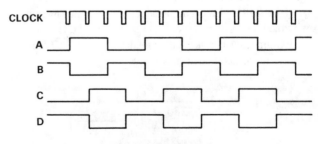

NORMAL DRIVE MODE

Provides, sequentially overlapping, 2 phases on at a time.

Figure 10-62. Half-Step, Wave and Normal-Drive Mode Waveforms

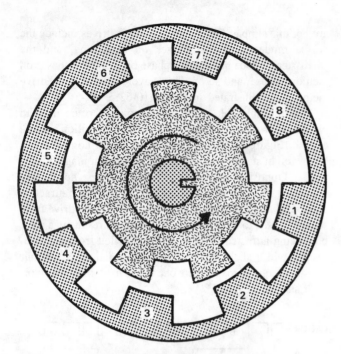

Figure 10-63. 8-Pole, 64-Steps per Revolution, Stepper Motor

The motor stator windings in this example are bridge driven with half "H" drivers at each end of the coils as illustrated in Figure 10-64. The drivers used are the SN75603 and SN75604. They provide complementary outputs allowing each set of coils to be controlled by the same data input pulse. The enables for each set are also paralleled and only four data inputs and four enable inputs are required. Driver data inputs (AD through DD) and enable inputs (AE through DE) provide a type of wave-mode drive that results in sequential stepping. Motor control waveforms are generated using the controller circuitry illustrated in Figure 10-65.

The clock generator shown in Figure 10-65 is a TLC555 timer connected as an astable oscillator operating at 400 Hz. Its output feeds an SN74198 8-bit shift register which generates eight waveforms, four data and four enable. These waveforms are combined with the aid of an SN7432 quadruple OR gate to form the correct drive patterns. Notice that simply changing the operating mode of the SN74198 register from shift left to shift right results in reversing the direction of motor rotation.

The shift register is automatically cleared during power-up by the resistor-capacitor (RC) network on the $\overline{\text{CLR}}$ input. The RC time constant needed to obtain the desired $\overline{\text{CLR}}$ pulse width may be calculated as follows:

$$RC = \frac{t_{wC}}{2.303 \left(-\log \dfrac{1 - V_T}{V_S} \right)}$$

where

t_{wC} is the clear pulse width; 1.0 ms for this application

V_T is the typical clear input threshold; 1.4 V

V_S is the supply voltage

In this example, the calculated value for RC is about 3.0 ms. Selecting a 0.1 μF capacitor for C, the resistor value will be 30 kΩ.

At power-up, the $\overline{\text{CLR}}$ pulse sets all shift register outputs low. The SN74198 outputs are connected to inputs of an SN74260 device that contains two 5-input NOR gates. With all of the SN74260 inputs low at turn-on, its outputs will be high. The outputs are fed through OR gates to the serial inputs (SR and SL) of the shift register. Thus, it provides the initial logic high-level input that will be shifted, QA through QH. When QH goes high, it is fed through OR gates back to the register's serial inputs to continue the cycle. During reverse rotation, the QA pulse is fed back to the serial inputs to continue the cycle.

Resulting data and enable control pulses are illustrated in Figure 10-66. Shift register outputs QA, QB, QC, and QD are the data pulses AD, BD, CD, and DD, respectively. An SN7432 device containing four 2-input OR gates is used to combine the QA with QE, QB with QF, QC with QG and QD with QH to form the enable pulses AE, BE, CE, and DE.

A 400 Hz clock input will result in 2.5 ms wide steps. This type of application requires 64 steps (0.16 seconds) per revolution of the motor. The motor speed will therefore be 375 rpm during continuous operation.

DRIVING DATA TRANSMISSION LINES

Because of their drive power capabilities, peripheral drivers are suitable for driving data transmission lines. The SN75450B type devices, for example, allow sink or source mode or differential line drive operation as illustrated in Figure 10-67. In Figure 10-67(a), the SN75450 is connected as a dual source-mode line driver and the outputs are terminated to ground with resistors equal to the line's characteristic impedance. For this illustration, the output transistor base bias resistor (R_b) was calculated using typical 25 °C values for the SN75450B output transistor characteristics and a V_O of 3.5 V. The line termination resistor is 120 Ω, and transistor h_{FE} is 50.

$$R_b = (V_{CC} - V_b)/I_b$$

where:

$V_{CC} = 5$ V
$V_b = V_O + V_{be} = 3.5$ V + 0.8 V = 4.3 V
$I_b = I_O/h_{FE} = 29.17$ mA/50 = 0.583 mA
$R_b = (5$ V − 4.3 V)/0.583 mA = 1.2 kΩ.

Figure 10-67(b) illustrates dual sink mode line drivers with the lines terminated in their characteristic impedance to the supply rail. Other peripheral drivers from the SN75451B, SN75461, and SN75471 families may be used as sink drivers. A combination of the two drive modes is shown in Figure 10-67(c) where the SN75450B is connected as a single balanced (differential) mode driver. In this example, the line termination resistors are equal to about one-half of the characteristic line impedance (or 62 Ω) and will

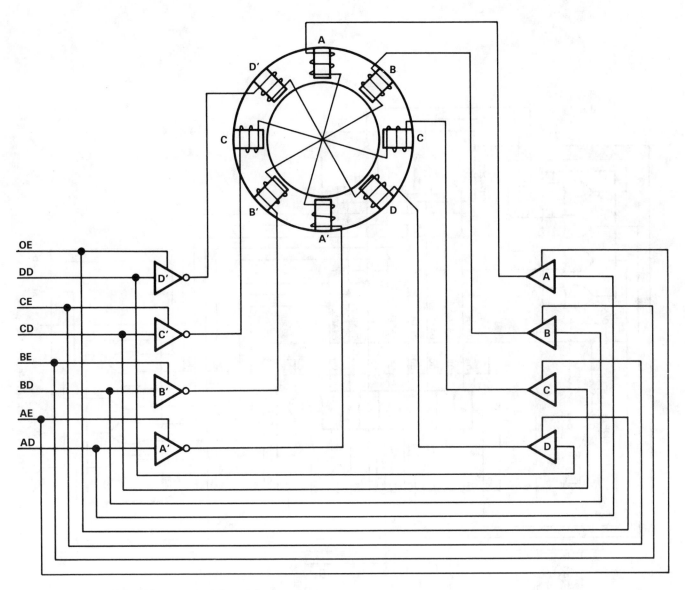

A, B, C and D are SN75603 drivers (non-inverted outputs).
A', B', C' and D' are SN75604 drivers (inverted outputs).

Figure 10-64. Bridge Driven 8-Pole Stepper Motor

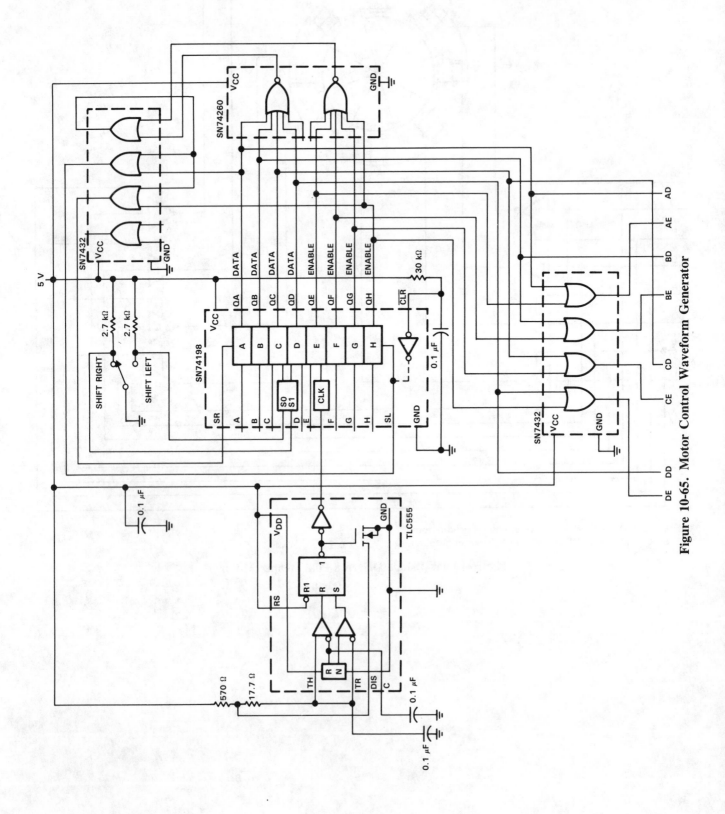

Figure 10-65. Motor Control Waveform Generator

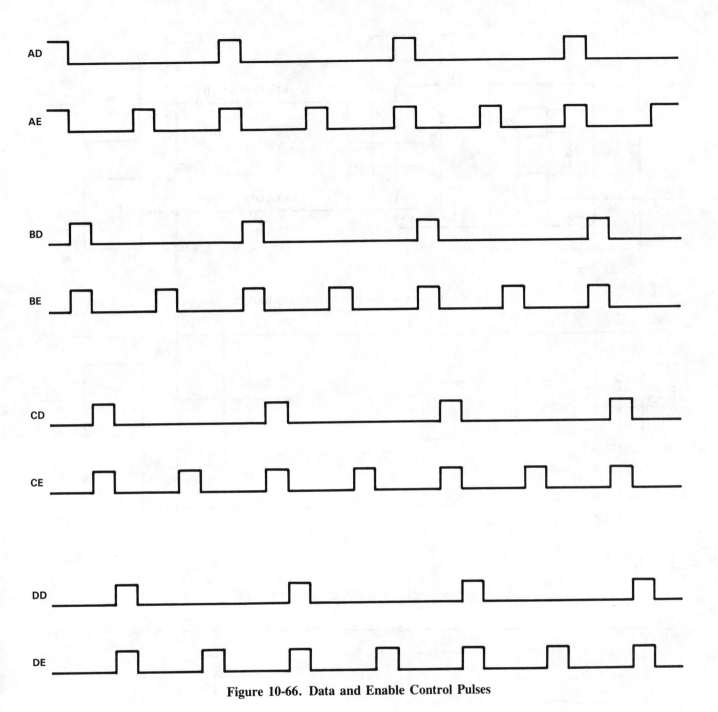

Figure 10-66. Data and Enable Control Pulses

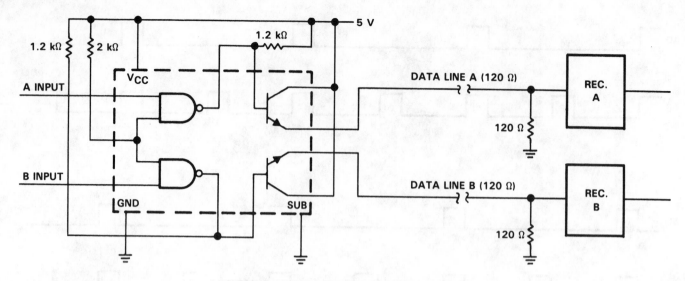

(a) SN75450B AS DUAL SOURCE-MODE LINE DRIVER

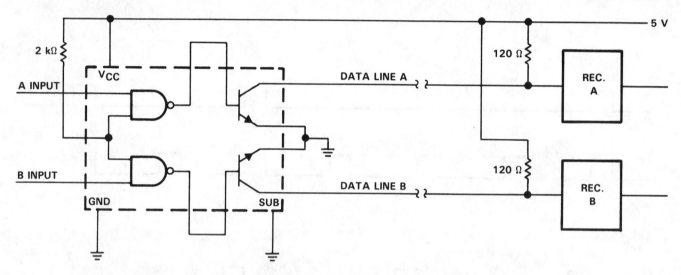

(b) SN75450B AS DUAL SINK-MODE LINE DRIVERS

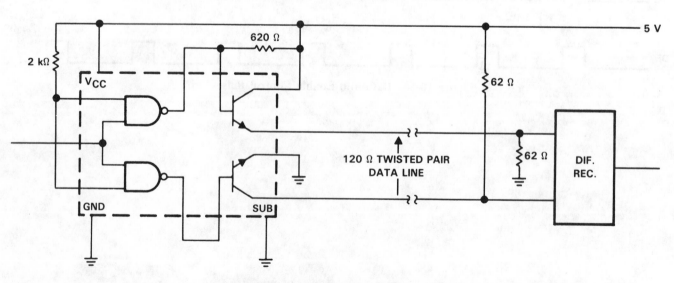

(c) SN75450B AS DIFFERENTIAL LINE DRIVER

Figure 10-67. SN75450B as Data Line Driver

result in a driver output current of 3.5 V/62 Ω or about 56.5 mA. Base current I_b is equal to 56.5 mA/50 or 1.13 mA. Base resistance R_b, for the source transistor, is (5 V - 4.3 V)/1.13 mA or about 620 Ω.

With any of these three drive modes, there are some limitations on speed due to the long RC time constants of line capacitance and termination resistance. Rather than using open-collector or emitter-follower output drive devices when high speed operation is desired, a totem-pole output driver is recommended. The SN75372 is a dual driver with totem-pole outputs. This device will drive single ended coaxial or twisted-pair lines at high data rates as illustrated in Figure 10-68. With transition times less than 12 ns, the SN75372 drivers can transmit data at frequencies up to 20 MHz.

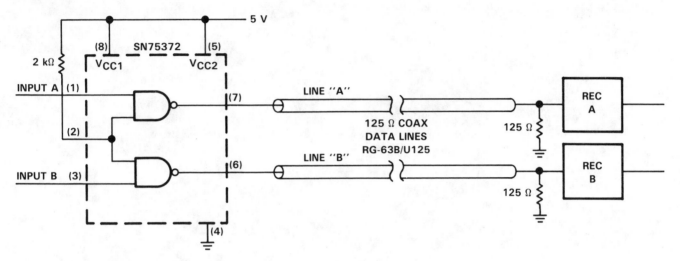

For Data Transmission Rates Up to 20 MHz

Figure 10-68. High-Speed Coaxial Data Line Transmission

Section 11

Data Acquisition Systems

A data acquisition system in the broad sense could mean any method used for obtaining or acquiring any type of data, however, a data acquisition system (DAS) usually denotes a group of electronic devices that are connected to perform the measurement and quantization of analog electrical signals for digital processing. Stated another way, the DAS is the interface between analog and digital electrical signals. Although most physical attributes are analog in nature, many are not electrical and must be converted to the electrical world using sensors or transducers. For the purposes of this text this conversion will be assumed and all inputs will be considered electrical.

TYPICAL SYSTEMS

A digital voltmeter is a DAS that simply displays its output as a numerical value. An automatic process control system containing a large computer and a DAS provides a greater range of data acquisition functions. The trend seems to be for a DAS to be coupled with a microprocessor or minicomputer in most applications today. A typical system with a minicomputer is illustrated in Figure 11-1.

Key parts of the system include:

Analog Multiplexer — This function allows the selection of any one of several analog signals.

Signal Conditioner — This function usually follows the multiplexer, but could be used on any or all input signal lines, to prepare the signals for conversion. The conditioning required might include one or more of the following: linear or logarithmic amplification, filtering, peak detection, sample-and-hold, or track-and-hold.

Analog-to-Digital (A/D Converter) — The A/D converter translates the analog signal to a digital format.

I/O Controller — The I/O controller generates the system timing and controls the memory READ/WRITE functions.

Minicomputer — An operational computer system based on a microprocessor chip. It contains memory (storage) and is controlled by software.

Output Buffer — The output buffer combines the data with the proper signal format for output to the specific peripheral devices. The output buffer is also called a peripheral controller.

Peripheral — Different microcomputer-type output devices such as a line printer, floppy-disk, or magnetic tape storage unit.

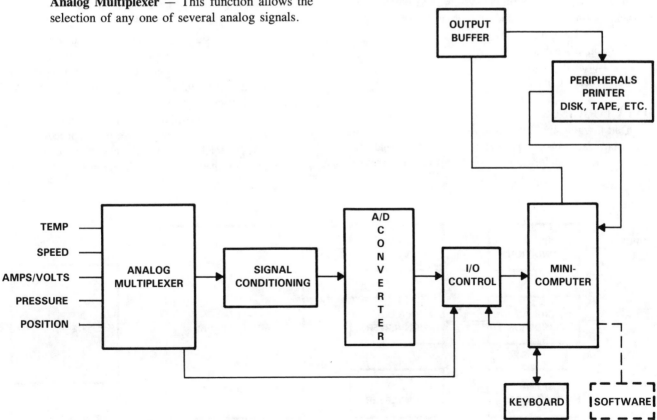

Figure 11-1. Typical Automated DAS

MICROPROCESSOR-CONTROLLED DAS

The latest trend in test instrumentation is to include a microprocessor in the equipment as shown in Figure 11-2. This system gives the user more flexibility and programming power.

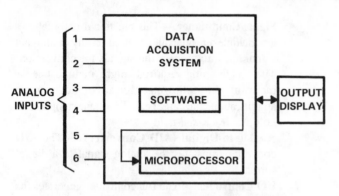

Figure 11-2. Microprocessor-Controlled DAS

A different configuration is shown in Figure 11-3. In this system each channel has its own signal conditioning and A/D converter. This allows processing of analog signals from very diverse sources or when high calibration accuracy is required.

BASIC USES OF A DAS

Any hardware system that is analog and digital in nature will have a DAS included in the system. The potential use of such a system will include four broad areas:
1. Data Logging
2. Signal Analysis
3. Automated Testing
4. Process Control

DATA LOGGING

Data logging involves not only measuring the analog inputs but also translating the results to digital signals and storing the data for further processing or analysis.

Figure 11-1 shows such a system. These functions may be performed at precise intervals and at accuracies beyond a person's capability of doing it manually. Present data logging equipment is highly portable and very automated. When in remote locations very little operator attention is required.

SIGNAL ANALYSIS

Many processes used today generate time-varying signals from which data must be extracted. Examples are radar tracking, air traffic control, seismic exploration, and patient monitoring in hospitals. The most common method of analyzing this information is to translate it into an appropriate Fourier series transform for analysis in the frequency domain. Low cost computer hardware coupled with fast Fourier transforms (FFT), makes this approach feasible. Growth in this area is limited only by the cost of computer storage, speed of the computer, and the data acquisition hardware.

PROCESS CONTROL

For many years process control systems were human-controlled. Today, the DAS is a way of life. The process monitoring systems presently in use may monitor hundreds of inputs, display the readings, make calculations, and provide instant data feedback to the operator. The operator takes appropriate action based on the computer output. As computer equipment becomes more complex, more decisions are made by the system. See Figure 11-4 for a typical computer-automated process control system.

BASIC SAMPLING CONCEPTS

For a data acquisition system to function, the input signal must be sampled. Sampling may be defined as measuring a continuous function at discrete time intervals. These sampled signals represent some analog parameter converted to a series of discrete values.

A typical data acquisition sampling system is shown in Figure 11-5 and is made up of these major processes:
1. SAMPLING — the act of measuring a continuous function at discrete time intervals

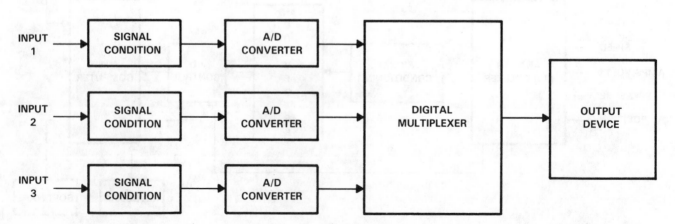

Figure 11-3. DAS with Conditioning and A/D Converter for Each Input

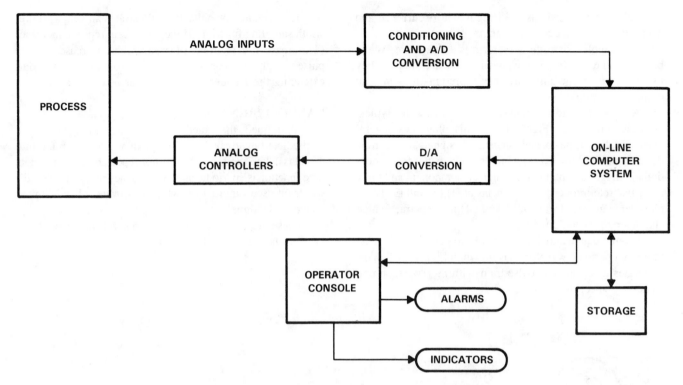

Figure 11-4. Computer-Automated Process Control

2. QUANTIZATION — approximating the linear curve by a series of stair-step values called levels
3. DIGITAL PROCESSING — evaluating the linear function as a series of discrete values represented by ones and zeros
4. RECOVERY — by the use of a D/A converter, returning the processed information to an analog form.

Besides the required number of bits (or quantization levels), the only other major problems are proper A/D conversion and the conversion of the digital signal back to analog.

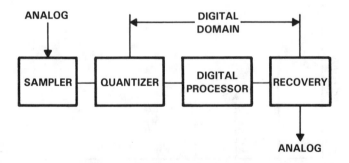

Figure 11-5. Data Acquisition Sampling System

TYPES OF SIGNALS

Before designing a data acquisition system and applying these sampling concepts, we must determine the nature of the signal to be processed. Is it dc or dynamic? Is it steady or random? What are the dynamic range, frequency range, and noise characteristics?

DC SIGNALS

While it may seem strange at first, dc signals too are sampled. In many systems the dc power supplies are monitored to keep them within tolerance. The system must be designed to compensate for undesirable noise signals sometimes present on the input being sampled.

Signal averaging is often a solution to this problem. In using this approach, a large number of measurements of the combined dc/noise signals is made and their sum is divided by the number of measurements. The sample rate should be relatively slow to minimize the effects of random noise signals. Another method used when sampling a dc signal containing noise would be to remove the noise with a low-pass filter. Often a combination of both averaging and filtering are used.

DYNAMIC SIGNALS

The majority of signals we are interested in are dynamic in nature. These signals are constantly changing, often in an unpredictable manner. Dynamic signals may be represented and analyzed in either the time domain or in the frequency domain.

The two major classes of dynamic signals are:
1. Deterministic signals — those that have known characteristics and can be described by mathematics.
2. Random Signals — those whose behavior is highly unpredictable.

DETERMINISTIC SIGNALS

Both periodic signals and transients are classified as deterministic signals. Periodic signals repeat at regular intervals, while transients are irregular in occurrence and decay to a zero value after a length of time.

The simplest periodic waveform is a sine wave. While both the time and frequency domain plots are simple, they prove that even the sine wave has representations in both domains (Figure 11-6).

A somewhat more complex plot is shown in the typical periodic function in Figure 11-7. This illustrates several harmonically, time-locked sinusoidal signals. The time domain plot is shown at (a) while (b) depicts the frequency-domain magnitude plot containing a dc component and three other discrete frequency components at f1, f2, and f3. These plots are in the form of discrete line spectra, which characterize periodic waveforms.

From a data acquisition point of view, one of the most important periodic waveforms is shown in Figure 11-8. This illustrates a rectangular pulse train with its corresponding frequency spectrum.

The rectangular pulse train illustrates an actual, finite-width sampling function. If the pulse train rep rate increases (T decreases), the number of spectral lines decreases. As the pulse width decreases, the zero crossings move out, extending the frequency content over a wider range.

RANDOM SIGNALS

As defined, the exact value of a random signal cannot be predicted in advance. Most signals we deal with fall into this class. The random signal is not repeatable, therefore its time function is aperiodic and its frequency plot is a "closed" spectrum. See Figure 11-9 which illustrates a band-limited nonperiodic signal.

These are the types of signals that will be encountered in everyday use of a data acquisition system.

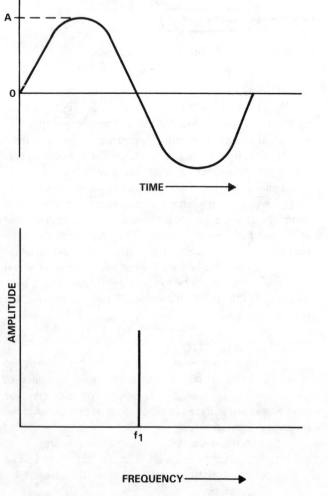

Figure 11-6. Time and Frequency Domain
of a Single Sine Wave

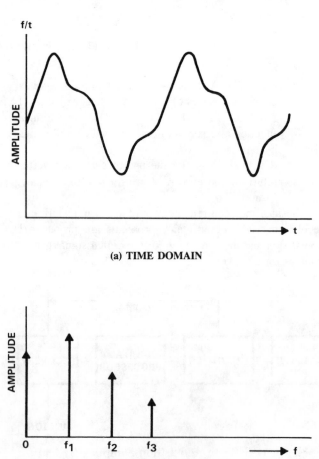

(a) TIME DOMAIN

(b) FREQUENCY DOMAIN

Figure 11-7. Typical Periodic Function

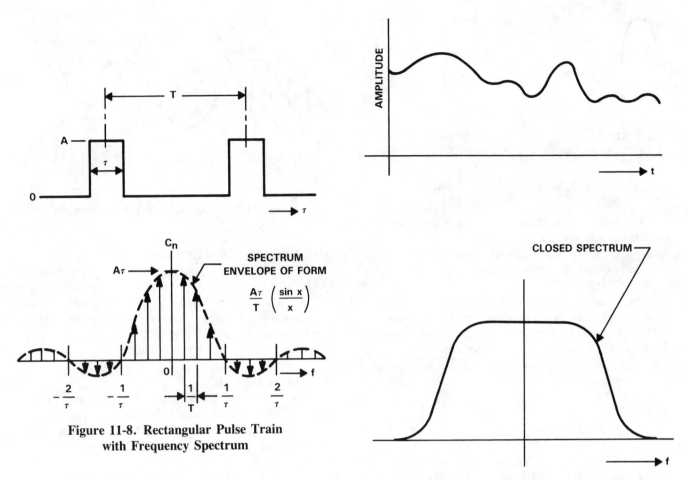

Figure 11-8. Rectangular Pulse Train
with Frequency Spectrum

Figure 11-9. Band-Limited Nonperiodic Signal

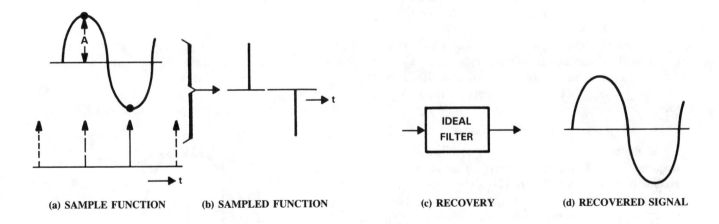

(a) SAMPLE FUNCTION   (b) SAMPLED FUNCTION   (c) RECOVERY   (d) RECOVERED SIGNAL

Figure 11-10. Basics of the Sampling Theorem

FREQUENTLY USED TERMS AND CONCEPTS

THE SAMPLING THEOREM

The sampling theorem states that signals with a finite bandwidth of f-hertz can be described by sampling the signals at instants separated by $T = 1/2f$ seconds. Basically, this means the signal must be sampled at a rate at least twice the frequency of the highest frequency in the spectrum being analyzed. This concept is shown in Figure 11-10.

Here, the sine wave is sampled at its positive and negative peaks. Part (b) shows the sampled function with respect to time. Part (d) shows the original sine wave reproduced after passing it through an ideal "box" filter (c).

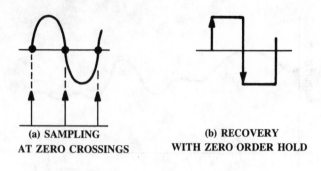

(a) SAMPLING
AT ZERO CROSSINGS

(b) RECOVERY
WITH ZERO ORDER HOLD

Figure 11-11. Sampling Theorem Problems

A number of difficulties can occur in practice when sampling at precisely the Nyquist rate. Two problems may occur. First, if the sampling points chosen fall precisely on the zero crossings, zero output would occur. See Figure 11-11(a).

The second difficulty arises in attempting to build an ideal filter that cuts perfectly above the highest frequency, f. Consider what happens if the reconstruction filter is a simple D to A converter (zero-order hold). As seen in Figure 11-11(b), the square wave has many more frequencies than the original single sine wave.

ALIASING

The dictionary defines the word alias as "an assumed name" or "to be called by another name." Likewise, the concept of aliasing means the same thing when applied to frequencies in a sampled data system. When a given frequency is sampled at a rate much lower in frequency than the original, the resultant frequency is the "alias" of the original. An example of aliasing in the time domain is shown in Figure 11-12.

The actual signal is being sampled at a rate that is somewhat less than one sample per cycle. The resultant frequency is about one-third the original signal. If we could accurately read the waveforms in Figure 11-12, we would find:

$$f(alias) = f(actual) - f(sample)$$

In the time domain, the pulses spread apart. In the frequency domain, the spectra move closer together and eventually overlap as in Figure 11-13.

It is important to note that the aliasing is centered about a point that is equal to one-half the sample frequency or $f_s/2$.

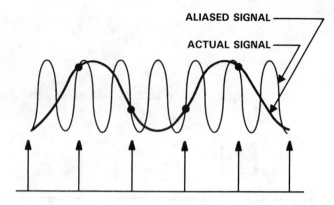

ALIASED SIGNAL
ACTUAL SIGNAL

Figure 11-12. Aliasing in the Time Domain

THE NYQUIST CRITERION

When sampling an analog signal, the Nyquist criterion must always be followed. The Nyquist criterion, named after the individual who discovered it, states that the sampling frequency must be greater than twice that of the highest frequency to be sampled ($f_s > 2f_h$). F_s is the sample frequency and f_h is the highest frequency to be sampled.

ALIASING FILTER

The signals of interest usually contain noise which is much wider in bandwidth than the frequencies of interest. Unless some type of filtering is used, the noise folds over and back into the useful spectrum. The obvious solution is then to add a presampling or "anti-aliasing" filter prior to sampling. Figure 11-14 (a) and (b) shows the removal of aliasing with a presampling filter.

To illustrate the point, a "box" filter is shown. Although this type of filter is impossible to construct, it shows that the spectra no longer overlap. In practice, complex filters introduce time delay and are expensive to build. In a multichannel system, each channel might have a filter prior to the multiplexer and the system cost would increase significantly. The final solution results from a combination of sample rate selection and presample filtering.

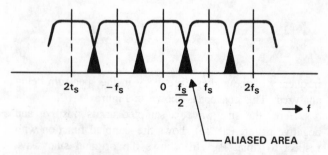

Figure 11-13. Aliasing in the Frequency Domain

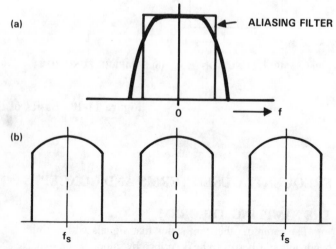

(a)

ALIASING FILTER

(b)

Figure 11-14. Removal of Aliasing with a Presampling or Aliasing Filter (a) and Result (b)

QUANTIZATION

In order for a computer or other digital equipment to process signals, they must first be converted from the analog domain to the digital domain. This process is called quantization. Quantization may be defined as the conversion of an input function which may have values in a continuous range to an output that has only discrete values.

Quantization is sometimes combined with sampling as the operations occur simultaneously. The transfer function for a typical quantizer or A/D converter is found in Figure 11-15.

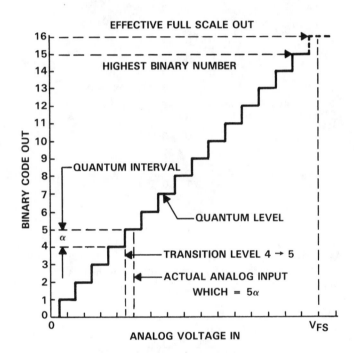

Figure 11-15. Transfer Function for 4-Bit Binary (16-Level) Quantizer

The horizontal scale represents the analog input voltage which may be any value between zero volts and the full-scale value, V_{FS}. The output can exist only as one of 16 discrete values from zero to 15. This "staircase" output has "flats" which represent the digital output code. These "flats" represent the quantum levels. For a binary-based (power-of-2) quantizer, the quantum interval is defined as:

$$\alpha = \frac{V_{FS}}{2^n}$$

where:

V_{FS} = full-scale voltage input
2 = binary number base
n = number of bits (binary)
2^n = number of quantizing intervals.

For the 4-bit quantizer shown in Figure 11-15, the quantum interval is:

$$\alpha = \frac{V_{FS}}{2^n} = \frac{V_{FS}}{2^4} = \frac{V_{FS}}{16}$$

The steps continue to double in number and halve in size as the number of quantizer bits increase, as shown in Table 11-1.

Table 11-1. Binary Steps Per Bit

n	Steps
1	2
2	4
3	8
4	16
5	32
6	64
7	128
8	256
9	512
10	1,024
11	2,048
12	4,096
13	8,182
14	15,384
15	32,768
16	65,536

BINARY CODES

Straight and Offset Binary Codes

Several binary codes are commonly used. The two most common are straight binary and offset binary as found in Table 11-2.

Table 11-2. Straight and Offset Binary Codes

	Bit Position				Level	
	2^3	2^2	2^1	2^0	Straight Binary	Offset Binary
Straight						
Binary Zero ⮞	0	0	0	0	0	−7
	0	0	0	1	1	−6
	0	0	1	0	2	−5
	0	0	1	1	3	−4
	0	1	0	0	4	−3
	0	1	0	1	5	−2
	0	1	1	0	6	−1
Offset	0	1	1	1	7	−0
Binary Zero ⮞						
	1	0	0	0	8	+0
	1	0	0	0	9	+1
	1	0	1	0	10	+2
	1	0	1	1	11	+3
	1	1	0	0	12	+4
	1	1	0	1	13	+5
	1	1	1	0	14	+6
	1	1	1	1	15	+7

Straight binary is a unipolar code where 0000 is equal to 0 (also analog zero) while 1111 is equal to level 15. Offset binary is a bipolar code which has 0000 equal to -7 and 1111 equal to $+7$. For this situation, the most significant bit is the sign bit, with 0 = minus($-$) and 1 = plus($+$). Actually, there are no "zero" levels in the offset binary code. Rather, there are two near-zero levels, each half a quantum interval from true analog zero.

One's and Two's Complement Codes

Two more codes that are widely used and are more compatible with digital computers are the one's complement and two's complement binary codes. Although both codes are bipolar, they differ chiefly in the positioning of the true analog zeros.

The one's complement code is identical to the offset binary code with the exception that the sign bit is inverted. In this case, 1 is negative and 0 is positive. Like offset binary, this system has two zero levels.

The two's complement code differs in that it has only one zero level and it is coincident with the 0000 code. There is also one more negative level (-8) than positive ($+7$). The sign is also the complement of that for offset binary. See Table 11-3.

Table 11-3. One's and Two's Complement Binary Codes

Bit Position				Level	
2^3	2^2	2^4	2^0	Two's	One's
1	0	0	0	-8	-7
1	0	0	1	-7	-6
1	0	1	0	-6	-5
1	0	1	1	-5	-4
1	1	0	0	-4	-3
1	1	0	1	-3	-2
1	1	1	0	-2	-1
1	1	1	1	-1	-0

One's Zero ➤

Two's Zero ➤

Bit Position				Level	
0	0	0	0	0	$+0$
0	0	0	1	$+1$	$+1$
0	0	1	0	$+2$	$+2$
0	0	1	1	$+3$	$+3$
0	1	0	0	$+4$	$+4$
0	1	0	1	$+5$	$+5$
0	1	1	0	$+6$	$+6$
0	1	1	1	$+7$	$+7$

All three bipolar codes discussed thus far are variations of straight binary and require offsetting of analog zero with respect to code levels or inversion of the sign bit.

Absolute-Value-Plus-Sign Code

Another binary code, popular in the 1970s, is the absolute-value-plus-sign code shown in Table 11-4.

In this code, the lower three bits are exactly the binary representation of the associated level, e.g., $111 = 7$. Therefore, the lower three bit codes reflect about zero level ($+3$ has the same code as -3). As in other bipolar codes, the most significant (2^3) bit is the sign designator.

Table 11-4. Absolute-Value-Plus-Sign Code

Bit Position				Level
2^3	2^2	2^1	2^0	
0	1	1	1	-7
0	1	1	0	-6
0	1	0	1	-5
0	1	0	0	-4
0	0	1	1	-3
0	0	1	0	-2
0	0	0	1	-1
0	0	0	0	-0

Zero ➤

Bit Position				Level
1	0	0	0	$+0$
1	0	0	1	$+1$
1	0	1	0	$+2$
1	0	1	1	$+3$
1	1	0	0	$+4$
1	1	0	1	$+5$
1	1	1	0	$+6$
1	1	1	1	$+7$

Gray Code

Another code used by some quantizers is the Gray code. See Table 11-5.

Table 11-5. The Gray Code

Bit Position				Level
2^3	2^2	2^1	2^0	
0	0	0	0	0
0	0	0	1	1
0	1	0	1	2
0	0	1	0	3
0	1	1	0	4
0	1	1	1	5
0	1	0	1	6
0	1	0	0	7

Midpoint ➤

Bit Position				Level
1	1	0	0	8
1	1	0	1	9
1	1	1	1	10
1	1	1	0	11
1	0	1	0	12
1	0	1	1	13
1	0	0	1	14
1	0	0	0	15

This code has two unique features. First, like the absolute-value-plus-sign code, the lower three bits reflect about the scale midpoint. Secondly, only one bit changes as the code progresses from one level to the next. The Gray code is used for most types of electromechanical quantizers and for cyclic-type A/D converters.

BCD Code

The final code to be discussed is the 8421 binary-coded decimal, which is commonly referred to as BCD. This code is used when direct decimal readouts are necessary. Although binary in form, the codes exist only in decades (0 to 9). When the decade limit is exceeded, the code carries to the next higher decade. See Table 11-6.

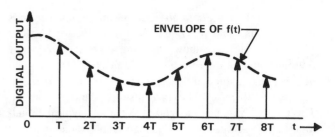

Figure 11-16. Typical Digital Processor Output, f(t) [Error of Omission]

Table 11-6. The 8421 Code

8	4	2	1	Level
0	0	0	0	0
0	0	0	1	1
0	0	1	0	2
0	0	1	1	3
0	1	0	0	4
0	1	0	1	5
0	1	1	0	6
0	1	1	1	7
1	0	0	0	8
1	0	0	1	9

Each increasing decade is identical to the figures shown in Table 6, but has a 10X magnifier associated with it. The quantum interval for 8421 BCD code is calculated as:

$$\alpha = \frac{V_{FS}}{10^m}$$

where:

V_{FS} = full scale voltage input
m = number of decades
10^m = number of quantum levels.

SIGNAL RECOVERY

The sample data process cannot be considered complete until a function called recovery is effected. The recovery function must accurately reconstruct the original time signal or a derivative of it. The two main sources of errors in recovery are errors of omission and errors of commission.

An error of omission is shown in Figure 11-16. The dashed line shows the sampled analog signal while the arrowheads depict the specific sampling points. A certain amount of signal trend information is missing in between these points. Once quantization has occurred and the samples have only discrete values, difficulties are compounded.

The second error problem is the error of commission. See Figure 11-17. The signal sampling is shown in Figure 11-17(a). When examining the frequency spectrum in Figure 11-17(b), not only is the original signal spectrum present, but many additional spectra are also present and centered about multiples of the sample frequency. If these extra outputs are not eliminated, erroneous results will occur.

The ideal recovery function would be a rectangular or "box" filter. Unfortunately, it is not physically possible to implement a filter with these ideal characteristics.

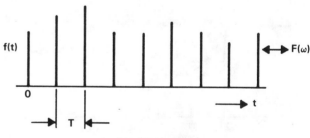

(a) SAMPLED INPUT

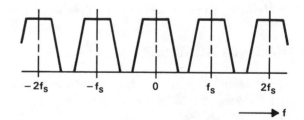

(b) FREQUENCY DOMAIN RESULT

Figure 11-17. Error of Commission

11-9

One of the most common methods of reconstruction is the zero-order hold function as shown in Figure 11-18.

Figure 11-18(a) shows that the time and frequency representation for the zero-order hold function is rectangular in the time domain and (sin x)/x in the frequency domain. Figure 11-18(b) shows the final result in the time domain. This output is instantly recognized as that of a typical digital-to-analog (D/A) converter. While this reconstruction process results in some error, the worst case error would occur at zero crossing of the sine wave at the highest frequency in the original signal.

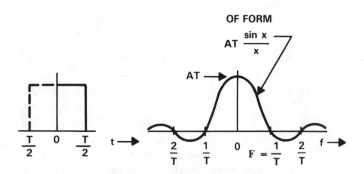

(a) RECOVERY FUNCTION, IN BOTH DOMAINS

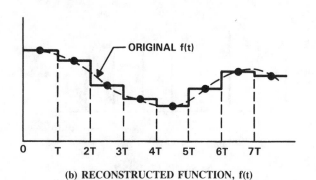

(b) RECONSTRUCTED FUNCTION, f(t)

Figure 11-18. Staircase or Zero-Order Hold Recovery

A DATA ACQUISITION EXAMPLE

The following example will help to tie together the concepts discussed thus far. A simple digital data system is shown in Figure 11-19(a).

Assume the analog input has the characteristics of a fourth-order Butterworth spectrum and the highest frequency of interest is at 20 kHz. The four initial requirements for our example are:

1. Dynamic range >70 dB
2. Aliasing error $<0.1\%$
3. Recovery error $<0.1\%$
4. Aperture error $<0.1\%$

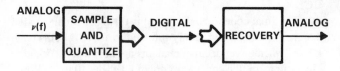

(a) EXAMPLE SAMPLE DATA SYSTEM

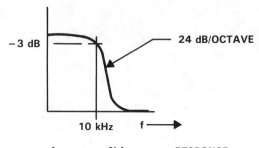

(b) CHARACTERISTICS OF FOURTH ORDER INPUT

f	%A	RESPONSE
10 kHz	70.7	-3 dB
20 kHz	4.4	-27 dB
40 kHz	0.28	-51 dB
80 kHz	0.018	-75 dB
120 kHz	0.0044	-87 dB
160 kHz	0.0011	-99 dB

Figure 11-19. Data Acquisition Example

The parameters to be determined are:
1. Number of quantization levels (bits)
2. Sample rate
3. Sampling aperture operator
4. Recovery operator

The first unknown to determine is the number of bits. The system dynamic range has been specified as 70 dB, which is equivalent to the ratio of 3162 to 1. The closest larger binary number is 4096, or 12 bits, which is equal to 72 dB. Thus, a 12-bit A/D converter would be chosen.

The next unknown to consider is the sampling rate. This becomes somewhat more complex as two requirements must be satisfied to determine this parameter; one for aliasing and one for recovery error. Consider the aliasing problem first. It requires the error due to folding to be less than 0.1% of the expected value at the highest frequency of interest, 20 kHz. Figure 11-19(b) shows that at 20 kHz, the signal levels are 4.4% of the dc value. The overlap of the first harmonic spectrum must not be greater than 0.1% of the 4.4% value. This then is the 0.0044% point on the chart in Figure 11-19(b) or -87 dB. This overlap occurs when the first spectrum contribution of $f_s - 120$ kHz coincides with the 20 kHz point on the original spectrum. This is shown in Figure 11-20.

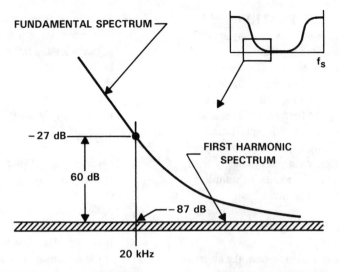

Figure 11-20. Effect of Aliasing, $f_S = 120$ kHz

This shows a 1000 to 1, or 60 dB, difference between the two spectra. The sample rate to satisfy this requirement must be equal to or greater than:

$$120 \text{ kHz} + 20 \text{ kHz} = 140 \text{ kHz}$$

Using this analogy, a 140 kHz sampling frequency will satisfy the aliasing error requirement of this example. Before selecting 140 kHz as the sampling rate, consider the impact of the recovery error requirement. Assume we choose the recovery filter to have the same characteristics as the input spectrum, namely a 4-pole Butterworth. To meet an rms recovery error of 0.1%, a 4-pole Butterworth filter spectrum must have 36 samples per cycle of a 3-dB bandwidth. Calculate as follows:

$$\frac{f_s}{f_{3dB}} = 36 \text{ samples per cycle}$$

where:

$$f_s = \text{sampling frequency}$$
$$f_{3dB} = 3 \text{ dB down point or 10 kHz}$$
$$f_s = 36 \times 10 \text{ kHz} = 360 \text{ kHz}$$

To meet the original requirements of our example, at a sampling rate of 360 kHz, aliasing is negligible. A suitable selection for this example would be a 12-bit, 500 kHz, successive approximation A/D converter.

The third consideration is aperture error and a decision must be made concerning whether to use a sample-and-hold function ahead of the converter. The basic aperture of an A/D converter is:

$$\text{Aperture} = \frac{1}{f_s} = \frac{1}{500 \text{ kHz}} = 2 \text{ } \mu s$$

An aperture like this would be equivalent to placing a (sin x)/x filter in the system, with the first zero crossing at 500 kHz. This would cause between one and 2% attenuation at 20 kHz. To achieve our design requirement of less than 0.1%, a sample-and-hold with an aperture of less than 100 ns would be required.

The final requirement to contend with is the recovery operator. The final recovery filter was previously chosen to be a 4-pole Butterworth. Before continuing, the data must first be converted from digital binary numbers to analog values. If we utilized a simple D/A converter (zero-order hold) with 500 kHz update rate ahead of the filter, we still have the (sin x)/x filtering problem. In fact, it will be more severe than the final recovery filter. One solution would be to utilize a D/A converter with return-to-zero outputs of less than 100 ns in width. The final system configuration is shown in Figure 11-21.

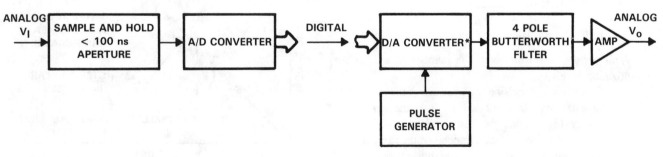

*100 ns return-to-zero

Figure 11-21. Complete Data Acquisition System

SUMMARY OF DESIGN CONSIDERATIONS

There is usually more than one solution for any specific A/D design situation. Filters other than the 4-pole Butterworth filter could be used in recovery. Tighter aliasing error requirements could have greatly affected the sample rate. Each problem must be considered individually, and then a decision made after considering the hardware available and the parametric requirements of the specific application.

To implement a successful design, you must know what the input signal spectrum looks like, what you ultimately wish to obtain from it, and how to apply the rules of sampling, quantization, and recovery on an individual basis.

SYSTEM PARAMETERS

To properly design a data acquisition system, the system parameters must be thoroughly defined. A large number of converter products are available today. Manufacturers may specify the parametric capabilities of devices in different ways. Two basic factors should be a key in choosing the right device for a particular application. The first is to completely define the design objectives and the second is to understand what the data sheet specifications mean. If there is any difficulty in understanding the specs, consult with the manufacturer before proceeding with the design.

UNDERSTANDING MAJOR SYSTEM PARAMETERS

Absolute Accuracy — This is an indication of the discrepancy between the data acquisition system measured value of a given input and the National Bureau of Standards measured volt. It implies that the system's voltage reference is traceable to the NBS standard and is periodically calibrated.

Relative Accuracy — This is a measure of the data acquisition system's ability to yield correct output codes, for all possible input voltages, relative to its full-scale range. It is a direct function of the system's linearity.

The most typical way of stating accuracy is:

$\pm 0.01\%$ of FSR (analog inaccuracies)

where:

FSR = full-scale reading

This term includes all the analog sources of error such as linearity, drift, and component error.

$\pm 1/2$ LSB (quantization error)

where

LSB = least significant bit

This term is determined by the number of available quantization levels.

Precision — Precision is a measure of the system's repeatability. In other words, it is the system's ability to produce the same output code when making successive measurements of the same input value. Internal system-generated noise is one cause of poor precision.

Resolution/Dynamic Range — This is the ability of a data acquisition system to distinguish between adjacent analog input levels. A 10-bit system would be capable of distinguishing between input levels that differ by 1/1024 of the full-scale range (10 bits $= 2^{10} = 1024$). Resolution also defines the system dynamic range which may be stated:

1024 to 1 or (dB $= 20 \log_{10} 1024$) $= 60$ dB

Overrange — When the normal full-scale voltage is exceeded on many data systems, the output continues to register the code assigned to the full-scale value.

Linearity — Two types of nonlinearity are involved. The first is integral nonlinearity, which is the maximum deviation from a best straight line drawn through the transfer curve. See Figure 11-22.

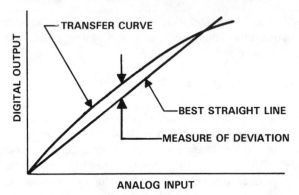

Figure 11-22. Integral Nonlinearity

Integral nonlinearity is usually specified as a percentage of full scale or in terms of quantum level (such as $\pm 1/2$ LSB).

The second type of nonlinearity is differential nonlinearity. This is the measure of the maximum deviation of any quantum interval in the system's transfer function from the theoretical value. Differential nonlinearity may be specified as a percentage of quantum interval or LSB. See Figure 11-23.

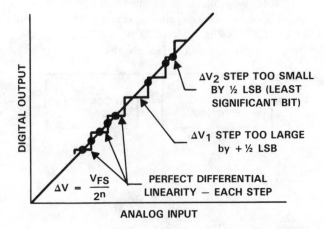

Figure 11-23. Examples of Differential Nonlinearity

Monotonicity — Monotonicity, in a data acquisition system, is a guarantee that the system's output code will continue to increase (or decrease) with a continuously increasing (or decreasing) input. This problem most likely occurs at the halfway or quarter-way points of full-scale readings.

Temperature Coefficients — Changes in ambient temperature affect a system's ability to produce accurate measurements. The two major problems are gain change and zero shift. See Figure 11-24.

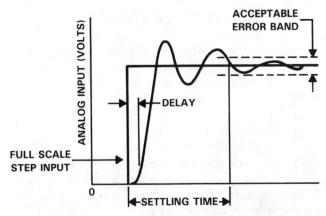

Figure 11-25. Example of Settling Time

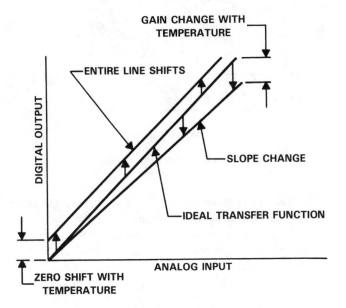

Figure 11-24. Temperature Effects on Accuracy

Temperature can also adversely affect integral and differential linearity. Temperature coefficients are factors which should not be overlooked.

   Noise — Noise is a constant problem in any acquisition system. Normal noise is that noise appearing at the two measurement terminals, along with the desired signal. Common-mode noise is interference that appears simultaneously on both measurement inputs. Most common-mode noise may be attributable to the 60 Hz power line frequency.

   Conversion Speed — Conversion speed provides an indication of system sampling rate, and is usually expressed in conversions per second. A communications type data acquisition system, for example, might have a conversion rate of 10 conversions per microsecond which equates to a conversion speed of 10M conversions per second.

   Settling Time — This is the time between the application of a full-scale input step and the stabilization of the output to within a specified error band. Once the ringing has settled to within the error band, an accurate conversion can occur. See Figure 11-25.

Aperture Time — Aperture time is the width of the sampling window or the actual time the system requires to obtain a sample.

Acquisition Time — Acquisition time applies to systems containing a sample-and-hold function. It defines the time required to sample an analog signal within the specified system accuracy. A typical acquisition time would be specified to 0.1% of full-scale = 4 μs.

Hold Time/Droop Rate — This parameter also concerns systems with a sample-and-hold function. Hold time defines how long the system can hold the stored value to a given accuracy. Droop rate indicates the loss of charge per unit of time. Although both specifications are still used, droop rate is more popular.

Cross Talk — This is very important in a multichannel system. It is an indication of signal leakage from one channel to any other.

DC Input Characteristics — These characteristics are offset voltages, offset currents, and bias currents. An offset of 10 mV will shift inputs of some systems by several significant bits. Although sometimes neglected, the dc input characteristics are important in some applications and should not be overlooked.

METHODS OF A/D CONVERSION

Several types of A/D converters are available. Some feature speed of conversion, while others are known for their higher resolution. The four most widely used methods are as follows:

1. Single slope
2. Dual slope
3. Successive approximation
4. Flash and semi-flash

Each of these methods has advantages and disadvantages. Some of the primary features of these conversion techniques are shown in the following chart:

TYPE	SPEED	ERROR	RESOLUTION
Single-slope	Slow (ms)	High ±1 LSB	Med-High (8—14 Bits)
Dual-slope	Slow (ms)	High ±1 LSB	High (10—18 Bits)
Successive-Approximation	Med (μs)	Low-Med ±0.5 LSB	Med-High (8—16 Bits)
Flash	Fast (μs)	Low-Med ±0.25 LSB	Low-Med (4—8 Bits)

KEY SELECTION CRITERIA FOR A/D CONVERTERS

NUMBER OF BITS

In a digital (binary) system, each bit of information is represented by a "1" or a "0". Four bits are required to represent the decimal numbers 0 through 15 (16 different states). An 8-bit number can represent the decimal numbers 0 through 255 (256 states). It follows that with more bits, or sampling points of the input signal, the output data will be more precise. For example, a 4-bit converter could specify values between 0 and 1.0 V in 1/16-V increments, but an 8-bit converter could specify values in this same range in 1/256-V increments. Thus, more bits of output data provide higher resolution.

CONVERSION SPEED

The process of sampling and converting an analog signal value to a digital number is not instantaneous. A finite amount of time is required for the conversion process to be completed. Faster individual conversions means more conversions per unit of time. For example, a conversion speed of 1.0 ms implies 1000 conversions per second and a conversion speed of 100 μs implies 10,000 conversions per second.

CONVERSION ACCURACY

There is a limit to how closely an A/D converter can approximate a given voltage level. The A/D converter approximates the linear curve by a series of stair-step values called quantization levels. See Figure 11-26.

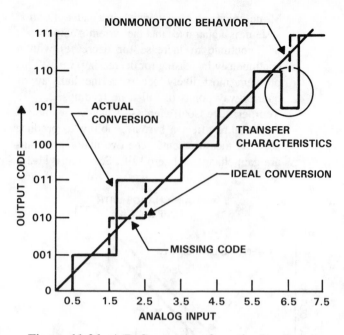

Figure 11-26. A/D Conversion Quantization Levels

Depending on when the sample is taken, the sample can be in error. The method for determining the error that can be expected from a given A/D converter is expressed relative to the number of quantization levels, specifically to the variation in LSB.

$$\text{Value of LSB} = \frac{\text{Full scale voltage}}{2n}$$

where:

LSB = least significant bit
n = number of bits of resolution

From this we see that a 10-bit A/D converter is not as good as an 8-bit converter if the 10-bit A/D converter has an error of ±4 LSB and the 8-bit A/D has an error of only ±1/2 LSB when sampling a full-scale voltage range of 4 V. As an example, assume a 4 V full-scale range and compare an 8-bit converter with a 10-bit converter.

8-Bit Converter $\quad \text{LSB} = \dfrac{4\,V}{2^8} = 15.6\,\text{mV per step}$

10-Bit Converter $\quad \text{LSB} = \dfrac{4\,V}{2^{10}} = 3.9\,\text{mV per step}$

If the 8-bit error is ±1/2 LSB, it is accurate to ±7.8 mV. The 10-bit converter with an error of ±4 LSB is accurate to only ±15.6 mV.

The accuracy of an analog-to-digital conversion may be defined as linearity. Errors are defined and measured in terms of the analog values at which the transitions occur, in relation to the ideal transition values. Figure 11-27 shows the transition errors including offset and range errors.

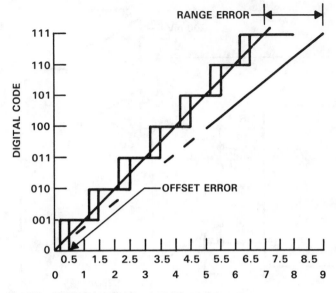

Figure 11-27. A/D Transition Errors

CONVERSION TECHNIQUES

SINGLE-SLOPE A/D CONVERTER

A single-slope integrator arrives at its digital output by comparing the unknown analog input signal to a ramp voltage generated from a reference. The ramp must be precisely controlled so that it is very linear, and takes exactly 2 raised to the nth number of clock pulses to go from 0 V to full scale. By counting the number of clock pulses required to build the ramp from 0 V to the value of the unknown analog signal, a digital value of the analog signal is obtained. For example, assume a 6-bit A/D converter and a full scale reading of 5 V. See Figure 11-28.

6 Bit A/D Converter = $2^{(6)}$ or 64 increments
Full scale resolution = 64 divisions
Full scale voltage = 5 V
Analog input voltage = 34/64 of full scale
Analog input voltage = 5 V × 34/64 = 2.6 V

The requirements for a good single-slope A/D converter are:

1. A stable reference voltage to build the ramp
2. A stable clock (oscillator) to count the ramp
3. A stable and linear integrator to control the accuracy of the ramp.

Figure 11-29 shows the functional block diagram of a single-slope A/D converter.

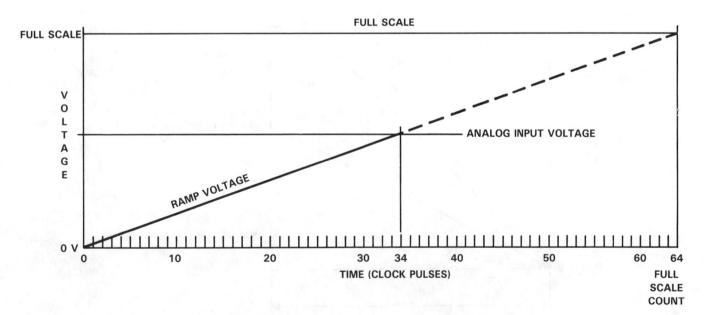

Figure 11-28. Ramp/Analog Voltage Comparison on a 6-Bit A/D Converter

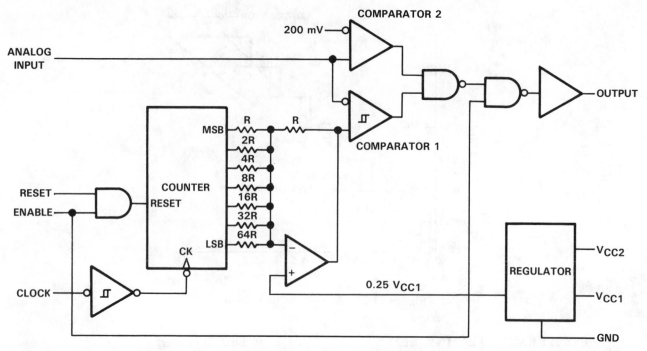

Figure 11-29. Single-Slope A/D Converter

DUAL-SLOPE A/D CONVERTER

Dual-slope conversion is an indirect method for A/D conversion where an analog voltage and a reference voltage are converted into time periods by an integrator, and then measured by a counter. The speed of this conversion technique is slow but the accuracy is high.

Like the single-slope A/D, the dual-slope integrator arrives at its digital output by building ramps. Its greater accuracy and lower conversion speed result from building two ramps instead of one. The first integration (ramp building) is accomplished using the unknown analog input and integrating for a fixed time (T). The peak value of this integration is proportional to the value of the unknown analog input. See Figure 11-30.

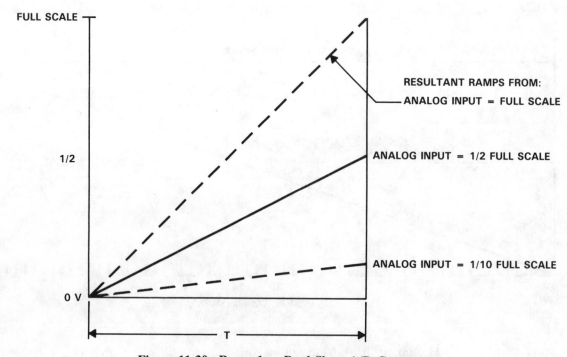

Figure 11-30. Ramp 1 — Dual-Slope A/D Converter

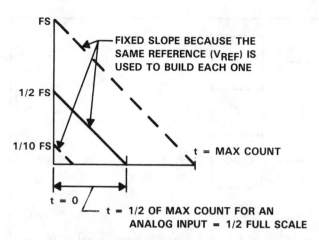

Figure 11-31. Ramp 2 — Dual-Slope A/D Converter

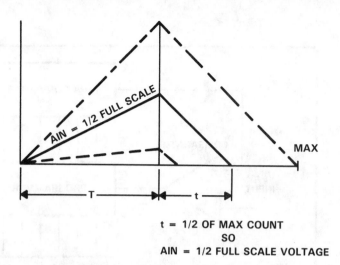

Figure 11-32. Ramps 1 and 2 — Dual-Slope A/D Converter

The resultant value is then integrated back down to 0 V at a fixed rate by using a reference voltage for a variable time (t). Since the reference causes this second ramp to have a fixed slope, the time (t) required to reach 0 V is proportional to the peak value of the first ramp. See Figure 11-31.

Putting these two ramps together, using the same clock to count (T) and (t), and using the same integration network to build both ramps allows errors in the clock frequency and linearity errors due to a nonlinear integrator to cancel. This leaves the stability of the reference voltage as the determining factor in overall accuracy. See Figure 11-32 for both ramps.

While the conversion speed of a dual-slope A/D converter is slow, resolution ranges of from 8 to 14 bits are available. Figure 11-33 shows a functional block diagram of a dual-slope A/D converter.

SUCCESSIVE-APPROXIMATION CONVERTERS

The successive-approximation converter has been used for a long time and is still the most common A/D converter for applications with sampling rates of about 1 MHz maximum. The name "successive approximation" comes from the fact that the converter arrives at a numerical value by making successively more precise observations until it

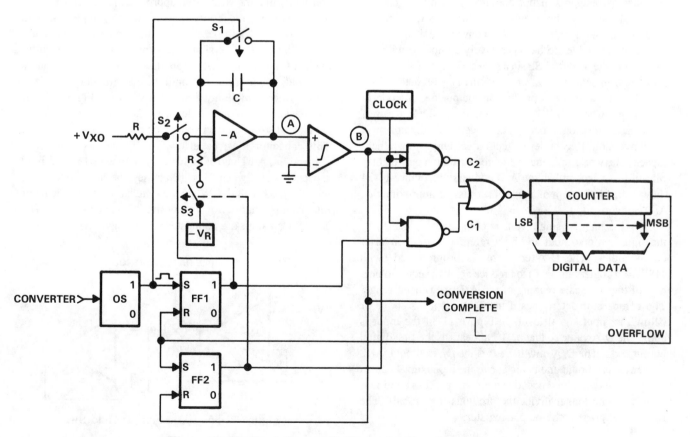

Figure 11-33. Dual-Slope A/D Converter Functional Diagram

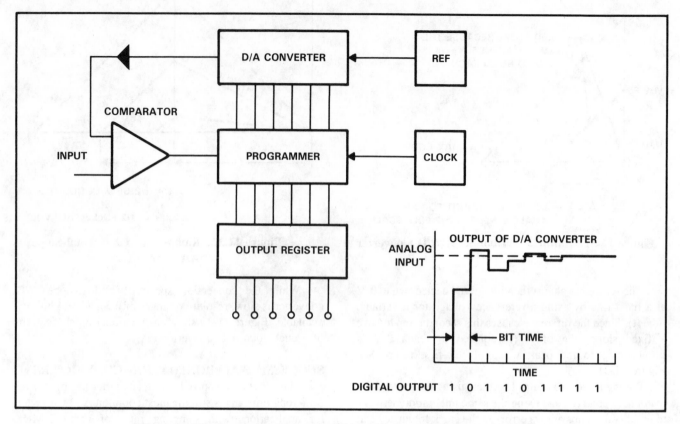

Figure 11-34. Successive-Approximation A/D Converter Block Diagram

has arrived at the closest answer.

Successive approximation does not use ramps to generate its digital output as the integration techniques did. Instead of ramps, it uses a digital-to-analog converter that allows the respective bit values to be successively compared with the input, from the MSB to LSB, in a series of steps. This method compares, in sequence, a series of binary-weighted values with the analog input to produce an output digital code in "n" steps, where "n" is the resolution of the converter. In the past, "resistor-ladder" designs were used to implement the conversion. Recent design techniques employ switched-capacitor networks utilizing charge redistribution on weighted capacitors to accomplish the A/D conversion. Figure 11-34 shows a block diagram of a typical successive-approximation A/D converter.

On the input, a single voltage comparator is driven by the signal on one input and by a reference D/A converter on the other. The D/A converter is programmed (with MSB = 1, other bits = 0) to produce a voltage equal to one-half of the full-scale reading of the A/D measuring range. The comparator determines if the D/A output is above or below the input signal level. In Figure 11-34 the input is above the D/A value so that a "1" appears as the first digital output code. The D/A now increases the D/A output by 1/2 the previous amount and being above the input signal causes a "0" to be the second coded value. This process continues until it can no longer divide the previous step in half. The result now goes to the output register.

While all this happens very fast, the signal must be held constant while the successive-approximation converter does its processing. This is accomplished with a sample-and-hold circuit which would be ahead of the comparator in Figure 11-34. Between conversions, the sample-hold circuit acquires the input signal, and, just before the conversion starts, places it in hold, where it remains until the conversion is complete. A simplified sample-hold circuit is shown in Figure 11-35.

Basically, a sample-hold circuit is a "voltage memory" device that stores a given voltage on a high-quality capacitor. This capacitor must have low leakage and low dielectric absorption. An electronic switch is connected to the hold capacitor so that when the switch closes, the capacitor charges with the input voltage. When the switch opens, the capacitor retains this charge for the desired period of time. The TLC271 op amps are connected as unity-gain buffer stages,

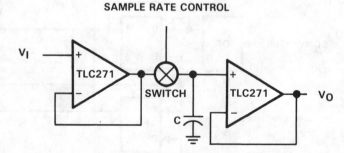

Figure 11-35. Basic Sample-Hold Switch

11-18

offering the high impedance needed in this application. This is very important when operating at the output of an analog multiplexer because the source should not be loaded. The output impedance of the buffer is low so that the sample-hold circuit can more easily drive the A/D converter input. Usually an FET type of device is used as the electronic sample rate control as it must be fast and have very low off-state leakage.

Binary Bit Weighting

In a binary number, the MSB (most significant bit) is always equal to 1/2 the full scale or maximum value. The next MSB is equal to 1/4 the maximum value, the next 1/8, the next 1/4, etc., until we arrive at the LSB (least significant bit) which has the value of the Maximum Value/$2^{(n)}$ (where "n" = number of bits).

As an example, assume we have a full scale (or maximum value) of 4 V. With a 4-bit converter, a 4-bit binary number would be arrived at as follows:

MSB (most significant bit) = 4 × 1/2 = 2 V
Next MSB   = 4 × 1/4 = 1 V
Next MSB   = 4 × 1/8 = 1/2 V
LSB (least significant bit) = 4 × 1/16 = 1/4 V
Full-Scale Reading   = 3 3/4 V

Thus, on a 4-bit converter the resolution is defined as:

$$\text{Resolution} = \frac{\text{Full Scale}}{2^4 \text{ bits}} = 1/4 \text{ V}$$

Therefore we can only determine the actual value in 1/4 V steps.

Creating a Digital Number

As stated previously, a successive-approximation converter has a built-in D/A converter which it uses to compare the output of the D/A with the analog input. It must then decide which binary number most closely represents the analog input. To do this the A/D starts with the MSB. See Figure 11-36.

As seen in the above example, the A/D compares its 1/2 full-scale value with the analog input and decides if it is at or above, or below, the 1/2 full scale value. If the input is below it sets the MSB to "0", if the input is at or above the 1/2 full scale value, it sets the MSB to "1". It then considers the next MSB in the same manner and continues until it has checked the LSB. After checking the LSB, the number remaining in the counter is the most accurate binary representation of the analog input voltage that is within the capability of the 4-bit A/D.

For the 4-bit converter, only 4 steps were required to arrive at the digital number. For an 8-bit converter, 8 steps would be required. The number of bits equals the number of steps. Each step is completed very fast, so even an 8-bit A/D requires much less time than would be required by a single ramp as in the case of a single-slope A/D converter.

The requirements for building a good successive-approximation converter are a stable reference voltage, an

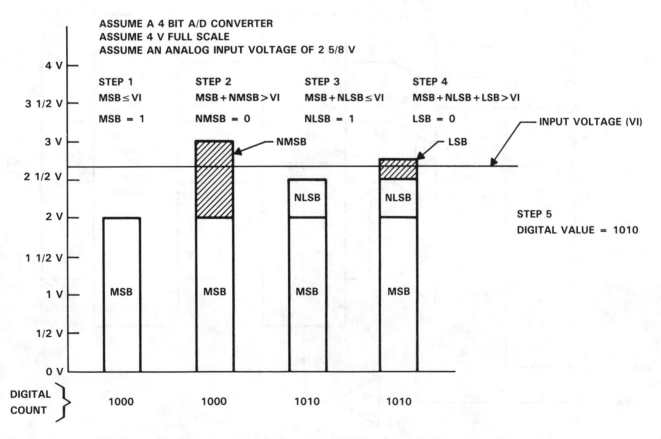

Figure 11-36. 4 V 4-Bit Successive Approximation A/D Converter Output

accurate D/A converter, and sufficient control logic to handle the counters, output registers, etc.

In summary, the successive-approximation A/D is faster but generally less accurate than a single- or dual-slope A/D.

FLASH A/D CONVERTERS

The flash A/D converter gets its name from its ability to do the conversion very rapidly. The flash converter gives an answer every time it receives a clock pulse.

As discussed previously, the successive-approximation A/D converter needs a clock pulse for each bit of resolution it provides. The flash A/D converter also does not need a sample-hold circuit or a D/A converter. The fast conversion speed is accomplished by providing a comparator for every quantization level except one ($2^n - 1$ comparators). The quantization level that does not require a comparator can be zero or full scale depending on how the reference is applied.

The power and speed of this technique lies in the fact that all comparators examine the input voltage simultaneously in parallel and make an immediate, total conversion. The limiting factor of this conversion system is the number of comparators required, which necessitates a larger bar of silicon material and increased cost. See Figure 11-37 for the block diagram of a flash converter.

As shown, one input of each comparator is connected to the input signal. The other input is connected to a voltage divider (resistor ladder network) chain that is fed by the reference. The reference is equal to the full-scale input signal voltage. The manner in which the flash A/D converter performs a quantization is relatively simple. When the rising edge of the clock arrives, the comparators quickly latch in a "1" or "0" state depending on whether the input signal is above or below the reference level at that instant. Those comparators referenced above the input signal remain turned off, representing a "0" state. The comparators at or below the input signal conversely become a "1" state. The code resulting from this comparison is converted to a binary code by the decoder network. The number of bits in the code is equal to the number of bits of resolution.

In summary, a flash converter can give an answer every time it receives a clock pulse, although $2^n - 1$ comparators are required to perform the conversion function. It is significant to note that each time the number of bits of resolution increases by one, the number of comparators essentially doubles. As an example:

A 6-bit resolution converter requires 63 comparators.

A 7-bit resolution converter requires 127 comparators.

An 8-bit resolution converter requires 255 comparators.

Presently the units on the market range from 4 bits at 50 MHz to 100 MHz to 6 bits at 100 MHz. These devices are built using the bipolar process.

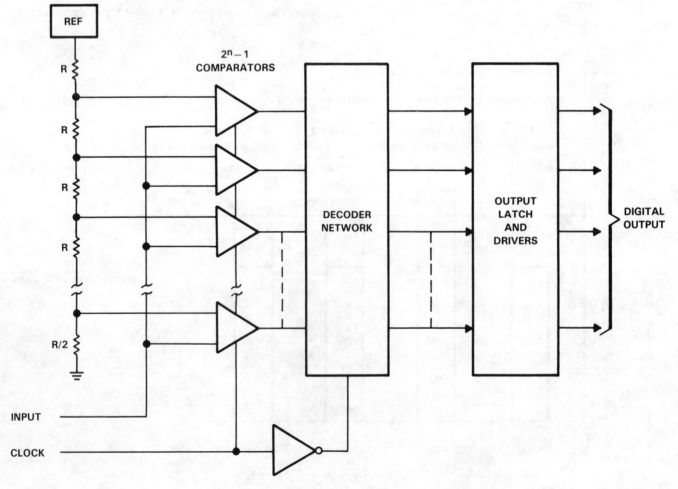

Figure 11-37. Flash A/D Converter Block Diagram

11-20

DEVICE TYPES

The use of digital techniques in measurement, communication, and control systems has increased rapidly in the past few years. Control systems use a combination of digital and analog (linear) circuit techniques. One of the key devices in these systems is the analog-to-digital (A/D) converter. Because of their inherent complexity, A/D converters have always been a challenge to IC designers. Through the continuous progress in design and process technology, economical monolithic devices are available to meet a wide range of application requirements. Today, the LinCMOS™ technology provides increased speeds, better input stability, lower power requirements, and increased thermal stability for demanding applications.

ADC0803 AND ADC0805 8-BIT SUCCESSIVE APPROXIMATION A/D CONVERTERS WITH DIFFERENTIAL INPUTS

The ADC0803 family of A/D converters are 8-bit devices which feature a conversion time of 100 μs and an access time of 135 ns. This converter requires no zero adjust, has an on-chip clock generator, and will operate from a single 5-V supply. As shown in Figure 11-38, the pinout allows for easy PC board layout. The data output pins are grouped together and the ground and V_{CC} pins are located at the package corners.

N DUAL-IN-LINE PACKAGE

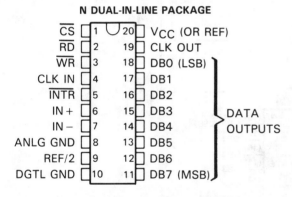

Figure 11-38. ADC0803 and ADC0805 Pinout (Top View)

Description

The ADC0803 and ADC0805 are CMOS 8-bit successive-approximation analog-to-digital converters that use a modified potentiometric 256-resistor ladder network. These devices are designed to operate from common microprocessor control buses, with the 3-state output latches driving the data bus. The devices can be made to appear to the microprocessor as a memory location or an I/O port. Figure 11-39 shows a functional block diagram and the timing diagrams are illustrated in Figure 11-40.

A differential analog voltage input allows increased common-mode rejection and eliminates offset due to the zero-input analog voltage value. Although a reference input

LinCMOS is a trademark of Texas Instruments.

(REF/2) is available to allow 8-bit conversion over smaller analog voltage spans or to make use of an external reference, ratiometric conversion is possible with the REF/2 input open. Without an external reference, the conversion takes place over a span from V_{CC} to analog ground (ANLG GND). The devices can operate with an external clock signal or, the on-chip clock generator can be used independently by adding an external resistor and capacitor to set the time period.

Principles of Operation

The ADC0803 and ADC0805 each contain a circuit equivalent to a 256-resistor network. Analog switches are sequenced by successive-approximation logic to match an analog differential input voltage ($V_{I+} - V_{I-}$) to a corresponding tap on the resistor network. The most significant bit (MSB) is tested first. After eight comparisons (64 clock periods), an 8-bit binary code (1111 1111 = full scale) is transferred to an output latch and the interrupt ($\overline{INTR}$) output goes low. The device can be operated in a free-running mode by connecting the $\overline{INTR}$ output to the write ($\overline{WR}$) input and holding the conversion start($\overline{CS}$) input at a low level. To ensure start-up under all conditions, a low-level $\overline{WR}$ input is required during the power-up cycle. Taking $\overline{CS}$ low any time after that will interrupt a conversion in process.

When the $\overline{WR}$ input goes low, the internal successive-approximation register (SAR) and 8-bit shift register are reset. As long as both $\overline{CS}$ and $\overline{WR}$ remain low, the analog-to-digital converter will remain in its reset state. One to eight clock periods after $\overline{CS}$ or $\overline{WR}$ makes a low-to-high transition, conversion starts. When the $\overline{CS}$ and $\overline{WR}$ inputs are low, the start flip-flop is set and the interrupt flip-flop and 8-bit register are reset. The next clock pulse transfers a logic high to the output of the start flip-flop. The logic high is ANDed with the next clock pulse, placing a logic high on the reset input of the start flip-flop. If either $\overline{CS}$ or $\overline{WR}$ have gone high, the set signal to the start flip-flop is removed, causing it to be reset. A logic high is placed on the D input of the 8-bit shift register and the conversion process is started. If the $\overline{CS}$ and $\overline{WR}$ inputs are still low, the start flip-flop, the 8-bit shift register, and the SAR remain reset. This action allows for wide $\overline{CS}$ and $\overline{WR}$ inputs with conversion starting from one to eight clock periods after one of the inputs goes high.

When the logic high input has been clocked through the 8-bit shift register, completing the SAR search, it is applied to an AND gate controlling the output latches and to the D input of a flip-flop. On the next clock pulse, the digital word is transferred to the 3-state output latches and the interrupt flip-flop is set. The output of the interrupt flip-flop is inverted to provide an $\overline{INTR}$ output that is high during conversion and low when the conversion is completed.

When a low is at both the $\overline{CS}$ and $\overline{RD}$ inputs, an output is applied to the DB0 through DB7 outputs and the interrupt flip-flop is reset. When either the $\overline{CS}$ or $\overline{RD}$ input returns to a high state, the DB0 through DB7 outputs are disabled (returned to the high-impedance state). The interrupt flip-flop remains reset.

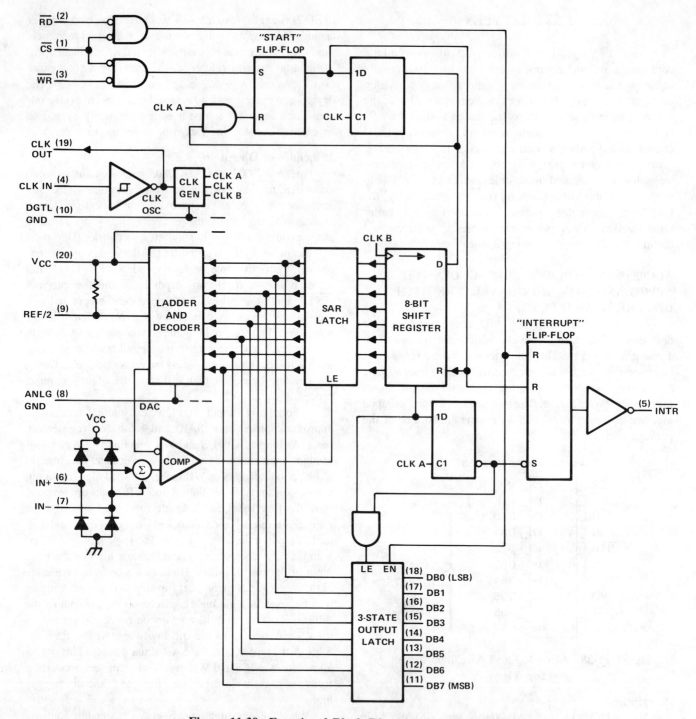

Figure 11-39. Functional Block Diagram (Positive Logic)

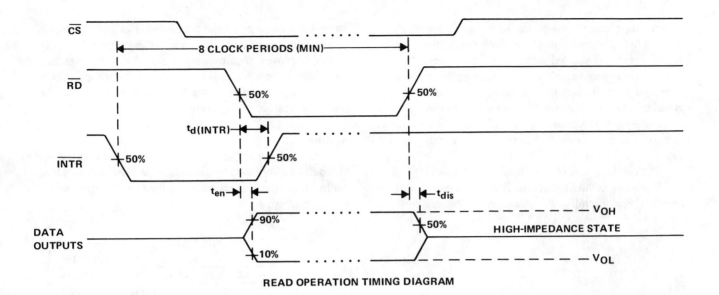

READ OPERATION TIMING DIAGRAM

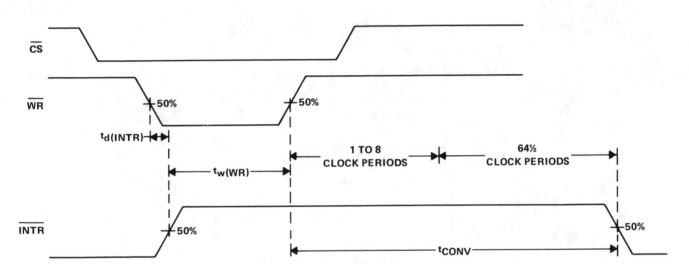

WRITE OPERATION TIMING DIAGRAM

Figure 11-40. Timing Diagrams

ADC0808, ADC0809 CMOS ANALOG-TO-DIGITAL CONVERTERS WITH 8-CHANNEL MULTIPLEXERS

The ADC0808 and ADC0809 are 8-bit A/D converters that offer latched address inputs as well as latched 3-state outputs. The total unadjusted error in the ADC0808 is $\pm 1/2$ LSB max and for the ADC0809 is ± 1.0 LSB max. These devices feature ratiometric conversion with a conversion time of 100 μs. They may be powered from a single 5-V supply and are easily connected to a microprocessor. Figure 11-41 shows how the inputs and address lines have been grouped to facilitate PC board layout.

Description

The ADC0808 and ADC0809 are monolithic CMOS devices with an 8-channel multiplexer, an 8-bit analog-to-digital A/D converter, and microprocessor-compatible control logic. The 8-channel multiplexer can be controlled

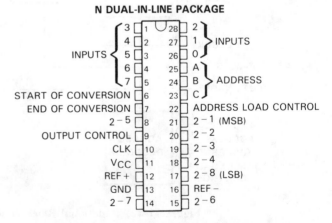

Filgure 11-41. ADC0808, ADC0809 Pinout (Top View)

by a microprocessor through the 3-bit address decoder with address load control to select any one of eight single-ended analog switches connected directly to the converter. The 8-bit A/D converter uses the successive-approximation conversion technique featuring a high-impedance threshold detector, a switched-capacitor array, a sample-and-hold, and a successive-approximation register (SAR). The functional block diagram is shown in Figure 11-42 and the operating sequence in Figure 11-43.

The comparison and conversion methods used eliminate the possibility of missing codes, nonmonotonicity, and the need for zero or full-scale adjustment. Also featured are latched 3-state outputs from the converter and latched inputs to the multiplexer address decoder. The single 5-V supply and low power requirements make the ADC0808 and ADC0809 especially useful for a wide variety of applications. Ratiometric conversion is made possible by access to the reference voltage input terminals.

Multiplexer

The analog multiplexer selects 1 of 8 single-ended input channels as determined by the address decoder. Address load control loads the address code into the decoder on a low-to-high transition.

MULTIPLEXER FUNCTION TABLE

INPUTS				SELECTED
ADDRESS			ADDRESS	ANALOG
C	B	A	STROBE	CHANNEL
L	L	L	↑	0
L	L	H	↑	1
L	H	L	↑	2
L	H	H	↑	3
H	L	L	↑	4
H	L	H	↑	5
H	H	L	↑	6
H	H	H	↑	7

H = high level, L = low level
↑ = low-to-high transition

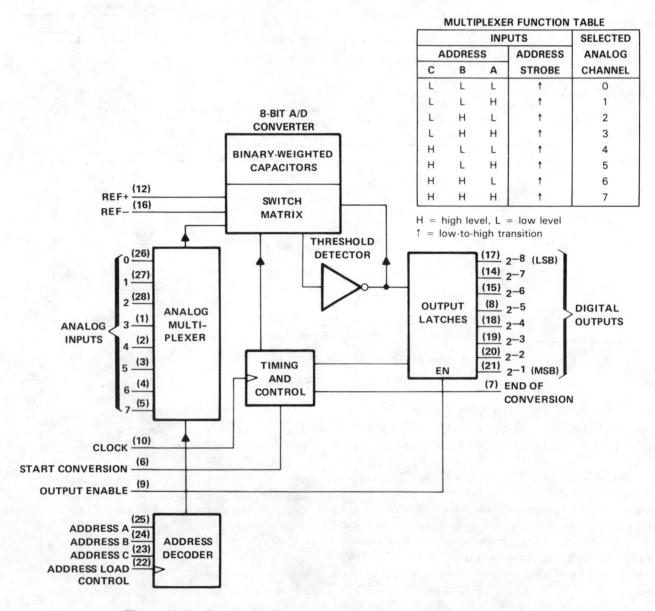

Figure 11-42. Functional Block Diagram (Positive) and Function Table

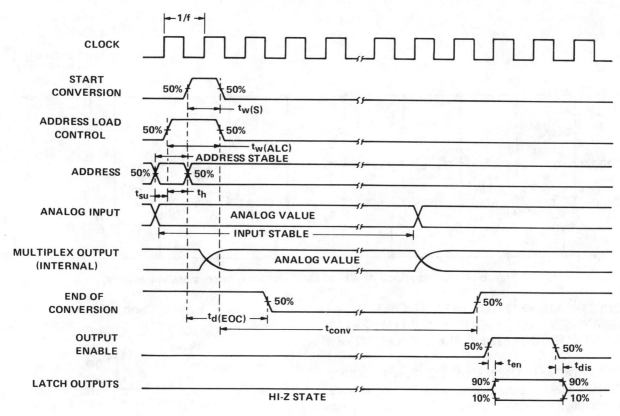

Figure 11-43. Operating Sequence

Converter

The CMOS threshold detector in the successive-approximation conversion system determines each bit by examining the charge on a series of binary-weighted capacitors (Figure 11-44).

In the first phase of the conversion process, the analog input is sampled by closing switch S_C and all S_T switches, and by simultaneously charging all the capacitors to the input voltage.

In the next phase of the conversion process, all S_T switches and the S_C switch are opened and the threshold detector begins identifying bits by identifying the charge (voltage) on each capacitor relative to the reference voltage. In the switching sequence, all eight capacitors are examined separately until all 8 bits are identified, and then the charge-convert sequence is repeated. In the first step of the conversion phase, the threshold detector looks at the first capacitor (weight = 128). Node 128 of this capacitor is

switched to the reference voltage, and the equivalent nodes of all the other capacitors on the ladder are switched to REF $-$. If the voltage at the summing node is greater than the trip point of the threshold detector (approximately one-half the V_{CC} voltage), a bit is placed in the output register, and the 128-weight capacitor is switched to REF $-$. If the voltage at the summing node is less than the trip point of the threshold detector, this 128-weight capacitor remains connected to REF $+$ through the remainder of the capacitor-sampling (bit-counting) process. The process is repeated for the 64-weight capacitor, then the 32-weight capacitor, and so on until all bits are counted.

With each step of the capacitor-sampling process, the initial charge is redistributed among the capacitors. This conversion process is successive approximation, but relies on charge redistribution rather than a successive-approximation register (and reference D/A) to count and weigh the bits from MSB to LSB.

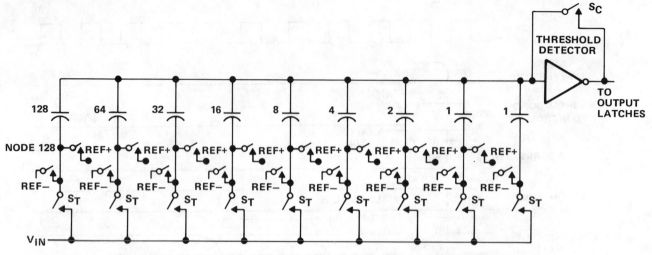

Figure 11-44. Simplified Model of the Successive-Approximation System

ADC0831, ADC0832, ADC0834, ADC0838 2-, 4-, 8-CHANNEL A/D PERIPHERALS WITH SERIAL CONTROL AND MULTIPLEXER OPTIONS

The ADC0831, ADC0832, ADC0834, and ADC0838 A/D converters are another family of A/D converters that may be easily used with a microprocessor or operated stand-alone. There is no zero or full-scale adjust. An on-chip shunt regulator on the ADC0834 and ADC0838 allows operation with high-voltage supplies. A conversion time of 32 μs is possible at a clock frequency of 250 kHz. The input voltage range is 0 to 5 V with a single 5 V supply. The systematic placing of data lines, V_{CC}, and ground terminals shown in Figure 11-45 allows efficient PC board layout.

Description

The ADC0831, ADC0832, ADC0834, and ADC0838 are 8-bit successive-approximation analog-to-digital converters each with a serial input/output and configurable input multiplexer with up to 8 channels on the ADC0838. The serial input/output can be used with standard shift registers or microprocessors. See Figure 11-46 for a functional block diagram of the ADC0831 and ADC0832 and Figure 11-47 for the functional block diagram of the ADC0834 and ADC0838.

The 2-, 4-, or 8-channel multiplexers are software configured for single-ended or differential inputs as well as channel assignment. Figures 11-48 and 11-49 show the ADC0832 and ADC0834—838 input and data output timing diagrams and Figure 11-50 shows the ADC0831 sequence of operation. The operating sequence for the ADC0832 is illustrated in Figure 11-51 while the ADC0834 sequence of operation is shown in Figure 11-52. The ADC0838 sequence of operation is shown in Figure 11-53.

The differential analog voltage input increases the common-mode rejection and eliminates the offset due to the analog zero input voltage value. In addition, the voltage reference input can be adjusted to allow encoding any smaller analog voltage span to the full 8 bits of resolution.

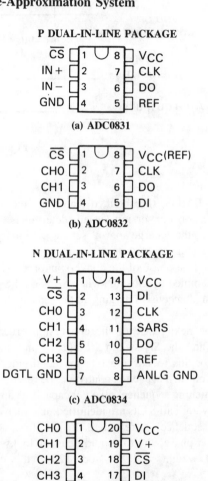

P DUAL-IN-LINE PACKAGE

(a) ADC0831

(b) ADC0832

N DUAL-IN-LINE PACKAGE

(c) ADC0834

(d) ADC0838

Figure 11-45. ADC0831, ADC0832, ADC0834, ADC0838 Pinouts (Top View)

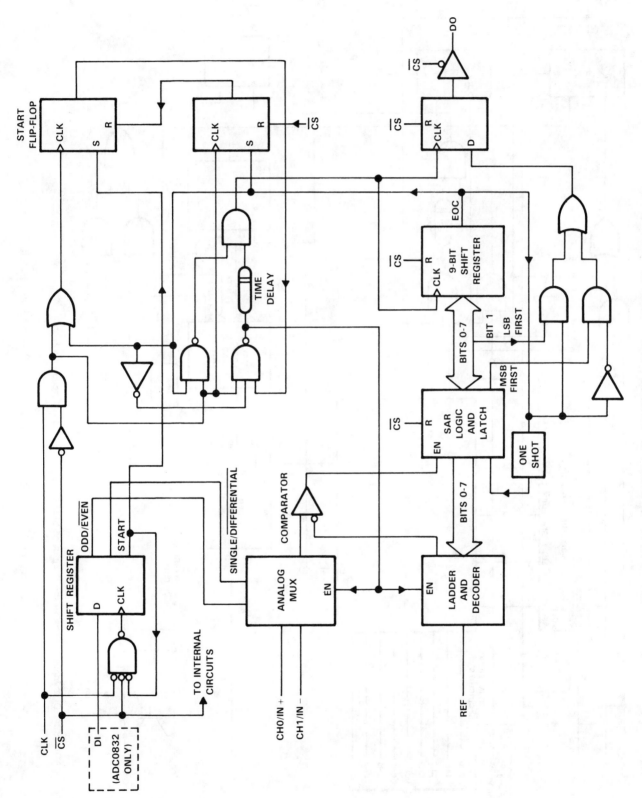

Figure 11-46. ADC0831A-B, ADC0832A-B Functional Block Diagram

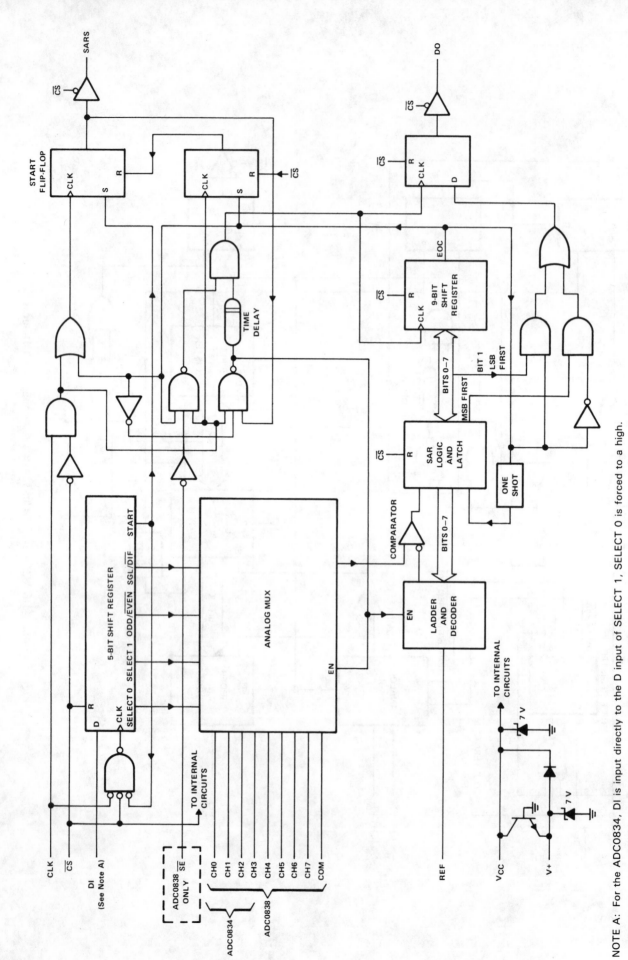

Figure 11-47. ADC0834A-B, ADC0838A-B Functional Block Diagram

NOTE A: For the ADC0834, DI is input directly to the D input of SELECT 1, SELECT 0 is forced to a high.

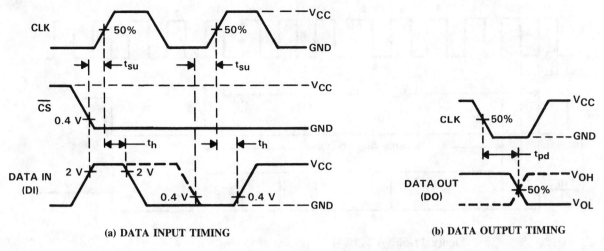

(a) DATA INPUT TIMING

(b) DATA OUTPUT TIMING

Figure 11-48. ADC0832 Data Input and Output Timing Diagram

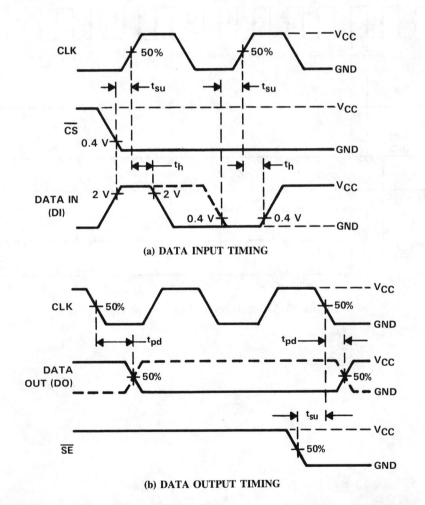

(a) DATA INPUT TIMING

(b) DATA OUTPUT TIMING

Figure 11-49. ADC0834, ADC0838 Data Input and Output Timing Diagram

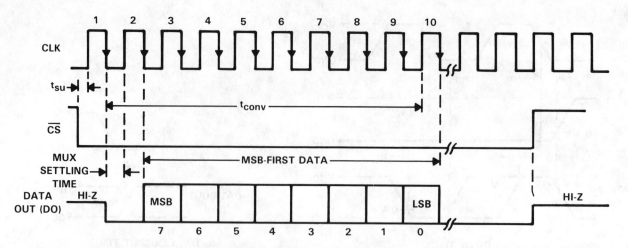

Figure 11-50. ADC0831 Sequence of Operation

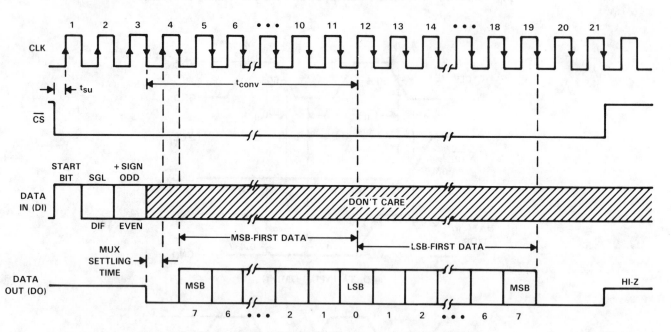

Figure 11-51. ADC0832 Sequence of Operation

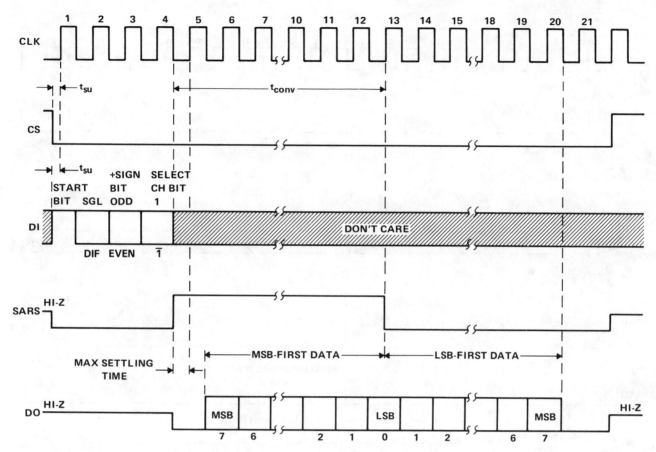

Figure 11-52. ADC0834 Sequence of Operation

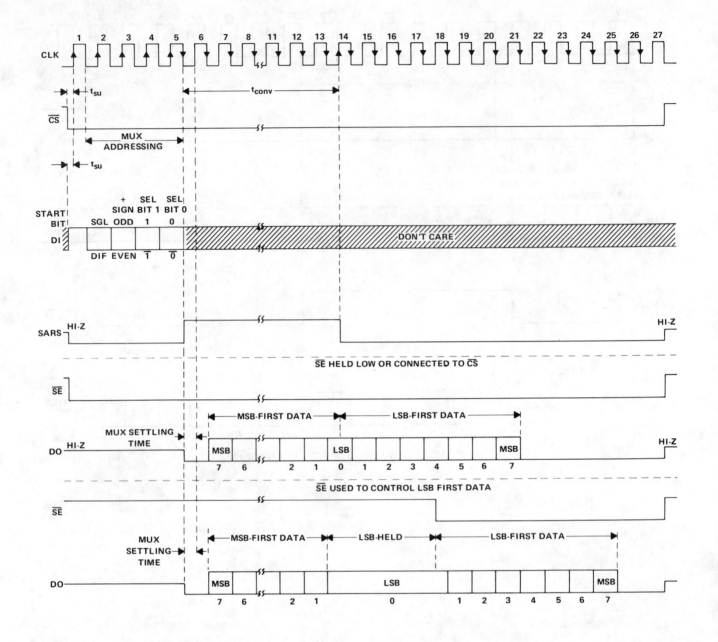

Figure 11-53. ADC0838 Sequence of Operation

Principles of Operation

The ADC0831, ADC0832, ADC0834, and ADC0838 use a sample data comparator structure that converts differential analog inputs by a successive-approximation routine. The input voltage to be converted is applied to a channel terminal and is compared to ground (single-ended), to an adjacent channel (differential), or to a common terminal (pseudo differential) that can be an arbitary voltage. The input terminals are assigned a positive (+) or negative (−) polarity. If the signal input applied to the assigned positive terminal is less than the signal on the negative terminal, the converter output is all zeros.

Channel selection and input configuration are under software control using a serial data link from the controlling processor. A serial communication format allows more functions to be included in a converter package with no increase in pin count. In addition, it eliminates the transmission of low-level analog signals by facilitating the remote location of the converter at the analog sensor. This process returns noise-free digital data to the processor.

A particular input configuration is assigned during the multiplexer addressing sequence. The multiplexer address is shifted into the converter through the data input (DI) line.

(The ADC0831, unlike the ADC0832, ADC0834, and ADC0838, contains only one differential input channel having a fixed polarity assignment and does not have multiplexer addressing.) The multiplexer address selects the analog inputs to be enabled and determines whether the input is single-ended or differential. When the input is differential, the polarity of the channel input is assigned. Differential inputs are assigned to adjacent channel pairs. For example, channel 0 and channel 1 may be selected as a differential pair. However, these channels cannot act differentially with any other channel. In addition to selecting the differential mode, the polarity may also be selected. Either channel of the channel pair may be designated as negative or positive.

As shown in Figure 11-54, the ADC0838 has several input multiplexer options. In Figure 11-54(a), it is used for 8 single-ended inputs. In Figure 11-54(b), it is being used for a pseudo differential input. In this mode, the voltage on the COM(−) input is negative compared to any other channel. This voltage can be any reference potential common to all channel inputs. This feature is useful in single-supply applications where all analog circuits are biased to a potential other than ground. Figure 11-54(c) shows an ADC0838 set up with four differential inputs. Figure 11-54(d) illustrates

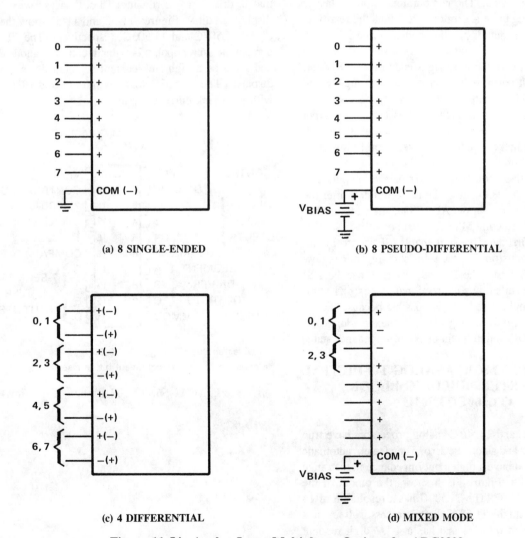

(a) 8 SINGLE-ENDED

(b) 8 PSEUDO-DIFFERENTIAL

(c) 4 DIFFERENTIAL

(d) MIXED MODE

Figure 11-54. Analog Input Multiplexer Options for ADC0838

the mixed mode option. Selection of this mode allows combinations of differential pairs and single-ended inputs.

A conversion is initiated by setting the chip select ($\overline{CS}$) input low, which enables all of the logic circuits. The $\overline{CS}$ input must be held low for the complete conversion process. On each low-to-high transition of the clock input received from the processor, the data on the DI input is clocked into the multiplexer address shift register. The first logic high on the input is the start bit. A 2-to-4 bit assignment word follows the start bit. On each successive low-to-high transition of the clock input, the start bit and assignment word are shifted through the shift register. When the start bit has been shifted into the start location of the multiplexer register, the input channel has been selected and conversion starts. The SAR status output (SARS) goes high to indicate that a conversion is in progress and the DI input to the multiplexer shift register is disabled for the duration of the conversion.

An interval of one clock period is automatically inserted to allow for the selected multiplexer channel to settle. The data output (DO) comes out of the high-impedance state and provides a leading low for this one clock period of multiplexer settling time. The SAR comparator compares successive outputs from the resistive-ladder with the incoming analog signal. The comparator output indicates whether the analog input is greater or less than the resistive-ladder output. This data is parallel loaded in to a 9-bit shift register that immediately outputs an 8-bit serial data word to the DO output with the most significant bit (MSB) first. After eight clock periods, the conversion is complete and the SAR status (SARS) output goes low. When $\overline{CS}$ goes high, all internal registers are cleared. At this time, the output circuits go to three state. If another conversion is desired, the $\overline{CS}$ line must make a high-to-low transition followed by address information.

In the ADC0831, only MSB-first data is output. The ADC0832 and ADC0834 output the LSB-first data after the MSB-first data stream. In the ADC0838, the programmer has the option of selecting MSB-first data or LSB-first data. To output LSB-first data, the shift enable ($\overline{SE}$) control input must go low. Data stored in the 9-bit shift register is now output with LSB first. The DI and DO pins can be tied together and controlled by a bidirectional processor I/O bit received on a single wire. This is possible because the DI input is examined only during the multiplexer addressing interval and the DO output is still in a high-impedance state.

TL500C THRU TL503C ANALOG-TO-DIGITAL CONVERTER BUILDING BLOCKS TL500C/TL501C

The TL500C and TL501C analog processors have true differential inputs and automatic zero. They feature automatic polarity status and have a high input impedance in the range of 10^9 Ω typically. Figure 11-55 shows the pinout of the TL500C and TL501C. The TL500C has a resolution of 14 bits when used with the TL502C digital processor. It features a linearity error of 0.001% and has a 4-1/2 digit readout

N DUAL-IN-LINE PACKAGE

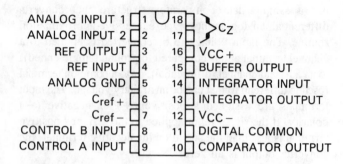

Figure 11-55. TL500C and TL501C Pinout (Top View)

accuracy when used with an external reference. The TL501C is capable of 10 to 13 bits of resolution when used with the TL502C processor. It has a linearity error of 0.01% and a 3-1/2 digit readout accuracy.

TL502C/TL503C

The TL502C and TL503C are digital processors which feature interdigit blanking as well as over-range blanking. They have fast display scan rates and the internal oscillator may be driven or free-running. These devices have 4-1/2 digit display circuitry. Figures 11-56 and 11-57 show the pinout of the TL502C and TL503C, respectively. The TL502C is compatible with popular 7-segment common-anode displays and features a high-sink-current segment driver for large displays. The TL503C features multiplexed BCD outputs with high sink current capability.

N DUAL-IN-LINE PACKAGE

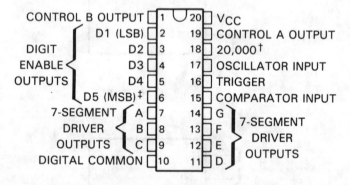

†Provides an output of $f_{osc} \div 20{,}000$.

‡Means D5, the most significant bit, is also the sign bit.

Figure 11-56. TL502 Pinout (Top View)

N DUAL-IN-LINE PACKAGE

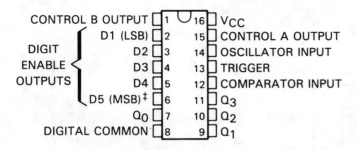

‡Means D5, the most significant bit, is also the sign bit.

Figure 11-57. TL503 Pinout (Top View)

General Overall Description

The TL500C and TL501C analog processors and TL502C and TL503C digital processors provide the basic functions for a dual-slope-integrating analog-to-digital converter.

The TL500C and TL501C contain the necessary analog switches and decoding circuits, reference voltage generator, buffer, integrator, and comparator. These devices may be controlled by the TL502C, TL503C, by discrete logic, or by a software routine in a microprocessor.

The TL502C and TL503C both include oscillator, counter, control logic, and digit enable circuits. The TL502C provides multiplexed outputs for 7-segment displays, while the TL503C has multiplexed BCD outputs.

When used in complementary fashion, these devices form a system that features automatic zero-offset compensation, true differential inputs, high input impedance, and capability for 4-1/2 digit accuracy. Applications include the conversion of analog data from high-impedance sensors of pressure, temperature, light, moisture, and position. Analog-to-digital-logic conversion provides display and control signals for weight scales, industrial controllers, thermometers, light-level indicators, and many other applications. See Figure 11-58.

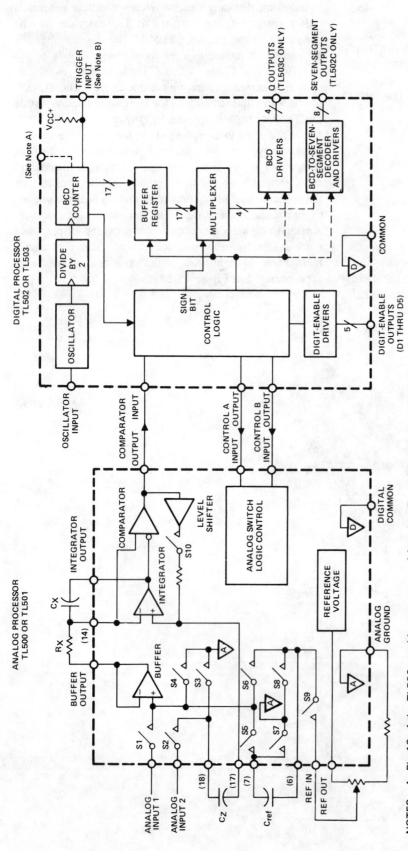

Figure 11-58. Block Diagram of Basic Analog-to-Digital Converter Using TL500C or TL501C and TL502C or TL503C

NOTES: A. Pin 18 of the TL502 provides an output of f_{osc} (oscillator frequency) $\div$ 20,000.
B. The trigger input assumes a high level if not externally connected.

MODE	ANALOG INPUT	COMPARATOR	CONTROLS A AND B	ANALOG SWITCHES CLOSED
Auto Zero	X	Oscillation	L L	S3, S4, S7, S9, S10
Hold[†]				
Integrate Input	Positive	H	H H	S1, S2
	Negative	L		
Integrate	X	L[‡]	L H	S3, S6, S7
Reference		H[‡]	H L	S3, S5, S8

H ≡ High, L ≡ low, X ≡ Irrelevant

[†] If the trigger input is low at the beginning of the auto-zero cycle, the system will enter the hold mode. A high level (or open circuit) will signal the digital processor to continue or resume normal operation.

[‡] This is the state of the comparator output as determined by the polarity of the analog input during the integrate input phase.

11-36

Description of TL500C and TL501C Analog Processors

The TL500C and TL501C analog processors are designed to compensate automatically for internal zero offsets, integrate a differential voltage at the analog inputs, integrate a voltage at the reference input in the opposite direction, and provide an indication of zero-voltage crossing. The external control mechanism may be a microcomputer and software routine, discrete logic, or a TL502C or TL503C digital processor. The TL500C and TL501C are designed primarily for simple, cost-effective, dual-slope analog-to-digital converters. Both devices feature true differential analog inputs, high input impedance, and an internal reference-voltage source. See Figure 11-59 for the input schematic and Figure 11-60 for the output schematic. The TL500C provides 4-1/2 digit readout accuracy when used with a precision external reference voltage. The TL501C provides 100-ppm linearity error and 3-1/2 digit accuracy capability. These devices are manufactured using TI's advanced technology to produce JFET, MOSFET, and bipolar devices on the same chip. The TL500C and TL501C are intended for operation over the temperature range of 0 °C to 70 °C.

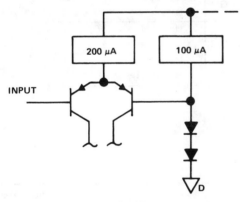

CONTROL A AND CONTROL B INPUTS

Figure 11-59. TL500C and TL501C Input Schematic

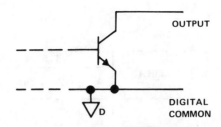

COMPARATOR OUTPUT

Figure 11-60. TL500C and TL501C Output Schematic

Description of TL502C/503C Digital Processors

The TL502C and TL503C are control logic devices designed to complement the TL500C and TL501C analog processors. They feature an internal oscillator, interdigit blanking, and fast display scan rate. The internal-oscillator input is a Schmitt trigger circuit that can be driven by an external clock pulse or provide its own time base with the addition of a capacitor. The typical oscillator frequency is 120 kHz with a 470-pF capacitor connected between the oscillator input and ground.

The TL502C provides 7-segment-display output drivers capable of sinking 100 mA and compatible with popular common-anode displays. The TL503C has four BCD output drivers capable of 100-mA sink current. The code for each digit is multiplexed to output drivers in phase with a pulse on the appropriate digit-enable line at a digit rate equal to f_{osc} divided by 200 (Table 11-7 and Figure 11-61). Each digit-enable output is capable of sinking 20 mA. Figure 11-62 shows input/output schematics of the TL502C and TL503C. Table 11-8 is the Table of Special Functions.

Table 11-7
(a) DIGITS 1 THRU 4 NUMERIC CODE

NUMBER	TL502C SEVEN-SEGMENT LINES							TL503C BCD OUTPUT LINES			
	A	B	C	D	E	F	G	Q3 8	Q2 4	Q1 2	Q0 1
0	L	L	L	L	L	L	H	L	L	L	L
1	H	L	L	H	H	H	H	L	L	L	H
2	L	L	H	L	L	H	L	L	L	H	L
3	L	L	L	L	H	H	L	L	L	H	H
4	H	L	L	H	H	L	L	L	H	L	L
5	L	H	L	L	H	L	L	L	H	L	H
6	L	H	L	L	L	L	L	L	H	H	L
7	L	L	L	H	H	H	H	L	H	H	H
8	L	L	L	L	L	L	L	H	L	L	L
9	L	L	L	L	H	L	L	H	L	L	H

H = high level, L = low level

(b) DIGIT 5 (MOST SIGNIFICANT DIGIT) CHARACTER CODES

CHARACTER	TL502C SEVEN-SEGMENT LINES							TL503C BCD OUTPUT LINES			
	A	B	C	D	E	F	G	Q3 8	Q2 4	Q1 2	Q0 1
+	H	H	H	H	L	L	L	H	L	H	L
+1	H	L	L	H	L	L	L	H	H	H	L
−	L	H	H	L	H	H	L	H	L	H	H
−1	L	L	L	L	H	H	L	H	H	H	H

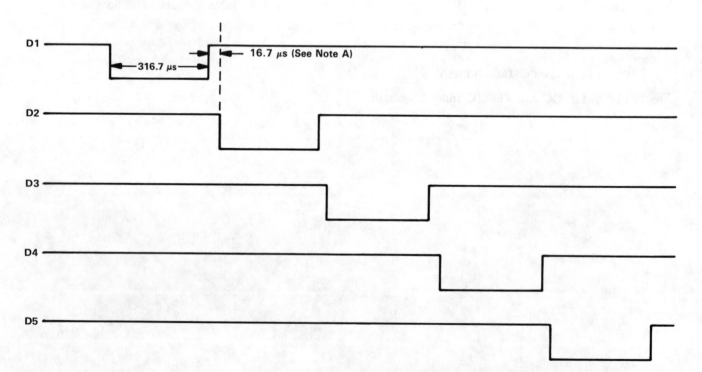

NOTE A: The BCD or seven-segment driver outputs are present for a particular digit slightly before the falling edge of that digit enable.

Figure 11-61. TL502 and TL503 Digit Timing with 120 kHz Clock Signal at Oscillator Input

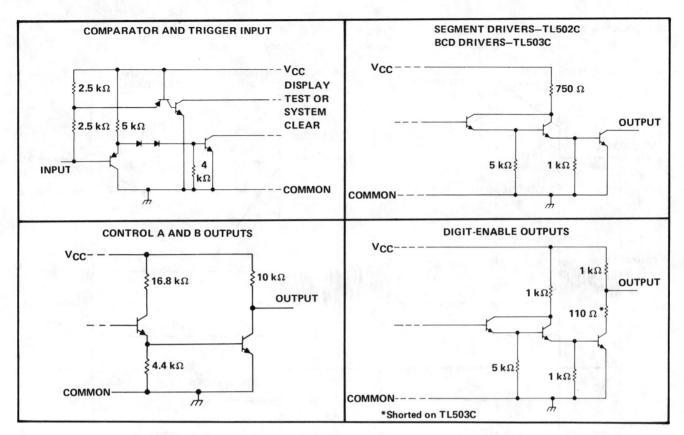

Figure 11-62. TL502C and TL503C Inputs and Outputs Schematics

Table 11-8. Table of Special Functions*

TRIGGER INPUT	COMPARATOR INPUT	FUNCTION
$V_I \le 0.8$ V	$V_I \le 6.5$ V	Hold at auto-zero cycle after completion of conversion
2 V $\le V_I \le 6.5$ V	$V_I \le 6.5$ V	Normal operation (continuous conversion)
$V_I \le 6.5$ V	$V_I \ge 7.9$ V	Display Test: All BCD outputs high
$V_I \ge 7.9$ V	$V_I \le 6.5$ V	Internal Test
Both inputs go to $V_I \ge 7.9$ V simultaneously		System Clear: Sets CBC counter to 20,000. When normal operation is resumed, cycle begins with Auto Zero.

*V_{CC} = 5 V ± 10%

The comparator input of each device, in addition to monitoring the output of the zero-crossing detector in the analog processor, may be used in the display test mode to check for wiring and display faults. A high logic level (2 to 6.5 V) at the trigger input with the comparator input at or below 6.5 V starts the integrate-input phase. Voltage levels equal to or greater than 7.9 V on both the trigger and comparator inputs clear the system and set the BCD counter

to 20,000. When normal operation resumes, the conversion cycle is restarted at the zero phase. These devices are manufactured using I^2L and bipolar technologies.

Principles of Operation

The basic principle of dual-slope-integrating converters is relatively simple (refer to Figure 11-58). The relationship of the charge and discharge values is shown in Figure 11-63.

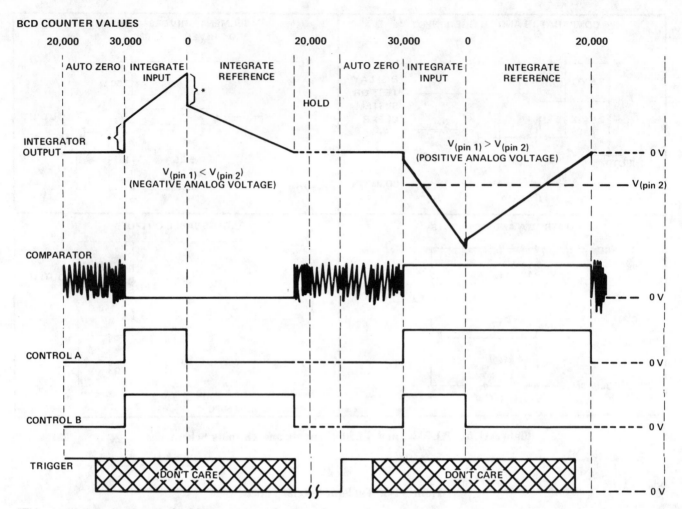

*This step is the voltage at pin 2 with respect to analog ground.

Figure 11-63. Voltage Waveforms and Timing Diagram

Capacitor, C_X, is charged through the integrator from V_{CT} for a fixed period of time at a rate determined by the value of the unknown voltage input. The capacitor is then discharged at a fixed rate determined by the reference voltage back to V_{CT} where the discharge time is measured precisely.

$$V_{CX} = V_{CT} - \frac{V_I\, t_1}{R_X C_X} \quad \text{charge} \tag{1}$$

$$V_{CT} = V_{CX}\, \frac{V_I}{V_{ref}} = -\frac{t_2}{t_1} \quad \text{discharge} \tag{2}$$

Combining equations 1 and 2 results in:

$$\frac{V_I}{V_{ref}} = -\frac{t_2}{t_1} \tag{3}$$

where:

V_{CT} = Comparator (offset) threshold voltage

V_{CX} = Voltage change across C_X during t_1 and t_2 (equal in magnitude)

V_I = Average value of input voltage during t_1

t_1 = Time period over which unknown voltage is integrated

t_2 = Unknown time period over which a known reference voltage is integrated

Equation (3) illustrates the major advantages of a dual-slope converter:

1. Accuracy is not dependent on absolute values of t_1 and t_2, but is dependent on their ratios. Long-term clock frequency variations will not affect the accuracy.
2. Offset values, V_{CT}, are not important.

The BCD counter in the digital processor (Figure 11-58) changes at a rate equal to one-half the oscillator frequency. The BCD counter and the control logic divide each measurement cycle into three phases.

Auto-Zero Phase

The cycle begins at the end of the integrate-reference phase when the digital processor applies low levels to inputs A and B of the analog processor. If the trigger input is at a high level, a free-running condition exists and continuous conversions are made. However, if the trigger input is low, the digital processor stops the counter at 20,000, entering a hold mode. In this mode, the processor samples the trigger input every 4000 oscillator pulses until a high level is detected. When this occurs, the counter is started again and is carried to completion at 30,000. The reference voltage is stored in the reference capacitor C_{ref}, comparator offset voltage is stored in the integration capacitor C_X, and the sum of the buffer and integrator offset voltages is stored on auto-zero capacitor C_Z. During the auto-zero phase, the comparator output is characterized by an oscillation (limit cycle) of indeterminate waveform and frequency that is filtered and dc shifted by the level shifter.

Integrate-Input Phase

The auto-zero phase is completed at a BCD count of 30,000, and high levels are applied to both control inputs to initiate the integrate-input phase. The integrator charges C_X for a fixed time of 10,000 BCD counts at a rate determined by the input voltage. Note that during this phase, the analog inputs see only the high impedance of the noninverting operational amplifier input. Therefore, the integrator responds only to the difference between the analog input terminals, providing true differential inputs.

Integrate-Reference Phase

At a BCD count of 39,999 + 1 = 40,000 or 0, the integrate-input phase is terminated and the integrate-reference phase is begun by sampling the comparator output. If the comparator output is low, corresponding to a negative average analog input voltage, the digital processor applies a low and a high to control inputs A and B, respectively, the positive side of the reference voltage stored on C_{ref} is applied to the buffer. If the comparator output is high, corresponding to a positive input, control inputs A and B are made high and low, respectively, and the negative side of the stored reference voltage is applied to the buffer. In either case, the processor automatically selects the proper logic state to cause the integrator to ramp back toward zero at a rate proportional to the reference voltage. The time required to return to zero is measured by the counter in the digital processor. The phase is terminated when either the integrator output crosses zero and the counter contents are transferred to the register, or the BCD counter reaches 20,000 and the overrange indication is activated. When activated, the overrange indication blanks all but the most significant digit and sign.

Seventeen parallel bits (4-1/2 digits) of information are strobed into the buffer register at the end of the integration phase. Information for each digit is multiplexed out to BCD outputs (TL503C) or the 7-segment drivers (TL502C) at a rate equal to the oscillator frequency divided by 400.

Capacitor Selection Guidelines

The auto-zero capacitor C_Z and reference capacitor C_{ref} should be within the recommended range of operating conditions and should have low leakage characteristics. Most film-dielectric capacitors and some tantalum capacitors provide acceptable results. Ceramic and aluminum capacitors are not recommended because of their relatively high leakage characteristics.

The integrator capacitor C_X should also be within the recommended range and must have good voltage linearity and low dielectric absorption. A polypropylene-dielectric capacitor similar to TRW's X363UW is recommended for 4-1/2 digit accuracy. For 3-1/2 digit applications, polyester, polycarbonate, and other film dielectrics are usually suitable. Ceramic and electrolytic capacitors are not recommended.

The time constant $R_X C_X$ should be kept as near the minimum value as possible and is given by the formula:

$$\text{Minimum } R_X C_X = \frac{V_{ID} \text{ (full scale) } t_1}{V_{OM-} - V_{I(pin\ 2)}}$$

where:

$$V_{ID} \text{ (full scale)} = \text{Voltage on pin 1 with respect to pin 2}$$
$$t_1 = \text{Input integration time in seconds}$$
$$V_{I(pin\ 2)} = \text{Voltage on pin 2 with respect to analog ground}$$

Bypassing and Stray Coupling

Stray coupling from the comparator output to any analog pin (in order of importance 17, 18, 14, 7, 6, 13, 1, 2, 15) must be minimized to avoid oscillations. In addition, all power supply pins should be bypassed at the package, for example, by a 0.01 μF ceramic capacitor.

Analog and digital common are internally isolated and may be at different potentials. Digital common can be within 4 V of the positive or negative supply with the logic decode still functioning properly.

TL505C ANALOG-TO-DIGITAL CONVERTER

The TL505 is a dual-slope A/D converter with 1-bit resolution and a 3-digit accuracy of 0.1%. It has an internal reference, automatic zero, and operates from a single supply. It features a high-impedance MOS input and consumes typically only 40 mW of power. It is designed chiefly for use with the TMS1000 microprocessor family for low-cost, high-volume applications. Figure 11-64 shows the package pin assignment.

Description

The TL505C is an analog-to-digital converter building block designed for use with TMS1000 type microprocessors. See Figure 11-65 for the functional block diagram. The TL505C contains the analog elements (operational amplifier, comparator, voltage reference, analog switches, and switch drivers) necessary for a unipolar automatic-zeroing dual-slope converter. The logic for the dual-slope conversion can be performed by the associated MPU as a software routine or it can be implemented with other components such as the TL502C logic-control device.

The high-impedance MOS inputs permit the use of less expensive, lower value capacitors, and conversion speeds from 20 per second to 0.05 per second.

N DUAL-IN-LINE PACKAGE

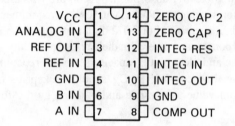

Figure 11-64. TL505C Pinout (Top View)

Definition of Terms

Zero Error — The intercept (b) of the analog-to-digital converter system transfer function $y = mx + b$, where y is the digital output, x is the analog input, and m is the slope of the transfer function, which is approximated by the ratiometric reading.

Linearity Error — The maximum magnitude of the deviation from a straight line between the end points of the transfer function.

Ratiometric Reading — The ratio of negative integration time (t_2) to positive integration time (t_1).

Principles of Operation

A block diagram of an MPU system utilizing the TL505C is shown in Figure 11-66. The TL505C operates in a modified positive-integration three-step dual-slope conversion mode. The A/D converter waveforms during the conversion process are illustrated in Figure 11-67.

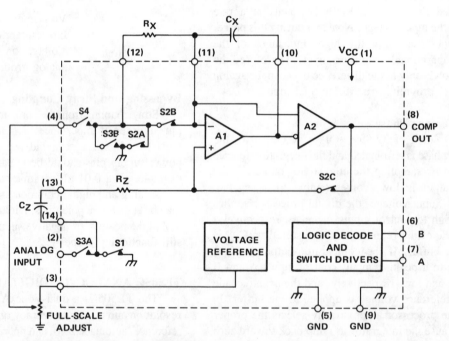

NOTE: Analog and digital GND are internally connected together.

Figure 11-65. Functional Block Diagram

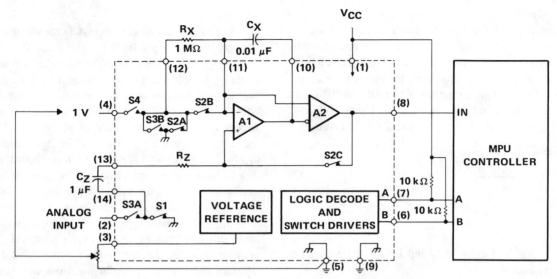

Figure 11-66. Functional Block Diagram of TL505C Interface with a Microprocessor System

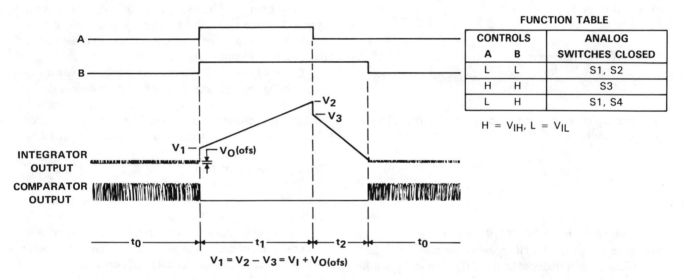

FUNCTION TABLE		
CONTROLS		ANALOG
A	B	SWITCHES CLOSED
L	L	S1, S2
H	H	S3
L	H	S1, S4

$$H = V_{IH}, \ L = V_{IL}$$

$$V_1 = V_2 - V_3 = V_I + V_{O(ofs)}$$

Figure 11-67. Conversion Process Timing Diagrams

The first step of the conversion cycle is the auto-zero period t_0 during which the integrator offset is stored in the auto-zero capacitor and the offset of the comparator is stored in the integrator capacitor. To accomplish this, the MPU takes both the A and B inputs low. This is decoded by the switch drivers, which close S1 and S2. The output of the comparator is connected to the input of the integrator through the low-pass filter consisting of R_Z and C_Z. The closed loop of A1 and A2 will seek a null condition where the offsets of the integrator and comparator are stored in C_Z and C_X, respectively. This null condition is characterized by a high-frequency oscillation at the output of the comparator. The purpose of S2B is to shorten the amount of time required to reach the null condition.

At the conclusion of t_0, the MPU takes both A and B inputs high. This closes S3 and opens all the other switches.

The input signal V_I is applied to the noninverting input of A1 through C_Z. V_I is then positively integrated by A1. Since the offset of A1 is stored in C_Z, the change in voltage across C_X will be due to only the input voltage. It should be noted that since the input is integrated in a positive integration during t_1, the output of A1 will be the sum of the input voltage, the integral of the input voltage and the comparator offset, as shown in Figure 11-67. The change in voltage across capacitor C_X during t_1 is given by:

$$V_{CX(1)} = \frac{V_I \, t_1}{R_1 \, C_X} \tag{1}$$

where

$R_1 = R_X + R_{S3B}$ and R_{S3B} is the resistance of switch S3B.

At the end of t_1, the MPU takes the A input low and the B input high. This turns on S1 and S4; all other switches are turned off. In this state, the reference is integrated by A1 in a negative sense until the integrator output reaches the comparator threshold. At this point, the comparator output goes high. This change in state is sensed by the MPU, which terminates t_2 by again taking the A and B inputs both low. During t_2, the change in voltage across C_X is given by:

$$V_{CX(2)} = \frac{V_{ref}\, t_2}{R_2} \qquad (2)$$

where

$R_2 = R_X + R_{S4} + R_{ref}$ and R_{ref} is the equivalent resistance divider.

Since $V_{CX1} = V_{CX2}$, equations (1) and (2) can be combined to give

$$V_I = V_{ref} \frac{R_1\, t_2}{R_2\, t_1} \qquad (3)$$

This equation is a variation on the ideal dual-slope equation, which is

$$V_I = V_{ref} \frac{t_2}{t_1} \qquad (4)$$

Ideally then, the ratio of R1/R2 would be exactly equal to one. In a typical TL505C system where $R_X = 1$ MΩ, the scaling error introduced by the difference in R_1 and R_2 is so small that it can be neglected, and equation (3) reduces to (4).

TL507I, TL507C
ANALOG-TO-DIGITAL CONVERTER

The TL507 is an economical single-slope, 7-bit resolution A/D converter. It features guaranteed monotonicity and ratiometric conversion. Operating from a single supply, it consumes only 25 mW of power at a supply voltage of 5 V and has a conversion speed of approximately 1 ms. See Figure 11-68 for the pin assignment of this converter.

P DUAL-IN-LINE PACKAGE

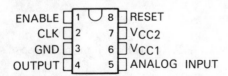

Figure 11-68. TL507 Pinout (Top View)

Description

The TL507 is a single-slope A/D converter designed to convert analog input voltages between 0.25 V_{CC} and 0.75 V_{CC} into a pulse width modulated output code. It contains a 7-bit synchronous counter, a binary weighted resistor ladder network, an operational amplifier, two comparators, a buffer amplifier, an internal regulator, and necessary logic circuitry. Integrated injection logic (I^2L) technology makes it possible to offer this complex circuit at low cost in a small dual-in-line 8-pin package.

In continuous operation, it is possible to obtain conversion speeds up to 1000 per second. The TL507 requires external signals for clock, reset, and enable. Versatility and simplicity of operation coupled with low cost, make this converter especially useful for a wide variety of applications. See Figure 11-69 for the functional block diagram and Figure 11-70 for input/output schematics.

Definition of Terms

Zero Error — The intercept (b) of the A/D converter system function y = mx + b, where y is the output of the ladder network, x is the analog input, and m is the slope of the transfer function.

Overall Error — The magnitude of the deviation from a straight line between the endpoints of the transfer function.

Differential Nonlinearity — Maximum deviation of an analog value change that is associated with a 1 bit code change (1 clock pulse) from its theoretical value of 1 LSB.

Principles of Operation

The TL507 like all single-slope converters is basically a voltage-to-time or current-to-time converter. An external clock signal is applied through a buffer to a negative edge triggered synchronous counter. Binary-weighted resistors from the counter are connected to an operational amplifier used as an adder. The operational amplifier generates a signal that ramps from 0.75 V_{CC1} down to 0.25 V_{CC1}. Comparator 1 compares the ramp signal to the analog input signal. Comparator 2 functions as a fault detector. With the analog input voltage within the 0.25 V_{CC1} to 0.75 V_{CC1} range, the duty cycle of the output signal is determined by the unknown analog input as shown in the waveform and function table in Figure 11-71.

For illustration, assume $V_{CC1} = 5.12$ V,

$0.25\ V_{CC1} = 1.28$ V

$1 \text{ binary count} = \frac{(0.75 - 0.25)\ V_{CC1}}{128} = 20$ mV

$0.75\ V_{CC1} - 1 \text{ count} = 3.82$ V

The output is an open-collector npn transistor capable of withstanding up to 18 V in the off state. The output is current limited within the 8 mA to 12 mA range; however, care must be taken to ensure that the output does not exceed 5.5 V in the on state. The voltage regulator section allows operation from either an unregulated 8 V to 18 V V_{CC2} supply source or a regulated 3.5 V to 6 V V_{CC1} supply. Regardless of which external power source is used, the internal circuitry operates at V_{CC1}. When operating from a V_{CC1} source, V_{CC2} may be connected to V_{CC1} or left open.

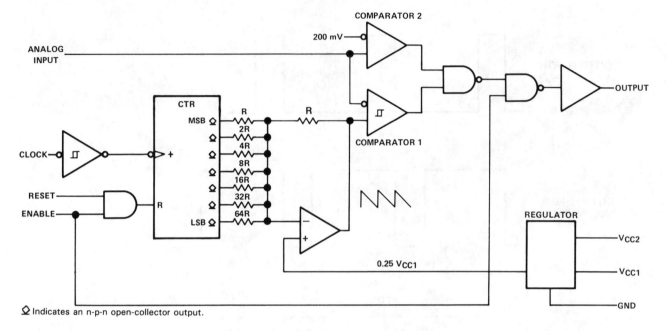

Figure 11-69. TL507 Functional Block Diagram

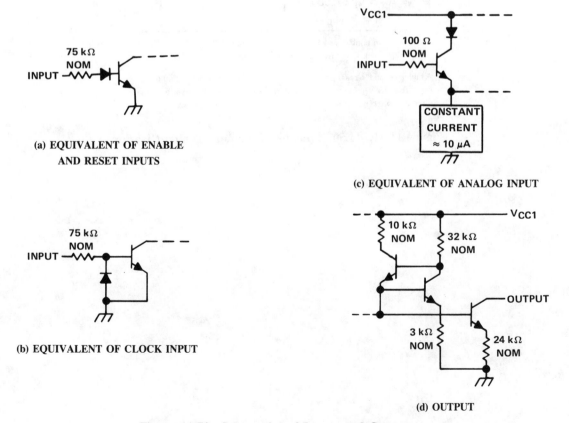

Figure 11-70. Schematics of Inputs and Outputs

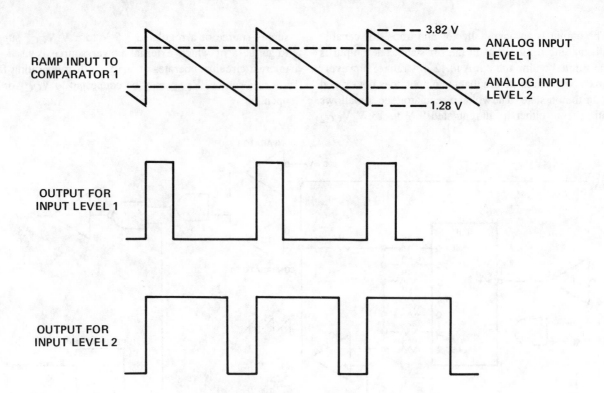

FUNCTION TABLE

ANALOG INPUT CONDITION	ENABLE	OUTPUT
X	L†	H
$V_I < 200$ mV	H	L
$V_{ramp} > V_I > 200$ mV	H	H
$V_I > V_{ramp}$	H	L

†Low level on enable also inhibits the reset function.

H = high level, L = low level, X = irrelevant

A high level on the reset pin clears the counter to zero, which sets the internal ramp to 0.75 V_{CC}. Internal pull-down resistors keep the reset and enable pins low when not connected.

Figure 11-71. TL507 Function Table and Timing Diagram

TLC532, TLC533 LinCMOS 8-BIT ANALOG-TO-DIGITAL PERIPHERALS WITH 5 ANALOG AND 6 MULTIPURPOSE INPUTS

The TLC532 and TLC533 are 8-bit A/D converters built with the LinCMOS technology. The unadjusted error is ±0.5 LSB and the devices feature ratiometric conversion. The I/O bus is three-state bidirectional. The access plus conversion time is 15 μs maximum for the TLC532 and 30 μs for the TLC533. There are 5 analog and 6 multipurpose inputs plus an analog on-chip 12-channel mixer. Other features are three on-chip sample-and-hold functions and 5 V single supply operation. Power consumption is low, typically 6.5 mW. Figure 11-72(a) and (b) show the device pinouts. The I/O data bus lines as well as the analog input lines are grouped together for efficient PC board layout.

Description

The TLC532A and TLC533A are monolithic LinCMOS peripheral intergrated circuits each designed to interface with a microprocessor for analog data acquisition. Figure 11-73 shows the function table and Figure 11-74 is the functional block diagram. These devices are complete peripheral data acquisition systems on a single chip that can convert analog signals to digital data from up to 11 external analog terminals. Each device features operation from a single 5-V supply. Each contains a 12-channel analog multiplexer, an 8-bit ratiometric analog-to-digital (A/D) converter, a sample-and-hold, three 16-bit registers, and microprocessor-compatible control logic circuitry. Additional features include a built-in self-test, six multipurpose (analog or digital) inputs, five external analog inputs, and an 8-pin input/output (I/O) data port. The three on-chip data registers store the control data, the conversion results, and the input digital data that can be accessed via the microprocessor data bus in two 8-bit bytes (most significant byte first). In this manner, a microprocessor can access up to 11 external analog inputs or six digital signals and the positive reference voltage that may be used for self-test.

The A/D conversion uses the successive-approximation technique and switched-capacitor circuitry. This method eliminates the possibility of missing codes, nonmonotonicity, and a need for zero or full-scale adjustment. Any one of 11 analog inputs (or self-test) can be converted to an 8-bit digital word and stored in 10 μs (TLC532A) or 20 μs (TLC533A) after instructions from the microprocessor have been recognized. The on-chip sample-and-hold functions automatically miniminze errors due to noise on the analog inputs. Furthermore, differential high-impedance reference inputs are available to help isolate the analog circuitry from the logic and supply noises while easing ratiometric conversion and scaling.

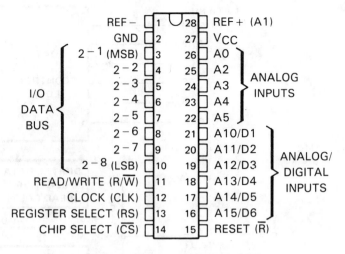

(a) J OR N DUAL-IN-LINE PACKAGE

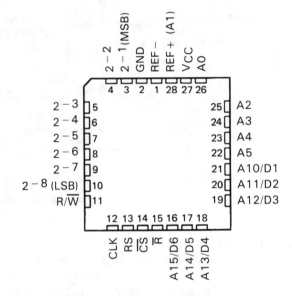

(b) FN CHIP-CARRIER PACKAGE

Figure 11-72. TLC532A and TLC533A Pinouts (Top View)

ADDRESS/CONTROL					DESCRIPTION
R/$\overline{W}$	RS	$\overline{CS}$	$\overline{R}$	CLK	
X	X	X	L†		Reset
L	H	L	H	↓	Write bus data to control register
H	L	L	H	↑	Read data from analog conversion register
H	H	L	H	↑	Read data from digital data register
X	X	H	H	X	No response

H = High-level, L = Low-level, X = Irrelevant
↓ = High-to-low transition, ↑ = Low-to-high transition
†For proper operation, Reset must be low for at least three clock cycles.

Figure 11-73. Function Table

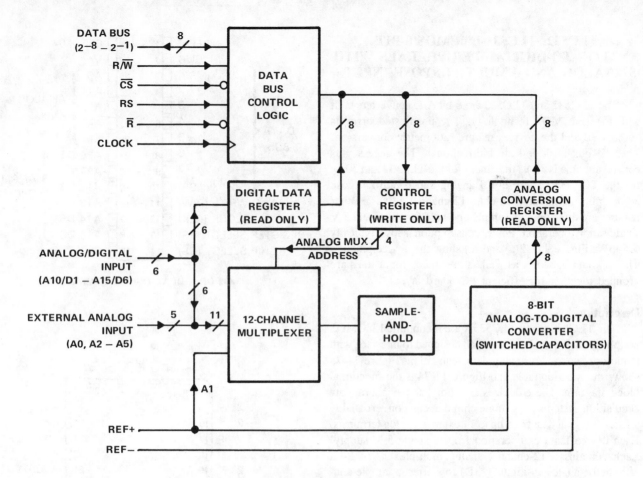

Figure 11-74. Functional Block Diagram

Principles of Operation

The TLC532A and TLC533A can be directly connected to a microprocessor-based system. Control of the TLC532A and TLC533A is handled via the 8-line TTL-compatible three-state data bus, the three control inputs (Read/Write, Register Select, and Chip Select), and the Clock input. Each device contains three 16-bit internal registers. These registers are the control register, the analog conversion data register, and the digital data register.

A high level at the Read/Write input and a low level at the Chip Select input set the device to output data on the 8-line data bus for the microprocessor. A low level at the Read/Write input and a low level at the Chip Select input set the device to receive instructions into the internal control register on the 8-line data bus from the microprocessor. When the device is in the read mode and the Register Select input is at a low level, the microprocessor reads the data contained in the digital data register.

The control register is a write-only register into which the microprocessor writes the command instructions for the device to start A/D conversion and to select the analog channel to be converted. The analog conversion data register is a read-only register that contains the current converter status and the most recent conversion results. The digital data register is also a read-only register that holds the digital input

Internally each device contains a byte pointer that selects the appropriate byte during two cycles of the clock input in a normal 16-bit microprocessor instruction. The internal pointer will automatically point to the most significant byte any time that the Chip Select is at a high level for at least one clock cycle. This causes the device to treat the next signal on the 8-line data bus as the most significant (MS) byte. A low level at the Chip Select input activates the inputs and outputs and an internal function decoder. However, no data is transferred until the Clock goes high. The internal byte pointer first points to the MS byte of the selected register during the first clock cycle. After the first clock cycle in which the MS byte is accessed, the internal pointer switches to the least significant (LS) byte and remains there for as long as Chip Select is low. The MS byte of any register may be accessed by either an 8-bit or 16-bit microprocessor instruction; however, the LS byte may only be accessed by a 16-bit microprocessor instruction.

Noramlly, a two-byte word is written into or read from the controlling microprocessor, but a single byte can be read by the microprocessor by proper manipulation of the Chip Select input. This can be used to read conversion status from the analog conversion data register or the digital multipurpose input levels from the digital data register.

A conversion cycle is started after a two-byte instruction is written into the control register and the start conversion (SC) bit is a logic high. This two-byte instruction also selects the input analog channel to be converted. The end-of-conversion (EOC) status bit in the analog conversion data register is reset and it remains at that level until the conversion is completed, when the status bit is set again. After conversion, the results are loaded into the analog conversion data register. These results remain in the analog conversion data register until the next conversion cycle is completed. If a new conversion command is entered into the coantrol register while the conversion cycle is in progress, the on-going conversion will immediately begin.

The Reset input allows the device to be externally forced to a known state. When a low level is applied to the Reset input for a minimum of three clock periods, the start conversion bit of the control register is cleared. The A/D converter is then idled and all the outputs are placed in the high-impedance off-state. However, the content of the analog conversion data register is not affected by the Reset input going to a low level. A typical operating sequence timing diagram is shown in Figure 11-75.

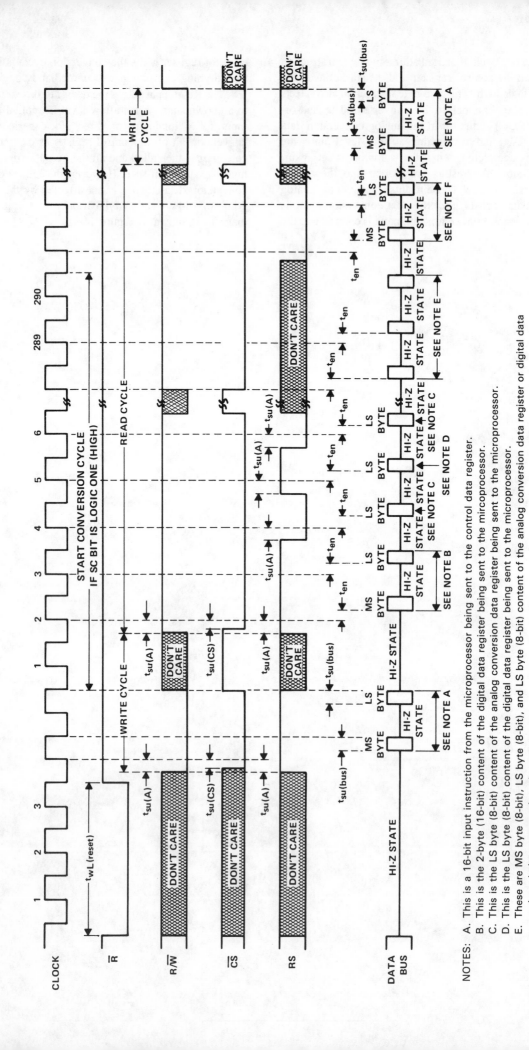

NOTES: A. This is a 16-bit input instruction from the microprocessor being sent to the control data register.
B. This is the 2-byte (16-bit) content of the digital data register being sent to the microprocessor.
C. This is the LS byte (8-bit) content of the analog conversion data register being sent to the mircoprocessor.
D. This is the LS byte (8-bit) content of the digital data register being sent to the microprocessor.
E. These are the MS byte (8-bit), LS byte (8-bit), and LS byte (8-bit) content of the analog conversion data register or digital data register being sent to the microprocessor.
F. This is the 2-byte (16-bit) content of the analog conversion data register being sent to the microprocessor.

Figure 11-75. Typical Operating Sequence

TLC540, TLC541 8-BIT ANALOG-TO-DIGITAL PERIPHERALS WITH SERIAL CONTROL AND 11 INPUTS

The TLC540 and TLC541 are LinCMOS A/D peripherials built around an 8-bit switched-capacitor successive-approximation A/D converter. Included is an on-chip 12-channel analog multiplexer and built-in self-test mode. The sample-and-hold is software controllable. Total unadjusted error is ± 0.5 LSB max. The pin layout is shown in Figures 11-76(a) and (b).

Typical Performance	TLC540	TLC541
Acquisition Time	2 μs	3.6 μs
Conversion Time	9 μs	17 μs
Sampling Rate	75×10^3	40×10^3
Power Dissipation	6 mW	6 mW

Description

The TLC540 and TLC541 are LinCMOS A/D peripherials built around an 8-bit switched-capacitor successive-approximation A/D converter. They are designed for serial interface to a microprocessor or peripheral via a three-state output with up to four control inputs (including independent System Clock, I/O Clock, Chip Select ($\overline{CS}$), and Address Input). The system clock is a 4-MHz clock for the TLC540 and a 2.1-MHz clock for the TLC541. Figure 11-77 shows a functional block diagram. Each device design features simultaneous read/write operation that allows high-speed data transfers and sample rates of up to 75,180 samples per second for the TLC540 and 40,000 samples per second for the TLC541. In addition to the high-speed converter and versatile control logic, there is an on-chip 12-channel analog multiplexer that can be used to sample any one of 11 inputs or an internal "self-test" voltage, and a sample-and-hold that can operate automatically or under processor control.

The converters incorporated in the TLC540 and TLC541 feature differential high-impedance reference inputs that facilitate ratiometric conversion, scaling, and analog circuitry isolation from logic and supply noises. A switched-capacitor design allows guaranteed low-error (± 0.5 LSB) conversion in 9 μs for the TLC540 and 17 μs for the TLC541 over the full operating temperature range.

Principles of Operation

The TLC540 and TLC541 are each complete data acquisition systems on a single chip. With judicious interface timing, a TLC540 conversion can be completed in 9 μs, while complete input-conversion-output cycles are being repeated every 14 μs. With a TLC541, a conversion can be completed in 19 μs, while complete input-conversion-output cycles are repeated every 35 μs. Furthermore, this fast conversion can be executed on any of 11 inputs or its built-in "self-test", and in any order desired by the controlling processor.

The System and I/O clocks are normally used independently and do not require any special speed or phase

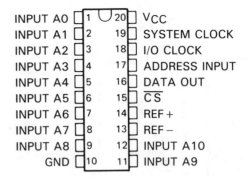

(a) N DUAL-IN-LINE PACKAGE

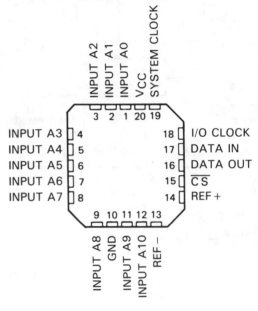

(b) FN CHIP-CARRIER PACKAGE

Figure 11-76. TLC540 and TLC541 Pinouts (Top View)

relationships between them. This independence simplifies the hardware and software control tasks for the device. Once a clock signal within the specification range is applied to the system clock input, the control hardware and software need only be concerned with addressing the desired analog channel, reading the previous conversion result, and starting the conversion by using the I/O clock. The System clock will drive the "conversion crunching" circuitry so that the control hardware and software need not be concerned with this task. See Figure 11-78 for Operating Sequence Timing.

When chip select ($\overline{CS}$) is high, the Data Out pin is in a three-state condition, and the Address Input and I/O Clock pins are disabled. This feature allows each of these pins, with the exception of $\overline{CS}$, to share a control logic point with their counterpart pins on additional A/D devices when additional TLC540 and TLC541 devices are used. This feature serves to minimize the required control logic pins when using multiple A/D devices.

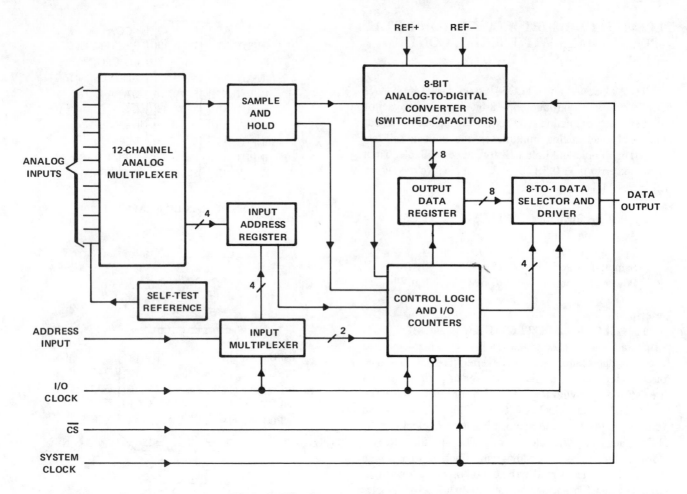

Figure 11-77. Functional Block Diagram

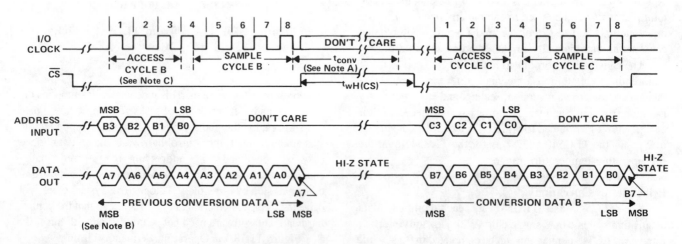

NOTES: A. The conversion cycle, which requires 36 internal system clock periods, is initiated with the 8th I/O clock pulse trailing edge after $\overline{CS}$ goes low for the channel whose address exists in memory at the time.

B. The most significant bit (MSB) will automatically be placed on the DATA OUT bus after $\overline{CS}$ is brought low. The remaining seven bits (A6-A0) will be clocked out on the first seven I/O clock falling edges.

C. To minimize errors caused by noise at the $\overline{CS}$ input, the internal circuitry waits for three internal system clock cycles (1.4 μs at 2 MHz) after a chip select transition before responding to control input signals. Therefore, no attempt should be fmade to clock-in address data until the minimum chip-select setup time has elapsed.

Figure 11-78. Operating Sequence

The control sequence has been designed to minimize the time and effort required to initiate conversion and obtain the conversion result. A normal control sequence is:

1. $\overline{CS}$ is brought low. To minimize errors by noise at the $\overline{CS}$ input, the internal circuitry waits for two rising edges and then a falling edge of the System clock after a $\overline{CS}$ (high-to-low) transition before the transition is recognized. This technique is used to protect the device against noise when the device is used in a noisy environment. The MSB of the previous conversion result will automatically appear on the DATA OUT pin.

2. A new positive-logic multiplexer address is shifted in on the first 4 rising edges of the I/O clock. The negative edges of these four I/O clocks shift out the 2nd, 3rd, 4th, and 5th most significant bits of the previous conversion result. The on-chip sample-and-hold begins sampling the newly addressed analog input after the 4th falling edge. The sampling operation basically involves the charging of internal capacitors to the level of the analog input voltage.

3. Three clock cycles are then applied to the I/O pin and the 6th, 7th, and 8th conversion bits are shifted out on the negative edges of these clock cycles.

4. The final (8th) clock cycle is applied to the I/O clock pin. The falling edge of this clock cycle completes the analog sampling process and initiates the hold function. Conversion is then performed during the next 36 System clock cycles.

$\overline{CS}$ can be kept low during periods of multiple conversion. Also, if $\overline{CS}$ is taken high, it must remain high until the end of conversion. Otherwise, a valid falling edge of $\overline{CS}$ will cause a reset condition, which will abort the conversion in progress.

A new conversion may be started and the ongoing conversion simultaneously aborted by performing steps 1 through 4 before the 36 System clock cycles occur. Such action will yield the conversion result of the previous conversion and not the ongoing conversion.

It is possible to connect the System and I/O clocks together in special situations where controlling circuitry must be minimized. In this case, the following special requirements must be taken into consideration in addition to the requirements of the normal control sequence, which was previously described.

1. When $\overline{CS}$ is recognized by the device to be at a low level, the common clock signal is used as an I/O clock. When the $\overline{CS}$ is recognized by the device to be at a high level, the common clock signal is used to drive the ''conversion crunching'' circuitry.

2. The device will recognize a $\overline{CS}$ transition only when the $\overline{CS}$ input changes and subsequently the System Clock pin receives two positive edges and then a negative edge. For this reason, after a $\overline{CS}$ negative falling edge, the first two clock cycles will not shift in the address because a low $\overline{CS}$ must be recognized before the I/O clock can shift in an analog channel

address. Also, upon shifting in the address, $\overline{CS}$ must be raised after the 6th I/O clock, so that a $\overline{CS}$ low level will be recognized upon the lowering of the 8th I/O clock signal. Otherwise, additional common clock cycles will be recognized as I/O clocks and will shift in an erroneous address.

For certain applications, such as strobing, conversion must be started at a specific point in time. These devices will accommodate these applications. Although the on-chip sample-and-hold begins sampling upon the negative edge of the 4th I/O clock cycle, the hold function is not initiated until the negative edge of the 8th I/O clock cycle. Thus, the control circuitry can leave the I/O clock signal in its high state during the 8th I/O clock cycle, until the moment at which the analog signal must be converted. The TLC540 and TLC541 will continue sampling the analog input until the 8th falling edge of the I/O clock. The control circuitry or software will then immediately lower the I/O clock signal and initiate the hold function to hold the analog signal at the desired point in time and start conversion.

TLC548 TLC549 LinCMOS 8-BIT ANALOG-TO-DIGITAL PERIPHERAL WITH SERIAL CONTROL

The TLC548 and TLC549 are LinCMOS A/D converters using the 8-bit switched-capacitor successive-approximation technology. The typical unadjusted error is ± 0.5 LSB maximum. Conversion time is 17 μs and total access plus conversion time is 45,500 conversions per second minimum for the TLC548 and 40,000 conversions per second minimum for the TLC549. The sample-and-hold is on-chip and is software controllable. Power consumption is a low 6 mW typical. Power supply range is from 3 V to 6 V. Figure 11-79 shows the pinout arrangement of the device.

P DUAL-IN-LINE PACKAGE

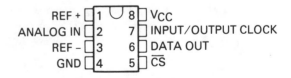

Figure 11-79. TLC548 and TLC549 Pinout (Top View)

Description

The TLC548 and TLC549 are LinCMOS A/D peripheral integrated circuits built around an 8-bit switched-capacitor successive-approximation ADC. See Figure 11-80 for the functional block diagram. They are designed for serial interface with a microprocessor peripheral through a three-state data output and an analog input. The TLC548 and TLC549 use only the Input/Output clock (I/O Clock) input along with the Chip Select ($\overline{CS}$) input for data control. The I/O clock input frequency of the TLC548 is guaranteed up to 2.048 MHz, and the I/O Clock input frequency of the TLC549 is guaranteed to 1.1 MHz.

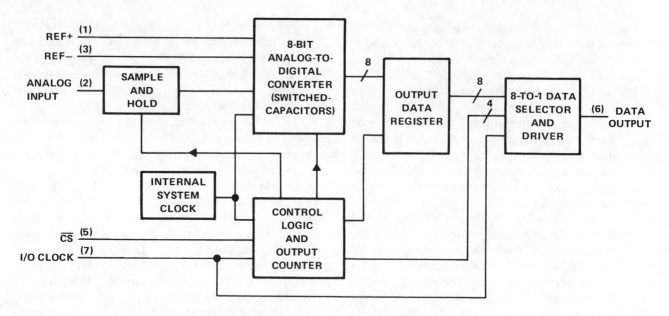

Figure 11-80. Functional Block Diagram

Operation of the TLC548 and TLC549 is very similar to that of the more complex TLC540 and TLC541 devices. However, the on-chip system clock of these devices allows internal device operation to proceed independently of serial input/output data timing and permits manipulation of the TLC548 and TLC549 as desired for a wide range of software and hardware requirements. The I/O Clock, together with the internal system clock, allows high-speed data transfer and minimum sample rates of 45,500 per second for the TLC548, and 40,000 per second for the TLC549.

Additional TLC548/549 features include versatile control logic, an on-chip sample-and-hold circuit that can operate automatically or under microprocessor control, and a high-speed converter with differential high-impedance reference voltage inputs that ease ratiometric conversion, scaling, and circuit isolation from logic and supply noises. Design of the totally switched-capacitor successive-approximation converter circuit allows guaranteed low-error conversion of ± 0.5 least significant bit (LSB) in less than 17 μs.

Principles of Operation

The TLC548 and TLC549 are each a complete data acquisition system on a single chip. Each includes such functions as internal system clock, sample-and-hold, 8-bit A/D converter, data register, and control logic. See Figure 11-81 for operating sequence timing. For flexibility and access speed, there are two control inputs: I/O Clock and Chip Select ($\overline{CS}$). These control inputs and a TTL-compatible three-state output facilitate serial communications with a microprocessor or minicomputer. A conversion can be completed in a maximum of 17 μs, while complete input-conversion-output cycles are repeated at a maximum of every 22 μs for the TLC548 and 25 μs for the TLC549, respectively.

The internal system clock and I/O clocks are used independently and do not require any special speed or phase relationships between them. This independence simplifies the hardware and software control tasks for the device. Due to this independence and the internal generation of the system clock, the control hardware and software need only be

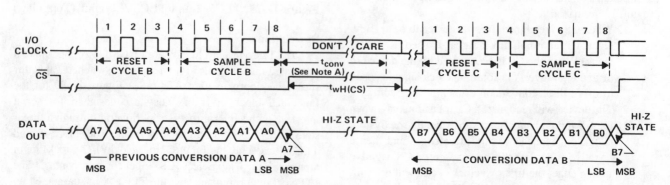

NOTE A: The conversion cycle, which requires 36 system clock periods, is initiated with the 8th I/O clock↓.

Figure 11-81. Operating Sequence

concerned with reading the previous conversion result and starting the conversion by using the I/O clock. The internal system clock will drive the "conversion crunching" circuitry so that the control hardware and software need not be concerned with this task.

When $\overline{CS}$ is high, the Data Out pin is in a high-impedance condition and the I/O clock pin is disabled. This feature allows each of these pins, with the exception of $\overline{CS}$, to share a control logic point with its counterpart pin when additional TLC548 or TLC549 devices are used. This serves to minimize the required control logic pins when using multiple TLC548 and TLC549 devices.

The control sequence has been designed to minimize the time and effort required to initiate conversion and obtain the conversion result. A normal control sequence is:

1. $\overline{CS}$ is brought low. To minimize errors caused by noise at the $\overline{CS}$ input, the internal circuitry waits for two rising edges and then a falling edge of the internal system clock after a $\overline{CS}$ (high-to-low) transition before the transition is recognized. This technique is used to protect the device against noise when the device is used in a noisy environment. The MSB of the previous conversion result will automatically appear on the data out pin.

2. The negative edges of the first four I/O clocks shift out the 2nd, 3rd, 4th, and 5th most significant bits of the previous conversion result. The on-chip sample-and-hold begins sampling the analog input after the 4th falling edge. The sampling operation basically involves the charging of internal capacitors to the level of the analog input voltage.

3. Three more clock cycles are then applied to the I/O pin and the 6th, 7th, and 8th conversion bits are shifted out on the negative edges of these clock cycles.

4. The final (8th) clock cycle is applied to the I/O Clock pin. The on-chip sample-and-hold begins the hold function upon the falling edge of this clock cycle. The hold function will continue for the next four internal system clock cycles. After these four system clock cycles, the hold function ends and conversion is performed during the next 32 system clock cycles, giving a total of 36 cycles. After the 8th I/O clock cycle, $\overline{CS}$ must go high or the I/O clock must remain low for at least 36 system clock cycles to allow for the hold and conversion functions.

$\overline{CS}$ can be kept low during periods of multiple conversion. Also, if $\overline{CS}$ is taken high it must remain high until the end of conversion. Otherwise, a falling edge of $\overline{CS}$ will cause a reset condition, which will abort the conversion in progress.

A new conversion may be started and the ongoing conversion simultaneously aborted by performing steps 1 through 4 before the 36 system clock cycles occur. Such action will yield the conversion result of the previous conversion and not the ongoing conversion.

For certain applications, such as strobing, it is necessary to start conversion at a specific point in time. These devices will accommodate these applications. Although the on-chip sample-and-hold begins sampling upon the negative edge of the 4th I/O clock cycle, the hold function does not begin until the negative edge of the 8th I/O clock cycle, until the moment at which the analog signal must be converted. The TLC548 and TLC549 will continue sampling the analog input until the 8th falling edge of the I/O clock. The control circuitry or software will then immediately lower the I/O clock signal and start the holding function to hold the analog signal at the desired point in time and start conversion.

TL0808, TL0809 LOW-POWER CMOS ANALOG-TO-DIGITAL CONVERTERS WITH 8-CHANNEL MULTIPLEXERS

The TL0808 and TL0809 are monolithic CMOS 8-bit A/D converters. They will operate with a 3-V supply making them ideal for battery type applications. Conversion time is 100 μs and they consume typically only 0.3 mW of power. They feature latched three-state outputs and latched address inputs. These A/D devices eliminate the possibility of missing codes, nonmonotonicity, and the need for zero or full-scale adjustment. The pin layout is shown in Figure 11-82. This grouping of input and address lines permits efficient pc board layout.

N DUAL-IN-LINE PACKAGE

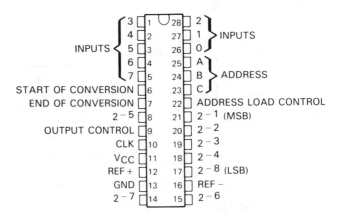

Figure 11-82. TL0808 and TL0809 Pinouts (Top View)

Description

The TL0808 and TL0809 are monolithic CMOS devices with a 8-channel multiplexer, an 8-bit analog-to-digital (A/D) converter, and microprocessor-compatible control logic. See Figure 11-83 for the functional block diagram and function table. The 8-channel multiplexer can be controlled by a microprocessor through a 3-bit address decoder with address load to select any one of eight single-ended analog switches connected directly to the comparator. The 8-bit A/D converter uses the successive-approximation conversion technique featuring a high-impedance threshold detector, a switched-capacitor array, a sample-and-hold, and a successive approximation register (SAR).

The comparison and conversion methods used eliminate the possibility of missing codes, nonmonotonicity, and the need for zero or full-scale adjustment. Also featured are three-state output latches from the SAR and latched inputs to the multiplexer address decoder. The single 3-V supply and extremely low power requirements make the TL0808 and TL0809 especially useful for a wide variety of applications including portable battery and LCD applications. Ratiometric conversion is made possible by access to the reference voltage input terminals.

Principles of Operation

The TL0808/0809 each consists of an analog signal multiplexer, an 8-bit successive-approximation converter, and related control and output circuitry.

MULTIPLEXER FUNCTION TABLE

INPUTS				SELECTED
ADDRESS			ADDRESS	ANALOG
C	B	A	STROBE	CHANNEL
L	L	L	↑	0
L	L	H	↑	1
L	H	L	↑	2
L	H	H	↑	3
H	L	L	↑	4
H	L	H	↑	5
H	H	L	↑	6
H	H	H	↑	7

H = high level, L = low level
↑ = low-to-high transition

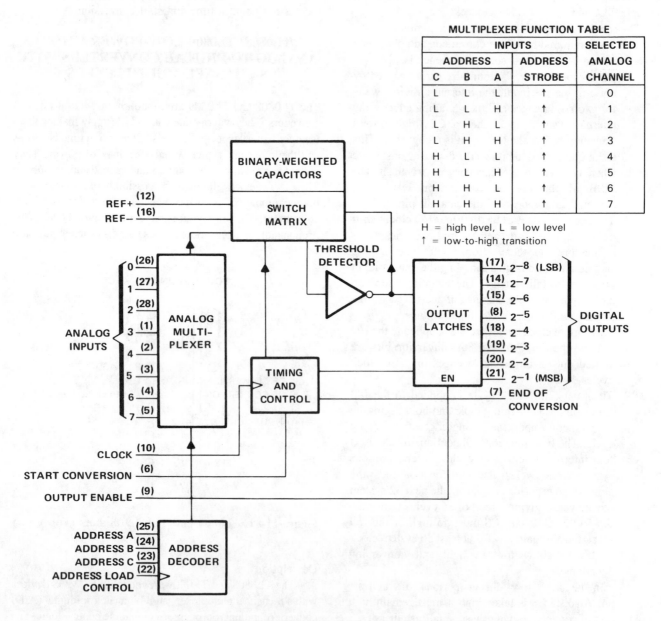

Figure 11-83. Functional Block Diagram (Positive Logic)

Multiplexer

See Figure 11-84 for the operating sequence timing. The analog multiplexer selects 1 of 8 single-ended input channels as determined by the address decoder. The Address Load Control clocks the address code into the decoder on a low-to-high transition. The output latch is reset by the positive-going edge of the start pulse. Sampling also starts with the positive-going edge of the start pulse and lasts for 32 clock periods. The conversion process may be interrupted by a new start pulse before the end of 64 clock periods. The previous data will be lost if a new start-of-conversion occurs before the 64th clock pulse. Continuous conversion may be accomplished by connecting the End-of-Conversion output to the start input. If used in this mode, an external pulse should be applied after power-up to assure start up.

Converter

The CMOS threshold detector in the successive-approximation conversion system determines each bit by examining the charge on a series of binary-weighted capacitors as shown in Figure 11-85. The conversion process uses successive approximation, but it relies on charge redistribution rather than a successive-approximation register (and reference DAC) to count and weight the bits from MSB to LSB.

In the first phase of the conversion process, the analog input is sampled by closing switch SC and all ST switches, and by simultaneously charging all the capacitors to the input voltage. In the next phase of the conversion process, all SC and ST switches are opened and the threshold detector begins identifying bits by identifying the charge (voltage) on each capacitor relative to the reference voltage. In the switching sequence, all eight capacitors are examined separately until all eight bits are identified, and then the charge-convert sequence is repeated.

In the first step of the conversion phase, the threshold detector looks at the first capacitor (weight = 128). Node 128 is switched to the reference voltage, and the equivalent nodes of all the other capacitors on the ladder are switched to REF −. If the voltage at the summing node is greater than the trip-point of the threshold detector (approximately one-half the V_{CC} voltage), a bit is placed in the output register, and the 128-weight capacitor is switched to REF −. If the voltage at the summing node is less than the trip point of the threshold detector, this 128-weight capacitor remains connected to REF + through the remainder of the capacitor-sampling (bit-counting) process. The process is repeated for the 64-weight capacitor, the 32-weight capacitor, and so on, until all bits are counted.

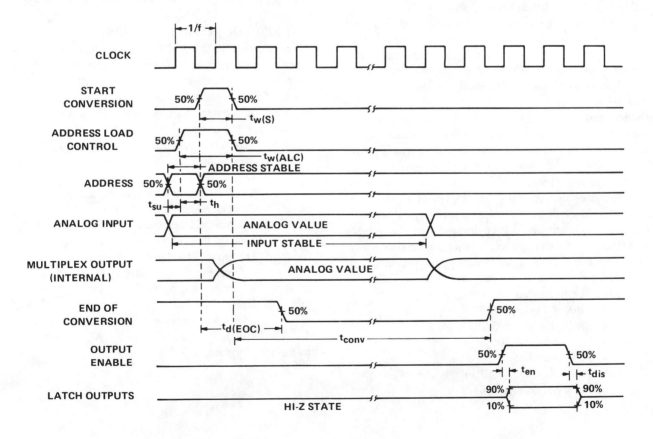

Figure 11-84. Operating Sequence

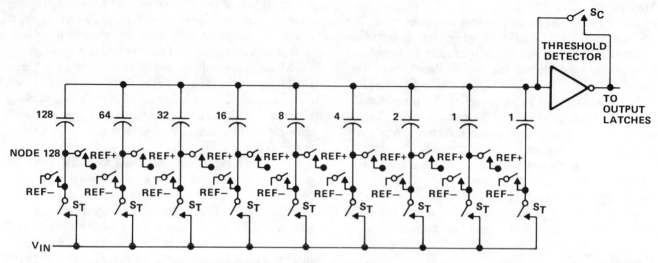

Figure 11-85. Simplified Model of the Successive-Approximation System

A/D CONVERTER APPLICATIONS

A/D converters are used in industrial control systems, automatic testing systems, communications and signal analysis systems, consumer appliance controls, displays, and automotive applications. Digital methods, especially with microprocessors, can provide powerful tools for dealing with analog functions in all of these applications.

An expanding use of A/D conversions is in digital signal processing. Television stations use this highly specialized technique to perform isolation and special effects seen on sports telecasts. However, digital signal processing is not limited to television. The medical community uses it to create and store visual images from computerized CAT scanners and ultrasound analysis equipment. Digital signal processing is also used by the military in radar, sonar, missile tracking, and secure communications. Geologists also use these techniques in analyzing earth science data, and astronomers apply the techniques in using radio telescopes.

Because the majority of applications for A/D converters involve the use of microprocessors, primary emphasis in this section has been directed toward interfaces for A/D converters and a variety of popular microprocessors. Most of the applications present hardware configurations and associated software that can be used with the following microprocessor integrated circuits:

1. Zilog Z80A and Z80
2. Intel 8051 and 8052
3. Intel 8048 and 8049
4. Motorola 6805
5. Motorola 6800, 6802, 6809, and 6809E
6. Rockwell 6502 and 6522 VIA

These microprocessors are representative of devices being used today and their inclusion in this book should not be implied as recommendation, or that other devices should not be considered for specific applications. Likewise, the circuits and software are presented only as working examples of each type of interface.

INTERFACE FOR ADC0803, ADC0804, AND ADC0805 CONVERTERS TO ZILOG Z80A AND Z80 MICROPROCESSORS

This application illustrates the circuit configuration and the associated software that can be used to operate the ADC080X family of A/D converters with the ZILOG Z80A and Z80. The A/D circuits are 8-bit successive-approximation A/D converters that feature microprocessor-compatible control logic and parallel communication with the microprocessor via the data bus. The configuration features are as follows:

1. Minimum circuitry
2. Low cost
3. Very fast communication between the microprocessor and A/D converter
4. Optional microprocessor-interrupt acknowledgement of the end of conversion
5. Differential analog voltage inputs, which reject both common-mode voltages and the offset of the zero-input analog voltage value
6. Optional on-board generation of the A/D clock signal with an external resistor and capacitor.

The basic differences between the A/D converters in this family are given in Table 11-9.

Table 11-9. Differences Between Devices in the ADC080X A/D Converter Family (Note 1)

	0803	0804	0805	UNIT
Total maximum adjusted error (with full-scale adjust)	± 1/2	—	—	LSB
Total maximum unadjusted error ($V_{ref}/2$ = 2.5 V)	—	—	—	LSB
Total maximum unadjusted error ($V_{ref}/2$ = open)	—	± 1	± 1	LSB
Operating free-air temperature range	−40 to 85	0 to 70	−40 to 85	°C

NOTE 1: Conversion accuracies listed are with V_{CC} at 5 V and the clock frequency at 640 kHz. If a faster clock is used, conversion time will decrease proportionately and the accuracy will tend to decrease slightly.

Circuitry

Figure 11-86 shows the interconnection between the microprocessor and the A/D converter. The A/D converter write $\overline{WR}$ and read $\overline{RD}$ signals, which are generated by the microprocessor, are not masked by an addressing scheme. However, if additional I/O devices were placed on the data bus, masking could be easily designed.

The SN74393 dual binary counter generates a 500 kHz clock signal for the A/D converter. Any binary counter may be used instead of the SN74393. In addition, any clock signal within the clock frequency specification may be used. Another way to generate the clock signal is to use an external resistor and capacitor in conjunction with the CLK IN and CLK OUT pins of the A/D converter. The configuration and frequency equation for this method are shown in Figure 11-87.

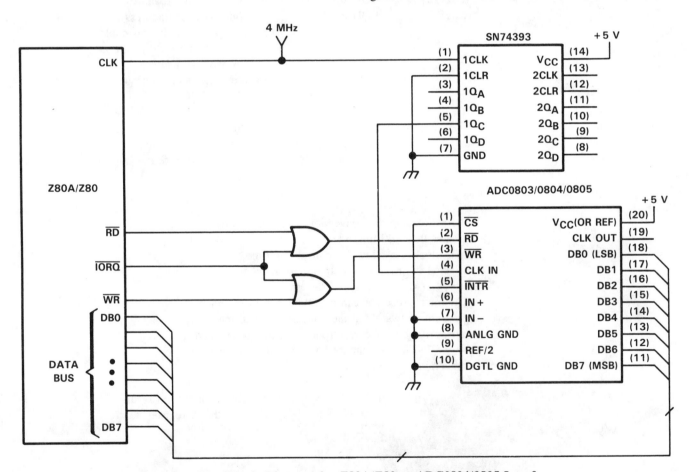

Figure 11-86. Circuit Diagram for Z80A/Z80 to ADC0804/0805 Interface

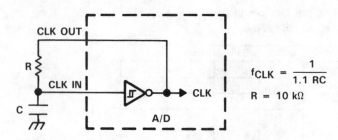

Figure 11-87. Configuration and Equations for On-Board Clock Generator

$$f_{CLK} = \frac{1}{1.1\ RC}$$

$$R = 10\ k\Omega$$

Timing Diagram

Figure 11-88 shows the timing diagram for the interface. With a 500 kHz clock, the conversion time is 140.5 μs and the A/D $\overline{INTR}$ pin is reset when the $\overline{RD}$ signal goes low.

Software

The following software listing presents the software for this interface. The software is minimal due to the microprocessor-compatibility of the A/D converter. In allowing time for conversion, the designer can use any delay that is convenient, including a timer function. Another method is to use the $\overline{INTR}$ signal on the A/D converter to start a microprocessor interrupt routine that reads the conversion result. If an on-board clock generator like the one shown in Figure 11-87 is used, the conversion time will vary with the tolerances of the resistor and capacitor. In this situation, the designer can allow a conservative delay to assure that the conversion is complete before reading the result. Another method might be to wait for an acknowledgement of the $\overline{INTR}$ signal and use this to inform the microprocessor of a completed conversion.

```
                        ;               Software for Z80/Z80A, ADC0803,
                        ;               ADC0804, ADC0805, Interface
                        ;
                        ;
        FF 00   WRITE:  EQU FFH         ;Interface write address
        FF 00   READ:   EQU FFH         ;Interface read address
                                        ;The above addresses could be
                                        ;anything unless they are used
                                        ;to drive an address decode
                                        ;circuit, which addresses
                                        ;the A/D IC

                        ;
0000    D3 FF   START:  OUT (WRITE),A   ;Start conversion of analog
                                        ;channel; the contents of the
                                        ;accumulator are not important;
                                        ;the write pulse however, is.


                        ;
                        ;Before reading the conversion result, a slightly
                        ;greater than 140 microsecond delay (with a 500 kHz
                        ;A/D clock), to allow for conversion time, can be
                        ;selected in any way which is convenient for the designer.
                        ;               OR
                        ;The microprocessor can continue to perform main
                        ;program software and the conversion result can be
                        ;retrieved by an interrupt routine, which is initiated
                        ;by the INTR(bar) signal, which signifies the end of
                        ;conversion, on pin 5.
                        ;
0002    DB FF           IN A, (READ)    ;Read conversion result
                                        ;into the A register

0004                    END
```

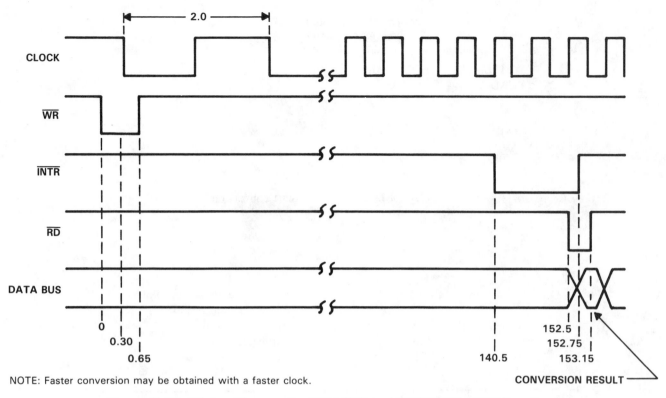

NOTE: Faster conversion may be obtained with a faster clock.

Figure 11-88. Timing Diagram for Z80A/Z80 to ADC0803/0804/0805 Interface

INTERFACE FOR ADC0803, ADC0804, AND ADC0805 CONVERTERS TO THE ROCKWELL 6502 MICROPROCESSOR

This application shows the circuit configuration and the associated software that can be used to operate this family of A/D converters with the Rockwell 6502 microprocessor. The interface circuit diagram is shown in Figure 11-89.

A data conversion cycle begins by performing a write to the ADC080X with a Start (STA) accumulator instruction. The conversion requires between 66 and 73 clock cycles. The ADC080X provides an interrupt request signal when the conversion is complete. The circuit timing is shown in Figure 11-90.

An alternate method for retrieving the conversion result would be to use a wait state in a software delay loop until the conversion results are read into the 6502 with an LDA instruction. A software listing for a typical interrupt service routine is shown. Note that the circuit employs a minimum of address decoding hardware; some applications may require additional decoding hardware.

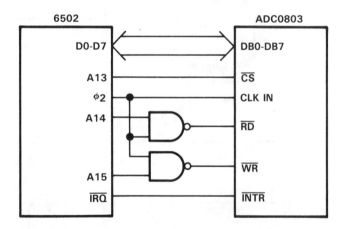

Figure 11-89. 6502 to ADC0803/0804/0805 Interface Circuit Diagram

```
;                      Register Assignments for ADC0803, ADC0804,
;                      and ADC0805 to 6502 Interface
;
WRITE:      .EQU 8800H
READ:       .EQU 4800H
;
;
MAIN:                  .
                       .
                       .


            STA WRITE              ;Start conversion

                       .
                       .
                       .

ISR:        PHA                    ;Save contents of accumulator
            LDA READ               ;Read conversion results
            STA DATA               ;Store results in memory
            PLA                    ;Restore accumulator
            RTI                    ;Return to main program
```

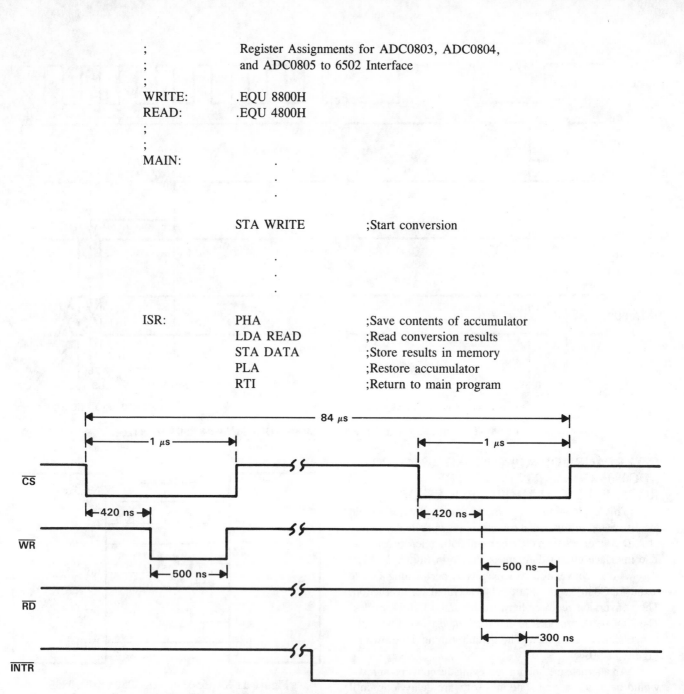

Figure 11-90. 6502 to ADC0803/0804/0805 Interface Timing Diagram

INTERFACE FOR ADC0808, ADC0809, TL0808, TL0809, TL520,TL521 AND TL522 CONVERTERS TO ZILOG Z80A AND Z80 MICROPROCESSORS

This application presents the circuit configuration and the associated software that can be used to operate this family of A/D converters with the ZILOG Z80A and Z80. These A/D circuits are 8-bit successive-approximation A/D converters that feature microprocessor-compatible control logic and parallel communication with the microprocessor via the data bus. This configuration features:

1. Minimum circuitry
2. Low cost

3. Very fast communication between the microprocessor and A/D converter
4. Optional microprocessor-interrupt acknowledgement of the end of conversion
5. Convenient control of the analog multiplexer.

Some of the basic differences between the A/D converters in these families are listed in Table 11-10. However, these are not the only differences. The TL0808 and TL0809 and ADC0808 and ADC0809 have the same pinout, but the TL520, TL521, and TL522 pinout is different from the 0808 and 0809.

Table 11-10. Differences Between ADC/TL0808-ADC/TL0809 and TL520-TL522 A/D Converter Families (Note 1)

	V$_{CC}$ SUPPLY (volts)	CONVERSION (μs)
TL0808/TL0809	2.75 V to 5.5 V	100 μs
ADC0808/ADC0809	4.5 V to 6 V	100 μs
TL520	3 V to 5.5 V	70 μs
TL521	3 V to 5.5 V	100 μs
TL522	2.75 V to 5.5 V	200 μs

NOTE 1: The conversion times require a 640 kHz clock. If faster conversion is desired, a faster clock is required. Both conversion time and conversion accuracy are specified in the data sheets when using a 640 kHz clock. If a faster clock is used, conversion time will decrease proportionately and the accuracy may stay within the 640 kHz specification or decrease slightly.

Figure 11-91 shows the interconnection points between the microprocessor and the A/D converter for a typical interface application.

Figure 11-92 shows the interconnection between the microprocessor and the TL0808 and TL0809 and ADC0808 and ADC0809 A/D converters. The interface will also work with the TL520, TL521, and TL522 if the different pinout is considered. The SN74393 dual binary counter is used to generate a 500 kHz clock for the A/D converter. Any binary counter may be used instead of the SN74393 or any clock signal within the A/D converter clock frequency specification may be used. The microprocessor-generated A/D control and address signals are not masked by an addressing scheme; however, if additional I/O devices are placed on the data bus, masking would be necessary.

Timing Diagram

Figure 11-93 is the timing diagram for the interface. With a 500 kHz A/D clock, the conversion time for the TL0808, TL0809, ADC0808, or ADC0809 devices is 130 μs with the EOC signal going low about 6 μs after activation of the output enable signal.

Software

The associated program listing presents the software routines for this interface. The coding is minimal due to the microprocessor-compatibility of the A/D converters. This code is written so that it may be easily incorporated into a subroutine if desired. In allowing time for conversion, the designer may use any delay which is convenient, including a timer function, or the End of Conversion (EOC) signal on the A/D converter to initiate a microprocessor interrupt routine that reads the conversion result.

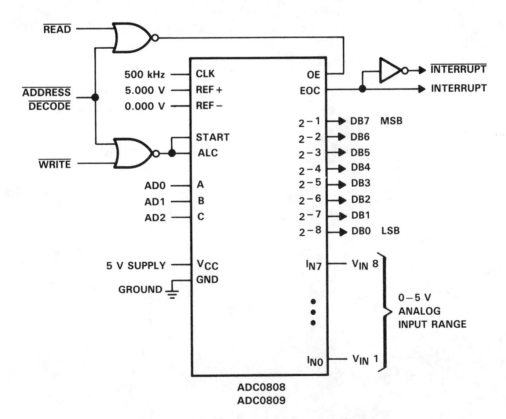

Figure 11-91. Typical Interface Application

```
                        ;Interface Software for Z80A/Z80-TL0808 / TL0809 ICs
                        ;                           -ADC0808 / ADC0809 ICs
                        ;                           -TL520 / TL521 / TL522 ICs
                        ;
                        ;
          FF 00         READ:        EQU FFH              ;Interface read address; this
                                                          ;address could be anything
                                                          ;unless it is used to drive
                                                          ;an address decode circuit,
                                                          ;which addresses the A/D IC

                        ;
0000      0E 00         START:       LD C,00H             ;Prepare to address analog
                                                          ;channel 0; 01 would address
                                                          ;channel 1, etc.

                        ;
0002      ED 49         LOD0808:     OUT (C),C            ;Start conversion of analog
                                                          ;channel per C register

                        ;
                        ;Before reading the conversion result, a slightly
                        ;greater than / microsecond delay (with a 500 kHz
                        ;A/D clock), to allow for conversion time, can be
                        ;selected in any way which is convenient for the
                        ;designer.
                        ;               OR
                        ;The microprocessor can continue to perform main
                        ;program software and the conversion result can be
                        ;retrieved by an interrupt routine, which is initiated
                        ;by the End of Conversion (EOC) signal on pin 7 of the
                        ;ADC0808/0809 and TL0808/0809 ICs and on pin 22 of the
                        ;TL520/521/522 ICs.
                        ;
0004      DB FF                      IN A, (READ)         ;Read conversion result
                                                          ;into the A register
0006                                 END
```

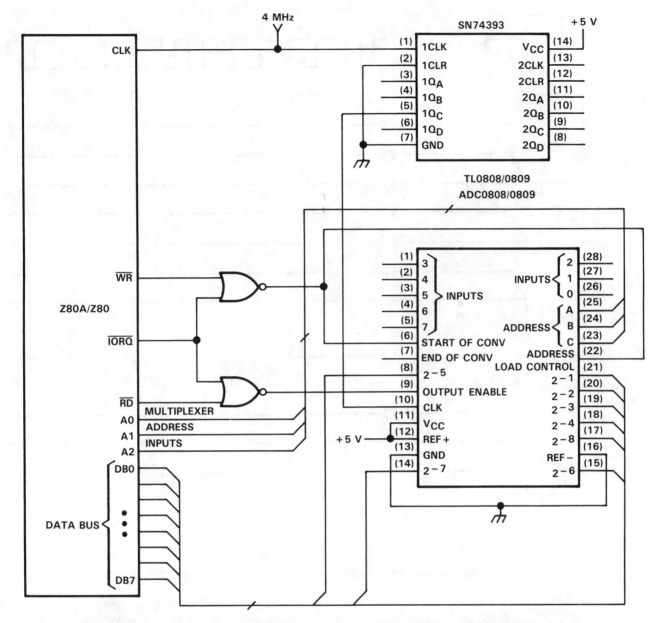

Figure 11-92. Circuit Diagram for Z80A/Z80 to TL0808/0809 ADC0808/0809 Interface

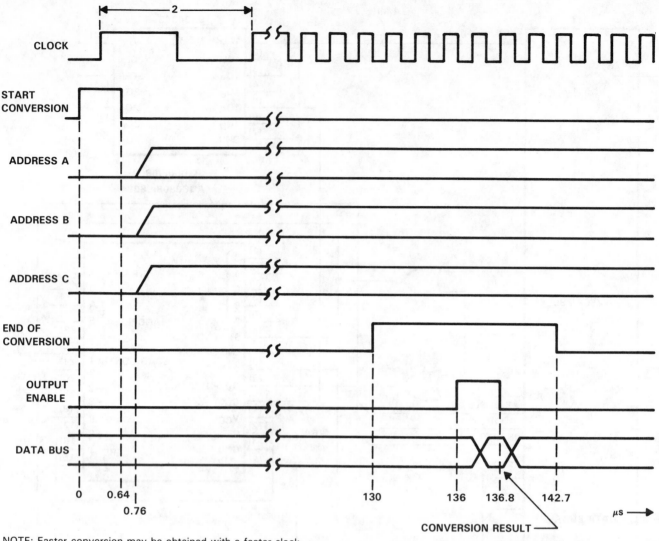

CLOCK

START
CONVERSION

ADDRESS A

ADDRESS B

ADDRESS C

END OF
CONVERSION

OUTPUT
ENABLE

DATA BUS

0   0.64
    0.76

130   136   136.8   142.7

μs ⟶

CONVERSION RESULT ⟶

NOTE: Faster conversion may be obtained with a faster clock.

Figure 11-93. Timing Diagram for Z80/Z80A to TL0808/0809 ADC0808/0809 Interface

INTERFACE FOR ADC0808 AND ADC0809 CONVERTERS TO THE ROCKWELL 6502 MICROPROCESSOR

This application shows the circuit configuration and the associated software that can be used to operate the ADC0808 and ADC0809 A/D converters with the Rockwell 6502 microprocessor. The interface circuit diagram is shown in Figure 11-94. A data conversion cycle is initiated by performing an STA instruction. This generates a start conversion pulse that clears the EOC line. The address in memory where the accumulator stores its contents determines which channel is selected for conversion. The conversion requires approximately 65 clock cycles. The clock frequency may vary from 10 kHz to 1.28 MHz, however, when it is above 640 kHz the accuracy may decrease slightly. The

circuit timing diagram is shown in Figure 11-95.

Upon completion of the conversion, the EOC signal is set and may be used to generate an interrupt signal for the Rockwell 6502. Also, a software delay loop may be used to achieve the proper delay until conversion is complete.

If the interrupt scheme is chosen, the interrupt service routine must either start another conversion cycle that will clear the EOC signal or disable the interrupts. If this is not done, the Rockwell 6502 will remain in the interrupt service routine loop. It may be desirable to add a flip-flop and an AND gate to control the enabling and disabling of the EOC signal. Structuring of the interrupt and the address decoding scheme are both flexible and should be optimized for the particular application. A software listing for an interrupt service routine follows.

```
;                    Register Assignments for ADC0808 and ADC0809
;                    Interface to the Rockwell 6502
ADC0808      .EQU 4000H
WRITE0       .EQU 8000H
WRITE1       .EQU 8001H
     .            .      .
     .            .      .
     .            .      .
WRITE7       .EQU 8007H
;
;
MAIN:            .
                 .
                 .
             STA WRITE0          ;Start conversion, clear interrupt
                 .
                 .
                 .


ISR:         PHA                 ;Save accumulator
             LDA ADC0808         ;Read conversion results into acc.
             STA DATA            ;Store conversion results
             PLA                 ;Restore accumulator
             PLP                 ;Pull status from stack
             SEI                 ;Disable interrupts
             PHP                 ;Push status back on stack
             RTI                 ;Return to main program
```

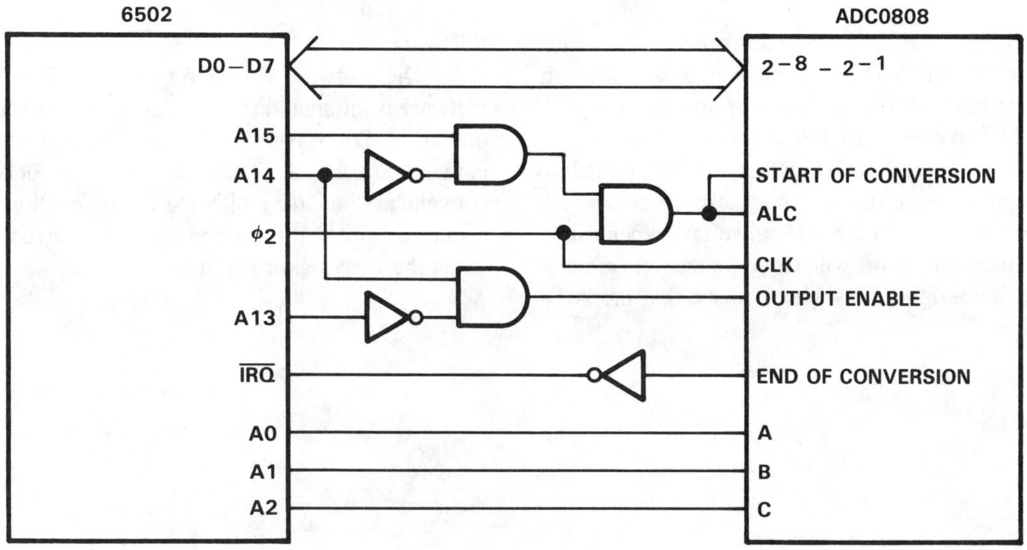

Figure 11-94. 6502 to ADC0808 Interface Circuit Diagram

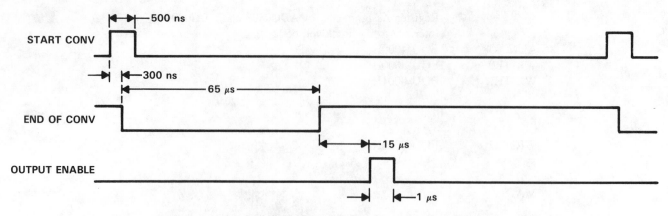

Figure 11-95. 6502 to ADC0808 Interface Timing Diagram

INTERFACE FOR ADC0808, ADC0809, TL0808, TL0809, TL520, TL521, AND TL522 CONVERTERS TO THE MOTOROLA 6800 MICROPROCESSORS

This application illustrates a circuit configuration and the software that can be used to operate this family of A/D converters with the Motorola 6800 family of microprocessors. Figure 11-91 shows a typical interface application. Figure 11-96 shows the interconnection between the Motorola 6802 microprocessor and the TL0808 or TL0809 A/D converters. Although Figure 11-96 shows circuitry for the TL0808/ADC0808 and TL0809/ADC0809 A/D converters, the interface will also work with the TL520, TL521, and TL522 devices if the different pinout is considered.

The SN74393 dual binary counter is used to generate a 500 kHz clock for the A/D converter. Any binary counter may be used in place of the SN74393 or any clock signal within the A/D converter clock frequency specification can be used. The 6800 E clock or E clock equivalent, if within the clock frequency specification, may also be connected directly to the clock input of the A/D converter. When using the Motorola microprocessors with clock frequencies greater than 1 MHz, the designer must check to see that the A/D

converter write and read timing specifications are met. The microprocessor-generated A/D control and address signals are masked by the 4000H-7FFFH addressing range. A more efficient addressing scheme can easily be implemented if desired. Table 11-11 provides information for adapting the circuit of Figure 11-96 for use with other members of the 6800 family of microprocessors.

Timing Diagram

Figure 11-97 is the timing diagram for this interface. With a 500 kHz A/D clock, the conversion time for the TL0808/TL0809 and ADC0808/ADC0809 A/D converters is 130 μs. The EOC signal falls approximately 6 μs after activation of the OUTPUT ENABLE signal.

Software

The software listing for this interface follows. The software is minimal due to the microprocessor-compatibility of the A/D converter devices. In allowing time for conversion, the designer can use any delay which is convenient, including either a timer function or the EOC signal to initiate the microprocessor interrupt routine that reads the conversion result.

```
;                        Software for Motorola 680X-TL0808 / TL0809 ICs
;                                              -ADC0808 / ADC0809 ICs
;                                              -TL520 / TL521 / TL522 ICs
;
;
        40 00       READ:       EQU 4000H           ;Interface read address; this
                                                    ;address is selected to ensure
                                                    ;that the read operation reads
                                                    ;the conversion result byte
                                                    ;and does not read the program
                                                    ;memory or on-board RAM (if
                                                    ;the microprocessor is the
                                                    ;6802) by mistake

                    ;
0000    B7 40 00    START:      STAA 4000H          ;Address analog channel 0 and
                                                    ;start conversion; 4001 would
                                                    ;address channel 1, etc.

                    ;
                    ;Before reading the conversion result, a slightly
                    ;greater than 130 microsecond delay (with a 500 kHz
                    ;A/D clock), to allow for conversion time, can be
                    ;selected in any way which is convenient for the
                    ;designer.
                    ;                   OR
                    ;The microprocessor can continue to perform main
                    ;program software and the conversion result can be
                    ;retrieved by an interrupt routine, which is initiated
                    ;by the End of Conversion (EOC) signal on pin 7 of the
                    ;ADC0808/0809 and TL0808/0809 ICs and on pin 22 of the
                    ;TL520/521/522 ICs.
                    ;
0003    B6 40 00                LDAA READ           ;Read conversion result
                                                    ;into the A accumulator
0006                            END
```

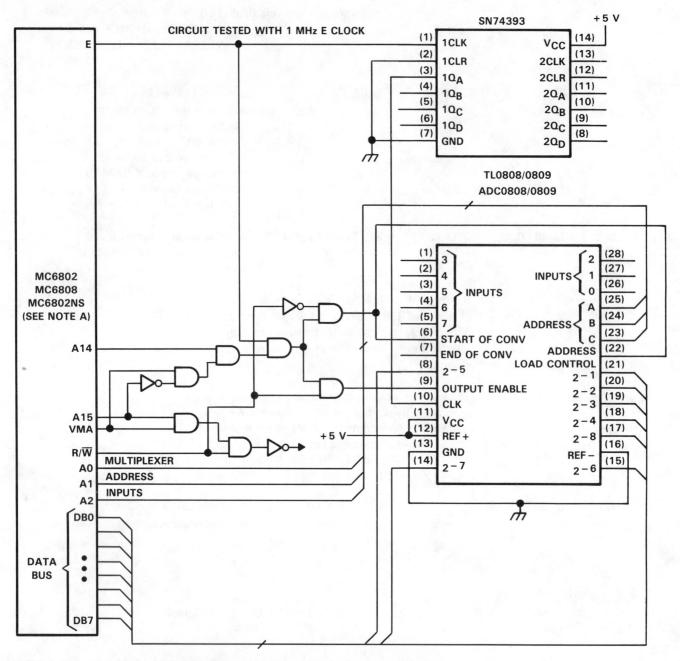

NOTE A: Refer to Table 11-11 for information about other Motorola microprocessors.

Figure 11-96. Circuit Diagram for TL0808/0809 Interface

**Table 11-11. Considerations for Using
Other Microprocessors**

MICROPROCESSOR	E CLOCK OR E CLOCK EQUIVALENT
6800	ϕ_2 is equivalent to the 6802/6809 E pin
6802/6808/6802ns	See Figure 11-96
6809	See Figure 11-96
6809E/68HC09E	See Figure 11-96 and ''Motorola 8-bit Microprocessor and Peripheral Data Book'' for clock generator — pg. 3-277

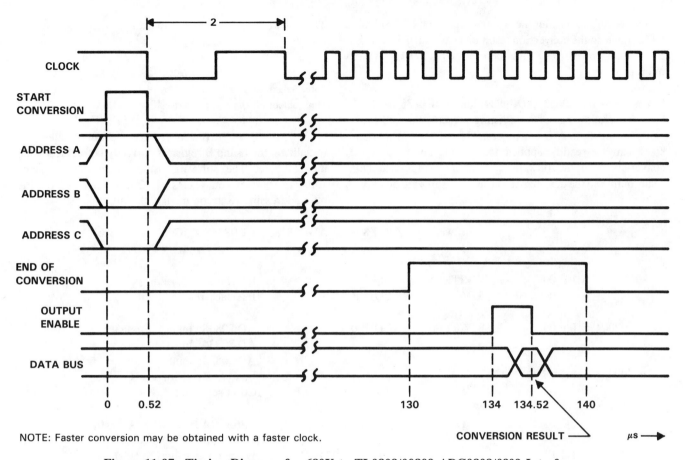

NOTE: Faster conversion may be obtained with a faster clock.

Figure 11-97. Timing Diagram for 680X to TL0808/00809 ADC0808/0809 Interface

INTERFACE FOR ADC0831, ADC0832, ADC0834, AND ADC0838 CONVERTERS TO ZILOG Z80A AND Z80 MICROPROCESSORS

The ADC0831, ADC0832, ADC0834, and ADC0838 devices are 8-bit successive-approximation analog-to-digital converters with serial input/output and configurable input multiplexers with up to 8 channels. These A/D converters are designed to easily communicate with microprocessors in a serial fashion. The hardware configurations and the associated software presented can be used to operate the ZILOG Z80A and Z80 microprocessors with the ADC0830 converters. The timing diagrams show the interaction between the microprocessor and the A/D converter.

The ADC0832, ADC0834, and ADC0838 A/D converters can be software configured in either the single-ended or differential input mode. Additionally, the differential ± inputs can be interchanged through software manipulation.

The serial interface features:
1. Low cost
2. Minimum circuitry
3. Fast conversion and communication between the A/D converter and the microprocessor
4. Remote control advantages of serial A/D converters.

Circuitry — ADC0832, ADC0834, and ADC0838

Figure 11-98 shows the interconnection between the microprocessor and the ADC0832, ADC0834, and ADC0838 A/D converters. The interconnection is identical for all three A/D converters. The microprocessor DO pin can be used to transmit to and receive digital data between the ADC083X DI and DO pins, respectively. The SN74126 3-state buffer output is in the high-impedance state except during a microprocessor read operation. The SN74174 quad D-type flip-flop is used to synchronize and slow down the write/read communication between the microprocessor and ADC083X so the ADC083X timing specifications are satisfied. Rather than NOR gates, OR gates may be used to activate the flip-flops and 3-state buffer on the positive transitions of the write WR and read RD strobes, instead of on the negative transitions. However, if OR gates are used, a SN74125 buffer must be used in place of the SN74126.

Timing Diagram — ADC0838 Device

Figure 11-99 is the timing diagram for the Z80A and Z80 to ADC0838 interface. Addressing the analog channel, performing conversion, and retrieving the conversion result requires 260 μs. The timing diagrams for the ADC0832 and ADC0834 converters are similar except fewer input bits are transmitted to the A/D converter. The 3-state buffer enable strobes occur 2.5 μs after the negative transition of the clock CLK signal so the conversion result bits have sufficient time to set up on the DO line before being read by the microprocessor.

Software — ADC0832, ADC0834, and ADC0838 Devices

The following software listing presents the interface software for these A/D converters. The software is written so it can be readily applied to any of the three A/D converters. Also, the software can be easily incorporated into a subroutine so the designer can access the software quickly.

The software differences for these A/D converters are as follows:

	ADC0832	ADC0834	ADC0838
Send Select Bit 1	No	Yes	Yes
Send Select Bit 0	No	No	Yes

The above differences are accommodated by initializing the C and B registers correctly before accessing the software. The software listing shows the interface software routines and Tables 11-12 through 11-17 list the mux addressing to accommodate the differences.

```
; Software for Z80/Z80A to ADC0838, ADC0834, ADC0832 Interface
;
;
          FF 00    WRITE:    EQU FFH          ;Interface write address
          FF 00    READ:     EQU FFH          ;Interface read address
                   ;
0000      06 05    START:    LD B,05H         ;ADC0838; Input bit counter = 5
                                              ;ADC0834; Input bit counter = 4
                                              ;ADC0832; Input bit counter = 3
0002      0E 15              LD C,15H         ;Load input address in C.
                                              ;C0 is sent first
                   ;
0004      3E 04    ADC83X    LD A,04H         ;CS(bar)(A2) = 1,CLK(A1) = 0,
                                              ;D1/D0(A0) = 0
0006      D3 FF              OUT (WRITE),A    ;Write these values to A/D
0008      3E 00              LD A,00H         ;Lower CS(bar)
000A      D3 F               OUT (WRITE),A
000C      CB 1F    ADC83X1:  RR A             ;Prepare to load next input bit in A
000E      CB 19              RR C             ;Put next input bit in carry
0010      CB 17              RL A             ;Load next input bit in A0 and realign
                                              ;A1 & A2 so they are correct
0012      D3 FF              OUT (WRITE),A    ;Set up next bit on D1 line
0014      F6 02              OR 02H           ;Raise CLK line and clock in,
0016      D3 FF              OUT (WRITE),A    ;next bit
0018      E6 FD              AND FDH          ;Lower CLK line
001A      D3 FF              OUT (WRITE),A
001C      10 EE              DJNZ ADC83X1     ;If B>0; branch
001E      06 08              LD B,08H         ;Conversion bit counter = 8
0020      3E 02    ADC83X0:  LD A,02H         ;Raise CLK & keep CS(bar) = 0
0022      D3 FF              OUT (WRITE),A
0024      3E 00              LD A,00H         ;Lower CLK line and clock out,
0026      D3 FF              OUT (WRITE),A    ;next conversion bit
0028      DB FF              IN A,(READ)      ;Put next conversion bit in A0
002A      CB 1F              RR A             ;Put conversion bit in carry
002C      CB 11              RL C             ;Put conversion bit in C0 and
                                              ;shift other conversion bits in C
002E      10 F0              DJNZ ADC83X0     ;If B>0; branch
0030      F6 04              OR 04H           ;Raise CS(bar)
0032      D3 FF              OUT (WRITE),A
0034                         END              ;Conversion result in register C
```

ADC0832 MUX ADDRESSING (5-BIT SHIFT REGISTER) (See Note 1)

Table 11-12. Single-Ended MUX Mode

START BIT	MUX ADDRESS SGL/$\overline{\text{DIF}}$	MUX ADDRESS ODD/SIGN	CHANNEL NO. 0	CHANNEL NO. 1	PUT DATA INTO REGISTER C
1	1	0	+		#03H
1	1	1		+	#0BH

Table 11-13. Differential MUX Mode

START BIT	MUX ADDRESS SGL/$\overline{\text{DIF}}$	MUX ADDRESS ODD/SIGN	CHANNEL NO. 0	CHANNEL NO. 1	PUT DATA INTO REGISTER C
1	0	0	+	−	#01H
1	0	1	−	+	#09H

NOTE 1: Internally, Select 0 is low. Select 1 is high, COMMON is internally connected to ANLG GND.

ADC0834 MUX ADDRESSING (5-BIT SHIFT REGISTER) (See Note 2)

Table 11-14. Single-Ended MUX Mode

START BIT	MUX ADDRESS SGL/$\overline{\text{BIT}}$	MUX ADDRESS ODD/SIGN	MUX ADDRESS SELECT 1	CHANNEL NO. 0	CHANNEL NO. 1	CHANNEL NO. 2	CHANNEL NO. 3	PUT DATA INTO REGISTER C
1	1	0	0	+				#03H
1	1	0	1			+		#0BH
1	1	1	0		+			#07H
1	1	1	1				+	#0FH

Table 11-15. Differential MUX Mode

START BIT	MUX ADDRESS SGL/$\overline{\text{BIT}}$	MUX ADDRESS ODD/SIGN	MUX ADDRESS SELECT 1	CHANNEL NO. 0	CHANNEL NO. 1	CHANNEL NO. 2	CHANNEL NO. 3	PUT DATA INTO REGISTER C
1	0	0	0	+	−			#01H
1	0	0	1			+	−	#09H
1	0	1	0	−	+			#05H
1	0	1	1			−	+	#0DH

NOTE 2: Internally, Select 0 is high, COMMON is internally connected to ANLG GND.

ADC0838 MUX ADDRESSING (5-BIT SHIFT REGISTER)

Table 11-16. Single-Ended MUX Mode

START BIT	MUX ADDRESS SGL/$\overline{\text{DIF}}$	MUX ADDRESS ODD/SIGN	MUX ADDRESS SELECT 1	MUX ADDRESS SELECT 0	ANALOG 0	ANALOG 1	ANALOG 2	ANALOG 3	ANALOG 4	ANALOG 5	ANALOG 6	ANALOG 7	ANALOG COM	PUT DATA INTO REGISTER C
1	1	0	0	0	+								−	#03H
1	1	0	0	1			+						−	#13H
1	1	0	1	0					+				−	#0BH
1	1	0	1	1							+		−	#1BH
1	1	1	0	0		+							−	#07H
1	1	1	0	1				+					−	#17H
1	1	1	1	0						+			−	#0FH
1	1	1	1	1								+	−	#1FH

Table 11-17. Differential MUX Mode

START BIT	MUX ADDRESS SGL/$\overline{\text{DIF}}$	ODD/SIGN	SELECT 1	SELECT 0	ANALOG DIFFERENTIAL CHANNEL-PAIR NO. 0: 0	1	1: 2	3	2: 4	5	3: 6	7	PUT DATA INTO REGISTER C
1	0	0	0	0	+	−							#011H
1	0	0	0	1			+	−					#11H
1	0	0	1	0					+	−			#09H
1	0	0	1	1							+	−	#19H
1	0	1	0	0	−	+							#05H
1	0	1	0	1			−	+					#15H
1	0	1	1	0					−	+			#0DH
1	0	1	1	1							−	+	#13H

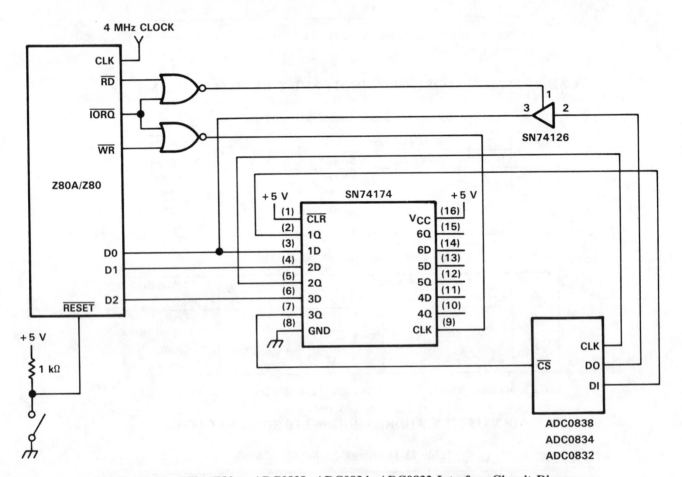

Figure 11-98. Z80A/Z80 to ADC0838, ADC0834, ADC0832 Interface Circuit Diagram

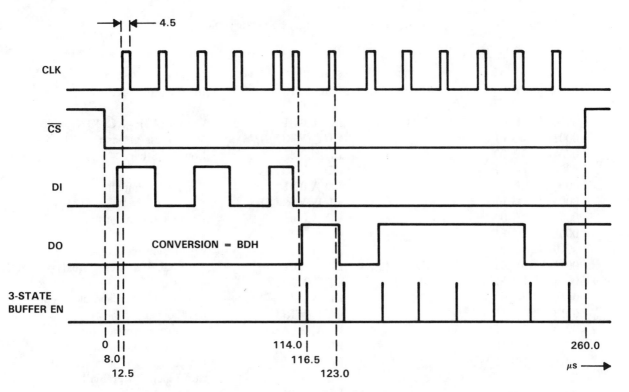

Figure 11-99. Timing Diagram for Z80A/Z80 to ADC0838 Interface
(Microprocessor Clock = 4 MHz)

Circuitry — ADC0831 Device

Figure 11-100 shows the interconnection between the microprocessor and the ADC0831 converter. The circuitry is basically the same as for the ADC0832, ADC0834, and ADC0838 devices except that the ADC0831 device does not have a DI line.

Timing Diagram — ADC0831 Device

Figure 11-101 is the timing diagram for the Z80A and Z80-ADC0831 interface. Performing conversion and retrieving the conversion result requires 181 μs. The 3-state buffer enable strobes occur 2.5 μs after the negative transition of the clock CLK signal so the conversion result bits have sufficient time to set up on the DO line before these bits are read by the microprocessor.

Software — ADC0831 Device

The software listing for the ADC0831 follows. The software can be easily incorporated into a subroutine so the designer can access the software quickly. An initial A/D clock cycle must occur before the eight subsequent clock cycles extract the conversion result bits from the A/D converter.

```
; Software for Z80/Z80A to ADC0831 Interface
;
;
         FF 00        WRITE:     EQU FFH              ;Interface write address
         FF 00        READ:      EQU FFH              ;Interface read address
                      ;
0000     3E 04        START:     LD A,04H             ;CS(bar)(A2) = 1,CLK(A1) = 40,
0002     D3 FF                   OUT (WRITE),A        ;Write these values to A/D
0004     3E 00                   LD A,00H             ;Lower CS(bar)
0006     D3 FF                   OUT (WRITE),A
0008     3E 02                   LD A,02H             ;Raise CLK initially
000A     D3 FF                   OUT (WRITE),A
000C     3E 00                   LD A,00H             ;Lower CLK initially
000E     D3 FF                   OUT (WRITE),A
0010     06 08                   LD B,08H             ;Conversion bit counter = 8
0012     3E 02        ADC8310:   LD A,00H             ;Raise CLK & keep CS(bar)=0
0014     D3 FF                   OUT (WRITE),A
0016     3E 00                   LD A,00H             ;Lower CLK line and clock out,
0018     D3 FF                   OUT (WRITE),A        ;next conversion bit
001A     DB FF                   IN A,(READ)          ;Put next conversion bit in A0
001C     CB 1F                   RR A                 ;Put conversion bit in carry
001E     CB 11                   RL C                 ;Put conversion bit in C0 and
                                                      ;shift other conversion bits in C
0020     10 F0                   DJNZ ADC8310         ;If B>0; branch
0022     3E 04                   LD A,04H             ;Raise CS(bar)
0024     D3 FF                   OUT (WRITE),A
0026                             END                  ;Conversion result in register C
```

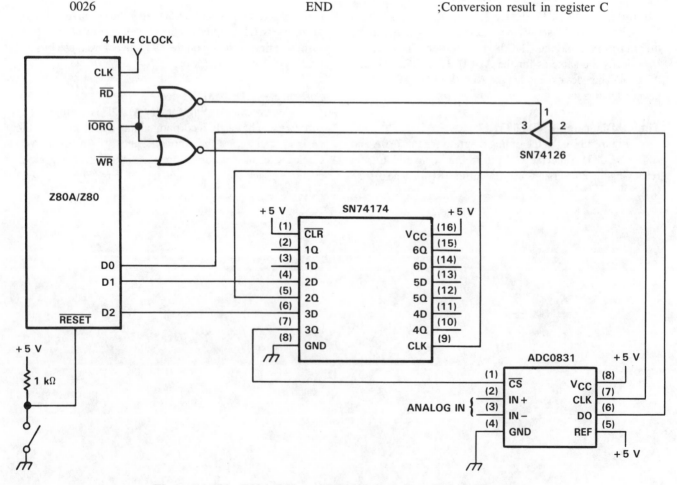

Figure 11-100. Z80A/Z80 to ADC0831 Interface Circuit Diagram

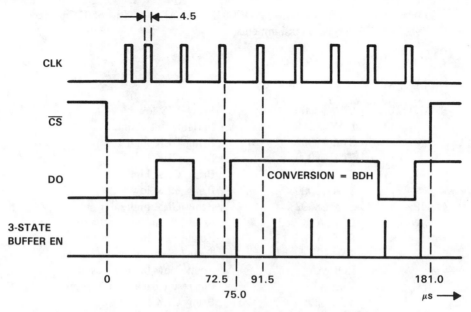

**Figure 11-101. Timing Diagram for Z80A/Z80 to ADC0831 Interface
(Microprocessor Clock = 4 MHz)**

INTERFACE FOR ADC0831, ADC0832, AND ADC0838 CONVERTERS TO THE ROCKWELL 6502 MICROPROCESSOR

The ADC083X family of successive-approximation A/D converters features 8-bit resolution, microprocessor-compatible control logic, serial data communication, and 1, 2, 4, or 8 analog inputs, which may be operated in the single-ended, differential, or pseudo-differential mode.

A circuit for an ADC0831 device interface to a 6502 microprocessor is shown in Figure 11-102. The timing diagram for this circuit is shown in Figure 11-103 and the interface software is shown as a listing. The interface circuit operates under complete control of the software. The system clock Phi 2 is used as the clocking signal to latch data from the address bus into the D-type flip-flops. By writing to the correct address locations, the CS and CLK signals are generated. The address decoding scheme is quite simple; therefore, a few additional gates may be required to provide proper address decoding in more complex applications. Since the ADC0831 device has only a single analog input, no multiplexer address is required which simplifies the timing.

A similar circuit for a 6502 to ADC0838 interface is presented in Figure 11-104. The timing diagram is shown in Figure 11-105 and the interface software listing follows. The operation of this circuit is basically the same as the above circuits with the addition of the DI input for the analog multiplexer address. Operation of the ADC0838 is the same as the ADC0831 converter with the exception of the multiplexer address which can be changed in software. With the software listings presented, one data acquisition cycle can be accomplished in 221 μs using the ADC0831 circuit and in 345 μs using the ADC0838 circuit.

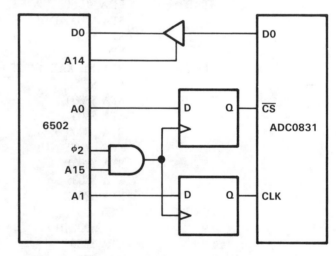

**Figure 11-102. 6502 to ADC0831
Interface Circuit Diagram**

```
; Software Listing for ADC0831, ADC0832, and ADC0838 to 6502 Interface
;                    Register Assignments
;
ADC0831        .EQU 4000H
;
;
START:         STA $8000          ; Bring /CS low
               LDY #$08           ; Initialize bit counter
               STA $8002          ; Bring CLK high
               STA $8000          ; Bring CLK low
               STA $8002          ; Bring CLK high
               STA $8000          ; Bring CLK low
LOOP:          STA $8002          ; Bring CLK high
               LDA ADC0831        ; Read bit into accumulator
               ROR A              ; Rotate new bit into carry
               TXA                ; Transfer result into accumulator
               ROL A              ; Rotate new bit into result
               TAX                ; Put result back into X register
               STA $8000          ; Bring CLK low
               DEY                ; Decrement bit counter
               BNE LOOP           ; Go back and get another bit
               STA $8001          ; Bring /CS high


;              Register Assignments
;
MUXADDRESS     .EQU 0000H
ADC0838        .EQU 4000H
;
;
START:         LDA #$C0           ;
               STA MUXADDRESS     ; Initialize muxaddress
               STA $8000          ; Bring /CS low
               LDY #$05           ; Initialize counter
LOOP1:         ROL MUXADDRESS     ; Rotate address bit into carry
               BCS SET            ; Is address bit set?
               STA $8000          ; Set up a low to be clocked in
               STA $8004          ; Bring CLK high write high to D1
               STA $8000          ; Bring CLK low
               JMP CONTINUE       ; Skip to continue
SET:           STA $8001          ; Set up a high to be clocked in
               STA $8005          ; Bring CLK high, write high to D1
               STA $8001          ; Bring CLK low
CONTINUE:      DEY                ; Decrement Counter
               BNE LOOP1          ; Go back to clock out next bit
               STA $8004          ; Bring CLK high
               STA $8000          ; Bring CLK low
               LDY #$08           ; Initialize counter
LOOP2:         STA $8004          ; Bring CLK high
               LDA ADC0838        ; Read bit in from ADC0838
               ROR A              ; Rotate new bit into carry
               TXA                ; Transfer result into accumulator
               ROL A              ; Rotate new bit into result
               TAX                ; Replace result back in X register
               STA $8000          ; Bring CLK low
               DEY                ; Decrement counter
               BNE LOOP2          ; Go back to clock next result bit
               STA $8002          ; Bring /CS high
```

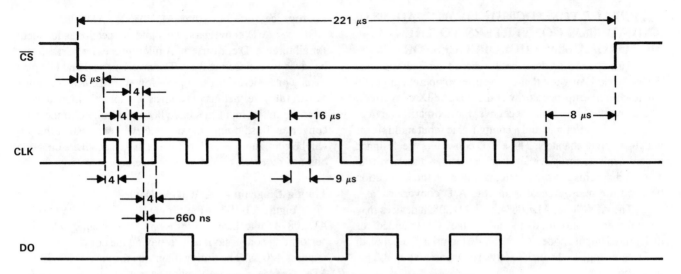

Figure 11-103. 6502 to ADC0831 Interface Timing Diagram

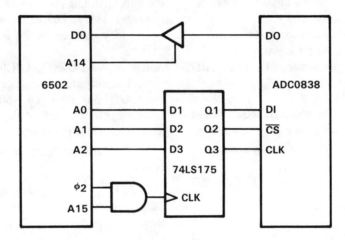

Figure 11-104. 6502 to ADC0838 Interface Circuit Diagram

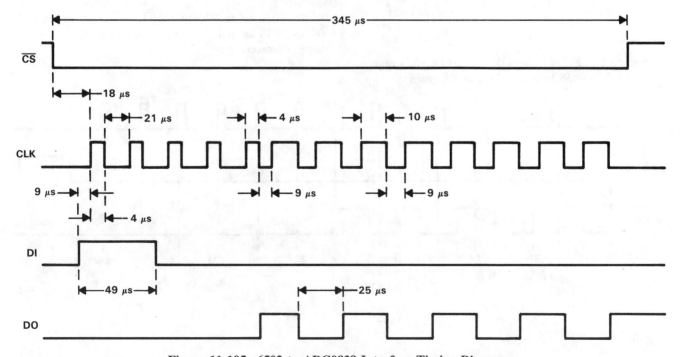

Figure 11-105. 6502 to ADC0838 Interface Timing Diagram

INTERFACE FOR ADC0831, ADC0832, ADC0834, AND ADC0838 CONVERTERS TO THE MOTOROLA 6805 MICROPROCESSOR

This application describes the circuit configuration and the associated software that can be used to operate the 6805 single-chip microprocessor with the ADC083X converters. The converters are 8-bit successive-approximation devices with serial input/output and a configurable input multiplexer with up to eight channels. These A/D converters are designed to communicate easily with microprocessors in a serial mode. Timing diagrams are presented that describe the interaction between the microprocessor and the A/D converter.

The ADC0832, ADC0834, and ADC0838 devices may be software configured in either the single-ended or differential input mode. Also, the differential [2] inputs can be interchanged through software manipulation. This interface features:

1. Low-cost A/D devices
2. Direct connection between the A/D converter and the microprocessor
3. Fast conversion and communication between the A/D converter and the microprocessor
4. Remote control advantage of serial A/D converters.

Circuitry — ADC0832, ADC0834, and ADC0838 Devices

Figure 11-106 shows the interconnection between the

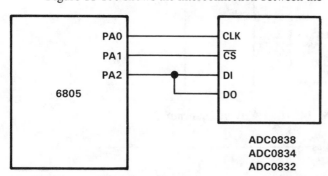

Figure 11-106. 6805 to ADC0832, ADC0834 and ADC0838 Interface Circuit Diagram

microprocessor and the ADC0832, ADC0834, and ADC0838 A/D converters. This interconnection is identical for all three A/D converters. A microprocessor port pin can be saved by connecting the A/D converters's DI and DO pins to only one microprocessor pin. However, the designer must be careful to define this port pin as an input pin immediately after the initial input bits have been transmitted to the A/D converter. In so doing, the designer will ensure that the pin will be configured as an input when the data conversion mode bits are received by the microprocessor.

Timing Diagram — ADC0838 Device

Figure 11-107 is the timing diagram for the 6805 to ADC0838 interface. Addressing the analog channel, performing conversion, and retrieving the conversion result requires 420 μs. The timing diagrams for the ADC0832 and ADC0834 devices are similar except fewer input bits are transmitted to the A/D converter. The CLC instruction in the software listing provides sufficient delay to allow the A/D converters to set up the conversion result bits on the DO line before they are read by the microprocessor.

Software — ADC0832, ADC0834, and ADC0838 Devices

The software listing describes the software routines for these A/D converters. The software is written so it can be readily applied to any of the three A/D converters as a subroutine so the designer can access the software quickly.

The software differences for these devices are listed below:

	ADC0832	ADC0834	ADC0838
Send Select Bit 1	No	Yes	Yes
Send Select Bit 0	No	No	Yes

The above differences are accommodated by initializing the accumulator and index register correctly before accessing the subroutine. The software listing and Tables 11-18 through 11-23 describe how to accommodate these differences.

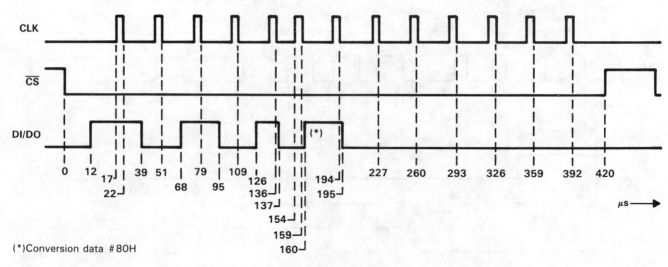

Figure 11-107. Timing Diagram for 6805 to ADC0838 Interface

```
                              ; Software for ADC0832, ADC0834, and ADC0838 to 6805 Interface
                              ;
            ORG  $100H        ; Initialize the starting address for the software
            LDA  #07H         ; Initialize Port A, I/O Pins
            STA  DDR A        ; Configures Port A.0 to A.2 as output pins
            LDA  #A8H         ; For MUX addressing, select appropriate START BIT,
                              ; SGL/DIF(bar), ODD/SIGN, SELECT BIT 1, SELECT BIT 0
                              ; & Load into Acc. Designer can select any MUX
                              ; addressing mode by changing the immediate data.
                              ; See Table 1 to 6 to select the desired MUX
                              ; addressing mode)
            LDX  #05H         ; Set bit counter to 5 or 4 or 3. Designer can
                              ; select any A/D chip out of ADC0838, ADC0834,
                              ; ADC0832.
                              ; Set the immediate date as follows:
                              ;  ADC0838 → Set bit counter to 5
                              ;  ADC0834 → Set bit counter to 4
                              ;  ADC0832 → Set bit counter to 3
            BSR  S83X         ; Load conversion mode bits into Data Input on
                              ; the A/D & Acquire Conversion result into Acc
                              ; from Data out.
                              ; Subroutine S83X
S83X    BCLR 0, Port A        ; Lower CLK
        BCLR 1, Port A        ; Lower Chip Select
S83XRO  ROL  A                ; Shift Conversion Mode bit into C
        BCS  S83XBC           ; If Carry is set ; BRANCH
        BCLR 2, Port A        ; Set 083X DI line to 0
        JMP  S83XJM           ; Go & Raise CLK
S83XBC  BSET 2, Port A        ; Set 083 DI line to 1
S83XJM  BSET 0, Port A        ; Raise CLK
        BCLR 0, Port A        ; Lower CLK
        DECX                  ; Decrement Counter
        BNE  S83XRO           ; Do 5 or 4 or 3 times
        BCLR 2, DDR A         ; Configures Port A.2 as an input pin
        LDX  RO, #08H         ; Set bit counter to 8
S83XI   BSET 0, Port A        ; Raise CLK
        BCLR 0, Port A        ; Lower CLK
        CLC                   ; Initialize C = 0
        BRCLR 2, Port A, S83X ; If 083X Data out = 0; BRANCH
        SEC                   ; 083X Data out = 1; Set C = 1
S83XB   LDA  $20H             ; Get Serial Buffer
        ROL  A                ; Shift Data out Bit into Serial Buffer
        STA  $20H             ; Store Serial Buffer
        DECX                  ; Decrement Counter
        BNE  S83XI            ; Do 8 times
        BSET 1, Port A        ; Raise Chip Select
        LDA  $20H             ; Conversion Data is in Accumulator
        RTS                   ; Return from Subroutine
        END                   ;
```

ADC0832 MUX ADDRESSING (5-BIT SHIFT REGISTER) (See Note 1)

Table 11-18. Single-Ended MUX Mode

START	MUX ADDRESS		CHANNEL NO.		PUT DATA
BIT	SGL/$\overline{DIF}$	ODD/SIGN	0	1	INTO A$_{CC}$
1	1	0	+		# C0H
1	1	1		+	# D0H

Table 11-19. Differential MUX Mode

START	MUX ADDRESS		CHANNEL NO.		PUT DATA
BIT	SGL/$\overline{DIF}$	ODD/SIGN	0	1	INTO A$_{CC}$
1	0	0	+	−	# 80H
1	0	1	−	+	# 90H

NOTE 1: Internally, Select 0 is low. Select 1 is high, COMMON is internally connected to ANLG GND.

ADC0834 MUX ADDRESSING (5-BIT SHIFT REGISTER) (See Note 2)

Table 11-20. Single-Ended MUX Mode

START	MUX ADDRESS			CHANNEL NO.				PUT DATA
BIT	SGL/$\overline{BIT}$	ODD/SIGN	SELECT 1	0	1	2	3	INTO A$_{CC}$
1	1	0	0	+				# C0H
1	1	0	1		+			# D0H
1	1	1	0		+			# E0H
1	1	1	1				+	# F0H

Table 11-21. Differential MUX Mode

START	MUX ADDRESS			CHANNEL NO.				PUT DATA
BIT	SGL/$\overline{BIT}$	ODD/SIGN	SELECT 1	0	1	2	3	INTO A$_{CC}$
1	0	0	0	+	−			# 80H
1	0	0	1		+	−		# 90H
1	0	1	0	−	+			# A0H
1	0	1	1			−	+	# B0H

NOTE 2: Internally, Select 0 is high, COMMON is internally connected to ANLG GND.

ADC0838 MUX ADDRESSING (5-BIT SHIFT REGISTER)

Table 11-22. Single-Ended MUX Mode

START BIT	SGL/DIF	ODD/SIGN	SELECT 1	SELECT 0	0	1	2	3	4	5	6	7	COM	PUT DATA INTO ACC
1	1	0	0	0	+								−	#C0H
1	1	0	0	1			+						−	#C8H
1	1	0	1	0					+				−	#D0H
1	1	0	1	1							+		−	#D8H
1	1	1	0	0		+							−	#E0H
1	1	1	0	1				+					−	#E8H
1	1	1	1	0						+			−	#F0H
1	1	1	1	1								+	−	#F8H

Table 11-23. Differential MUX Mode

START BIT	SGL/DIF	ODD/SIGN	SELECT 1	SELECT 0	0 (0)	0 (1)	1 (2)	1 (3)	2 (4)	2 (5)	3 (6)	3 (7)	PUT DATA ACC
1	0	0	0	0	+	−							#80H
1	0	0	0	1			+	−					#88H
1	0	0	1	0					+	−			#90H
1	0	0	1	1							+	−	#98H
1	0	1	0	0	−	+							#A0H
1	0	1	0	1			−	+					#A8H
1	0	1	1	0					−	+			#B0H
1	0	1	1	1							−	+	#C8H

Circuitry — ADC0831 Device

Figure 11-108 shows the interconnection between the microprocessor and the ADC0831 converter. To assure that the conversion result bits can be read by the microprocessor, the port pin which is assigned to the DO line of the A/D converter must be configured as an input.

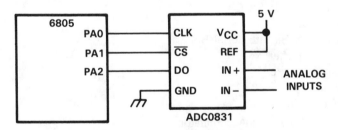

Figure 11-108. 6805 to ADC0831 Interface Circuit Diagram

Timing Diagram — ADC0831 Device

The timing diagram for the 6805 to ADC0831 interface is shown in Figure 11-109. Performing conversion and retrieving the conversion result requires 265 μs. The CLC instruction in the software listing provides sufficient delay to allow the A/D converters to set up the conversion result bits on the DO line before they are read by the microprocessor.

Software — ADC0831 Device

The following software listing presents the routines for this converter. The software is written as a subroutine so the designer can access the software quickly. An initial A/D clock cycle must occur before the eight subsequent clock cycles can be used to extract the conversion result bits from the A/D converter.

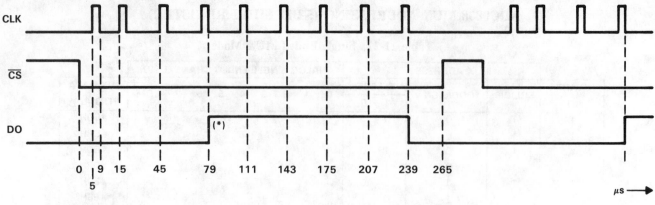

(*)Conversion data #3EH

Figure 11-109. Timing Diagram for 6805 to ADC0831 Interface

```
                              ; Software for 6805 to ADC0831 Interface
                              ;
          ORG  $100H          ; Initialize the starting address for the software
          BSR  S831           ; Branch to the Subroutine
                              ; Load 9 clocks to CLK input on the A/D [
                              ; Acquire the Conversion Result into Accumulator.
                              ;
                              ; Subroutine S831
S831      LDA  #03H           ; Initialize Port A, I/O pins
          STA  DDR A          ; Configures Port A.0 and A.1 as output pins
                              ; and Port A.2 as an input pin
          BCLR 0, Port A      ; Lower CLK
          BCLR 1, Port A      ; Lower Chip Select
          BSET 0, Port A      ; Raise CLK
          BCLR 0, Port A      ; Lower CLK
          LDX  #08H           ; Set bit counter to 8
S831I     BSET 0, Port A      ; Raise CLK
          BCLR 0, Port A      ; Lower CLK
          CLC                 ; Initialize C = 0
          BRCLR 2, Port A, S831B  ; If 0831 Data out = 0; BRANCH
          SEC                 ; 0831 Data out = 1; Set C = 1
S831B     LDA  $20H           ; Get Serial Buffer
          ROL  A              ; Shift Data out Bit into Serial Buffer
          STA  $20H           ; Store Serial Buffer
          DECX                ; Decrement Counter
          BNE  S831I          ; Do 8 times
          BSET 1, Port A      ; Raise Chip Select
          LDA  $20H           ; Conversion Data is in Accumulator
          RTS                 ; Return from Subroutine
          END                 ;
```

INTERFACE FOR ADC0831, ADC0832, ADC0834, AND ADC0838 CONVERTERS TO MOTOROLA 6800, 6802, 6809, AND 6809E MICROPROCESSORS

The ADC083X family of 8-bit successive-approximation A/D converters is designed to communicate easily with microprocessors in a serial mode. This application shows the circuit configurations and the software that can be used to operate this family of microprocessors with the ADC083X converters. Timing diagrams show the interaction between the microprocessor and the A/D converter.

The ADC0832, ADC0834, and ADC0838 A/D converters can be software configured in either the single-ended or differential input mode. Further, the differential inputs can be interchanged through software manipulation. This interface features:

1. Low-cost A/D converter
2. Minimum circuitry
3. Fast conversion and communication between the A/D converter and the microprocessor
4. Remote control advantages of serial A/D converters.

Circuitry — ADC0832, ADC0834, and ADC0838 Devices

Figure 11-110 shows the interconnection between the microprocessor and the ADC0832, ADC0834, and ADC0838 A/D converters. The interconnection is identical for all three converters. The microprocessor DO pin can be used to transmit to and receive digital data from the ADC083X DI and DO pins, respectively. The SN74126 3-state buffer output is in the high-impedance state except during a microprocessor read operation. The SN74174 quad D-type flip-flop is used to synchronize and slow down the write/read communication between the microprocessor and ADC083X converter family so the timing requirements are satisfied. The configuration of AND gates and inverters assures that the flip-flops receive the A/D input data on the negative edge of the E clock. Thus, the flip-flops will receive the correct data. Also, this configuration enables the 3-state buffer and allows the microprocessor to read the A/D DO line when the read/write and address lines are correctly activated. Table 11-24 provides information for adapting the circuit of Figure 11-110 for use with other members of the 6800 family of microprocessors.

Timing Diagram — ADC0838 Device

Figure 11-111 presents the timing diagram for the 6802 to ADC0838 interface. Addressing the analog channel, performing conversion, and retrieving the conversion result requires 449 μs. The timing diagrams for the ADC0832 and ADC0834 converters are similar except fewer input bits are transmitted to the A/D converter. The 3-state buffer enable strobes occur 3 μs after the negative transition of the CLK signal so the conversion result bits have sufficient time to set up on the DO line before these bits are read by the microprocessor.

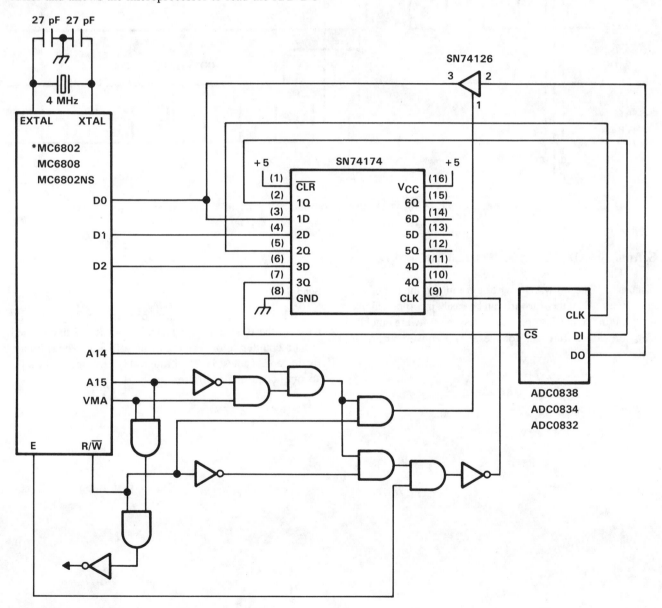

*See Table 11-24 for information about other Motorola microprocessors.

Figure 11-110. Circuit Diagram for the ADC0832, ADC0834, and ADC0838 Interface

11-85

Table 11-24. Adapting Figure 11-121 and Figure 11-124 to Other 6800 Family Microprocessors

MICROPROCESSOR	CRYSTAL OSCILLATOR AND E PIN
6800	ϕ_2 is equivalent to the 6802/6809 E pin. See Figure 11-121 and Figure 11-124.
6802/6808/6802ns	See Figure 11-121 and Figure 11-124
6809	See Figure 11-121 and Figure 11-124
6809E/68HC09E	See Figure 11-121. MC6809 Data sheet for Clock Generator information.

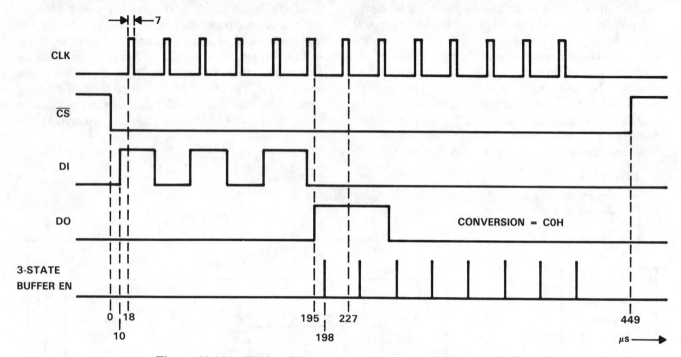

Figure 11-111. Timing Diagram for 6802 — ADC0838 Interface
(1 MHz Microprocessor Clock Cycle)

Software — ADC0832, ADC0834, and ADC0838 Devices

The software listing for these converters follows this discussion. The software is written for any of the three A/D converters. Also, the software can be easily incorporated into a subroutine so the designer can access the software quickly. The software differences for these A/D converters are as follows:

	ADC0832	ADC0834	ADC0838
Send Select Bit 1	No	Yes	Yes
Send Select Bit 0	No	No	Yes

The above differences are accommodated by correctly clearing the B accumulator and a byte of RAM that is used as a bit counter before accessing the software. The software listing and Tables 11-25 through 11-30 describe how to accommodate these differences.

```
; Software for 680X Family to ADC0838, ADC0834, ADC0832 Interface
;
;
                40 00      WRITE:      EQU 4000H          ;Interface write address
                40 00      READ:       EQU 4000H          ;Interface read address
                           ;
0000    C6 15   START:     LDAB #15H              ;Load input address in B, See
                                                  ;Tables 11-25 to 11-30, B0 is sent first
0002    86 0               LDAA #10H              ;ADC0838; Input bit counter = 10H
                                                  ;ADC0834; Input bit counter = 08H
                                                  ;ADC0832; Input bit counter = 04H
0004    97 00   STAA       0000H                  ;Store bit counter in RAM
                ;
0006    86 04   ADC83X     LDAA #04H              ;CS(bar)(A2) = 1, CLK(A1) = 0,
                                                  ;DI/D0(A0) = 0
0008    B7 40 00 STAA      WRITE                  ;Write these values to A/D
000B    86 00   LDAA       #00H                   ;Lower CS(bar)
000D    B7 40 00           STAA WRITE
0010    46      ADC83XI:   RORA                   ;Prepare to load next bit in A
0011    56                 RORB                   ;Put next input bit in carry
0012    49                 ROLA                   ;Load next input bit in A0 and re-
                                                  ;align A1 & A2 so they are correct
0013    B7 40 00           STAA WRITE             ;Set up next bit on DI line
0016    8A 02              ORAA #02H              ;Raise CLK line and clock in,
0018    B7 40 00           STAA WRITE             ;next bit
001B    84 FD              ANDA #FDH              ;Lower CLK line
001D    B7 40 00           STAA WRITE
0020    76 00 00           ROR 0000H              ;Rotate bit counter
0023    24 EB              BCC ADC83XI            ;If carry = 0; branch
0025    86 80              LDAA #80H              ;Conversion bit counter = 8
0027    97 00              STAA 000H              ;Store bit counter in RAM
0029    86 2    ADC83XO:   LDAA #02H              ;Raise CLK & keep CS(bar) = 0
002B    B7 40 00           STAA WRITE
002E    86 00              LDAA #00H              ;Lower CLK line and clock out,
0030    B7 40 00           STAA WRITE             ;next conversion bit
0033    B6 40 00           LDAA READ              ;Put next conversion bit in A0
0036    46                 RORA                   ;Put conversion bit in carry
0037    59                 ROLB                   ;Put conversion bit in B0 and
                                                  ;shift other conversion bits in B
0038    76 00 00           ROR 0000H              ;Rotate bit counter
003B    24 EC              BCC ADC83XO            ;If carry = 0; branch
003D    86 04              LDAA #04H              ;Raise CS(bar)
003F    B7 40 00           STAA WRITE
0042                       END                    ;Conversion result is in B accumulator
```

ADC0832 MUX ADDRESSING (5-BIT SHIFT REGISTER) (See Note 1)

Table 11-25. Single-Ended MUX Mode

| START BIT | MUX ADDRESS | | CHANNEL NO. | | PUT DATA INTO |
	SGL/$\overline{\text{DIF}}$	ODD/SIGN	0	1	ACCUMULATOR B
1	1	0	+		#03H
1	1	1		+	#0BH

Table 11-26. Differential MUX Mode

| START BIT | MUX ADDRESS | | CHANNEL NO. | | PUT DATA INTO |
	SGL/$\overline{\text{DIF}}$	ODD/SIGN	0	1	ACCUMULATOR B
1	0	0	+	−	#01H
1	0	1	−	+	#05H

NOTE 1: Internally, Select 0 is low. Select 1 is high, COMMON is internally connected to ANLG GND.

ADC0834 MUX ADDRESSING (5-BIT SHIFT REGISTER) (See Note 2)

Table 11-27. Single-Ended MUX Mode

| START BIT | MUX ADDRESS | | | CHANNEL NO. | | | | PUT DATA INTO |
	SGL/$\overline{\text{BIT}}$	ODD/SIGN	SELECT 1	0	1	2	3	ACCUMULATOR B
1	1	0	0	+				#03H
1	1	0	1		+			#0BH
1	1	1	0			+		#07H
1	1	1	1				+	#0FH

Table 11-28. Differential MUX Mode

| START BIT | MUX ADDRESS | | | CHANNEL NO. | | | | PUT DATA INTO |
	SGL/$\overline{\text{BIT}}$	ODD/SIGN	SELECT 1	0	1	2	3	ACCUMULATOR B
1	0	0	0	+	−			#01H
1	0	0	1			+	−	#09H
1	0	1	0	−	+			#05H
1	0	1	1			−	+	#0DH

NOTE 2: Internally, Select 0 is high, COMMON is internally connected to ANLG GND.

ADC0838 MUX ADDRESSING (5-BIT SHIFT REGISTER)

Table 11-29. Single-Ended MUX Mode

START BIT	SGL/DIF	ODD/SIGN	SELECT 1	SELECT 0	0	1	2	3	4	5	6	7	COM	PUT DATA INTO ACCUMULATOR B
							ANALOG SINGLE-ENDED CHANNEL NO.							
1	1	0	0	0	+								−	#03H
1	1	0	0	1			+						−	#13H
1	1	0	1	0					+				−	#0BH
1	1	0	1	1							+		−	#1BH
1	1	1	0	0		+							−	#07H
1	1	1	0	1				+					−	#17H
1	1	1	1	0						+			−	#0FH
1	1	1	1	1								+	−	#1FH

Table 11-30. Differential MUX Mode

START BIT	SGL/DIF	ODD/SIGN	SELECT 1	SELECT 0	0	1	2	3	4	5	6	7	PUT DATA INTO ACCUMULATOR B
					0		1		2		3		
1	0	0	0	0	+	−							#01H
1	0	0	0	1			+	−					#11H
1	0	0	1	0					+	−			#09H
1	0	0	1	1							+	−	#19H
1	0	1	0	0	−	+							#05H
1	0	1	0	1			−	+					#15H
1	0	1	1	0					−	+			#0DH
1	0	1	1	1							−	+	#13H

Circuitry — ADC0831 Device

Figure 11-112 shows the interconnection between the microprocessor and the ADC0831 converter. The circuitry is basically the same as for the ADC0832, ADC0834 and ADC0838 interface except the ADC0831 does not have a DI line. Table 11-24 provides information that will help the designer adapt the circuit of Figure 11-112 to other members of the 6800 family of microprocessors.

Timing Diagram — ADC0831 Device

Figure 11-113 shows the timing diagram for the 6802 to ADC0831 interface. Performing conversion and retrieving the conversion result requires 283 μs. The 3-state buffer enable strobes occur 3 μs after the negative transition of the CLK signal so the conversion result bits have sufficient time to set up on the DO line before these bits are read by the microprocessor.

Software — ADC0831 Device

The following software listing provides the software routines for this converter. The software can be easily incorporated into a subroutine so the designer can access the software quickly. An initial A/D clock cycle must occur before the eight subsequent clock cycles can be used to extract the conversion result bits from the A/D converter.

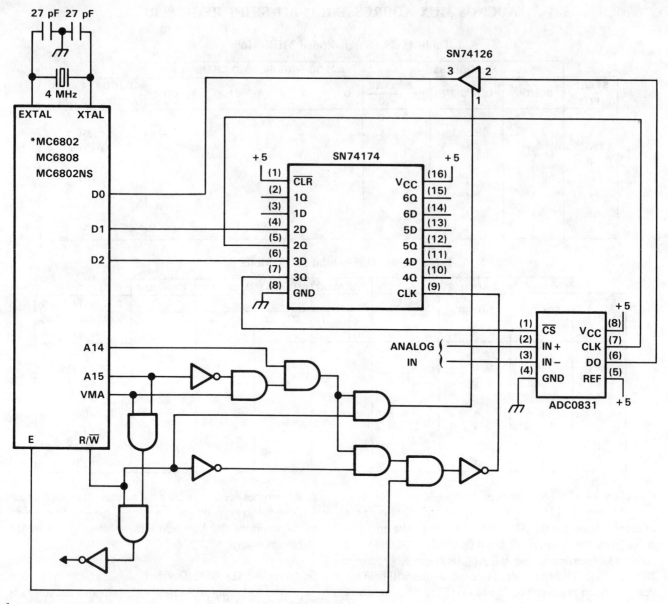

Figure 11-112. Circuit Diagram for the ADC0831 Interface

*See Table 11-24 for information about other Motorola microprocessors.

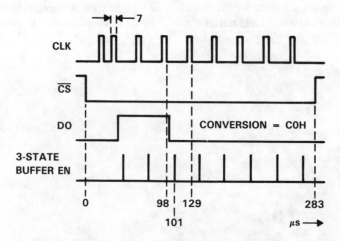

Figure 11-113. Timing Diagram for 6802 to ADC0831 Interface (1 MHz Microprocessor Clock Cycle)

; Software for 680X Family to ADC0831 Interface
;
;

```
                40 00       WRITE:      EQU 4000H          ;Interface write address
                40 00       READ:       EQU 4000H          ;Interface read address
                            ;
    0000        86 04       START:      LDAA #04H          ;CS(bar)(A2)=1, CLK(A1)=0
    0002        B7 40 00                STAA WRITE         ;Write these values to A/D
    0005        86 00                   LDAA #00H          ;Lower CS(bar)
    0007        B7 40 00                STAA WRITE
    000A        86 02                   LDAA #02H          ;Raise CLK initially
    000C        B7 40 00                STAA WRITE
    000F        86 00                   LDAA #00H          ;Lower CLK initially
    0011        B7 40 00                STAA WRITE
    0014        86 80                   LDAA #80H          ;Conversion bit counter = 8
    0016        97 00                   STAA 0000H         ;Store bit counter in RAM
    0018        86 02       ADC8310:    LDAA #02H          ;Raise CLK and keep CS(bar)=0
    001A        B7 40 00                STAA WRITE
    001D        86 00                   LDAA #00H          ;Lower CLK line and clock out,
    001F        B7 40 00                STAA WRITE         ;next conversion bit
    0022        B6 40 00                LDAA READ          ;Put next conversion bit in A0
    0025        46                      RORA               ;Put conversion bit in carry
    0026        59                      ROLB               ;Put conversion bit in B0 and
                                                           ;shift other conversion bits in B
    0027        76 00 00                ROR 0000H          ;Rotate bit counter
    002A        24 EC                   BCC ADC8310        ;If carry=0; branch
    002C        86 04                   LDAA #04H          ;Raise CS(bar)
    002E        B7 40 00                STAA WRITE
    0031                                END                ;Conversion result is in B accumulator
```

INTERFACE FOR ADC0831, ADC0832, ADC0834, AND ADC0838 CONVERTERS TO INTEL 8051 AND 8052 MICROPROCESSORS

This appplication presents the circuit configurations and the associated software that can be used to operate the 8051 and 8052 family of microprocessors with the ADC083X converters. Timing diagrams show the interaction between microprocessor and the A/D converter.

The ADC0832, ADC0834, and ADC0838 converters can be software configured in either the single-ended or differential input mode. Also, the differential ± inputs may be interchanged through software manipulation. This serial interface features:

1. Low-cost A/D converter
2. Direct connection between the A/D converter and the microprocessor
3. Fast conversion and communication between the A/D converter and the microprocessor
4. Remote control advantages of serial A/D converters.

The 8051 and 8052 microprocessor family consists of the following:

8031AH	8052AH
8032AH	80C51
8051AH	8751H

Circuitry — ADC0832, ADC0834, and ADC0838 Devices

Figure 11-114 shows the interface between the microprocessor and the ADC0832, ADC0834, and ADC0838 A/D converters. The interconnection is identical for these converters. A microprocessor port pin can be saved by connecting the A/D converter's DI and DO pins to one microprocessor port pin. However, the designer must be careful to define this port pin as an input pin immediately after the initial input bits have been transmitted to the A/D converter. By doing this, the designer will ensure that the pin will be configured as an input when the data conversion bits are received by the microprocessor.

Timing Diagram — ADC0838 Device

Figure 11-115 presents the timing diagram for the 8051 and 8052 to ADC0838 interface. Addressing the analog channel, performing conversion, and retrieving the conversion result requires only 150 μs. The timing diagrams for the ADC0834 and ADC0832 converters are similar except fewer input bits are transmitted to the A/D converter. The CLR instruction in the software listing provides sufficient delay to allow the A/D converter to set up the conversion result bits on the DO line before they are read by the microprocessor.

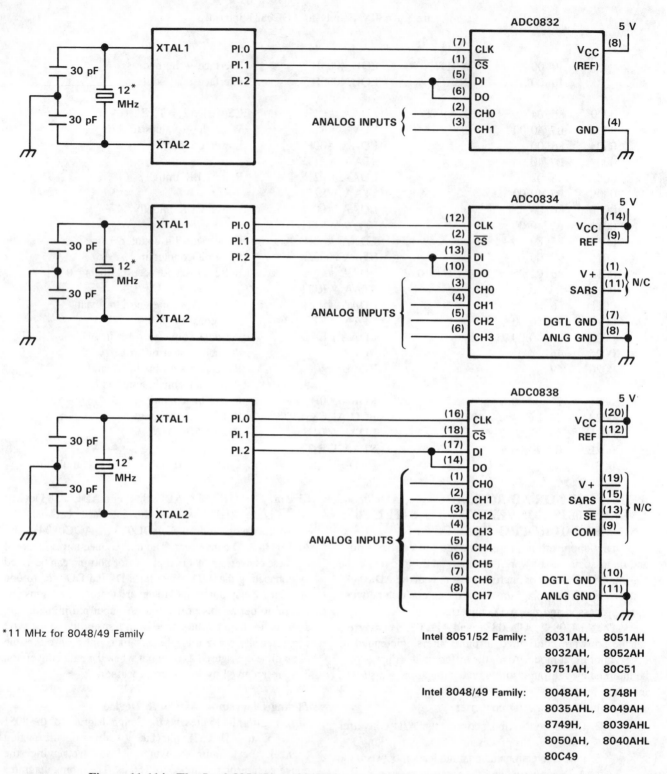

Figure 11-114. The Intel 8051/52 or 8048/49 to ADC0832, ADC0834, and ADC0838 Interface Circuit Diagram

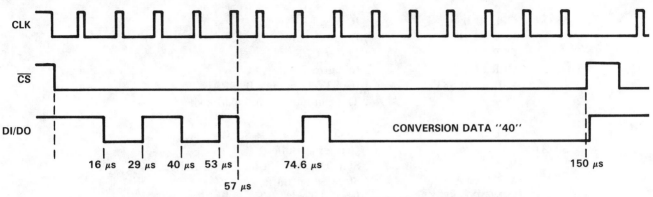

CLK

$\overline{\text{CS}}$

DI/DO

CONVERSION DATA "40"

16 μs 29 μs 40 μs 53 μs 74.6 μs 150 μs

57 μs

Figure 11-115. Timing Diagram for the Intel 8051/52 to ADC0838 Interface

Software

The following software listing provides the software routines for these A/D converters. The software is written so it can be readily applied to any of the three A/D converters and as a subroutine so the designer can access the software quickly.

The software differences for these A/D converters are as follows:

	ADC0832	ADC0834	ADC0838
Send Select Bit 1	No	Yes	Yes
Send Select Bit 0	No	No	Yes

The above differences are accommodated by initializing the accumulator and the R6 register correctly before accessing the subroutine. The software listing and Tables 11-31 through 11-36 describe how to accommodate these differences.

```
                              ; Software for Intel 8051/52 or 8048/49 to ADC0832, 0834, 0838 Interface
                              ;
            MOV A, #A8H       ; For MUX addressing, select appropriate
                              ; START BIT, SGL/DIF(bar), ODD/SIGN, SELECT
                              ; BIT 1, SELECT BIT 0 and Load into Acc.
                              ; Designer can select any MUX addressing
                              ; mode by changing the immediate data
                              ; (See Table 1 to 6 to select the desired
                              ; MUX addressing mode)
            MOV R6, #05H      ; Set bit counter to 5
                              ; Designer can select any A/D chip out of
                              ; ADC0838, ADC0834, ADC0832
                              ; Set the immediate data as follows:
                              ;   ADC0838 → Set bit counter to 5
                              ;   ADC0834 → Set bit counter to 4
                              ;   ADC0832 → Set bit counter to 3
            ACALL S83X        ; Load conversion mode bits into Data Input
                              ; on the A/D chip & Acquire Conversion
                              ; Result into Accumulator from Data Output
                              ;
                              ; Subroutine S83X
            CLR P1.0          ; Lower CLK
S83X        CLR P1.1          ; Lower Chip Select
S83XRL      RLC A             ; Shift Conversion Mode bit into C
            JC S83XJC         ; If Carry is set ; BRANCH
            ANL P1, #FBH      ; Set 083X DI line to 0
            SJMP S83XSJ       ; Go & Raise CLK
S83XJC      ORL P1, #04H      ; Set 083X DI line to 1
S83XSJ      NOP               ; Delay to Set up Mode bit
            CPL P1.0          ; Raise CLK
            NOP               ; Delay to slow CLK
            CLR P1.0          ; Lower CLK
            DJNZ R6, 83XRL    ; Do 5, 4, or 3 times
            ORL P1, #04H      ; Configure P1.2 as an input pin
            MOV R0, #08H      ; Set bit counter to 8
```

```
S83XI    CPL P1.0               ; Raise CLK
         NOP                    ; Delay to slow CLK
         CLR P1.0               ; Lower CLK
         CLR C                  ; Initialize C = 0
         JNB P1.2, S83XJ        ; If 083X Data out = 0; BRANCH
                                ; This pin must be in input mode


         CPL C                  ; 083X Data out = 1; Set C = 1
S83XJ    MOV A,R1               ; Get Serial Buffer
         RLC A                  ; Shift Data out Bit into Serial Buffer
         MOV R1,A               ; Store Serial Buffer
         DJNZ R0, S83XI         ; Do 8 times
         CPL P1.1               ; Raise Chip Select
         MOV A,R1               ; Conversion Data is in Accumulator
         RET                    ;
         END;
```

ADC0832 MUX ADDRESSING (5-BIT SHIFT REGISTER) (See Note 1)

Table 11-31. Single-Ended MUX Mode

START	MUX ADDRESS		CHANNEL NO.		PUT DATA
BIT	SGL/$\overline{\text{DIF}}$	ODD/SIGN	0	1	INTO A$_{CC}$
1	1	0	+		#C0H
1	1	1		+	#D0H

Table 11-32. Differential MUX Mode

START	MUX ADDRESS		CHANNEL NO.		PUT DATA
BIT	SGL/$\overline{\text{DIF}}$	ODD/SIGN	0	1	INTO A$_{CC}$
1	0	0	+	−	#80H
1	0	1	−	+	#A0H

NOTE 1: Internally, Select 0 is low. Select 1 is high, COMMON is internally connected to ANLG GND.

ADC0834 MUX ADDRESSING (5-BIT SHIFT REGISTER) (See Note 2)

Table 11-33. Single-Ended MUX Mode

START	MUX ADDRESS			CHANNEL NO.				PUT DATA
BIT	SGL/$\overline{\text{BIT}}$	ODD/SIGN	SELECT 1	0	1	2	3	INTO A$_{CC}$
1	1	0	0	+				#C0H
1	1	0	1		+			#D0H
1	1	1	0		+			#E0H
1	1	1	1				+	#F0H

Table 11-34. Differential MUX Mode

START	MUX ADDRESS			CHANNEL NO.				PUT DATA
BIT	SGL/$\overline{\text{BIT}}$	ODD/SIGN	SELECT 1	0	1	2	3	INTO A$_{CC}$
1	0	0	0	+	−			#80H
1	0	0	1			+	−	#90H
1	0	1	0	−	+			#A0H
1	0	1	1			−	+	#B0H

NOTE 2: Internally, Select 0 is high, COMMON is internally connected to ANLG GND.

ADC0838 MUX ADDRESSING (5-BIT SHIFT REGISTER)

Table 11-35. Single-Ended MUX Mode

START BIT	MUX ADDRESS				ANALOG SINGLE-ENDED CHANNEL NO.									PUT DATA INTO A_CC
	SGL/DIF	ODD/SIGN	SELECT 1	0	0	1	2	3	4	5	6	7	COM	
1	1	0	0	0	+								−	#C0H
1	1	0	0	1		+							−	#C8H
1	1	0	1	0			+						−	#D0H
1	1	0	1	1				+					−	#D8H
1	1	1	0	0	+								−	#E0H
1	1	1	0	1			+						−	#E8H
1	1	1	1	0						+			−	#F0H
1	1	1	1	1								+	−	#F8H

Table 11-36. Differential MUX Mode

START BIT	MUX ADDRESS				ANALOG DIFFERENTIAL CHANNEL-PAIR NO.								PUT DATA A_CC
	SGL/DIF	ODD/SIGN	SELECT 1	0	0		1		2		3		
					0	1	2	3	4	5	6	7	
1	0	0	0	0	+	−							#80H
1	0	0	0	1			+	−					#88H
1	0	0	1	0					+	−			#90H
1	0	0	1	1							+	−	#98H
1	0	1	0	0	−	+							#A0H
1	0	1	0	1			−	+					#A8H
1	0	1	1	0					−	+			#B0H
1	0	1	1	1							−	+	#B8H

Circuitry — ADC0831 Device

Figure 11-116 shows the interconnection between the microprocessor and the ADC0831 converter. To assure that the conversion result bits can be read by the microprocessor, the designer must configure the microprocessor port pin which is assigned to the DO line of the A/D converter as an input.

Timing Diagram — ADC0831 Device

Figure 11-117 shows the timing diagram for the 8051 and 8052 to ADC0831 interface. Performing conversion and retrieving the conversion result requires only 91 μs. The CLR C instruction in the software listing provides sufficient delay to allow the A/D converter to set up the conversion result bits on the D line before they are read by the microprocessor.

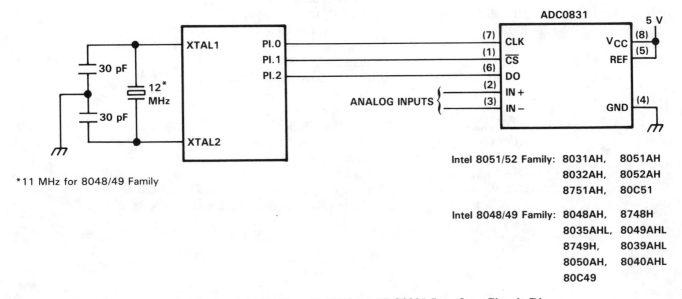

Figure 11-116. The Intel 8051/52 or 8048/49 to ADC0831 Interface Circuit Diagram

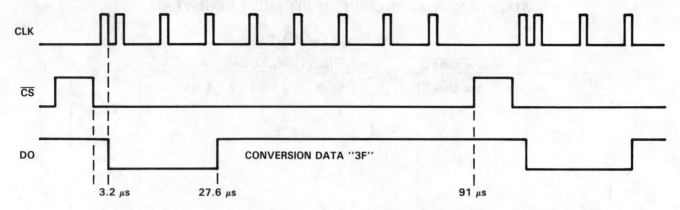

Figure 11-117. Timing Diagram for the Intel 8051/52 to ADC0831 Interface

Software — ADC0831 Device

The following software listing provides the software routines for this A/D converter. The software is written as a subroutine so the designer can access the software quickly.

An initial A/D clock cycle must occur before the eight subsequent clock cycles can be used to extract the conversion result bits from the A/D converter.

```
                              ; Software for Intel 8051/52 Interface to ADC0831
                              ;
                              ;
                              ;
                              ;
              ACALL S831      ; Load 9 clocks to CLK input on the A/D
                              ; chip & Acquire Conversion result
                              ; into Accumulator
                              ;
                              ; Subroutine S831
       S831   ORL P1,04H      ; Configure P1.2 as an input pin
              CLR P1.0        ; Lower CLK
              CLR P1.1        ; Lower Chip Select
              CPL P1.0        ; Raise CLK
              NOP             ; Delay to slow CLK
              CLR P1.0        ; Lower CLK
              MOV R0, #8H     ; Set bit counter to 8
       S831I  CPL P1.0        ; Raise CLK
              NOP             ; Delay to slow CLK
              CLR P1.0        ; Lower CLK
              CLR C           ; Initialize C = 0
              JNB P1.2,S831J  ; If 0831 Data out = 0 ; BRANCH
                              ; This pin must be in input mode
              CPL C           ; 0831 Data out = 1 ; Set C = 1
       S831J  MOV A,R1        ; Get Serial Buffer
              RLC A           ; Shift Data out Bit into Serial Buffer
              MOV R1,A        ; Store Serial Buffer
              DJNZ R0,S831I   ; Do 8 times
              CPL P1.1        ; Raise Chip Select
              MOV A,R1        ; Conversion Data is in Accumulator
              RET             ;
              END             ;
```

INTERFACE FOR ADC0831, ADC0832, ADC0834, AND ADC0838 CONVERTERS TO INTEL 8048 AND 8049 MICROPROCESSORS

Refer to the INTERFACE FOR ADC0831, ADC0832, ADC0834, AND ADC0838 CONVERTERS TO INTEL 8051 AND 8052 MICROPROCESSORS application for additional information regarding these devices. The 8048 and 8049 microprocessor family consists of the following:

8035AHL	8045AH	8748AH
8039AHL	8049AH	8749H
8040AHL	8050AH	80C49

Circuitry — ADC0838, ADC0834, and ADC0832

See Figure 11-114 in the previous application for circuitry also used with this family of converters.

Timing Diagram — ADC0838 Device

Figure 11-118 presents the timing diagram for the 8048 and 8049 to ADC0838 interface. Addressing the analog channel, performing conversion, and retrieving the conversion result requires 309 μs. The timing diagrams for the ADC0834 and ADC0832 converters are similar except fewer input bits are transmitted to the A/D converter.

The NOP and CLR C instructions in the software listing provide sufficient delay to allow the A/D converter to set up the conversion result bits on the DO line before they are read by the microprocessor. Also, these instructions guarantee that occasional spikes on the DI/DO line during clock edges will not cause erroneous readings. These spikes appeared during our testing and may have been due to our test circuit layout. However, these spikes are harmless because they appear during clock edges and the delay time of the NOP and CLR C instructions is longer than the decay time of the spikes. Therefore, the DO line can be read accurately. If the designers feel uncomfortable with these spikes, they may connect the DI/DO line to a microprocessor port other than the port to which the A/D clock line is connected. For example, the A/D converter chip select and clock lines might be connected to port 1 of the microprocessor, while the DI/DO line might be connected to port 2. This use of two different ports eliminates the occasional spike action.

Software — ADC0838, ADC0834, and ADC0832 Devices

See the INTERFACE FOR ADC0831, ADC0832, ADC0834, AND ADC0838 CONVERTERS TO INTEL 8051 AND 8052 MICROPROCESSORS application for more information on these A/D converters. When reading that application, substitute the software listing for this application.

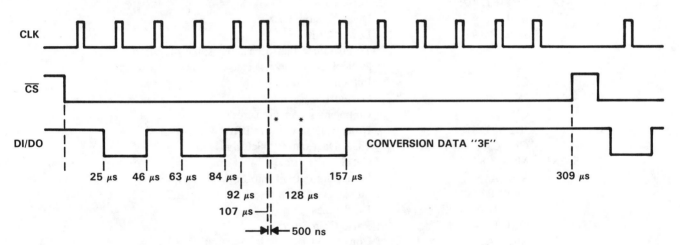

*Designer may see some spikes on the Data Output that do not effect the reading of the conversion results, since the microprocessor reads the DO pin after the spikes occur (see description of timing diagram).

Figure 11-118. Timing Diagram for the Intel 8048/49 to ADC0838 Interface

```
                    ; Software for Intel 8051 and 8052 to ADC8032,
                      ADC8034 and ADC8038 Interface
      MOV A, #A8H     ; For MUX addressing, select appropriate
                    ; START BIT, SGL/DIF(bar), ODD/SIGN, SELECT
                    ; BIT 1, SELECT BIT 0 and Load into Acc.
                    ; Designer can select any MUX addressing

                    ; mode by changing the immediate data
                    ; (See Table 1 to 6 to select the desired
                    ; MUX addressing mode)
```

```
                MOV R6, #05H         ; Set bit counter to 5
                                     ; Designer can select any A/D chip out of
                                     ; ADC0838, ADC0834, ADC0832
                                     ; Set the immediate data as follows:
                                     ; ADC0838 → Set bit counter to 5
                                     ; ADC0834 → Set bit counter to 4
                                     ; ADC0832 → Set bit counter to 3
                CALL S83X            ; Load conversion mode bits into Data Input
                                     ; on the A/D chip & Acquire Conversion
                                     ; Result into Accumulator from Data Output
                                     ;
                                     ;
                                     ; Subroutine S83X
                ANL P1, #FEH         ; Lower CLK
     S83X       ANL P1, #FDH         ; Lower Chip Select
     S83XRL     RLC A                ; Shift Conversion Mode bit into C
                JC S83XJC            ; If Carry is set: BRANCH
                ANL P1, #FBH         ; Set 83X DI line to 0
                JMP S83XSJ           ; Go & Raise CLK
     S83XJC     ORL P1, #04H         ; Set 83X DI line to 1
     S83XSJ     NOP                  ; Delay to Set up Mode bit
                ORL P1, #01H         ; Raise CLK
                NOP                  ; Delay to slow CLK
                ANL P1, #FEH         ; Lower CLK
                DJNZ R6,S83XRL       ; Do 5, 4, or 3 times
                MOV R0, #08H         ; Set bit counter to 8
     S83XI      ORL P1, #01H         ; Raise CLK
                NOP                  ; Delay to slow CLK
                ANL P1, #FEH         ; Lower CLK
                NOP                  ; To guarantee that occasional spike on DI/DO
                                     ; at clock edges do not cause erroneous readings
                CLR C                ; Initialize C = 0
                IN A,P1              ; Get Port 1
                CPL A                ; Complement Accumulator
                JB2 S83XJ            ; If 083X Data out = 1: BRANCH
                CPL C                ; 083X Data out = 1: Set C = 1
     S83XJ      MOVA,R1              ; Get Serial Buffer
                RLC A                ; Shift Data out Bit into Serial Buffer
                MOV R1,A             ; Store Serial Buffer
                DJNZ R0,S83XI        ; Do 8 times
                ORL P1, #02H         ; Raise Chip Select
                MOV A,R1             ; Conversion Data is in Accumulator
                RET                  ;
                END                  ;
```

Circuitry — ADC0831 Device

See the INTERFACE FOR ADC0831, ADC0832, ADC0834, AND ADC0838 CONVERTERS TO INTEL 8051 AND 8052 MICROPROCESSORS application.

Timing Diagram — ADC0831 Device

Figure 11-119 shows the timing diagram for the 8048 and 8049 to ADC0831 interface. Performing conversion and retrieving the conversion result requires only 207 μs. The CLR instruction in the software listing provides sufficient delay to allow the A/D converter to set up the conversion

result bits on the DO line before they are read by the microprocessor.

Software — ADC0831 Device

The following software listing provides the software routines for this A/D converter. The software is written as a subroutine so the designer can access the software quickly. An initial A/D clock cycle must occur before the eight subsequent clock cycles can be used to extract the conversion result bits from the A/D converter.

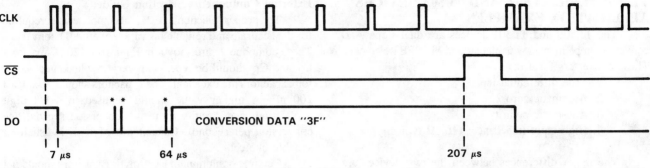

*Designer may see some spikes on the Data Output that do not effect the reading of the conversion results, since the microporcessor reads the DO pin after the spikes occur (see description of timing diagram).

Figure 11-119. Timing Diagram for the Intel 8048/49 to ADC0831 Interface

```
                                    ; Software for Intel 8048/49 Interface to ADC0831
                                    ;
                                    ;
                                    ;
                CALL S831           ; Load 9 clocks to CLK input on the A/D
                                    ; chip & Acquire Conversion result
                                    ; into Accumulator
                                    ;
                                    ; Subroutine S831
        S831    ORL P1,#04H         ; Configure P1.2 as an input pin
                ANL P1,#FEH         ; Lower CLK
                ANL P1,#FDH         ; Lower Chip Select
                ORL P1,#01H         ; Raise CLK
                NOP                 ; Delay to slow CLK
                ANL P1,#FEH         ; Lower CLK
                MOV R0,#08H         ; Set bit counter to 8
        S831I   ORL P1,#01H         ; Raise CLK
                NOP                 ; Delay to slow CLK
                ANL P1,#FEH         ; Lower CLK
                NOP                 ; To guarantee that occasional spikes on
                                    ; DI/DO line at clock edges do not cause
                                    ; erroneous readings
                CLR C               ; Initialize C = 0
                IN A,P1             ; Get Port 1
                CPL A               ; Complement Accumulator
                JB2 S831J           ; If 0831 Data out = 1 ; BRANCH
                CPL C               ; 0831 Data out = 1 ; Set C = 1
        S831J   MOV A,R1            ; Get Serial Buffer
                RLC A               ; Shift Data out Bit into Serial Buffer
                MOV R1,A            ; Store Serial Buffer
                DJNZ R0,S831I       ; Do 8 times
                ORL P1,#02H         ; Raise Chip Select
                MOV A,R1            ; Conversion Data is in Accumulator
                RET                 ;
                END                 ;
```

TL500, TL501, TL502, AND TL503 DEVICES APPLICATION EXAMPLES

The TL500 and TL501 devices are dual-slope A/D converters implemented with bipolar and MOSFET elements. These converters feature:

1. True differential inputs
2. Automatic zero
3. Automatic polarity
4. High input impedance: (10^9 Ω typically).

The major differences between the two devices are given in Table 11-37.

Table 11-37. Major Differences Between the TL500 and the TL501

	TL500	TL501
Resolution	14 bits (with TL502)	10-13 bits (with TL502)
Linearity Error	0.001%	0.01%
Readout Accuracy	4-1/2 Digits (with external ref.)	3-1/2 Digits

The TL502 and TL503 devices are digital processors, or control circuits, that contain an oscillator circuit and output devices. These are monolithic circuits using I^2L and bipolar techniques. These digital processing devices feature:

1. Fast display scan rates
2. Internal oscillator (free-running or driven)
3. Interdigit blanking
4. Overrange blanking
5. 4-1/2 digit display circuitry
6. High sink-current digit drivers.

The major differences between the two digital processors are given in Table 11-38.

Table 11-38. Major Differences Between the TL502 and the TL503

TL502	TL503
Compatible with Seven-Segment Common-Anode Displays	Multiplexed BCD Outputs
High Sink-Current Segment Driver for Large Displays	High Sink-Current BCD Outputs

The dual-slope technique used on the TL500 and TL501 devices results in a high degree of noise rejection due to input voltage integration. This means an expensive sample-and-hold circuit can be avoided, even when the input has a high noise content.

When compared to successive-approximation A/D converters, the speed of the dual-slope technique (less than 150 conversions per second) seems slow. However, this speed is generally adequate for most industrial applications such as temperature measurement or digital panel meters that depend upon the human eye for reading speed.

External Components Selection Guide

The proper selection of external components is required for the accurate and reliable operation of an A/D converter. These components are shown in Figure 11-120. Capacitors C_Z and C_R should be types with very low leakage; plastic foil capacitors are excellent. The capacitors should be at least 200 nF. A higher value should be used in 4-1/2 digit applications and circuits with low conversion speeds (one conversion per second). Capacitance of 1 μF is usually a suitable choice.

Ceramic or aluminum electrolytic capacitors should not be used because of their high leakage. This also applies when choosing C_X, particularly in 4-1/2 digit applications when a capacitor with very low dielectric loss should be used. Polypropylene capacitors are well suited to this application.

The value of resistor R_X should be in the range of 15 kΩ to 100 kΩ; a carbon film resistor would be suitable. The minimum RC time constant for the integrator can be calculated using the following equation:

$$TC = R_X \times C_X \times \frac{V_{EMAX} \times t_1}{(V_{OMAX}) - V_{CM}}$$

In the above equation, V_{EMAX} represents the maximum input voltage between pins 1 and 2. When using the TL502 or TL503 device, this value will be 200 mV. V_{OMAX} is the maximum output voltage of the integrator at pin 13. With a supply voltage of ± 12 V, this should be +8 V or −5 V. V_{CM} is the common-mode voltage between pin 2 and the analog ground as indicated in Figure 11-121.

If a differential input is not required, connecting pin 2 to analog ground will set V_{CM} to zero. The duration of the V_E integration phase is equal to t_1. When using the TL502 or TL503 device, t_1 can be calculated from the oscillation frequency:

$$t_1 = \frac{20,000}{f_{osc}}$$

Example: A value of 470 pF for C_{osc} results in a frequency of approximately 160 kHz. Thus, the period of t_1 is 125 ms. (The oscillator circuit is part of the logic control circuit in TL502 or TL503, and is discussed later.)

Resistor R_E and capacitor C_E are not strictly necessary to the operation of the converter as resistor R_E constitutes a protection circuit for the analog inputs. If the input voltages cannot be guaranteed to remain below the value of the supply voltage V_{CC}, then R_E should be included to limit the current which will flow through the TL500 and TL501 internal clamp diodes. The value of R_E should not exceed 100 kΩ.

Each analog input is connected through an analog switch to the high-impedance buffer or integrator inputs. These switches are MOSFET devices having fixed drain-gate capacitance. If the inputs to the switches are also high impedance, then significant feed-through or interaction may occur between the switch control signals and the input voltages. Eventual converter errors can be avoided if the input impedance is decreased using capacitor C_E with a capacitance of 10 to 100 nF.

The reference voltage at pin 4, which is 100 mV for an input voltage range of 200 mV, may be derived from the internal reference voltage (REF OUT) through a voltage divider. The total resistance of the divider should be within the range of 1 kΩ to 10 kΩ. For applications that require high precision, a temperature-compensated reference voltage source is recommended. If the value of R_X is within the recommended range of 15 kΩ to 100 kΩ, the output impedance of the reference voltage source should not exceed 2 kΩ.

Supply voltages for the TL500 and TL501 devices may be selected within the range of 5 V to 18 V and -8 V to -18 V. Operation at voltages approaching the lower limits is not recommended and should only be considered for applications which do not require high resolution.

Both the common-mode range and the maximum output voltage of the integrator are approximately 3 V. In the case of 4-1/2 digit applications, a power supply of ± 12 V minimum is recommended because of the eventual increase in the dynamic range of the integrator.

The analog and digital ground connections of the TL500 and TL501 devices are internally isolated from each other. The digital ground connection serves as a reference point for the A and B control inputs and the comparator output. The operation of the TL500 and TL501 devices is guaranteed if the potential of the digital ground is within the range of $-V_{CC}$ to $+V_{CC}$ minus 4 V.

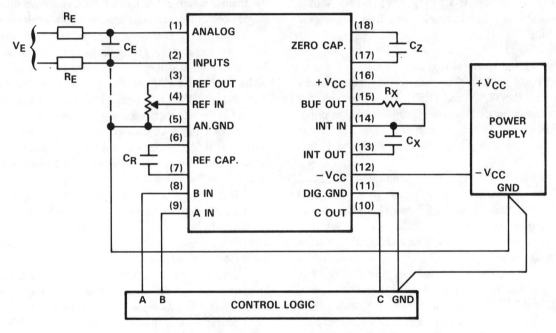

Figure 11-120. TL500/501 Wiring Diagram

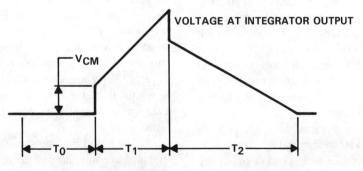

V_{CM} = VOLTAGE BETWEEN ANALOG GROUND (PIN 5) AND ANALOG INPUT (PIN 2).

Figure 11-121. Example With Constant Common-Mode Voltage

Printed Circuit Board Layout Notes

When constructing an A/D circuit on a printed circuit board, the layout has a significant effect on accuracy. Coupling of the digital control signals to the analog components can often be the cause of unexplainable errors. The following points should be observed.

It is essential that the TL500 and TL501 supply voltages be adequately decoupled. Tantalum capacitors are recommended and should be located as close as possible to the device. The analog and digital grounds should be connected together at the power supply as illustrated in Figure 11-120. This ensures that no current from the digital circuitry can flow along the analog ground connections. The ground connection of the internal reference voltage-divider network or an external reference voltage should be connected directly to the analog ground pin of the TL500 and TL501 devices. The connections to C_R, C_Z, C_X, and R_X should be kept as short as possible and isolated from any digital control wiring.

TL502 and TL503 Control Circuits

The TL502 and TL503 contain not only the necessary logic to control the TL500 and TL501 A/D converters, but also have outputs for driving a 4-1/2 digit display in either the 7-segment or BCD code respectively. The TL502 and TL503 devices are also suitable for controlling the TL505 device.

Figure 11-122 shows the block diagram of the TL502 or TL503 device connected to a TL500 or TL501 A/D converter. The oscillator in the TL502 and TL503 can operate either as a free-running oscillator with an external capacitor connected between the oscillator input and digital ground or may be driven by an external TTL signal. The oscillation frequency when operating in the free-running mode is approximately 160 kHz with a 470 pF external capacitor. The TL502 and TL503 devices require 80,000 cycles from the oscillator for a complete conversion. If a 470 pF capacitor is used, the result is a conversion speed of about 2 measurements per second.

For 4-1/2 digit applications, an external oscillator is recommended. This oscillator should not have any short-term frequency variations, although long-term variations do not have a direct effect on the conversion result. To obtain a stable display, it is necessary to reject the power line frequency. It is desirable that the oscillation frequency be an even multiple of the power line frequency. The external oscillator should be synchronized.

The oscillation frequency range is approximately 10 kHz to 1 MHz. Below 10 kHz the display begins to flicker, and because of the long conversion time, this low frequency is not recommended. Above 1 MHz, reliable operation of the TL502 and TL503 devices is no longer guaranteed for all operating conditions. Higher conversion speeds can be achieved with the A/D converters if a fast TTL or microprocessor control is used.

Logic Inputs and Outputs

The TL502 and TL503 devices have two inputs, trigger and comparator, that can control a total of five different functions.

The conversion may be interrupted using the trigger input; a logic low (0) at the trigger input will cause the converter to stop at the beginning of the next Auto Zero phase. Under these conditions, logic outputs A and B are reset low and the previous conversion result is displayed. However, the internal operation of the device continues, and if the trigger input receives another high level, a new conversion begins with the next Auto Zero phase.

If the trigger input rises more than approximately 2.5 V above V_{CC}, the device enters a test mode in which the operation of the multiplexer is inhibited and all the segments or BCD outputs are activated. Certain decades of the main counter are bypassed relative to the digit output under control to accelerate the operation of the controller.

The comparator input receives information from the A/D converter regarding the polarity of the unknown voltage. This information determines whether the A or B output is set low in the following phase and when a positive or negative reference voltage is integrated. The control logic recognizes the end of the integration phase when the switching threshold of the comparator is crossed. The value from the main counter is then transferred to a register, and the A and B logic outputs are set low. Figure 11-123 shows a typical circuit with both comparator and trigger inputs.

If the comparator input rises more than approximately 2.5 V above V_{CC}, a display test begins but the control logic continues to operate. However, normal conversion is not guaranteed in this case because comparator information cannot be communicated.

If signals are applied simultaneously to both the trigger and comparator inputs, the logic is reset. When this mode ends, control begins again with the start of the next Auto Zero phase; until then, the outputs display 19999.

The TL502 device has an oscillator output that is labelled 20,000. The frequency of this output is the oscillator frequency divided by a factor of 20,000 and has a mark-space ratio of 1:4. This output is suitable for controlling a phase locked loop (PLL) circuit so the oscillator frequency is locked to an even multiple of the line frequency. Figure 11-124 shows the block diagram of a typical PLL circuit. This type of circuit is recommended when a stable display is required at relatively high conversion rates that will remain stable, even when the input voltage has superimposed line frequency hum.

The A and B outputs, as well as the 20,000 oscillator output, have internal pull-up resistors; therefore, they are TTL compatible and the A and B outputs can be connected directly to the relevant inputs of the TL500 and TL501 devices.

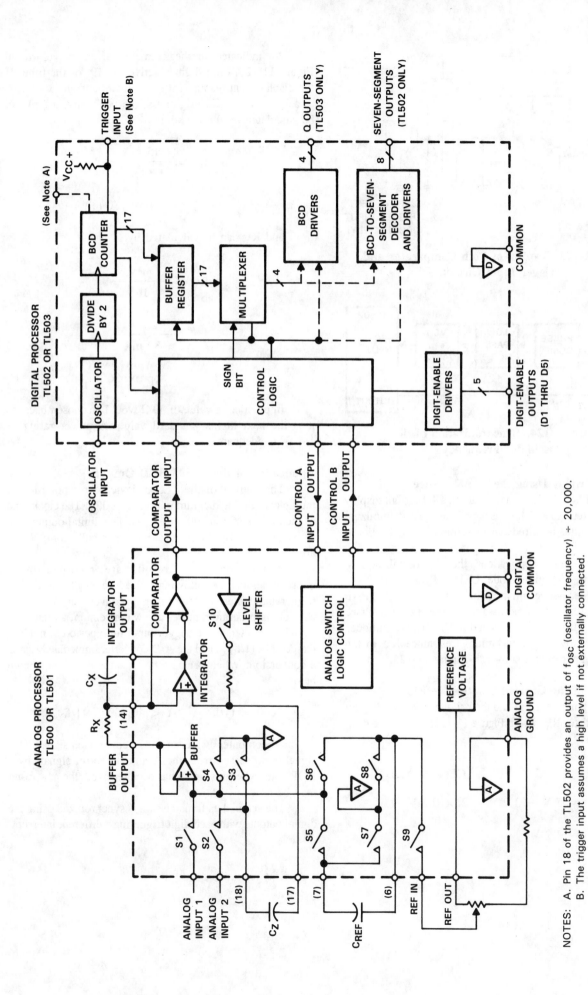

Figure 11-122. Block Diagram of Basic Analog-to-Digital Converter Using TL500C or TL501C and TL502C or TL503C

NOTES: A. Pin 18 of the TL502 provides an output of f_{OSC} (oscillator frequency) ÷ 20,000.
B. The trigger input assumes a high level if not externally connected.

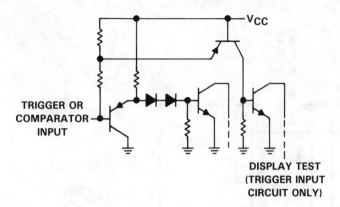

Figure 11-123. Typical of Both Comparator and Trigger Input Circuits

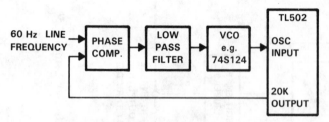

Figure 11-124. Synchronization of the Oscillator Frequency

Driving a Display Using the TL502 Device

The TL502 device can drive up to a 4-1/2 digit display in a multiplexed mode. This has the advantage of minimizing the device outputs required and also simplifies the PC board layout.

In most simple applications, the display will consist of 7-segment, common-anode LEDs. Because the I^2L technology does not permit the manufacture of suitable PNP drive transistors, it is necessary to use external drive transistors to drive each digit anode. The display cathodes are connected by a current-limiting resistance R_V to the relevant segment output shown in Figure 11-125.

Calculation of the Current-Limiting Resistor

The segment current and the brightness of the display are set by resistor R_V. From Figure 11-125, the voltage drop across R_V is:

$$V_{RV} = V_{CC} - V_F - V_{CEsat} - V_{OL}$$

$$= 5\ V - 1.7\ V - 2 \times 0.2\ V$$

$$= 2.9\ V$$

As indicated in the segment current waveform in Figure 11-126, each digit is active for 19ᶜ of the time. If the display requires an average segment current of 15 mA to achieve a specific brightness, the peak current (Ip) can be calculated by the equation:

$$I_P = \frac{15\ mA}{19\%}$$

$$= \text{approx. } 80\ mA$$

and R_V can be calculated by:

$$R_V = \frac{V_{RV}}{I_P}$$

$$= \frac{2.9\ V}{80\ mA}$$

$$= \text{approx. } 36\ \Omega$$

In practice, a value of 39 Ω would be chosen because it is the next higher standard value. A power rating of 1/4 W is adequate.

Application of the TL503 BCD Outputs

The outputs of the TL503 device correspond to the multiplexer timing diagram in Figure 11-126. The clock time of the multiplexer is 200 x $(1/f_{osc})$. The digital outputs are active low. The BCD outputs are the open-collector type and are active high.

If the conversion results are recorded by a microprocessor, the following procedure may be used. The trigger input must be low prior to conversion. When the controlling program requires a new conversion, a logic high signal is directed to the trigger input for a period of not less than 2000 x $(1/f_{osc})$ and the A/D converter immediately starts a conversion. When period t satisfies the condition shown below

$$2000 \times (1/f_{osc}) < t < 40,000 \times (1/f_{osc}),$$

the microcomputer awaits an end of conversion signal. This signal is processed from the A and B control signals by a NOR gate. After this signal has been received, the new result may be used.

The actual data transmission is synchronized using the digital outputs with each digit triggering a different interrupt.

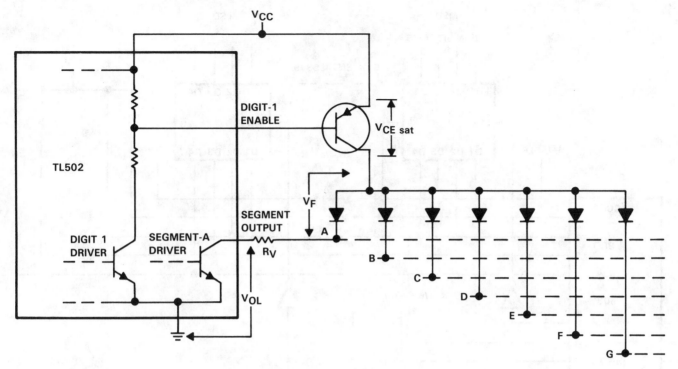

Figure 11-125. Display Driving

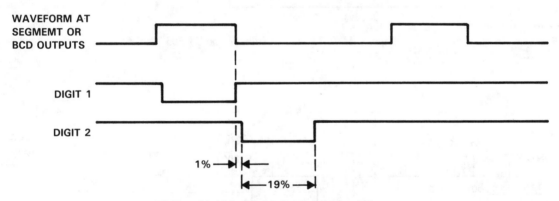

Figure 11-126. Multiplexer Timing Diagram

This guarantees the integrity of the BCD information. The minimum time required for a complete transfer, therefore, corresponds to the duration of the multiplexer period. This time can be decreased if two or more bits of digital information are combined. It is possible to combine all 18 bits using the demultiplexer shown in Figure 11-127. The BCD information from each digit is transmitted to the quadruple D-type flip-flop using the negative edge of the digital outputs. Figure 11-126 indicates that this negative edge is available after a time delay of two clock cycles. If required, the inputs to the flip-flops may be inhibited during the read process by using the enable inputs.

Digital Panel Meter Using the TL501 and TL502 Devices

The input voltage range of the digital panel meter shown in Figure 11-128 is ±2.0 V and when using the component values shown, a conversion rate of approximately 2 conversions per second is produced. Other conversion rates can be achieved by using different size capacitors for C_O and C_X. Pins 2 and 5 on the TL501 device may be jumpered together when no common-mode input voltage exists. The equation for selecting values for R_X and C_X is given in the External Components Selection Guide section. When necessary, an RC time constant protection network may be connected to the input.

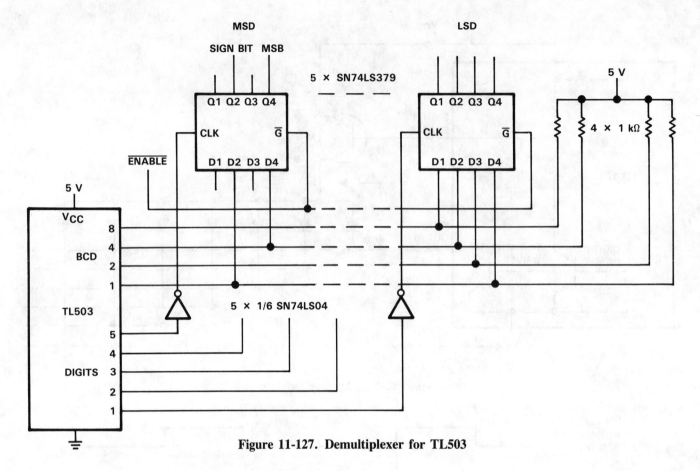

Figure 11-127. Demultiplexer for TL503

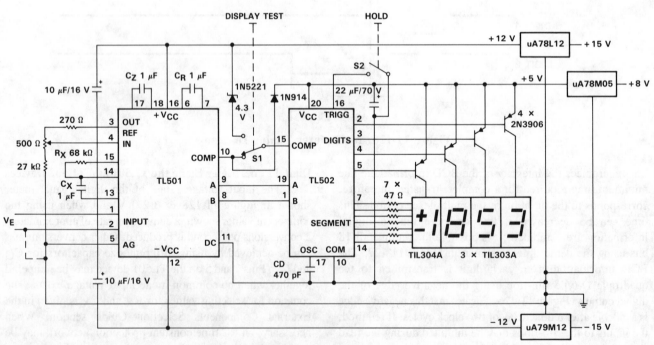

Figure 11-128. Digital Panel Meter Using TL501

Note that the recommended use of separate grounds is shown in Figure 11-128. It is essential that the analog and digital grounds be connected only at the power supply.

Switch S1 provides a means of activating the display test mode. The required 7.5-V supply is provided from the 12 V line using a 1N5221 zener diode to provide a 4.3-V drop. A changeover switch is required because the comparator output of the TL501 device is not short circuit protected. Remember that after a display test, the display contains an invalid result.

By operating the HOLD switch, S2, conversion results can be stored or readings can be manually observed. The different pin-out of the TIL304A device (sign and MSB digits) means that pins 1 and 14 must be connected to the digit driver. Pin 7 connects to the limiting resistor of segment E or F and pin 8 connects to the limiting resistor of segment G.

Digital Thermometer

The digital thermometer diagram shown in Figure 11-129 is an application that uses the differential inputs of the TL500 or TL501 device to measure a voltage from a bridge circuit. When the circuit is calibrated, the voltage that represents an actual temperature is indicated on the LED readout.

The temperature sensor is a silicon thermistor with a positive temperature coefficient and a low time constant (less than 1.5 seconds). Excluding linearity of the sensor, the linearity error of the circuit is less than 0.5% over the temperature range of $-55\,°C$ to $+125\,°C$.

The calibration of the circuit is straightforward. First, potentiometer P0 is adjusted for the zero point. This is accomplished by adjusting P0 to a preset voltage of 0.258 x V_{REF} V (approximately 0.31 V) at the wiper of P0. Second, the thermometer is calibrated by applying a known temperature to the thermistor and adjusting potentiometer P1

for about 0.153 x V_{REF} V (approximatley 0.19 V) at the wiper of P1.

The decimal point is fixed between digits 2 and 3 by connecting the relevant cathode of the display to digital ground through a 56-Ω resistor.

Precision Panel Meter Application

Figure 11-130 shows a 4-1/2 digit precision panel meter using the TL500 and TL502 devices. The input voltage range is ± 200 mV. This means that current measurements are possible with only an additional shunt resistor.

The use of a precision reference voltage source is recommended so the circuit may be used over a wide temperature range. The 100 mV reference voltage is obtained from a TL431 shunt regulator and a voltage divider comprised of a 250-Ω cermet trimmer potentiometer and a 2500-Ω metal film resistor. No allowance has been made for a common-mode input voltage. If this is required, the connection from pin 2 to pin 5 should be opened and a larger capacitor inserted.

The TL502 device is controlled by a NE555 external oscillator. Its frequency is adjusted to an even multiple of the power line frequency to guarantee a high degree of power line frequency suppression.

The latch, implemented with an SN74LS02 device, increases the performance of the circuit at input voltages of almost zero. If the input voltage is very low, the integrator of the TL500 device can only produce a small ramp signal. In this case, low-level interfering signals are sufficient to cause an incorrect comparator switching sequence. This result is especially noticeable during the switch-over from the up-integration phase to the down-integration phase. The comparator signal, delayed by one-half of a clock cycle, is directed to the control logic through the latch. Thus, interference impulses caused by crosstalk from the A and B control signals no longer have any influence on the comparator input.

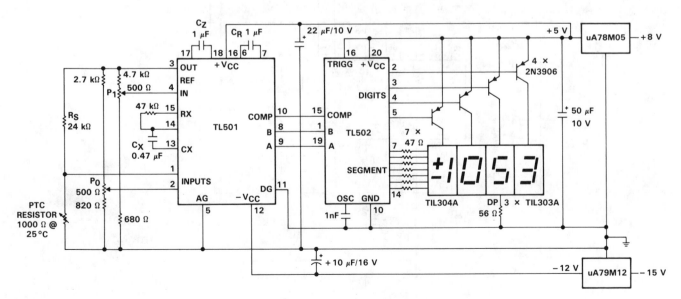

Figure 11-129. Digital Thermometer

11-107

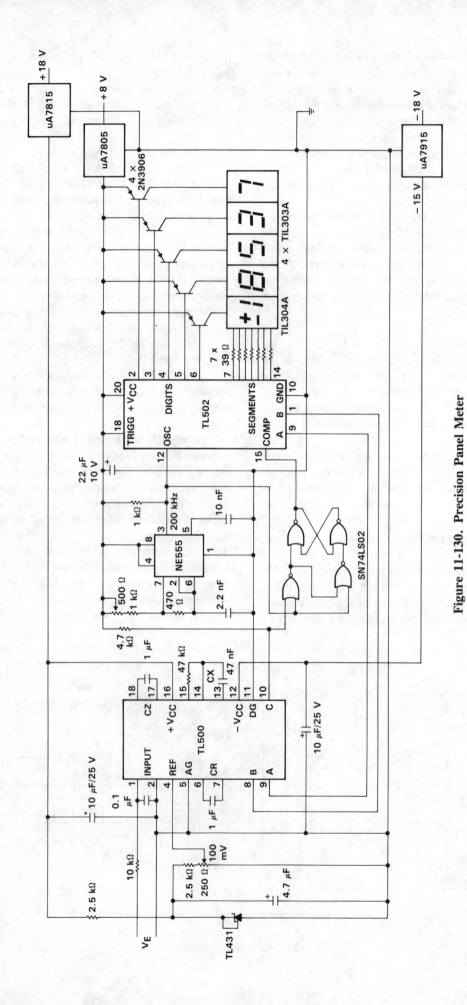

Figure 11-130. Precision Panel Meter

The suggested display is composed of four TIL303A devices and a single TIL304A device. When connecting the TIL304A device, the minus sign should be connected to segment outputs A and D. All other connections are identical to the TIL303A device.

TL505 A/D CONVERTER APPLICATIONS

The TL505 device is a dual-slope A/D converter with 10-bit resolution and 3-digit accuracy. It has a high-impedance MOSFET input. It was developed for use with a microcomputer, but can be used in combination with discrete control logic such as the TL502 or TL503 devices. The TL505 device requires only a single supply voltage in the 7- to 15-V range.

Functional Description

Figure 11-131 shows a simplified functional circuit diagram of the TL505 device. The dual-slope principle has already been described using the TL500 device; therefore, only the functional differences of the TL505 device will be described. Figure 11-132 describes the function of the A and B control inputs of the TL505 device with reference to the switches in Figure 11-131.

The TL505 device, like the TL500 device, starts a conversion during the Auto Zero phase when all offset error voltages are stored on capacitors C_X and C_Z. However, the operation of the TL505 device differs from the TL500 device in the integration phase that follows the Auto Zero phase. During the integration phase, operational amplifier A1 functions as a noninverting integrator. The input voltage is

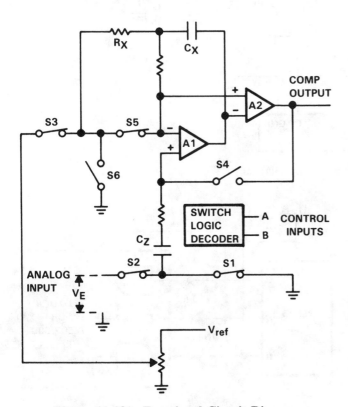

Figure 11-131. Functional Circuit Diagram for the TL505

CONTROL INPUTS		SWITCHES CLOSED	FUNCTION
A	**B**		
0	0	1,4,5,6	Auto Zero
0	1	1,3	Integration of reference voltage
1	0	1	Hold
1	1	2,4	Integration of input voltage

Figure 11-132. Function of the A and B Control Inputs of the TL505

switched through S2 and C_Z to the high-impedance input; therefore, an additional buffer stage is not required. In the following down-integration phase, A1 is used as an inverting integrator. A positive reference voltage is connected to the input through S3 so that a negative-going waveform appears at the output. This phase lasts until the switching threshold of the comparator A2 is reached and the control supplies a logic low signal to both the A and B inputs of the TL505 device. Figure 11-133 shows the conversion process timing diagram and its function table.

External Component Selection Guide

The suggestions given on component selection for the TL500 and TL501 circuits are also valid for the TL505 A/D converter. Figure 11-134 shows the TL505 device wiring diagram. Capacitor C_Z must be a low-leakage type with a minimum capacitance of 150 nF. Capacitor C_X must also be a low-leakage type. Ceramic or electrolytic capacitors should not be used. Resistance R_X is not critical and a carbon film resistor in the range of 0.5 to 2.0 MΩ is adequate. The value of capacitor C_X may be calculated using the following formula:

$$C_X \geq \frac{V_{EMAX} \times t_1}{R_X(V_{CC} - V_{EMAX} - 2\text{ V})}$$

V_{EMAX} is the maximum input voltage and a value of 2 V should be assumed if the TL502 or TL503 devices are used as the control logic. For safe operation, the maximum input voltage must not exceed 4 V when the TL505 device is used in other applications not involving the TL502 and TL503 devices. The factor $(V_{CC} - V_{EMAX} - 2\text{ V})$ represents the maximum voltage variation at the integrator output and time t_1 represents the duration of the V_E integration phase. The internal resistance of the reference voltage source connected to pin 4 should be as low as possible because this resistance will influence the conversion result due to the operation of the TL505 device.

If the value of R_X lies within the range of 0.5 to 2.0 MΩ, the output impedance of the reference voltage source should not exceed 2 kΩ. In most cases when the TL502 or TL503 device is used as the control logic, a reference voltage will be 1 V and the total resistance of the reference voltage divider must not exceed 10 kΩ.

Resistor R_E and capacitor C_E constitute a protection circuit for the input similar to the TL500 and TL501 devices.

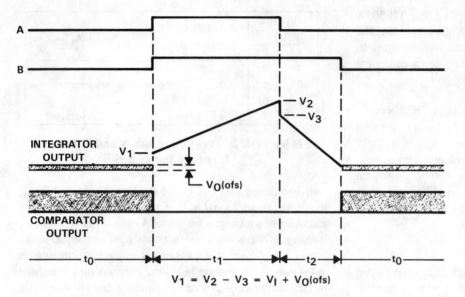

FUNCTION TABLE

CONTROLS		ANALOG
A	B	SWITCHES CLOSED
L	L	S1, 4, 5, 6
H	H	S2, 4
L	H	S1, 3

$H = V_{IH}, L = V_{IL}$

$$V_1 = V_2 - V_3 = V_I + V_{O(ofs)}$$

Figure 11-133. Conversion Process Timing Diagrams

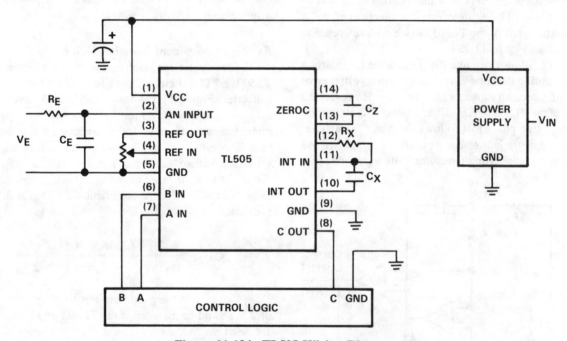

Figure 11-134. TL505 Wiring Diagram

It should be noted that the logic inputs of the TL505 device are not TTL compatible. Figure 11-135 shows a simple interface circuit which should be used with the TL505, TL502, and TL503 device combination.

Digital Panel Meter Using the TL505 and TL502 Devices

A typical application of the TL505 device is given in Figure 11-136 and shows a digital panel meter for positive input voltages. The single-point ground method is the best way to ensure acceptable results.

The values shown are for a maximum input voltage of 2 V and a conversion rate of approximately 2.5 conversions per second. The RC time constant circuit protection network connected to the input is optional. The

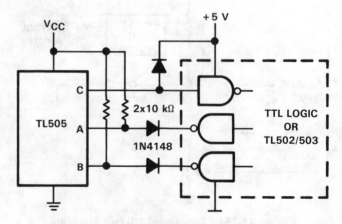

Figure 11-135. TL505C — TTL Interface

diodes in the A and B control lines ensure that the input voltage limits on the A and B inputs of the TL505 device are not exceeded. The diode at the comparator input of the TL502 device prevents any spurious initiation of the display test mode.

TL505 to TMS1000 Interface

Figure 11-137 shows an application of a TL505 device to TMS1000 device series interface. The microcomputer may be any one of the TMS1000 series requiring a 15-V power supply. A 6.8-V zener diode is used as an interface between the comparator output and the TMS1000 input. This results in the minimum possible standing current in the low state while maintaining a high noise immunity in the high state. The two 1N914 diodes prevent the TL505 device inputs from going below 0.2 V and may be omitted when using the TL505A device. The 2N2222 transistor pulls the inputs of the TMS1000 low during the start period.

The selection of external components for the TL505 device is dependent on the required input voltage range. The calculations are shown in the data sheet.

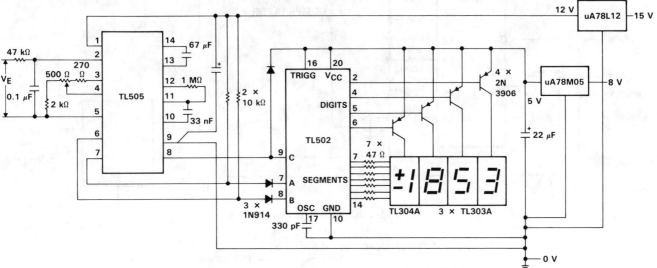

Figure 11-136. Digital Panel Meter Using TL505C

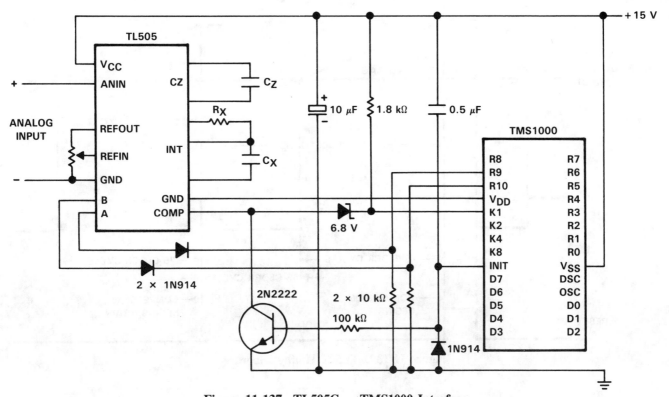

Figure 11-137. TL505C — TMS1000 Interface

11-111

TL507 A/D CONVERTER APPLICATIONS
TL507 Device Description

The TL507 device is a low-cost single-slope A/D converter which converts analog input voltages into a pulse-width-modulated (PWM) output code. In other words, the TL507 device is a voltage or current-to-time converter and the user can select the time reference by selection of the clock frequency. The operation of the TL507 device is explained with the block diagram in Figure 11-138 and the timing diagram in Figure 11-139.

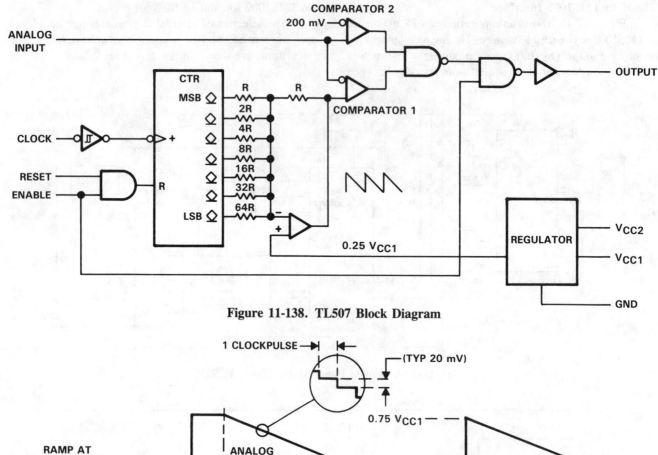

Figure 11-138. TL507 Block Diagram

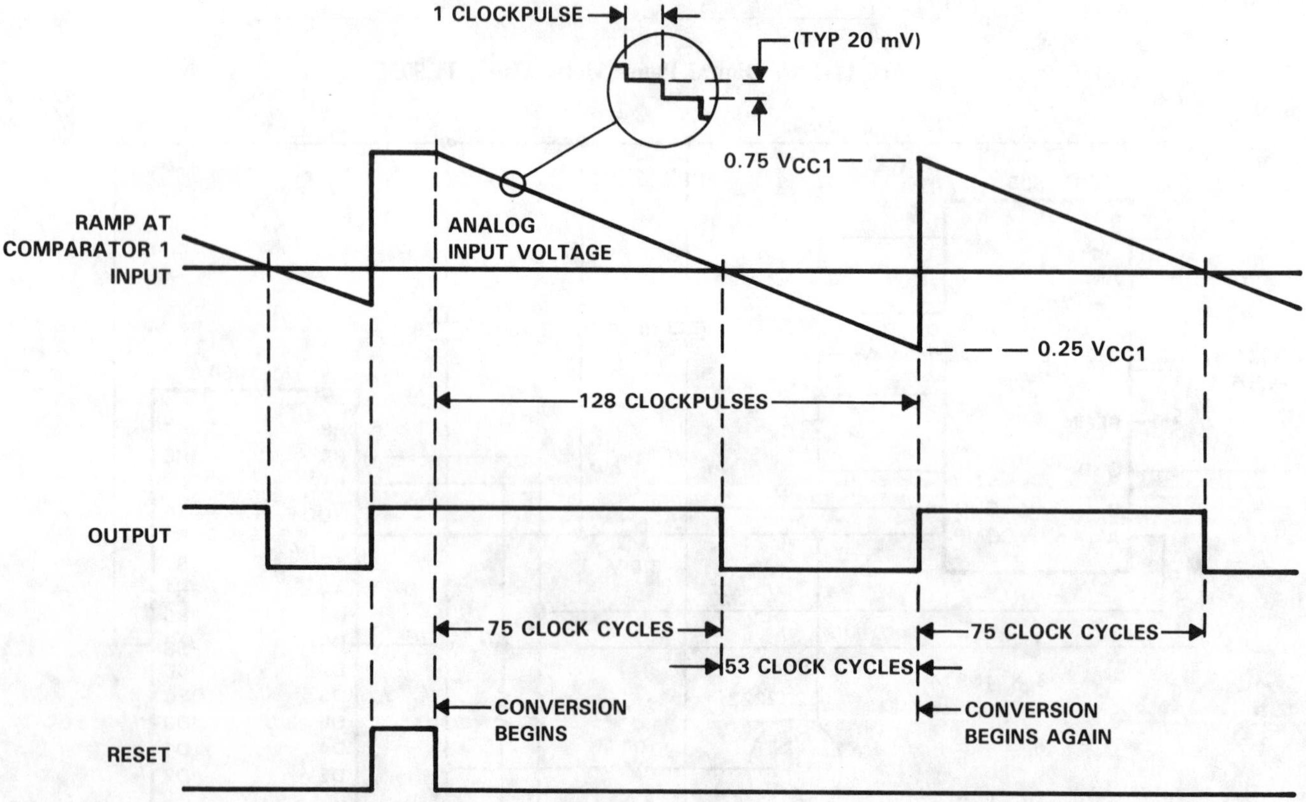

Figure 11-139. TL507 Timing Diagram

The internal 7-bit counter drives a binary-weighted resistor network to produce a monotonically negative-going ramp as the circuit is clocked. Comparator 1 compares the analog input voltage with the ramp and produces a low state at the output when the ramp voltage is less than the analog input voltage. This function happens when the enable input is high and the reset input is low.

The reset and enable inputs are inputs to an AND gate and are TTL voltage-level compatible. When the reset and enable inputs are at a high level, the counter is reset and the ramp is at its maximum value ($0.75 V_{CC1}$). When the reset input pin goes low, the TL507 device starts the A/D conversion cycle.

A conversion can begin by taking the reset input low or by clocking the circuit after the end of the previous conversion cycle as shown in Figure 11-139. The ramp input to the comparator is always at its highest voltage at the beginning of a conversion and the TL507 device output is high. The ramp voltage decreases with each succeeding clock cycle and the TL507 device output remains high until the ramp voltage becomes less than the analog input voltage; then, the output goes low. Thus, the number of clock cycles that occur during the period when the TL507 device output is low is directly proportional to the analog input voltage. A convenient method for obtaining a clock count, which is directly proportional to the analog input voltage, can be obtained with the simple NOR gate circuit in Figure 11-140.

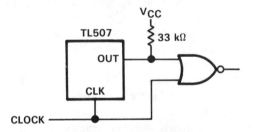

Figure 11-140. Circuit for Obtaining Clock Count Proportional to the Analog Input Voltage

The lower limit of the ramp is set by an internal voltage divider at $0.25 V_{CC1}$. This lower limit corresponds to the highest possible count which is $2^7 - 1 = 127$. Since the counter outputs are open-collector transistors that are off at the highest count, no current flows through the binary-weighted resistor network. Thus, the output of the operational amplifier is equal to the voltage ($0.25 V_{CC1}$) at its positive input. The upper limit of the ramp corresponds to a count of zero when the binary-weighted resistor network is grounded. The output of the operational amplifier is calculated as follows:

$$V_{RAMPmax} = 0.25 V_{CC1} (1 + R [1/R + - - - 1/64R])$$

$$\approx 0.75 V_{CC1}$$

The counter is decreased by one least significant bit for each clock pulse; thus, the ramp voltage decreases by $V_{CC1}/256$ as illustrated in Figure 11-139 and as calculated below:

$$\Delta V_{RAMP} = 0.25 V_{CC1} (R) (1/64R)$$

$$= \frac{V_{CC1}}{256} \approx 20 \text{ mV, typically.}$$

The input voltage, as illustrated in Figure 11-139, resulted in the device output being low for 53 clock cycles. The value of this input voltage is therefore

$$V_{IN} = \frac{\text{low clock count}}{\text{total clock count}} \times \frac{V_{CC1}}{2}$$

$$(\text{where } V_{CC1} = 5 \text{ V})$$

$$= \frac{53}{128} \times 2.5 \text{ V} \approx 1 \text{ V.}$$

TL507 Device Inputs and Outputs

When the enable input, pin 1, is held low, the TL507 device output is forced high regardless of all other input states. Since the TL507 device output is an open-collector transistor, multiple TL507 outputs can be wire ORed together, and the enable input functions as a multiplexer. When the enable input of one of the wire ORed TL507s goes high, the circuit performs the conversion operation and only its output controls the bus. The enable input is TTL compatible and is active high.

The clock input, pin 2, is connected to the counter input through a Schmitt-trigger gate that provides a large amount of hysteresis for noise immunity; therefore, the clock input can accept inputs with slow rise and fall times. Because of the hysteresis, the clock input is not compatible with TTL voltage levels and a pull-up resistor of approximately 4.7 kΩ must be used when this input is driven from a TTL output. The counter is clocked on the negative edge of the clock waveform as shown in Figure 11-141. The propagation delay time t_{pdCO}, as illustrated in Figure 11-141, is from the clock's negative edge to an output level change and is made up of two parts. One is the shift register delay (t_{pdCR}) and is from the clock input's negative edge to the analog input-ramp voltage crossing. The other (t_{pdRO}) is from the analog input-ramp voltage crossing to the output changing state. Although t_{pdRO} is a fixed value of about 3 μs, the value of t_{pdCR} may vary, depending on the value of the analog input voltage, from 1 μs to about 3 μs. Total delay, $t_{pdCO} = t_{pdCR} + t_{pdRO}$, must be less than one clock period. The recommended operating clock frequency is 125 kHz for an 8 μs time period. The maximum recommended clock frequency of 150 kHz yields a 6.67 μs time period.

The TL507 output at pin 4 is an open-collector transistor, and it must be pulled up with a suitable resistor determined by the load current and voltage requirements. The typical output characteristic of the TL507 device is shown in Figure 11-142 to aid in selecting the pull-up resistor

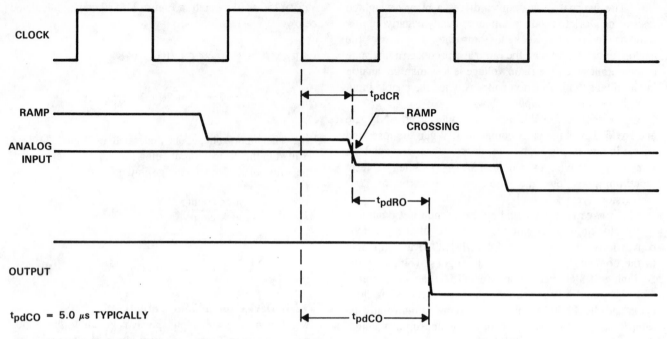

$t_{pdCO} = 5.0\ \mu s$ TYPICALLY

t_{pdCR} = Delay from the negative clock edge to the ramp crossing.

t_{pdRO} = Delay from the analog input-ramp crossing to the output changing state.

t_{pdCO} = Total delay time from the negative clock edge to the output changing state.

Figure 11-141. Typical TL507 Propagation Delay

value. Although the output acts like a current source at loads of about 4 mA, this characteristic should not be used to limit the output current when driving a low impedance load such as an optical coupler input diode. Also, the low state voltage should not exceed 5.5 V to keep from exceeding the power dissipation limits.

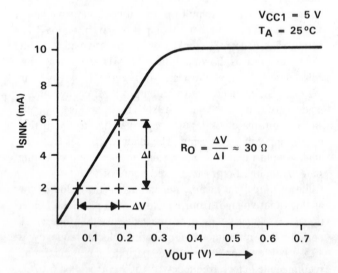

Figure 11-142. Typical TL507 Output Characteristic

The analog input, pin 5, is buffered by an emitter follower prior to being connected to the inputs of the comparators, and the buffer reduces the worst-case analog input current to 300 nA. The circuit or transducer, which drives the analog input, must have a low enough input impedance to minimize the effects of the input current or it will create an input offset. Comparator 2 has a nominal offset voltage of 200 mV as indicated in Figure 11-138. When the analog input voltage is less than 200 mV, the output is forced to a low state to indicate that the analog input voltage is out of range. Since the offset voltage has a ±50% tolerance, it is not suitable for measurement purposes. When the analog input exceeds the upper ramp voltage of 0.75 V_{CC1}, the output goes low again as an indication of overrange. Since the analog input current of the TL507 device is typically 10 nA, it can be used in a sample-hold configuration as shown in Figure 11-143.

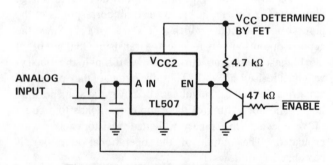

Figure 11-143. Sample Hold Circuit

The TL507 device has an internal regulator that converts an unregulated 8 V to 18 V input into a regulated 5 V. The unregulated input, pin 7, is called V_{CC2}. The regulated voltage is available on pin 6 as V_{CC1}. If more current is needed, pin 7 can be left open and a regulated voltage between 3.5 Vand 6 V can be applied to pin 6. The typical output characteristic for the regulator is shown in Figure 11-144. Additional loading on the regulated V_{CC1} should not exceed 5 mA. When the internal regulator is used, both the V_{CC1} and V_{CC2} pins should be decoupled with a 10 μF capacitor to ground.

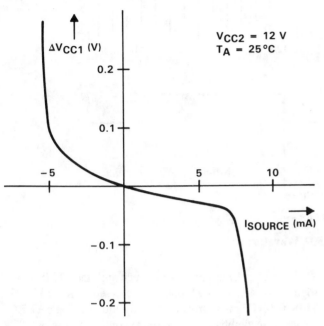

Figure 11-144. Typical V_{CC1} Characteristic

Single-Wire Power, Data, and Clock Cycle Transmitter

In this application, a remote sensor and the TL507 are used to provide a pulse-width-modulated signal representing a sensed analog input level. Connection between the remote location and a central control logic unit is over a single wire and common ground. This wire carries the clock pulses and unregulated power to the remote location and also returns the PWM signal.

Figures 11-145 and 11-146 show the basic circuitry and resulting waveforms. Q1 is used to chop the supply voltage and provide clock pulses to the remote location. The clock signal is fed directly to the TL507 clock input. It is also connected to diode D1 and capacitor C1 for filtering to provide the dc power for the TL507 A/D converter and the TLC271 signal conditioner. Adjusting the TLC271 gain provides the proper input levels for conversion by the TL507, allowing calibration of the output in degrees Celsius or Fahrenheit. Since the TL507 is continuously clocked, without any reset or enable, it continuously converts the analog input signal as illustrated in Figure 11-146. The pulse-width-modulated output signal from the TL507 is fed back over the wire and through a filter (R1 and C2) to remove the clock pulses. A TL331 comparator detects the PWM signal produced by the TL507. The comparator output, along with SN7400 and SN74ALS109A control logic, controls the TIL307 intelligent display.

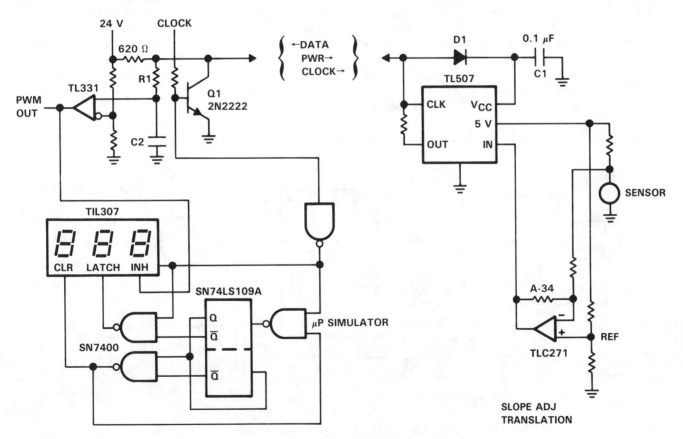

Figure 11-145. Remote-Sensor, Single-Wire A/D Converter

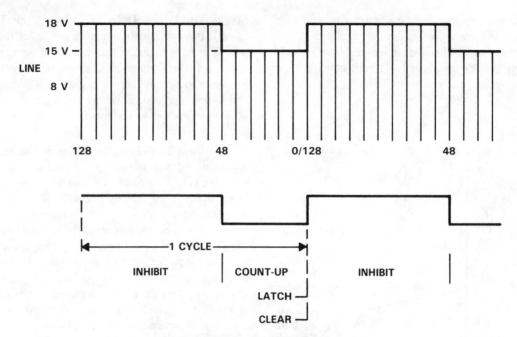

Figure 11-146. TL507 Waveforms

TL507 High Voltage Isolation Application and Signal Coupling

The circuit in Figure 11-147 illustrates the use of an opto-coupler interface between a microprocessor and the TL507 device. In this circuit, the TL507 may be located at a high voltage point while the TIL125 devices provide isolation for the microprocessor. This design is intended to minimize the opto-coupler switching times as much as possible without sacrificing conversion performance.

TL507 to TMS1000 Interface

In some applications, the TL507 device is used with a single-chip microprocessor such as the TMS1000. Figure 11-148 shows a TL507 device used with the PMOS TMS1000NLP microprocessor. Control of the single TL507 device is accomplished using serial control. Using parallel control would allow direct control of 4 TL507 devices. If necessary, up to 15 TL507s may be controlled if the 4 kΩ inputs of the TMS1000 are expanded using an external TMS1025 integrated circuit. The organization of the inputs and outputs of the TMS1000 microprocessor shown in Figure 11-148 is given only as an example and may be changed as required.

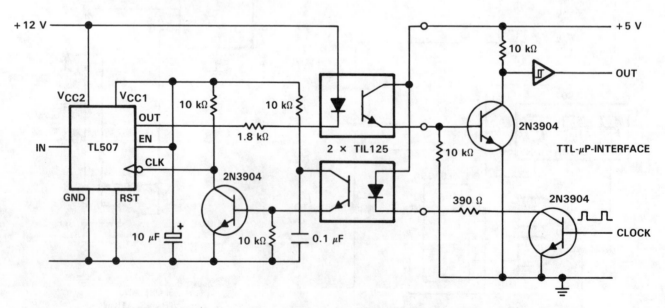

Figure 11-147. Example of High Voltage Isolation and Signal Coupling

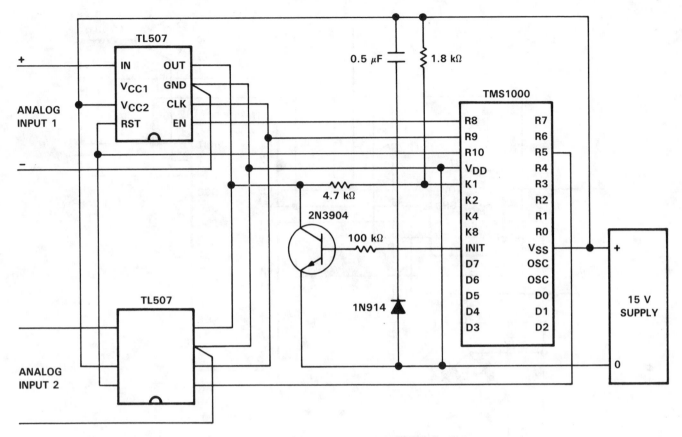

Figure 11-148. TL507 — TMS1000 Interface

TL507 to TMS1000C Interface

Figure 11-149 shows the interface between a TL507 device and a TMS1000C microprocessor. This is the CMOS version of the single-chip TMS1000 microprocessor. This circuit is particularly relevant for applications requiring low power consumption. Since the current requirement of the TMS1000C microprocessor is less than 3 mA, it may be supplied from the V_{CC1} regulator of the TL507 device. Since the total current requirement of the circuit is less than 10 mA, it can be powered from a single 9 V battery. The transistor in parallel with the TL507 output may be omitted if the undefined state of the TL507 output is taken into account during initialization.

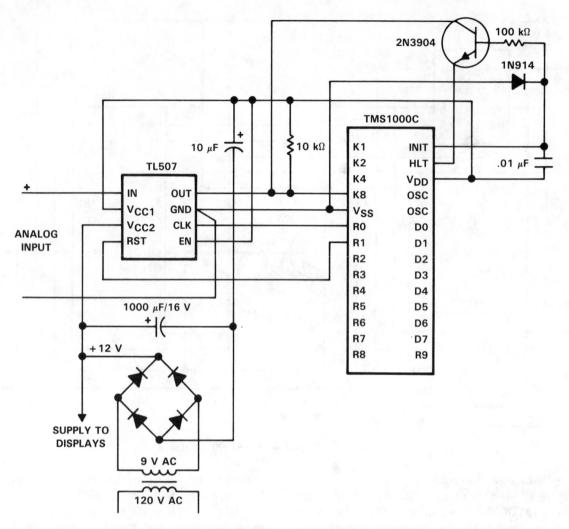

Figure 11-149. TL507 — TMS1000C Interface

Summary

The low cost of the TL507 device allows it to be used in almost any 7-bit application. By converting the analog signal to a stream of digital pulses, where the number of pulses is directly proportional to the analog input, the TL507 device can be used with many different data transmission schemes. Simple coding of the clock frequency or ac power line polarity can be combined with the TL507 device to achieve unique, noise-immune, inexpensive data transmission systems.

INTERFACE FOR TLC532A, TLC533A, TL530, AND TL531 DEVICES TO ZILOG Z80A AND Z80 MICROPROCESSORS

The TLC532A device provides a direct interface to a microprocessor-based system. It contains three 16-bit internal registers: control register, analog conversion register, and digital data register. Control signals enter on an 8-line, TTL compatible, 3-state data bus with read/write, register select, and chip select control inputs.

The interface circuitry is shown in Figure 11-150. This circuitry is powerful since it allows the microprocessor to write to and read from the A/D converters quickly and efficiently via the bus. Since these devices are bus-oriented, the interface circuitry uses them in an optimum fashion by providing microprocessor bus write-and-read capability. This interface circuitry allows the microprocessor to fully use all of the A/D converter's write-read instructions. One-byte instructions can be used to initiate the A/D conversion by setting the SC bit and to determine the end of conversion by monitoring the EOC bit.

The circuitry uses a Z80A microprocessor. However, the same circuitry could use a Z80 microprocessor by changing the clock frequency from 4 MHz to 2 MHz. Such a change will double the A/D conversion time since the frequency of the clock signal into the A/D converter is halved. The 4 MHz clock signal can be changed to any frequency with the provision that the resulting A/D converter clock frequency, whose frequency is the microprocessor clock frequency divided by 4, does not exceed the A/D upper clock frequency limit. Also, a microprocessor clock signal whose frequency is greater than 4 MHz may be used. In this case the resulting A/D converter clock frequency must not exceed the upper frequency limit for the A/D converter.

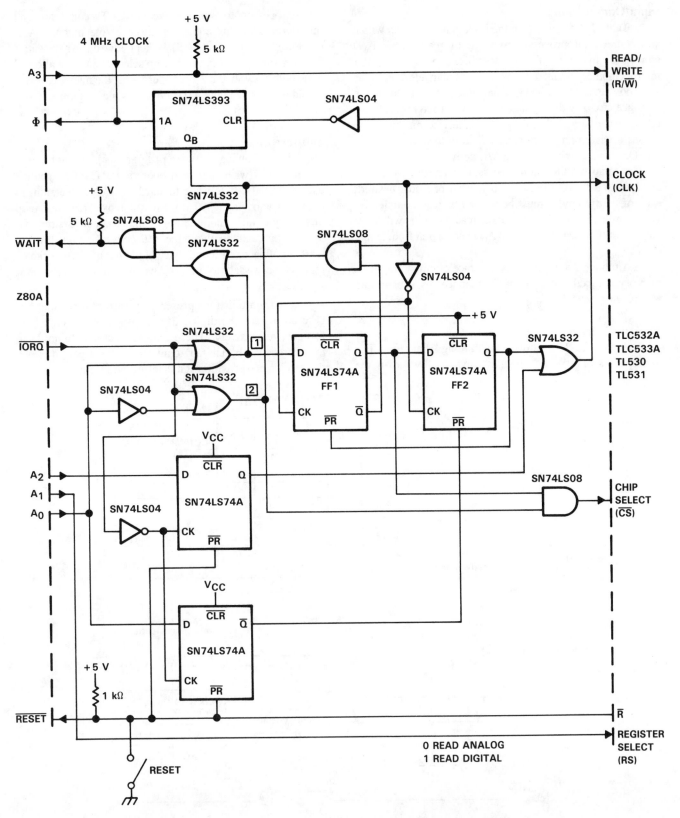

NOTE: Interface has been tested with TTL-LS parts only, however interface should work with HCMOS ICs.

Figure 11-150. Z80A Microprocessor Interface Diagram, TLC532A, TLC533A, TL530, and TL531

Input/Output Mapping

Table 11-39 shows the Input/Output (I/O) map which allows the microprocessor to write or read to the A/D converter in any of the possible modes. If other microcomputer I/O interfaces are required in addition to the A/D converter, the $\overline{IORQ}$ signal must be blocked in the interface circuitry during non-A/D write or read operations. $\overline{IORQ}$ is the signal that drives the interface circuitry and causes a write or read to occur. Failure to mask or block the $\overline{IORQ}$ signal will allow the A/D converter and other I/O devices to access the microprocessor bus at the same time and cause a data collision. If a 2-byte write or read is desired, each of the two bytes must be written or read completely. Partial execution of a 2-byte write or read operation will leave chip select (CS) low and prevent A/D conversion from taking place.

Although not tested, this interface circuitry might be memory mapped, rather than I/O mapped, by substituting MREQ in place of $\overline{IORQ}$.

Timing Diagram

Figure 11-151 shows the timing diagram. The microprocessor's WAIT input is used to slow down the microprocessor to synchronize the A/D converter and microprocessor bus communications. Note that the A/D converter's chip select (CS) is kept low during an entire 2-byte write or read cycle. If chip select (CS) were allowed to go high between the 2-byte operations, the microprocessor would write or read the A/D converters most significant byte twice.

Software

The following software program listings can be used for both A/D write and read operations. The first routine uses an initial 2-byte write to start the A/D conversion, to address analog input A0, to select a 1-byte read test loop or a delay to allow completion of the A/D conversion, and uses a 2-byte read to obtain the A/D conversion result in the least significant or second byte. Note that a 2-byte write or read is necessary to access the A/D converter's least significant byte.

Two more listings present software routines that can be used to read the digital data registers in 2- and 1-byte formats, respectively. These read instructions do not have to be preceded by an A/D conversion delay period, since these registers provide digital information only.

Table 11-39. TLC532A I/O Map

		A_3	A_2	A_1	A_0	NO. OF BYTES COMMUNICATED
Control Register Write	MSB	0	0	1	0	2
	LSB	0	0	1	1	
	MSB	0	1	1	0	1
Analog Data Register Read	MSB	1	0	0	0	2
	LSB	1	0	0	1	
	MSB	1	1	0	0	1
Digital Data Register Read	MSB	1	0	1	0	2
	LSB	1	0	1	1	
	MSB	1	1	1	0	1

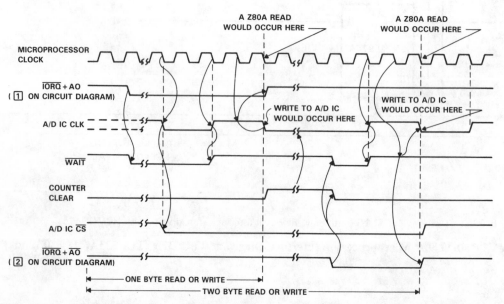

Figure 11-151. Timing Diagram for the Z80A/Z80 Interface to the TLC532A, TLC533A, TL530, and TL531

```
                              ; Software for A/D Conversion Using the TLC532A,
                                TLC533A, TL530 and TL531
                              ;
0000                          ORG 00H                   ;(used by cross-assembler)
0000    3E 01                 LD A,01H                  ;01H into accumulator
0002    03 02                 OUT (02H),A               ;write MSB (STRT CONV = 1)
0004    3E 00                 LD A,00H                  ;00H into accumulator
0006    D3 03                 OUT (03H),A               ;write LSB (ADDR = 0)
                              ;
0008    DB 0C        TEST:    IN A,(OCH)                ;read MSB analog register
000A    F2 08 00              JP P,TEST                 ;branch if not END OF CONVERSION = 0
                              ;
                              ;The above test loop can be substituted by any delay
                              ;which meets the A/D IC's time requirements for A/D
                              ;conversion.
                              ;
0000    DB 08                 IN A,(08H)                ;read analog register MSB
000F    DB 09                 IN A,(09H)                ;read analog register LSB,
                                                        ;conversion result is in accumulator
                              ;
                              ;
--------------------------------------------------------------------------------------------------
                              ;
                              ; Software for Digital Data Register 2-Byte Read
                              ;
0100                          ORG 100H                  ;(used by cross assembler)
0100    DB 0A                 IN A,(0AH)                ;fetch MSB
0102    47                    LD B,A                    ;store MSB
0103    DB 0B                 IN A,(0BH)                ;fetch LSB, MSB is in B and
                                                        ;LSB is in A
                              ;
                              ;
--------------------------------------------------------------------------------------------------
                              ;
                              ; Software for Digital Data Register 1-Byte Read
                              ;
0200                          ORG 200H                  ;(used by cross assembler)
0200    DB 0E                 IN A,(OEH)                ;fetch MSB, MSB is in A
0202    47                    END                       ;(used by cross-assembler)
```

Additional Comments

The time required for analog-to-digital conversion for the A/D converters in this application are presented in Table 11-40. It is necessary that the microprocessor allow these time periods for A/D conversion, either through a test loop, where the microprocessor reads the end of conversion (EOC) bit, or a delay period that may be obtained in any convenient manner by the microprocessor.

Table 11-40. A/D Conversion Times for the Z80A/Z80 Interface to the TLC532A, TLC533A, TL530 and TL531

	A/D IC CLOCK CYCLE'S REQUIRED FOR CONVESION (See Note 1)	CONVERSION TIME (μs) USING AN A/D IC CLOCK FREQUENCY OF 1 MHz (See Note 2)
TLC532A	29½	29.5
TLC533A	29½	29.5
TL530	290½	290.5
TL531	290½	290.5

NOTES: 1. Time 0 is the first rising edge of the first write cycle of a 2-byte write operation.
2. The interface circuitry will generate an A/D IC clock signal of 1 MHz when using a 4 MHz microprocessor clock signal.

INTERFACE FOR TLC532A, TLC533A, TL530, AND TL531 DEVICES TO INTEL 8048 AND 8049 AND INTEL 8051 AND 8052 MICROPROCESSORS

This application covers all the possible interface combinations for these parts. These interfaces offer the following advantages:

1. Fast complete cycle times for loading the A/D address, performing conversion, and retrieving the conversion
2. Flexible use of A/D converter pins for either A/D conversion or input of digital data
3. Low cost.

These Intel microprocessor families consist of the following:

INTEL 8048 and 8049 Family

8035AHL	8048AH	8748H
8039AHL	8049AH	8749H
8040AHL	8050AH	80C49

INTEL 8051 and 8052 Family

8031AH	8051AH	8751H
8032AH	8052AH	80C51

Because of differences in timing, two different circuit configurations are required for interfacing these A/D converters and microprocessor families. For reference, these two configurations are called Interface 1 and Interface 2.

Interface 1 is more desirable than Interface 2 since it does not constrain the microprocessor's data bus operations. It also allows the microprocessor to use the A/D converter software in external program memory. Unfortunately, different timing specifications prevent Interface 1 from being used for Interface 2 (8051 and 8052 to TL530, TL531, and TLC533A interface).

Hardware — Interface 1

Figure 11-152 shows the circuit configuration for Interface 1. This circuit can be used for the 8051 and 8052 devices interface to the TLC532A device, and the 8048 and 8049 devices interface to the TL530, TL531, and TLC533A devices.

The system clock for the A/D converter is obtained from the microprocessor crystal oscillator. To assure proper operation of the crystal oscillator, a high impedance buffer must be used to prevent overloading the oscillator. An important design detail is that the low and high level input requirements of the buffer must lie within the range of the oscillator signal. This compatibility will prevent missing edge transitions in the A/D converter's system clock signal. The buffered crystal oscillator signal must be frequency divided to assure that the resulting system clock frequency does not exceed the upper frequency limit of the A/D converter. Any convenient divider circuitry may be used to accomplish this task.

Timing Diagram — Interface 1

The timing diagrams for the 8051, 8052 and 8048, 8049 devices to TLC532A converter interface are presented in Figures 11-153 and 11-154, respectively. The timing diagrams cover a complete conversion cycle and the reading of the digital data registers. In Figure 11-154, the required number of conversion clocks will increase for the TL530, TL531, and TLC533A devices due to a lower clock frequency operating range and the higher number of conversion clock cycles that are required by the TL530 and TL531 converters.

Software — Interface 1

The software listing for the interface between the 8051 and 8052 devices and the TLC532A device follows. The software listing for the 8048 and 8049 to the TL530, TL531, TLC532A, and TLC533A interface is also included. The conversion software program or the digital data register software program can be incorporated into a subroutine so the designer can easily access the software with a simple subroutine call. Also, the conversion software assumes that the A/D converter address has been placed in register R2 and, upon completion of A/D conversion, stores the conversion result in register R2. The digital data register software stores the most significant (MS) byte and the least significant (LS) byte in registers R2 and R3, respectively.

```
                              ; Software for Intel 8051/52 Family to
                              ; TLC532A Interface 1
                              ;
                              ;               Software for Conversion
                              ;
0000                          ORG 000H
0000    43 90 0C              ORL P1, #0CH           ;CS, RS = 1
0003    C2 90                 CLR P1.0               ;Counter clear = 0
0005    7F 01                 MOV R7, #01H           ;Do delay to allow 3 system clocks
0007    DF FE     D1CSEQ1:    DJNZ R7,D1CSEQ1        ;to occur so A/D IC will
                                                     ;recognize CS = 1
0009    B2 90                 CPL P1.0               ;Counter clear = 1
000B    53 90 FS              ANL P1, #F5H           ;CS, R/W(bar) = 0
000E    74 01                 MOV A, #01H            ;Write MSByte (Set SC = 1)
0010    F2                    MOVX @R0,A
0011    EA                    MOV A,R2               ;Get A/D IC address, which is
                                                     ;assumed to be in R2
0012    F2                    MOVX @R0,A             ;Write LSByte or A/D IC address
0013    43 90 0A              ORL P1, #0AH           ;CS, R/W(bar) = 1
0016    C2 90                 CRL P1.0               ;Counter clear = 0
                        ;
                        ;       A delay must occur here to allow the A/D IC to
                        ;       complete conversion. The delay must allow at least
                        ;       29.5 A/D IC system clock cycles to occur.
                        ;
0018    B2 90                 CPL P1.0               ;Counter clear = 1
001A    53 90 F3              ANL P1, #F3H           ;CS, RS = 0
001D    E2                    MOVX A,@R0             ;Read Analog Conversion Register MSByte
001E    E2                    MOVX A,@R0             ;Read Analog Conversion Register LSByte
                                                     ;or conversion result
001F    FA                    MOV R2,A               ;Store conversion result
0020    B2 93                 CPL P1.3               ;CS = 1
0022    C2 90                 CLR P1.0               ;Counter clear = 0
                        ;
--------------------------------------------------------------------------------------------------
                              ; Software for Reading Digital Data Registers
                              ;
0100                          ORG 0100H
0100    43 90 0E              ORL P1, #0EH           ;CS, RS, R/W(bar) = 1
0103    C2 90                 CLR P1.0               ;Counter clear = 0
0105    7F 01                 MOV R7, #01            ;Do delay to allow 3 system clocks
0107    DF FE     D2CSEQ1:    DJNJ R7,D2CSEQ1        ;to occur so A/D IC will
                                                     ;recognize CS = 1
0109    B2 90                 CPL P1.0               ;Counter clear = 1
010B    C2 93                 CLR P1.3               ;CS = 0
010D    E2                    MOVX A,@R0             ;Read Digital Data Register MSByte
010E    FA                    MOV R2,A               ;Store Digital Data Register MSByte
010F    E2                    MOVX A,@R0             ;Read Digital Data Register LSByte
0110    FB                    MOV R3,A               ;Store Digital Data Register LSByte
0111    B2 93                 CPL P1.3               ;CS = 1
0113    C2 90                 CLR P1.0               ;Counter clear = 0
0115                          END
```

```
; Software for Intel 8048/49 Family —
; TL530, TL531, TLC532A, and TLC533A Interface 1
;
;                                Software for Conversion
;
                                 ORG 000H
0000    89 0C              ORL P1, #0CH            ;CS, RS = 1
0002    99 FE              ANL P1, #FEH            ;Counter clear = 0
0004    BF 01              MOV R7, #01H            ;Do delay to allow 3 system clocks
0006    EF 06    D1CSEQ1:  DJNZ R7,D1CSEQ1         ;to occur so A/D IC will
                                                   ;recognize CS = 1
0008    89 01              ORL P1, #01H            ;Counter clear = 1
000A    99 F5              ANL P1, #F5H            ;CS, R/W(bar) = 0
000C    23 01              MOV A, #01H             ;Write MSByte (Set SC = 1)
000E    90                 MOVX @R0,A
000F    FA                 MOV A,R2                ;Get A/D IC address, which is
                                                   ;assumed to be in R2
0010    90                 MOVX @R0,A              ;Write LSByte or A/D IC address
0011    89 0A              ORL P1, #0AH            ;CS, R/W(bar) = 1
0013    99 FE              ANL P1, #FEH            ;Counter clear = 0
                 ;
                 ;        A delay must occur here to allow the A/D IC to
                 ;        complete conversion. The delay must allow at least
                 ;        29.5 A/D IC system clock cycles to occur for the
                 ;        TLC532A and TLC533A. The TL530 and TL531 require
                 ;        290.5 system clock cycles.
                 ;
0015    89 01              ORL P1. #01H            ;Counter clear = 1
0017    99 F3              ANL P1, #F3H            ;CS, RS = 0
0018    80                 MOVX A,@R0              ;Read Analog Conversion Register MSByte
0019    80                 MOVX A,@R0              ;Read Analog Conversion Register LSByte
                                                   ;or conversion result
001A    AA                 MOV R2,A                ;Store conversion result
001B    89 08              ORL P1, #08H            ;CS = 1
001D    99 FE              ANL P1, #FEH            ;Counter clear = 0
                 ;
-------------------------------------------------------------------------------------------------
                 ;        Software for Reading Digital Data Registers
                 ;
                                 ORG 0100H
0100    89 0E              ORL P1, #0EH            ;CS, RS, R/W(bar) = 1
0102    99 FE              ANL P1, #FEH            ;Counter clear = 0
0104    BF 01              MOV R7, #01             ;Do delay to allow 3 system clocks
0106    EF 06    D2CSEQ1:  DJNZ R7,D2CSEQ1         ;to occur so A/D IC will
                                                   ;recognize CS = 1
0108    89 01              ORL P1, #01H            ;Counter clear = 1
010A    99 F7              ANL P1, #F7H            ;CS = 0
010C    80                 MOVX A,@R0              ;Read Digital Data Register MSByte
010D    AA                 MOV R2,A                ;Store Digital Data Register MSByte
010E    80                 MOVX A,@R0              ;Read Digital Data Register LSByte
010F    AB                 MOV R3,A                ;Store Digital Data Register LSByte
0110    89 08              ORL P1, #08H            ;CS = 1
0112    99 FE              ANL P1, #FEH            ;Counter clear = 0
                                 END
```

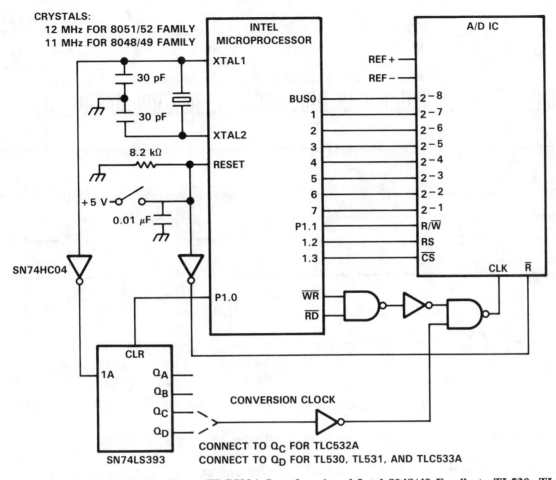

Figure 11-152. Intel 8051/52 Family to TLC532A Interface 1 and Intel 8048/49 Family to TL530, TL531, TLC532A, and TLC533 Interface 1

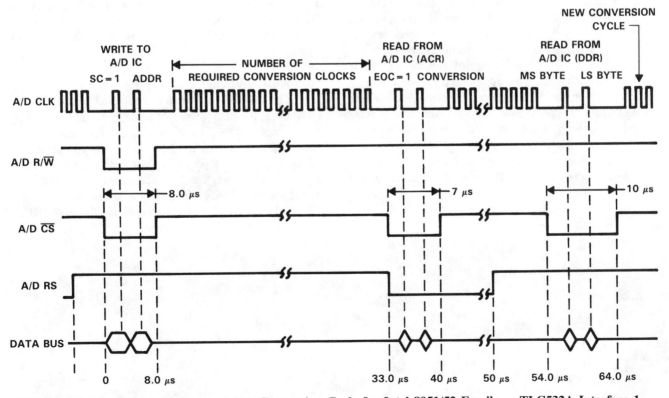

Figure 11-153. Timing Diagram of A/D Conversion Cycle for Intel 8051/52 Family — TLC532A Interface 1

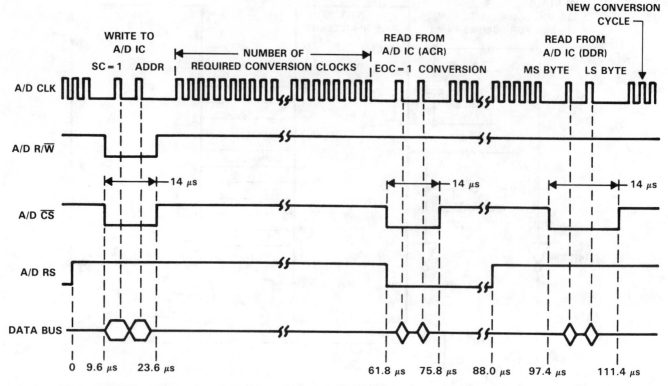

Figure 11-154. Timing Diagram of A/D Conversion Cycle for Intel 8048/49 Family — TLC532A

Hardware — Interface 2

Figure 11-155 presents the circuit configuration for Interface 2. This circuit can be used for the 8051 and 8052 devices interface to the TL530, TL531, and TLC533A devices.

The A/D converter's data port can be connected to any of the microprocessor's ports. However, if the connection is made to Port 0, the interface circuitry prevents the incorporation of the A/D software into the external program memory. As seen in the following software listings, three microprocessor instructions are required to read a byte of information from the A/D converter. Unfortunately, the first of the three instructions causes the A/D converter to drive the microprocessor bus in a non-3-state mode that prevents the microprocessor from retrieving any subsequent instructions. Refer to the Hardware — Interface 1 paragraph for a description of how the A/D converter system conversion clock is generated.

Timing Diagram — Interface 2

The timing diagram for the 8051 and 8052 devices interface to the TLC533A device is presented in Figure 11-156. The timing diagram covers a complete conversion cycle and the reading of the digital data registers. The required number of conversion clocks will increase for the TL530 and TL531 devices due to the higher number of conversion clock cycles that are required by the TL530 and TL531 converters.

Software — Interface 2

The software listings for Interface 2 follow. Unlike Interface 1, the microprocessor write (WR) and read (RD) pins are not used to generate the A/D converter clock during write and read operations. Instead, the clock is generated by toggling the P1.4 pin. This change between Interface 1 and Interface 2 was made to accommodate the different timing specifications of the 8051 and 8052 devices and the TL530, TL531, and TLC533A converters. (See the Interface 1 paragraph for further description of the software listings.)

```
; Software Conversion for Intel 8051/52
; Family — TL530, TL531, TLC533A Interface 2
;

0000                              ORG 000H
0000    43 90 0C                  ORL P1,#0CH        ;CS, RS = 1
0003    53 90 EE                  ANL P1,#EEH        ;Counter clear, Slow clock = 0
0006    7F 02                     MOV R7,#02H        ;Do delay to allow 3 system clocks
0008    DF FE       D1CSEQ1:      DJNZ R7,D1CSEQ1    ;to occur so A/D IC will
                                                     ;recognize CS = 1

000A    B2 90                     CPL P1.0           ;Counter clear = 1
000C    53 90 F5                  ANL P1,#F5H        ;CS, R/W(bar) = 0
000F    75 B0 01                  MOV P3,#01H        ;Write MSByte (Set SC = 1) to Port 3
0012    B2 94                     CPL P1.4           ;Raise slow clock
0014    B2 94                     CPL P1.4           ;Lower slow clock
0016    8A B0                     MOV P3,R2          ;Write A/D IC address, which is
                                                     ;assumed to be in R2, to Port 3

0018    B2 94                     CPL P1.4           ;Raise slow clock
001A    B2 94                     CPL P1.4           ;Lower slow clock
001C    75 B0 FF                  MOV P3,#FFH        ;Put Port 3 pins to input mode
001F    43 90 0A                  ORL P1,#0AH        ;CS, R/W(bar) = 1
0022    C2 90                     CLR P1.0           ;Counter clear = 0

            ;
            ;                     A delay must occur here to allow the A/D IC to
            ;                     complete conversion. The delay must allow at least
            ;                     29.5 A/D IC system clock cycles to occur for the
            ;                     TLC533A. Similarly, at least 290.5 clock cycles
            ;                     must occur for the TL530 and TL531. Further the
            ;                     frequency of the clock signal must not exceed
            ;                     the specification for the A/D IC.
            ;

0024    B2 90                     CPL P1.0           ;Counter clear = 1
0026    53 90 F3                  ANL P1,#F3H        ;CS, RS = 0
0029    B2 94                     CPL P1.4           ;Raise slow clock
002B    AA B0                     MOV R2,P3          ;Read MSByte of the analog
                                                     ;conversion register
002D    B2 94                     CPL P1.4           ;Lower slow clock
002F    B2 94                     CPL P1.4           ;Raise slow clock
0031    AA B0                     MOV R2,P3          ;Read and store LSByte of the analog
                                                     ;conversion register or conversion
0033    B2 94                     CPL P1.4           ;Lower slow clock
0035    B2 93                     CPL P1.3           ;CS = 1
0037    C2 90                     CLR P1.0           ;Counter clear = 0

            ;
            ;
```

```
                    ;
                    ;
                    ; Software for Reading Digital Data Registers
                    ;
    0100                              ORG 0100H
    0100    43 90 0E                  ORL P1, #0EH          ;CS, RS, R/W(bar) = 1
    0103    53 90 EE                  ANL P1, #EEH          ;Counter clear, Slow clock = 0
    0106    7F 02                     MOV R7, #02           ;Do delay to allow 3 system clocks
    0108    DF FE        D2CSEQ1:     DJNZ R7,D2CSEQ1       ;to occur so A/D IC will
                                                            ;recognize CS = 1
    010A    B2 90                     CPL P1.0              ;Counter clear = 1
    010C    75 B0 FF                  MOV P3, #FFH          ;Put Port 3 pins to input mode
    010F    C2 93                     CLR P1.3              ;CS = 0
    0111    B2 94                     CPL P1.4              ;Raise slow clock
    0113    AA B0                     MOV R2,P3             ;Read and store MSByte of the digital
                                                            ;data register
    0115    B2 94                     CPL P1.4              ;Lower slow clock
    0117    B2 94                     CPL P1.4              ;Raise slow clock
    0119    AB B0                     MOV R3,P3             ;Read and store LSByte of the digital
                                                            ;data register
    011B    B2 94                     CPL P1.4              ;Lower slow clock
    011D    B2 93                     CPL P1.3              ;CS = 1
    011F    C2 90                     CLR P1.0              ;Counter clear = 0
    0121                              END
```

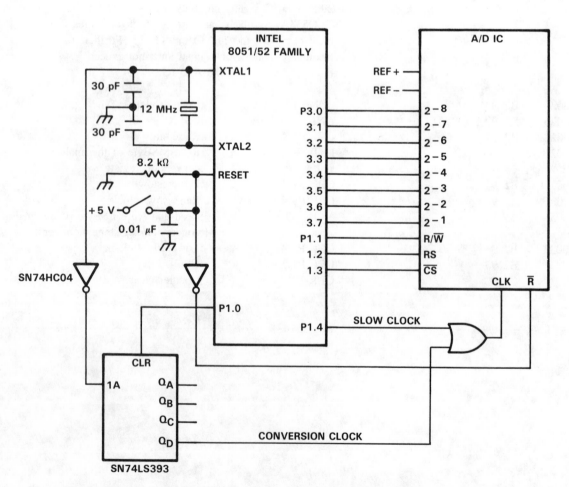

Figure 11-155. Intel 8051/52 Family — TL530, TL531, TLC533A Interface 2

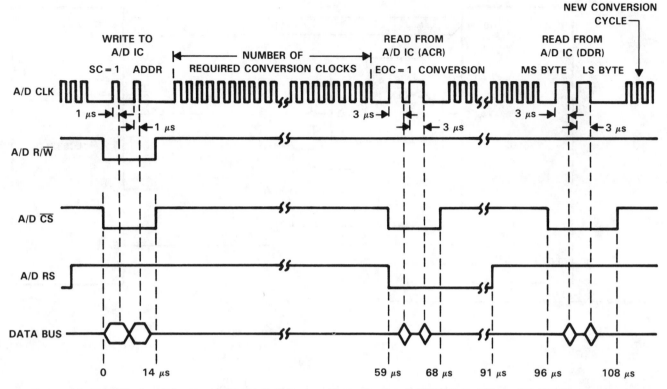

Figure 11-156. Timing Diagram of A/D Conversion Cycle for Intel 8051/52 Family to TLC533A Interface 2

Additional Comments

The A/D converter clock signal should not be interrupted during A/D conversion because of the possible loss of electrical charge on the A/D converter's internal capacitors. Interruption of the clock signal during an A/D converter write or read operation presents no problem. All of the software offered for use with these interfaces complies with the above requirement since the conversion clock is not stopped until after the A/D conversion is completed.

INTERFACE FOR TLC532A, TLC533A, AND TL530 DEVICES TO THE ROCKWELL 6502 MICROPROCESSOR

This application describes techniques for interfacing the TLC532A converter device and compatible peripheral chips to the Rockwell 6502 microprocessor. The TLC532A family of A/D converters consists of chips offering combinations of conversion speed and number of analog inputs. The interface presented deals specifically with the TLC532A, TLC533A, and the TL530 devices. Other members of the TL/TLC53x family are not included because the bus timing specification cannot be guaranteed. It is important that SN74S00 or SN74AS00 NAND gates be used to guarantee operation of this interface circuit.

Principles of Operation

The TLC532A device to 6502 device interface circuit and timing diagrams are shown in Figures 11-157 and 11-158, respectively. A listing of the control software necessary to operate the interface follows.

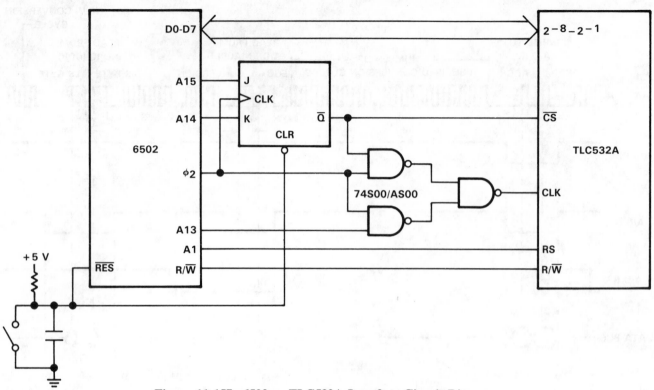

Figure 11-157. 6502 — TLC532A Interface Circuit Diagram

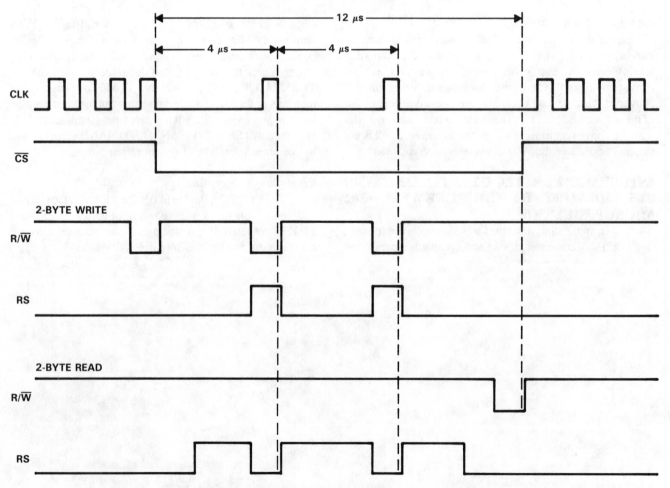

Figure 11-158. 6502 — TLC532A Interface Timing Diagram

```
        ;
        ;ADDRESS ASSIGNMENTS
        ;
        ANALYZER:   .EQU 0000H
        ACMSB:      .EQU 2400H
        ACLSB:      .EQU 2401H
        CRMSB:      .EQU 2402H
        CRLSB:      .EQU 2403H
        DDMSB:      .EQU 2502H
        DDLSB:      .EQU 2503H
        SETCS:      .EQU 4400H
        CLRCS:      .EQU 8400H
        ;
        ;
        ;
                    LDX # $01        ; SET SC BIT
                    LDA # $00        ; SELECT MUXADDRESS
                    STA CLRCS        ; BRING CHIP SELECT LOW
                    STX CRMSB        ; LOAD CONTROL REG MSB
                    STA CRLSB        ; LOAD CONTROL REG LSB, START CONV.
                    LDA SETCS        ; BRING CHIP SELECT HIGH
                    LDY # $07        ;
        LOOP:       DEY              ; DELAY UNTIL CONVERSION COMPLETE
                    BNE LOOP         ;
                    LDA CLRCS        ; BRING CHIP SELECT LOW
                    LDA ACMSB        : READ CONV. REG MSB (EOC BIT)
                    LDA ACLSB        ; READ CONV. REG LSB (RESULT)
                    STA SETCS        ; BRING CHIP SELECT HIGH
```

A conversion cycle begins by bringing chip select ($\overline{CS}$) low. This is accomplished by clocking a high into the J input of the flip-flop on the trailing edge of PH 2, bringing $\overline{Q}$ low. A low on $\overline{Q}$ also inhibits the PH 2 clock from reaching the TLC532A device. When the first byte of a 2-byte write cycle is being written, address bit A13 goes high. This allows one PH 2 pulse to reach the TLC532A device and clock in the byte. The write cycle for the second byte is exactly the same as the first. Chip select ($\overline{CS}$) is then brought high by clocking a high into the K input of the flip-flop to bring $\overline{Q}$ high again. Analog-to-digital conversion requires 29 clock cycles. Therefore, a delay loop is included in the software listing to allow for this conversion time.

The 2-byte read cycle operation is similar to that of the write cycle since it uses address bit A13 as a gating signal for the clock input to the TLC532A device. One- or 2-byte reads of either the analog conversion registers or the digital data registers may be performed with the proper register select signals. The software listing shows the necessary software for a 2-byte read of the analog conversion register, and leaves the A/D conversion result in the accumulator. One complete data acquisition cycle can be performed every 66 microseconds.

INTERFACE FOR TLC532A, TLC533A, TL530, AND TL531 DEVICES TO THE MOTOROLA 6800, 6802, 6809, AND 6809E MICROPROCESSORS

Since these A/D converters are bus oriented, the interface circuitry uses these converters in an optimum way by providing microprocessor bus write and read capability. Two interface circuits are presented: Interface 1 in Figure 11-159 and Interface 2 in Figure 11-160 that provide 1- and 2-byte write/read communications capability between the microprocessor and the A/D converter. Only the 2-byte capability is demonstrated in this application. To use the 1-byte capability of the A/D converter, 1-byte external memory write/read instructions can be used.

Input/Output Mapping

Table 11-41 presents the I/O map that allows the microprocessor to write or read to the A/D converter. This I/O map is intended as a guide and does not preclude other mapping possibilities.

Table 11-41. I/O Map

	A15	A14	A13*	R/W̄
Control Register Write	0	1	1	0
Analog Data Register Read	0	0	1	1
Digital Data Register Read	0	1	1	1
ROM Program	1	X	X	X

*A13 = 1 to prevent the 6802 from reading its internal RAM and the A/D IC simultaneously.

Timing Diagram

A timing diagram for Interface 1 with a 1 MHz microprocessor clock cycle is shown in Figure 11-161. The microprocessor write and read timing diagrams for Interface 2 are shown in Figures 11-162 and 11-163.

Software

The following software listings can be used for both interface circuits. These listings show the software for conversion and for reading the digital data registers, respectively. For conversion, the software must determine the end of conversion by monitoring the end of conversion bit (EOC) in the most significant (MS) byte of the analog conversion register or by allowing a delay so the A/D converter can complete conversion. The length of the delay can be computed by referring to Table 11-42. This table shows the number of clock cycles required for each A/D converter to complete conversion.

Table 11-42. A/D Conversion Time

A/D IC	A/D IC CLOCK CYCLES REQUIRED FOR CONVERSION (See Note)
TLC532A	29½
TLC533A	29½
TL530	290½
TL531	290½

NOTE: Time 0 is the first rising edge of the first write cycle of a 2-byte write operation.

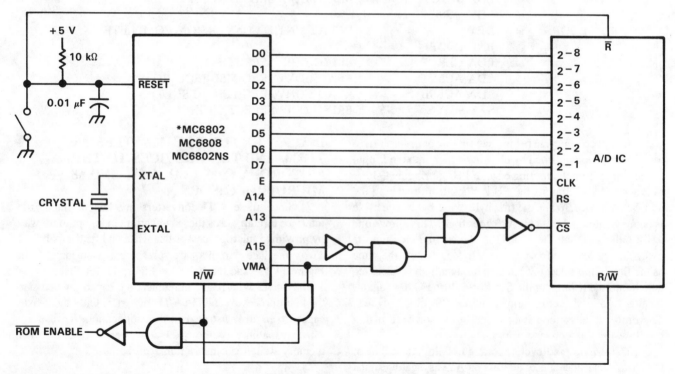

*See Table 11-43

Figure 11-159. Interface 1 Hardware Circuitry

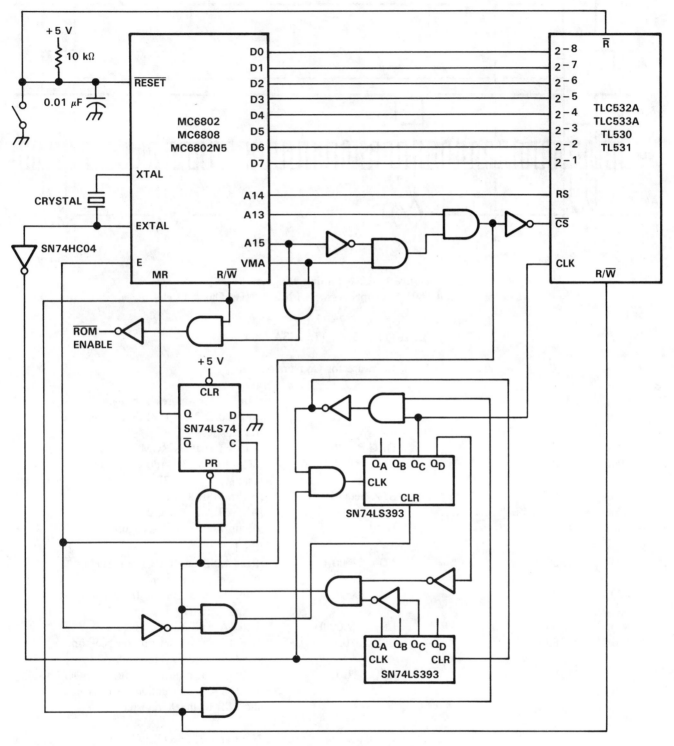

*See Table 11-43 for information about other microprocesor configurations.

Figure 11-160. Interface 2 Hardware Circuitry

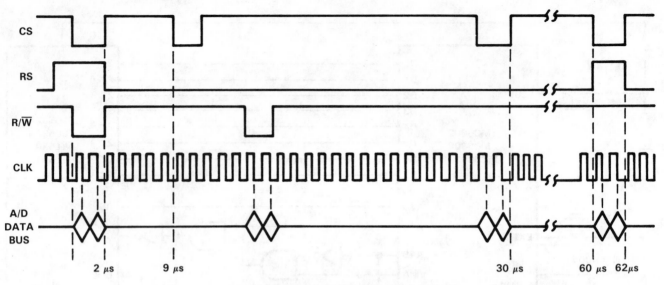

Figure 11-161. Timing Diagram
(Interface 1 and 6802 Microprocessor with a 1 MHz Clock Cycle)

```
                    ; Software for 6800 Family —
                    ; TL530, TL531, TLC532A, TLC533A Interface
                    ;
                    ;               Software for Conversion
                    ;
07FE    80                                          ;Microprocessor reset, branch
07FF    00                                          ;to 8000H. 2716 UVPROM was used
                                                    ;for this test.
8000                              ORG 8000H         ;A15 = 1 enables ROM memory
8000    CE 01 05                  LDX #0105H        ;SC = 1 in MSB, Address analog
                                                    ;channel #5 in LSB
8003    FF 60 00                  STX $6000H        ;Write MSB and LSB to A/D IC
8006    01            STILLC:     NOP               ;Allow I A/D clock cycle to
8007    01                        NOP               ;occur so that A/D recognizes
                                                    ;CS⁵1
8008    FE 20 00                  LDX $2000H        ;Read MSB and LSB from Analog
                                                    ;Conversion Register
800B    DF 00                     STX $0000H        ;Store MSB & LSB in 0000H & 0001H
                                                    ;RAM memory
800D    96 00                     LDAA $0000H       ;Load A with MSB
800F    85 80                     BITA $80H         ;If EOC bit = 0; set microprocessor
                                                    ;condition code Z bit
8011    27 F3                     BEQ STILLC        ;If EOC = 0; branch and re-test EOC bit.
                                                    ;If EOC = 1; conversion result is in
                                                    ;the LSB of the X register.
                    ;
                    ;The loop within the statements STILLC: NOP and BEQ STILLC can
                    ;be replaced with a delay, which can be obtained in any way
                    ;which is convenient to the designer. After the delay, the
                    ;software must execute a LDX $2000H instruction and the con-
                    ;version result will lie in the LSB of the X register. Table 1142
                    ;shows the delays which are required for each A/D IC.
                    ;               Software for Reading Digital Data Registers
                    ;
8013    FE 60 00                  LDX $6000H        ;Read MSB and LSB from Digital Data
                                                    ;Register. These bytes are now in the
                                                    ;X register.

8016                              END
```

Considerations for Different Microprocessors

Table 11-43 lists the factors to be considered when using the Motorola microprocessors. The crystal oscillator and E pin considerations in column 1 apply to both Interface 1 and 2. Considerations such as slowing the microprocessor (in column 2) apply only to Interface 2. All of the microprocessors can be easily adapted to Interface 1. Table 11-43 and Figure 11-162 enable the designer to readily adapt all of the microprocessors except the 6800 to Interface 2. For the 6800 device with a clock frequency greater than 1 MHz to the A/D converter, an external clock generator which can be slowed down must be used.

Table 11-43. Considerations for Using Other Microprocessors

MICROPROCESSOR	CRYSTAL OSCILLATOR AND E PIN	SLOWING THE MICROPROCESSOR*
6800	ϕ_2 is equivalent to the 6802/6809 E pin	Need an external circuit to slow the microprocessor
6802/6808/6802ns	See Figure 11-159	See Figure 11-159
6809	See Figure 11-159	See Figure 11-159 Substitute MRDY for MR
6809E/68HC09E	See Figure 11-159	Need an external circuit to slow the microprocessor

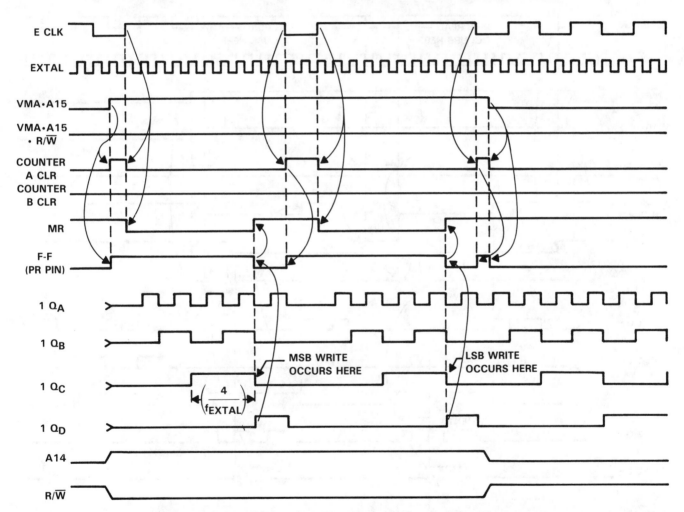

Figure 11-162. Interface 2 Microprocessor Write Timing Diagram

Interface Selection

In general, microprocessors whose clock frequency is less than or equal to 1 MHz can use Interface 1. The TL530, TL531, and TLC533A device specifications are very closely matched but not completely with 1 MHz microprocessor operation and Interface 1. The TLC532A device works satisfactorily with 1 MHz microprocessor operation and Interface 1, although even faster microprocessor clock speeds can be used. If the 68AOX and 68BOX microprocessors are run at full speed, Interface 2 must be used to assure matching between A/D converter and microprocessor specifications. Interface 2 will work with a microprocessor clock frequency that is greater than 2 MHz; however, 2 MHz is generally the upper limit for Motorola microprocessors. Figure 11-164 shows the pertinent timing considerations.

Additional Comments

Although Interface 2 appears complicated, an examination of its write and read timing diagrams in Figures 11-162 and 11-163 reveal the simple strategy of the interface design. This strategy is twofold. First, the lowering of the clock signal during an A/D write causes the microprocessor to leave its temporarily frozen or halted state. This action assures that the necessary data is written into the A/D converter before the microprocessor continues. Thus, a write is guaranteed. Second, the microprocessor's reading of the A/D converter must occur before the clock signal is lowered. Thus, a read is guaranteed. This inexpensive interface allows the designer to take full advantage of these flexible A/D devices.

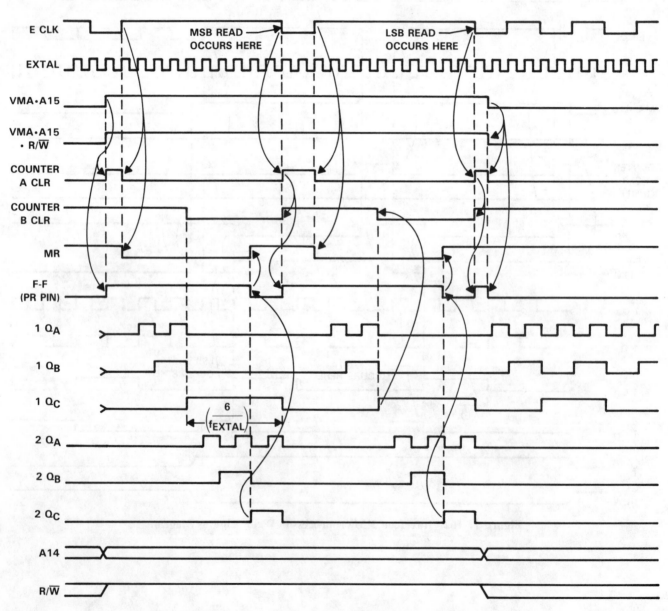

Figure 11-163. Microprocessor Read Timing Diagram

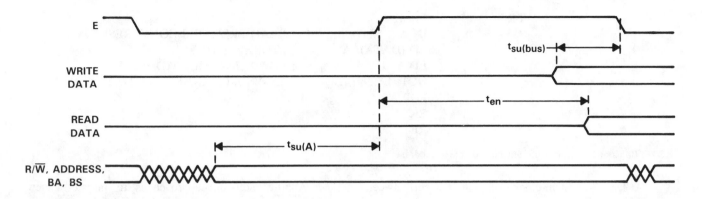

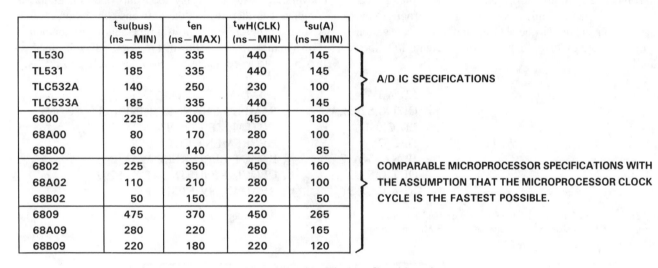

	$t_{su(bus)}$ (ns—MIN)	t_{en} (ns—MAX)	$t_{wH(CLK)}$ (ns—MIN)	$t_{su(A)}$ (ns—MIN)	
TL530	185	335	440	145	⎫ A/D IC SPECIFICATIONS
TL531	185	335	440	145	
TLC532A	140	250	230	100	
TLC533A	185	335	440	145	⎭
6800	225	300	450	180	⎫
68A00	80	170	280	100	
68B00	60	140	220	85	
6802	225	350	450	160	⎬ COMPARABLE MICROPROCESSOR SPECIFICATIONS WITH THE ASSUMPTION THAT THE MICROPROCESSOR CLOCK CYCLE IS THE FASTEST POSSIBLE.
68A02	110	210	280	100	
68B02	50	150	220	50	
6809	475	370	450	265	
68A09	280	220	280	165	
68B09	220	180	220	120	⎭

Figure 11-164. Timing Considerations

INTERFACE FOR THE TLC540 DEVICE TO ZILOG Z80A AND Z80 MICROPROCESSORS

The TLC540 is a complete data acquisition system on a chip. It includes an analog multiplexer, sample-and-hold, 8-bit A/D converter, three 16-bit registers, and microprocessor control inputs. These inputs are chip select ($\overline{CS}$), address in, system clock, and I/O clock. The system clock and I/O clock do not require any special speed or phase relationship, and are normally used independently. This allows the system clock to run at up to 4 MHz, which ensures a conversion time of less than 9 μs, while the I/O clock runs at up to 2.048 MHz to allow a maximum data transfer rate.

Two different interfaces are shown in this application. While both are TLC540 to Z80 interfaces, the two differ as follows:

1. The control signals are generated by additional hardware for Interface 1, but the control signals for Interface 2 are generated through software
2. Interface 1 is about five times faster than Interface 2
3. Interface 1 has less complicated software than Interface 2

4. Interface 1 costs more than Interface 2, but Interface 2 requires less PC board space.

Interface 1

The circuit shown in Figure 11-165 initiates conversion with an OUT instruction. The timing diagram for this circuit is shown in Figure 11-166. When $\overline{IORQ}$ and $\overline{WR}$ go low, $\overline{CS}$ is brought low while the universal shift register is placed in the load data mode and the system clock is enabled to the clock input to latch in the multiplexer address. The rising edge of $\overline{IORQ}$ enables the I/O clock to the shift register to shift the multiplexer address out while shifting the previous conversion result in. Sampling begins at the falling edge of the fourth I/O clock pulse and continues until the eighth falling edge. At this time the I/O clock is disabled and $\overline{CS}$ is brought high to ensure that the TLC540 device will remain undisturbed during the conversion. Conversion of the addressed analog input requires 36 system clock cycles. During this time, previous conversion results may be read by using an IN instruction. A typical interrupt service routine is shown in the software listing that follows.

```
ISR        PUSH AF
           IN A,(LS299)        ; READ PREVIOUS CONV. RESULT
           LD (DATA), A        ; STORE RESULT
           LD A,D              ; LOAD NEW MUX ADDRESS
           OUT (LS299),A       ; SAMPLE CHANNEL [ INITIATE CONV.
           POP AF
           EI
           RETI
```

NOTE: Register D contains the analog multiplexer address.

At least 36 system clock cycles must occur between interrupts to ensure proper operation. If a new multiplexer address is shifted in while a conversion is in progress, the ongoing conversion will be aborted and a new conversion will be initiated by the falling edge of the eighth I/O clock pulse.

Another software approach is to initiate a new conversion, wait in a delay loop until the conversion is complete and then read in the previous conversion results. A simple program segment using the delay loop method is shown in the software listing that follows. Using this method it is possible to complete a conversion cycle in 21.5 μs.

```
           LD A,D              ; LOAD MUX ADDRESS
           OUT (LS299),A       ; SAMPLE CHANNEL [ INITIATE CONV.
           LD C,03H            ; INITIALIZE COUNTER
WAIT:      DEC C               ; DECREMENT COUNTER
           JP NZ, WAIT         ; IF NOT ZERO KEEP WAITING
           OUT (LS299),A       ; CLOCK RESULTS INTO LS299
           IN A,(LS299)        ; READ CONV RESULTS
```

NOTE: A count of 03H will produce a delay of 10.50 microseconds, suitable for 4 MHz operation, while a count of 05H produces a delay of 17.50 microseconds, suitable for 2.5 MHz operation.

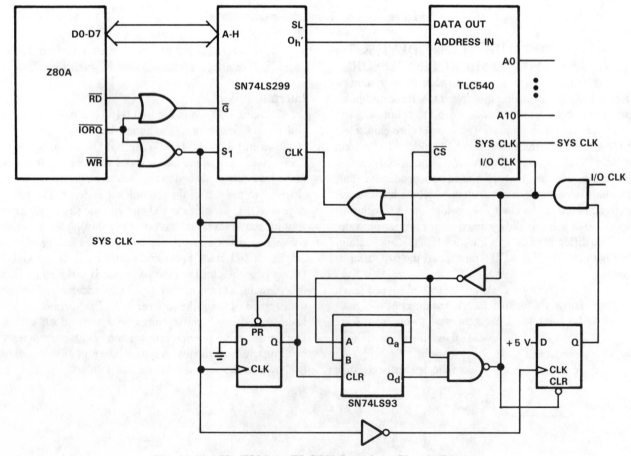

Figure 11-165. Z80A to TLC540 Interface Circuit Diagram

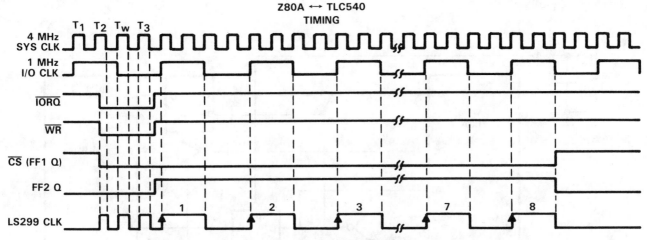

Figure 11-166. Z80A to TLC540 Interface Timing Diagram

Interface 2

The circuit diagram for Interface 2 shows the software controlled TLC540 to Z80A interface and is shown in Figure 11-167. Circuit timing is shown in Figure 11-168.

Execution of an IN instruction causes the $\overline{RD}$ line and the $\overline{IORQ}$ line to become active to shift an address bit into the TLC540 device and a data bit out of the TLC540 device. $\overline{CS}$ is brought low by latching in a low from address bit A0 on the positive going edge of the $\overline{WR}$ signal. A simple program segment that shifts out a new analog multiplexer address while also shifting in previous conversion results is shown in the software listing that follows.

This program segment uses the B register to store the conversion result, the C register as a bit counter, and the D register to hold the analog multiplexer address. The analog multiplexer address is shifted left out of the D register; therefore, the 4-bit address must be placed in the most

significant 4-bits of the byte. Conversion results are read in one bit at a time and then shifted left to the proper position in the B register. Sampling of the addressed input begins at the falling edge of the fourth I/O clock pulse and continues until the falling edge of the eighth I/O clock pulse starts the time conversion. Conversion requires 36 system clock cycles; therefore, an appropriate software delay must be included. The amount of the delay depends on the system clock frequency. If a new multiplexer address is shifted in before a conversion has been completed, the ongoing conversion will be aborted and a new conversion cycle will begin at the eighth falling edge of the I/O clock. $\overline{CS}$ is brought high after the eighth falling edge of the I/O clock to ensure that extraneous noise or glitches on the I/O clock line are not interpreted as the beginning of a new cycle. Using this program segment with the system clock at 4 MHz, it is possible to initiate a new conversion cycle and read the results of the previous conversion in 138 μs.

```
             LD C,08          ; INITIALIZE BIT COUNTER
             LD B,00          ; CLEAR RESULT REGISTER
             OUT (CSLOW),A    ; BRING CHIP SELECT LOW
      LOOP   RLC B            ; ROTATE RESULT LEFT
             LD (HL),D        ; LATCH ADDRESS BIT INTO D FF
             IN A. (BIT)      ; READ IN DATA BIT
             AND 01H          ; MASK OFF BIT 0
             OR B             ; OR NEW BIT INTO RESULT REG.
             LD B,A           ; STORE IN RESULT REGISTER
             RLC D            ; SHIFT ADDRESS LEFT
             DEC C            ; DECREMENT BIT COUNTER
             JP NZ,LOOP       ; GET ANOTHER BIT IF NOT ZERO
             OUT (CSHIGH),A   ; BRING CHIP SELECT HIGH
```

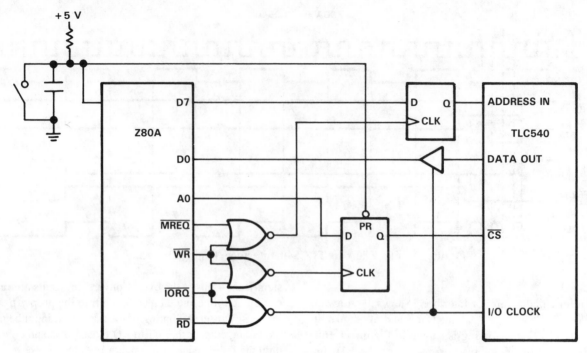

Figure 11-167. Z80A to TLC540 Interface Circuit Diagram

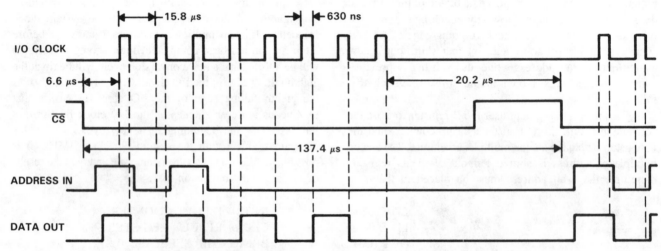

Figure 11-168. Z80A to TLC540 Interface Timing Diagram

INTERFACE FOR THE TLC540 DEVICE TO THE ROCKWELL 6502 MICROPROCESSOR USING THE 6522 VIA

Interfacing the TLC540 A/D converter to the 6502 microprocessor can be accomplished by several methods. This application presents the design of an interface using the 6522 VIA. A method using TTL gates is described in the next application. Cost and performance are the basic trade-offs between the two designs. The 6522 device interface is faster, but the TTL method costs less.

Principles of Operation

The interface circuit diagram is shown in Figure 11-169. Timing for a data read cycle and an address write cycle is shown in Figures 11-170 and 11-171, respectively. A software listing that initializes the 6522 device is shown below. Interface control software is also listed below.

The interface makes use of the serial port available on port B pins CB1 and CB2. Since the serial port is not capable of full duplex communication, the port is configured to function as an output port for the address write cycle and as an input port during the data read cycle. This requires the use of a SN74LS126 3-state buffer. The D-type flip-flops are used to delay the I/O clock to ensure that the set up and hold times for shifting data in and out are met. Port B pins PB0 and PB1 are used to generate $\overline{CS}$ and the output enable signal for the 3-state buffer.

A data conversion cycle begins by bringing $\overline{CS}$ low. This is accomplished by writing a low to PB0. The analog multiplexer address is shifted out by writing to the SR register of the 6522. A delay loop is inserted to wait until the

11-140

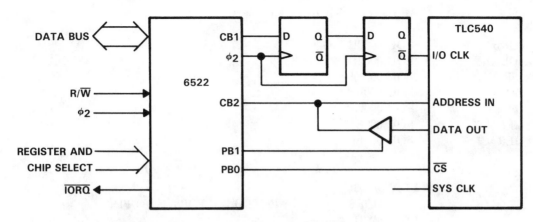

Figure 11-169. 6502 to TLC540 Interface Circuit Diagram (6522 VIA)

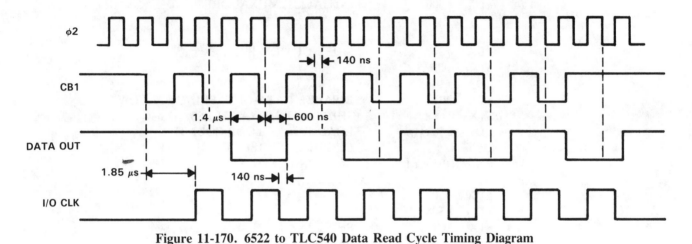

Figure 11-170. 6522 to TLC540 Data Read Cycle Timing Diagram

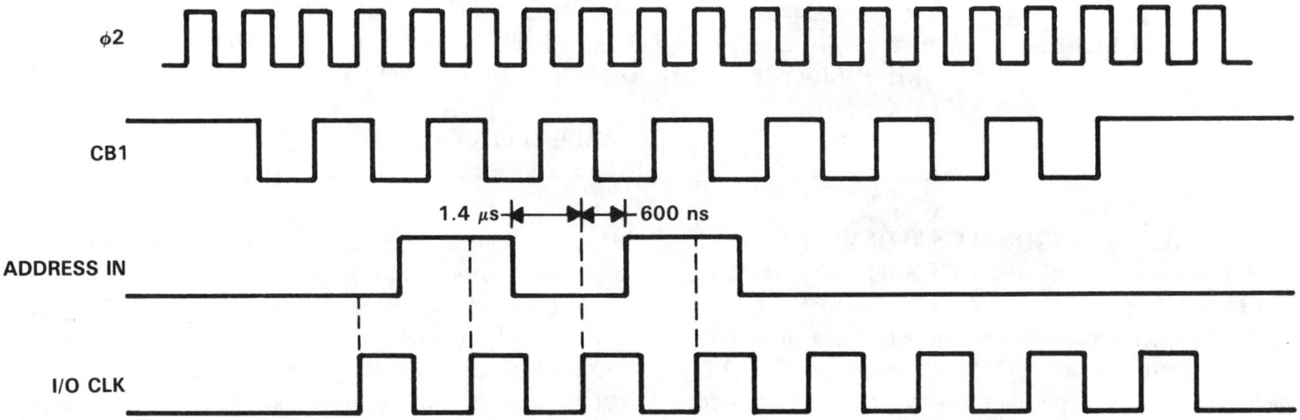

Figure 11-171. 6502 to TLC540 Address Write Cycle Timing Diagram (6522 VIA)

multiplexer address has been shifted out. The serial port is then configured as an input port in order to shift in the A/D conversion results. The output of the 3-state buffer is enabled by writing a high to PB1, and data is shifted into the SR register of the 6522. Again, a delay loop is included to wait until the data is shifted in. $\overline{CS}$ is then brought high, and the 3-state buffer is disabled, completing one data acquisition cycle in 55 μs.

```
;
;
ORB             .EQU 0000H
DDRB            .EQU 0002H
SR              .EQU 000AH
ACR             .EQU 000BH
PCR             .EQU 000CH
IFR             .EQU 000DH
IER             .EQU 000EH
;
;

        LDA   # $03       ;
        STA   DDRB        ; INITIALIZE PORT B I/O PINS
        LDA   # $01       ;
        STA   ORB         ; BRING CHIP SELECT HIGH
        LDX   # $00       ; INITIALIZE MUXADDRESS


        LDA   # $18       ; SHIFT OUT ON PHI 2
        STA   ACR         ;
        LDA   # $00       ;
        STA   ORB         ; DISABLE DATA OUT, BRING /CS LOW
        STX   R           ; SHIFT OUT MUXADDRESS TO 540
        LDY   # $02       ; LOAD DELAY LOOP COUNTER
DELAY1  DEY               ; DECREMENT DELAY COUNTER
        BNE   DELAY1      ; BRANCH IF NOT ZERO
        NOP               ;
        LDX   # $02       ;
        LDA   # $08       ;
        STA   ACR         ; SHIFT IN ON PHI 2
        STX   ORB         ; ENABLE OUTPUT OF 74LS126
        LDA   SR          ; DUMMY LOAD TO SHIFT RESULTS IN
        LDY   # $03       ; LOAD DELAY LOOP COUNTER
DELAY2  DEY               ; DECREMENT DELAY LOOP COUNTER
        BNE   DELAY2      ; BRANCH IF NOT ZERO
        LDA   # $01       ;
        STA   ORB         ; DISABLE DATA OUT, BRING /CS HIGH
        LDA   SR          ; READ CONVERSION RESULTS INTO 6502
```

INTERFACE FOR THE TLC540 DEVICE TO THE ROCKWELL MICROPROCESSOR USING TTL GATES

This application shows an interface between the 6502 microprocessor and the TLC540 A/D converter using TTL gates. The previous application showed an interface between these same two parts using the 6522 VIA. Cost and performance are the basic trade-offs between the two designs. The 6522 device interface is faster, but the TTL method costs less.

Principles of Operation

The basic premise of the interface circuit shown in Figure 11-172 is that all timing control signals are generated under the control of software. Circuit timing is shown in Figure 11-173 and interface control software is listed below.

A data conversion cycle is initiated by bringing $\overline{CS}$ low. This is accomplished by latching a low into the D-type flip-flop from address line A1 on the positive edge of system clock Phi 2. Address bit A14 is used as a gating signal to prevent the TLC540 device from being inadvertently selected during normal program execution. After $\overline{CS}$ is brought low, I/O clock pulses shift in the multiplexer address while shifting out the previous conversion results. The I/O clock is enabled by gating the positive going pulse of Phi 2 to the TLC540 I/O CLK input. This gating occurs by addressing a location so A15 is high. The high on A15 also enables the 3-state buffer output onto the data bus. Address bit A0 determines whether the multiplexer address bit being written is a high or a low. This multiplexer address bit is shifted into the TLC540 device on the positive edge of Phi 2.

Once the data is loaded into the accumulator, it is rotated into the carry bit and then rotated into memory. $\overline{CS}$ is brought high again by writing a high into the D-type flip-flop by placing a high on address bit A1. This cycle can be completed every 172 μs.

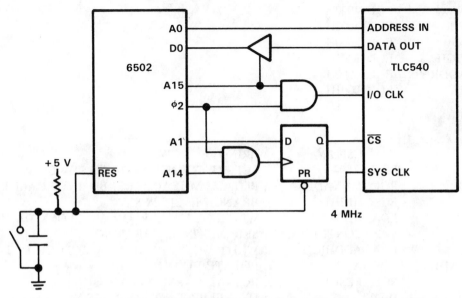

Figure 11-172. 6502 to TLC540 Interface Circuit Diagram (TTL Gates)

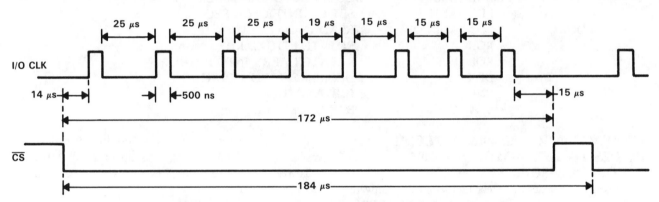

Figure 11-173. 6502 to TLC540 Interface Timing Diagram (TTL Gates)

This interface circuit uses an address decoding scheme that requires a minimum of decoding hardware. Small modifications of the address decoding scheme may be necessary in order to fit the interface to a particular application.

```
CSLOW           .EQU 4000H
CSHIGH          .EQU 4002H
ADDLOW          .EQU 8000H
ADDHIGH         .EQU 8001H
MUXADDRESS      .EQU 0000H
RESULT          .EQU 0001H
;
;
;
                STA  CSLOW         ; BRING /CS LOW
                LDX  #$04          ; SET BIT COUNTER FOR FIRST 4 BITS
LOOP1:          ROL  MUXADDRESS    ; ROTATE MUXADDRESS BIT INTO CARRY
                BCS  HIGH          ; BRANCH IF BIT IS SET
                LDA  ADDLOW        ; WRITE OUT A LOW ON A0, CLOCK DATA IN
                JMP  CONTINUE      ; SKIP NEXT INSTRUCTION
HIGH:           LDA  ADDHIGH       ; WRITE OUT A HIGH ON A0, CLOCK DATA IN
CONTINUE:       ROR  A             ; ROTATE DO INTO CARRY
                ROL  RESULT        ; ROTATE CARRY INTO RESULT
                DEX                ; DECREMENT BIT COUNTER
                BNE  LOOP1         ; GO BACK FOR ANOTHER BIT
                LDX  #$04          ; SET COUNTER FOR SECOND 4 BITS
LOOP2:          LDA  ADDLOW        ; READ IN DATA BIT
                ROR  A             ; ROTATE INTO CARRY
                ROL  RESULT        ; ROTATE CARRY INTO RESULT
                DEX                ; DECREMENT BIT COUNTER
                BNE  LOOP2         ; GET ANOTHER BIT
                STA  CSHIGH        ; BRING /CS HIGH
```

INTERFACE FOR TLC540 AND TLC541 DEVICES TO THE MOTOROLA 6805 MICROPROCESSOR

This application describes techniques for using software controlled interfaces between the TLC540 and TLC541 devices and the 6805 microprocessor. Interfaces for the 6805 microprocessor to the TLC540 and TLC541 devices may be accomplished using either of two methods:

1. Generating all necessary control signals under software control by toggling the output port pins
2. Using the serial peripheral interface (SPI) to generate necessary control signals for data transfer.

The TLC540 and TLC541 devices are particularly well suited for use with the SPI, however, not all 6805 microprocessors include the SPI on the chip. Therefore, the software controlled interface has the advantage although it is less efficient.

Both the TLC540 and TLC541 devices comprise a complete data acquisition system on a chip. The system includes functions such as an analog multiplexer, sample-and-hold, 8-bit A/D converter, data and control registers, and microprocessor-compatible control logic. The four control inputs are $\overline{CS}$, I/O clock, system clock, and address input.

The I/O clock and system clock need no special speed or phase relationship, and are normally used independently. This allows the system clock to run up to 4 MHz for the TLC540 device and up to 2.1 MHz for the TLC541 device.

This ensures conversion times of less than 9 and 17 μs, respectively. The I/O clock runs up to 2.048 MHz and 1.1 MHz, respectively, to allow a maximum data transfer rate.

Principles of Operation

The circuit diagram for the software controlled 6805 microprocessor to TLC540 and TLC541 device interface is shown in Figure 11-174. Circuit timing is shown in Figure 11-175. The software controlled interface makes use of four pins on port A of the 6805 microprocessor. Three of these pins are used as outputs to generate the $\overline{CS}$, I/O clock and, multiplexer address inputs. The remaining pin is used as an input to receive the conversion results from the

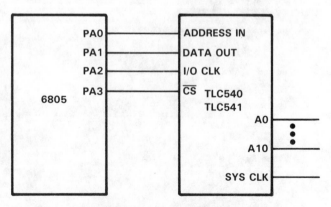

Figure 11-174. 6805 to TLC540 and TLC541 Interface Circuit Diagram

TLC540 and TLC541 devices. A short program segment which can be used to initialize the input/output pins is listed below.

A program listing which controls the actual transfer of data is also given below. This program block sends out the 4-bit analog multiplexer address that is shifted into the TLC540 and TLC541 devices on the first four I/O clock rising edges. Sampling of the addressed input begins at the falling edge of the fourth I/O clock and continues until the falling edge of the eighth I/O clock occurs. Conversion requires 36 clock cycles of the system clock, which can run at up to 4 MHz to allow conversion in only 9 μs. Conversion results are shifted out of the TLC540 and TLC541 devices on the negative edge of the I/O clock with the program block leaving these results in the accumulator. With the 6805 device running at 5 MHz and the TLC540 device system clock at 4 MHz, one conversion cycle can load an analog multiplexer address and read the results from the previous conversion. This cycle can be completed in 270 μs. All 11 analog inputs can be consecutively converted in 3.25 ms.

```
PORTA       .EQU   0000H
DDRA        .EQU   0004H
MUXADDRESS  .EQU   0010H
            .ORG   0100H

START:      LDA    #0DH           ;
            STA    DDRA           ; INITIALIZE PORT A I/O PINS
            BSET   3,PORTA        ; BRING CHIP SELECT HIGH
            LDA    #0DH           ;
            STA    MUXADDRESS     ; INITIALIZE MULTIPLEXER ADDRESS


            LDX    #04            ; LOAD COUNTER FOR FIRST 4 BITS
            BCLR   3,PORTA        ; BRING CHIP SELECT LOW
;
LOOP1:      ROL    MUXADDRESS     ; ROTATE MUXADD BIT INTO CARRY BIT
            BCS    SET            ; GO TO SET IF MUXADD BIT IS 1
            BCLR   0,PORTA        ; WRITE OUT A 0 TO TLC540 ADDRESS IN
            JMP    SKIP           ; SKIP NEXT INSTRUCTION
SET:        BSET   0,PORTA        ; WRITE OUT A 1 TO TLC540 ADDRESS IN
SKIP:       BSET   2,PORTA        ; BRING I/O CLOCK HIGH
            BRSET  1,PORTA,LABEL1 ; READ DATA BIT INTO CARRY BIT
LABEL1:     ROLA                  ; ROTATE NEW DATA BIT INTO ACCUM
            BCLR   2,PORTA        ; BRING I/O CLOCK LOW
            DECX                  ; DECREMENT COUNTER
            BNE    LOOP1          ; CONTINUE IF COUNTER IS NOT ZERO
;
            LDX    #04            ; LOAD COUNTER FOR LAST 4 DATA BITS
;
LOOP2:      BSET   2,PORTA        ; BRING I/O CLOCK HIGH
            BRSET  1,PORTA,LABEL2 ; READ DATA BIT INTO CARRY BIT
LABEL2:     ROLA                  ; ROTATE NEW DATA BIT INTO RESULT
            BCLR   2,PORTA        ; BRING I/O CLOCK LOW
            DECX                  ; DECREMENT COUNTER
            BNE    LOOP2          ; CONTINUE IF COUNTER IS NOT ZERO
;
            BSET   3,PORTA        ; BRING CHIP SELECT HIGH
```

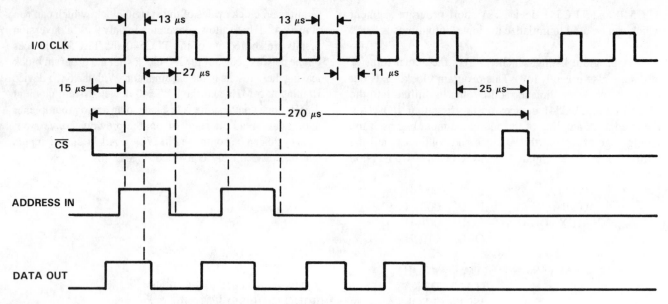

Figure 11-175. 6805 to TLC540 and TLC541 Interface Timing

SOFTWARE INTERFACE FOR TLC540 AND TLC541 DEVICES TO INTEL 8051 AND 8052 MICROPROCESSORS

Two types of interfaces are presented. Figures 11-176 and 11-177, and the following software listing supports Interface 1. In this interface, the A/D converter clock signal is derived from the microprocessor ALE clock. In Interface 2, the A/D converter clock signal is derived from the microprocessor crystal oscillator. This interface is supported by Figures 11-178 and 11-179, and a software listing.

These interfaces minimize the hardware and rely extensively on software techniques. Although the amount of hardware and associated cost is reduced, the use of more software increases the time required to load the address into and retrieve the conversion data from the A/D converter. However, the trade-off of minimum hardware versus longer conversion time may benefit many designs.

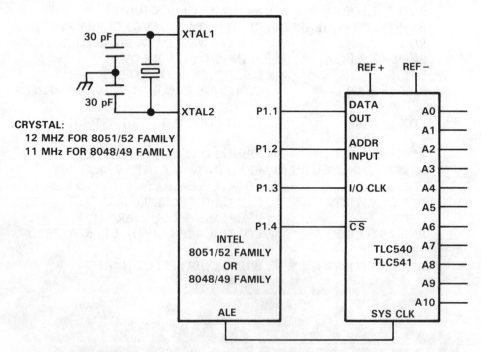

INTEL 8051/52 FAMILY: 8031AH/8051AH/8032AH/8052AH/8751H/80C51
INTEL 8048/49 FAMILY: 8048AH/8748H/8035AHL/8049AH/8749H/8039AHL/8050AH/8040AHL/80C49

Figure 11-176. TLC540 and TLC541 — Intel 8051/52 and 8048/49 Microprocessor Interface 1 Diagram

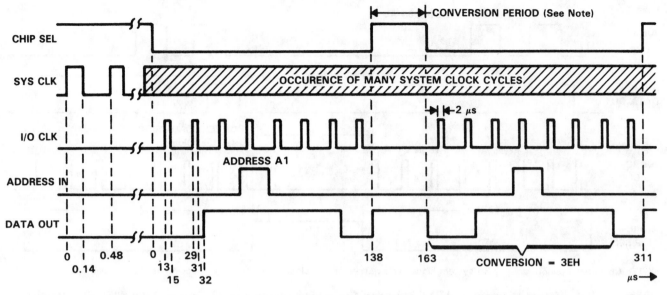

NOTE: Conversion period requires 36 system clock cycles after 8th I/O clock goes low upon clocking in the address.

Figure 11-177. Timing Diagram of an A/D Conversion Cycle for TLC540 and TLC541 to Intel 8051/52 Interface 1

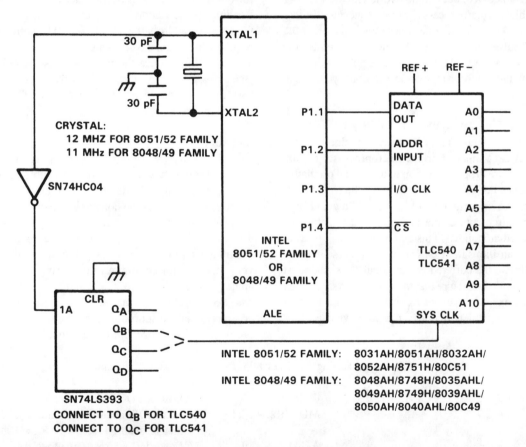

Figure 11-178. TLC540 and TLC541 — Intel 8051/52 and 8048/49 Microprocessor Interface 2 Diagram

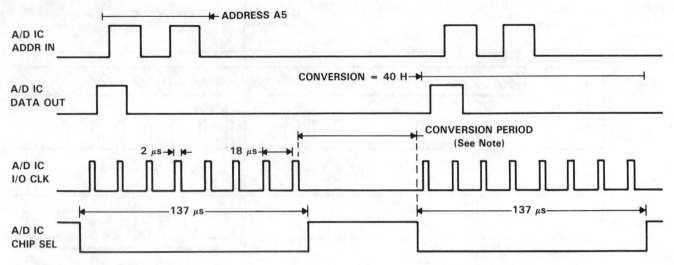

NOTE: Conversion period requires 36 system clock cycles after 8th I/O clock goes low upon clocking in the address.

Figure 11-179. Timing Diagram of A/D Conversion Cycle for TLC540 and TLC541 — Intel 8051/52 Interface 2

Circuit — Interface 1 (ALE CLOCK)

The interface shown in Figure 11-176 will always work with the TLC540 device, but will not work with the TLC541 device at the higher microprocessor clock frequencies. Before using the TLC541 device, the designer must verify that the high and low pulse widths of the ALE clock signal meet the requirements of the TLC541 device. These pulse widths are dependent on the microprocessor instruction cycle clock frequency.

Circuit — Interface 2 (CRYSTAL CLOCK)

As shown in Figure 11-178, the system clock for the A/D converter is obtained from the microprocessor crystal oscillator. To assure proper operation of the crystal oscillator, a high impedance buffer must be used to tap the signal from the oscillator. An important detail is that the low and high level input requirements of the buffer must lie within the range of the oscillator signal. This compatibility will prevent missing edge transitions in the A/D converter's system clock signal. In addition, the buffered crystal oscillator signal must be frequency divided to assure that the resulting system clock signal does not exceed the upper frequency limit of the A/D converter. Any convenient divider circuitry may be used for this task.

Timing Diagrams

Figures 11-177 and 11-179 show the timing diagrams for the A/D conversions of Interfaces 1 and 2, respectively. Loading the new address and retrieving the conversion for the previous address requires 137 μs. The conversion period requires 36 system clocks. This time period is longer for the TLC541 device since the microprocessor clock frequency (see software listing) or A/D clock frequency (see software listing) must be lower to satisfy the TLC541 device specifications.

Software

The software listings for Interface 1 and Interface 2 follow. These software programs use a subroutine, ACALL, that simultaneously loads the A/D converter with a new address and retrieves the conversion result for the previous address. This subroutine can be used any time the designer desires to load a new address and retrieve a conversion value. Simultaneous loading and retrieving makes this subroutine very effective for continuous monitoring of A/D converter analog inputs. The subroutine assumes that the new address has been previously placed in the R2 register. Upon completion of the subroutine, the conversion result for the previous address is left in the accumulator.

```
;   Software Listing for Intel 8051/52 - TLC540/541 Interface 1
;
0000    C2 93           CLR P1.3            ;Lower I/O clock
0002    7A 10           MOV R2, #10H        ;Load A/D analog input address. Note
                                            ;that to send address = 1, R2 = 10h.
0004    11 0E           ACALL S540D         ;Load 540 address, assumes the address
                                            ;is currently in R2.
;
```

```
0006    7B 09                   MOV R3,#09H     ;This software loop allows a
                                                ;conservative 40 A/D system clocks,
                                                ;since only 36
0008    DB FE       DELAY:      DJNZ R3,DELAY   ;clocks are required to perform
                                                ;conversion, to be emitted from the
                                                ;microprocessor ALE pin

            ;
000A    7A 10                   MOV R2,#10H     ;Load A/D analog input address. Note
                                                ;that to send address = 1, R2 = 10h.
000C    11 0E                   ACALL S540D     ;Load new 540 address, assumes this
                                                ;address is in R2; leaves the
                                                ;conversion result for the previous
                                                ;address in A

            ;
            ;                   Subroutine ACALL
            ;
000E    7E 08       S540D:      MOV R6,#08H     ;Set bit counter to 8
0010    C2 94                   CLR P1.4        ;Lower chip select
0012    C3          S540DL:     CLR C           ;Initialize C = 0
0013    30 91 01                JNB P1.1,S540DI ;If 540 Data Out = 0; branch
0016    B3                      CPL C           ;540 Data Out = 1; set C = 1
0017    EA          S540DI:     MOV A,R2        ;Get serial buffer
0018    33                      RLC A           ;Shift Data Out bit into serial
                                                ;buffer and shift 540 address
0019    FA                      MOV R2,A        ;Store serial buffer
001A    50 05                   JNC S540DWO     ;If 540 address bit = 0; branch
001C    43 90 04                ORL P1,#04H     ;Set 540 branch line to 1
001F    80 03                   SJMP S540DWE    ;Go and raise the I/O clock
0021    53 90 FB    S540DWO:    ANL P1,#FBH     ;Set 540 address line to 0
0024    00          S540DWE:    NOP             ;Allow address line to setup
0025    B2 93                   CPL P1.3        ;Raise I/O clock
0027    00                      NOP             ;Delay to slow I/O clock
0028    C2 93                   CLR P1.3        ;Lower I/O clock
002A    DE E6                   DJNZ R6,S540DL  ;Do all 8 bits
002C    B2 94                   CPL P1.4        ;Raise chip select
002E    EA                      MOV A,R2        ;Get serial buffer

            ;
0023    33                      RLC A           ;6543210C 7; b7 is now in carry
0024    33                      RLC A           ;543210C7 6; b6 is now in carry
0025    92 E1                   MOV ACC.1,C     ;54321067 6; put b6 into ACC.1
0027    A2 E2                   MOV C,ACC.2     ;54321067 0; put b0 into C
0029    33                      RLC A           ;43210670 5; b5 is now in carry
002A    92 E3                   MOV ACC.3,C     ;43215670 5; put b5 into ACC.3
002C    A2 E4                   MOV C,ACC.4     ;43215670 1; put b1 into C
002E    33                      RLC A           ;32156701 4; b4 is now in carry
002F    92 E5                   MOV ACC.5,C     ;32456701 4; put b4 into ACC.5
0031    A2 E6                   MOV C,ACC.6     ;32456701 2; put b2 into C
0033    33                      RLC A           ;24567012 3; b3 is now in carry
0034    92 E7                   MOV ACC.7,C     ;34567012 3; put b3 into ACC.7
0036    23                      RL A            ;45670123  ; prepare for SWAP A
0037    C4                      SWAP A          ;01234567  ; bits are ordered
                                                ;correctly
                                                ;conversion result is in accumulator

0038                            END
```

```
;   Software Listing for Intel 8051/52 Family -
;   TC540/541 Serial Interface 2
;
0000    74 0A       SR540L:     MOV A,#0AH          ;A/D IC address of 5 in this example
;
;The serial port will send 0A(00001010) with the least
;significant bit first. Therefore, the A/D IC will see
;(01010000), which will load an address of 5 into the control
;register.
;
0002    C2 97                   CLR P1.7            ;Disable 125
0004    B2 97                   CPL P1.7
0006    C2 96                   CLR P1.6            ;Lower chip select
0008    53 98 ED                ANL SCON,#EDH       ;Reset REN & TI flags
000B    F5 99                   MOV SBUF,A          ;Send 540 address (LSB FIRST)
000D    30 99 FD    SNDTST:     JNB SCON.1,SNDTST   ;TI flag not set; branch
                                                    ;until transmission is complete
;
;A delay must occur here to allow the A/D IC to complete
;conversion. The delay must llow 36 A/D IC system clock
;cycles to occur.
;
0010    B2 97                   CPL P1.7            :Enable 125
0012    43 98 10                ORL SCON,#10H       :Set REN
0015    53 98 FE                ANL SCON,#FEH       :Reset RI
0018    30 98 FD    RCVTST:     JNB SCON.0,RCVTST   :RI FLAG not set; branch
                                                    :until reception is complete
001B    B2 96                   CPL P1.6            ;Raise chip select
001D    E5 99                   MOV A,SBUF          ;Get SBUF
;
;The serial port read reverses the data conversion bits coming
;to the microprocessor so that they are in the following order:
;b0(1sb),b1,b2,b3,b4,b5,b6,b7(msb). These bits (01234567) along
;with the carry bit (C) in the following instruction comments
;are presented so that the reader will understand the technique,
;which is used to place the bits in their proper order.
;
001F    33                      RLC A               ;6543210C 7; b7 is now in carry
0020    33                      RLC A               ;543210C7 6; b6 is now in carry
0021    92 E1                   MOV ACC.1,C         ;54321067 6; put b6 into ACC.1
0023    A2 E2                   MOV C,ACC.2         ;54321067 0; put b0 into C
0025    33                      RLC A               ;43210670 5; b5 is now in carry
0026    92 E3                   MOV ACC.3,C         ;43215670 5; put b5 into ACC.3
0028    A2 E4                   MOV C,ACC.4         ;43215670 1; put b1 into C
002A    33                      RLC A               ;32156701 4; b4 is now in carry
002B    92 E5                   MOV ACC.5,C         ;32456701 4; put b4 into ACC.5
002D    A2 E6                   MOV C,ACC.6         ;32456701 2; put b2 into C
002F    33                      RLC A               ;24567012 3; b3 is now in carry
0030    92 E7                   MOV ACC.7,C         ;34567012 3; put b3 into ACC.7
0032    23                      RL A                ;45670123 ; prepare for SWAP A
0033    C4                      SWAP A              ;01234567 ; bits are ordered
                                                    ;correctly
                                                    ;conversion result is in accumulator

0034                            END
```

Hardware

The circuit for Interface 1, which obtains the clock signal from the ALE pin, is shown in Figure 11-180. The circuit for Interface 2, which obtains the clock signal from the crystal oscillator, is illustrated in Figure 11-182. The signal at the microprocessor TXD pin must be inverted so the communication protocols for the microprocessor and the A/D converter are compatible. Use of a SN74LS125 3-state buffer allows the microprocessor serial port to be used for both transmission and reception. Although this method does not permit simultaneous loading of a new address while retrieving the conversion of a previously loaded address, the time loss is small. This is because the serial port can quickly load an address and retrieve the resulting conversion.

Timing Diagrams

Figures 11-181 and 11-183 show the timing diagrams for an A/D conversion for each of the two circuits. Loading the address, waiting for conversion, and retrieving the conversion result requires 56 μs for Interface 1 and 48 μs for Interface 2 when using the TLC540 device. This time period is longer for the TLC541 device because the system clock frequency must be lower for this device.

Software

The software routines for Interface 1 and Interface 2 were listed earlier. The serial port mode 0 is used. If desired, the software can be incorporated into a subroutine so the designer can address the software with a simple subroutine

call. Careful attention must be exercised when placing the address bits in the serial buffer since the serial port sends the least significant bit first and the A/D converter accepts this bit as the most significant bit of the address. A similar process occurs when the serial port receives the conversion result.

SOFTWARE INTERFACE FOR TLC540 AND TLC541 DEVICES TO INTEL 8048 AND 8049 MICROPROCESSORS

Hardware

Figures 11-176 and 11-178 show the circuit diagrams for this application. Also refer to the SOFTWARE INTERFACE FOR TLC540 and TLC541 DEVICES TO INTEL 8051 AND 8052 MICROPROCESSORS application.

Timing Diagram

Figure 11-184 of this application shows the timing of an A/D conversion. Loading the new address and retrieving the conversion for the previous address requires 250 μs. The conversion period requires 36 system clocks, however, this period is longer for the TLC541 device because the system clock frequency must be lower for this device.

Software

Software listings are attached below. Also refer to the SOFTWARE INTERFACE FOR TLC540 AND TLC541 DEVICES TO INTEL 8051 AND 8052 MICRO-PROCESSORS application.

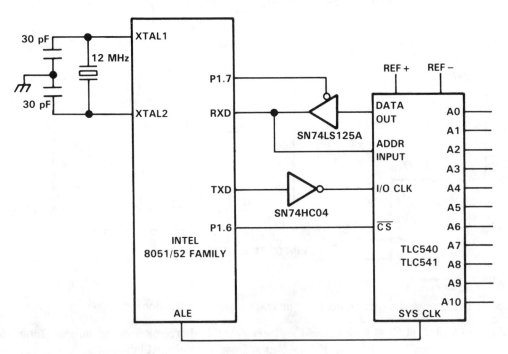

INTEL 8051/52 FAMILY: 8031AH/8051AH/8032AH/8052AH/8751H/80C51
CONNECT TO Q$_B$ FOR TLC540
CONNECT TO Q$_C$ FOR TLC541

Figure 11-180. TLC540 and TLC541 — Intel 8051/52 Microprocessor Interface 1 Diagram (Clock Signal from ALE)

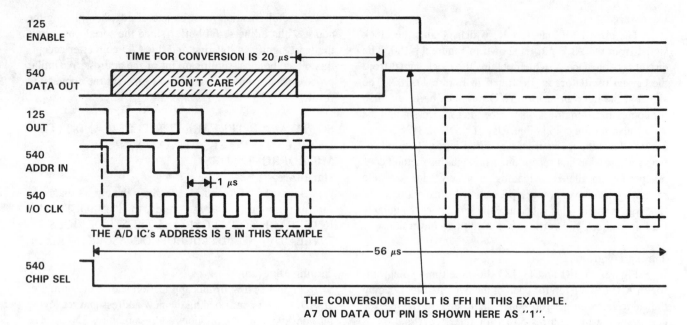

Figure 11-181. Timing Diagram for an A/D Conversion for the TLC540 and TLC541 — Intel 8051/52 Microprocessor Serial Port Interface 1

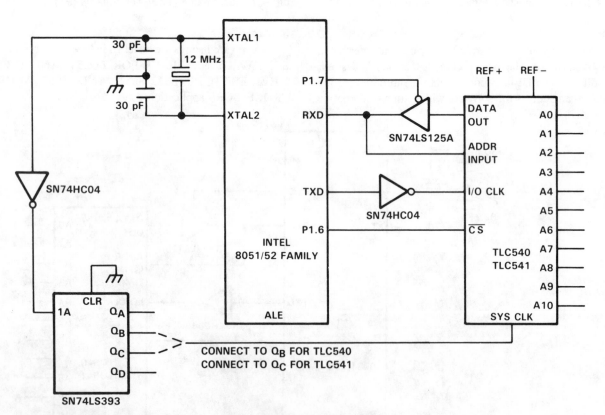

INTEL 8051/52 FAMILY: 8031AH/8051AH/8032AH/8052AH/8751H/80C51

Figure 11-182. TLC540 and TLC541 — Intel 8051/52 Microprocessor Serial Port Interface 2 Diagram (Clock Signal from Crystal Oscillator)

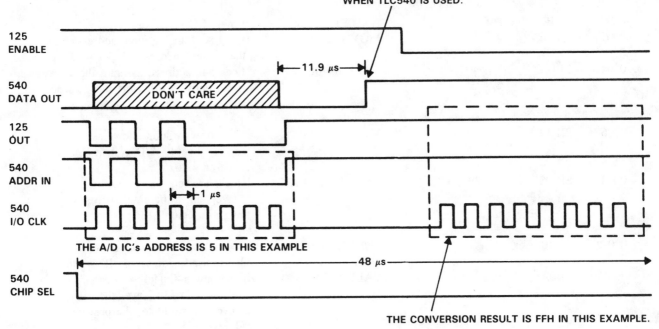

IN THIS EXAMPLE, THE CONVERSION RESULT IS FFH. A7 ON DATA PIN IS SHOWN HERE AS "1". CONVERSION TAKES 11.9 μs WHEN TLC540 IS USED.

125 ENABLE

540 DATA OUT — DON'T CARE — ⟵ 11.9 μs ⟶

125 OUT

540 ADDR IN

540 I/O CLK — 1 μs

THE A/D IC's ADDRESS IS 5 IN THIS EXAMPLE

540 CHIP SEL — 48 μs

THE CONVERSION RESULT IS FFH IN THIS EXAMPLE.

Figure 11-183. Timing Diagram for an A/D Conversion for the TLC540 and TLC541 — Intel 8051/8052 Microprocessor Serial Port Interface 2

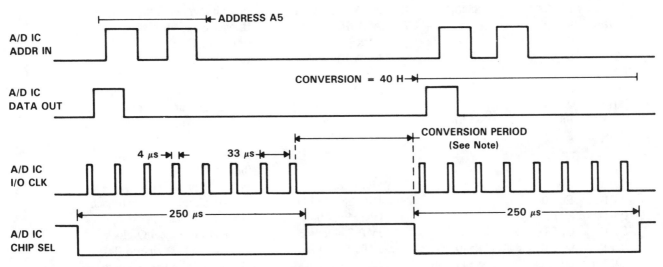

A/D IC ADDR IN — ADDRESS A5

A/D IC DATA OUT — CONVERSION = 40 H

CONVERSION PERIOD (See Note)

A/D IC I/O CLK — 4 μs — 33 μs

A/D IC CHIP SEL — 250 μs — 250 μs

NOTE: Conversion period requires 36 system clock cycles after 8th I/O clock goes low upon clocking in the address.

Figure 11-184. Timing Diagram of A/D Conversion Cycle for TLC540 and TLC541 — Intel 8048/49 Microprocessor Interface 2

```
0000   99 F7                   ANL P1, #F7H        ;Lower I/O clock
0002   B8 10                   MOV R0, #10H        ;Load A/D analog input address. Note
                                                   ;that to send address = 1, R2 = 10h.
0004   14 0A                   CALL S540D          ;Load 540 address, assumes the address
                                                   ;is currently in R0. Note that to send
                                                   ;address = 1, R0 = 10h.

;

0006   BB 13                   MOV R3, #13H        ;This software loop allows a
                                                   ;conservative 40 A/D system clocks,
                                                   ;since only 36
0008   EB 08       DELAY:      DJNZ R3,DELAY       ;clocks are required to perform
                                                   ;conversion, to be emitted from the
                                                   ;microprocessor ALE pin

;

000A   B8 10                   MOV R0, #10H        ;Load A/D analog input address. Note
                                                   ;that to send address = 1, R2 = 10h.
000C   14 0A                   CALL S540D          ;Load new 540 address, assumes this
                                                   ;address is in R0; leaves the
                                                   ;conversion result for the previous
                                                   ;address in A

;
;                                      Subroutine S540D
;

000E   BA 08       S540D:      MOV R2, #08H        ;Set bit counter to 8
0010   99 EF                   ANL P1, #EFH        ;Lower chip select
0012   97          S540DL:     CLR C               ;Initialize C = 0
0013   09                      IN A,P1             ;Get Port 1
0014   37                      CPL A               ;Complement accumulator
0015   32 17                   JB1 S540DI          ;If 540 Data Out = 0; branch
0017   A7                      CPL C               ;540 Data Out = 1; set C = 1
0018   F8          S540DI:     MOV A,R0            ;Get serial buffer
0019   F7                      RLC A               ;Shift Data Out bit into serial
                                                   ;buffer and shift 540 address
                                                   ;bit into C
001A   A8                      MOV R0,A            ;Store serial buffer
001B   E6 20                   JNC S540DWO         ;If 540 address bit = 0; branch
001D   89 04                   ORL P1, #04H        ;Set 540 branch line to 1
001F   04 21                   JMP S540DWE         ;Go and raise the I/O clock
0021   99 FB       S540DWO:    ANL P1, #FBH        ;Set 540 address line to 0
0023   00          S540DWE:    NOP                 ;Allow address line to setup
0024   89 08                   ORL P1, #08H        ;Raise I/O clock
0026   00                      NOP                 ;Delay to slow I/O clock
0027   99 F7                   ANL P1, #F7H        ;Lower I/O clock
0029   EA 11                   DJNZ R2,S540DL      ;Do all 8 bits
002B   89 10                   ORL P1, #10H        ;Raise chip select
002D   F8                      MOV A,R0            ;Get serial buffer
002E   83                      RET
```

```
;   Software Listings for Intel 8048/49 Family -
;   TLC540/541 Interface 2
;

0000    99 F7                       ANL P1, #F7H        ;Lower I/O clock
0002    B8 10                       MOV R0, #10H        ;Load A/D analog input address. Note
                                                        ;that to send address = 1, R2 = 10h.
0004    14 06                       CALL S540D          ;Load 540 address, assumes the address
                                                        ;is currently in R0. Note that to send
                                                        ;address = 1, R0 = 10h.

;
;A delay must occur here to allow the A/D IC to complete
;conversion. The delay must allow 40 A/D IC system clock
;cycles to occur.
;

0006    B8 10                       MOV R0, #10H        ;Load A/D analog input address. Note
                                                        ;that to send address = 1, R2 = 10h.
0008    14 06                       CALL S540D          ;Load new 540 address, assumes this
                                                        ;address is in R0; leaves the
                                                        ;conversion result for the previous
                                                        ;address in A

;
;                   Subroutine S540D
;

000A    BA 08       S540D:          MOV R2, #08H        ;Set bit counter to 8
000C    99 EF                       ANL P1, #EFH        ;Lower chip select
000E    97          S540DL:         CLR C               ;Initialize C = 0
000F    09                          IN A,P1             ;Get Port 1
0010    37                          CPL A               ;Complement accumulator
0011    32 14                       JB1 S540DI          ;If 540 Data Out = 0; branch
0013    A7                          CPL C               ;540 Data Out = 1; set C = 1
0014    F8          S540DI:         MOV A,R0            ;Get serial buffer
0015    F7                          RLC A               ;Shift Data Out bit into serial
                                                        ;buffer and shift 540 address
                                                        ;bit into C
0016    A8                          MOV R0,A            ;Store serial buffer
0017    E6 1D                       JNC S540DWO         ;If 540 address bit = 0; branch
0019    89 04                       ORL P1, #04H        ;Set 540 branch line to 1
001B    04 1F                       JMP S540DWE         ;Go and raise the I/O clock
001D    99 FB       S540DWO:        ANL P1, #FBH        ;Set 540 address line to 0
001F    00          S540DWE:        NOP                 ;Allow address line to setup
0020    89 08                       ORL P1, #08H        ;Raise I/O clock
0022    00                          NOP                 ;Delay to slow I/O clock
0023    99 F7                       ANL P1, #F7H        ;Lower I/O clock
0025    EA 0E                       DJNZ R2,S540DL      ;Do all 8 bits
0027    89 10                       ORL P1, #10H        ;Raise chip select
0029    F8                          MOV A,R0            ;Get serial buffer
002A    83                          RET
```

INTERFACE FOR THE TLC549 DEVICE TO THE ZILOG Z80A MICROPROCESSOR

This application describes a technique for interfacing the TLC549 A/D converter to the Z80A microprocessor. The TLC549 device is a complete data acquisition system on a chip. It differs from the TLC540 device since the TLC549 device has one analog input and an on-chip system clock while the TLC540 device has 11 analog inputs and requires an external system clock input. These differences allow the TLC549 device to be packaged in an 8-pin DIP to reduce size and cost. Timing is the same as the TLC540 device except that it does not require an analog multiplexer address since it is a single channel data acquisition system. The I/O clock may run at any frequency up to 2.048 MHz.

Principles of Operation

The circuit diagram for the TLC549 device to Z80A microprocessor interface is shown in Figure 11-185. The circuit timing is shown in Figure 11-186.

$\overline{CS}$ is brought low by latching a low from address bit A0. Execution of an IN instruction causes $\overline{RD}$ and $\overline{IORQ}$ to become active and generate one I/O clock pulse. A data bit is read before the falling edge of the I/O clock and the falling edge shifts out the next data bit. Sampling of the analog input begins at the falling edge of the fourth I/O clock and continues until the falling edge of the eighth I/O clock. At that time, conversion begins and $\overline{CS}$ is brought high. Since $\overline{CS}$ high disables all inputs and outputs, conversion may proceed. Conversion requires 17 μs. A simple program segment that starts a conversion and reads in previous conversion results is shown in the software listing. If this program segment is placed in a loop, it is possible to initiate a conversion and read previous results in 111 μs.

Figure 11-185 presents a software controlled interface for the Z80A and the TLC549 devices. It is possible to increase performance by introducing additional hardware into the interface. For more description of such an interface, refer to the TLC540 DEVICE INTERFACE FOR THE TLC540 DEVICE TO ZILOG Z80A AND Z80 MICRO-PROCESSORS application.

```
            LD  C,08H          ; Load bit counter
            LD  B,00H          ; Initialize result register
            OUT (CSLOW),A      ; Bring chip select low
LOOP        RLC B              ; Rotate result reg. left
            IN  A, (BIT)       ; Read in a bit & clock next
            AND 01H            ; Mask off bit 0
            OR  B              ; Or new bit with result
            LD  B,A            ; Store in results register
            DEC C              ; Decrement bit counter
            JP  NZ,LOOP        ; Get another bit if not zero
            OUT (CSHIGH),A     ; Bring chip select high
```

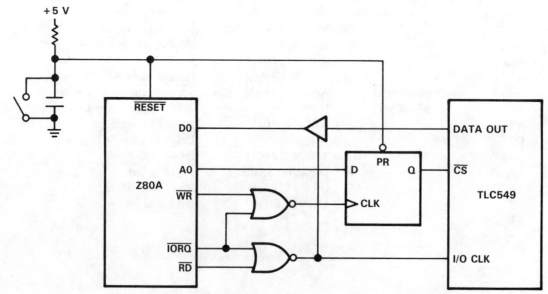

Figure 11-185. Z80A to TLC549 Interface Circuit Diagram

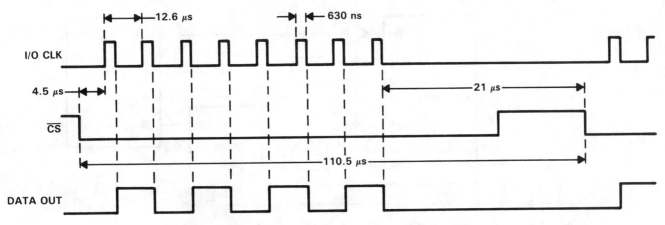

Figure 11-186. Z80A to TLC549 Interface Timing Diagram

INTERFACE FOR THE TLC549 DEVICE TO THE ROCKWELL 6502 MICROPROCESSOR USING TTL GATES

This application presents the design of an interface for the TLC549 A/D converter to the 6502 microprocessor. Interfacing techniques for the TLC549 device are similar to those of the TLC540 A/D peripheral chip. The interface uses a small number of TTL gates similar to the INTERFACE FOR THE TLC540 DEVICE TO THE ROCKWELL 6502 MICROPROCESSOR USING TTL GATES presented in a previous application. Since the TLC549 device converts only one analog input channel, no multiplexer address is required. This reduces the amount of software code and execution time required.

Principles of Operation

The TLC549 device to 6502 microprocessor interface circuit is shown in Figure 11-187. The timing diagram is shown in Figure 11-188. A data conversion cycle is initiated when $\overline{CS}$ is brought low. This occurs when a low on address line A0 is latched into the D-type flip-flop on the positive edge of clock Phi 2. Address bit A14 is used as a gating signal to prevent the TLC549 device from inadvertently being selected during normal program execution.

Data is shifted out of the TLC549 device on the positive edge of the I/O clock which is generated by gating the positive going edge of clock Phi 2 with address line A15. When A15 goes high, the output of the 3-state buffer is enabled onto the data bus and the conversion result bit is shifted into the 6502 device on the trailing edge of clock Phi 2. Once the data is in the accumulator, it is rotated into the carry bit, and then rotated into memory. $\overline{CS}$ is then brought high by latching a high into the D-type flip-flop. This is done by placing a high on address bit A0. This cycle can be completed every 133 μs.

It should be noted that this circuit uses an address decoding scheme that requires a minimum of decoding hardware. Small modifications to the decoding scheme may be necessary to match the interface properly to a particular application. The software listing for this application follows.

```
CSHIGH      .EQU 4001H
CSLOW       .EQU 4000H
RESULT      .EQU 0000H
CLOCK       .EQU 8000H
;
;
;
            STA   SLOW         ; Bring /CS low
            LDX   #$08         ; Initialize bit counter
LOOP        LDA   LOCK         ; Clock in data bit
            ROR   A            ; Rotate do into carry bit
            ROL   RESULT       ; Rotate carry bit into result
            DEX                ; Decrement bit counter
            DNE   LOOP         ; Get another bit if not zero
            STA   CSHIGH       ; Bring /CS high
```

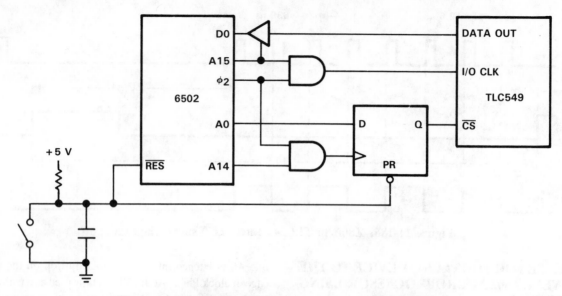

Figure 11-187. 6502 to TLC549 Interface Circuit Diagram

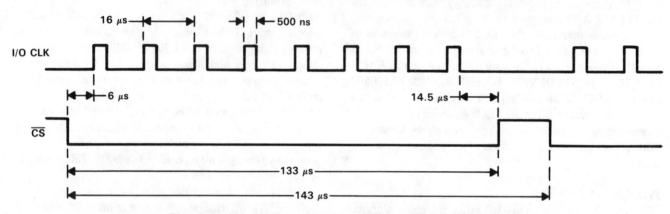

Figure 11-188. 6502 to TLC549 Interface Timing Diagram

INTERFACE FOR THE TLC549 DEVICE TO THE ROCKWELL 6502 MICROPROCESSOR USING THE 6522 VIA

This application presents the design of an interface circuit for the TLC549 A/D to the 6502 microprocessor through the 6522 VIA. Interfacing techniques for the TLC549 are very similar to those for the TLC540. The interface presented in this application utilizes the 6522 VIA, and is similar to the INTERFACE FOR THE TLC549 DEVICE TO THE ROCKWELL 6502 MICROPROCESSOR USING THE 6522 VIA presented in a previous application. Since the TLC549 converts only one analog input, no multiplexer address is required, thus, less software is required to control the interface.

Principles of Operation

The TLC549 to 6522 interface circuit is shown in Figure 11-189. The timing diagram is shown in Figure 11-190. An initialization software listing and an interface control software listing are included below.

A data conversion cycle begins by bringing $\overline{CS}$ low. This is accomplished by writing a low to port B output pin PBO. Previous conversion results are shifted in by reading the SR (Figure 11-188). The D-type flip-flops (Figure 11-186) are included to effectively delay the I/O clock in order to meet all of the set up and hold time requirements for shifting the data. A delay loop is included (Figure 11-188) to allow for the data to be shifted in. On the eighth falling edge of the I/O clock, conversion of the analog input begins, and is completed in 17 μs. $\overline{CS}$ is then brough high by writing a high to PBO. Previous conversion results can then be read into the 6502 from the 6522 while a new conversion is in progress. One cycle as described here can be completed in 48 μs.

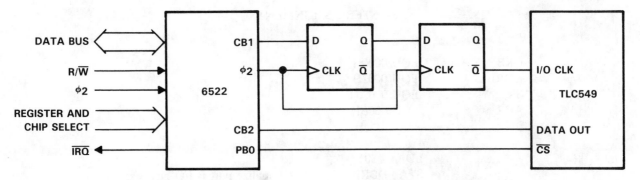

Figure 11-189. 6522 to TLC549 Interface Circuit Diagram

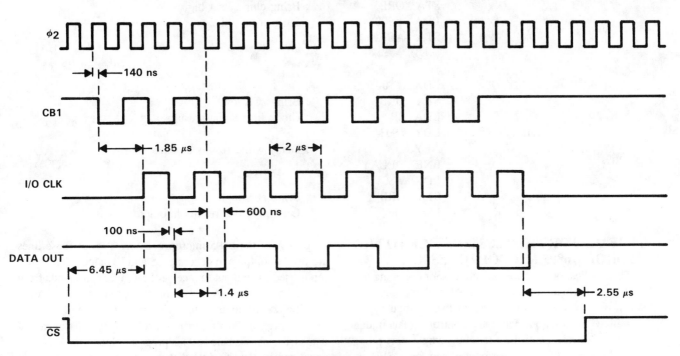

Figure 11-190. 6522 to TLC549 Interface Timing Diagram

```
;
;
ORB              .EQU 0000H
DDRB             .EQU 0002H
SR               .EQU 000AH
ACR              .EQU 000BH
;
;
                 LDA  # $01     ;
                 STA  DDRB      ; Initialize port B I/O pins
                 LDA  # $01     ;
                 STA  ORB       ; Bring chip select high
                 LDA  # $08     ; Shift in on phi 2
                 STA  ACR       ;


                 LDX  # $00     ;
                 STX  ORB       ; Bring /CS low
                 LDA  SR        ; Clock in previous results
DELAY:           LDY  # $03     ;
                 DEY            ; Delay while results clocked in
                 BNE  DELAY     ;
                 LDX  # $01     ;
                 STX  ORB       ; Bring /CS high
                 LDA  SR        ; Load previous results into acc
```

INTERFACE FOR THE TLC549 DEVICE TO THE MOTOROLA 6805 MICROPROCESSOR

This application describes techniques for operating the TLC549 A/D peripheral chip with the 6805 microprocessor. This can be accomplished by either of two methods:

1. Generating all necessary control signals under software control by toggling output port pins
2. Using the serial peripheral interface (SPI) to generate necessary control signals for data transfer.

The TLC549 device is well suited for use with the SPI, however, not all 6805 microprocessors include the SPI on the chip. Therefore, the software controlled interface has the advantage although it is less efficient.

The TLC549 device is a complete data acquisition system on a chip. The system includes functions such as sample-and-hold, 8-bit A/D converter, on-chip oscillator, and microprocessor-compatible control logic. Data can be acquired with only three I/O lines: $\overline{CS}$, I/O clock, and a serial data line. Conversion requires 17 μs, and the I/O clock can run up to 2.048 MHz.

Principles of Operation

The circuit diagram for the software controlled TLC549 device to 6805 microprocessor interface is shown in Figure 11-191. Circuit timing is shown in Figure 11-192. The software controlled interface makes use of three pins on port A of the 6805 microprocessor. Two of these lines are used as outputs to generate $\overline{CS}$ and I/O clock. The other pin is used as an input to receive the conversion results from the TLC549 device. A short program listing that can be used to initialize the input/output pins is given below.

A program listing that controls the actual transfer of data is also shown below. This program block brings $\overline{CS}$ low and generates the I/O clock that shifts the previous conversion results out of the TLC549 device. Sampling of the analog input begins at the falling edge of the fourth I/O clock and continues until the falling edge of the eighth I/O clock occurs. When sampling ends, conversion begins and is completed in 17 μs. With the 6805 device running at 5 MHz, one conversion cycle that initiates a new conversion and reads the results of the previous conversion can be completed in 204 μs.

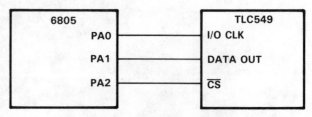

Figure 11-191. 6805 to TLC549 Interface Circuit Diagram

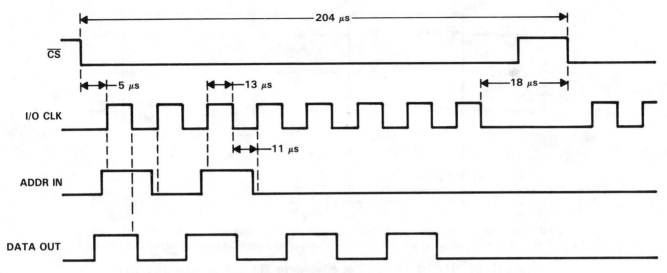

Figure 11-192. 6805 to TLC549 Interface Timing Diagram

```
PORTA       .EQU   0000H
DDRA        .EQU   0004H
            .ORG   0100H
START:      LDA    #05H         ;
            STA    DDRA         ; Initialize port a I/O pins
            BSET   2,PORTA      ; Bring chip select high




            LDX    #08          : Initialize counter
            BCLR   2,PORTA      ; Bring chip select low
LOOP:       BSET   0,PORTA      ; Bring I/O clock high
            BRSET  1,PORTA,LABEL ; Read data bit into carry bit
LABEL:      ROLA                ; Rotate carry into accumulator
            BCLR   0,PORTA      ; Bring I/O clock low
            DECX                ; Decrement counter
            BNE    LOOP         ; Continue if counter is not zero
            BSET   2,PORTA      ; Bring chip select high
```

SOFTWARE INTERFACE FOR THE TLC549 DEVICE TO INTEL 8048 AND 8049 MICROPROCESSORS

The operation of the TLC549 device is similar to the more complex TLC540 and TLC541 devices. The TLC549 device provides an on-chip system clock that operates typically at 4 MHz. The on-chip system clock allows internal device operation to proceed independently of serial input/output data timing. This independence permits manipulation of the TLC549 device for a wide range of software and hardware requirements.

This software interface application features:

1. A low-cost A/D converter
2. Minimum hardware requirement
3. Remote control of serial A/D converters.

Hardware

Figure 11-193 shows the circuit for the 8048 and 8049 microprocessor to TLC549 device software interface. This interface uses only three port pins of the microprocessor. While the TLC549 device internal sample-and-hold circuitry is accessing and holding the new analog signal, Port 1.3 and $\overline{CS}$ must be low. I/O clock is transmitted through port 1.2 to pin 7 of the TLC549 device. Port 1.1 acquires the conversion data from the data output pin of the TLC549 device.

Timing Diagram

Figure 11-194 shows the timing diagram for the interface. The timing diagram indicates that one complete conversion cycle requires 454 μs.

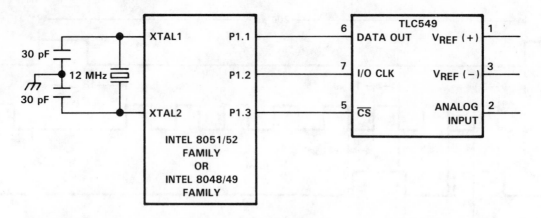

INTEL 8051/52 FAMILY: 8031AH/8051AH/8032AH/8052AH/8751AH/80C51
INTEL 8048/49 FAMILY: 8048AH/8748H/8035AHL/8049AH/8749H/8039AHL/8050AH/8040AHL/80C49

Figure 11-193. The Intel 8051/52 or 8048/49 to TLC549 Software Interface

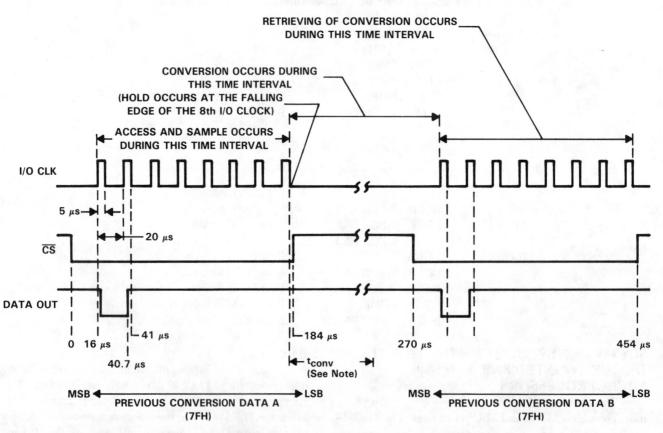

NOTE; t$_{conv}$: The conversion cycles, which requires 36 internal system clock periods, is initiated with the 8th I/O clock↓ after $\overline{CS}$↓. Conversion time requires a minimum of 17µs.

Figure 11-194. Timing Diagram for the Intel 8048/49 — TLC549 Software Interface

After $\overline{CS}$ goes low, eight I/O clocks access and sample the new analog signal from the analog input. At the same time, I/O clock falling edges shift out the previous conversion result. This conversion result is only valid for the previous analog signal sample.

Conversion occurs in the time interval starting when the eighth I/O clock goes low. The conversion time interval requires 36 internal system clock cycles. The designer cannot see the system clock, but the designer can obtain the conversion time from the data sheets. Typically, a time interval of 17 µs is required for conversion.

See the SOFTWARE INTERFACE FOR TLC540 AND TLC541 DEVICES TO INTEL 8048 AND 8049 MICROPROCESSORS application for more information.

Software

The interface software program listing follows. Also see the SOFTWARE INTERFACE FOR THE TLC549 DEVICE TO INTEL 8051 AND 8052 MICRO-PROCESSORS application. There are differences between the software for the 8051 and 8052 devices and for the 8048 and 8049 devices. The JB1 instruction is used to detect the polarity of the conversion result bits for the 8048 and 8049 device interface. If desired, the software can be incorporated into a subroutine so the designer can access the software with a simple subroutine call.

```
; Software Listing for Intel 8048/49
; Interface — TLC549
;
;
;
                CALL S549D      ; Access, sample and hold the new
                                ; analog signal.
                                ;
                                ; A delay must occur here to allow the A/D
                                ; chip to complete conversion. The delay
                                ; must allow 36 A/D chip internal system
                                ; clock cycles to occur. Conversion time
                                ; requires a minimum of 17 microseconds.
                                ;
                CALL S549D      ; Access, sample and hold the new
                                ; analog signal. Bring out the previous
                                ; conversion result.
                                ;
                                ; Subroutine CALL
                                ;
S549D           ANL P1, # FBH   ; Lower I/O clock
                MOV R0, # 08H   ; Set bit counter to 8
                ANL P1, # F7H   ; Lower Chip Select
S549H           CLR C           ; Initialize C = 0
                IN A,P1         ; Get Port 1
                CPL A           ; Compliment Accumulator
                JB1 S549J       ; If 549 data out = 0; branch
                CPL C           ; 549 data out = 1; Set C = 1
S549J           MOV A,R1        ; Get serial buffer
                RLC A           ; Shift data out bit into serial buffer
                MOV R1,A        ; Store serial buffer
                ORL P1, # 04H   ; Raise I/O clock
                NOP             ; Delay to slow I/O clock
                ANL P1, # FBH   ; Lower I/O clock
                DJNZ R0,S549H   ; Do 8 times
                ORL P1, # 08H   ; Raise Chip Select
                MOV A,R1        ; Conversion data in A
                RET             ;
                END             ;
```

SERIAL PORT INTERFACE FOR THE TLC549 DEVICE TO INTEL 8051 AND 8052 MICROPROCESSORS

The operation of the TLC549 device is similar to that of the more complex TLC540 and TLC541 devices. The TLC549 device provides an on-chip system clock that operates typically at 4 MHz. The on-chip system clock allows internal device operation to proceed independently of serial input/output data timing to permit manipulation of the TLC549 device for a wide range of software and hardware requirements.

Since the TLC549 device needs only the I/O clock input and $\overline{CS}$ input for data control, this serial port interface offers the simplest interface example. It has excellent performance and uses a most economical LinCMOS converter chip. This serial interface application features:

1. Low-cost A/D converter
2. Minimum hardware requirement
3. Fast conversion and communication speed between the A/D converter and the microprocessor
4. Remote control of the A/D converters.

This is an excellent interface if the Intel microprocessor's serial port does not have to be used for another purpose. Even if another purpose is required, the serial port can be multiplexed through good design so both the A/D converter and the additional purpose can be accommodated by the serial port.

Hardware

Figure 11-195 shows the 8051 and 8052 device serial port to the TLC549 device interface circuit. By using the inverted TXD shift clock as an I/O clock for the A/D converter, previous conversion data can be transferred serially from the TLC549 device to the RXD pin of the microprocessor.

Timing Diagram

Figure 11-196 shows the timing diagram for a complete A/D conversion. The timing diagram shows that one complete conversion cycle requires 78.98 μs.

After $\overline{CS}$ goes low, eight I/O clocks access and sample the new analog input. At the same time, I/O clock falling edges shift out the previous conversion result. This conversion result is valid only for the previous analog signal sample.

Conversion occurs in the time interval starting when the eighth I/O clock goes low. The conversion time interval requires 36 internal system clock cycles. The designer cannot see the system clock, but can obtain the conversion time from the data sheets. Typically, a time interval of 17 μs is required for conversion.

See the SOFTWARE INTERFACE FOR TLC540 AND TLC541 DEVICES TO INTEL 8051 AND 8052 MICROPROCESSORS application for more information.

Software

The interface software listing follows. The serial port Mode 0 state is used to permit 8-bit transmission and reception. Note that the A/D converter device sends the most significant bit of the conversion result first and the serial buffer receives this bit as the least significant bit. The latter part of the software program is responsible for reversing the conversion bits and placing them in the proper order.

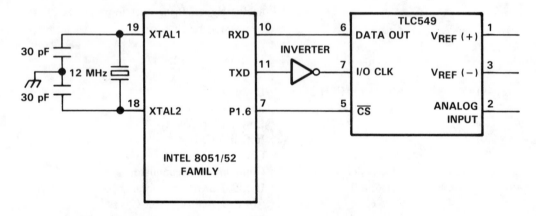

INTEL 8051/52 FAMILY: 8031AH/8051AH/8032AH/8052AH/8751H/80C51
INVERTER: 74HC04, 74LS04

Figure 11-195. The Intel 8051/52 Serial Port to TLC549 Interface

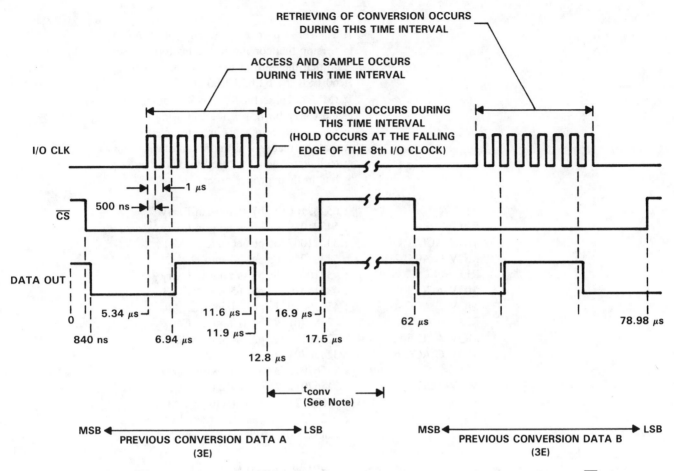

Figure 11-196. Timing Diagram for the Intel 8051/52 Serial Port to TLC549 Interface

NOTE; t_{conv}: The conversion cycles, which requires 36 internal system clock periods, is initiated with the 8th I/O clock↓ after $\overline{CS}$↓. Conversion time requires a minimum of 17µs.

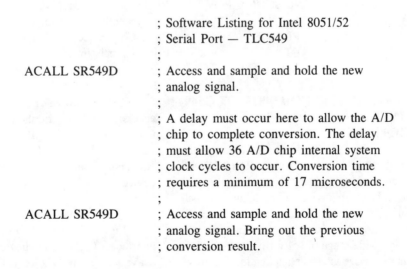

```
                              ; Software Listing for Intel 8051/52
                              ; Serial Port — TLC549
                              ;
              ACALL SR549D    ; Access and sample and hold the new
                              ; analog signal.
                              ;
                              ; A delay must occur here to allow the A/D
                              ; chip to complete conversion. The delay
                              ; must allow 36 A/D chip internal system
                              ; clock cycles to occur. Conversion time
                              ; requires a minimum of 17 microseconds.
                              ;
              ACALL SR549D    ; Access and sample and hold the new
                              ; analog signal. Bring out the previous
                              ; conversion result.
```

```
                                      ;
                                      ; The serial port read reverses the data con-
                                      ; version bits coming to the microprocessor
                                      ; so that they are in the following order:
                                      ; b0(LSB),b1,b2,b3,b4.b5.b6.b7
                                      ; (MSB). These bits (01234567) along with
                                      ; the Carry bit (C) in the following
                                      ; instruction comments are presented so
                                      ; that the reader will understand the
                                      ; technique, which is used to place the
                                      ; bits in their proper order.
                                      ;
                RLC A                 ; 6543210C 7; b7 is now in Carry
                RLC A                 ; 543210C7 6; b6 is now in Carry
                MOV ACC.1,C           ; 54321067 6; put b6 into ACC.1
                MOV C,ACC.2           ; 54321067 0; put b0 into C
                RLC A                 ; 43210670 5; b5 is now in Carry
                MOV ACC.3,C           ; 43215670 5; put b5 into ACC.3
                MOV C,ACC.4           ; 43215670 1; put b1 into C
                RLC A                 ; 32156701 4; b4 is now in Carry
                MOV ACC.5,C           ; 32456701 4; put b4 into ACC.5
                MOV C,ACC.6           ; 32456701 2; put b2 into C
                RLC A                 ; 24567012 3; b3 is now in Carry
                MOV ACC.7,C           ; 34567012 3; put b3 into ACC.7
                RL A                  ; 45670123   ; prepare for SWAP A
                SWAP A                ; 01234567   ; bits are ordered correctly
                                      ; Conversion result is in Accumulator
                                      ;
                                      ; Subroutine ACALL
SR549D          CLR P1.6              ; Lower Chip Select
                ORL SCON, #10H        ; Set REN
                ANL SCON, #FEH        ; Reset RI
                JNB SCON.0,RCV        ; RI flag not set; BRANCH
                                      ; until reception is complete
                CPL P1.6              ; Raise Chip Select
                RET                   ; Conversion is in SBUF
                END                   ;
```

SOFTWARE INTERFACE FOR THE TLC549 DEVICE TO INTEL 8051 AND 8052 MICROPROCESSORS

The operation of the TLC549 device is similar to the more complex TLC540 and TLC541 devices. The TLC549 device provides an on-chip system clock that operates typically at 4 MHz. The on-chip system clock allows internal device operation to proceed independently of serial input/output data timing to permit manipulation of the TLC549 device for a wide range of software and hardware requirements. This software interface application features:

1. Low-cost A/D converter
2. Minimum hardware requirement
3. Remote control of serial A/D converters.

Hardware

Figure 11-197 shows the circuit for the 8051 and 8052 device to TLC549 device software interface. This application requires only three port pins of the microprocessor. While the TLC549 device internal sample-and-hold circuitry is accessing and holding the new analog signal, Port 1.3 and $\overline{CS}$ must be low. The I/O clock is transmitted through port 1.2 to pin 7 of the TLC549 device. Port 1.1 acquires the conversion data from the data output pin of the TLC549 device.

Timing Diagram

Figure 11-198 shows the timing diagram for the interface. The timing diagram shows that one complete conversion cycle requires 245 μs.

After $\overline{CS}$ goes low, eight I/O clocks access and sample the new analog input. At the same time, I/O clock falling edges shift out the previous conversion result. This conversion result is valid only for the previous analog signal sample.

Conversion occurs in the time interval starting when the eighth I/O clock goes low. The conversion time interval requires 36 internal system clock cycles. The designer cannot see the system clock, but can obtain the conversion time from the data sheets. Typically, a time interval of 17 μs is required for conversion.

See the SOFTWARE INTERFACE FOR TLC540 AND TLC541 DEVICES TO INTEL 8051 AND 8052 MICROPROCESSORS application for more information.

Software

The internal software program follows. As shown in the SOFTWARE INTERFACE FOR TLC540 AND TLC541 DEVICES TO INTEL 8051 AND 8052 MICROPROCESSORS application, the instruction RLC A (Rotate Accumulator Left through the Carry flag) loads each conversion bit successively into the accumulator from the data output of the TLC549 device. If desired, the software can be incorporated into a subroutine so the designer can access the software with a simple subroutine call.

```
                                        ; Software Listing for Intel 8051/52
                                        ; Interface — TLC549
                                        ;
                                        ;
                ACALL S549D             ; Access, sample and hold the new
                                        ; analog signal.
                                        ;
                                        ; A delay must occur here to allow the A/D
                                        ; chip to complete conversion. The delay
                                        ; must allow 36 A/D chip internal system
                                        ; clock cycles to occur. Conversion time
                                        ; requires a minimum of 17 microseconds.
                                        ;
                ACALL S549D             ; Access, sample and hold the new analog
                                        ; signal. Bring out the previous
                                        ; conversion result.
                                        ;
                                        ; Subroutine ACALL
S549D           CLR P1.2                ; Lower I/O Clock
                MOV R0, #08H            ; Set bit counter to 8
                CLR P1.3                ; Lower Chip Select
S549I           CLR C                   ; Initialize C = 0
                JNB P1.1,S549J          ; If 549 data out = 0; BRANCH
                CPL C                   ; 549 data out = 1; set C = 1
S549J           MOV A,R1                ; Get serial buffer
                RLC A                   ; Shift data out bit into serial buffer
                MOV R1,A                ; Store serial buffer
                CPL P1.2                ; Raise I/O clock
                NOP                     ; Delay to slow I/O clock
                CLR P1.2                ; Lower I/O clock
                DJNZ R0,S549I           ; Do 8 times
                CPL P1.3                ; Raise Chip Select
                MOV A,R1                ; Conversion data in A
                RET                     ;
                END                     ;
```

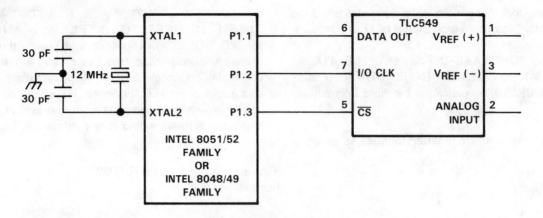

INTEL 8051/52 FAMILY: 8031AH/8051AH/8032AH/8052AH/8751AH/80C51
INTEL 8048/49 FAMILY: 8048AH/8748H/8035AHL/8049AH/8749H/8039AHL/8050AH/8040AHL/80C49

Figure 11-197. The Intel 8051/52 or 8048/49 to TLC549 Software Interface

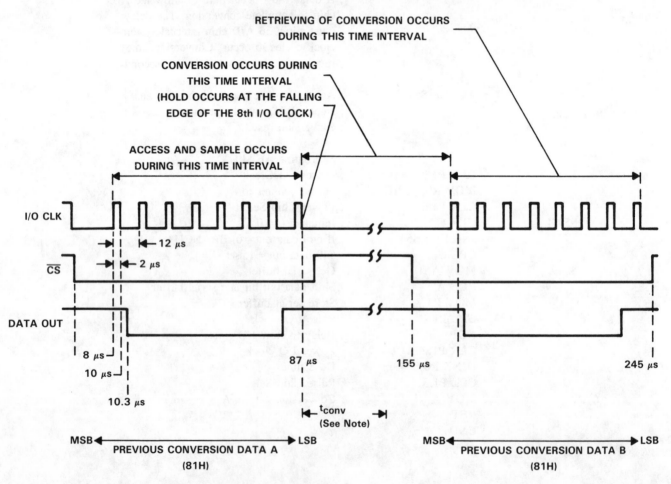

NOTE; t_{conv}: The conversion cycles, which requires 36 internal system clock periods, is initiated with the 8th I/O clock↓ after $\overline{CS}$↓. Conversion time requires a minimum of 17μs.

Figure 11-198. Timing Diagram for the Intel 8051/52 — TLC549 Software Interface

INTERFACE FOR TLC1540 AND TLC1541 DEVICES TO THE ZILOG Z80A MICROPROCESSOR

This application describes a technique for interfacing the Z80A microprocessor to the TLC1540 10-bit A/D converter using software control. The TLC1540 device is a complete data acquisition system on a chip. It includes an analog multliplexer, sample-and-hold, 10-bit successive-approximation A/D converter, and microprocessor-compatible control inputs. These inputs are chip select, address input, system clock, and I/O clock. The system clock and I/O clock need no special speed or phase relationship and are normally operated independently. This allows the system clock to run up to 2.1 MHz to ensure conversion in 21 μs while the I/O clock runs up to 1.1 MHz to allow a maximum data transfer rate.

Principles of Operation

A circuit diagram for the software controlled Z80A microprocessor to TLC1540 device interface is shown in Figure 11-199. A timing diagram for the interface is shown in Figure 11-200. The control software routine listing follows the illustrations.

A data conversion cycle is initiated by bringing $\overline{CS}$ low. This is accomplished by latching a low on address line A0 into the D-type flip-flop. Execution of an IN instruction causes the RD line and the $\overline{IORQ}$ line to become active and generate one I/O clock pulse.

A multiplexer address bit is shifted in on the rising edge of the I/O clock pulse and a conversion result data bit is shifted out on the trailing edge of the pulse. The I/O clock pulse also connects the output of the SN74126 3-state buffer onto the data bus that allows a data bit to be read into the Z80A accumulator. Once in the accumulator, the data bit is rotated through the carry bit location and into register B or C. Register B contains the eight most significant bits while register C contains the two least significant bits so the data is left justified. Register D contains the multiplexer address in the 4 most significant bit positions and is shifted out to the left. On the falling edge of the fourth I/O clock pulse, sampling of the addressed analog input begins and continues until the falling edge of the tenth I/O clock pulse occurs. At that time, $\overline{CS}$ is brought high and conversion begins. Conversion requires 44 system clock cycles, and $\overline{CS}$ should remain high until conversion is complete. If these conditions change, the ongoing conversion will be aborted and a new conversion cycle will begin. It is possible to read conversion results every 141 μs.

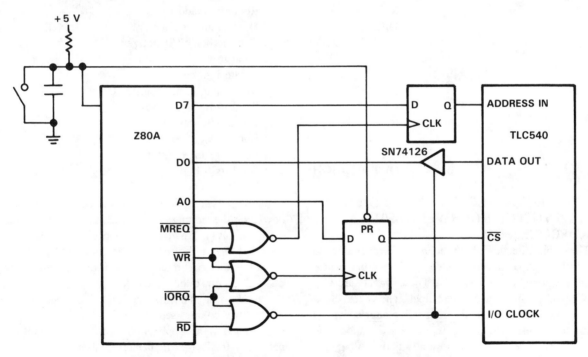

Figure 11-199. Z80A to TLC1540 Interface Circuit Diagram

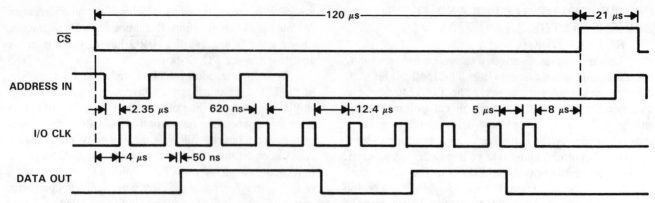

Figure 11-200. Z80A to TLC1540 Interface Timing Diagram

Software for Z80A to TLC1540
and TLC1541 Interface

```
        LD    E,08H        ; Load bit counter
        LD    D,50H        ; Initialize muxaddress to Ch 5
        OUT   (CSLOW),A     ; Bring /CS low
LOOP:   LD    (HL),D        ; Write out muxaddress bit
        IN    A,(BIT)       ; Read in a bit & clock next bit
        RRA                 ; Rotate bit into carry
        RL    B             ; Rotate into result register B
        RLC   D             ; Rotate muxaddress left
        DEC   E             ; Decrement bit counter
        JP    NZ,LOOP       ; Get another bit if not zero
        IN    A,(BIT)       ; Read in bit 1 of result
        RRA                 ; Rotate into carry
        RL    C             ; Rotate into result register C
        IN    A,(BIT)       ; Read in bit 0 of result
        RRA                 ; Rotate into carry
        RR    C             ; Rotate result register C
        RR    C             ; Rotate result register C
        OUT   (CSHIGH),A    ; Bring /CS high
```

INTERFACE FOR TLC1540 AND TLC1541 DEVICES TO THE ROCKWELL 6502 MICROPROCESSOR

This application describes a technique for operating the TLC1540 10-bit A/D converter with the 6502 microprocessor software generated control signals. These signals are $\overline{CS}$, I/O clock, address input, and system clock. The system clock signal is required to drive the successive-approximation conversion process. Reference inputs allow ratiometric conversion, however, they are normally tied to V_{CC} and ground. One of 11 analog inputs or a self-test mode is selected with the on-board analog multiplexer. The TLC1540 device is pin-for-pin compatible with the TLC540 device, allowing easy system upgrade.

Hardware

A circuit diagram for the 6502 microprocessor to TLC1540 device interface is shown in Figure 11-201. I/O clock pulses are generated by enabling the positive going pulse of the $\phi2$ clock to the TLC1540 device. Address input bits are latched into the TLC1540 device on the leading edge

of the $\phi2$ pulse and data bits are shifted out of the TLC1540 device on the trailing edge of the pulse. Address lines A1, A10, and A11 are decoded to enable the output of the SN74126 3-state buffer onto the data bus and the data bit is read by the 6502 microprocessor on the negative edge of $\phi2$. $\overline{CS}$ is controlled by clocking the level of A2 into the D-type flip-flop. Thus, by writing to the proper address, $\overline{CS}$ can be lowered or raised. A simple address decoding scheme is presented in this application. Small modifications may be necessary to fit a particular application.

Timing Diagram

A conversion cycle is initiated by bringing $\overline{CS}$ low. This is accomplished by writing to an address so that A2 is low. The multiplexer address should be placed in the most significant 4 bits of the MUXADDRESS memory location. A multiplexer address is rotated left into the carry, and then tested and written out. The LDA instruction causes one I/O clock pulse to be generated and one data bit to be read. Once the data bit is read in, it is rotated into the carry bit location, and then into RAM. The conversion result is stored in a left justified format in memory locations 0001H and 0002H. After the tenth I/O clock pulse, $\overline{CS}$ is brought high and conversion starts. Conversion requires 44 system clock cycles, and a delay loop may be needed to allow enough time for conversion when using low system clock frequencies. When conversion is complete, $\overline{CS}$ is brought low. The time for one conversion cycle, 209 µs, is indicated in Figure 11-202.

Software

All control signals are directly or indirectly under control. A control software listing follows.

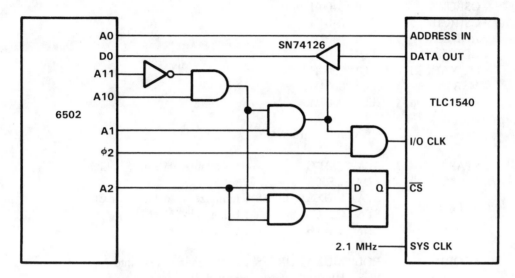

Figure 11-201. 6502 to TLC1540 Interface Circuit Diagram

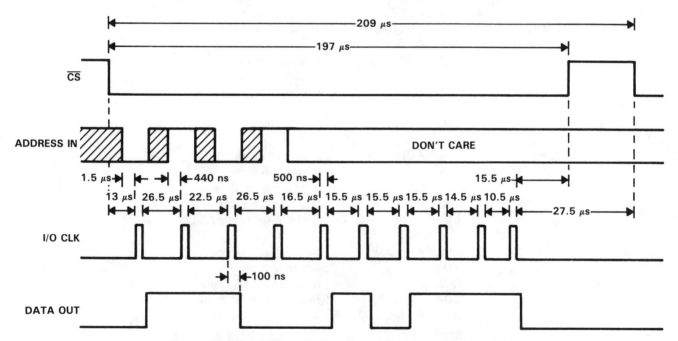

Figure 11-202. 6502 to TLC1540 Interface Timing Diagram

```
;                         *** Register Assignments ***
;
CSLOW          .EQU 0400H
CSHIGH         .EQU 0404H
ADDLOW         .EQU 0402H
ADDHIGH        .EQU 8403H
MUXADDRESS     .EQU 0000H
MSB            .EQU 0001H
LSB            .EQU 0002H
;
;                         *** Main Program ***
;
START:    STA  CSHIGH       ; Make sure chi select is high
          LDA  #$50         ;
          STA  MUXADDRESS   ; Initialize muxaddress to Ch 5
          STA  CSLOW        ; Bring chip select low
          LDX  #$04H        ; Load counter

LOOP1:    ROL  MUXADDRESS   ; Rotate muxaddress into carry
          BCS  HIGH         ; Branch if bit is set
          LDA  ADDLOW       ; Write out A low on A0, clock data in
          JMP  OVER         ; Skip next instruction
HIGH:     LDA  ADDHIGH      ; Write out A high on A0, clock data in
OVER:     ROR  A            ; Rotate do into carry
          ROL  MSB          ; Rotate carry into MSB result
          DEX               ; Decrement counter
          BNE  LOOP1        ; Go back for another bit

          LDX  #$04H        ; Load counter

LOOP2:    LDA  ADDLOW       ; Read in data bit
          ROR  A            ; Rotate into carry
          ROL  MSB          ; Rotate into MSB result
          DEX               ; Decrement counter
          BNE  LOOP2        ; Go back for another bit

          LDA  ADDLOW       ; Read in bit 1 of result
          ROR  A            ; Rotate into carry
          ROL  LSB          ; Rotate into LSB result
          LDA  ADDLOW       ; Read in bit 0 of result
          ROR  A            ; Rotate into carry
          ROR  LSB          ; Rotate LSB result
          ROR  LSB          ; Rotate LSB result
          STA  CSHIGH       ; Bring chip select high
```

INTERFACE FOR TLC1540 AND TLC1541 DEVICES TO THE MOTOROLA 6802 MICROPROCESSOR

This application describes a technique that uses software control in a 6802 microprocessor to TLC1540 10-bit A/D converter interface. The TLC1540 device is a complete data acquisition system on a chip. It contains an analog multiplexer, sample-and-hold, 10-bit successive-approximation A/D converter, and microprocessor-compatible control inputs. These inputs are $\overline{CS}$, I/O clock, address input, and system clock. The system clock and the I/O clock need no special speed or phase relationship and are normally operated independently. This allows the system clock to run up to 2.1 MHz to ensure conversion in 21 μs, while the I/O clock runs up to 1.1 MHz to allow a maximum data transfer rate.

Principles of Operation

A circuit diagram for the software controlled TLC1540 device to 6802 microprocessor interface is shown in Figure 11-203. The timing diagram is shown in Figure 11-204. A control software routine listing follows.

A data conversion cycle is initiated by bringing $\overline{CS}$ low. This is accomplished by latching a low from data bus line DO into the D-type flip-flop. Executing a STAA instruction latches a multiplexer address bit into the D-type flip-flop to ensure that set-up time requirements are met. Execution of an LDAB instruction enables the E clock to the TLC1540 device to generate one I/O clock pulse. It also enables the output of the 3-state buffer onto the data bus to read in a data bit on the trailing edge of the E clock. The trailing edge of E clock also shifts out the next bit of the conversion result. Once a data bit is read into accumulator B, it is rotated into the carry bit and then rotated into accumulator A. This rotates

the multiplexer address one bit to the left. Since the eight most significant bits are stored in the first byte of RAM and the two least significant bits are stored in the second byte of RAM, the data is left justified. Conversion of the addressed channel begins on the falling edge of the tenth I/O clock pulse and requires 44 system clock cycles. $\overline{CS}$ is brought high at the beginning of conversion and should remain high until conversion is complete. A 2 MHz system clock can be obtained by tapping off the EXTAL pin into an SN74HC04 inverter and frequency divide-by-two circuit. The high impedance of the HCMOS device prevents the oscillator from being loaded excessively. It should be noted that many of the gates in this interface are used for the address decoding scheme for the control software presented. One data conversion cycle may be completed in 225 μs as indicated in Figure 11-204.

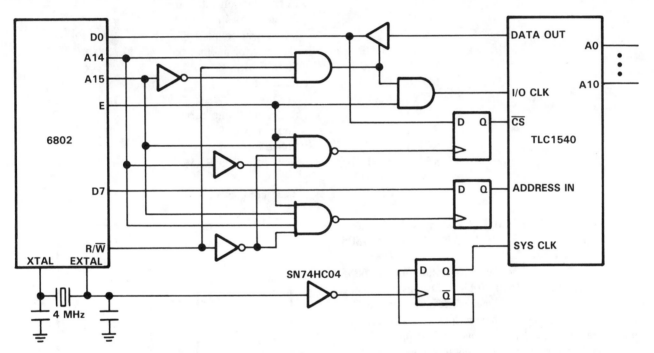

Figure 11-203. 6802 to TLC1540 Interface Circuit Diagram

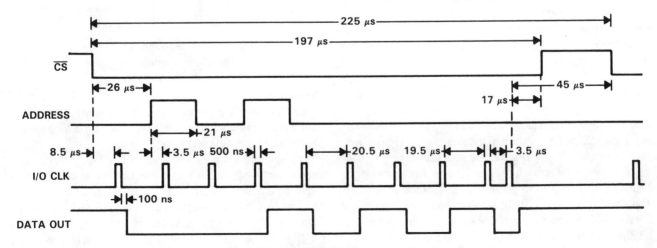

Figure 11-204. 6802 to TLC1540 Interface Timing Diagram

11-173

```
;

ADDRESS     .EQU    $C000H      ; Address to select address FF
DATA        .EQU    $4000H      ; Address to select 3-state buff
CS          .EQU    $8000H      ; Address to select CS FF
;
;
;
START:      LDAA    #50H        ; Initialize muxaddress to channel 5
            LDX     #08H        ; Load counter
            LDAB    #00H        ;
            STAB    CS          ; Bring chip select low
LOOP:       STAA    ADDRESS     ; Write out muxaddress bit
            LDAB    DATA        ; Read data bit and clock
            RORB                ; Rotate data bit into carry
            ROLA                ; Rotate carry into result
            DEX                 ; Decrement counter
            BNE     LOOP        ; Go back for another bit
            STAA    $0000H      ; Store 8 MSB'S in RAM
            LDAB    DATA        ; Read bit 1
            LDAA    DATA        ; Read bit 0
            RORA                ; Rotate bit 0 into carry
            RORB                ; Rotate bits 0 & 1
            RORB                ; Rotate bits 0 & 1
            STAB    $0001H      ; Store LSB'S in RAM
            LDAA    #01H        ;
            STAA    CS          ; Bring chip select high
```

SOFTWARE INTERFACE FOR TLC1540 AND TLC1541 DEVICES TO THE MOTOROLA 6805 MICROPROCESSOR

This application describes an interface for the 6805 microprocessor to TLC1540 10-bit A/D converter using software generated control signals. These signals are $\overline{CS}$, I/O clock, address input, and system clock. The system clock signal is required to drive the successive-approximation conversion process. Reference inputs allow ratiometric conversion, however, they are normally tied to V_{CC} and ground. One of 11 analog inputs or a self-test mode is selected with the on-board analog multiplexer. The TLC1540 device is pin-for-pin compatible with the TLC540 device to allow easy system upgrade.

Hardware

A circuit diagram for the 6805 microprocessor to TLC1540 device interface is shown in Figure 11-205. Four port pins in PORTA are used to transfer control signals and data. A system clock signal may be obtained by tapping off the oscillator input into a high impedance buffer or inverter, such as a SN74HC04, to prevent loading of the oscillator. The signal can then be divided to the required frequency. Care should be used to ensure that system clock pulse width requirements of the TLC1541 device are met.

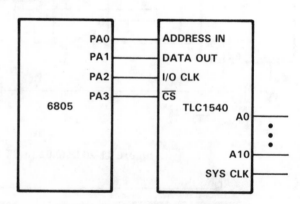

**Figure 11-205. 6805 to TLC1540
Interface Circuit Diagram**

Software

All interface control signals are generated through software manipulation of port pins. The single loop control software listing follows.

The multiplexer address should be loaded in the most significant 4 bits of the accumulator during initialization. A data bit is read into the carry bit location, rotated into the accumulator, which then rotates a multiplexer address bit out. One I/O clock pulse is then generated by toggling bit 2 of PORTA. The multiplexer address bit is latched in on the leading edge of the I/O clock, and a data bit is shifted out on the trailing edge of the I/O clock. This procedure is placed in a loop and repeated eight times. The most significant bits are then stored in the first byte of the on-board RAM. The last two result bits are read in and stored in the second byte of RAM causing the data to be stored in a left justified format. After the results are read in, $\overline{CS}$ is brought high and should remain high until conversion is complete. Conversion requires 44 system clock cycles and a possible time delay depending on the system clock frequency. Using this program, a conversion cycle can be completed in 360 μs.

A timing diagram for a 2-loop control program is shown in Figure 11-206. The 2-loop control program listing follows Figure 11-206. This program has a loop to shift out the four multiplexer address bits and has another loop to shift in the next four bits. There are two individual clock cycles for the last two bits. Although this program uses a few more bytes of program memory, cycle time can be reduced to 312 μs. It is also possible to write a brute-force routine that uses no loops and generates 10 individual clock pulses. This routine reduces cycle time to 260 μs but requires more than twice as much program memory space.

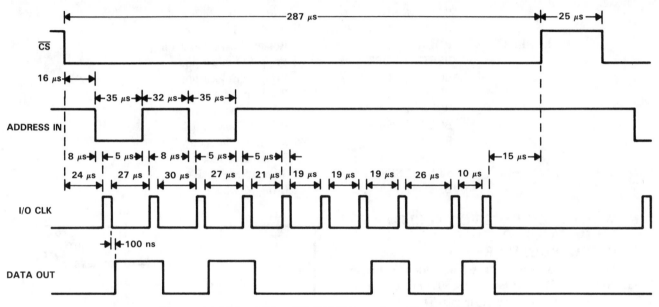

Figure 11-206. 6805 to TLC1540 Interface Timing Diagram

```
;                              Software Listing for TLC1540 and TLC1541 to
                                       Motorola 6805 Interface
                       *** Register Assignments***
;
PORTA        .EQU   0000H        ; Port A I/O pins
DDRA         .EQU   0004H        ; Data direction register A
MSBS         .EQU   0010H        ; Conversion result MSB'S
LSBS         .EQU   0011H        ; Conversion result LSB'S
;
;                     *** MAIN PROGRAM ***
;
START:       LDA    #0DH         ;
             STA    DDRA         ; Initialize port A I/O pins
             BSET   3,PORTA      ; Make sure /CS is high
             LDA    #50H         ; Initialize muxadd to channel 5
             LDX    #08H         ; Initialize counter
             BCLR   3,PORTA      ; Bring /CS low
```

```
;
LOOP:            BRSET  1,PORTA,LABEL1 ; Read data bit into carry
LABEL1           ROLA               ; Rotate into accumulator
                 BCS    HIGH        ; Go to high if muxadd bit is 1
                 BCLR   0,PORTA     ; Write a 0 to TLC1540 address in
                 JMP    CLOCK       ; Skip next instruction
HIGH:            BSET   0,PORTA     ; Write a 1 to TLC1540 address in
CLOCK:           BSET   2,PORTA     ; Bring I/O clock high
                 BCLR   2,PORTA     ; Bring I/O clock low
                 DECX               ; Decrement counter
                 BNE    LOOP        ; Go back for another bit
;
                 STA    MSBS        ; Store MSB'S in RAM
                 BRSET  1,PORTA,LABEL2 ; Read bit 1 into carry
LABEL2           ROLA               ; Rotate into accumulator
                 BSET   2,PORTA     ; Bring I/O clock high
                 BCLR   2,PORTA     ; Bring I/O clock low
                 BRSET  1,PORTA,LABEL3 ; Read bit 0 into carry
LABEL3           BSET   2,PORTA     ; Bring I/O clock high
                 BCLR   2,PORTA     ; Bring I/O clock low
                 RORA               ; Rotate accumulator
                 RORA               ; Rotate accumulator
                 STA    LSBS        ; Store LSB'S in RAM
                 BSET   3,PORTA     ; Bring /CS high
```

SOFTWARE INTERFACE FOR TLC1540 AND TLC1541 DEVICES TO THE INTEL 8051 AND 8052 MICROPROCESSORS

This application describes a technique for operating the TLC1540 10-bit A/D converter with the 8051 microprocessor using software generated control signals. These signals are $\overline{CS}$, I/O clock, address input, and system clock. The system clock signal is required to drive the successive-approximation conversion process. Reference inputs allow ratiometric conversion, however, they are normally tied to V_{CC} and ground. One of 11 analog inputs or a self-test mode is selected with the on-board multiplexer. The TLC1540 device is pin-for-pin compatible with the TLC540 device to allow easy system upgrade.

Hardware

The system clock is derived from the ALE signal of the 8051 device. Another method uses a signal that is tapped off the oscillator through a high impedance buffer or inverter and divided down to the appropriate frequency. Care should be taken when using the ALE signal for the TLC1541 system clock to ensure that the high and low pulse widths are within specifications. The ALE signal is dependent upon the oscillator frequency and may not meet pulse width specifications at high oscillator frequencies. A circuit diagram is shown in Figure 11-207.

Timing Diagram

A timing diagram for the interface is shown in Figure 11-208. The subroutine can be executed in 79 μs. With a system clock of 2.1 MHz from the ALE pin, conversion results may be read every 101 μs.

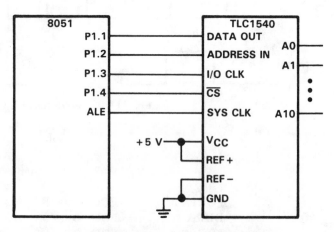

Figure 11-207. 8051 to TLC1540 Interface Circuit Diagram

Software

All interface control signals are generated through software manipulation of port pins. A subroutine is used to load a new multiplexer address and retrieve a previous conversion result. A listing of the subroutine follows the discussion.

A multiplexer address should be loaded into the most significant 4 bits of the accumulator before calling the subroutine. Previous conversion results are returned left justified, with R2 holding the eight most significant bits and R3 holding the two least significant bits. After returning from the subroutine, a delay loop is executed to allow time for the conversion. Conversion requires 44 system clock cycles, therefore, delay loops of appropriate length should be included according to the system clock frequency.

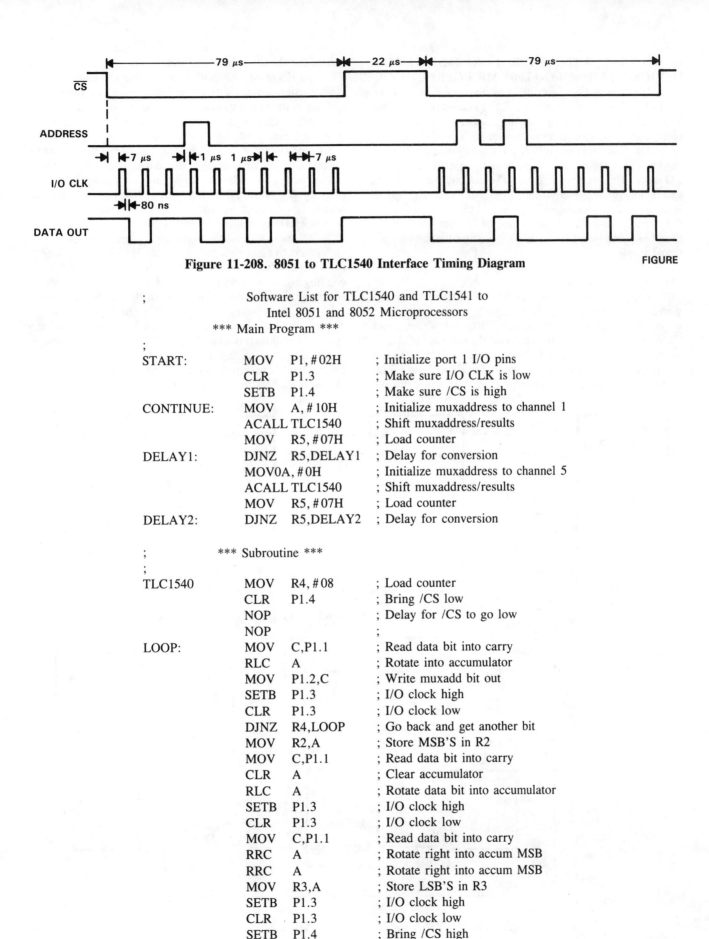

Figure 11-208. 8051 to TLC1540 Interface Timing Diagram

```
;                    Software List for TLC1540 and TLC1541 to
                           Intel 8051 and 8052 Microprocessors
                  *** Main Program ***
;
START:      MOV   P1,#02H      ; Initialize port 1 I/O pins
            CLR   P1.3         ; Make sure I/O CLK is low
            SETB  P1.4         ; Make sure /CS is high
CONTINUE:   MOV   A,#10H       ; Initialize muxaddress to channel 1
            ACALL TLC1540      ; Shift muxaddress/results
            MOV   R5,#07H      ; Load counter
DELAY1:     DJNZ  R5,DELAY1    ; Delay for conversion
            MOV0A,#0H          ; Initialize muxaddress to channel 5
            ACALL TLC1540      ; Shift muxaddress/results
            MOV   R5,#07H      ; Load counter
DELAY2:     DJNZ  R5,DELAY2    ; Delay for conversion


;               *** Subroutine ***
;
TLC1540     MOV   R4,#08       ; Load counter
            CLR   P1.4         ; Bring /CS low
            NOP                ; Delay for /CS to go low
            NOP                ;
LOOP:       MOV   C,P1.1       ; Read data bit into carry
            RLC   A            ; Rotate into accumulator
            MOV   P1.2,C       ; Write muxadd bit out
            SETB  P1.3         ; I/O clock high
            CLR   P1.3         ; I/O clock low
            DJNZ  R4,LOOP      ; Go back and get another bit
            MOV   R2,A         ; Store MSB'S in R2
            MOV   C,P1.1       ; Read data bit into carry
            CLR   A            ; Clear accumulator
            RLC   A            ; Rotate data bit into accumulator
            SETB  P1.3         ; I/O clock high
            CLR   P1.3         ; I/O clock low
            MOV   C,P1.1       ; Read data bit into carry
            RRC   A            ; Rotate right into accum MSB
            RRC   A            ; Rotate right into accum MSB
            MOV   R3,A         ; Store LSB'S in R3
            SETB  P1.3         ; I/O clock high
            CLR   P1.3         ; I/O clock low
            SETB  P1.4         ; Bring /CS high
            RET                ; Return to main program
```

HARDWARE INTERFACE FOR THE TLC545 DEVICE TO THE ZILOG Z80A MICROPROCESSOR

This application describes a technique for interfacing the TLC545 A/D converter to the Z80A microprocessor. The TLC545 device is a complete data acquisition system on a chip. The system includes an analog multiplexer, sample-and-hold, 8-bit A/D converter, and microprocessor control inputs. These inputs are $\overline{CS}$, address in, system clock, and I/O clock. The system clock and I/O clock require no special speed or phase relationship, and are normally used independently. This allows the system clock to run up to 4 MHz to ensure a conversion time of less than 9 μs while the I/O clock runs up to 2.1 MHz to allow a maximum data transfer rate.

Principles of Operation

The circuit shown in Figure 11-209 initiates conversion with an OUT instruction. The timing diagram for this circuit is shown in Figure 11-210. When $\overline{IORQ}$ and $\overline{WR}$ go low, $\overline{CS}$ is brought low while the universal shift register is placed in the load data mode and the system clock is enabled to the clock input to latch in the multliplexer address. The multiplexer address should occupy the five most significant

bits of the shift register since the data is shifted out to the left. The rising edge of $\overline{IORQ}$ enables the I/O clock to the shift register, which shifts the multiplexer address out while shifting in the previous conversion result. Sampling begins at the falling edge of the fourth I/O clock cycle and continues until the eighth falling edge occurs. At this time the I/O clock is disabled and $\overline{CS}$ is brought high to ensure that the TLC545 device will remain undisturbed during the conversion. Conversion of the addressed analog input requires 36 system clock cycles. During conversion time, previous conversion results may be read by using an IN instruction. A typical interrupt service routine is shown below.

At least 36 system clock cycles must be used for conversion to ensure proper operation. If a new multiplexer address is shifted in while a conversion is in progress, the ongoing conversion will be aborted and a new conversion will be initiated at the falling edge of the eighth I/O clock cycle. Another software approach is to initiate a new conversion, wait in a delay loop until the conversion is complete, and then read in the previous conversion results. A sample program segment using the delay loop method follows this discussion. Using this method, a conversion cycle can be completed in 21.5 μs.

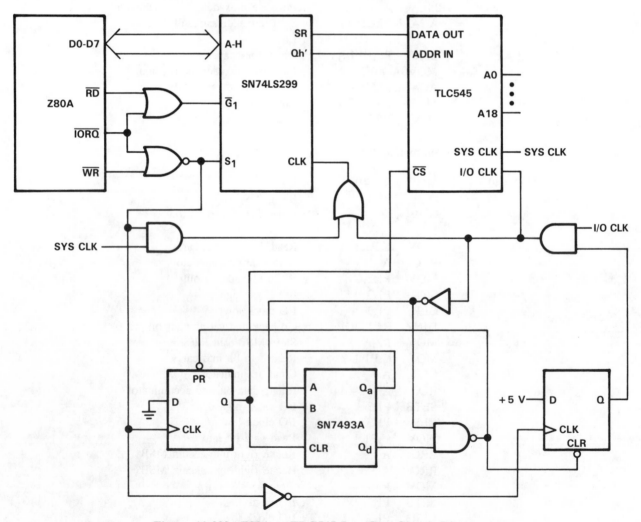

Figure 11-209. Z80A to TLC545 Interface Circuit Diagram

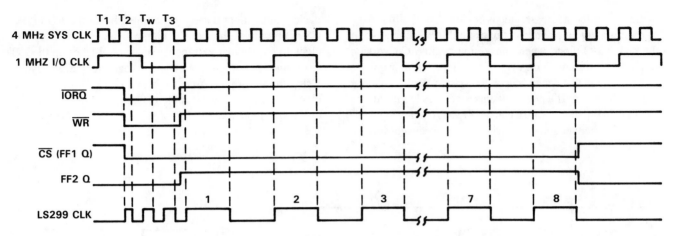

Figure 11-210. Z80A to TLC545 Interface Timing Diagram

```
ISR        EX AF,AF              ; Save accumulator
           IN A,(LS299)          ; Read previous conv. result
           LD (DATA), A          ; Store result
           LD A, (MUXADDRESS)    ; Load new mux address from RAM
           OUT (LS299),A         ; Initiate new conversion cycle
           EX AF,AF'             ; Restore accumulator
           EI
           RETI

           LD A, (MUXADDRESS)    ; Load mux address from RAM
           OUT (LS299),A         ; Sample channel & initiate conv.
           LD C,03H              ; Initialize counter
WAIT:      DEC C                 ; Decrement counter
           JP NZ, WAIT           ; If not zero keep waiting
           OUT (LS299),A         ; Shift results into LS299
           IN A,(LS299)          ; Read conv results
```

NOTE: A count of 03H will produce a delay of 10.50 μs, suitable for 4 MHz operation, while a count of 05H produces a delay of 17.50 μs, suitable for 2.5 MHz operation.

SOFTWARE INTERFACE FOR THE TLC545 DEVICE TO THE ZILOG Z80A MICROPROCESSOR

This application describes a technique for operating the TLC545 A/D converter with the Z80A microprocessor using software generated control signals. These signals are $\overline{CS}$, address in, I/O clock, and system clock. The system clock is needed to drive the successive-approximation conversion process. The system clock and the I/O clock require no special speed or phase relationship and are normally used independently. This allows the system clock to run up to 4 MHz to ensure conversion in less than 9 μs while the I/O clock runs up to 2.1 MHz to allow a maximum data transfer rate. Reference inputs allow ratiometric conversion; however, they are normally tied to V_{CC} and ground. One of 19 analog inputs or a self-test mode is selected with the on-board analog multiplexer. Timing for the TLC545 device is identical to the TLC540 device with the exception that one additional multiplexer address bit must be shifted out to address the additional analog inputs.

Principles of Operation

The circuit diagram for the software controlled Z80A microprocessor to TLC545 device interface is shown in Figure 11-211. Circuit timing is shown in Figure 11-212.

Execution of an IN instruction causes the $\overline{RD}$ line and the $\overline{IORQ}$ line to become active and shift an address bit in and a data bit out of the TLC545 device. $\overline{CS}$ is brought low by latching in a low from address bit A0 on the positive going edge of the $\overline{WR}$ signal. A simple program segment listing that shifts out a new analog multiplexer address while also shifting in previous conversion results follows the discussion.

This program segment uses the B register to store the conversion result, the C register as a bit counter, and the D register to hold the analog multiplexer address. The analog multiplexer address is shifted left out of the D register, therefore, the 5-bit address must be placed in the five most significant bits of the byte.

Conversion results are read in one bit at a time and then shifted left to the proper position in the B register.

Sampling of the addressed input begins at the falling edge of the fifth I/O clock and continues until the falling edge of the eighth I/O clock when conversion begins. Conversion requires 36 system clock cycles, therefore, an appropriate software delay that depends upon the system clock frequency must be included. If a new multiplexer address is shifted into the TLC545 device before a conversion has been completed, the ongoing conversion will be aborted and a new conversion cycle will begin at the eighth falling edge of the I/O clock. $\overline{CS}$ is brought high after the eighth falling edge of the I/O clock to ensure that extraneous noise or glitches on the I/O clock line are not interpreted as the begining of a new cycle. Using the program segment just presented with the system clock at 4 MHz, it is possible to initiate a new conversion cycle and read the results of the previous conversion in 138 μs as indicated in Figure 11-212.

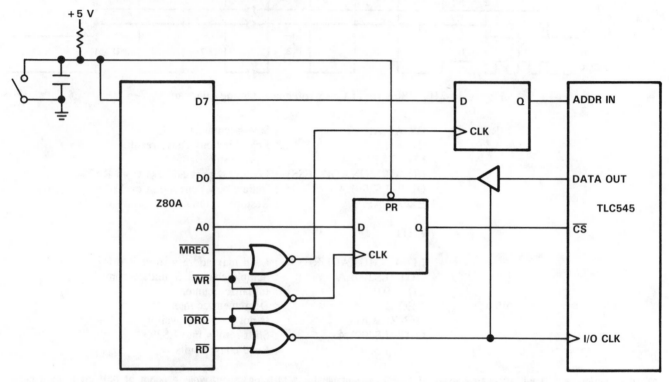

Figure 11-211. Z80A to TLC545 Software Interface Circuit Diagram

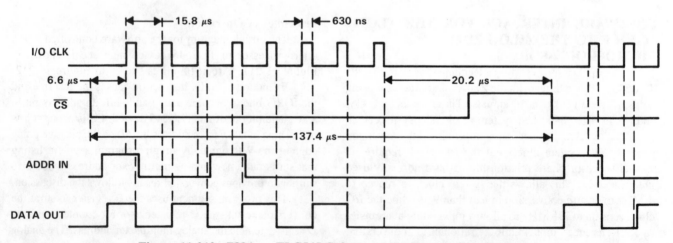

Figure 11-212. Z80A to TLC545 Software Interface Timing Diagram

```
;               *** REGISTER ASSIGNMENTS ***
;
;
CSLOW           .EQU 0100H          ; A0 = 0
CSHIGH          .EQU 0101H          ; A0 = 1
BIT             .EQU 00FFH          ; Arbitrary address
;
;               *** MAIN PROGRAM ***
;
                LD C,08             ; Initialize bit counter
                LD B,00             ; Clear result register
                OUT (CSLOW),A       ; Bring chip select low
LOOP            RLC B               ; Rotate result left
                LD (HL),D           ; Latch address bit into D FF
                IN A, (BIT)         ; Read in data bit
                AND 01H             ; Mask off bit 0
                OR B                ; Or new bit into result reg.
                LD B,A              ; Store in result register
                RLC D               ; Shift address left
                DEC C               ; Decrement bit counter
                JP NZ,LOOP          ; Get another bit if not zero
                OUT (CSHIGH),A      ; Bring chip select high
```

SOFTWARE INTERFACE FOR THE TLC545 AND TLC546 DEVICES TO INTEL 8051 AND 8052 MICROPROCESSORS

Two interface circuits are discussed in this application. The Interface 1 circuit diagram and associated timing diagram are shown in Figure 11-213 and Figure 11-214, respectively. A software listing of Interface 1, in which the A/D clock signal is derived from the microprocessor ALE clock, follows the illustrations.

Figure 11-215 and Figure 11-216 show the Interface 2 circuit diagram and associated timing diagram. The software listing for this interface follows the illustrations.

These interfaces minimize the amount of hardware and rely mainly upon software techniques. Although the amount of hardware and its cost are reduced, the use of more software increases the time required to load the address into, and retrieve the conversion data from, the A/D converter. However, the trade-off of minimum hardware versus longer conversion time may benefit many designs.

Hardware

The interface shown in Figure 11-213 will always work with the TLC545 device, but will not work with the TLC546 device at the higher microprocessor instruction cycle frequencies. Before using the TLC546 device, the designer must verify that the high and low pulse widths of the ALE clock signal meet the specifications of the TLC546 device. These pulse widths are dependent upon the microprocessor instruction cycle frequency.

In the A/D converter of Figure 11-215, the system clock for the A/D converter is obtained from the microprocessor crystal oscillator. To assure proper operation of the crystal oscillator, a high impedance buffer must be used to tap the signal from the oscillator. An important detail is that low and high level input buffer requirements must lie within the range of the oscillator signal. This compatibility will prevent missing edge transitions in the A/D converter's system clock signal. Subsequently, the buffered crystal oscillator signal must be frequency divided to assure that the resulting system clock signal does not exceed the upper frequency specification of the A/D converter. Any convenient divider circuitry may be used to accomplish this task.

Timing Diagrams

Figures 11-214 and 11-216 show the timing diagrams for the A/D conversions of Interfaces 1 and 2, respectively. Loading the new address and retrieving the conversion for the previous address requires 137 μs maximum. The conversion period requires 36 system clocks. This period is longer for the TLC546 device since the microprocessor clock frequency (see Figure 11-213) or the A/D clock frequency (see Figure 11-215) must be lower to satisfy the TLC546 device specifications.

Software

Refer to the software listings for Interfaces 1 and 2 for the following discussion. The software uses a subroutine, ACALL that simultaneously loads the A/D converter with a new address and retrieves the conversion result for the previous address. This subroutine can be used any time the designer desires to load a new address and retrieve a conversion value. This simultaneous loading and retrieving makes the subroutine very appropriate for continuous monitoring of several A/D converter analog inputs. The subroutine assumes that the new address has been prevously placed in the five most significant bits of register R2. Upon completion of the subroutine, the conversion result for the previous address is left in the accumulator.

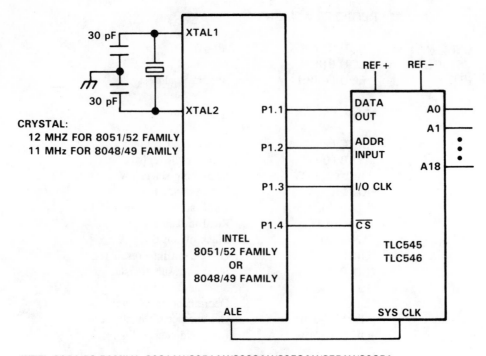

CRYSTAL:
12 MHZ FOR 8051/52 FAMILY
11 MHz FOR 8048/49 FAMILY

INTEL 8051/52 FAMILY: 8031AH/8051AH/8032AH/8052AH/8751H/80C51
INTEL 8048/49 FAMILY: 8048AH/8748H/8035AHL/8049AH/8749H/8039AHL/8050AH/8040AHL/80C49

Figure 11-213. TLC545 and TLC546 to Intel Microprocessor Interface 1 Diagram

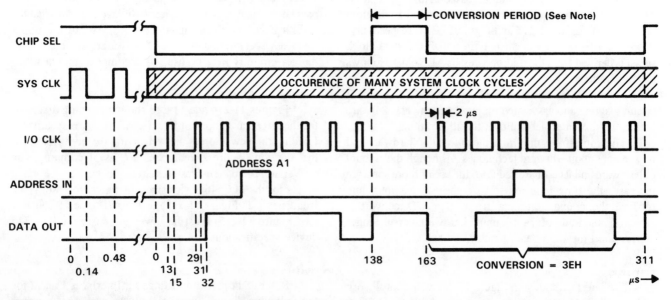

NOTE: Conversion period requires 36 system clock cycles after 8th I/O clock goes low upon clocking in the address.

Figure 11-214. Timing Diagram of an A/D Conversion Cycle for TLC545 and TLC546 to Intel 8048/8049 and 8051/8052

```
; Software Listing for Intel 8051/52 to TLC545/546 Interface 1
;
                CLR P1.3              ;Lower I/O clock
                MOV R2, #10H         ;Load A/D analog input address. Note
                                     ;that to send address = A2, R2 = 10H.
                ACALL S545D          ;Load 545 address, assumes the address
                                     ;is currently in R2.
;
                MOV R3, #09H         ;This software loop allows a
                                     ;conserative 40 A/D system clocks,
                                     ;since only 36
DELAY:          DJNZ R3,DELAY        ;clocks are required to perform
                                     ;conversion, to be emitted from the
                                     ;microprocessor ALE pin
;
                MOV R2, #10H         ;Load A/D analog input address. Note
                                     ;that to send address = A2, R2 = 10H.
                ACALL S545D          ;Load new 545 address, assumes this
                                     ;address is in R2; leaves the
                                     ;conversion result for the previous
                                     ;address in A
;
;               Subroutine ACALL
;
S545D           MOV R6, #08H         ;Set bit counter to 8
                CLR P1.4             ;Lower chip select
S545DL:         CLR C                ;Initialize C = 0
                JNB P1.1,S545DI      ;If 545 Data Out = 0; branch
                CPL C                ;545 Data Out = 1; set C = 1
S545DI:         MOV A,R2             ;Get serial buffer
                RLC A                ;Shift Data Out bit into serial
                                     ;buffer and shift 545 address
                MOV R2,A             ;Store serial buffer
                JNC S545DWO          ;If 545 address bit = 0; branch
                ORL P1, #04H         ;Set 545 address line to 1
                SJMP S545DWE         ;Go and raise the I/O clock
S545DWO:        ANL P1, #FBH         ;Set 545 address line to 0
S545DWE:        NOP                  ;Allow address line to setup
                CPL P1.3             ;Raise I/O clock
                NOP                  ;Delay to slow I/O clock
                CLR P1.3             ;Lower I/O clock
                DJNZ R6,S545DL       ;Do all 8 bits
                CPL P1.4             ;Raise chip select
                MOV A,R2             ;Get serial buffer
                RET
                END
```

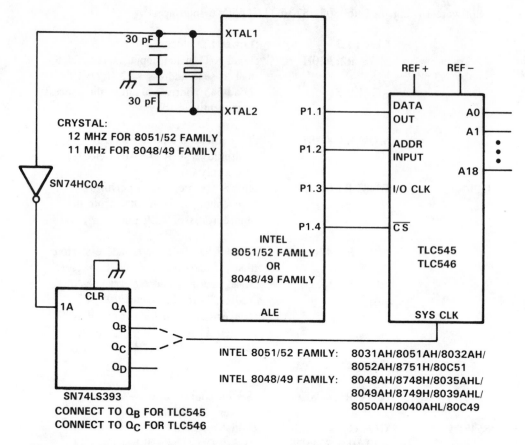

Figure 11-215. TLC545 and TLC546 to Intel Microprocessor Interface 2 Diagram

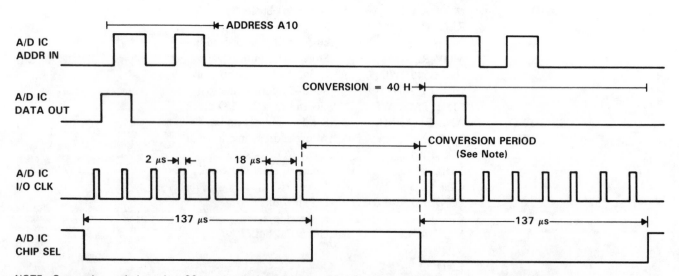

NOTE: Conversion period requires 36 system clock cycles after 8th I/O clock goes low upon clocking in the address.

Figure 11-216. Timing Diagram of an A/D Conversion Cycle for TLC545 and TLC546 to Intel 8048/8049 and 8051/8052

; Software Listing for Intel 8051/31 - TLC545/546 Interface 2
;

```
                    CLR P1.3            ;Lower I/O clock
                    MOV R2, #50H        ;Load A/D analog input address. Note
                                        ;that to send address = A10, R2 = 50H.
                    ACALL S545D         ;Load 545 address, assumes the address
                                        ;is currently in R2.
;
;              A delay must occur here to allow the A/D IC to complete
;              conversion. The delay must allow 36 A/D IC system clock
;              cycles to occur.
;
                    MOV R2, #50H        ;Load A/D analog input address. Note
                                        ;that to send address = 10, R2 = 50H.
                    ACALL S545D         ;Load new 545 address, assumes this
                                        ;address is in R2; leaves the
                                        ;conversion result for the previous
                                        ;address in A
;
;              Subroutine ACALL
;
S545D               MOV R6, #08H        ;Set bit countr to 8
                    CLR P1.4            ;Lower chip select
S545DL:             CLR C               ;Initialize C = 0
                    JNB P1.1,S545DI     ;If 545 Data Out = 0; branch
                    CPL C               ;545 Data Out = 1; set C = 1
S545DI:             MOV A,R2            ;Get serial buffer
                    RLC A               ;Shift Data Out bit into serial
                                        ;buffer and shift 545 address
                    MOV R2,A            ;Store serial buffer
                    JNC S545DWO         ;If 545 address bit = 0; branch
                    ORL P1, #04H        ;Set 545 address line to 1
                    SJMP S545DWE        ;Go and raise the I/O clock
S545DWO:            ANL P1, #FBH        ;Set 545 address line to 0
S545DWE:            NOP                 ;Allow address line to setup
                    CPL P1.3            ;Raise I/O clock
                    NOP                 ;Delay to slow I/O clock
                    CLR P1.3            ;Lower I/O clock
                    DJNZ R6,S545DL      ;Do all 8 bits
                    CPL P1.4            ;Raise chip select
                    MOV A,R2            ;Get serial buffer
                    RET
                    END
```

SERIAL PORT INTERFACE FOR TLC545 AND TLC546 DEVICES TO INTEL 8051 AND 8052 MICROPROCESSORS

Two interface circuits are presented for this application. Figures 11-217, 11-218, and a software listing support Interface 1 in which the A/D clock signal is derived from the microprocessor ALE clock. Figures 11-219, 11-220, and a software listing support Interface 2 in which the A/D clock signal is derived from the microprocessor crystal oscillator. These interfaces not only use a low cost A/D converter, but also require minimal hardware and provide quick conversion and A/D microprocessor communications. They also provide the remote location possibilities of a serial A/D converter.

This application is an excellent one if the Intel microprocessor's serial port does not have to be used for another purpose. Even if another purpose is required, the serial port may be multiplexed through good design so both the A/D converter and the additional purpose may be accommodated by the serial port.

Hardware

The circuit is presented in Figures 11-217 and 11-219. The signal at the microprocessor's TXD pin must be inverted so the communication protocols for the microprocessor serial port and the A/D converter are compatible. Use of an SN74125 3-state buffer allows the microprocessor serial port to be used for both transmission and reception. Use of the 3-state buffer prevents the simultaneous loading of a new address while retrieving the conversion of a previously loaded address. However, this time loss is small because the serial port can quickly load an address and retrieve the resulting conversion.

Timing Diagram

Figures 11-218 and 11-220 are the timing diagrams for an A/D conversion for each of the two circuits. With the TLC545 device, loading the address, waiting for conversion, and retrieving the conversion result requires 56 μs for Interface 1 and 48 μs for Interface 2. This time period is longer for the TLC546 device since the system clock frequency must be lower for this device.

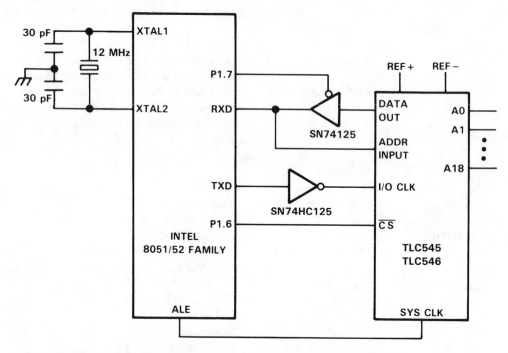

INTEL 8051/52 FAMILY: 8031AH/8051AH/8032AH/8052AH/8751AH/80C51

Figure 11-217. TLC545 and TLC546 to Intel Microprocessor Interface 1 Diagram

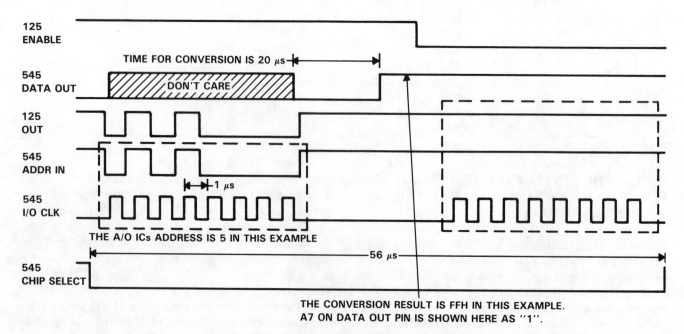

Figure 11-218. Timing Diagram for an A/D Conversion

Software

Refer to the interface software listings presented earlier. The serial port mode 0 is used. If desired, the software can be incorporated into a subroutine so the designer can address the software with a simple subroutine call. Particular attention must be exercised when placing the address bits in the serial buffer since the serial port sends the least significant bit of the address first and the A/D converter accepts this bit as the most significant bit of the address. A similar process occurs when the serial port receives the conversion result.

```
; Software Listing for Intel 8051/52 Family - TLC545/546 Interface 1
;
SR545L;          MOV A, #0AH          ;A/D IC address of A16 in this example
;
;The serial port will send 0A(00001010) with the least significant
;bit first. Therefore, the A/D IC will see (01010000), which
;will load an address of A16 into the control register.
;
                 CLR P1.7             ;Disable 125
                 CPL P1.7
                 CLR P1.6             ;Lower chip select
                 ANL SCON,4EDH        ;Reset REN & TI flags
                 MOV SBUF,A           ;Send 545 address (LSB FIRST)
SNDTST:          JNB SCON.1,SNDTST    ;TI flag not set; branch
                                      ;until transmission is complete
;
;A delay must occur here to allow the A/D IC to complete
:conversion. The delay must allow 36 A/D IC system clock
;cycles to occur.
;
                 CPL P1.7             ;Enable 125
                 ORL SCON, #10H       ;Set REN
                 ANL SCON, #FEH       ;Reset RI
RCVTST:          JNB SCON.0,RCVTST    ;RI FLAG not set; branch
                                      ;until reception is complete
                 CPL P1.6             ;Raise chip select
                 MOV A,SBUF           ;Set SBUF
;
;The serial port read reverses the data conversion bits coming
;to the microprocessor so that they are in the following order:
;b0(lsb),b1,b2,b3,b4,b5,b6,b7(msb). These bits (01234567) along
;with the carry bit (C) in the following instruction comments are
;presented so that the reader will understand the technique, which
;is used to place the bits in their proper order.
;
                 RLC A                ; 6543210C 7; b7 is now in carry
                 RLC A                ; 543210C7 6; b6 is now in carry
                 MOV ACC.1,C          ; 54321067 6; put b6 into ACC.1
                 MOV C,ACC.2          ; 54321067 0; put b0 into C
                 RLC A                ; 43210670 5; b5 is now in carry
                 MOV ACC.3,C          ; 43215670 5; put b5 into ACC.3
                 MOV C,ACC.4          ; 43215670 1; put b1 into C
                 RLC A                ; 32156701 4; b4 is now in carry
                 MOV ACC.5,C          ; 32456701 4; put b4 into ACC.5
                 MOV C,ACC.6          ; 32456701 2; put b2 into C
                 RLC A                ; 24567012 3; b3 is now in carry
                 MOV ACC.7,C          ; 34567012 3; put b3 into ACC.7
                 RL A                 ; 45670123   ; prepare for SWAP A
                 SWAP A               ; 01234567   ; bits are ordered correctly
                                      ; conversion result is in accumulator
```

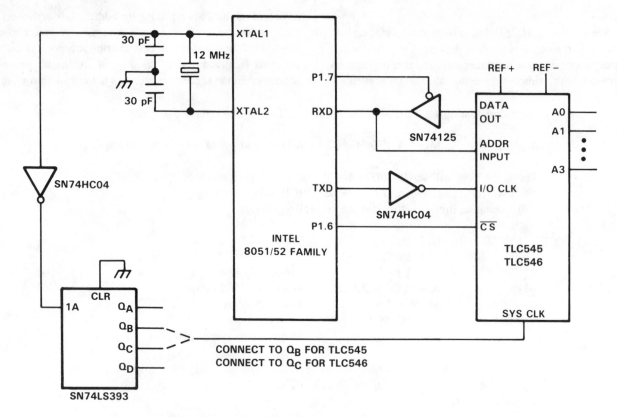

INTEL 8051/52 FAMILY: 8031AH/8051AH/8032AH/8052AH/8751AH/80C51

Figure 11-219. TLC545 and TLC546 to Intel Microprocessor Interface 2 Diagram

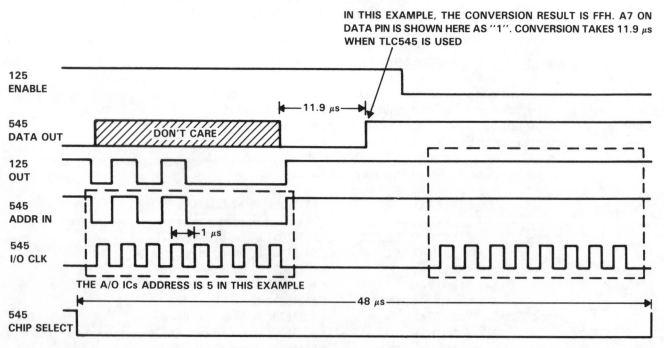

Figure 11-220. Timing Diagram for an A/D Conversion

; Software Listing for Intel 8051/52 Family - TLC545/546 Interface 2
;
SR545L;   MOV A, #0AH   ;A/D IC address of A16 in this example
;
;The serial port will send 0A(00001010) with the least significant bit first. Therefore, the A/D
;IC will see (01010000), which will load an address of A16 into the control register.
;

```
                          CLR P1.7                 ;Disable 125
                          CPL P1.7
                          CLR P1.6                 ;Lower chip select
                          ANL SCON, #EDH           ;Reset REN & TI flags
                          MOV SBUF,A               ;Send 545 address (LSB FIRST)
SNDTST:                   JNB SCON.1,SNDTST        ;TI flag not set; branch
                                                   ;until transmission is complete
;
                          MOV R3, #09H             ;This software loop allows a
                                                   ;conservative 40 A/D system clocks,
                                                   ;since only 36
DELAY:                    DJNZ R3,DELAY            ;clocks are required to perform
                                                   ;conversion, to be emitted from the
                                                   ;microprocessor ALE pin
;
                          CPL P1.7                 ;Enable 125
                          ORL SCON, #10H           ;Set REN
                          ANL SCON, #FEH           ;Reset RI
RCVTST:                   JNB SCON.0,RCVTST        ;RI FLAG not set; branch
                                                   ;until reception is complete
                          CPL P1.6                 ;Raise chip select
                          MOV A,SBUF               ;Set SBUF
```

;
;The serial port read reverses the data conversion bits coming
;to the microprocessor so that they are in the following order:
;b0(lsb),b1,b2,b3,b4,b5,b6,b7(msb). These bits (01234567) along
;with the carry bit (C) in the following instruction comments are
;presented so that the reader will understand the technique, which
;is used to place the bits in their proper order.
;

```
                          RLC A                    ; 6543210C 7; b7 is now in carry
                          RLC A                    ; 543210C7 6; b6 is now in carry
                          MOV ACC.1,C              ; 54321067 6; put b6 into ACC.1
                          MOV C,ACC.2              ; 54321067 0; put b0 into C
                          RLC A                    ; 43210670 5; b5 is now in carry
                          MOV ACC.3,C              ; 43215670 5; put b5 into ACC.3
                          MOV C,ACC.4              ; 43215670 1; put b1 into C
                          RLC A                    ; 32156701 4; b4 is now in carry
                          MOV ACC.5,C              ; 32456701 4; put b4 into ACC.5
                          MOV C,ACC.6              ; 32456701 2; put b2 into C
                          RLC A                    ; 24567012 3; b3 is now in carry
                          MOV ACC.7,C              ; 34567012 3; put b3 into ACC.7
                          RL A                     ; 45670123   ; prepare for SWAP A
                          SWAP A                   ; 01234567   ; bits are ordered correctly
                                                   ; conversion result is in accumulator
```

INTERFACE FOR THE TLC545 DEVICE TO THE ROCKWELL 6502 MICROPROCESSOR USING TTL GATES

The TLC545 A/D converter can be operated with the 6502 microprocessor using several methods. This application presents the design of an interface using TTL gates. Another method using the 6522 VIA is described in the next application. Cost and performance are the basic trade-offs between the two designs. The 6522 device interface is faster, but the TTL method costs less.

Principles of Operation

The basic premise of the interface circuit shown in Figure 11-221 is that all timing control signals are generated under software control. Circuit timing is shown in Figure 11-222. This is followed by the control software listing.

A data conversion cycle is initiated by bringing $\overline{CS}$ low. This is accomplished by latching a low into the D-type flip-flop from address line A1 on the positive edge of system clock $\phi2$. Address bit A14 is used as a gating signal to prevent the TLC545 device from being inadvertently selected during normal program execution. After $\overline{CS}$ is brought low,

I/O clock pulses shift in the multiplexer address that is stored in the five most significant bits of a byte in RAM while shifting out the previous conversion results. The I/O clock is enabled by gating the positive going pulse of $\phi2$ to the TLC545 device I/O CLK input. The gating occurs by addressing a location so A15 is high. The high on A15 also enables the output of the 3-state buffer onto the data bus. Address bit A0 determines whether the multiplexer address bit is a high or a low. This multiplexer address bit is shifted into the TLC545 device on the positive edge of clock $\phi2$ and a data bit is placed onto the data bus at this time by the TLC545 device. The data bit is latched into the 6502 device on the negative edge of clock $\phi2$.

Once the data is loaded into the accumulator, it is rotated into the carry bit and then into memory. $\overline{CS}$ is brought high again by writing a high into the D-type flip-flop. This is done by placing a high on address bit A1. This cycle can be completed every 176 μs.

This interface circuit uses an address decoding scheme that requires a minimum of decoding hardware. Small modifications to the address decoding scheme may be necessary to fit the interface to a particular application.

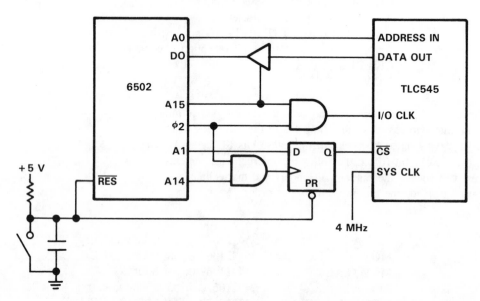

Figure 11-221. 6502 to TLC545 Interface Circuit Diagram

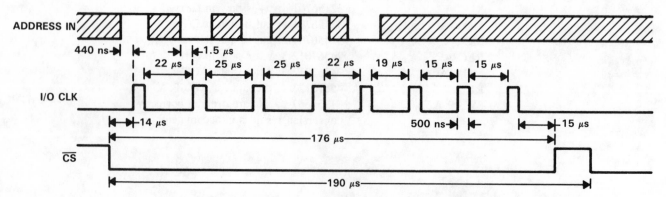

Figure 11-222. 6502 to TLC545 Interface Timing Diagram

```
CSLOW              .EQU 4000H
CSHIGH             .EQU 4002H
ADDLOW             .EQU 8000H
ADDHIGH            .EQU 8001H
MUXADDRESS         .EQU 0000H
RESULT             .EQU 0001H
;
;
;
                   LDA    #$90          ; Initialize muxaddress
                   STA    MUXADDRESS    ; To 10010 = channel 18
                   STA    CSLOW         ; Bring /CS low
                   LDX    #$05          ; Set bit counter for 5 MSB'S

LOOP1:             ROL    MUXADDRESS    ; Rotate muxaddress bit into carry
                   BCS    SET           ; Branch if bit is set
                   LDA    ADDLOW        ; Write out A low on A0, clock data in
                   JMP    CONTINUE      ; Skip next instruction
SET:               LDA    ADDHIGH       ; Write out A high on A0, clock data in

CONTINUE:          ROR    A             ; Rotate do into carry
                   ROL    RESULT        ; Rotate carry into result
                   DEX                  ; Decrement bit counter
                   BNE    LOOP1         ; Go back for another bit

                   LDX    #$03          ; Set counter for 3 LSB'S

LOOP2:             LDA    ADDLOW        ; Read in data bit
                   ROR    A             ; Rotate into carry
                   ROL    RESULT        ; Rotate carry into result
                   DEX                  ; Decrement bit counter
                   BNE    LOOP2         ; Get another bit

                   STA    CSHIGH        ; Bring /CS
```

INTERFACE FOR THE TLC545 DEVICE TO THE ROCKWELL 6502 MICROPROCESSOR USING THE 6522 VIA

This application presents the design of an interface using the 6522 VIA. Another method using TTL gates is described in the previous application. Cost and performance are the basic trade-offs between the two designs. The 6522 device interface is faster, but the TTL method costs less.

Principles of Operation

The interface circuit diagram is shown in Figure 11-223. Timing for a data read cycle and an address write cycle is shown in Figures 11-224 and 11-225, respectively. Software listings for the initialization of the 6522 device and interface control follow the timing diagrams.

The interface makes use of the serial port available on port B pins CB1 and CB2. Since the serial port is not capable of full duplex communication, previous conversion results cannot be read in while a new multiplexer address is being shifted out. Thus, the interface can be used only for individual conversion cycles where the port is configured dynamically; that is, as an output port for the address write cycle, and as an input port during the data read cycle. This requires the inclusion of an SN74LS126 3-state buffer. The D-type flip-flops are used to effectively delay the I/O clock to ensure that the set up and hold times for shifting data in and out are met. Port B pins PB0 and PB1 are used to generate $\overline{CS}$ and the output enable signal for the 3-state buffer.

A data conversion cycle begins by bringing $\overline{CS}$ low. This is accomplished by writing a low to PB0. The analog multiplexer address, which is stored in the five most significant bits of the X index register, is shifted out by writing to the SR register of the 6522 device. A delay loop is inserted to wait until the multiplexer address has been shifted out. The serial port is then configured as an input port to shift in the A/D conversion results. Conversion requires 36 system clock cycles, therefore, an appropriate delay loop dependent upon system clock frequency may be required. The output of the 3-state buffer is enabled by writing a high to PB1, and data is shifted into the SR register of the 6522 device. Again, a delay loop is included to wait until the data is shifted in. $\overline{CS}$ is then brought high, and the 3-state buffer is disabled to complete one data acquisition cycle. A data acquisition cycle can be completed in 55 μs.

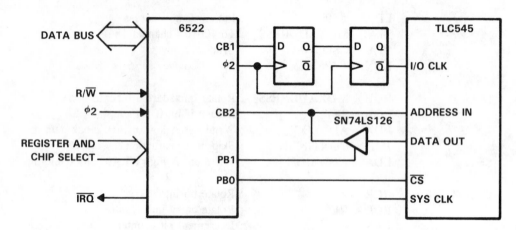

Figure 11-223. 6522 to TLC545 Interface Circuit Diagram

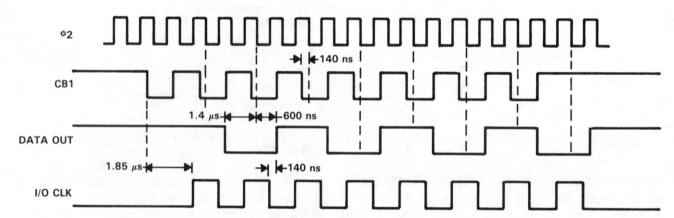

Figure 11-224. 6522 to TLC545 Data Read Cycle Timing Diagram

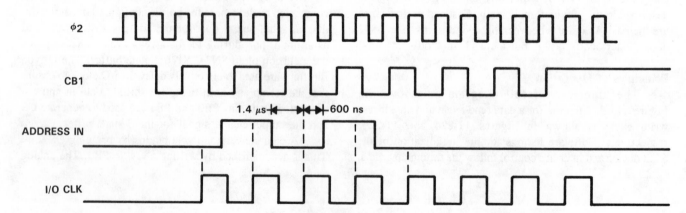

Figure 11-225. 6522 to TLC545 Address Write Cycle Timing Diagram

```
;
;
ORB          .EQU 0000H
DDRB         .EQU 0002H
SR           .EQU 000AH
ACR          .EQU 000BH
PCR          .EQU 000CH
IFR          .EQU 000DH
IER          .EQU 000EH
;
;

              LDA  #$03          ;
              STA  DDRB          ; Initialize port B I/O pins
              LDA  #$01          ;
              STA  ORB           ; Bring chip select high
              LDX  #$50          ; Initialize muxaddress = 01010

              LDA  #$18          ; Shift out under phi 2 control
              STA  ACR           ;
              LDA  #$00          ;
              STA  ORB           ; Disable data out, bring /CS low
              STX  SR            ; Shift out muxaddress to 545
              LDY  #$02          ; Load delay loop counter
DELAY1        DEY                ; Decrement delay counter
              BNE  DELAY1        ; Branch if not zero
              NOP                ;
              LDX  #$02          ;
              LDA  #$08          ;
              STA  ACR           ; Shift in under phi 2 control
              STX  ORB           ; Enable output of 74LS126
              LDA  SR            ; Dummy load to shift results in
              LDY  #$03          ; Load delay loop counter
DELAY2        DEY                ; Decrement delay loop counter
              BNE  DELAY2        ; Branch if not zero
              LDA  #$01          ;
              STA  ORB           ; Disable data out, bring /CS high
              LDA  SR            ; Read conversion results into 6502
```

INTERFACE FOR THE TLC545 AND TLC546 DEVICES TO THE MOTOROLA 6805 MICROPROCESSOR

This application describes techniques for operating the TLC545 and TLC546 A/D converters with the 6805 microprocessor using software control. This can be accomplished by two methods:

1. Generating all necessary control signals under software control by toggling output port pins
2. Using the serial peripheral interface (SPI) to generate necessary control signals for data transfer.

The TLC545 and TLC546 devices are particularly well suited for interfacing to the SPI, however, not all 6805 microprocessors include the SPI on chip. Thus, the software controlled interface has the advantage although it is less efficient.

Both the TLC545 and TLC546 devices comprise a complete data acquisition system on a chip. The system includes such functions as analog multiplexer, sample-and-hold, 8-bit A/D converter, and microprocessor-compatible control logic. The four control inputs are $\overline{CS}$, I/O clock, system clock, and address input.

The I/O and system clock require no special speed or phase relationship, and are normally used independently. This allows the system clock to run up to 4 MHz for the TLC545 device and up to 2.1 MHz for the TLC546 device to ensure conversion times of less than 9 and 17 μs, respectively. The I/O clock runs up to 2.1 MHz and 1.1 MHz, respectively, to allow a maximum data transfer rate.

Principles of Operation

Circuit diagram for the software controlled 6805 microprocessor to TLC545 and TLC546 device interface is shown in Figure 11-226. Circuit timing is shown in Figure 11-227. The software controlled interface makes use of four pins on port A of the 6805 microprocessor. Three of these pins are used as outputs to generate the $\overline{CS}$, I/O clock, and multiplexer address input. The remaining pin is used as an input to receive the conversion results from the TLC545 and TLC546 devices. A software program segment illustrates a method that can be used to initialize the input/output pins.

A program listing which controls actual transfer of data follows the initialization method. This program block sends out the 5-bit analog multiplexer address that is shifted into the TLC545 and TLC546 devices on the first five I/O clock rising edges. Sampling of the addressed input begins at the falling edge of the fifth I/O clock and continues until the falling edge of the eighth I/O clock occurs. Conversion requires 36 clock cycles of the system clock, which can run up to 4 MHz to allow conversion in as little as 9 μs. Conversion results are shifted out of the TLC545 and TLC546 devices on the negative edge of the I/O clock and the program block leaves these results in the accumulator. With the 6805 device running at 5 MHz and the TLC545 device system clock at 4 MHz, one conversion cycle loads an analog multiplexer address and reads the results from the previous conversion. This can be completed in 248 μs.

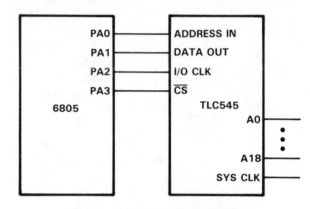

Figure 11-226. 6805 to TLC545 Interface Circuit Diagram

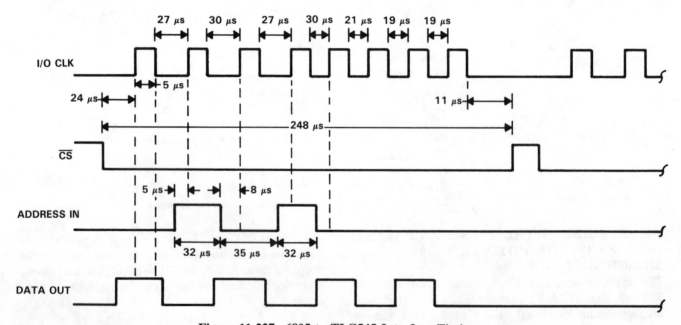

Figure 11-227. 6805 to TLC545 Interface Timing

```
PORTA       .EQU 0000H
DDRA        .EQU 0004H
            .ORG 0100H

START:      LDA   #0DH      ;
            STA   DDRA      ; Initialize port A I/O pins
            BSET  3,PORTA   ; Bring chip select high
            LDA   #50       ; Initialize multiplexer address


            LDX   #05       ; Load counter for first 5 bits
            BCLR  3,PORTA   ; Bring chip select low

LOOP1:      BRSET 1,PORTA,LABEL1 ; Read data bit into carry bit
LABEL1:     ROL   A         ; Rotate muxadd bit and result bit
            BCS   SET       ; Go to set if muxadd bit is 1
            BCLR  0,PORTA   ; Write out A 0 to TLC545 address in
            JMP   CLOCK     ; Skip next instruction
SET:        BSET  0,PORTA   ; Write out A 1 to TLC545 address in
CLOCK:      BSET  2,PORTA   ; Bring I/O clock high
            BCLR  2,PORTA   ; Bring I/O clock low
            DECX            ; Decrement counter
            BNE   LOOP1     ; Continue if counter is not zero
;
            LDX   #03       ; Load counter for last 3 data bits
;
LOOP2:      BRSET 1,PORTA,LABEL2 ; Read data bit into carry bit
LABEL2:     ROL   A         ; Rotate new data bit into result
            BSET  2,PORTA   ; Bring I/O clock high
            BCLR  2,PORTA   ; Bring I/O clock low
            DECX            ; Decrement counter
            BNE   LOOP2     ; Continue if counter is not zero
;
            BSET  3,PORTA   ; Bring chip select high
```

INTRODUCTION TO DATA ACQUISITION FOR DIGITAL SIGNAL PROCESSORS

Digital Signal Processing (DSP) involves the representation, transmission, and manipulation of signals using numerical techniques and digital processors. It has been a fast growing technology during the past few years, with the impact of programmability and VLSI broadening the range of problems that can be served with DSP. Its applications have been expanded to encompass not only traditional radar signal processing but also today's image processing, speech processing, and telecommunications.

Both the theoretical and practical aspects of DSP have made tremendous progress. While more DSP algorithms are being discovered, better tools are also being developed to implement these algorithms. One of the most important recent breakthroughs in electronic technology is the high-speed digital signal processor. These single-chip processors are now commercially available in very large-scale integration (VLSI) circuits from semiconductor suppliers. Digital signal processors are essentially high-speed microprocessors/microcomputers, designed specifically to perform the computation-intensive algorithms. By taking advantage of the advanced architecture, parallel processing, and dedicated DSP instruction sets, these devices can execute millions of DSP operations per second.

With the advantages offered by VLSI, innovative engineers are discovering more and more applications where digital signal processors efficiently provide better solutions than their analog counterparts for reasons of reliability, flexibility, repeatibility, compactness, and long-term stability. Digital signal processors do, however, require accurate analog-to-digital (A/D) and digital-to-analog (D/A) converters. These converters interface the inputs and outputs of the DSP to the analog world.

Texas Instruments has developed a state-of-the-art analog interface IC, the TLC32040, to interface Texas Instruments TMS320 family of digital signal processors to the analog world. In addition to the A/D and D/A converters, the TLC32040 also contains filters for reducing noise and aliasing, which is an undesired sampled-data phenomenon. The TLC32040 also contains a serial port, which allows direct interface to most TMS320 digital signal processors. More information on the TLC32040 is presented in the following pages.

INTERFACING THE TLC32040 TO THE TMS320 FAMILY OF DIGITAL SIGNAL PROCESSORS

Description

The TLC32040 is a complete analog-to-digital and digital-to-analog input/output system on a single monolithic CMOS chip. This device integrates a band-pass, switched-capacitor, antialiasing input filter, a 14-bit resolution A/D converter, four microprocessor-compatible serial port modes, a 14-bit resolution D/A converter, and a low-pass, switched-capacitor, output-reconstruction filter. The device offers numerous combinations of Master Clock input frequencies and conversion/sampling rates, which can be changed via digital processor control.

Typical applications for this IC include modems (7.2-, 8-, 9.6-, 14.4-, and 19.2-kHz sampling rates), analog interface for digital signal processors, speech recognition/storage systems, industrial process control, biomedical instrumentation, acoustical signal processing, spectral analysis, data acquisition, and instrumentation recorders. Four serial modes, which allow direct interface to the TMS32011, TMS32020, and TMS320C25 digital signal processors, are provided. Also, when the transmit and receive sections of the Analog Interface Circuit (AIC) are operating synchronously, it will interface to two SN54299 or SN74299 serial-to-parallel shift registers. These shift registers can then interface in parallel to the TMS32010, other digital signal processors, or to external FIFO circuitry. Output data pulses are emitted to inform the processor that data transmission is complete, or to allow the DSP to differentiate between two transmitted bytes. A flexible control scheme is provided so that the functions of the IC can be selected and adjusted coincidentally with signal processing via software control.

The antialiasing input filter comprises seventh-order and fourth-order CC-type (Chebyshev/elliptic transitional) low-pass and high-pass filters, respectively, and a fourth order equalizer. The input filter is implemented in switched-capacitor technology and is preceded by a continuous time filter to eliminate any possibility of aliasing caused by sampled data filtering. When no filtering is desired, the entire composite filter can be switched out of the signal path. A selectable, auxiliary, differential analog input is provided for applications where more than one analog input is required.

The A/D and D/A converters each have 14 bits of resolution with 10 bits of integral linearity guaranteed over any 10-bit range. Currently the AIC is being evaluated from the perspective of offering several different versions of the AIC. These AICs would be screened at the factory for enhanced performance in areas such as total harmonic distortion or A/D and D/A accuracy. One of these devices may provide integral linearity greater than 10 bits.

The A/D and D/A architectures guarantee no missing codes and monotonic operation. An internal voltage reference is provided to ease the design task and to provide complete control over the performance of the IC. The internal voltage is brought out to a pin and is available to the designer.

Separate analog and digital voltage supplies and grounds are provided to minimize noise and ensure a wide dynamic range. Also, the analog circuit path contains only differential circuitry to keep noise to an absolute minimum. The only exception is the DAC sample-and-hold, which utilizes pseudo-differential circuitry.

The output-reconstruction filter is a seventh-order CC-type (Chebyshev/elliptic transitional low-pass filter with a fourth-order equalizer) and is implemented in switched-capacitor technology. This filter is followed by a continuous-time filter to eliminate images of the digitally encoded signal.

Principles of Operation

As shown in Figures 11-228 and 11-229, the AIC is easily interfaced to the TMS32020, TMS320C25, TMS320C17 and TMS32011 serial ports. The TMS32020/C25 can communicate with the AIC either synchronously or asynchronously depending on the information in the control register. If d5 in the AIC control register is a 0, the transmit and receive sections of the AIC will operate asynchronously; if d5 is a 1, these sections will operate synchronously. The operating sequence for synchronous communication with the TMS32020 and TMS320C25 is shown in Figure 11-230. For asynchronous communication, the operating sequence is similar but $\overline{FSX}$ and $\overline{FSR}$, in general, do not occur at the same time (see Figure 11-231). For proper operation, the TXM bit in the TMS32020/C25 control register should be set to 0 so that the $\overline{FSX}$ pin of the TMS32020/C25 is configured as an input, the format status bit (FO) of the TMS32020/C25 should be set to 0, and the AIC WORD/$\overline{BYTE}$ pin should be at a logic high. After each receive and transmit operation, the TMS32020/C25 asserts an internal receive (RINT) and transmit (XINT) interrupt which may be used to control program execution.

The interface to the TMS32011 and TMS320C17 also requires no external circuitry and, like the TMS32020 interface, can operate either synchronously or asynchronously. The operating sequence for synchronous communication is shown in Figure 11-232. In this case, the AIC must transmit data in two 8-bit bytes and therefore the WORD/$\overline{BYTE}$ pin must be held low. As with the TMS32020/C25, the TMS32011/C17 control register should be configured to respond to external framing pulses ($\overline{FSX}$ and $\overline{FSR}$) and an external Shift Clock. The $\overline{FSX}$ and $\overline{FSR}$ interrupts may be used to control program execution.

Since the TMS32010 and the TMS320C15 have no serial port, a direct connection to the AIC is not possible. However, as shown in Figure 11-233, the AIC can interface to two 74LS299 Serial-to-Parallel shift registers which can then be connected to the data bus of the TMS32010 or the TMS320C15. The AIC must be operated synchronously and in 16-bit (word) mode. The operating sequence is shown in Figure 11-234. In this configuration, the device labeled A1 represents a data-delay. The data-delay may be realized with three cascaded AND gates connected between "I" and "O" ("C" is not connected), or a single D-type flip-flop with the

D-input connected to "I", the Q-output connected to "O" and the noninverting clock input connected to "C". All of the logic circuitry excluding the two 74LS299s and the 74LS138 3-to-8-line decoder may be replaced with a single

PAL. Data is written to the AIC with an OUT instruction, specifying port address 1; a read is accomplished with an IN instruction specifying port address 0.

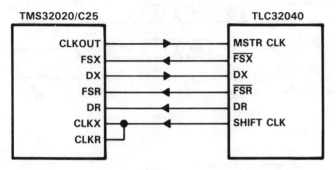

Figure 11-228. AIC Interface to TMS32020/C25

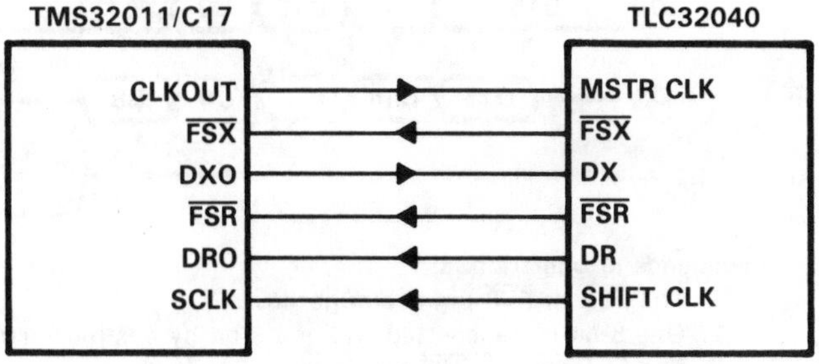

Figure 11-229. AIC Interface to TMS32011/C17

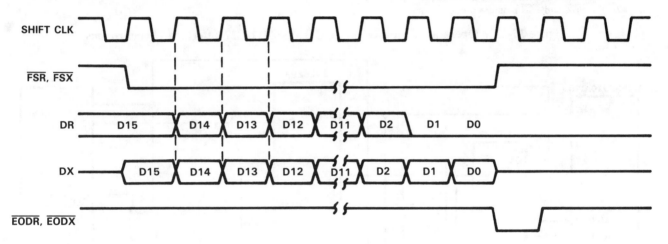

The sequence of operation is:
1. The $\overline{FSX}$ or $\overline{FSR}$ pin is brought low.
2. One 16-bit word is transmitted or one 16-bit byte is received.
3. The $\overline{FSX}$ or $\overline{FSR}$ pin is brought high.
4. The $\overline{EODX}$ or $\overline{EODR}$ pin emits a low-going pulse as shown.

Figure 11-230. Operating Sequence for AIC-TMS32020/C25 Interface—Synchronous

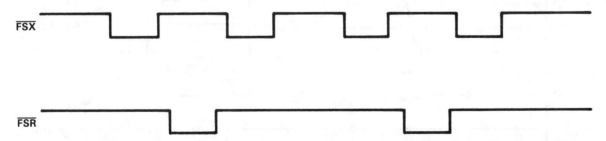

Figure 11-231. Asynchronous Communication: AIC-TMS32020/C25 Interface

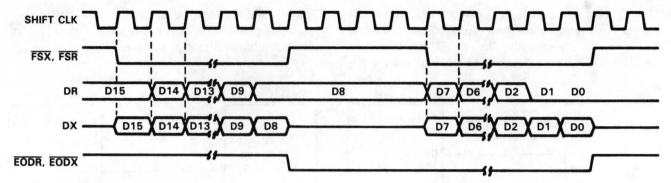

The sequence of operation is:
1. The $\overline{FSX}$ or $\overline{FSR}$ pin is brought low.
2. One 8-bit is transmitted and one 8-bit byte is received.
3. The $\overline{EODX}$ and $\overline{EODR}$ pins are brought low.
4. The $\overline{FSX}$ and $\overline{FSR}$ emit a positive frame-sync pulse that is four shift clock cycles wide.
5. One 8-bit byte is transmitted and one 8-bit byte is received.
6. The $\overline{EODX}$ and $\overline{EODR}$ pins are brought high.
7. The $\overline{FSX}$ and $\overline{FSR}$ pins are brought high.

Figure 11-232. Operating Sequence for AIC-TMS32011/C17 Interface—Synchronous

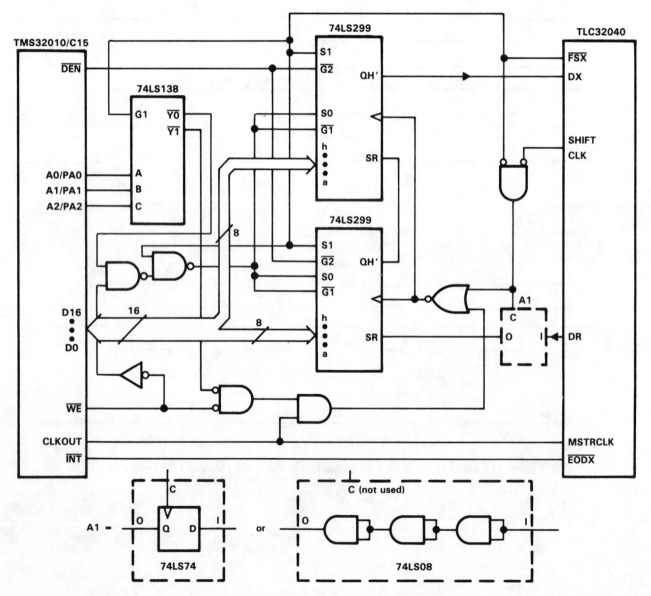

Figure 11-233. AIC Interface to TMS32010/C15

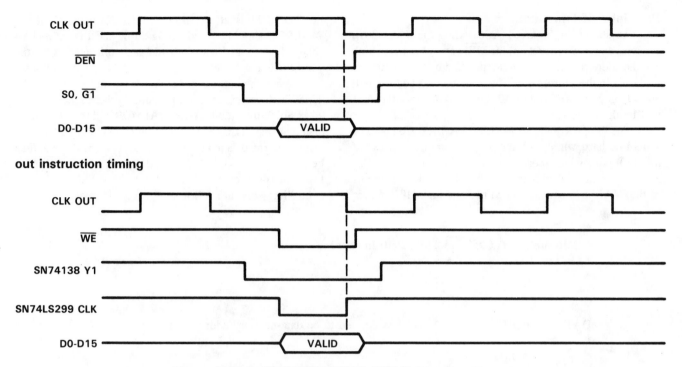

out instruction timing

Figure 11-234. TMS32010/C15-TLC32040 Interface Timing

INTERFACING THE TLC7524 TO THE TMS32010

Hardware

Due to the high-speed operation of the internal logic circuitry of the TLC7524 8-bit digital-to-analog converter, the interface to the TMS32010 Digital Signal Processor requires a minimum of external circuitry. As shown in Figure 11-235, the interface circuitry consists of logic to decode the address of the peripheral. Here we have used one SN74ALS679 12-bit Address Comparator.

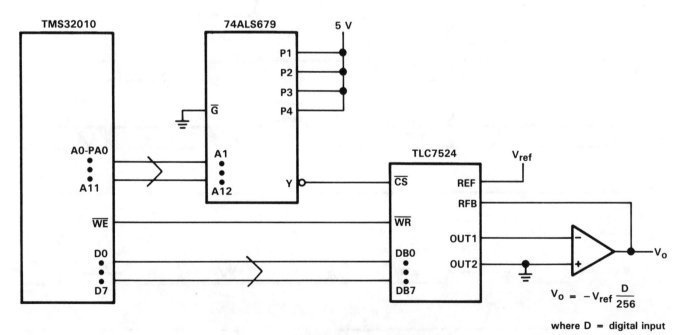

$$V_o = -V_{ref} \frac{D}{256}$$

where D = digital input

Figure 11-235. TLC7524 to TMS32010 Interface

Principles of Operation

As shown in Figure 11-236, when the TMS32010 executes an OUT instruction, the $\overline{\text{MEN}}$ output remains high and the address of the peripheral specified by the instruction is placed on the address bus. The three-bit port address will appear on pins A0 through A2, and pins A3 through A11 will be driven low.

The P-inputs of the SN74ALS679 Address Comparator should be hard wired with the preprogrammed address of the I/O port which corresponds to the TLC7524. For example, if the TLC7524 is at I/O port 0, all of the P-inputs of the 74ALS679 should be tied to logic high. If however the TLC7524 is at I/O port address 1, input P2 of the 74ALS679 should be tied to logic low and P1, P3 and P4 should be tied to logic high (detailed information on the 74ALS679 may be found in the ALS/AS Logic Data Book).

When the address appearing at the address bus of the TMS32010 corresponds to the address programmed into the 74ALS679, the output of the 74ALS679 will be driven low, enabling the TLC7524 (see Figure 11-236, the timing diagram). The data to be written to the TLC7524 should then be on the data bus of the TMS32010 so that when $\overline{\text{WE}}$ goes low, the data can be latched into the TLC7524. The controlling software follows.

```
*
* Software for TLC7524 to TMS32010 Interface
*
*
*
*
DAT       EQU     0          DAT is defined to be data-memory address zero.
*
          AORG    0
          LDPK    0          Set Data-Memory Page Pointer to 0.
*
START     OUT     DAT,0      Place the contents of data-memory address 0 in
*                            the data latch of the DAC. The TLC7524 is at port address 0.
*         B       START
          END
```

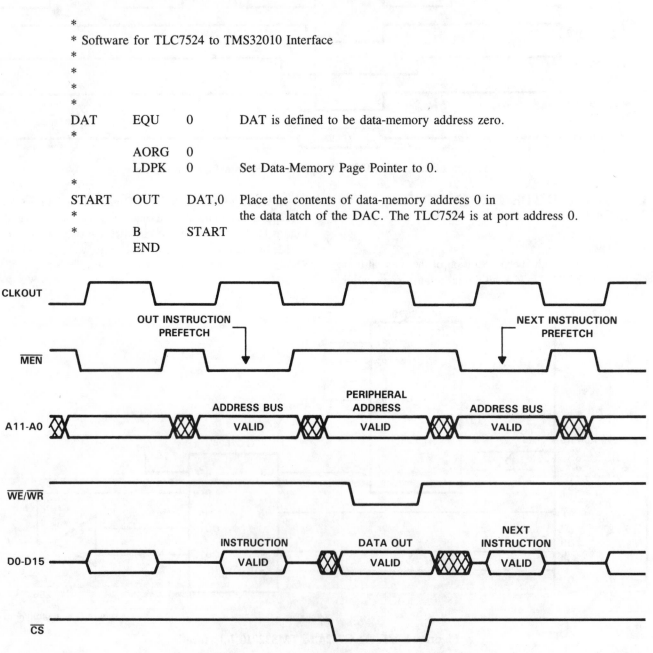

Figure 11-236. OUT Instruction Timing for TMS32010

INTERFACING THE TLC0820 TO THE TMS32010

Hardware

Because the control circuitry of the TLC0820 8-bit A/D converter operates much more slowly than the TMS32010, it cannot be directly interfaced. The following describes in detail the circuit shown in Figure 11-237. All of the logic functions are implemented with 74ALS and 74LS Low-power Schottky Logic. The devices used are:

1	SN74ALS679	12-bit Address Comparator
1	SN74LS74	Dual Positive-edge Triggered D-type Flip-flops
1	SN74ALS465	Octal Buffer with 3-state Output
1	SN74LS32	Quad 2-Input OR-Gate

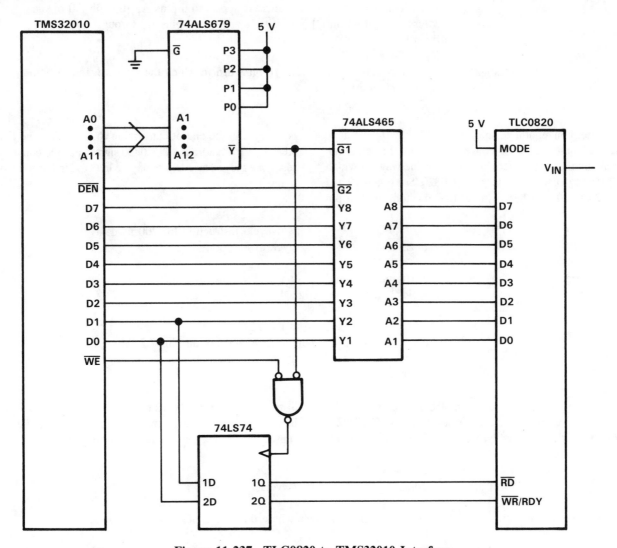

Figure 11-237. TLC0820 to TMS32010 Interface

Principles of Operation

The TMS32010 Digital Signal Processor operates the TLC0820 by writing to the D flip-flops the necessary states of the $\overline{WR}$ and $\overline{RD}$ pins. The $\overline{WR}$ and $\overline{RD}$ signals must initially be at a logic high level, corresponding to a high level at the output of both flip-flops. To begin conversion, $\overline{WR}$ must be lowered for a minimum of 600 nanoseconds and then raised. The conversion begins on the rising edge of the $\overline{WR}$ pulse, and approximately 600 nanoseconds later, the conversion is complete. Since the OUT instruction requires two CLKOUT cycles (each of which lasts 200 ns when the TMS32010 is running at 20 MHz), the write pulse may be implemented with the following code (also shown in the software listing).

OUT	<dma1>, <PA>	
NOP		or any 1-cycle instruction
OUT	<dma2>, <PA>	2-cycle instruction

where   <dma1>   is a data-memory address of a 16-bit word in which the bit assigned to $\overline{WR}$ is a logic-0 and the bit assigned to $\overline{RD}$ is a logic-1.

                        <PA>   is the port address assigned to the TLC0820 and programmed into the 12-bit address comparator. The port address appears on pins A2 through A0 of the address bus during an OUT or an IN operation, pins A3 through A11 are driven low.

                        <dma2>   is a data-memory address of a 16-bit word in which the bits assigned to both $\overline{WR}$ and $\overline{RD}$ are logic 1s.

After conversion has begun, the TLC0820 must have time to complete the A/D conversion before the result can be read by the DSP. There are two possible ways in which the TMS32010 may operate the TLC0820 (See Figures 2 and 3 of TLC0820 datasheet). In the first method, the $\overline{RD}$ pin is lowered at the end of conversion (600 ns after the rising edge of $\overline{WR}$), before the TLC0820 signals that the conversion result is ready. The $\overline{RD}$ pin may be lowered in the following manner:

NOP		any 1-cycle instruction for 200 ns delay
OUT	<dma3>, <PA>	2-cycle instruction

where   <dma3>   is the data-memory address of a 16-bit word in which the bit assigned to $\overline{RD}$ is a logic 0 and the bit assigned to $\overline{WR}$ is a logic 1.

                        <PA>   is the port address assigned to the TLC0820 and programmed into the 12-bit address comparator. The port address appears on pins A2 through A0 of the address bus during an OUT or an IN operation, pins A3 through A11 are driven low.

No more than 300 ns after the $\overline{RD}$ pin is lowered, the conversion result will appear on the data bus of the TLC0820, where it can then be read by the TMS32010. Since the IN instruction requires 400 ns to execute, no additional delay is necessary. The data may be read and the $\overline{RD}$ pin subsequently raised in the following manner:

IN	<dma4>, <PA>	2 cycles
OUT	<dma2>, <PA>	2 cycles

where   <dma4>   is the data-memory address where the conversion result will be stored.

                        <PA>   is the port address of the TLC0820.

                        <dma2>   is the data-memory address of a 16-bit word in which both the bits assigned to $\overline{RD}$ and $\overline{WR}$ are logic 1s.

After the conversion result has been read it will be necessary to mask the data to obtain the 8-bits of interest. The interface timing is shown in Figure 11-238.

INTERFACING THE TLC7524 TO THE TMS32020

Hardware

Due to the high-speed operation of the internal logic circuitry of the TLC7524 8-bit digital-to-analog converter, the interface to the TMS32020 Digital Signal Processor requires a minimum of external circuitry.

As shown in Figure 11-239, the interface circuitry consists of logic to decode the address of the peripheral. Here we have used one SN74ALS138 3-to-8-Line Decoder.

Principles of Operation

As shown in Figure 11-240, when the TMS32020 executes an OUT instruction, the peripheral address is placed on the address bus and the $\overline{IS}$ line goes low, indicating that

```
*
*
* Software for TLC0820 Interface to TMS32010
*
*
           AORG    0
           LDPK    0          Set Data-Memory Page-Pointer to 0.
*
* Data memory locations 1, 2, and 3 should be initialized as follows:
*
*     ADDRESS    DATA
*        1       0001H
*        2       0002H
*        3       0003H
*
* This is necessary because the LSB of the data will be written to the /WR line of the TLC0820
* and the next-to-LSB will be written to the /RD line of the TLC0820.
*
* The following program segment manipulates the /WR and /RD control inputs of the TLC0820.
*
*
START      OUT     3,0        Raise /WR and /RD.
           NOP
           OUT     2,0        Lower /WR.
           NOP
           OUT     3,0        Raise /WR.
           NOP
           OUT     1,0        Lower /RD.
           IN      0,0        Read conversion result and store in data-memory location 0.
           OUT     3,0        Raise /RD.
           B       START
           END
```

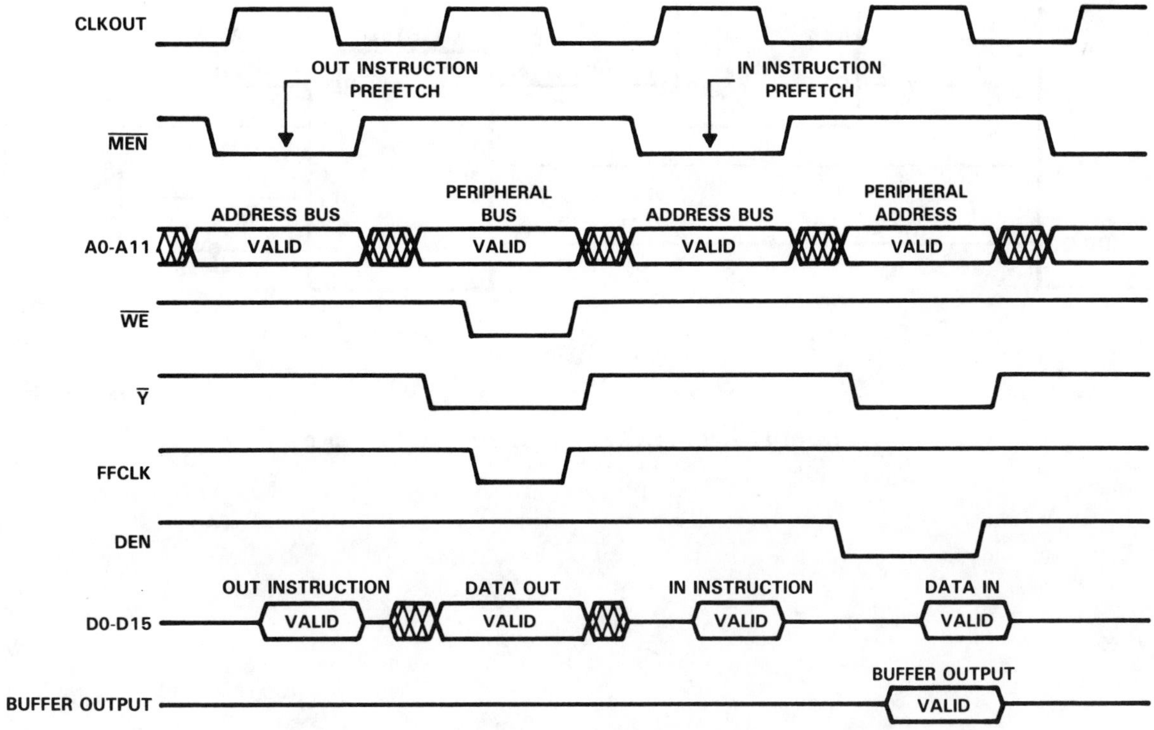

Figure 11-238. TLC0820 to TMS32010 Interface Timing

the address on the bus corresponds to an I/O port and not external program or data memory. A low level at $\overline{\text{IS}}$ will enable the 74ALS138 decoder, and the Y-output, corresponding to the address on the bus, is brought low.

When the Y-output is brought low, the TLC7524 is enabled and the data appearing on the databus will be latched into the D/A converter by $\overline{\text{STRB}}$. The controlling software is shown in the following listing.

```
*
*
* Software for TLC7524 to TMS32020 Interface
*
*
            AORG    >0
            LDPK    >0          Set data memory page pointer to 0.
*
* The following program segment transfers the data in the lower 16 bits of the accumulator to the
* data latches of the TLC7524 digital-to-analog converter.

*
START       SACL    >60         Save low accumulator in >60.
            OUT     >60,0       Move contents of >60 to TLC7524.
            B       START
            END
```

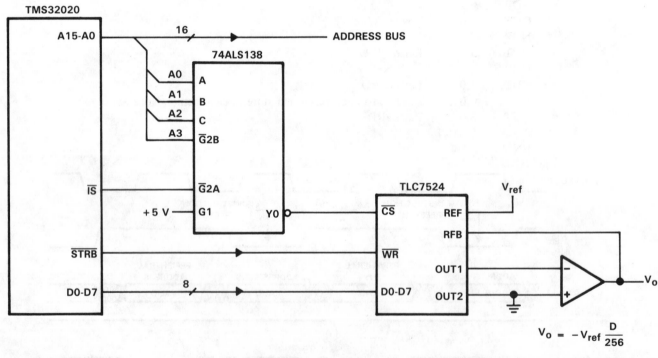

$$V_o = -V_{ref}\frac{D}{256}$$

where D = digital input

Figure 11-239. TLC7524 to TMS32020 Interface

INTERFACING THE TLC0820 TO THE TMS32020

Hardware

Because the control circuitry of the TLC0820 8-bit A/D converter operates much more slowly than the TMS32020, it cannot be directly interfaced. The following describes in detail the interface circuit shown in Figure 11-241. As drawn, the interface circuitry employs the following logic devices.

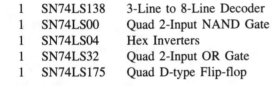

1	SN74LS138	3-Line to 8-Line Decoder
1	SN74LS00	Quad 2-Input NAND Gate
1	SN74LS04	Hex Inverters
1	SN74LS32	Quad 2-Input OR Gate
1	SN74LS175	Quad D-type Flip-flop

Principles of Operation

The 74LS138 decodes the addresses assigned to the TLC0820. One of the addresses is used when performing a write operation; the other, for a read. The two different addresses are necessary to ensure that the correct number of wait states is provided for the write and read operations.

The controlling software is shown in the following listing. To begin conversion, the TMS32020 must supply the TLC0820 with a write pulse ($\overline{WR}$) of at least 600 ns. With the TMS32020 running at 20 MHz and the TLC0820 configured as slow memory, three wait states are necessary to provide a write pulse of sufficient length.

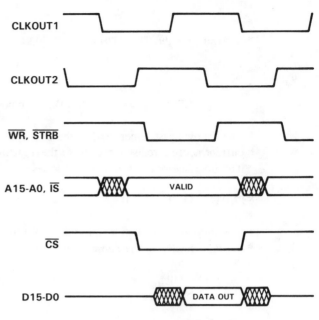

Figure 11-240. TMS32020 OUT Instruction Cycle

```
OUT     *,0     output the data word specified by the current auxiliary register
                to peripheral on port address 0. A low at Y0 of the
                74ALS138 will allow 3 wait states.
```

After conversion has begun (with the rising edge of the $\overline{WR}$ signal), the TMS32020 must wait at least 600 ns before the conversion result can be read. Sufficient delay should be provided in software.

```
IN      <dma>,1     read a data word from peripheral on port address 1. Store in
                    data memory location <dma>. A low at Y1 of the
                    74ALS138 will provide 2 wait states.
```

To read the conversion result, sufficient wait states should be provided to allow for the data access time (320 ns minimum) of the TLC0820. As shown in the timing diagram of Figure 11-242, two wait states are provided when accessing port 1.

```
*
*
* Software for the TLC0820 to TMS32020 Interface
*
*
          AORG    >0
          LDPK    >0          Set Data-Memory Page Pointer to >0.
*
* The following statement writes the data contained in the data-memory location specified by the
* current address-register pointer to the TLC0820. The data is not important* the write pulse is.
* The interface provides for 3 wait states.
*
START     OUT     *,0
*
* The following instructions provide a 600 ns delay between the rising edge of the write pulse to
* the falling edge of the subsequent read.
*
          NOP
          NOP
*
* The following IN instruction reads the conversion result and stores the result in data-memory
* location >60. The interface provides for 2 wait states.
*
          IN      >60,1
          B       START
          END
```

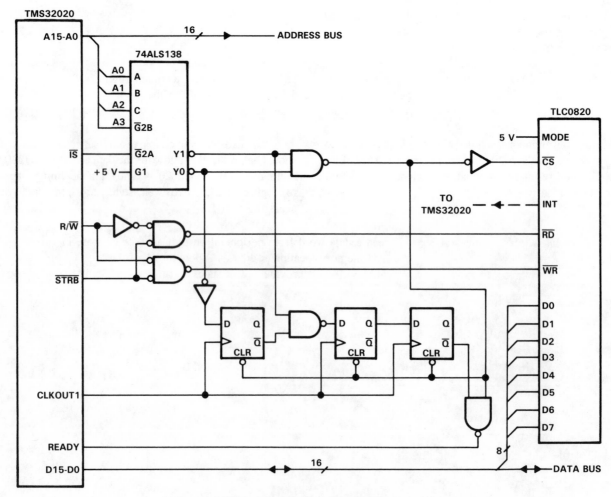

Figure 11-241. TLC0820 to TMS32020 Interface

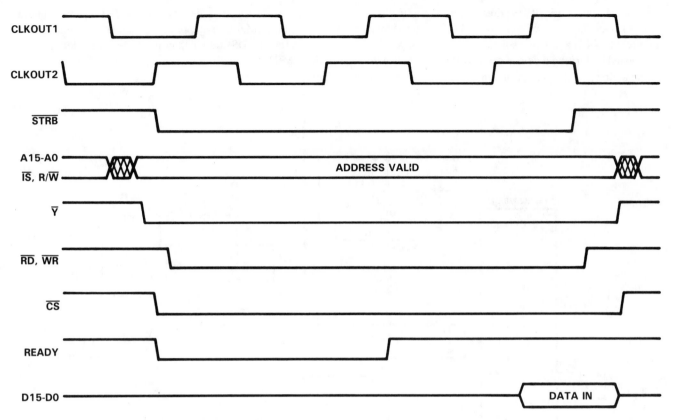

Figure 11-242. TLC0820 IN Instruction Timing — Two Wait States

INTERFACING THE TLC1540, TLC1541, TLC540 AND TLC541 TO THE TMS32020 AND TMS320C25

The TMS32020/C25 can be easily interfaced to the TLC154X and TLC54X family of data acquisition devices via the serial port. The TLC1540/1 and TLC540/1 are multiple-analog-input devices that are useful in cases where more than one analog signal needs to be monitored. The TLC540/1 contains an 8-bit ADC with 11 external analog inputs. If more precision is required, the TLC1540/1 contains a 10-bit ADC with 11 external analog inputs. These devices contain separate input and output serial ports. The input port is used for selecting which analog input is to be monitored. The host processor simply sends the address of the desired analog input. The output port is used for transmitting the digital representation of the processor.

Hardware

As shown in Figure 11-243, glue logic is required to generate the clock for the serial interface and the chip select for the TLC54X/154X. A total of three packages is required: two 74HC163 synchronous counters and two inverters from a 74HC04. Timing is not critical, therefore other technologies such as S, LS, AS, ALS, F, and ACl can be used. In this application, the glue logic can be replaced by one PAL. A '16R8 would be appropriate.

Please refer to Figure 11-243 for the following explanation. U1 divides the CLKOUT signal from the TMS32020/C25 to provide the clock for the serial interface. The TMS32020/C25 internally divides the input crystal clock by four. In the case of a TMS320C25 using a 40 MHz crystal, U1 divides by eight and by sixteen to provide a 1.25 MHz and 625 kHz clock, respectively. The 1.25 MHz clock can be used for the TLC540, and the 625 kHz clock can be used for the TLC541 and TLC1540/1. The TMS32020/C25 generates a framing signal on pin FSX that is used to trigger a chip-select from U2. When triggered, U2 provides a chip-select 8 and 10 clocks wide for the TLC540/1 and TLC1540/1, respectively. The ripple carry signal from U2 is used as the chip-select signal. Triggering occurs when the load input is driven low. This occurs when the FSX signal from the TMS32020/C25 goes high. On the next rising clock edge, U2 is loaded with the values on its preset inputs. When U2 is loaded, ripple carry goes low until a count of 15 is reached. Then, ripple carry goes high, disabling the count. The count disable is accomplished by connecting the inverted ripple carry to ENABLE P input. The preset input values are 7 and 5 to generate a chip-select 8 and 10 clocks wide, respectively.

If several devices need to be networked together, multiple TLC54X and TLC154Xs can be connected together on the same serial bus. This is possible because the serial output ports are in a high-impedance state when not selected.

Additional glue logic is required for this application. A latched decoder such as a 74HCT137 (U4) as shown in Figure 11-244 is a possible solution. The appropriate device is selected by writing a device value to U4. The device value is the address of the output pin on U4 that routes the chip-select to the desired device. U4 is I/O mapped in this application. Memory mapping is possible by using the data-space select (-DS) rather than the I/O-space select (-IS) on the TMS32020/C25.

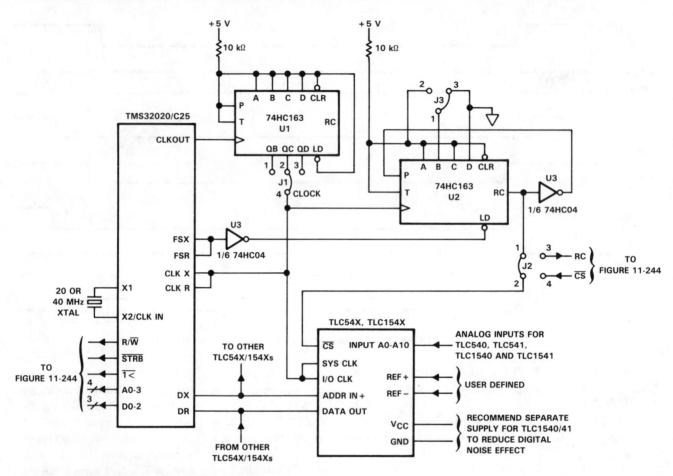

TMS32020/C25-TLC54X, TLC154X INTERFACE:
- Data transfer software initiated on TMS32020/C25
- TMS32020/C25 using burst mode for transfers on serial port
- 3 packages used for glue logic: 74HC04 and 2X 74HC163
- Glue logic can be replaced with one PAL, 16R8

J1 — TMS32020 operation:
    Connect 1 and 4 for 1.25 MHz clock
    Connect 2 and 4 for 625 kHz clock
    TMS320C25 operation:
    Connect 2 and 4 for 1.25 MHz clock
    Connect 3 and 4 for 625 kHz clock
J2 — Connect 1 and 2 for single TLC54X/154 X operation
    Connect 1 and 3, 2 and 4 for multiple TLC54X/154X operation
J3 — Connect 1 and 3 for TLC154X
    Connect 1 and 2 for TLC540/1.

Figure 11-243. TLC54X, TLC154X to TMS32020/C25 Interface

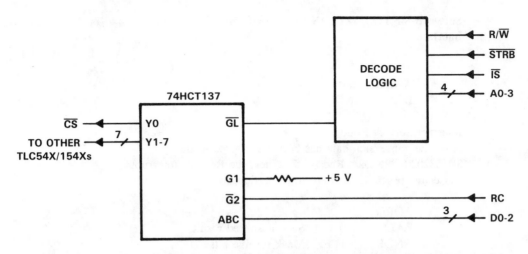

Figure 11-244. Expansion Circuitry for TMS32020/C25 to TLC154X/54X Interface

Software

For proper operation of the TMS32020/C25 with the hardware, two of its status bits, TXM and FO, must be correctly set as shown in the following software listing. TXM is the transmit mode bit that configures the FSX pin as an input or an output. This application requires FSX to be an output. Consequently, the TXM bit must be set high. The value of the TXM bit can be modified by using the STXM or RTXM instructions. Note that TXM is set low upon reset. The TXM bit should be set high using the STXM instruction during the reset initialization routine. FO is the serial I/O format bit that defines the data transfer to be either 8 or 16-bit formats. This application requires data transfers to be in a 16-bit format. Therefore, the FO bit must be set low. The FO bit can be modified using the FORT instruction. Note that the FO bit is correctly set low at reset. TLM and FO can also be modified using the LST1 instruction.

Attention must be paid to the bit positions of the data in the 16-bit word on the TMS32020/C25. Bit positions are skewed as a result of the 2.5 system clock delay in the TLC54X/154X to recognize a chip-select. The TMS32020/C25 must embed the analog input address at the proper bit locations to be correctly recognized by the TLC540/1 and TLC1540/1. Similarly, the TMS32020/C25 must correctly interpret the proper bit locations of the data sent by the TLC54X/154X. The timing diagrams (see Figure 11-245) illustrate the skew in the bit locations. Bit manipulations are easily performed using logical operators and bit shifting such as the AND, ANDK, OR, ORK, XOR, XORK, LAC and LACT instructions. Table 11-44 shows the proper bit locations.

Table 11-44. TMS320/C25 Data Bit Positions

BIT POSITIONS	15	14	13	12	11	10	9	8	7	6	5	4	3	2	1	0
TLC540/1 AND TLC1540/1 ADDRESS IN			B3	B2	B1	B0										
TLC154X DATA OUT				A9	A8	A7	A6	A5	A4	A3	A2	A1	A0			
TLC54X DATA OUT				A7	A6	A5	A4	A3	A2	A1	A0					

Other Considerations

Data transfers between the TMS32020/C25 and the TLC54X/154X are initiated under software control. External stimulus for these transfers is possible by using the TMS32020/C25 external interrupt inputs (INT) and the branch control input (BIO).

The TLC54X/154X require a minimum number of system clock cycles to occur between valid transfers to allow for ADC conversion. Conversion time periods are specified on the TLC54X/154X data sheets. Be sure to wait the conversion time period between transfers to obtain valid data. Wait periods can be timed by using the timer function in the TMS32020/C25.

```
* Software for TLC54X/154X to TMS32020/C25
* Interface
*
*
             AORG    >0
             B       INIT
*
* Interrupt Service Routine:
* After the conversion data has been transferred to the
* TMS32020/C25 Data Receive Register, the Receive Interrupt
* causes the processor to branch to location 26.
*
             AORG    26
             ZALS    >0       Save unshifted result.
             ANDK    >1FFF    Mask receive bits.
             RPTK    2        Shift receive bits.
             SFR
             SACL    >60      Save conversion result in >60.
             RPTK    127      Delay for conversion.
             NOP
             RPTK    127
             NOP
             RPTK    127
             NOP
             B       TXRX     Get new conversion result.
*
* Initialization Routine:
*
INIT         DINT             Disable interrupts.
             LDPK    >0       Set Data-Memory Page Pointer to >0.
             LACK    >10      Enable (RINT) interrupt.
             SACL    >4
             STXM             Set transmit mode (FSX is an output).
             FORT    0        Set data format to 16 bits.
*
* Data Transfer Routine:
*
TXRX         LACK    >5       Load Accumulator with address (channel 5).
             RPTK    9        Shift address 10 places to the left.
             SFL
             SACL    >1       Place channel address in Transmit Register.
             EINT             Enable Interrupts.
WAIT         B       WAIT     Wait for receive interrupt.
```

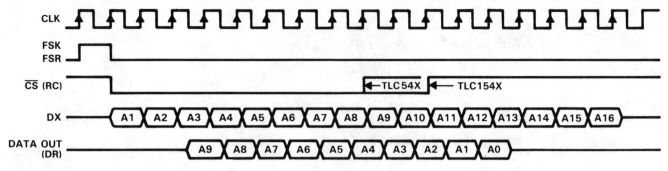

CLK
FSK / FSR
CS (RC) ← TLC54X ← TLC154X
DX: A1 A2 A3 A4 A5 A6 A7 A8 A9 A10 A11 A12 A13 A14 A15 A16
DATA OUT (DR): A9 A8 A7 A6 A5 A4 A3 A2 A1 A0

TMS32020/C25
NOTES: ● Burst mode for serial data transfer
● FSX is an output, TXM status bit set to "1"
● 16-bit word transfers required, FO status bit set to "0"

Figure 11-245. Timing Diagram for TLC154X

INTERFACING THE TLC548/9 TO THE TMS32020

Hardware

The interface circuit employs three packages:

1	SN74LS175	Quad Positive-edge Triggered D-type Flip-flop with Common Clock and Clear
1	SN74LS02	Quad Positive-NOR Gate
1	SN74LS74	Dual Positive-edge Triggered D-type Flip-flop with Independent Clocks, Clear and Preset.

Note: One 74LS175 may be substituted for the 74LS74.

The 74LS74 (or, 1/2 of a 74LS175) is used to implement a ring counter which divides the frequency of the CLKOUT signal, provided by the TMS32020, by four and yields four separate output phases. For example, if the frequency of CLKOUT is 5 MHz, then the output frequency of the counter is 1.25 MHz. If the 74LS74 is used, the clocks should be tied together and all PRESET and CLEAR inputs should be tied to the positive supply voltage. If the 74LS175 is used, only two of the four flip-flops are needed and the common CLEAR should be tied to the positive supply. The complete interface circuit is shown in Figure 11-246.

Principles of Operation

As shown in Figure 11-247, the timing diagram, XF is normally high and is brought low under software control (an RXF instruction as shown in the software listing). Since XF is tied to CS of the TLC549, CS is also brought low. The TLC548/9 requires a maximum of 1.4 microseconds of delay from the falling edge of CS until valid data appears at the output. This is implemented with a 74LS175 which provides 2 microseconds of delay between the falling edge of CS and the falling edge of FSR, which begins data transfer.

After the falling edge of FSR, the first data bit is clocked into the DRR (Data Receive Register) of the TMS32020 on the falling edge of CLKR. Two-hundred nanoseconds later, the TLC548/9 sees a falling edge of I/O CLOCK and places the next bit on the DATA OUT pin. As shown in Figure 11-247, the phase relationship between CLKR and I/O CLOCK allows 600 ns from the falling edge of I/O CLOCK (which clocks-out the next data bit) and the next falling edge of CLKR (which clocks the bit into the DRR register of the TMS32020).

Since the TLC548/9 is an 8-bit A/D converter, the serial port of the TMS32020 should be configured to read 8-bits at a time. This is accomplished by setting FO, the internal format bit, equal to 1. On the eighth falling edge of CLKR after FSR goes low, the last data bit (the LSB) is clocked into the DRR register, and the TMS32020 generates RINT, the serial port read interrupt. If the interrupt is enabled (i.e. the TMS32020 has been instructed to recognize the interrupt), the TMS32020 executes the instruction residing in memory location 26, which should be a branch to a routine which reads the data in the DRR register and raises XF (with an SXF instruction).

If the interrupt (RINT) has been disabled, XF should be raised (with an SXF instruction) after the eighth falling edge of I/O CLOCK (i.e. a minimum of 44 CLKOUT cycles after it was initially lowered).

Because the eighth falling edge of I/O CLOCK initiates the hold-mode of the TLC548/9's internal sample-and-hold and starts the next conversion, it is important that the eighth falling edge of I/O CLOCK occur before XF is raised. XF should be kept high for a minimum of 17 microseconds (85 CLKOUT cycles) before the next RXF instruction, thus allowing the TLC548/9 enough time to complete the next conversion.

If XF is lowered before the conversion is complete, the TLC548/9 will halt conversion and the results of the previous conversion will be transferred to the DRR register of the TMS32020.

```
*
*
* Software for TLC548/9 to TMS32020 Interface
*
*
            AORG    >0
            B       INIT      Branch to initialization routine
*
* Interrupt Service Routine:
* After the conversion data has been transferred to the
* TMS32020 Data Receive Register, the Receive Interrupt
* causes the processor to branch to location 26.
*
            AORG    26
            SXF               Raise /CS (XF pin of TMS32020).
            LAC     >0        DRR is data-memory location 0.
            SACL    >60       Save result in data-memory address >60.
            RPTK    85        Delay for conversion.
            NOP
            B       GETDAT
*
* Initialization Routine:
*
INIT        DINT              Disable interrupts.
            SXF               Raise /CS.
            LDPK    >0        Set data-memory page pointer.
            LACK    >10       Load Interrupt Mask Register
            SACL    >4        with correct data to enable RINT.
            FORT    1         Set data format to 8 bits.
*
* Data Transfer Routine:
* This routine lowers /CS of the TLC548 and begins data transfer.
*
GETDAT      RXF               Lower /CS.
            EINT              Enable Interrupts.
WAIT        B       WAIT      Wait for Receive Interrupt.
            END
```

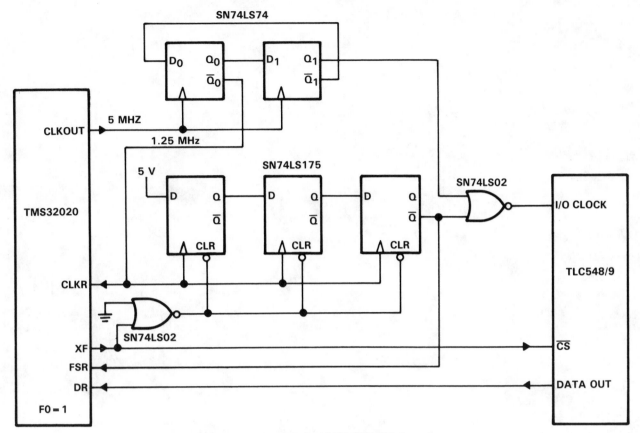

Figure 11-246. TLC549-TMS32020 Interface

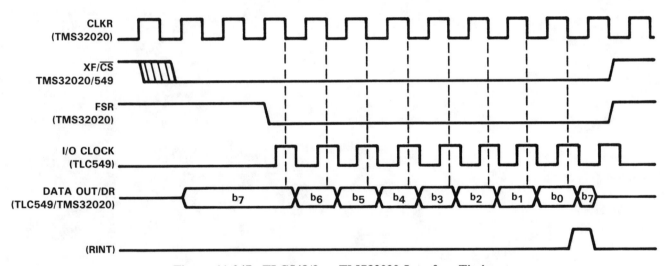

Figure 11-247. TLC548/9 to TMS32020 Interface Timing

Section 12

Special Functions

THE HALL EFFECT

Edwin Hall first noted the effect that bears his name at Johns Hopkins University in 1879. He was investigating the effects of a steady magnetic field on current in a thin gold foil. He observed a small voltage at the edges of a current-carrying gold foil when a magnetic field was applied that was perpendicular to the foil.

THE HALL EFFECT IN SILICON

The Hall effect occurs as current flows through a semiconductor material in the presence of a magnetic field. As electrons or holes flow through the material, their path depends on the charge density and velocity of the majority carriers as well as the magnetic flux. In Figure 12-1, if current (I) flows from left to right with magnetic flux (B) in the direction shown, the force (F) applies downward on the majority carriers, whether holes or electrons. Charge carriers collect near the bottom surface of the semiconductor material and generate the Hall voltage. In n-type material, the majority carriers are electrons and the polarity of the Hall voltage becomes negative on the bottom surface with respect

majority carriers and the charge density. The Hall effect ideally produces a repeatable Hall voltage that is linearly proportional to the external magnetic field.

However, in addition to magnetic-field intensity, there are other semiconductor-material factors that govern the Hall voltage, including temperature, mechanical stress, and current. Both mechanical stress and temperature changes affect mobility of the majority carriers. Also changes in current flow cause nonlinear fluctuations in the Hall voltage. A constant-current generator eliminates current changes, and temperature-compensation circuits offset the thermal effects. The nonlinearity caused by mechanical stress is not easily corrected. The most desirable approach for minimizing the effects of mechanical stress utilizes the architecture of the sensor. Figure 12-2 shows the geometric layout of the orthogonal cross-coupled Hall cell. The cell with four sensors

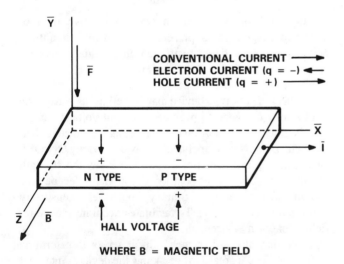

WHERE B = MAGNETIC FIELD

Figure 12-1. Hall-Effect Current Flow

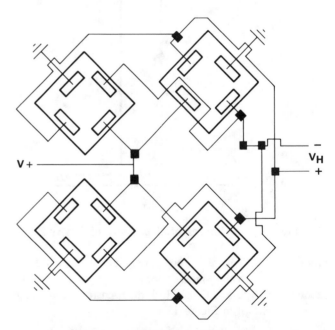

Figure 12-2. Hall-Effect Sensor Layout

connected in an orthogonal manner, reduces the effect of mechanical stress and also improves sensitivity to magnetic fields. The cross coupling of devices also reduces process-related variations and reduces the offset by a factor of 16.

to the top. The converse is true for p-type material because holes are the majority carriers. Accurate diffusion of a specific impurity into silicon determines the mobility of the

HALL-EFFECT DEVICES

A Hall-effect device is a circuit consisting of a Hall-effect cell, signal conditioning functions which may include hysteresis, and an output transistor integrated into a monolithic chip. The three basic types of Hall-effect devices are the switch, the latch, and the linear device. The switch and latch are digital devices while the linear Hall device provides a voltage output that is linear with respect to changes in magnetic flux density. The unit of magnetic flux density in the International System of Units is the tesla (T). The tesla is equal to one weber per square meter. Values expressed in milliteslas may be converted to gauss by multiplying by ten, e.g., 50 millitesla = 500 gauss.

SWITCHING DEVICES

The typical switching device is used in applications calling for a normally OFF Hall-effect switch. This device turns ON in the presence of a positive magnetic field and turns OFF when the field is removed. A typical transition diagram for this type of device is shown in Figure 12-3. Note that both the operate point (B_{OP}) and the release point (B_{RP}) are positive values. The hysteresis provides stable switching characteristics. The operating and release values, and the

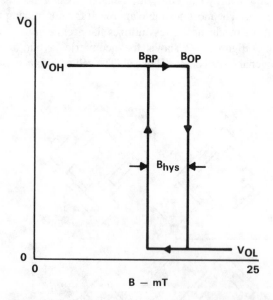

Figure 12-3. Representative Curve of V_O vs B

width of the hysteresis, are parameters that should be considered when choosing a device and magnet for a specific application. The functional block diagram for a switching device is shown in Figure 12-4.

LATCHING DEVICES

A Hall-effect latch is a switch that turns on in the presence of a positive magnetic field and off in the presence of a negative magnetic field. The maximum positive field strength required to turn ON a typical latch is 25 mT. The minimum field to turn a typical latch OFF is –25 mT. See Figure 12-5 for a hysteresis-loop characteristic diagram. The

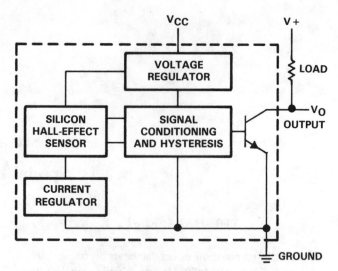

Figure 12-4. Functional Block Diagram of Hall-Effect Switch

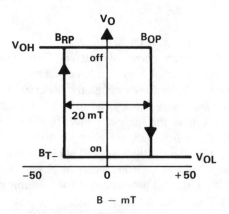

Figure 12-5. Representative Curve of V_O vs B

output transistor is turned on when a positive field of sufficient magnitude is present and remains on until a negative magnetic field of sufficient magnitude is present.

LINEAR HALL DEVICES

A linear Hall-effect device may be defined as a magnetic field sensor designed to provide an output voltage change that is linearly proportional to a change in the applied magnetic field. Not all applications involve strictly ON/OFF switch conditions; sometimes you must know the strength of a magnetic field and its polarity. Linear Hall-effect devices contain no hysteresis circuitry, but their sensitivity (approximately 16 mV/mT) facilitates accurate magnetic-field-strength measurement.

You may also utilize such a linear sensor to determine a magnetic field's polarity if you know the device's intercept value (the point at which the sensor's output voltage characteristic crosses the zero magnetic field strength line). An output voltage greater than the intercept value indicates the presence of a north magnetic pole, while a smaller output denotes a south pole.

Figure 12-6 shows the functional block diagram of a TL3103 linear Hall-effect device. This circuit incorporates a Hall element as the primary sensor along with a voltage reference and a precision amplifier.

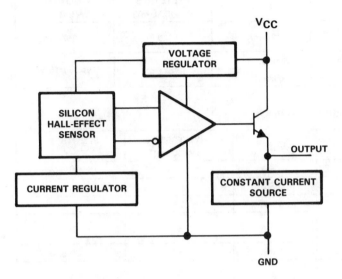

Figure 12-6. TL3103 Functional Block Diagram

Temperature stabilization and internal trimming circuitry provide a device that features high overall sensitivity accuracy with less than 5% error over its operating temperature range. The Hall voltage is amplified to provide a convenient voltage level that is proportional to the magnetic field sensed. The nominal output voltage in the presence of a zero magnetic field is 6 V. The output voltage increases 16 mV/mT with a positive magnetic field and decreases 16 mV/mT with a negative magnetic field as shown in Figure 12-7.

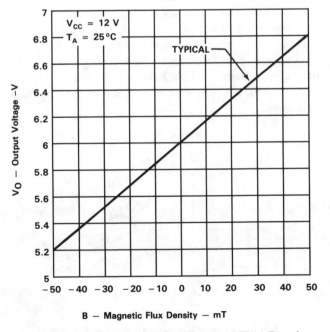

Figure 12-7. TL3103 V_O vs Magnetic Flux Density

ADVANTAGES OF HALL-EFFECT DEVICES

Hall-effect devices give distinct advantages over mechanical and optoelectronic switches. For example, switching thresholds in Hall-effect devices do not degrade with time, as they do in emitter/sensor pairs. Additionally, while stray light affects photosensors in some applications, stray magnetic fields do not generally trigger Hall-effect sensors. As another advantage, you can isolate magnetically activated Hall-effect devices from environmental hazards such as dirt, dust, light, water, or vapor. You cannot as readily seal mechanical and optical switches and relays against contamination. Where moisture is a problem or where a switch spark might ignite explosive vapors, a specially sealed mechanical switch entails high cost. Further Hall-effect advantages over mechanical switches and relays include no contacts to wear, pit, or weld. No-contact switching implies a low failure rate and no maintenance. An internal hysteresis circuit in Hall-effect switches also eliminates contact bounce, a serious problem where mechanical switches must interface directly to a microprocessor.

HALL-EFFECT DEVICE SELECTION

It is helpful to understand that two conventions are used by manufacturers in specifying the magnetic properties of today's devices. The definitions "into" and "out of" the cell cause confusion because both are defined in terms of the orientation of a bar magnet relative to the cell. If the magnet is perpendicular to the cell and the north pole is closest to the sensor, the field is "into the cell" when using the north-pole positive convention. If the south pole is closest to the sensor, the field is "out of" the cell. (See Figure 12-8.) The south-pole positive convention reverses this definition. It is important to take note of the orientation of the magnet with respect to the face of the Hall-effect unit when selecting a device for a specific application.

The TL31xx series uses the north-pole positive convention, as opposed to other similar devices that use the south-pole positive convention. Also, TI second sources UGN30xx parts with the TL30xx series, and these devices are south-pole positive. Therefore, you should check the specifications carefully before selecting a device. Table 12-1 shows a list of Hall-effect devices with respect to their electrical and magnetic properties.

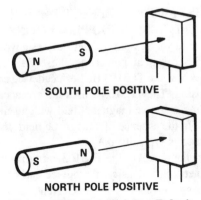

SOUTH POLE POSITIVE

NORTH POLE POSITIVE

Figure 12-8. Determining Polarity

Table 12-1. Electrical and Magnetic Properties
of Hall-Effect Devices

NORTH POLE POSITIVE DEVICES

DEVICE	TYPE	V_{CC} SUPPLY	V_{OUT}	I_{OUT}	HYSTERESIS (TYPICAL mT)	SENSITIVITY (mT)
TL3101	General-purpose switch	5 V	30 V	20 mA	20	$+25/-25$
TL3103	Linear	12 V	5.8 to 6.2 V (B = 0)	Sink 0.5 mA / Source 2 mA	$-----$	(Typical) 16 V/T

SOUTH POLE POSITIVE SWITCHES

DEVICE	TYPE	V_{CC} SUPPLY	V_{OUT}	I_{OUT}	HYSTERESIS (TYPICAL mT)	SENSITIVITY (mT)
TL3013	N.O.	4.5 to 40 V	40 V	30 mA	7.5	45/25
TL3019	N.O.	4.5 to 40 V	40 V	30 mA	12	50/12.5
TL3020	N.O.	4.5 to 40 V	40 V	30 mA	5.5	35/5
TL3030	General Purpose	4.5 to 40 V	40 V	30 mA	5	$+25/-25$
TL3040	N.O.	4.5 to 40 V	40 V	30 mA	5	20/5

N.O. = Normally open.

HALL-EFFECT APPLICATIONS

Although the key feature of a silicon Hall-effect device is its ability to sense magnetic fields, applications are not limited to magnetic-field-related uses. You can utilize them to sense virtually any type of movement by incorporating magnetic material in the moving object. While environmental conditions such as moisture and vibration can adversely affect optical and mechanical devices, Hall-effect units are immune to most environmental conditions.

Traditionally, engineers have not used Hall-effect devices because their cost was much higher than opto or mechanical components. The cost of Hall components has dropped significantly in the past five years so this is not a significant factor in most designs. Designers can now consider using Hall sensors in many applications where mechanical or optical sensors have been used.

The following applications demonstrate methods of using Hall-effect sensors in isolated feedback applications and to sense motion or position.

TL3103 LINEAR HALL-EFFECT DEVICE IN ISOLATED SENSING APPLICATIONS

For several years, opto coupler devices have been used for isolated sensing in power supplies. This application demonstrates how the TL3103 Hall-effect device can provide isolated sensing. The TL3103 senses the presence of either a positive or a negative magnetic field with a sensitivity of 16 mV/mT. In the absence of a magnetic field, the TL3103 output voltage is typically 6 V.

Because the output of the TL3103 varies proportionally to the magnetic flux density, the device can be used with a toroid to sense current or voltage. Current can be measured

as shown in Figure 12-9. When the toroid terminals A and B are connected in series with the circuit to be measured, a change in current changes the magnetic flux density in the toroid gap and causes a voltage change in the TL3103 output. Whether the change is an increase or decrease depends on the direction of current flow through the toroid. The output of the TL3103 can be used to drive an amplifier as shown in Figure 12-9. The TL3103 output is 6 V in the absence of toroid current and flux. One input of the amplifier is therefore referenced to 6 V, allowing a level shift to 0 V at its output for zero sense current.

A similar arrangement provides an isolated voltage measurement. This is accomplished by connecting a resistor in series with the toroid as illustrated in Figure 12-10. The voltage to be measured is between terminals C and D. In this configuration, the current in the toroid is determined by the output voltage (Is = Vout/Rs). Thus, the output variations of the TL3103 are proportional to the sensed output voltage variations.

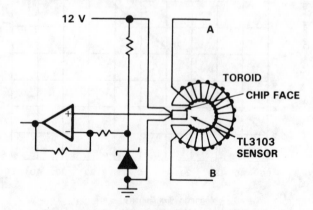

Figure 12-9. Current Measuring Circuit

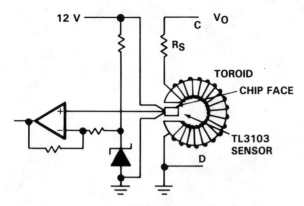

Figure 12-10. Voltage Measuring Circuit

Toroid Design

A variety of soft magnetic materials can be used in manufacturing cores for use with Hall-effect devices. These include manganese-zinc based ferrites, Molypermalloy powder cores, high flux powder cores, and strip wound tape cores.

The choice of the core to be gapped and used with a Hall-effect device depends upon the core characteristics required, cost, and stability over temperature. The effective permeability of the material is a function of the air gap made in the core and the initial permeability of the starting material. Permeability (μ) is defined as the ratio of magnetic flux density (B) in gauss to magnetic field intensity (H) in oersteds. (μ = B/H).

When the gap in a core exceeds a few thousandths of an inch, the effective permeability is determined essentially by the air gap. The magnetic field intensity (H) of a toroid is given by the expression NI/L where:

$$NI = \text{number of turns (N) x current (I)}$$
$$L = \text{mean length of the toroid}$$

The expression for the magnetic flux density then becomes:

$$B = (\mu)\,(NI/L)$$

With an air gap, the expression is altered to:

$$Bgap = \mu\,(NI/L+Kg)\,K$$
$$\mu_o = \text{permeability of air}$$
$$\quad (12.57 \times 10^{-7}\ \text{W/Am})$$
$$K = \text{relative permeability of the toroid}$$
$$\quad (\mu/\mu_o)$$
$$g = \text{length of the air gap}$$
$$W/Am = \text{Webers/amp-meter}$$

As previously discussed, the output of the TL3103 is:

$$V_{sense} = 6\ \text{volts} + (16mV/mT)(\mu_o\ NI/g)$$

This shows how the output of the TL3103 varies with the amp-turns of the toroid. The applications in Figures 12-9

and 12-10 use an Arnold toroid #A393163-2 with a 0.165 inch air gap which is sufficient for an LU package. The magnetic flux density in the air gap is:

$$B(gauss) = 1.92\ NI\ (\text{amp-turns})$$

Thus, the variation in the output of the TL3103 is:

$$V\ sense\ (mV) = (1.6)(1.92\ NI)$$

Therefore, the sensitivity of the TL3103 device to the current in the winding is determined by the number of turns in the winding.

V sense	N	I
614 mV	20	10 A
614 mV	200	1 A
614 mV	2000	100 mA

The features of this approach are:
1. Minimum power loss in the sensing element
   P loss = I^2R in the toroid
   (R < 0.01 for 20 turns)
2. Isolated feedback, no passive connection required.

TL594 ISOLATED FEEDBACK POWER SUPPLY

Figure 12-11 is a power supply circuit using the isolated feedback capabilities of the TL3103 for both current and voltage sensing. (See Figures 12-9 and 12-10). This supply is powered from the ac power line and has an output of 5 V at 1.5 A. Both output voltage and current are sensed and the error voltages are applied to the error amplifiers of the TL594 PWM control IC.

The 24 V transformer produces about 35 V at the 1000 μF filter capacitor. The 20 kHz switching frequency is set by the 6 kΩ resistor and the 0.01 μF capacitor on pins 6 and 5, respectively. The TL594 is set for push-pull operation by tying pin 13 high.

The 5 V reference on pin 14 is tied to pin 15 which is the reference for the current error amplifier. The 5 V reference is also tied to pin 2 which is the reference for the output voltage error amplifier. The output voltage and current limit are set by adjustment of the 10 kΩ pots in the TL3103 error sensing circuits. A pair of TIP31E npn transistors are used as switching transistors in a push-pull circuit. The transformer design information is given in Figure 12-12.

TACHOMETER AND DIRECTION OF ROTATION CIRCUIT

In machine and equipment design, some applications require measurement of both the shaft speed and the direction of rotation. Figure 12-13 shows the circuit of a tachometer which also indicates the direction of rotation.

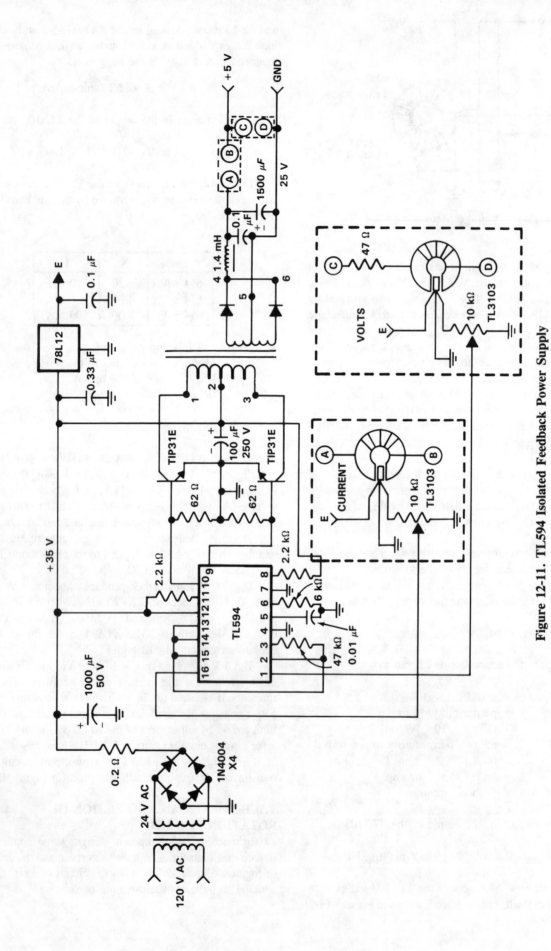

Figure 12-11. TL594 Isolated Feedback Power Supply

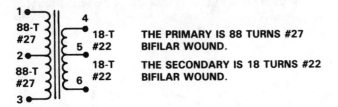

CORE IS FERROXCUBE #846T250/3C8 (BLACK)

THE PRIMARY IS 88 TURNS #27 BIFILAR WOUND.

THE SECONDARY IS 18 TURNS #22 BIFILAR WOUND.

Figure 12-12. Transformer Design Data

Tachometer Operation

The flywheel sensor is a TL3101 Hall-effect switch selected for latch operation. The flywheel has two magnets embedded in the outer rim about 45° apart. One magnet has the north pole toward the outside and the other magnet has the south pole toward the outside rim of the flywheel. Due

to the magnet spacing, a short ON pulse is produced by the TL3101 in one direction and a long ON pulse in the other direction. A 0–50 μA meter is used to monitor the flywheel speed while the LEDs indicate the direction of rotation.

The output from the TL3101 is applied to pin 3 (CLR 1) of an SN74LS123 one-shot multivibrator. The one-shot output pulse, at pin 13 (Q), goes high with a high input pulse. The output pulse duration is set to about 3 μs with the 1 μF capacitor and the 10 kΩ resistor on pins 14 and 15. The pulses at the output go through a low-pass filter, through diode D2, and to the 0–50 μA meter which is the tachometer. Since the TTL low-level output for the SN74LS123 is still about 0.3 V above ground, the meter would read slightly above zero when the flywheel is not moving. The purpose of the germanium diode, D2, is to produce a 0.3 V drop to correct for the 0.3 V low-level output. The meter may be calibrated in revolutions per minute (rpm), or as the user desires.

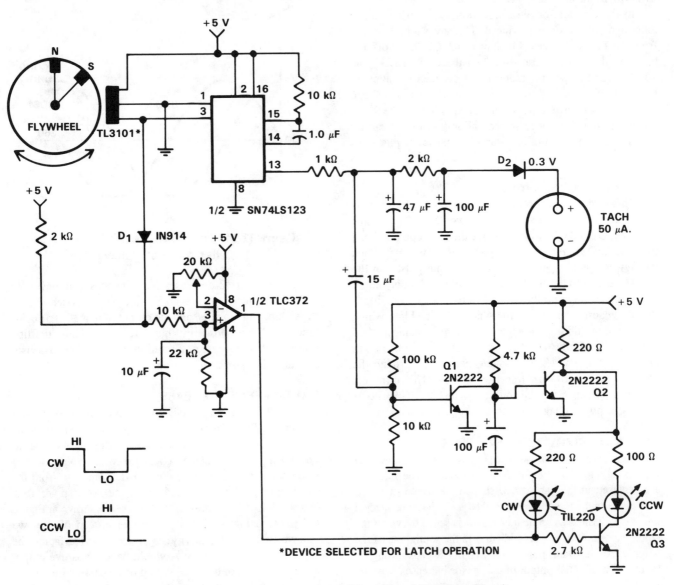

Figure 12-13. Tachometer and Direction of Rotation Circuit

12-7

Direction of Rotation Circuit

The direction of rotation circuit can be divided into three parts:

1. TLC372 device for input conditioning and reference adjustment
2. Two 2N2222 transistors which apply the V_{CC} to the two LEDs when needed.
3. The two TIL220 LEDs which indicate clockwise (CW) or counterclockwise (CCW) direction of rotation.

The input pulses going to the SN74LS123 one-shot multivibrator are also applied through diode D1 to the noninverting input of the TLC372 device. The inverting input is connected to a 20 kΩ potentiometer, providing an adjustable reference. When the flywheel is rotating in a clockwise direction, a short-high and a long-low pulse train is produced. This causes the output of the TLC372 device to be low, turning on the CW LED and holding the CCW LED off by turning the 2N2222, Q3, off.

In the counterclockwise direction, a short-low and a long-high pulse train is produced. This gives a high output from the TLC372 device which turns the CW LED off and turns on the 2N2222 transistor and the CCW LED. The 20 kΩ reference potentiometer is set between the average voltage level of the high-level pulses and the average voltage level of the low-level pulses. The speed of the flywheel will not make any difference in the calibration of the circuit because the ratio of the high- and low-pulse lengths stays the same.

LED V_{CC} Supply Control

This circuit keeps the LEDs off when the flywheel is not rotating. The circuit consists of two 2N2222 transistor switches, Q1 and Q2, which remove the 5 V supply voltage from the LED circuit when the flywheel is not turning. When the flywheel is stopped, there are no output pulses on the SN74LS123 output pin 13. At this time, there is only about 0.5 V bias voltage on the base of transistor Q1, holding it off. This causes the base of transistor Q2 to be high, which turns it on, prohibiting the supply voltage from being applied to the LEDs. With the wheel in motion, a train of pulses appears at the SN74LS123 output that passes through the 15 μF capacitor to turn transistor Q1 on. This turns off transistor Q2, allowing the collector to go high. This allows the 5 V supply voltage to activate the LED circuit.

ANGLE OF ROTATION DETECTOR

Figure 12-14(a) shows two TL3103 linear Hall-effect devices used for detecting the angle of rotation. The TL3103s are centered in the gap of a U-shaped permanent magnet. The angle that the south pole makes with the chip face of unit #1 is defined as angle θ. Angle θ is set to 0° when the chip face of unit #1 is perpendicular to the south pole of the magnet. As the south pole of the magnet sweeps through a 0° to 90° angle, the output of the sensor increases from 0° to

value of V_{OQ} to a peak value of $+Vp$ at 90°. As the magnet continues to rotate to 180°, the output of the sensor retraces its path to V_{OQ}.

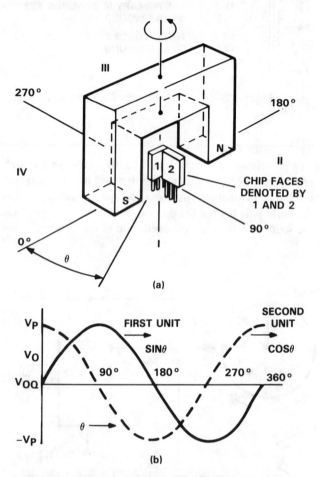

Figure 12-14. Two Linear Hall-Effect Devices Detect Angle of Rotation

Sensor unit #2 decreases from its peak value of $+Vp$ at 0° to a value V_{OQ} at 90°. So, the output of sensor unit #1 is a sine function of θ and the output of unit #2 is a cosine function of θ as shown in Figure 12-14(b). Thus, the first sensor yields the angle of rotation and the second sensor indicates the quadrant location.

HALL-EFFECT COMPASS

The TL3103 linear Hall-effect device may be used as a compass. By definition, the north pole of a magnet is the pole that is attracted by the magnetic north pole of the earth. The north pole of a magnet repels the north-seeking pole of a compass. By convention, lines of flux emanate from the north pole of a magnet and enter the south pole. (See Figure 12-15.) The circuit of the compass is shown in Figure 12-16. By using two TL3103 devices instead of one, we achieve twice the sensitivity. With each device facing the opposite direction, device A would have a positive output while the output of device B would be negative with respect to the zero magnetic field level. This gives us a differential signal to apply to the TLC251 op amp. The op amp is connected as a difference amplifier with a gain of 20. Its output is applied to a null meter or a bridge balance indicator circuit.

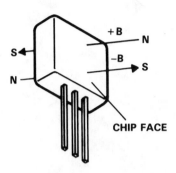

Figure 12-15. Definition of Magnetic Flux Polarity

SECURITY DOOR AJAR ALARM

In security systems for buildings, a switch of some type is installed on each door to be monitored. These may be mechanical switches or reed relay type switches operated by a permanent magnet placed in the door. A TL3019 is a normally open, south-pole-positive Hall-effect device, which may be used in this type of application. Figure 12-17 shows the basic circuit.

In operation, the TL3019 device will activate, or go low, when a south pole of a magnet comes near the chip face of the device. The example shows five doors. Each door has a magnet embedded in its edge with the south pole facing the outer surface. At the point where the magnet is positioned with the door closed, a TL3019 sensor is placed in the door jamb. With the door closed, the Hall devices will be in a logic low state. This design has five doors and uses five TL3019 devices. Each TL3019 has a 4 kΩ resistor in series with it and all door sensor and resistor sets are in parallel

and connected to the inverting input of an LM393 comparator. With all doors closed, the effective resistance will be about 800 Ω and produce 2.2 V at the inverting input. The noninverting input goes to a voltage divider network which sets the reference voltage. The 1.5 kΩ potentiometer is adjusted so the indicator goes out with all doors closed. This will cause 2.35 V to appear at the noninverting input of the comparator. When a door opens, the voltage at the inverting input will go to 2.5 V which is greater than Vref, and the LED will light.

A large number of doors and windows may be monitored with this type of circuit. Also, it could be expanded to add an audible alarm in addition to the visual LED.

MULTIPLE POSITION CONTROL SYSTEM

In machine equipment design, it is sometimes necessary to select an operation or perform steps of another operation in a specific sequence. This application allows a drill press operator to choose a certain size bit, use it, and go to another selection.

Figure 12-18 shows the circuit of a multiple position sensor system. Eight TL3019 Hall devices are positioned outside the rim of a rotating disc for use in a drill-turret control application. Each of the normally OFF switches activates when aligned with the south (S) magnetic pole of the magnet, allowing a computer to stop the turret at any switch location. For example, if the computer chooses drill bit 4, and the turret reaches the bit 4 position, the output of sensor 4 turns on and sets an output latch. This latch output stops the turret motor. When an operator depresses the START pushbutton, the output latch resets and the turret rotates to the next selected drill bit.

There are many variations of this application that may be used in other designs. A comparatively new application is the brushless dc motor. Brushless dc motors are essentially brush-type dc motors turned inside out. Power is fed directly to the armature windings while a permanent magnet field is

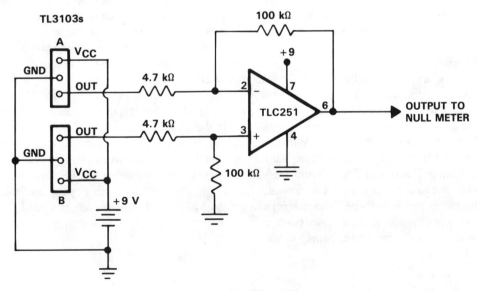

Figure 12-16. Linear Hall-Effect Compass

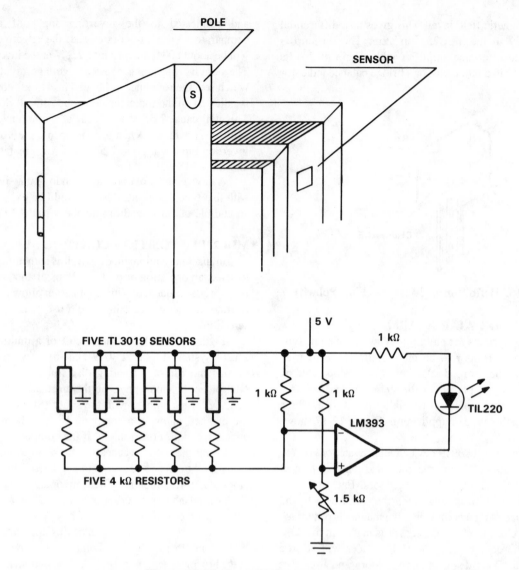

Figure 12-17. Security Door Ajar Alarm

the rotating member. In this type of motor, a Hall-effect sensor senses the position of the rotating magnet and excites the proper windings using a logic and driver circuit.

DOOR OPEN ALARM

Door open alarms are used chiefly in automotive, industrial, and appliance applications. This type of circuit can sense the opening of a refrigerator door. When the door opens, a triac could be activated to control the inside light.

Figure 12-19 shows a door position alarm. When the door is opened, an LED turns on and the the piezo alarm sounds for approximately 5 seconds. This circuit uses a TL3019 Hall-effect device for the door sensor. This normally open switch is located in the door frame. The magnet is mounted in the door. When the door is in the closed position, the TL3019 output goes to logic low, and remains low until the door is opened.

This design consists of a TLC555 monostable timer circuit. The 1 μF capacitor and 5.1 MΩ resistor on pins 6 and 7 set the monostable RC time constant. These values allow the LED and piezo alarm to remain on about 5 seconds when triggered. One unusual aspect of this circuit is the method of triggering. Usually a 555 timer circuit is triggered by taking the trigger, pin 2, low which produces a high at the output, pin 3. In this configuration with the door in the closed position, the TL3019 output is held low. The trigger, pin 2, is connected to 1/2 the supply voltage V_{CC}. When the door opens, a positive high pulse is applied to control pin 5 through a 0.1 μF capacitor and also to reset pin 4. This starts the timing cycle. Both the piezo alarm and the LED visual indicator are activated.

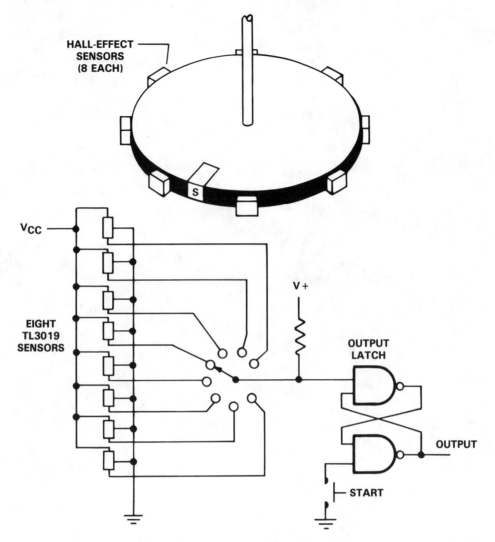

Figure 12-18. Multiple Position Sensor System

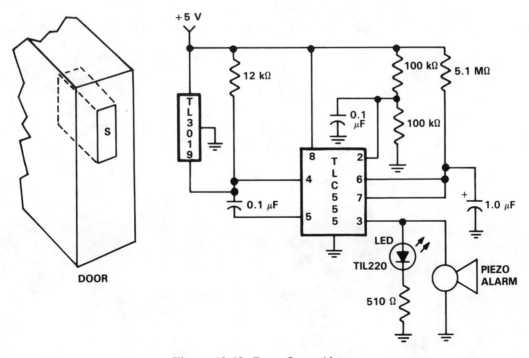

Figure 12-19. Door Open Alarm

APPENDIX

Device Numbering System

Table A-1, an overview of the device numbering system, shows the meaning of the various characters in Texas Instruments Linear and Interface circuit device numbers. Texas Instruments devices that are direct alternate sources for other manufacturers' parts carry the original part number including its prefix. Alteration in device characteristics from the original data sheet specifications, generally to improve performance, results in a new number with the appropriate SN55, SN75, TL, or TLC prefix. The type of package is also included in the device number. Table A-2 lists the package suffixes and their definitions.

Table A-1. Linear and Interface Circuits

XXX XXXXX XX

└─ Package Type (See Table A-2)

ORIGINAL MANUFACTURER	TI PREFIX	DEVICE NUMBER	TEMP* RANGE
TI	TL or TLC	XXXC	COM
		XXXI	IND
		XXXM	MIL
	SN	75XXX	COM
		55XXX	MIL
NATIONAL	LM	1XXX	MIL
		2XXX	IND
		3XXX	COM
	ADC	XXXX	COM
	DS	78XX	MIL
		88XX	COM
RAYTHEON	RC	4XXX	COM
	RM	4XXX	MIL
SIGNETICS	NE	5/55XX	COM
	SA	5/55XX	AUTO
	SE	5/55XX	MIL
	N8T	XX	COM
FAIRCHILD	uA	7XXXC	COM
		7XXXI	IND
		7XXXM	MIL
	uA	9XXX	COM
MOTOROLA	MC	13/33XX	IND
		14/34XXX	COM
		15/35XX	MIL
SPRAGUE	UCN	XXX	COM
	UDN	XXXX	COM
	ULN	XXXX	COM
AMD	AM	XXXXXM	MIL
		XXXXXC	COM
SILICON GENERAL	SG	15XX	MIL
		25XX	IND
		35XX	COM
PMI	OP-	XX	COM

Table A-2. Packages

TYPE	PACKAGE DESCRIPTION
N	Plastic DIP
NE, NG	Plastic DIP, copper lead frame
NF	Plastic DIP, 28 pin, 400 mil
NT	Plastic DIP, 24 pin, 300 mil
P	Plastic DIP, 8 pin
D	Plastic SO, small outline
J	Ceramic DIP
JD	Ceramic DIP, side braze
JG	Ceramic DIP, 8 pin
FE, FG	Ceramic chip carrier, rectangular
FH, FK	Ceramic chip carrier, square
FN	Plastic chip carrier, square
KA	TO-3 metal can
KC	TO-220 plastic, power tab
LP	TO-226 plastic
U	Ceramic flatpack, square
W, WC	Ceramic flatpack, rectangular

*Temperature ranges:

COM = 0°C to 70°C

IND = −25°C to 85°C

AUTO = −40°C to 85°C

MIL = −55°C to 125°C

INDEX

Absolute-value-plus-sign code, 11-8
AC plasma display drivers, 8-16 to 8-29
AC Thin Film Electroluminescent display (TFEL) drivers, 8-42 to 8-50
Accuracy in DAS systems, 11-12, 11-14
Active filters, 2-10 to 2-13, 4-3
  applications for, 3-6 to 3-11
Active tone controls, 3-11 to 3-13
A/D converters (*see* Analog-to-digital converters in DAS)
ADC0803 and ADC0805, 11-21 to 11-23
  interfacing of: to Rockwell 6502, 11-61 to 11-62
    to Z80, 11-58 to 11-61
ADC0804, interfacing of:
  to Rockwell 6502, 11-61 to 11-62
  to Z80, 11-58 to 11-61
ADC0808 and ADC0809 A/D converters, 11-23 to 11-26
  interfacing of: to Motorola 6800, 11-68 to 11-71
    to Rockwell 6502, 11-66 to 11-68
    to Z80, 11-62 to 11-66
ADC0831, ADC0832, ADC0834, and ADC0838 A/D peripherals, 11-26 to 11-34
  interfacing of: to Intel 8048 and 8049, 11-97 to 11-99
    to Intel 8051 and 8052, 11-91 to 11-96
    to Motorola 6800, 6802, and 6809(E), 11-84 to 11-91
    to Motorola 6805, 11-80 to 11-84
    to Rockwell 6502, 11-77 to 11-79
    to Z80, 11-71 to 11-77
Alarm, door ajar, 12-9 to 12-11
Alarm driver, 10-21
Aliasing, 11-6, 11-10 to 11-12
AM26LS31 line driver, 9-42
AM26LS32A line receiver, 9-42
AM26S10/S11 transceiver, 9-8, 9-9
Amplifiers:
  audio, 3-11 to 3-17
  video, 4-1 to 4-6
  (*See also* Operational amplifiers)
Analog-to-digital (A/D) converters in DAS, 11-1, 11-2, 11-10 to 11-12
  criteria for, 11-14 to 11-15
  devices for, 11-21 to 11-58
  and quantization, 11-7
  techniques for, 11-15 to 11-20
Aperture error in DAS, 11-11
Astable timer operating mode, 7-2, 7-4 to 7-6, 7-10, 7-12
Audio op amp applications, 3-11 to 3-17
Automobiles, circuits for, 3-5, 6-35 to 6-36, 10-21

Balanced modulator, 4-3, 4-5

Balanced transmission systems, 9-2 to 9-4, 9-12, 9-13, 9-17
Band-gap regulator references, 5-4 to 5-5
Band-pass active filters, 2-12 to 2-13, 3-7 to 3-10, 4-4
Band-reject active filters (*see* Notch filters)
Bandwidth:
  of band-pass filters, 2-12, 2-13
  and comparators, 2-16
  of LinCMOS op amps, 2-14 to 2-15
  op amp, 2-1 to 2-3, 2-7
  and video amplifiers, 4-1
Battery charger, 6-24 to 6-27
Baxandall tone control, 3-11 to 3-13
BCD (binary-coded decimal), 11-9
BIDFET ICs, 8-1 to 8-2
BIFET op amps, 2-4, 2-14
Binary-coded decimal code, 11-9
Binary codes, 11-7 to 11-9
Binary-to-frequency conversion, 4-5
Bipolar op amps, 2-13 to 2-14
Bits, binary:
  number of, in DAS, 11-10, 11-14
  weighting of, in A/D converters, 11-19
Boost regulators, 6-2 to 6-3, 6-14 to 6-17, 6-20
Bridge-balance indicator, 3-1 to 3-2
Bridge motor drive, 10-21 to 10-22
Buck regulators, 6-2, 6-5 to 6-7, 6-14 to 6-16, 6-18 to 6-20
Buffers:
  in DAS, 11-1
  data, 8-28 to 8-29
  voltage, 3-6
Bulk sustain, 8-19 to 8-20
Bus interfacing for IEEE 488 standard, 9-51 to 9-56
Butterworth filters, 3-8, 11-11
Byte-serial bit-parallel data transfer method, 9-46, 9-47

Capacitance-to-voltage meter, 7-20 to 7-21
Capacitors and capacitance:
  and AC TFEL displays, 8-44, 8-50
  in A/D converters, 11-41, 11-100, 11-102, 11-109
  and cables, 3-11
  with differentiators, 2-9
  filters, 5-19 to 5-21
  and op amp speed, 2-5 to 2-6
  for switching power supplies, 6-12 to 6-13
Closed-loop gain, op amp, 2-3
CMRR (*see* Common-mode rejection ratio)
CMVR (common-mode voltage range), 9-15 to 9-16
Coaxial cable, 8-8, 9-4
Column drivers, 8-44, 8-46 to 8-48

Common-mode rejection ratio (CMRR):
  in differential amplifiers, 3-3
  and error amplifiers, 5-6
  for LinCMOS op amps, 2-15
  op amp, 2-2, 2-5
Common-mode voltage range (CMVR), 9-15 to 9-16
Comparators, 2-15 to 2-18
  applications with, 3-21 to 3-23
  Schmitt triggers as, 3-4
  (*See also* Operational amplifiers)
Compass, Hall-effect, 12-8 to 12-9
Compensation, op amp, 2-1, 2-4, 2-16
Connectors, IEEE 488, 9-48 to 9-49
Constant-current limiting, 5-14
Control circuit for switching power supplies, 6-1, 6-7 to 6-12
Control element for voltage regulators, 5-1 to 5-2
Control system, multiple position, 12-9 to 12-11
Coupling:
  with A/D converters, 11-116
  and display drivers, 8-27 to 8-29, 8-47, 8-49
  of video amplifiers, 4-2
Crowbar circuits, 6-23, 6-38
Crystal filters, 4-4
Current limiting, 6-23 to 6-26
  in display drivers, 11-104
  in lamp circuits, 3-22, 10-14 to 10-17
  in rectifier circuits, 5-20 to 5-21
  for regulators, 5-13 to 5-15
  for switching power supplies, 6-9 to 6-11
Current-mode line drivers, 9-5, 9-13 to 9-15
Current regulator, 5-8

D/A (digital-to-analog) converters, 11-10, 11-11
Darkroom enlarger timer, 7-14
Data acquisition systems (DAS), 11-1 to 11-14
  A/D converters for, 11-14 to 11-20, 11-58 to 11-195
  devices for, 11-21 to 11-58
  for digital signal processors, 11-195 to 11-213
Data logging with DAS, 11-2
Data transmission systems, 9-1 to 9-22
  EIA standards, 9-22 to 9-46
  IBM System 360/370, 9-58 to 9-70
  IEEE 488-1978, 9-46 to 9-58, 9-61
  line drivers for, 10-26, 10-30 to 10-31
  (*See also* Drivers; Receivers; Transceivers)
DC motor drives, 10-22 to 10-23
DC output offsets, 2-8 to 2-10

DC plasma display drivers, 8-9 to 8-11
Dead time and switching power supplies, 6-8 to 6-10, 6-12
Delay timers, 7-1 to 7-2, 7-25 to 7-26
Deterministic signals, 11-3 to 11-4
Difference amplifiers, op amp, 2-9
Differential amplifier, 3-2 to 3-3
Differential transmissions, 9-11 to 9-15
   line receivers for, 9-24, 9-26, 9-28
   party-line systems, 9-17 to 9-18
   voltage-mode, 9-19 to 9-22
Differentiators, op amp, 2-9 to 2-10
Digit scan display method, 8-4, 8-6
Digital interfaces, 9-33, 9-35, 9-39 to 9-41
Digital logic (see Logic circuits)
Digital numbers, creation of, by A/D converter, 11-19 to 11-20
Digital panel meters, 11-105 to 11-107, 11-110 to 11-111
Digital signal processors (DSP), 11-195 to 11-213
Digital thermometer, 11-107
Digital-to-analog (D/A) converters, 11-10, 11-11
Diode limiters, 3-3
Direction of rotation circuit, 12-5, 12-7 to 12-8
Disk power supply, 6-4 to 6-5
Display drivers, 8-1 to 8-7
   AC plasma, 8-16 to 8-29
   AC TFEL, 8-42 to 8-50
   DC plasma, 8-9 to 8-11
   gas discharge, 8-11 to 8-13
   high-voltage, 8-11, 8-13 to 8-16
   LED, 8-3 to 8-8
   using TL502 A/D converter, 11-104, 11-105
   vacuum fluorescent, 8-29 to 8-42
Distortion:
   and active tone controls, 3-11
   and bandwidth, 2-3
   with BIFETS, 2-14
   and transmission lines, 9-3
Distribution amplifiers, 3-13 to 3-14
DM256X64A display, 8-37, 8-39
DMOS (Double-Diffused MOS) structures, 8-1 to 8-2
Door ajar alarms, 12-9 to 12-11
Dot matrix display, 8-30, 8-35 to 8-42
Double-Diffused MOS (DMOS) structures, 8-1 to 8-2
Drift, op amp, 2-5
Drivers, transmission line, 9-1, 9-4 to 9-6
   differential, 9-11, 9-19 to 9-20
   EIA standards, 9-22 to 9-24, 9-33 to 9-36, 9-41
   IBM System 360/370, 9-59 to 9-63
   IEEE 488 standard, 9-52
   single-ended, 9-6
DS3680 telephone relay driver, 10-9 to 10-11
DSP (digital signal processors), 11-195 to 11-213
Dual polarity power supply, 6-36 to 6-38
Dual-slope A/D converters, 11-16 to 11-17, 11-34 to 11-44
Dual-tracking voltage regulators, 5-7, 5-16

Duty cycle:
   for capacitance-to-voltage meter, 7-21
   and inverting regulators, 6-3
   in motor control circuit, 7-22 to 7-23
   and switching supplies and regulators, 5-2 to 5-3, 5-9 to 5-12, 6-2, 6-13
   with timers, 7-5, 7-17
   varying of, 7-17
   with VFD drivers, 8-32 to 8-34
Dynamic signals in DAS, 11-3

EIA transmission standards:
   RS-232-C, 9-22 to 9-33, 9-49
   RS-422-A, 9-37 to 9-40
   RS-423-A, 9-33 to 9-38
   RS-485, 9-38, 9-40 to 9-46
Equal-component low-pass filters, 3-8
Erase operations with display drivers, 8-17, 8-19, 8-20
Error contributors and amplifiers for power supplies, 5-1, 5-3 to 5-6, 6-1, 6-9 to 6-11
Errors, sampling, 11-9, 11-11, 11-15

Fanout of comparators, 2-16 to 2-18
Feed-forward compensation, 2-6 to 2-7
Feedback loops, op amp, 2-8 to 2-9
Filters:
   in DAS, 11-5 to 11-6, 11-12
   for regulators, 5-18 to 5-21
   (See also Active filters)
Flash A/D converters, 11-20
Flasher, LED, 7-17 to 7-18
Flip-flops, 7-2, 7-9
Floating voltage regulators, 5-8
Floppy disk power supply, 6-4 to 6-5
Flyback switching regulator, 6-3
Fold-back current limiting, 5-14 to 5-15, 6-33, 6-37
Forward converters, 6-3 to 6-4
Free-running timer mode, 7-2, 7-4 to 7-6, 7-10, 7-12
Frequency and frequency response:
   and AC TFEL brightness, 8-43, 8-44
   of active filters, 2-11, 2-12, 3-7 to 3-9
   conversion of, from binary, 4-5
   in differentiator circuits, 2-10
   of function generator, 3-20
   in integrator circuits, 2-10
   of microphone preamplifier, 3-17
   of notch filters, 3-10 to 3-11
   and op amp gain, 2-2 to 2-3
   and op amp phase margin, 2-6
   oscillator, 3-16, 7-15
   of ramp generator, 7-16
   and sampling theorem, 11-5 to 11-6
   of switching power supplies, 6-2, 6-7 to 6-8, 6-10
   in switching regulators, 5-10 to 5-12
   synthesizer of, 7-25, 7-26
   timer, 7-4 to 7-6, 7-10, 7-25
   and twisted-pair transmission lines, 9-20, 9-22
   and VCOs, 7-20
Frequency counters, preamplifier for, 4-2 to 4-3
FSK (frequency shift keyer), 4-3 to 4-6

Full bridge converters, 6-4
Full-wave rectifiers, 5-18, 5-20
Function generator, 3-20 to 3-21

Gain:
   and active filters, 3-7
   of comparators, 2-16
   and frequency, 2-2 to 2-3, 2-6
   and noise, 2-6
   in op amp circuits, 2-8
   and video amplifiers, 4-1
Gain-bandwidth product, 2-3, 2-5
Gas discharge display drivers, 8-11 to 8-13, 8-16 to 8-29
Gating with display drivers, 8-27 to 8-29
Generators, 3-16 to 3-17
Gray binary code, 11-8
Ground loops, 5-17, 9-2
Grounding:
   in motor control circuits, 7-22
   of transmission lines, 9-1, 9-2, 9-4, 9-62, 9-63
   of video amplifiers, 4-2
   and voltage regulators, 5-6 to 5-7

H-bridge converters, 6-4
H motor drive, 10-21 to 10-22
Half bridge converters, 6-4
Half-wave rectifiers, 5-18, 5-20
Hall effect, 12-1
   applications with, 12-4 to 12-11
   devices for, 12-2 to 12-4
Hewlett Packard Company and IEEE 488-1978 standard, 9-47
High-pass active filters, 2-10 to 2-11, 3-8 to 3-9, 4-4
High voltages:
   display drivers using, 8-9, 8-11 to 8-13
   isolation of, 11-116
   regulators for, 5-8, 6-33
   VFD drivers for, 8-33 to 8-35, 8-38
Hysteresis with comparators, 2-18, 2-19
Hysteresis modes, SN75154, 9-28 to 9-30

IBM System 360/370 interface circuits, 9-58 to 9-70
IEEE 488-1978 standard, 9-46 to 9-58, 9-61
Inductors for switching power supplies, 6-12
Input offset voltage, 2-1, 2-3 to 2-4, 2-14 to 2-15, 5-5
Inrush current, 3-22
Instrumentation meter driver, 3-18 to 3-19
Integration phases in A/D converter, 11-41
Integrators, op amp, 2-10
Intel 8048 and 8049, interfacing of:
   to ADC0831, ADC0832, ADC0834, and ADC0838, 11-97 to 11-99
   to TL530, TL531, TLC532A, and TLC533A, 11-122 to 11-129
   to TLC540 and TLC541 (software), 11-151 to 11-155
   to TLC549, 11-161 to 11-163
Intel 8051 and 8052, interfacing of:
   to ADC0831, ADC0832, ADC0834, and ADC0838, 11-91 to 11-96

Intel 8051 and 8052, interfacing of (*Cont.*):
  to TL530, TL531, TLC532A, and
    TLC533A, 11-122 to 11-129
  to TLC540 and TLC541, 11-146 to
    11-151
  to TLC545 and TLC546 (serial port),
    11-185 to 11-189
  to TLC545 and TLC546 (software),
    11-181 to 11-185
  to TLC549 (serial port), 11-163 to 11-166
  to TLC549 (software), 11-166 to 11-168
  to TLC1540 and TLC1541 (software),
    11-176 to 11-177
Intelligent switches, 10-21
Interference:
  and microphone cables, 3-11
  and switching power supplies, 6-2
  and transmission lines, 9-3
Inverting op amps, 2-8, 3-4
Inverting regulators, 6-3, 6-14 to 6-21
I/O controller in DAS, 11-1
Isolated feedback power supply, 12-5, 12-6
Isolation:
  and active filters, 3-7
  and Hall effect, 12-3 to 12-5
  high voltage, 11-116

Kick circuit with AC plasma display
    drivers, 8-24 to 8-25
Kilby, Jack, and op amps, 2-1

Lamps:
  driver circuits, 3-22, 8-7, 10-14 to 10-17
  in oscillator circuits, 3-16
Latch-up, 2-5, 5-16
Latches, Hall-effect, 12-2
Layout:
  for A/D converters, 11-102
  and switching power supplies, 6-18
  for video amplifiers, 4-2
  for voltage regulators, 5-17 to 5-18
Lead-acid battery charger, 6-24 to 6-27
LED display drivers, 3-22, 8-3 to 8-8,
  10-17, 10-19
LED flasher, 7-17 to 7-18
"Lights on" alarm, 10-21
Limiting circuits, voltage, 3-3 to 3-4
  (*See also* Current limiting)
LinCMOS operational amplifiers, 2-4, 2-14
  to 2-15, 3-17 to 3-21
  A/D converters, 11-21
  timers, 7-5 to 7-6
Line-to-ground and line-to-line termina-
    tion, 9-4, 9-12, 9-13, 9-43, 9-44
Linear ramp generator, 7-15 to 7-16
Linearity in DAS, 11-12 to 11-14, 11-42
LM311 comparator, 2-18, 3-22
LM317 regulator, 3-19, 5-16 to 5-17
LM318 op amp, 3-11, 3-12
LM393 comparator, 3-23 to 3-24, 12-9,
  12-10
LM741 op amp, 3-16
LM2904 op amp, 3-5
Logic circuits:
  and comparators, 3-22 to 3-23
  and Schmitt triggers, 3-4
  supervisor devices for, 6-27 to 6-31

Logic gate arrays, power supply for, 3-19
Logic levels, conversion and shifting of,
  3-4, 8-7, 10-1
Low-pass active filters, 2-10 to 2-11, 3-8 to
  3-9, 4-4

Magnetic fields and Hall effect, 12-1 to
  12-2
MC1445 op amp, 4-3, 4-5, 4-6
MC1488 line driver, 9-24, 9-26
MC1489(A) line receivers, 9-29
MC3423, 6-38 to 6-39
MC3446 bus transceiver, 9-56, 9-60
MC3486 line driver, 9-42, 9-43
MC3487 receiver, 9-42, 9-43
Memory, computer, supervisor devices
  for, 6-27 to 6-31
Messages for IEEE 488 standard, 9-50 to
  9-51
Metal migration in ICs, 10-2
Meters:
  capacitance-to-voltage, 7-20 to 7-21
  drivers for, 3-18 to 3-19
  panel, 11-105 to 11-111
  scaler for, 6-24
Microphone preamplifiers, 3-11, 3-17
Microprocessors:
  in DAS, 11-1, 11-2
  supervisor devices for, 6-27 to 6-31
  (*See also* Intel; Motorola; Rockwell; *and*
    Zilog *entries*)
MIL-STD-188C standard, 9-24, 9-26 to
  9-29
Missing pulse detector, 7-11 to 7-13
Modulation input of uA2240, 7-8, 7-23,
  7-25
Monostable timer operating mode, 7-2 to
  7-4, 7-9 to 7-10, 7-13
Motorola 6800, interfacing of:
  to ADC0808 and ADC0809, 11-68 to
    11-71
  to ADC0831, ADC0832, ADC0834, and
    ADC0838, 11-84 to 11-91
  to TL520, TL521, and TL522, 11-68 to
    11-71
  to TL530, TL531, TLC532A, and
    TLC533A, 11-131 to 11-137
  to TL0808 and TL0809, 11-68 to 11-71
Motorola 6802, interfacing of:
  to ADC0831, ADC0832, ADC0834, and
    ADC0838, 11-84 to 11-91
  to TL530, TL531, TLC532A, and
    TLC533A, 11-131 to 11-137
  to TLC1540 and TLC1541, 11-172 to
    11-174
Motorola 6805, interfacing of:
  to ADC0831, ADC0832, ADC0834, and
    ADC0838, 11-80 to 11-84
  to TLC540 and TLC541, 11-144 to 11-146
  to TLC545 and TLC546, 11-193 to
    11-195
  to TLC549, 11-160 to 11-161
  to TLC1540 and TLC1541 (software),
    11-174 to 11-176
Motorola 6809(E), interfacing of:
  to ADC0831, ADC0832, ADC0834, and
    ADC0838, 11-84 to 11-91

Motorola 6809(E), interfacing of (*Cont.*):
  to TL530 and TL531, 11-131 to 11-137
  to TLC532A and TLC533A, 11-131 to
    11-137
Motors:
  control circuit for, 7-22 to 7-23
  driver circuits for, 10-21 to 10-26
Multiple-feedback band-pass filters, 3-9 to
  3-10
Multiplex, digital display, 8-4, 8-9, 11-104
  to 11-105
Multipoint communications, 9-41, 9-45 to
  9-46
Multivibrators, 3-16 to 3-17

NE555 timer, 7-2 to 7-6, 7-13 to 7-18
NE592A video amplifier, 4-2 to 4-3
NE5534 op amp, 3-14 to 3-15
Negative voltage regulators, 5-6 to 5-7,
  5-15 to 5-16, 5-24, 6-33
Night light, telephone controlled, 7-23,
  7-24
Noise:
  in DAS, 11-13
  and differentiator circuits, 2-10
  and grounding, 9-62, 9-63
  op amp, 2-6
  and receivers, 9-7, 9-8
  and SN55107A receiver, 9-17, 9-18
  and switching power supplies, 6-18
  and transmission lines, 9-2 to 9-4, 9-10
  to 9-12
Noninverting op amps, 2-7 to 2-8
Notch filters, 2-11 to 2-12, 4-4
  twin-T, 3-10 to 3-11, 3-20
Nulling of offset voltages, 2-4, 2-9
Numbering systems, device, 1-1 to 1-2
Nyquist criterion and sampling, 11-6

Offset binary codes, 11-7 to 11-8
Offset voltages and error amplifiers, 5-5 to
  5-6
One-channel balanced transmission
  system, 9-17
One-shot timer mode, 7-2 to 7-4, 7-9 to
  7-10, 7-13
One's complement binary code, 11-8
OP-07 op amp, 3-1, 3-2, 3-5 to 3-6
Open-loop gain and mode:
  and comparators, 2-16
  for op amps, 2-2 to 2-3
  and video amplifiers, 4-1
Operational amplifiers, 2-1, 2-7 to 2-9,
  3-1 to 3-5
  active filters using, 2-10 to 2-13, 3-6 to 3-11
  audio applications, 3-11 to 3-15
  choosing, 2-13 to 2-14
  LinCMOS, 2-4, 2-14 to 2-15, 3-17 to 3-21
  oscillators and generators using, 3-15 to
    3-17
  performance characteristics of, 2-2 to 2-6
  (*See also* Comparators)
Optical sensors for TTL interface circuit,
  3-1, 3-2
Opto isolated solenoid driver, 10-19 to
  10-21
Oscillation and comparators, 2-18 to 2-19

Oscillators:
    audio, 3-15 to 3-17
    square wave, 7-15, 7-16
    in switching power supplies, 6-7 to 6-8,
        6-10
    in timers, 7-9
    variable-duty-cycle, 7-17
    voltage-controlled, 7-18 to 7-20
Oscilloscopes:
    calibration circuit for, 7-13 to 7-14
    preamplifier for, 4-2 to 4-3
Output offset voltage, 2-4, 4-1 to 4-2
Overvoltage sensing, 6-38 to 6-39

Panel meters, 11-105 to 11-111
Panels, VFD, 8-29 to 8-31
Party-line systems, 9-8 to 9-10, 9-12, 9-13,
        9-17 to 9-20
Pass transistors, power supply, 3-19, 5-12
        to 5-14, 6-26
Passive filters, 2-11 to 2-12
Peak detectors, 3-17 to 3-18
Periodic signals in DAS, 11-3 to 11-4
Peripheral drivers:
    applications with, 10-14 to 10-21
    devices for, 10-1 to 10-14
    for motors, 10-21 to 10-29
    for transmission lines, 10-26, 10-30 to
        10-31
Phase and phase margin, 2-6, 2-8
Polarity:
    magnetic, and Hall effect, 12-2 to 12-4
    op amp output voltage, 2-1
Positive trigger pulse for timers, 7-18, 7-19
Positive voltage regulators, 5-6 to 5-7,
        5-16, 6-33
Power and transmission lines, 9-3
Power amplifiers, audio, 3-14 to 3-15
Power supplies:
    for data transmissions, 9-1
    dual polarity, 6-36 to 6-38
    5 volt/10 amp, 6-10 to 6-13
    general purpose, 6-34 to 6-35
    isolated feedback, 12-5, 12-6
    for mobile equipment, 6-35 to 6-36
    and supervisor devices, 6-27 to 6-31
    for timers, 7-10 to 7-11
    two-volt logic array, 3-19
    for video amplifiers, 4-2
    (See also Switching power supplies;
        Voltage regulators)
Preamplifiers:
    microphone, 3-11, 3-17
    oscilloscope/counter, 4-2 to 4-3
    using TL080, 3-11 to 3-13
Print hammer driver, 10-18, 10-19
Programmable timer, uA2240, 7-6 to 7-10,
        7-23 to 7-29, 10-20
Propagation delays, timer, 7-5 to 7-6
Pulse train for DAS, 11-4, 11-5
Pulse-width modulation (PWM):
    motor control circuit using, 7-22 to 7-23
    for motor drives, 10-22 to 10-23
    in switching power supplies, 6-1 to 6-2
    with TL594, 6-7 to 6-9
Push-pull converters, 6-3 to 6-4

PWM (see Pulse-width modulation)

Q of active filters, 2-11 to 2-13, 3-9 to 3-11
Quantization in sampling, 11-3, 11-7, 11-12,
        11-14
Quantum interval, 11-7

Ramp generator, 7-15 to 7-16
Random signals in DAS, 11-3 to 11-5
Receivers, transmission line, 3-4, 9-1, 9-15
        to 9-17
    for current-mode drivers, 9-14
    differential, 9-12, 9-19 to 9-20
    EIA standard, 9-24 to 9-32, 9-36, 9-41
    IBM System 360/370, 9-62, 9-63
    IEEE 488 standard, 9-52
    single-ended, 9-6
Rectifiers, 5-18, 6-1, 6-25
Reference, 10-volt, 3-5 to 3-6
Reference elements for voltage regulators,
        5-1, 5-3 to 5-5, 6-7
Refreshing of AC TFEL panels, 8-44, 8-47,
        8-48, 8-50
Relay driver circuits, 3-22, 7-14, 8-7, 10-9
        to 10-11, 10-17 to 10-21
Remote LEDs, driving, 10-17 to 10-19
Remote monitors, 9-40
Remote overvoltage protection circuit,
        6-38
Remote sensor, one-wire A/D converter,
        11-115 to 11-116
Remote voltage sensing and regulators,
        5-17
Repeaters, transmission line, 9-18, 9-21
Resistors:
    for A/D converters, 11-100, 11-109
    for current limiting, 5-13 to 5-14
Reverse bias protection, regulator, 5-13
Reversible motor and solenoid drivers,
        10-18, 10-20 to 10-23
RF amplifier, interface for, 8-8
Ribbon cable, 9-2, 9-4
Ripple, power supply, 5-21 to 5-22
    and switching power supplies, 6-13
    and timers, 7-11
Rise time, comparator, 2-18 to 2-19
Rockwell 6502, interfacing of:
    to ADC0803, ADC0804, and ADC0805,
        11-61 to 11-62
    to ADC0808 and ADC0809, 11-66 to
        11-68
    to ADC0831, ADC0832, ADC0834, and
        ADC0838, 11-77 to 11-79
    to TL530, TLC532A, and TLC533A,
        11-129 to 11-131
    to TLC540, 11-140 to 11-144
    to TLC545, 11-190 to 11-193
    to TLC549, 11-157 to 11-160
    to TLC1540 and TLC1541, 11-170 to
        11-172
Rockwell 6522 VIA, interfaces using,
        11-140 to 11-142, 11-158 to 11-160,
        11-191 to 11-193
Row drivers, 8-44, 8-45, 8-47 to 8-50
RS-232-C standard, 9-22 to 9-33, 9-38, 9-49
RS-422-A standard, 9-37 to 9-40

RS-423-A standard, 9-33 to 9-38
RS-485 standard, 9-38, 9-40 to 9-46

Safe operating area (SOA), regulator, 5-12
        to 5-13
Sample-and-hold circuits, 11-18, 11-114
Sampling concepts for DAS, 11-2 to 11-3,
        11-5 to 11-6
Sampling elements for voltage regulators,
        5-1, 5-5
Sampling rate, 11-10 to 11-11, 11-13
Scaler, voltmeter, 6-24
Schade, O. H., and filter design, 5-19
Schmitt triggers, 3-4
SCRs in crowbar circuits, 6-23, 6-38
Security alarm, 12-9, 12-10
Serial binary interfacing, 9-22
Series regulators, 5-2, 5-7 to 5-8, 5-16 to
        5-17, 6-23, 6-26
    terms for, 5-21 to 5-24
Shielding of coaxial lines, 9-4
Shunt limiter, 3-3 to 3-4
Shunt regulators, 3-19, 5-2, 5-9, 6-21 to
        6-23
Signals in DAS, 11-1 to 11-5
    recovery of, 11-9 to 11-10, 11-13
Silicon, Hall effect in, 12-1
Sine waves in DAS, 11-4
Single-ended coupling, 4-2
Single-ended data transmissions, 9-4 to
        9-10, 9-22, 9-25
Single-slope A/D converters, 11-15 to
        11-16, 11-44 to 11-46
Single wire transmission line, 9-1
Slew rate, 2-2, 2-5 to 2-7, 2-16
SN74ALS679 address comparator,
        11-199 to 11-201
SN74LS123, 12-7, 12-8
SN74LS138 line decoder, 11-197, 11-202 to
        11-204
SN7432 OR gate, 10-26, 10-28
SN55107A receiver, 9-17, 9-18
SN74198 shift register, 10-26, 10-28
SN74260 NOR gate, 10-26, 10-28
SN74299 shift register, 11-196, 11-197
SN74393 binary counter in A/D converter,
        11-59, 11-63, 11-68
SN75064, SN75065, SN75066, and
        SN75067 peripheral drivers, 10-10
        to 10-11
SN75068 and SN75069 switches, 10-12
SN75074 driver, 10-12, 10-20 to 10-21
SN75075 driver, 10-12
SN75107A receiver, 9-15 to 9-19
SN75108A receiver, 9-15, 9-18
SN75110A driver, 9-14
SN75113 and SN75114 line drivers, 9-19,
        9-22
SN75115 receiver, 9-19, 9-22
SN75121 driver, 9-8 to 9-9
SN75122 receiver, 9-8 to 9-9
SN75123 line driver, 9-63 to 9-64
SN75124 line receiver, 9-65 to 9-67
SN75125 receiver, 9-64, 9-66 to 9-68
SN75126 line driver, 9-64 to 9-65, 9-67 to
        9-68, 9-70

SN75127 receiver, 9-64, 9-66 to 9-68
SN75128 receiver, 9-64, 9-67 to 9-70
SN75129 receiver, 9-64, 9-67, 9-69
SN75130 line driver, 9-65, 9-66
SN75150 line driver, 9-22 to 9-24, 9-28, 9-32 to 9-33
SN75152 line receiver, 9-7 to 9-10, 9-24, 9-26 to 9-28
SN75154 line receiver, 9-9, 9-28 to 9-30, 9-32 to 9-33
SN75156 line driver, 9-22, 9-25, 9-33 to 9-34, 9-36 to 9-37
SN75157 receiver, 9-36, 9-37
SN75160A transceiver, 9-53 to 9-58
SN75161A and SN75162A transceivers, 9-54 to 9-59
SN75163A transceiver, 9-8, 9-9, 9-54
SN75172B driver, 9-41 to 9-44
SN75173A receiver, 9-41 to 9-45
SN75174B driver, 9-41 to 9-44
SN75175A receiver, 9-41 to 9-45
SN75176B, SN75177B, SN75178B, and SN75179B transceivers, 9-45 to 9-47
SN75188 line driver, 9-24, 9-26, 9-33, 9-34
SN75189(A) line receivers, 9-29, 9-31 to 9-34
SN75361A TTL to MOS driver, 9-7
SN75365 driver, 9-9, 9-10
SN75372 driver, 10-8 to 10-10, 10-31
    LED driver using, 10-17, 10-19
    motor drive using, 10-23, 10-24
SN75374 FET driver, 10-8 to 10-10
SN75431 peripheral driver series, 10-4, 10-5
SN75435 peripheral driver, 10-6 to 10-7
SN75436, SN75437, and SN75438 peripheral drivers, 10-5 to 10-6
SN75440 peripheral driver, 10-7, 10-8
SN75446 peripheral driver series, 10-4
SN75447 peripheral driver, 10-18, 10-19
SN75450B peripheral driver, 10-2 to 10-3, 10-26, 10-30
SN75451B peripheral driver, 10-3, 10-26
SN75461 peripheral driver series, 10-4, 10-5, 10-26
SN75465, SN75466, SN75467, SN75468, and SN75469 transistor arrays, 10-12 to 10-13
SN75471 peripheral driver series, 10-4, 10-5, 10-26
SN75476 peripheral driver series, 10-4
SN75480 DC plasma driver, 8-9 to 8-11
SN75491(A) and SN75492(A) LED segment drivers, 8-3, 8-6 to 8-8
SN75494 LED digit driver, 8-4
SN75497 and SN75498 LED drivers, 8-4, 8-5
SN75500A display driver, 8-17 to 8-29
SN75501C display driver, 8-17 to 8-20, 8-24 to 8-26
    as high-voltage VFD driver, 8-35, 8-37, 8-38
SN75512A VFD driver, 8-33, 8-34, 8-36, 8-37, 8-39, 8-40
SN75513A VFD driver, 8-33 to 8-37
SN75514 VFD driver, 8-33 to 8-34, 8-36

SN75518 VFD driver, 8-34, 8-37
SN75551 and SN75552 row drivers, 8-44, 8-45, 8-47, 8-48
SN75553 and SN75554 column drivers, 8-44, 8-46 to 8-48
SN75581 display driver, 8-11 to 8-13
SN75584A display driver, 8-11, 8-13 to 8-16
SN75603 and SN75604 peripheral drivers, 10-7 to 10-9
    motor drives using, 10-22 to 10-23, 10-26, 10-27
    solenoid driver using, 10-21
SN75605 peripheral drivers, 10-7 to 10-9, 10-21
SOA (safe operating area), regulator, 5-12 to 5-13
Soft-start protection circuits, 6-9, 6-12
Solenoid drivers, 10-18 to 10-21
Solid state displays, 8-42 to 8-50
Speaker amplifiers, 3-14 to 3-15
Speed:
    in DAS, 11-13, 11-14
    for data transmissions, 9-1, 9-53
    of LinCMOS, 2-15
    op amp, and bandwidth, 2-5, 2-6
    of peripheral devices, 10-2
Split axis sustain, 8-21, 8-22
Square waves:
    oscillators for, 7-15, 7-16
    and Schmitt triggers, 3-4
Staircase generator, 7-28 to 7-29
Staircase signal recovery, 11-10
Step-down regulators, 6-2, 6-5 to 6-7, 6-14 to 6-16, 6-18 to 6-20
Step-up regulators, 6-2 to 6-3, 6-14 to 6-17, 6-20
Stepper motor drive circuits, 10-23, 10-25 to 10-26
Strobing, display driver, 8-24 to 8-25
Successive-approximation A/D converter, 11-11, 11-17 to 11-34, 11-47 to 11-58
Summing amplifiers, op amp, 2-9
Supervisor devices, 6-27 to 6-31
Surge currents, 5-19
    and lamps, 3-22, 10-14 to 10-17
Sustain operations, 8-17, 8-19, 8-24 to 8-25
Switches:
    Hall-effect, 12-2, 12-3
    intelligent, 10-21
    peripheral drivers for, 10-2
    touch, 7-15
Switching power supplies, 6-18 to 6-24, 6-34 to 6-39
    applications using, 6-24 to 6-27, 6-31 to 6-34
    5 volt/10 amp, 6-10 to 6-13
    operation of, 6-1 to 6-5
    step-down, 6-2, 6-5 to 6-10, 6-14 to 6-16, 6-18 to 6-20
    step-up, 6-2 to 6-3, 6-14 to 6-17, 6-20
    and supervisor devices, 6-27 to 6-31
    with TL497A, 6-13 to 6-18
Switching regulators, 5-2 to 5-3, 5-9 to 5-12
Switching times, comparator, 2-18 to 2-19
Synthesizer, frequency, 7-25, 7-26

Tachometer, 12-5, 12-7 to 12-8
Telephone controlled night light, 7-23, 7-24
Telephone relay driver, 10-9 to 10-11
Temperature:
    and DAS, 11-13
    indicator for, automobile, 3-5
    and noise, 2-6
    and op amps, 2-4, 2-5
    and oscillator circuits, 3-16
    and regulator design, 5-4, 5-5, 5-17 to 5-18, 5-22, 5-23
    and timers, 7-11
Termination of transmission lines, 9-2 to 9-4, 9-12 to 9-13, 9-43, 9-44, 9-62
Tesla (unit of magnetic flux), 12-2
TFEL (see AC Thin Film Electroluminescent display drivers)
Thermistors, 3-5, 11-107
Thermometer, digital, 11-107
Three-terminal voltage regulators, 3-19, 5-15 to 5-17
Threshold comparators, 7-2, 7-8 to 7-9
Threshold sensing, 9-24, 9-27, 9-28
Threshold-voltage shifts, op amp, 2-14
TIL125 optocoupler, 11-116
TIL154 optocoupler, 10-20 to 10-21
TIL307 display, 11-115
TIL406 optical sensor, 3-1, 3-2
TIL804 digital display, 8-4, 8-6
Time-base generators, 7-1, 7-8
Time controlled solenoid driver, 10-19 to 10-21
Time delays, 7-1 to 7-2, 7-25 to 7-27
Time-multiplex techniques, digital display, 8-4, 8-9, 11-104 to 11-105
Timers, 7-1, 7-10 to 7-29
    (See also NE555 timer; TLC555 timer; uA2240 programmable timer)
TL061 and TL066 op amps, 2-14
TL070 op amp, 3-4
TL071 op amp, 3-16 to 3-17
TL080 IC preamplifier, 2-6 to 2-7, 3-11 to 3-13
TL081 op amp, 2-6 to 2-7, 3-7 to 3-10
TL082 and TL083 op amps, 2-6 to 2-7
TL084 op amp, 2-6 to 2-7, 3-13 to 3-14
TL271 op amp, 3-18 to 3-20
TL321 op amp, 2-2, 2-3
TL331 comparator, 11-115
TL430 shunt regulator, 6-21 to 6-23
TL431 shunt regulator, 3-19, 6-10, 6-21 to 6-23, 6-25 to 6-26
TL497A regulator, 6-13 to 6-21
TL500, TL501, TL502, and TL503 A/D converters, 11-34 to 11-37, 11-100 to 11-111
TL505 A/D converter, 11-41 to 11-44, 11-109 to 11-111
TL507 A/D converter, 11-44 to 11-46, 11-112 to 11-116
    interfacing of: to TMS1000, 11-116 to 11-117
        to TMS1000C, 11-117 to 11-118
TL520, TL521, and TL522, interfacing of: to Motorola 6800, 11-68 to 11-71
    to Z80, 11-62 to 11-66

TL530 and TL531, interfacing of:
  to Intel 8048, 8049, 8051, and 8052,
    11-122 to 11-129
  to Motorola 6800, 6802, and 6809(E),
    11-131 to 11-137
  to Rockwell 6502, 11-129 to 11-131
  to Z80, 11-118 to 11-121
TL593 floppy disk power supply, 6-4 to 6-5
TL594 regulator, 6-4 to 6-13, 12-5, 12-6
TL710 comparator, 2-18 to 2-19
TL0808 and TL0809 A/D converters, 11-55
    to 11-58
  interfacing of: to Motorola 6800, 11-68 to
    11-71
    to Z80, 11-62 to 11-66
TL810 comparator, 2-16 to 2-19
TL3019 Hall-effect device, 12-3, 12-9 to
    12-10
TL3103 Hall-effect device, 12-3 to 12-5,
    12-7 to 12-9
TL4810A VFD driver, 8-32 to 8-33
TL7700 supervisor ICs, 6-28 to 6-31
TLC251 op amp, 2-14 to 2-15, 3-17, 3-18,
    12-9
TLC271 op amp, 2-14 to 2-15
  in A/D converter, 11-18
  in function generator and notch filter,
    3-20
  in logic array power supply, 3-19
  as signal conditioner, 11-115
TLC311 comparator, 2-18
TLC372 op amp, 3-21, 12-8
TLC532 and TLC533 A/D peripherals,
    11-47 to 11-50
  interfacing of: to Intel 8048, 8049, 8051,
    and 8052, 11-122 to 11-129
    to Motorola 6800, 6802, and 6809(E),
      11-131 to 11-137
    to Rockwell 6502, 11-129 to 11-131
    to Z80, 11-118 to 11-121
TLC540 A/D peripherals, 11-51 to 11-53
  interfacing of: to Intel 8048 and 8049
    (software), 11-151 to 11-155
    to Intel 8051 and 8052, 11-146 to
      11-151
    to Motorola 6805, 11-144 to 11-146
    to Rockwell 6502, 11-140 to 11-144
    to TMS32020 and TMS320C25, 11-207
      to 11-211
    to Z80, 11-137 to 11-140
TLC541 A/D peripherals, 11-51 to 11-53
  interfacing of: to Intel 8048 and 8049
    (software), 11-151 to 11-155
    to Intel 8051 and 8052, 11-146 to
      11-151
    to Motorola 6805, 11-144 to 11-146
    to TMS32020 and TMS320C25, 11-207
      to 11-211
TLC545, interfacing of:
  to Intel 8051 and 8052 (serial port),
    11-185 to 11-189
  to Intel 8051 and 8052 (software), 11-181
    to 11-185
  to Motorola 6805, 11-193 to 11-195
  to Rockwell 6502, 11-190 to 11-193
  to Z80A (hardware), 11-178 to 11-179

TLC545, interfacing of (Cont.):
  to Z80A (software), 11-179 to 11-181
TLC546, interfacing of:
  to Intel 8051 and 8052 (serial port),
    11-185 to 11-189
  to Intel 8051 and 8052 (software), 11-181
    to 11-185
  to Motorola 6805, 11-193 to 11-195
TLC548 A/D peripherals, 11-53 to 11-55
  interfacing of: to TMS32020, 11-211 to
    11-213
TLC549 A/D peripherals, 11-53 to 11-55
  interfacing of: to Intel 8048 and 8049,
    11-161 to 11-163
    to Intel 8051 and 8052 (serial port),
      11-163 to 11-166
    to Intel 8051 and 8052 (software),
      11-166 to 11-168
    to Motorola 6805, 11-160 to 11-161
    to Rockwell 6502, 11-157 to 11-160
    to TMS32020, 11-211 to 11-213
    to Z80, 11-156 to 11-157
TLC555 timer, 7-5 to 7-6, 7-18 to 7-24,
    10-26, 10-28, 12-10, 12-11
  in counting circuit, 8-15 to 8-16
TLC556/NE556 dual timers, 7-6
TLC0820, interfacing of:
  to TMS32010, 11-201 to 11-202
  to TMS32020, 11-205 to 11-207
TLC1540 and TLC1541, interfacing of:
  to Intel 8051 and 8052 (software), 11-176
    to 11-177
  to Motorola 6802, 11-172 to 11-174
  to Motorola 6805 (software), 11-174 to
    11-176
  to Rockwell 6502, 11-170 to 11-172
  to TMS32020 and TMS320C25, 11-207 to
    11-211
  to Z80, 11-169 to 11-170
TLC7524, interfacing of:
  to TMS32010, 11-199 to 11-200
  to TMS32020, 11-202 to 11-204
TLC32040 analog interface IC and TMS320
    digital signal processor, 11-195
  interfacing of, 11-196 to 11-199
TMS9914A bus controller, 9-53, 9-56 to
    9-57, 9-61
TMS9940 microcomputer system, 6-30
TMS1000 microprocessor:
  interfacing of: to TL505, 11-41, 11-111
    to TL507, 11-116 to 11-117
  undervoltage protection for, 6-30
TMS1000C, interfacing of, to TL507,
    11-117 to 11-118
TMS32010, interfacing of:
  to TLC0820, 11-201 to 11-202
  to TLC7524, 11-199 to 11-200
TMS32020, interfacing of:
  to TLC540, TLC541, TLC1540, and
    TLC1541, 11-207 to 11-211
  to TLC548/9, 11-211 to 11-213
  to TLC0820, 11-205 to 11-207
  to TLC7524, 11-202 to 11-204
  to TLC32040, 11-196 to 11-197
Tone control circuits, 3-11 to 3-13
Toroids, 12-3, 12-4

Touch switch circuit, 7-15
Transceivers, transmission line:
  differential, 9-12
  RS-485, 9-45 to 9-46
  single-ended, 9-6
Transformers:
  and regulators, 5-11, 5-18
  for switching power supplies, 6-1, 6-2,
    6-5, 6-10
Transients:
  in DAS, 11-3 to 11-4
  and power dissipation, 10-2, 10-3
Transistor drivers, 8-7
Transmission lines, 9-1 to 9-4
  drive circuits for, 10-22 to 10-23, 10-26,
    10-30 to 10-31
  (See also Data transmission systems)
Trigger circuits, 7-2 to 7-4, 7-18, 7-19
TTL circuits, optical sensor interface for,
    3-1, 3-2
Twin-T notch filters, 3-10 to 3-11, 3-20
Twisted-pair transmission lines, 9-2 to 9-4,
    9-20, 9-22
Two's complement binary code, 11-8

uA723 voltage regulator, 6-31 to 6-36
uA741 in band-reject active filter, 2-11 to
    2-12
uA2240 programmable timer, 7-6 to 7-10,
    7-23 to 7-29, 10-20
uA7815 and uA7915 regulators, 5-16
uA9636 dual line driver, 9-22, 9-25, 9-33 to
    9-34
uA9637A receiver, 9-36, 9-37
UCN4810A VFD driver, 8-31 to 8-33
UDN2841 driver, 10-13 to 10-14
UDN2845 driver, 10-13 to 10-14, 10-18,
    10-20, 10-22
UGN30xx Hall-effect devices, 12-3
ULN2001A, ULN2002A, ULN2003A,
    ULN2004A, and ULN2005A
    transistor arrays, 10-12 to 10-13
ULN2064, ULN2065, ULN2066, and
    ULN2067 peripheral drivers,
    10-10 to 10-11
ULN2068 and ULN2069 switches, 10-12
ULN2074 and ULN2075 drivers, 10-12
Undervoltage protection, 6-30
Unit load concept, 9-41
Unity-gain bandwidth, 2-3, 2-7
Unity-gain filter, 2-10 to 2-11

Vacuum fluorescent displays, 8-29 to 8-42
VCO (voltage-controlled oscillators), 7-18
    to 7-20
VCVS (voltage-controlled-voltage-source)
    circuit, 3-8
VFD (vacuum fluorescent displays), 8-29
    to 8-42
Video amplifiers, 4-1 to 4-6
Voltage-controlled oscillators (VCO), 7-18
    to 7-20
Voltage-controlled timer, 7-23 to 7-25
Voltage-controlled-voltage-source (VCVS)
    circuit, 3-8
Voltage followers, 2-9, 3-4, 3-6

Voltage-mode line drivers, 9-5, 9-19 to
    9-22
Voltage regulators:
    adjustable, 5-7, 5-12, 5-16 to 5-17, 5-24,
        6-21 to 6-23
    components for, 5-1 to 5-3
    error contributors to, 5-3 to 5-6
    fixed, 5-7, 5-16, 5-23, 6-23
    layout and design of, 5-6 to 5-12, 5-17 to
        5-21
    safe operating area of, 5-12 to 5-15
    terms and definitions for, 5-21 to 5-23
    three-terminal, 3-19, 5-15 to 5-17
    uA723, 6-31 to 6-36
    (*See also* Power supplies)
Voltages:
    isolated measurements of, 12-4 to 12-5
    monitoring of, 6-27 to 6-31
    (*See also* Voltage regulators)
Voltmeter scaler, 6-24

Volume compressor circuits, 3-3

Waveform generator, 3-20 to 3-21
Wein bridge oscillator, 3-15 to 3-16
Wheatstone bridge, 3-1, 3-2
Window comparators, 3-21
Write operations, 8-17, 8-18, 8-20 to 8-22

X-axis drivers, 8-19 to 8-20
X-Y matrix displays and panels, 8-16 to
    8-29, 8-42 to 8-50

Y-axis drivers, 8-19 to 8-20

Z80 microprocessor (*see* Zilog Z80)
Zener diodes:
    adjustable op amps as, 3-19
    as regulator reference, 5-3 to 5-4
    shunt regulators as, 6-21 to 6-23
Zero-crossing detector, 3-23 to 3-24

Zero-order hold function, 11-10
Zilog Z80, interfacing of:
    to ADC0803, ADC0804, and ADC0805,
        11-58 to 11-61
    to ADC0808 and ADC0809, 11-62 to
        11-66
    to ADC0831, ADC0832, ADC0834, and
        ADC0838, 11-71 to 11-77
    to TL520, TL521, and TL522, 11-62 to
        11-66
    to TL530 and TL531, 11-118 to 11-121
    to TL0808 and TL0809, 11-62 to 11-66
    to TLC532A and TLC533A, 11-118 to
        11-121
    to TLC540, 11-137 to 11-140
    to TLC545 (hardware), 11-178 to 11-179
    to TLC545 (software), 11-179 to 11-181
    to TLC549, 11-156 to 11-157
    to TLC1540 and TLC1541, 11-169 to
        11-170